Forest Health

Its Assessment and Status

Forest Health
Its Assessment and Status

John L. Innes

Swiss Federal Institute for Forest, Snow and Landscape Research (Formerly: Research Division, Forestry Commission of Great Britain)

CAB INTERNATIONAL

CAB INTERNATIONAL
Wallingford
Oxon OX10 8DE
UK

Tel: Wallingford (0491) 832111
Telex: 847964 (COMAGG G)
Telecom Gold/Dialcom: 84: CAU001
Fax: (0491) 833508

ISBN 0 85198 793 1

Phototypeset from disc by Fleetlines Typesetters, Southend-on-Sea
Printed and bound in the UK at the University Press, Cambridge

Contents

Preface

This is not another book about the effects of air pollution on trees. Instead, it attempts to cover the subject of forest decline. Forest decline is frequently assumed to be inextricably linked with air pollution, but this is not so. There are some situations where air pollution has caused the decline and even death of forests, but it is also quite clear that forests can decline in the absence of air pollution. There has been a huge amount of experimental research on the effects of air pollutants on trees under artificial conditions. This has yielded valuable information on the mechanisms of the biological effects of pollutants and the concentrations of specific pollutants that are likely to cause injury. However, applying these results to trees grown in the uncontrolled conditions represented by the natural environment is extremely difficult. Consequently, very few studies (if any) have succeeded in bridging the gap between experimental environments and the real world. Because of this, experimental research on the effects of air pollution on trees has not been reviewed. Where appropriate, reference is made to some more relevant experiments, particularly if the experimental design and analysis of the experiments are known to have been valid. Regrettably, many studies had to be excluded either on the grounds of irrelevance or of poor scientific practice.

Acknowledgements

Several people kindly reviewed chapters in this book. Joseph Barnard, Peter Blaser, Otto-Ulrich Braker, Bill Burkman, Frank Haemmerli, Theodore Keller, Werner Landolt, Andy Moffat, David Peterson, Rodolph Schlaepfer, John Skelly and Kenneth Stolte all provided extremely valuable comments.

The material for chapters five to nine was largely collected by staff from the Forestry Commission's Forest Surveys Branch: Peter Charlesworth, Meirion Davies, Bert Dover, Dave Dymond, Nick Evans, Nigel Fearis, Keith Fletcher, Peter Hutchinson, Graham MacBryer, Ian McNicoll, Colin Olsson, John Proudfoot, John Read, Chris Ross, Nick Salter, Kenneth Thompson and Kenneth Williams all took part in at least one assessment. In addition, Alan Betts, Alex Morris and Bob Selmes helped with the co-ordination of the annual programmes. Lisa Smith and Joan Lees undertook many of the laboratory analyses described in Chapter 8. Eleanor Harland tracked down many of the more obscure references cited in the text. All of their hard work and dedication is gratefully acknowledged.

A final point

I have made every effort to standardize the terminology used. Two areas defeated me. Country names were correct at the time of writing but some changes (e.g. Czechoslovakia) have since occurred. Measurement units in the source material have been retained since while this may result in some confusion, complete standardization would have been impossible.

John Innes

1 Forest Decline

We should note that forests have always 'declined'.
Treshow and Anderson, 1989

Introduction

During the 1980s, considerable concern developed over the condition of trees in many parts of the world. Visitors to forests saw dead and dying trees and foresters described the increasing need to fell trees prematurely to prevent the spread of secondary organisms, particularly bark beetles. The story was picked up by an increasingly environmentally aware media and by politicians. As a result, the issue of forest decline rapidly became one of the most widely discussed environmental topics of the 1980s. Since then, public interest in the subject has waned and it no longer draws the same level of political interest. This was partly because of an improvement in the health of forests in the late 1980s (which was subsequently reversed). An additional factor was the small attention span that most issues receive from the media. For example, in the Federal Republic of Germany, reports in the media steadily declined over the period 1985–87. The reduction occurred as it became apparent that the massive decline on forest ecosystems predicted by some scientists was not going to materialize (Krämer and Suda, 1987). In this book, the whole question of forest health is examined, with particular reference to the five years of detailed monitoring undertaken in Great Britain. However, as a global problem, it would be unwise to restrict the scope of the book to Great Britain, and many examples from Europe, North America and elsewhere have been examined.

The reported deterioration in the condition of some forests prompted much research into cause–effect mechanisms. This research was primarily experimental in nature, using seedlings and young saplings grown under strictly controlled conditions. Very few studies have been based on trees grown in their natural forest habitat, although where such studies have been undertaken, the value of the results far exceeds those obtained from controlled-environment studies. The careful work undertaken in the Erzgebirge Mountains of Germany by E.-D. Schulze and co-workers (Schulze *et al.*, 1989) illustrates this well. Field situations are normally extremely complex, and strictly controlled laboratory conditions may be necessary to examine mechanisms of pollution impact. In the research on forest decline, by far the greatest emphasis to date has been placed on the role of air pollution. Despite this, there are many case studies where the nature and causes of tree and forest decline have been investigated without reference to air pollution. However, the scientific and popular literature now abounds with references to the widespread and increasing 'forest decline' that is purported to be occurring and which is attributed to the adverse effects of air pollution. So much emphasis has been placed on this that someone unfamiliar with the subject might think that the combined effects of acidic rainfall and other forms of pollution will result in the imminent death of the forests of the world.

The views of many were summarized in several international reports (Anon., 1987, 1988, 1989a, 1990b, 1991a). These stated, for example, that 'research has indicated that air pollution in many cases plays a significant role in forest decline. In many cases the existence and extent of forest damage cannot be explained without considering an influence of air pollution' (Anon., 1990b). This view, with its use of the qualification 'in many cases', represents a modification of earlier statements that 'research results obtained so far clearly indicate that air pollution is an essential, causal factor in the destabilization of forests' (Anon., 1987). Regrettably, these claims are not supported by the available data on forest decline. Not only can forest declines occur without any involvement of air pollution, but the declines that were previously attributed wholly or partly to air pollution are now being reconsidered in the light of new evidence on cause–effect mechanisms.

In this chapter, natural declines in several forest types are examined together with some cases of studies of the impact of pollution around strictly delineated sources of pollution. Forests occur throughout the world and, in most areas where they have been studied, declines have been reported. Forest declines are not restricted to the industrialized areas of the northern hemisphere as some have suggested, but occur in all types of forest, from tropical rain forest to boreal coniferous forests. Some of these are described in the remainder of this chapter.

Definitions

Before embarking on any discussion of forest decline, some of the terms that are widely used, together with their origin, should be considered. The issue of forest decline, particularly as it relates to the possible effects of air pollution, originated in Germany in the late 1970s and early 1980s, when the term 'Waldsterben' was used to describe the decline and death of trees at several high-altitude sites in the south and east of the country. The literal meaning of this term is 'forest death', although it was primarily used to distinguish the decline from the well-known damage caused by sulphur dioxide and smoke ('klassische Rauchschäden'). The term was later modified to 'neuartige Waldschäden', meaning new forest damages. In the latter case, the use of the plural term is significant. 'New' was used to try to distinguish the declines from similar events that had occurred in the past. Schütt and Cowling (1985) modified the above definition of 'Waldsterben' to 'the widespread and substantial decline in growth and the change in behaviour of many softwood and hardwood ecosystems in central Europe'. This definition, with its emphasis on forest ecosystems as opposed to tree health has subsequently been widely used in the literature. It is particularly useful as it implicitly acknowledges that changes in forest ecosystems can occur without coincident changes in tree health.

Schütt (1988) subsequently modified the definition even further and now sees Waldsterben as a 'disease of forest ecosystems'. Interestingly, this definition suggests a single disease, whereas many scientists believe that a variety of phenomena are affecting forest ecosystems throughout Europe. The problem here seems to lie in what forms a definite decline and what is assumed to be a decline: Schütt (1988) appears to consider that the reports of crown transparency from several countries show the spread of the problem across the entire northern hemisphere within a period of eight years. However, this 'spread' merely represented the start of national inventories of forest condition in individual countries. For example, trees with transparent crowns were present in Great Britain long before the start of formal surveys in 1984. Photographic evidence for persistent thin-crowned trees exists in many other countries (see, for example, Schweingruber, 1989), and there is no justification for viewing transparency as a new phenomenon.

The term 'forest decline' is now firmly established in the literature related to air pollution. This is unfortunate for several reasons. In many reported cases, the decline does not involve all species present in the forest; rather, only one or two species are involved. In such circumstances, it would be more accurately termed tree decline (Skelly, 1989). It has been clear for some time (Forschungsbeirat Waldschäden, 1986; Rehfuess, 1987) that even within individual species, several different types of decline exist and that these can be distinguished by symptomatology and ecology. A further problem with the use of the term 'forest decline' is that in many of the cases cited in the literature, a decline as such has not been shown. Rather, if the state of a forest was considered to have deteriorated over a period of time, then a decline was inferred. Such a procedure fails to consider the negligible amount of information available on the 'normal' condition of forests throughout the world. Lastly, the term 'forest decline' has a well-established meaning in the ecological literature, referring to the noticeable reduction in vigour of many trees growing together at the same site (Mueller-Dombois, 1988).

The term decline is used frequently in English translations, but it is also common to find the phenomenon called a disease. This is consistent with the concept of a disease that progresses slowly over time and is due to a combination of abiotic and biotic factors (Manion, 1981; 1989). In some cases concept below, the problem is a true decline; in others, a specific single cause can be identified and the case is not a true decline in the current phytopathological sense.

Forest decline should be clearly distinguished from forest dieback. The latter refers to the unseasonable partial or complete loss of foliage on many trees within a forest (Mueller-Dombois, 1988). Usually, it is a long-term phenomenon, with dieback progressing from the edges of the tree inwards and from the top downwards. Frequently, it is reversible, with recovery being possible. Dieback can be brought about by several factors, with water stress being a particularly frequent cause. In certain circumstances, dieback in individual trees should be seen as a natural adaptation of the tree to changed environmental circumstances (e.g. a lower water table) rather than the onset of a major deterioration in the health of the tree. Dieback may be a symptom of decline but the reverse is not necessarily the case.

Many more recent studies in the literature compare one or more aspects of full-crowned trees with trees having more transparent crowns. The former are usually described as healthy and the latter as declining or diseased (the cause often being assumed to be air pollution). The inference that trees with transparent crowns are declining is difficult to substantiate, particularly in view of the work relating growth to transparency (see Chapter 10). A tree cannot be classed as healthy simply because it has not lost any of its foliage. Foliage loss is often the end-result of a series of changes in the tree's condition. Similarly, the presence of a transparent crown (which may or may not be attributable to foliage loss) does not necessarily reveal a state of decline. This problem is particularly important when interpreting national surveys, as some assess transparency (crown density) whereas others assess needle or leaf loss. The two indices are different, but they are rarely distinguished. Below, the term transparency is used to denote a crown with progressively less foliage. This term does not imply a state of health, nor does it imply a change in crown condition.

Another area requiring some definition is the different types of stress that can affect trees. One way is to consider stresses as being of three types (Sinclair, 1964; Houston, 1973, 1981, 1984). The terminology of these groups has become confused with different authors using the same term to describe different types of stress; they are best described as predisposing, triggering and accelerating. Predisposing stresses are those that contribute to the general weakening of a tree. They include a variety of types, ranging from long-term climatic change and long-term low levels of air pollution to the general reduction in vigour that accompanies the ageing of a tree. Triggering stresses are those that move the decline of a tree from the long-term to the short-term. They usually involve a specific event such as a drought or severe insect infestation, which severely reduces the vigour of the tree, and their effects can often be classed as damage. Finally, there are the accelerating stresses which cause the death of the tree. These are normally biotic in origin, involving either insects or fungi. The category that a particular stress is placed in may vary, depending on its nature and the nature of any other stresses present. Consequently, it is sometimes extremely difficult to place a particular stress in one of the three categories.

Natural forest declines

There are many documented cases of natural forest declines. Sometimes, these may be attributable to a specific pathogen. An example is the decline of patches of rainforest in Queensland, Australia, and jarrah (*Eucalyptus marginata* Sm.) in western Australia caused by the root rotting fungus *Phytophthora cinnamomi* Rands (Podger, 1972; Brown, 1976). Often, no single pathogen is the clear cause, and the reasons for the decline are much less clear. In many cases it can be linked to stand dynamics (e.g. Lawesson, 1988; Loehle, 1988) and the frequency with which natural competition and/or successional processes cause the decline of individual species within forests, or the decline of the entire forest appears to have been underestimated in the past. Several case studies involving tree or forest declines are presented below.

Tropical and sub-tropical declines

Sri Lanka

Much of the natural forest of Sri Lanka has been destroyed, but remnants of natural forest occur in the upper montane zone (above 1500 m) in a few areas. The patches of woodland are typical of cloud forests throughout the world, with stunted growth and dense coverings of mosses, lichens and filmy ferns. The dominant species are usually *Calophyllum walkeri*, *Calophyllum trapezifolium* and *Calophyllum cuneifolium*, with several other species, such as *Syzygium* sp., *Eugenia* sp., *Litsea* sp., *Actinodaphne* sp., *Cinnamomum* sp., *Symplocos* sp. and *Elaeocarpus* sp., also being important. The forests typically have little or no regeneration possibly because of suppression by the understorey of *Strobilanthes* sp. and *Indocalamnus* sp., both of which have 6–12-year flowering cycles.

Declining *Calophyllum* and *Syzygium* trees were first reported from the slopes of Thotupolakanda, above the Horton Plains (Perera, 1978; Werner, 1984). Perera described widespread flowering of *Strobilanthes* and investigated the area as he was unable to explain why an understorey should be so visible from a distance. He reported over 50% mortality of the dominant *Calophyllum* and *Syzygium* canopy trees and the complete absence of regeneration of these species. The extensive lichen flora, primarily *Usnea barbata* (L.) Wigg. em. Mot., was described as 'dead, black and charred'. Interestingly, Perera hypothesized that local air pollution from agricultural activities on the nearby Horton Plains might be involved, although he also speculated about the role of increased rates of drainage.

Later, Werner (1984) described dense, undisturbed montane rainforest below the area with dieback. However, he noted that the lichen growth was only very local. Subsequently, Werner (1988) described the extent of the forest mortality as being more widespread in the Horton Plains area, occurring on Kirigalpota, along the Pidurutalagala ridge and on Kobonilgala, in the Knuckles Mountains. The dieback occurred only in low (3–5 m) single-storied upper montane rainforest on shallow soil.

Werner (1988) firmly linked the dieback to drought conditions, particularly those associated with periods of desiccating winds. An especially dry year was 1976, shortly before the dieback was first observed and another very dry period occurred during the highly anomalous El Niño event of 1982–83. When the area was revisited in 1986, following a series of wet years, some trees had recovered substantially and the lichen flora seemed to be increasing. However, there were few signs of any recovery in the original patch of damaged forest on Thotupolakanda (Werner, 1988).

Werner (1988) attributed to the decline to a combination of factors. He considered the gradual ageing of the homogeneously aged canopy to be a predisposing factor, encouraging cohort senescence (whereby a large group of even-aged trees decline and die simultaneously) when a particularly important stress occurred. This stress took the form of severe drought conditions in 1976, and was probably responsible for triggering the dieback.

Hawaii

The occurrence of cohort senescence and mortality has been studied in detail on some of the Hawaiian islands, although it also appears to occur on several other Pacific islands (Mueller-Dombois, 1982). The decline of ohi'a (*Metrosideros polymorpha* Gaud.) forest in Hawaii aroused considerable concern in the 1960s and 1970s in view of the importance of this species in ecosystems that support a variety of endemic flora and fauna. On Hawaii, approximately 97% of the 2200 native plant species are

Figure 1.1 Dieback of ohi'a on the island of Maui.

believed to be endemic. The original flora, before colonization by man, may have involved as many as 50 000 species (Kimura and Nagata, 1980). The ohi'a forms a closed tree canopy, with an understorey of mixed tree species and tree ferns (mainly *Cibotium glaucum*), shrubs, ferns, lobeliads and bryophytes. Consequently, any decline of the ohi'a radically affects the entire ecosystem.

In several areas, individual trees or entire stands have declined (Fig. 1.1). Records of dieback extend throughout the 20th century (Mueller-Dombois, 1983; Lewton-Brain, 1909; Lyon, 1909; Fosberg, 1961; Mueller-Dombois and Krajina, 1968; Burgan and Nelson, 1972; Jacobi *et al.*, 1988). The dieback has not taken the form of a single event, rather deterioration and recovery phases of stands have been asynchronous. The scale of the dieback has been substantial. For example, an increase from 120 ha in 1954 to 50 000 ha of affected forest occurred in 1977 on the windward side of the main island of Hawaii (Jacobi *et al.*, 1988). Most of the dieback has been reported from the windward slopes of Mauna Kea and Mauna Loa volcanoes, on the main island of Hawaii. However, it is also present on other islands in the archipelago (Oahu, Kauai, Maui and others) (Jacobi *et al.*, 1988).

Mueller-Dombois (1980, 1983) described five different types of dieback: wetland, dryland, bog-formation, displacement and gap-formation, and each is likely to be brought about by a slightly different combination of factors. Dryland dieback occurs in forests that have developed on young substrates (usually lava flows or coarse volcanic ash), characterized by shallow, well-drained soils. Small groups of trees show significant dieback, the groups being scattered among healthy trees. In marked contrast, wetland dieback normally involves the dieback of all trees within an area. The soil type is also different from sites with dryland dieback, consisting of a shallow layer of organic soil above an impermeable substrate. The substrates tend to be older than in cases of dryland dieback and frequently consist of the pahoehoe lava type, which is smooth and unbroken. In areas where pahoehoe is mixed with 'a'a (a well-drained, clinker-type lava) wetland dieback may progress through the forest until it reaches the 'a'a, where it will form a sharp boundary (Jacobi, 1983; Mueller-Dombois, 1985). Bog-formation dieback occurs on even older surfaces, with deep soils and poor drainage, developed on old, nutritionally poor ash. In the displacement dieback type, ohi'a is replaced by a closed tree fern forest (Mueller-Dombois, 1985). This dieback type occurs on moderately well-drained, relatively recent and nutritionally rich ash soils (Mueller-Dombois *et al.*, 1983). Gap formation dieback consists of the death of small groups of ohi'a on knolls and ridges (Mueller-Dombois, 1985) located in bogs. Sometimes, the ohi'a is mixed with koa trees (*Acacia koa* Gray) and these too may show signs of dieback and mortality.

Figure 1.2 Severe erosion on Kauai following loss of the forest cover.

Following the dieback, rapid regeneration of the canopy normally occurs, and this characteristic of the ohi'a decline points strongly toward a stand dynamics explanation for the phenomenon. Occasionally, the loss of the forest cover may trigger soil erosion (Fig. 1.2). However this appears to be restricted to certain soil types and the erosion is not a normal adjunct to the different declines.

Possible causes of the dieback

To explain the dieback, it is necessary to look at the ecology of the species. Ohi'a is a shade-intolerant species with no further recruitment into the stand population once canopy-closure has occurred (Friend, 1980; Burton and Mueller-Dombois, 1984). Figure 1.3 gives some examples of the population structures of developing ohi'a stands, with the point at which a marked decline occurs being apparent. Although ohi'a seedlings may be present in the closed forest, the low light conditions normally prevent them from developing into saplings (Mueller-Dombois *et al.*, 1984; Burton and Mueller-Dombois, 1984). This led Lyon (1909) to believe that ohi'a was a pioneer species incapable of maintaining a long-term dominance in the forest. Subsequently, the United States Forest Service introduced several forest species (e.g. the Australian swamp mahogany *Eucalyptus robusta* Sm. and the paperbark *Melaleuca quinquenervia*) in an attempt to ensure catchment protection. Interestingly, later research (Doty, 1983)

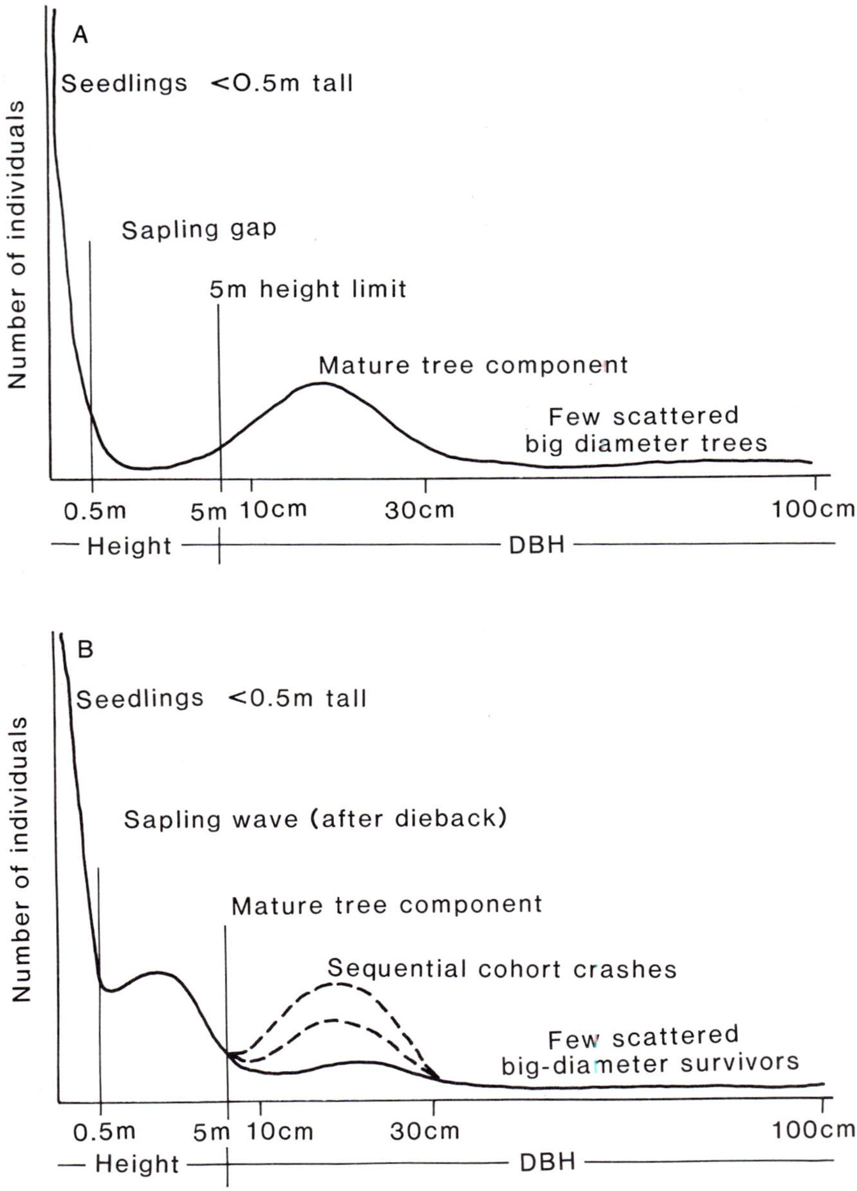

Figure 1.3 Cohorts of ohi'a before and after dieback. On the left-hand part of the *x*-axis, the trees are measured by height. At 5 m height, diameter at breast height (DBH) has been used. From: Mueller-Dombois (1987).

showed that there was no difference in the water quantity or quality draining areas with dieback and areas without dieback.

The air quality of the islands in the Hawaiian archipelago is normally considered to have been in pristine condition until quite recently and, although there are sulphur dioxide emissions from volcanoes which result in acidified rainfall in the area (Siegel *et al.*, 1990), the extent of the dieback bears no temporal or spatial relationship to the emissions of sulphur dioxide (Gerrish *et al.*, 1988) and the ohi'a can tolerate relatively high concentrations of sulphur dioxide (Winner and Mooney, 1985).

Nitrogen deficiency is thought to be involved with dryland dieback (Vitousek *et al.*, 1983) and, sometimes, fertilization of stunted trees led to their recovery (Kliejunas and Ko, 1974). Foliage of young trees growing in this habitat type (shallow, well-drained soils) had lower nitrogen contents than trees on older, more fertile sites (Balakrishnan and Mueller-Dombois, 1983). However, despite the work of Kliejunas and Ko (1974), nutrient deficiency has been ruled out as a primary cause of the dieback as nitrogen-phosphorus-potassium fertilizer treatments did not reduce mortality compared to unfertilized control plots (Mueller-Dombois, 1981; Gerrish and Bridges, 1984; Gerrish *et al.*, 1988). The most likely explanation for the problem is a combination of causes, with stand age predisposing trees to decline and the root pathogen *Phytophthora cinnamomi* and the bark beetle *Plagithmysus bilineatus* sometimes being involved as secondary factors (Mueller-Dombois, 1987). *Phytophthora cinnamomi* was particularly suspected of causing root dieback on trees growing in poorly drained sites, making them more susceptible to attack by *Plagithmysus bilineatus* (Papp *et al.*, 1979). However, both agents were ruled out as primary factors as they are not always associated with declining stands (Kliejunas and Ko, 1973; Hwang, 1977; Papp *et al.*, 1979). Seasonal soil moisture stress may act as an inciting stress in certain circumstances (Mueller-Dombois, 1980; Doty, 1982; Evenson, 1983).

One of the most widely accepted explanations for the dieback is through natural processes associated with stand dynamics. Gerrish (1990) used pipe-model theory to look at the relationships between sapwood, inner bark and leaf mass to test the hypothesis that the ratio of photosynthesis to respiration is lower in mature trees than in juvenile ones. He found that the sapwood to leaf-area ratios and the inner bark to leaf-area ratios were higher in mature trees and suggested that mature trees have a lower surplus carbohydrate production and were therefore less resistant to most stresses.

Changes in vigour through time were investigated by Jacobi *et al.* (1988). In a study area covering 100 000 ha on the windward side of Mauna Kea and Mauna Loa on the main island of Hawaii, they recorded significant changes in the vigour of stands between 1977 and 1986. Several stands with no or light dieback in 1977, had heavy to severe dieback when revisited in 1986. In addition, some stands with heavy dieback in 1977 were classified as severe in 1986. This suggested a lack of density-dependence and confirmed that neither natural nor anthropogenic stand thinning necessarily induces recovery (Gerrish and Bridges, 1984; Gerrish *et al.*, 1988). However, six of the 25 stands showed no progression of dieback. This may mean that under certain circumstances, the degree of stress on individual trees may be reduced through time or that some individuals are more resistant to decline than others. Two stands showed an improvement in vigour. The variation suggests a lack of a widespread triggering factor, with stand changes being dependent on the nature of the stands themselves. However, this may reflect the heterogeneity of the environment. Mueller-Dombois (1980) proposed that the dieback was initiated through a climatic instability that affected the soil moisture status. Any such effect would be dependent on a variety of factors including the soil type and the vegetation structure on it. Consequently, the synchronous onset of dieback across all habitat types would not be expected.

Further evidence that the process may be a natural one comes from studies of the vegetation history of Hawaii, which revealed that the amounts of ohi'a pollen preserved in bog sediments fluctuated over the last 10 000 years (Selling, 1948; Mueller-Dombois *et al.*, 1984). Fluctuations appear to have a 3–400 year periodicity, possibly coinciding with the average lifespan of an ohi'a generation. The pollen cores revealed no evidence of a decline of ohi'a during the period.

Conclusion

The study of the dieback bears some striking similarities to the study of forest decline in Europe. For example, Burgan and Nelson (1972) referred to the decline as a rapidly spreading epidemic, which they attributed to the recent introduction of the root pathogen *Armillaria mellea.* Similarly, Petteys *et al.* (1975) believed that the dieback was a new phenomenon that would lead to the destruction of the ohi'a forest in 15–25 years and described it as a severe epidemic. However, the documented records of dieback extending back to the turn of the century suggest that dieback of ohi'a has been present much longer than documented by Petteys *et al.* (1975). Had Petteys *et al.* (1975) given a better description of the symptomatology, it would have been apparent that only the upper storey was subject to dieback and that vigorous regeneration was present. This would have suggested that the disappearance of the forest was extremely unlikely. The most likely explanation for the phenomenon lies in stand dynamics, with a variety of factors determining the precise nature of the dieback in specific areas.

Decline in the productivity of Monterey pine plantations

Monterey pine (*Pinus radiata* D.Don) is the dominant plantation species in areas such as southern Australia, southern Africa and Chile and it is now of major economic importance. In recent years, many scientists have drawn attention to apparent declines in the productivity of this species. They have suggested that the fast growth may exhaust the mineral resources of some soils in which the trees were planted. This problem is likely to be particularly important in tropical and subtropical areas due to the generally impoverished nature of many soils. Evans (1990) reviewed the use of fast-growing plantations. He concluded that there was little evidence that the cultivation of the tree crop itself was the cause of the loss of productivity observed in many areas. Instead, he blamed several factors, including climatic fluctuations, specific nutrient deficiencies, weed competition effects, severe site disturbance at harvesting and poor species selection for the loss of productivity. Nonetheless, he admitted that there was a lack of good information in many areas.

Some of the most detailed reports of declines in productivity have come from southern Australia (Keeves, 1966; Boardman, 1978; Boardman and McGuire, 1990a, b; Woods 1990/1991) and New Zealand (Whyte, 1973). In Australia, reductions in productivity and deformation of the crown were first identified in the late 19th century. Simultaneously, growth reductions were seen in South Africa (Boardman and McGuire, 1990b). The extent of the problem was not fully recognized until the 1930s (Swain, 1935) and in 1946, 24 000 ha were affected in south-east Australia alone. The problem in this area seemed to be specifically related to zinc deficiency

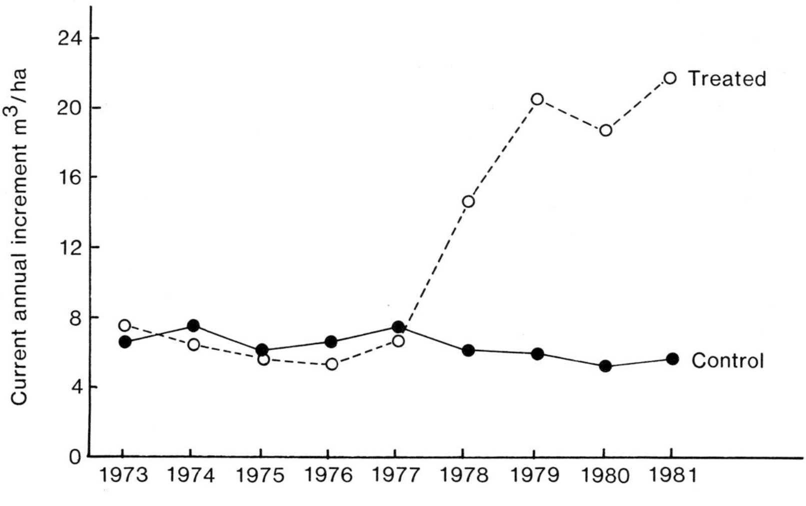

Figure 1.4 Response to nitrogen fertilization of a mid-rotation stand of maritime pine at Penola Forest, Australia. From: Woods (1990/1991).

Figure 1.5 Death of cipres de las Guaitecas (*Pilgerodendron uviferum* (D. Don) Florin) in the Golfo Elefantes, southern Chile caused by salt water flooding.

(Hearman, 1938; Smith and Bayliss, 1942; Boardman and McGuire, 1990a, b). The deficiency arises because of the high productivity of the pines, which depletes existing reserves of zinc in the soil. Elsewhere, boron deficiency is involved (Stone, 1990). For example, boron deficiency has been documented in Monterey pine stands in eastern Australia (Lambert and Turner, 1976, 1977), New Zealand (Will, 1978) and Chile (Schlatter and Gerding, 1985). Copper deficiency may also be involved at some sites (e.g. Ruiter, 1969; Turvey, 1984; Turvey *et al.*, 1992). Conversely, recent trials in Tantanoola Forest, Australia, suggested that the loss in productivity can be reversed by weed control after planting and application of N-based fertilizers (Fig. 1.4) (Woods, 1990/1991). Here the problem appears to be related to the loss of nitrogen during and following the burning of slash after the first rotation crop is felled.

Temperate forests

Chile

A similar, but longer term, phenomenon has been reported from the temperate rain forests of southern Chile. Veblen *et al.* (1983) recognized 17 different types of temperate broad-leaved forest in southern Chile, with *Nothofagus* sp. forming the dominant or co-dominant species in 13. Declines were recognized in several of these forest types, mostly related to stand dynamics rather than to any complex sequence of stresses. The declines described here are different from the spectacular declines that are present in southern Chile associated with flooding caused by land surface adjustments during earthquakes (Fig. 1.5).

Coigüe (*Nothofagus dombeyi* (Mirb.) Oerst.) has been reported to be in decline in many Chilean forests. Unlike many other cases described in this chapter, the decline is characterized by a reduction in the abundance of coigüe in forests. Cases where tree condition is declining can also occur (Fig. 1.6), but have not been reported in any detail. The decline in abundance is usually apparent through a lack of regeneration in undisturbed forests and a gradual ageing of the coigüe population in particular

Figure 1.6 Declining coigüe de Magallanes (*Nothofagus betuloides* (Mirbel) Oersted) in a high-altitude stand in southern Chile (Taitao Peninsula). Cause unknown.

stands, as opposed to a decline in either crown condition or radial increment. For example, high basal areas but very low densities were recorded in forests on the island of Talcán in the Chiloé archipelago (Armesto and Figueroa, 1987), suggesting a failure to recruit into the population. The lack of regeneration is widespread, as shown by the forest structure in the Valdivian Andes. In the past, the decline was widely attributed to supposed increasing aridity in southern Chile (e.g. Kalela, 1941; Auer, 1958, 1966; Pisano, 1975; Rey Balmaceda, 1985) and this line of argument has been uncritically accepted in the biogeographical literature (e.g. Holloway, 1954; Udvardny, 1969). According to Veblen and Lorenz (1988), the assumption that the native forests are inherently unstable has even been used to justify the logging of native forests on the grounds that they are incapable of regenerating.

The survival of coigüe within a given area appears to be determined by the incidence of large-scale allogenic disturbances (such as earthquake-induced landslides); where these are absent, large areas without coigüe occur (Veblen *et al.*, 1980, 1981; Donoso *et al.*, 1984, 1985; Armesto and Figueroa, 1987; Armesto and Fuentes, 1988). The connection appears to lie in the presence of a closed canopy. Coigüe is relatively intolerant of shade (e.g. Muller-Using and Schlegel, 1980; Read and Hill, 1985) and recruitment of young individuals into the canopy population appears to

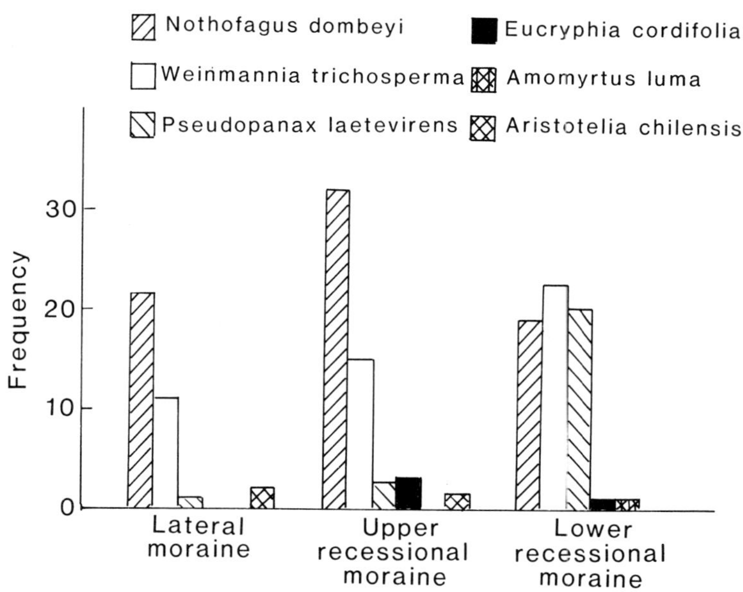

Figure 1.7 Tree species composition on moraines at the Casa Pangue glacier, Chile. The lateral moraine is the youngest surface and the lower recessional moraine the oldest. From: Veblen *et al.* (1989).

depend on the creation of substantial canopy gaps. Where gaps are small, shade-tolerant species take over.

Studies of succession on recently disturbed surfaces have shown that coigüe will rapidly colonise newly exposed surfaces (Uebelhör, 1984). Even-aged coigüe forests develop on these sites and the plots of the stem diameters in such stands follow a normal distribution. Where coigüe deaths occur, replacement is either by coigüe or by more shade-tolerant species, such as tepa or huahuán (*Laurelia philippiana* Looser) and trevo or palo santo (*Dasphyllum diacanthoides* (Less.) Cabr.). This successional stage is followed by a gradual decline in the density of coigüe, with the proportion of shade-tolerant species gradually increasing. There are several documented examples of this stage of forest, such as at San Pablo de Tregua, 37° 07′S, 72° 02′W (Veblen and Ashton, 1978; Veblen *et al.*, 1980). Eventually, the coigüe are eliminated from the forest if no disturbance occurs to restart the succession (Veblen *et al.*, 1981; Uebelhör, 1984; Veblen, 1989a). Rauli (*Nothofagus alpina* (Poepp. et Endl.) Oerst.) has very similar dynamics, with successful regeneration only occurring following large-scale disturbances (Veblen *et al.*, 1980). The pattern was nicely illustrated by examining the succession of trees on moraines of different ages at the Casa Pangue glacier (Veblen *et al.*, 1989) and the changes are shown in Fig. 1.7. The frequency of the coigüe increases from the lateral to the upper recessional moraine but then decreases on the lower recessional moraine as other species become more important.

Other types of *Nothofagus* are also known to have regeneration problems. Lenga (*Nothofagus pumilio* (Poepp. et Endl.) Krasser) and coigüe de Magallanes (*Nothofagus betuloides* Blume) are both failing to regenerate in a mixed species stand near Antillanca (Veblen *et al.*, 1977b; Veblen, 1979). Here, regeneration appears to be being prevented by a dense understorey of bamboo (*Chusquea tenuiflora*). This species has an irregular flowering pattern, with all individuals in an area flowering simultaneously and then dying. Veblen (1979) argued that dying off of the bamboo was necessary to allow regeneration of the *Nothofagus*, so that even-aged stands will tend to develop. In southern Patagonia, bamboo is absent and I have seen regeneration of lenga forest stands where grazing pressure is not too heavy. However, Armesto *et al.* (1992) suggested that lenga required substantial canopy gaps before regeneration could occur, with undisturbed stands reverting to a mixture of coigüe de Magallanes, ñirre and lenga.

Ñirre (*Nothofagus antarctica* Oerst.) is known to have problems regenerating in certain forest types (Veblen, 1989a, b). In the rain shadow area of the Andes, this species occurs with the Chilean incense cedar, *Austrocedrus chilensis* (D.Don) Florin and Boutelje, and with cöigue. However, without large-scale disturbances, particularly fire, the proportion of ñirre gradually declines and it is eventually replaced by cöigue after about 100 years following the initial disturbance. North-east Patagonia has a long history of fire disturbance (Veblen and Lorenz, 1988) and the presence of ñirre in the forests today appears to be the direct result of late 19th and 20th century fires.

The type of forest decline illustrated in this example is common in natural forest ecosystems. There are many forest communities that depend on disturbances for their survival, with fire-climax ecosystems being good examples. However, many advocates of forest decline have made the implicit assumption that forests should be in a steady state. The important point to emphasise here is that declines may occur without a triggering stress, although most declines described in this book can be related to specific suites of stresses. Succession is a widely recognized concept, and it plays an important part in determining forest dynamics. In such a context, the decline of an individual species is a normal process and is consistent with ecological theory.

Declines in the natural forests of New Zealand

Several different types of forest decline have been reported from New Zealand. Forest decline in mature southern beech (*Nothofagus*) forests is particularly frequent, with a variety of species being involved, including red beech (*Nothofagus fusca* Oerst.) (Hosking and Kershaw, 1985), silver beech (*Nothofagus menziesii* Oerst.) (Ogden, 1988; Hosking, 1989), clinker beech (*Nothofagus truncata* Cockayne) (Hosking and Hutcheson, 1986) and mountain beech (*Nothofagus solandri* var. *cliffortioides* Oerst.) (Skipworth, 1981; Grant, 1984; Hosking and Hutcheson, 1988; Ogden, 1988).

Various stress combinations appear to cause several types of decline in New Zealand's southern beech species and more than one explanation can be offered for each type. Ogden (1988) described the dynamics of different types of stands of red beech, mountain beech and silver beech. Each shows different types of decline, depending on the nature of the stands that they are growing in, and these are described below.

In pure stands of mountain beech, decline frequently occurs following the opening up of the canopy by storm damage (Wardle and Allen, 1983). Regeneration in pure stands is dependent on canopy gaps and normally occurs rapidly following disturbance. The result is a mosaic of even-aged patches within a heterogeneous stand. These pass through juvenile, mature and senescent phases, with a simultaneous decline in vigour occurring in all trees within a given patch. An example is provided by Hosking and Hutcheson (1988), who divided individual sample plots into stable, transitional and unstable types. The basal area of live trees progressively declined through these three categories. While there were few seedlings and saplings in stable plots, unstable plots had vigorous regeneration. The age range of trees was 50–120 years for stable plots, 100–130 years for transitional plots and >170 years for unstable plots.

Pure stands of red beech were examined in detail at Mount Colenso, North Island (Ogden, 1970; June and Ogden, 1975, 1978). Investigations in the area started following suggestions by various scientists that red beech was undergoing a climate-driven decline (Holloway, 1954; Elder, 1963; Franklin, 1967). The population structures of red beech in four stands suggested that a gradual downhill migration of red beech was occurring (Ogden, 1970). However, Ogden did not rule out the possibility that the lack of regeneration at higher altitudes was attributable to the absence of a suitable opportunity for regeneration, since regeneration was apparently favoured by large-scale disturbances. This was confirmed by a later study by June and Ogden (1978), who considered that the high-altitude stand had developed following a catastrophic disturbance and that the even-aged population was still actively undergoing competitive thinning. Because of this work, they suggested that following an initial major disturbance, a series of progressively less synchronised generations will occur, resulting in areas of increasingly heterogeneous forest.

In mixed stands of mountain beech and silver beech, stand dynamics controls the decline of mountain beech. Mountain beech is a light-demanding species which is short-lived. Consequently, it relies on large gaps in the canopy for regeneration. If these do not occur regularly, the species is eliminated from the forest, being replaced by silver beech. The initial establishment of mountain beech is dependent on the creation of a large gap, such as that caused by a catastrophic event. The occurrence of mixed mountain and red beech stands therefore suggests a major disturbance at some stage in the recent past. Population studies again provide evidence for this. For example, Wardle (1983) found that in mixed forests, the proportion of silver beech stems increased with increasing stand age, and this trend was also identified by Ogden (1985a, b).

Ogden (1988) provided an example of a decline in a mixture between red beech and silver beech. These stands are also inherently unstable, with silver beech gradually attaining dominance over the red beech, provided that no major disturbance occurs in the stand. The transition occurs because of the better shade-tolerance of silver beech and because of its lower susceptibility to attack by the scale insect, *Inglisia fagi*, and its greater resistance to attack by pin-hole borers (*Platypus* spp.) (Ogden, 1988).

Possible causes

As with many other cases, several stresses are involved in the decline of the southern beech stands. Competition alone may be the dominant factor, but usually one or more stresses either start or accelerate the changes. Drought has been widely cited as a triggering stress in the stands (Hocking, 1946; Elder, 1962; Grant, 1984; Hosking and Kershaw, 1985; Hosking and Hutcheson, 1986, 1988; Jane and Green, 1986). It is also involved in the dieback of other forest types in New Zealand (Atkinson and Greenwood, 1972; Ogden, 1976). In coastal locations, damage by salt-laden winds can trigger the decline of several forest types (Veblen and Stewart, 1980; Shaw, 1983; Campbell and Rudge, 1984) and physical damage by wind or snow has also initiated dieback (Wardle and Allen, 1983; Jane, 1986). Rawlings (1953) attributed the occurrence of dieback to the critical role of earthquakes. These trigger landslides which create bare surfaces, encouraging the development of even-aged stands. Surface exposure caused by earthquakes occurs frequently in New Zealand (Pain, 1971) and, in Papua New Guinea, land surfaces were estimated to be exposed by landslides on average every 200 years (Simonett, 1967; Pain and Bowler, 1973; Pain, 1972). Earthquake-induced landslides are also important in the creation of bare ground in Chile, where similar stand dynamics have been described (see the previous section).

Individual declines in New Zealand are frequently accelerated by a fungus (*Sporothrix* spp.) transmitted by pinhole borers (Faulds, 1977; Milligan, 1974; Wardle and Allen, 1983). As in many other cases, the beetles are attracted to damaged trees and then spread to other trees showing reduced vigour. Other insects, particularly defoliators, can also be involved, including *Neomycta pulicaris* Pascoe, *Proteodes carnifex* Butler and *Inglisia fagi* Maskell (Conway, 1952; Hosking and Kershaw, 1985; Hosking and Hutcheson, 1986).

The role of adverse climatic conditions, as opposed to catastrophic ones, seems uncertain. The annual variations in growth of both mountain and silver beech are partly controlled by summer temperature (Norton, 1984a, 1984b, 1984c), particularly at higher altitudes. However, Norton did not find a decrease in mean ring width with altitude. This might have been expected, given the estimated average summer temperature difference of 5°C between the highest and lowest sites used in his studies. Wardle (1970) also found no relationship between mean ring width and altitude in a study of mountain beech in the same area. However, the standard deviation and the common variance of the chronologies increased with altitude, suggesting altitude-dependent effects on the growth series.

Several other tree species in New Zealand are susceptible to decline. Decline of rata–kamahi (*Metrosideros umbellata* Cav.–*Weinmannia racemosa* L. f.) forests was described in detail by Stewart and Rose (1988) and Stewart (1989). The composition of these forests is very variable, creating a mosaic of different forest types. The decline of trees in these forests can be directly related to browsing stress caused by the bush-tailed possum (*Trichosurus vulpecula*). This animal has a distinct preference for stands with large numbers of old trees and stands with a high-proportion of seral shrub hardwood species (such as kotukutuku (*Fuchsia excorticata* Linn. f.), wineberry (*Aristotelia serrata*), mahoe (*Melicytus ramiflorus*), *Pseudopanax* spp. and pate (*Schefflera digitata*)) (Fitzgerald and Wardle, 1979; Veblen and Stewart, 1982a; Reif and Allen, 1988; Coleman *et al.*,

1980). Dieback increases as the possum populations build up (Pekelharing, 1979; Pekelharing and Reynolds, 1983; Batcheler, 1983). The stands most susceptible to feeding damage and therefore dieback are those consisting of large trees with even-sized diameter distributions (Allen and Rose, 1983; Stewart and Rose, 1988); similar but younger stands are not affected to the same extent, possibly because of a lack of understorey food species and the increased vigour of young trees (Allen and Rose, 1983; Stewart and Rose, 1988). As with *Nothofagus* stands, the even-aged stands seem to be the result of large-scale disturbances, particularly landslide-induced earthquakes. A further factor is the soil type, with dieback being more frequent on younger soils than on well-developed soils (Reif and Allen, 1988). Again, as with stands of southern beech, insects may act as accelerating stresses in the decline (Hoy, 1958; Payton, 1989).

Another type of decline that is thought to be occurring in New Zealand is associated with regeneration gaps. These are population structures which appear to show a lack of regeneration, with fewer than expected seedlings and saplings, similar to those found in the southern beech forests of Chile. Regeneration gaps occur with several species, including kahikatea (*Podocarpus dacrydioides* A. Rich), rimu (*Dacrydium cupressinum* Solander), cedar (*Librocedrus bidwillii* Hook. f.), totara (*Podocarpus totara* D.Don ex Lamb), matai (*Podocarpus spicatus* R.Br.) and kauri (*Agathis australis* Stend.) (Ogden, 1985a, b). The gaps occur when closed forest is examined; there is usually good regeneration of the species in canopy gaps. This strongly suggests a successional process.

Veblen and Stewart (1982b) made a detailed investigation of the regeneration gap in cedar stands on South Island. They explained the gap as a function of the regeneration dynamics of cedar. In closed canopy stands, there was a lack of seedlings and saplings. This was attributed to the slow growth of cedar and its relatively high tolerance to shade. However, rapid regeneration occurred in canopy gaps or gaps created by large-scale disturbances, such as landslides, windthrow and fire. This, again, reflects a situation where regeneration is primarily through the occurrence of gaps in closed-canopy forest. It provides another example of the questionable validity of the assumption that the regeneration of existing canopy species occurs without disturbance.

Forest declines in Tasmania

Rather similar problems to those in New Zealand also occur in Tasmania. In the 1960s, alpine ash (*Eucalyptus delegatensis* R.T.Baker), in the area between Mount Maurice and Ben Nevis in north-east Tasmania was recorded as declining (Ellis, 1964). Trees of all ages were affected in mixed age stands, predominantly of this one species. The decline also occurred elsewhere in Tasmania (Needham, 1960) and was not restricted to the Ben Nevis area. However, at Ben Nevis, the dieback was restricted to altitudes above 800 m above sea level, becoming severe at about 1000 m ASL (Ellis, 1964). The dieback took various forms, depending on the nature of the stand. In mixed-age stands, young trees were affected first and the decline then affected progressively older trees. The progression was gradual and mixed-age stands could take up to 60 years to die (Ellis, 1964). The health of the stands appeared to deteriorate as the age of the understorey increased, with this pattern being modified by soil drainage conditions at each site. The understorey always consisted of the original rain forest species, strongly suggesting a successional element in the development of the dieback.

Visible symptoms started with the thinning of foliage, followed by the dieback of shoots and the failure of new growth to replace the lost foliage (Ellis *et al.*, 1980). Leaf discoloration was not recorded before shedding. A period of epicormic development occurred on the main stem and branches, but this was only temporary. There was no evidence that *Phytophthora cinnamomi* was associated with the decline (Ellis *et al.*, 1980), although this finding was predictable in view of the altitudes of the sites involved and their ground conditions.

Burning, which was abruptly reduced between about 1830 and 1840, may be involved in the decline. The existing stands of alpine ash became established among grass following the destruction by burning of the original rain forest (Ellis, 1971). Dieback was particularly severe in stands with an understorey dominated by myrtle (*Nothofagus cunninghamii* Oerst.).

Ellis (1971) found an inverse relationship between the average annual soil temperature and the degree of dieback. He argued that a stand with a dense understorey of rainforest species would intercept more incoming solar radiation and would also increase the input of water through the more efficient condensation of fog particles. He found no relationship between the degree of dieback and either the soil pH or the amount of organic matter, despite both these variables changing with increasing understorey age. Soil moisture levels were also ruled out as a cause of the dieback. Instead, Ellis (1971) suggested that decreasing soil temperatures could trigger the dieback in the alpine ash. However, when this hypothesis was tested by clearing the understorey by cutting, the alpine ash failed to show any improvement in growth. Conversely, felling and burning the understorey resulted in marked improvements in the long-term growth of badly affected trees (Ellis *et al.*, 1980). Some of this recovery may have been due to the rejuvenation of the crown following crown scorch, as has been documented for jarrah (Kimber, 1976).

In southern Tasmania, decline of stringybark (*Eucalyptus obliqua* L'Herit.) and mountain ash (*Eucalyptus regnans* F. Muell.) was reported by Bowling and McLeod (1968) and Podger *et al.* (1980). The decline was first identified early in 1964 from a few tree mortalities in some permanent plots. At the time, the mortalities were tentatively attributed to *Armillaria* infection (Bowling and McLeod, 1968). Since then, the decline has been given the name 'regrowth dieback' as it is restricted to second-growth stands. By 1980, approximately 16 000 ha were affected (Podger *et al.*, 1980), although the problem appears to have stabilized since then (Wardlaw, 1989). As with the stands of alpine ash in the north, the area affected in south Tasmania has a long history of fire disturbance and the eucalypt stands have probably replaced rainforest. The dieback primarily affects stringybark and mountain ash, but may also affect peppermint (*Eucalyptus nitida* Hook.f.) trees growing in the area. It develops in regenerating stands more than 30–40 years old (Wardlaw, 1989). The dieback appears to occur on a variety of site types, being present in eucalypts growing above an understorey of myrtle, *Atherosperma moschatum* Labill. and *Dicksonia antarctica* Labill. on moist, well-drained sites, above *Bedfordia salicina* DC and *Pomaderris apetala* Labill. on dry, well-drained sites and above *Melaleuca squarrosa* Sm. on poorly drained sites (Podger *et al.*, 1980).

The decline is characterized by a reduction in the height and diameter growth rate of the trees and the dieback of branches in the primary crown (Podger *et al.*, 1980; Palzer, 1985). Within any affected stand, the distribution of affected trees seems to be random, although co-dominant trees appear to be more frequently affected than dominants (Podger *et al.*, 1980). Normally, epicormic shoots develop on the stem and along the main branches before the foliage dies in the primary crown and subsequent branch dieback occurs (Podger *et al.*, 1980). The size and colour of the leaves are unaffected. Mortality of the tree may occur up to 8–10 years after the onset of the decline (Kile *et al.*, 1981) and, in contrast to the decline of alpine ash, there is no evidence that once affected, a tree can recover. The properties of the timber are unaffected by the decline (Kile *et al.*, 1981), although the area affected is such that economic losses caused by a reduction of volume increment are likely (West and Podger, 1980).

The dieback is not related to thinning practices and it is not restricted to stands that have regenerated following logging. Fire does not appear to prevent the development of the dieback (Podger *et al.*, 1980). Podger *et al.* (1980) also ruled out competition and cohort senescence as a cause of the decline.

Analysis of basal area growth rates suggested that dieback may have started in 1959–60, a period which followed a series of unusually dry summers (West, 1979). The problem was aggravated by *Armillaria* sp. (Kile, 1980) and the paropsine beetle *Chrysophtharta bimaculata* Ol. (Kile, 1974; West and Podger, 1980). However, the *Armillaria* species that is present in trees in stands affected by the decline is a weak pathogen, incapable of being the primary agent in the decline (Kile, 1980).

The declines of these eucalypt stands are genuine cases of complex declines caused by a variety of factors. However, the importance of past management practices suggests that the problem may not be as complex as thought. Successional processes may play an important part in determining the occurrence of the declines. In the southern declines, external factors are more important than internal ones, and the link made by West (1979) between the decline and drought may be significant.

Boreal and sub-boreal declines

Alaska yellow-cedar

The Alaska yellow-cedar (*Chamaecyparis nootkatensis* (D.Don) Spach) is a widely distributed species in north-western America, extending from the south of Oregon northwards to Prince William Sound in Alaska. It shows a range of ecological tolerances, extending from sea-level to the tree-line in the centre of its range (Harris and Farr, 1974). Reports of decline in Alaska yellow-cedar go back over 100 years (Sheldon, 1912; Laurent, 1982; Frear, 1982; Shaw *et al.*, 1985; Hennon *et al.*, 1990a, b) and currently, more than 150 000 ha in south-eastern Alaska and British Columbia are thought to be affected (Hennon *et al.*, 1990a; Auclair *et al.*, 1990). The decline is believed to have started in about 1880 (Hennon *et al.*, 1990b). Mortality is currently continuing, although the rate of tree death is slow. Most mortality occurs on bogs and other damp sites (Downing, 1960; Laurent, 1982), although mortality has been spreading outward to better-drained areas (Hennon *et al.*, 1990a).

The decline of Alaska yellow-cedar is typified by reduced radial growth, fine root mortality, thinning and yellowing of foliage and invasion by secondary pests (Hennon *et al.*, 1990b). Occasionally, regeneration has occurred. However, careful dating suggested that the trees currently filling the spaces created by mortality are older than 100 years and must therefore have existed as understorey trees during the initial period of mortality (Hennon *et al.*, 1990b). In other cases, continued mortality has occurred without any evidence of regeneration.

Various investigations have been undertaken on the possible causes of the decline. So far, no single biotic or abiotic factor has been shown to be the exclusive cause (Hennon *et al.*, 1990a), suggesting that the decline is caused by a complex of interacting stresses. Hennon *et al.* (1990a) argued that the spread of the decline occurred along pre-existing environmental gradients rather than being a positive feedback process, and that the bogs existed before the onset of the decline. Furthermore, comparisons of air photographs taken over several years showed that whereas the extent of mortality at some sites had increased, the area of bog had not. This suggests that the formation of bogs is not the cause of the decline. Basal scars, caused by the teeth of brown bears (*Ursus arctos*) are widespread on the trees, affecting more than 50% of the trees on the Baranof and Chichagof Islands of southeast Alaska (Hennon *et al.*, 1990c). However, no relationship between the occurrence of the scars and the incidence of decline was observed. Various other explanations have been put forward, including climatic change and cohort senescence, but the matter remains unresolved.

This is a typical example of a species decline rather than a forest decline in that a disproportionate amount of mortality has occurred in the cedars in comparison to other conifers. However, mortality in other species, principally hemlocks (mountain hemlock and hemlock *Tsuga heterophylla* (Raf.) Sarg.) and Sitka spruce (*Picea sitchensis* (Bong.) Carr.), has been recorded, although it is uncertain if their mortality has any relationship with the mortality of the cedars. Hennon *et al.* (1990a) suggested that mountain hemlocks were currently stable, with few signs of decline in the living trees. The absence of a correlation between the timing of mortality in hemlocks and Alaska yellow-cedar may indicate that the presence of mortality in both species is coincidental. Clearly, this decline warrants further investigation.

Decline of trees in southern and central Sweden

Decline of Norway spruce and Scots pine has been reported along the entire range of the Swedish Scandes Mountains by Kullman and Högberg (1989), with particular damage occurring in 1987. Damage took the form of yellowing of needles that later turned into reddish-brown discoloration of the crowns of both Scots pine and Norway spruce. There was no evidence of needle pathogens, although dead needles were colonized by *Phacidium infestans* Karst. by 1988. On pines, almost all the older foliage was killed, but current year buds and shoots were not affected. The damage was not repeated in 1988. Damage increased with tree height, with little or no damage below 0.2 m. Damage also occurred on birches (*Betula* spp.) and trembling aspen (*Populus tremuloides* Mich.) as well as on shrubs, notably juniper *Juniperus communis* L.), crowberry (*Empetrum hermaphroditum* L.) and heather (*Calluna vulgaris* L.). In 1987, the damage was principally confined to 565–580 m ASL, with less damage both above and below this altitudinal belt.

The damage was restricted to peat soils which remained frozen until late in the spring. It was therefore attributed to winter desiccation associated with the exceptionally late thawing of seasonal ground frost in 1987 (Kullman and Högberg, 1989). On its own, this record of climatic damage would not form an instance of forest decline. However, in the literature, there are several records of decline and damage from the region. Henning (1895) reported dying pines in the area in the late 19th century. Kullman (1988, 1989) has since reported that both Norway spruce and Scots pine in the area were undergoing a period of retrogression throughout the 1960s and 1970s. This is quite surprising in view of the widespread evidence of rising tree-lines in many alpine areas (Innes, 1991a). The records of rising tree-lines are supported by numerous records

Table 1.1 Documented increases in growth since AD 1850

Species	Location	Starting	Reference
Fir species	South Europe	~ 1880	Briffa 1991
Silver fir	Vosges, France	~ 1850	Becker 1989
	South Germany	~ 1985	Kenk & Fischer 1988
Red fir	Washington	Before 1880	Graumlich & Brubaker 1987
Downy birch*	Greenland	~ 1880	Kuivinen & Lawson 1982
Birch species	N. Sweden	?	Eckstein *et al.* 1991
Beech	Germany	~ 1955	Kenk & Fischer 1988
European larch*	France	~ 1820	Tessier 1986
Tamarack*	Quebec Province	~ 1860	Arquillière *et al.* 1990
	Alaska/Canada	~ 1840	D'Arrigo *et al.* 1987
Norway spruce	Germany	~ 1955	Kenk & Fischer 1988
	North Europe	~ 1910	Briffa 1991
	Central Europe	~ 1870	Briffa 1991
	Germany	~ 1890	Eichkorn 1987
	Germany	~ 1920	Spelsberg 1987
White spruce*	Alaska/Canada	~ 1840	Jacoby & D'Arrigo 1989
	Quebec Province	~ 1860	Enright 1984
Black spruce*	Quebec Province	~ 1890	Payette *et al.* 1985
Pine species	South Europe	~ 1960	Briffa 1991
White pine*	California	~ 1870	Peterson & Arbaugh 1990
Lodgepole pine	California	~ 1850	Peterson & Arbaugh 1990
	California	~ 1950	Jacoby 1986
Rocky Mountain white pine*	Nevada	~ 1850	LaMarche *et al.* 1984
Bristlecone pine*	California	~ 1850	LaMarche *et al.* 1984
	New Mexico	~ 1850	LaMarche & Stockton 1974
	Colorado	~ 1850	LaMarche & Stockton 1974
			Graybill 1987
Scots pine	Finland	~ 1940	Hari & Arovaara 1988
	N. Europe	~ 1880	Briffa 1991
	N. Sweden	?	Eckstein *et al.* 1991
Douglas fir	South Germany	?	Kenk & Fischer 1988
	British Columbia	~ 1920	Parker 1987
White cedar*	Alaska/Canada	~ 1840	Jacoby & D'Arrigo 1989
Western hemlock	Washington	Before 1880	Graumlich & Brubaker 1987
Mountain hemlock	Washington	~ 1900	Heikkinen 1984a
	Washington	Before 1880	Graumlich & Brubaker 1987

From Innes (1991)

*Downy birch: *Betula pubescens* Ehrh.; European larch: *Larix decidua* Mill.; Tamarack: *Larix laricina* (Du Roi) K. Koch; White spruce: *Picea glauca* (Moench) Voss; Black spruce: *Picea mariana* (Mill.) Britten, Sterns & Poggenberg; White pine: *Pinus albicaulis* Engelm.; Rocky Mountain white pine: *Pinus flexilis* James; Bristlecone pine: *Pinus longaeva* Bailey or *Pinus aristata* Engelm.; White cedar: *Thuja occidentalis* L.

of increases in high-altitude tree growth (Table 1.1). These suggest that the warming trend recorded globally during the 20th century may be reflected in changes in tree-lines. However, the evidence from the Scandes

Figure 1.8 Cross-section through a fir wave. From: Sprugel (1976).

Mountains shows that global trends may be confounded regionally and that particular seasonal events may have a profound influence on trees and forest stands.

Wave phenomena

A particularly interesting type of decline occurs in several mountain areas. It takes the form of the wave-like progression of dieback through stands (Fig. 1.8). In the subalpine fir forests in Japan, the waves involve Veitch's silver fir (*Abies veitchii* Lindl.) and Maries's fir (*Abies mariesii* Mast.) (Iwaki and Totsuka, 1959; Kohyama, 1988). The waves occur as a striped-pattern of stand-level decline, known locally as Shimagare, after Mount Shimagare where frequent reports of the phenomenon have been made (Okubo, 1933; Iwaki and Totsuka, 1959; Kohyama and Fujita, 1981). The dieback fronts typically move gradually (ca 1.3 m per annum) through forests, usually in an up-slope direction and are up to 800 m long. They are separated in an up-slope direction by about 100 m of healthy forest. Regeneration occurs behind the fronts, developing from a dense understorey of seedlings present under the canopy in the dieback zone. In Japan, waves usually occur on the south slopes of mountains close to the summits, with the dieback fronts being exposed to the south. The dieback has been attributed to desiccation stress and mechanical damage, accelerated by the prevailing wind. It is particularly pronounced in overcrowded and even-aged stands.

The dieback is not restricted to Mt. Shimagare, occurring in several areas of central Honshu (Kohyama, 1988) and elsewhere. However, in Japan, it is restricted to certain fir species (*Abies veitchii*, *Abies mariesii* and *Abies sachalinensis* Mast.). Where the latter occur in mixed forests, dieback has not been recorded, nor has it been recorded in *Abies firma* Sieb. et Zucc. or *Abies homolepis* Sieb. et Zucc., which occur in mixed *Abies*-broadleaved forests. The susceptibility to dieback does not appear to bear any relation to soil type (Yoshino, 1976). The strips are restricted to near-summit areas, with the phenomenon being present in fir forests

Figure 1.9 Fir waves at Camel's Hump, Vermont.

of mountains with summits at 2300–2700 m ASL but not in higher mountains (Oka, 1983).

Progressive dieback of a similar nature has occurred along artificial edges, such as those created by road construction (Takahashi and Takahashi, 1974) and clear-felling (Takahashi, 1974). Such a pattern suggests a link between tree exposure and dieback. Further evidence for a link between exposure and decline is the tendency for a coincidence between the slope aspect and direction of the dieback edge of the stands (Iwaki and Totsuka, 1959). The mechanism of damage is believed to be desiccation, both in winter and summer (Kashimura, 1981; Kohyama, 1988), exacerbated by root damage caused by wind-rocking. Kohyama (1988) also proposed intraspecific competition as an additional stress factor, arising from the tendency for even-aged cohorts to develop in the wake of dieback fronts. These cohorts suffer from intense competition after about 20 years following release, resulting in heavy mortality and the development of small crowns in the remaining trees, which are then more susceptible to environmental stresses.

Similar wave-phenomena have been observed in North America. Waves occur in *Abies* forests in the Northern Appalachian Mountains (e.g. Sprugel, 1976; Reiners and Lang, 1979; Sprugel and Bormann, 1981; Marchand, 1984; Marchand and Goulet, 1984) (Fig. 1.9), in mountain hemlock stands in the Cascade Mountains (McCauley and Cook, 1980; Matson and Boone, 1984) and in balsam fir forests on Newfoundland

Table 1.2 Mean velocities of fir waves in the White Mountains of New Hampshire

Site and number	Distance travelled	Period (years)	Velocity (m yr^{-1})
Garfield Ridge			
G1	97.7±6.3	35	2.9±0.2
G2	29.3±4.1	35	0.9±0.1
G3	35.8±2.4	12	3.1±0.2
G4	42.3±4.5	35	1.3±0.1
Mount Lafayette			
L1	47.2±4.8	12	4.0±0.4
L2	17.2±4.5	12	1.6±0.4
L3	47.2±4.8	34	1.4±0.1
L4	34.3±4.7	12	2.9±0.4
L5	68.7±5.1	34	2.1±0.2
L6	64.4±5.0	12	6.0±0.5
Mount Moosilauke			
M1	84.5±7.4	35	2.7±0.2

From Boyce (1988)

(Robertson, 1987). Robertson (1987) distinguished two types of waves, one consisting of a regular event, the other being random. As in Japan, the waves of mortality are followed by vigorous regeneration of the species involved in the decline. Dieback can occur in various ages of forest, ranging from 35 to 75 years old (Marchand and Goulet, 1984), although Sprugel (1976) considered that the prevalence of waves in 50–60 year old stands suggested a role for natural cohort senescence. In view of the age range of dieback fronts reported in recent studies, this factor is only likely to be relevant to certain cases.

Details of waves, derived from aerial photographs taken by the U.S. Forest Service, are presented in Table 1.2 (Boyce, 1988). Velocities ranged from 0.9 m yr^{-1} to 6.0 m yr^{-1}, with the majority travelling at less than 3.0 m yr^{-1}. The estimated directions of the waves correspond closely with flagging directions, showing the importance of a wind-related process in determining the speed, occurrence and distribution of waves. Boyce argued that the wind velocity was critical, waves only occurring where the mean annual wind velocity was 7.7–8.6 m s^{-1}.

In the fir dieback, the primary mechanism involved is mechanical damage. Marchand (1984) and Marchand *et al.* (1986) described growth changes associated with waves in fir forests. Very often, the changes in growth occur before any visible signs of dieback. One characteristic of trees in dieback zones was the absence of several annual growth rings (usually 1977, 1979 and 1981), suggesting severe reductions in carbohydrate production in specific years due to loss of foliage. Exposure to wind along edges results in loss of foliage, often through rime-ice build-up (Marchand and Goulet, 1984; Marchand *et al.*, 1986) and breakage of fine

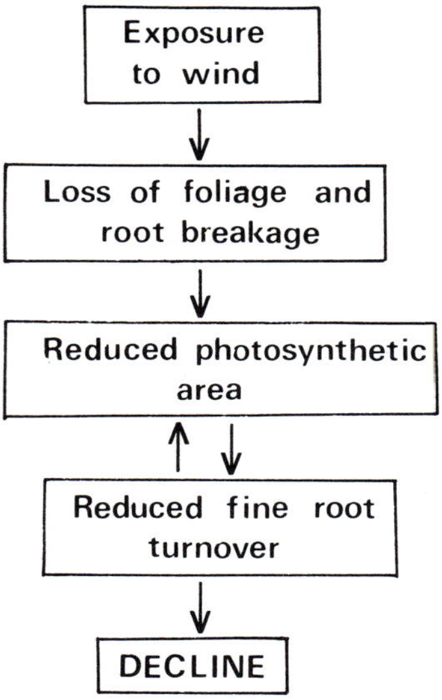

Figure 1.10 Hypothetical pathway for decline of fir trees in a fir wave (based on Marchand *et al.*, 1986).

roots. The rime-ice occurs when super-cooled water droplets impact on needles and immediately freeze, the increased weight leading to mechanical damage (Sprugel, 1976; Boyce, 1988). The reduced photosynthetic area means that insufficient resources are available to maintain the fine root turnover, and decline and mortality occurs eventually through a positive feedback process (Fig. 1.10). This hypothesis is supported by estimates of projected crown foliage, which show a significant reduction in the dieback front, whatever the stand age. The presence of needle loss and root breakage due to physical stress further support the hypothesis shown in Fig. 1.10 (Marchand, 1984; Marchand and Goulet, 1984). Root breakage is accompanied by the increased incidence of a root pathogen *Resinicium (Odontia) bicolor*, with substantial levels of infection in dieback front sites. Winter desiccation was largely ruled out as a cause of the decline as trees with similar levels of exposure near the tree-line are not particularly affected by dessication (Marchand and Chabot, 1978; Kincaid and Lyons, 1981).

Sprugel (1976) described several instances of wave mortality in the north-east USA. In all but one case, the mortality was present in pure fir forest, the remaining instance being spruce-fir forest. This is similar to the situation in Japan, where mortality is predominantly restricted to pure fir forests. In the cases described by Sprugel, the distribution of waves may reflect the requirement of red spruce (*Picea rubens* Sarg.)for disturbance-free conditions (Sprugel, 1976). If the wave mortality represents a repetitive situation which has been maintained for a long period, the frequent periods of exposure (at least once every 100 years) may have eliminated red spruce from the ecosystem. If correct, this hypothesis has important bearings on the purported red spruce decline described in the following chapter.

Waves also occur in stands of mountain hemlock (*Tsuga mertensiana* (Bong.) Carr.) (McCauley and Cook, 1980; Matson and Waring, 1984; Matson and Boone, 1984). In view of these reports, waves may be considerably more widespread than the literature currently suggests. In the mountain hemlock case, the dieback occurs in 300 year old stands, possibly because of a cumulative soil nitrogen deficit (Matson and Boone, 1984). Matson and Boone argued that the continued sequestration of nitrogen by the growing trees eventually reduced the available nitrogen to a point where the trees were unable to synthesize defensive compounds against pathogenic stresses, in this case the root fungus *Phellinus weirii.*

Summary

Wave mortality is a widespread phenomenon occurring in coniferous forests in a variety of situations. The process is linked to a combination of increasing susceptibility to stress as an even-aged stand gets older, and positive feedback processes once the initial opening up of the stand occurs. These feedback processes are relevant to several other types of decline described in the following two chapters although they have not been given sufficient attention in the literature.

Tree and forest declines around sources of pollution

The above examples all involved the decline or dieback of forests in areas where air pollution is not an important factor influencing the health of trees. At the opposite end of the spectrum, extensive damage to forests can occur around specific sources of pollution. These cases invariably involve direct injury to the foliage of trees. Locations such as Copper Basin in Tennessee and Sudbury in Ontario are well-documented (see below). In contrast, information is only just emerging about modern analogues in the USSR, China and Korea, such as the copper-nickel smelters of the Kola Peninsula (Alexeyev, 1989; Kryuchkov, 1991) and the severe pollution at Mount Chongqing Nanshan in south-west China (Yu *et al.*, 1988; Lee and Kim, 1990; Bian and Yu, 1992). Although attention has been given to the more famous sites, extensive mortality also occurred in Europe around certain smelters, refineries and other sources of pollution. For example, detailed studies have been made of the damage caused by fluorides from the aluminium smelters in the Arc Valley, Maurienne, France (Figs 1.11 and 1.12), and similar damage is apparent near other aluminium smelters (for example, at Øvre Årdal in southern Norway (Braanas, 1970) and at

Leanachan Forest near Fort William, Scotland). Fertilizer factories are also important sources of pollution and forest damage around them may be substantial (e.g. Laamanen *et al.*, 1972; Soikkeli and Paakkunainen, 1981; Patel and Devi, 1986; Väisänen, 1986) (Fig. 1.13). Miller and Mcbride (1975), Kozlowski and Constantinidou (1986a,b) and Smith (1990) provide full reviews of early work on the effects of air pollution on trees around point sources.

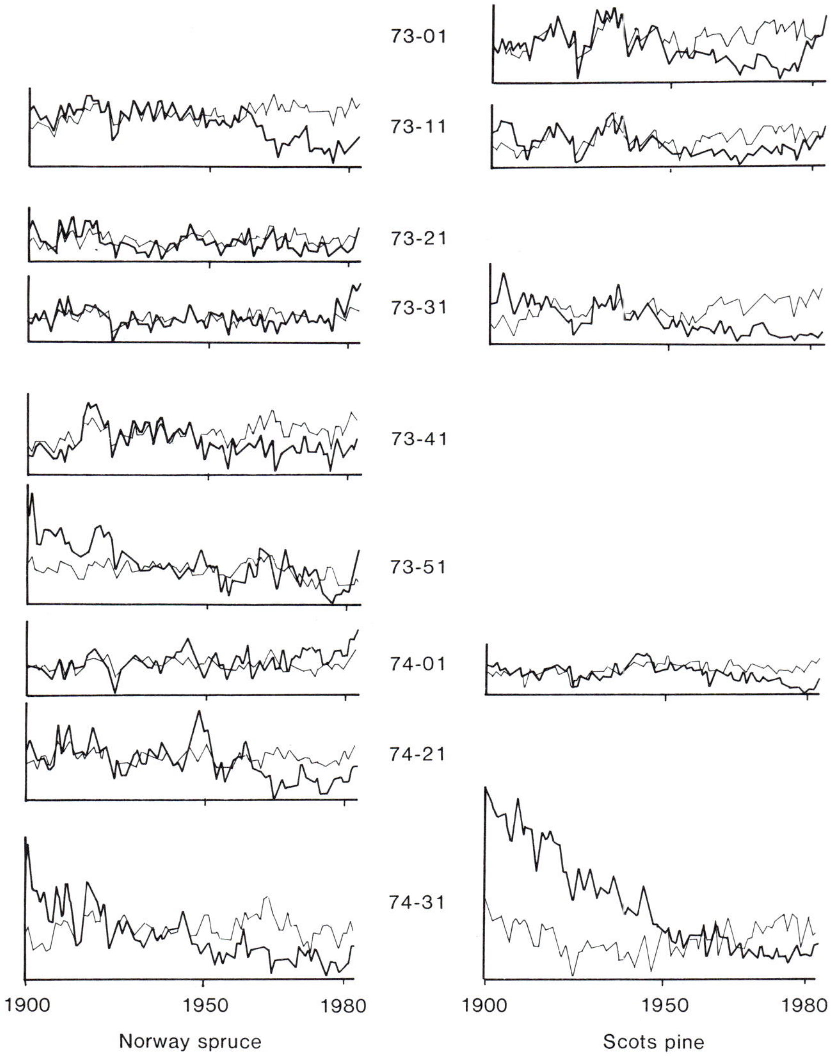

Figure 1.11 Observed (thick line) and predicted (thin line) growth of Norway spruce and Scots pine at several sites affected by fluorine pollution in the Arc Valley, France. From: Tessier *et al.* (1990).

The damage by pollution around specific sources can frequently be divided into two types. Close to the source, acute injury is present. This is caused by concentrations of particular pollutants that are directly toxic to plants. Sulphur dioxide and hydrogen fluoride are two of the best-known examples, but there are many other pollutants that will, in

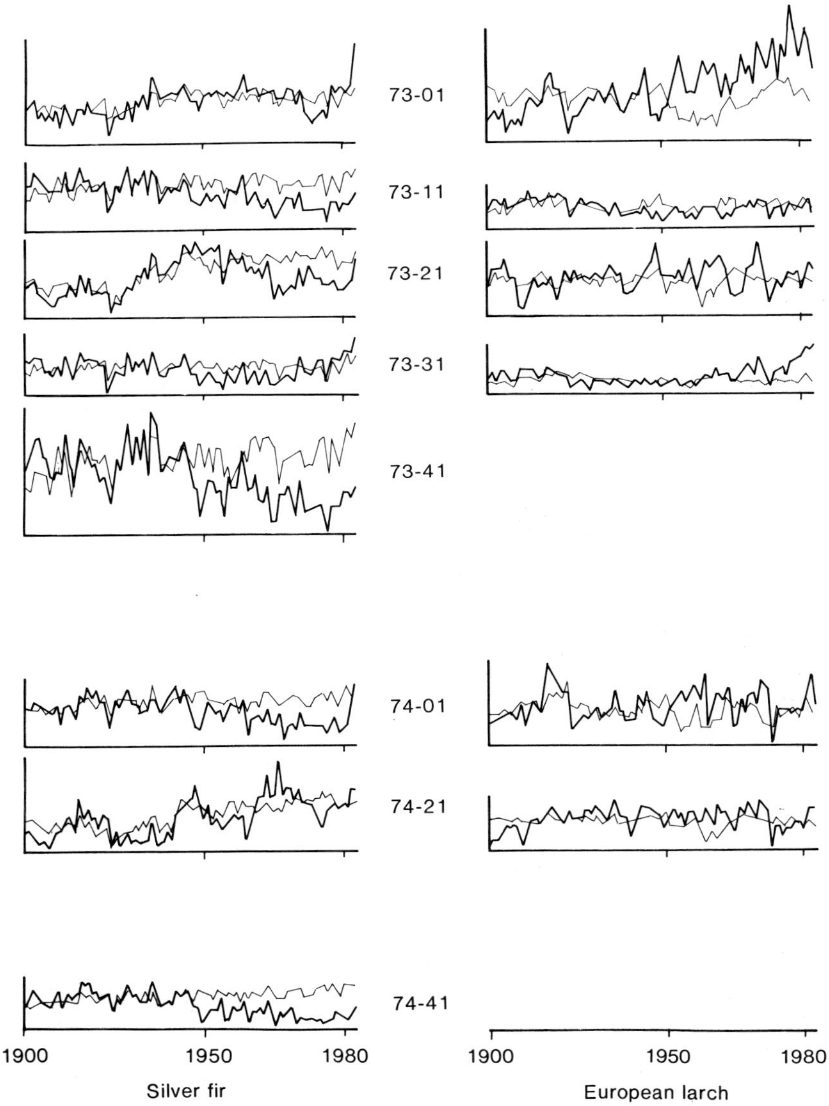

Figure 1.12 Observed (thick line) and predicted (thin line) growth of silver fir and European larch at several sites affected by fluorine pollution in the Arc Valley. From: Tessier *et al.* (1990).

Figure 1.13 A stand of Scots pine close to a fertilizer factory near Katowice, Poland. The trees are over 100-years-old, but radial growth has been neglible. The condition of the foliage was relatively good.

sufficient concentrations, injure or kill plants. The damage frequently takes the form of discoloured foliage, with tip-burn of needles and marginal necrosis of leaves being frequently reported.

Further away from the source, chronic injury may occur. This may take the form of direct effects resulting from, for example, the uptake of gases through the stomata on leaves. Alternatively, indirect effects may occur, such as changes in the vigour of plants as a result of the effects of pollution on some factor influencing the plants' vigour. Chronic injury is frequently very difficult to diagnose and it is often necessary to establish whether a particular stress would have been so severe in the absence of pollution. The complexity of the issue becomes apparent when it is considered that the response of a plant to a specific stress (e.g. susceptibility to frost damage) may change because of direct chronic injury.

Cases of acute injury are ones in which the role of air pollution has been conclusively identified. Consequently, they provide a useful basis for examining the much more complex issue of chronic pollution damage in forests. Several such case studies are described below.

Copper Basin, Tennessee, USA

Mining at Copper Basin (Polk County, Tennessee) started in 1850, although the peak activity occurred between 1890 and 1895 (Hursh, 1948). Initial deforestation of the mixed hardwood forests occurred in response to the demand for wood to fuel the open-pit ore-roasting process. Subsequently, in a zone close to the plants, about 2800 ha of forest was damaged by sulphur dioxide and failed to regenerate (Haywood, 1910; Hedgecock, 1914; Hursh, 1948). Immediately outside this intensely damaged area, an area of approximately 6900 ha of grassland (mainly broomsedge (*Andropogon scoparius*) occurred, bordered by a third zone of mixed grassland and pollution-resistant trees, approximately 12 000 ha in area. Damage to vegetation extended approximately 19–24 km to the north of the plant and 19 km to the west, with damage to eastern white pine (*Pinus strobus* L.), a particularly sensitive species, extending up to 32 km from the plant (Hursh, 1948). In the outer transition zone, the main tree species present were various oaks (*Quercus* spp.), hickory (*Carya* spp.), sourwood (*Oxydendrum arboreum* (L.) DC), dogwood (*Cornus* spp.), tupelo (*Nyssa sylvatica* Marsh.) and a few eastern white pine.

Sudbury, Ontario, Canada

The Sudbury copper and nickel smelters opened shortly after the ore bodies were first discovered in 1883. The main metals involved were nickel and copper. The smelting of these produced sulphur dioxide, sulphuric acid mist, acidic precipitation and contamination by heavy metals (Freedman and Hutchinson, 1980a, c; Gorham and Gordon, 1960a, b; Watson and Richardson, 1972). As with Copper Basin, early deforestation occurred through felling of adjacent forests to fuel the roastbeds. Open roastbeds were banned by 1928 and, subsequently, roasting was undertaken at the nearby smelters of Copper Cliff, Coniston and Falconbridge. Peak emissions occurred in the period 1965–1970, with emission rates of ca. 2.7 million tonnes per year, although the use of tall chimneys resulted in lower concentrations of sulphur dioxide and other gases in the immediate vicinities of the smelters (Chan and Lusis, 1985; Freedman and Hutchinson, 1980b).

Approximately 10 000 ha close to the smelters were classified as barren and a further 36 000 ha were considered to have an impoverished flora (Watson and Richardson, 1972; Gignac and Beckett, 1986; Winterhalder, 1988). Eastern white pine was killed over an area extending for 1860 km^2 and up to 40 km from the smelters. Foliar injury occurred over a further area of about 4100 km^2 (Linzon, 1971; Smith, 1981). The stratification of vegetation around the plant was examined in detail by Gorham and Gordon (1960a). They found that only the most tolerant taxa (including *Sambucus pubens*, *Polygonum cilinode*, red maple (*Acer rubrum* L.) and red oak (*Quercus rubra* L.) could survive within about 6.4 km of the plant. Freedman and Hutchinson (1980a, c) described an innermost damage zone that could only support a very limited vascular flora, predominantly consisting of the grasses *Deschampsia caespitosa*, *Agrostis hyemalis* and *Agrostis gigantea*, although red maple was also present.

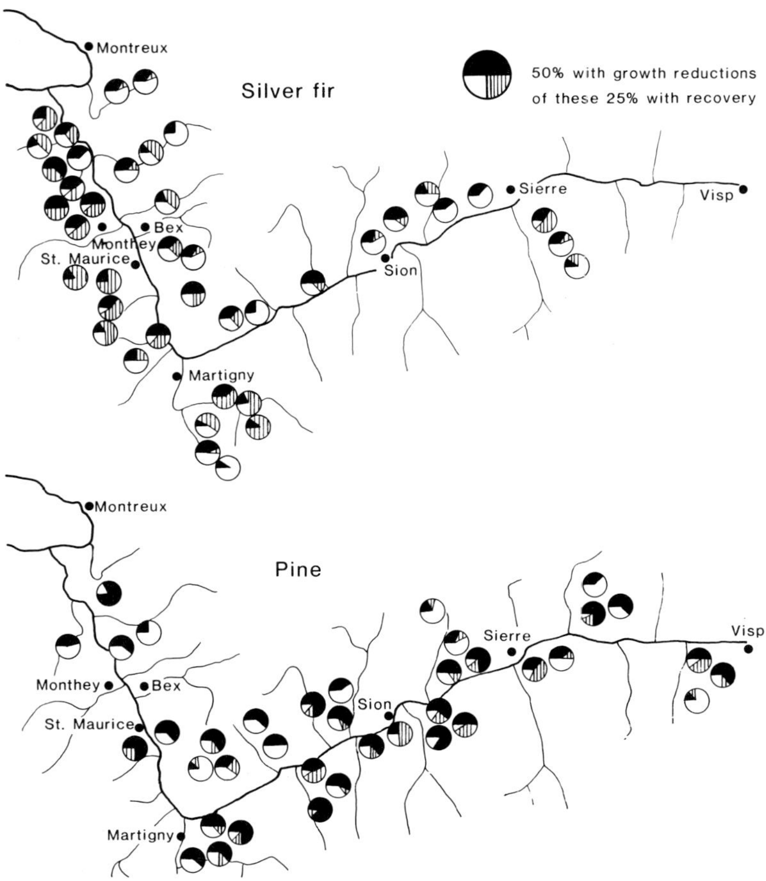

Figure 1.14 Proportion of silver fir (above) and pine (below) showing abrupt growth reductions in the Rhône Valley, Switzerland. The area of solid black indicates the proportion of trees showing a growth decline with no subsequent recovery. The proportion that declined and subsequently recovered is shown by shading. From: Kienast (1985).

Radford, Virginia, USA

The damage around the US Army Munitions plant at Radford, Virginia is of interest as the main form of pollution consists of the oxides of nitrogen. It was described by Skelly *et al.* (1972). The plant released both oxides of nitrogen and sulphur dioxide, although the latter was in low concentrations. The highest concentration of oxides of nitrogen recorded in the study was 0.0585 ppm hr^{-1} at a point about 600 m from the source. The highest sulphur dioxide concentrations were 0.69 ppm over a two-hour period, at a site 400 m from the source.

Damage to eastern white pine was recorded up to 3 km from the source. As with many other studies of this species, substantial tree-to-tree variations in susceptibility occurred. Typical symptoms included tufted needle growth and acute tip burn close to the source, and chlorotic mottling of needles further away.

The Rhône valley, Switzerland

Many forests in the Rhône Valley were severely damaged during the first half of the 20th century. The greatest damage was in the regions of Charrat/Saxon, Pfynwald and Visp (Flühler, 1981). Fluorine emissions in this area started in 1908, when two aluminium smelters were built, and were not reduced until 1981 (Flühler, 1983). Damage occurred from the early 1970s onwards and, in 1976, it covered about 130 km^2 (Kienast, 1985). Growth reductions (Fig. 1.14) were found throughout the valley and increment analysis suggested that damage to pines occurred in the period 1930–40 as well as in the 1970s (Kienast *et al.*, 1981; Kienast, 1985), whereas damage to silver fir was only recorded in the 1970s. Severe increment reductions in 1976 in the Rhône Valley were also documented by Kontic *et al.* (1986). Based on changes in annual radial increment, Schweingruber *et al.* (1985) described damage as occurring in several years, but it is unclear whether this was exclusively due to a combination of drought and fluorine pollution or whether other factors were involved. Damage was greatest downwind of sources, with no damage being recorded outside the main wind trajectories (Flühler, 1983). The slopes above Saxon and Charrat showed the greatest levels of damage: these lie directly in the path of the prevailing winds from the smelter at Martigny. Although not described, effects on certain Norway spruce provenances probably occurred, as this species is also sensitive to fluorine (Robak, 1969; Greve *et al.*, 1983, 1985).

Fluorine contents of needles were related to the proximity to specific sources at Chippis, Steg and Martigny (Flühler *et al.*, 1981) with concentrations of 1000 ppm and above being recorded. Sulphur pollution was also involved as needle chemical analyses revealed that sulphur contents were above background levels for Switzerland (Flühler, 1983). Chlorine contents were high, particularly close to sources of pollution in Visp and Monthey.

Tree ring analysis (Kienast *et al.*, 1981; Kienast, 1982) has shown that most damage occurred regardless of tree age between 1941 and 1947. Both healthy and declining trees were cored, with 80% of trees showing a sharp discontinuity in growth. The period of damage corresponds with the first occurrence of pronounced droughts after a second smelter was built at Martigny.

Spokane-Mead, Washington, USA

Damage to ponderosa pine was first noticed in the Franklin Park in about

1943. The damage extended along a valley, extending for approximately 18 km. Over the next ten years, virtually all the trees in a 800 ha area had died and trees over a further 12 950 ha were damaged (Adams *et al.*, 1952; Miller and McBride, 1975). Elevated fluorine contents in pine needles from the damaged area were found, with concentrations ranging up to 600 ppm (dry weight). This, together with the absence of any other factor that might have caused the damage, strongly implicated fluoride releases from the Kaiser aluminium ore reduction plant (Shaw *et al.*, 1951).

The role of hydrogen fluoride was confirmed by laboratory fumigations, which produced the same symptoms as seen in the field. Foliar damage, particularly to new needles, was accompanied by increment reductions (Lynch, 1951; Shaw *et al.*, 1951). The field evidence for damage, being supported by experimental evidence, provides a good example of the approach necessary to attribute damage to a specific cause.

Other sources of fluorine

Fluorine damage around aluminium smelters has been reported from a variety of other locations throughout the world. For example, Bunce (1989) described extensive damage around Kitimat in British Columbia. The damage occurred in forests consisting of western hemlock, western red cedar (*Thuja plicata* D.Don), Pacific silver fir (*Abies amabilis* Doug. ex Forbes) and Sitka spruce and took the form of marked reductions in growth. Fluorine damage in North America has also been reported at Columbia Falls, Montana (Carlson and Dewey, 1971; Carlson, 1972), the Dalles, Oregon (Compton *et al.*, 1961) and Georgetown Canyon, Idaho (Treshow *et al.*, 1967).

Similar damage was reported at Vettismorki in south-west Norway (Braanas, 1970), where both growth reductions (up to 72% in diameter increment) and mortality (ca. 15% of trees <60 cm diameter) of Scots pine and birch have occurred. Fluorine damage is not restricted to this site in Norway: Robak (1969) describes five Norwegian sites where severe damage to trees has occurred and further damage has been reported by Gilbert (1975). Fluorine damage has also been reported from Hungary (Keller, 1973) and Austria (Jung, 1968). In China, Ma Guangjing (1989) reported heavy mortality (46%) from hydrogen fluoride in a 1800 ha area of Masson pines (*Pinus massoniana* Lambert) around Chongquing, with the surviving trees being in very poor condition. Fluorine damage can usually be attributed to direct foliar injury, but secondary effects, such as the increased leaching from soils of organic carbon, iron and aluminium can be caused by fluorine (Häni, 1975; Polomski *et al.*, 1982a, b). In addition, there are many reports of more subtle effects, such as adverse changes on pollen germination and pollen tube growth (Facteau *et al.*, 1973) and cone and seed production (Sidhu and Staniforth, 1986).

In most of the above cases, the damage was accentuated by the location of the aluminium smelters. The process of aluminium production requires large amounts of electricity and smelters tend to be located close to sources of cheap electricity. The cheapest form of electricity is usually hydroelectric, but this requires a suitable topography for the generation of the head of water needed to drive the turbines. Consequently, the plants tend to be located in mountainous areas. These are also areas where marked temperature inversions occur. Inversions can prevent pollution from being dispersed, enabling the build-up of high concentrations of pollution. This is readily visible in the Rhône Valley in winter, when visitors to ski resorts are frequently above a marked zone of haze.

Elsewhere, fluorine pollution is associated with industrial complexes also producing a variety of other pollutants. For example, Kreutzer *et al.* (1983) described effects around a complex near Nuremburg and Huttunen (1973) found various effects around Oulu in Finland. Huttunen showed that damage was greatest in the winter months and she attributed this to the stress that the trees were already under. The study by Kreutzer *et al.* (1983) is difficult to interpret because of the large numbers of toxic chemicals involved. Król (1983) recorded damage in an area surrounding the Tarnobrzeg sulphur complex in south-east Poland, which was emitting a combination of sulphur dioxide, fluorine and dust. The study looked at the insect populations in the area, finding unusually high levels of *Xiphydria longicollis*, suggesting that fluorine may also have indirect effects on trees.

Conclusion

Several different types of decline have been described in this chapter. Clearly, various factors can cause the decline of a stand of trees or a forest. The natural declines described here can mostly be attributed to a complex suite of factors that operate together to weaken and eventually kill the trees. In cases where the decline takes the form of a progressive loss of basal area for a given species through time, stand dynamics may play an important part. Most forests should be viewed as highly dynamic ecosystems, with a range of successional species all vying for space. Some species are more successful at colonising at an early stage, and these species may also take advantage of small gaps. Other species are better adapted to colonizing and regenerating under an established forest cover.

Acute levels of pollution tend to directly affect trees and frequently appear to influence healthy trees as much as trees already under stress. While air pollution can be confidently implicated in the decline of forests at some sites, there are also many examples of declines in pristine areas where the symptoms cannot be attributed to pollution. Some of the most difficult cases in cause–effect studies lie between these two extremes, and these are examined in the following two chapters.

2 Forest Decline and Air Pollution in North and Central America

> Repetition of experiments is a key element in establishing scientific truth, and this must eventually be addressed in all fields of science including forestry.
> *Schreuder and Thomas, 1991*

Regional air pollution

There are relatively few cases of well-documented regional declines in forest health that can be directly attributed to the effects of air pollution. One reason for this is the complexity of the interrelationships between air pollution and forest health (e.g. Cowling, 1989). Air pollution rarely causes injury to trees on a regional scale, its effects over large areas usually being apparent through secondary processes (e.g. Alstad *et al.*, 1982). The possibility that air pollution might be affecting trees on a regional scale in North America has been recognized for some time. In some cases described in the previous chapter, point sources (e.g. Sudbury and Copper Hills) have produced so much pollution that regional-scale effects have occurred. In other cases, the effects are less apparent, but are spread over a much wider area. Two examples, described below, are the decline of trees around Mexico City and the decline of conifers inland of the Los Angeles conurbation.

Desierto de los Leones, Mexico

The metropolitan district of Mexico City covers an area of 800 km^2 in a basin surrounded by mountains about 600–800 m above the valley floor (Fig. 2.1). Within the city, de Bauer and Krupa (1990) found extensive foliar injury to blue gum (*Eucalyptus globulus* Labillardère). This was probably the result of acute injury by high concentrations of pollution. To the south and south-west of the city, the dominant direction of the wind, chlorotic banding and mottling, premature needle loss and branch mortality were recorded on Hartweg's pine (*Pinus hartwegii* Lindl.), Montezuma pine (*Pinus montezumae* Lamb.) and smooth-leaved pine

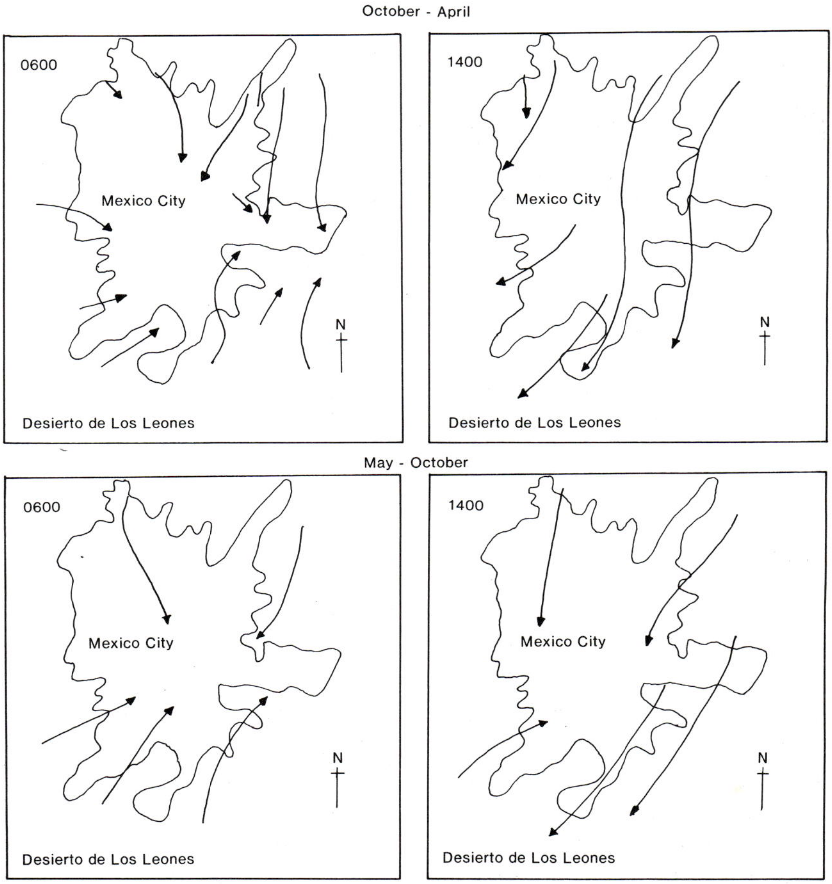

Figure 2.1 Diurnal and seasonal variations in the wind flow over Mexico City. From: Tovar (1989).

(*Pinus leiophylla* Schlecht. & Cham.) (Hernandez Tejeda *et al.*, 1982). The symptoms were more similar to ozone damage than damage by either sulphur dioxide or hydrogen fluorides.

One of the most affected species is the oyamel (*Abies religiosa* Schl. et Cham.). It is found at altitudes above 2700 m on the northern and north-eastern slopes of mountains to the south of the Valley of Mexico. Decline in this species was first recognized in 1981 and, since then, hundreds of thousands of trees have died (Ciesla and Macias Samano, 1987; Tovar, 1989). For example, in the Desierto de los Leones alone, there was 278 000 m^3 dead wood present in 1987. The decline initially took the form of small groups of trees dying (Tovar, 1989). These groups gradually expanded until, in 1983, there was 103 000 m^3 standing dead timber. Trees died from the base upwards, with the upper whorls being the last to die. During the decline of the trees, needle retention generally fell. Needles developed white spots which later turned brown and then coalesced, resulting in uniform brown discoloration on the upper sides of needles (Alvarado Rosales *et al.*, 1985). As in many other instances, a pest (in this case the bark beetle *Pseudohylesinus variegatus*) was ultimately responsible for the death of the trees (Tovar, 1989).

The decline of trees near Mexico City has important implications for many other expanding cities. Problems associated with photochemical oxidants are also apparent around Santiago in Chile and several other South American cities. Traffic controls are already being introduced in many of these locations. However, the strict emission controls in, for example, California, are a luxury that developing countries can ill afford. Consequently, further reports of forest decline around certain cities should be anticipated.

Decline of trees in western north America

Some of the best-documented examples of air pollution damage on a regional scale come from California and adjacent states. Here, the combination of a large urban population and meteorological conditions and relief that favours the formation of an inversion layer has resulted in the frequent occurrence of photochemical smogs. The persistent temperature inversions play a crucial part in the build-up of photochemical smogs on the west coast of the United States. Inversion conditions exist for 83–6% of the time in coastal California, but their frequency declines northwards, to a value of 30% in coastal Washington (Holzwirth and Fischer, 1979).

During summer, the average daylight concentration of ozone in urban areas is 40–60 ppb (Miller and McBride, 1989), with 7-hour concentrations regularly exceeding 100 ppb. Higher concentrations have been recorded in the Riverside area, and maximum hourly concentrations reach 2–300 ppb. The city pollution is frequently drawn inland because of thermal low pressure created by upward moving hot air in the desert interiors. Daytime up-slope winds subsequently transport the ozone into forested areas in the mountain ranges (Miller *et al.*, 1972; Pedersen and Cahill, 1989) (Fig. 2.2). Consequently, pollutant concentrations in areas such as the San Bernardino Mountains, the San Gabriel Mountains and the Santa Rosa Mountains of southern California (Fig. 2.3) can be relatively high. Outside these particular areas, concentrations of ozone can also be high, with similar concentrations occurring in the south of the Sierra Nevada. Reports from forest areas in Utah, Arizona, the northern Front Range of the Colorado Rocky Mountains and the western slopes of the northern Cascades (Böhm, 1989) reveal similar maximum hourly average concentrations of ozone, although cumulative exposures are not nearly as high. High peak concentrations (maximum hourly concentrations in excess of 120 ppb)

Figure 2.2 Poor valley air quality in the San Bernadino Mountains (upper) and Palomar Mountain (lower), California.

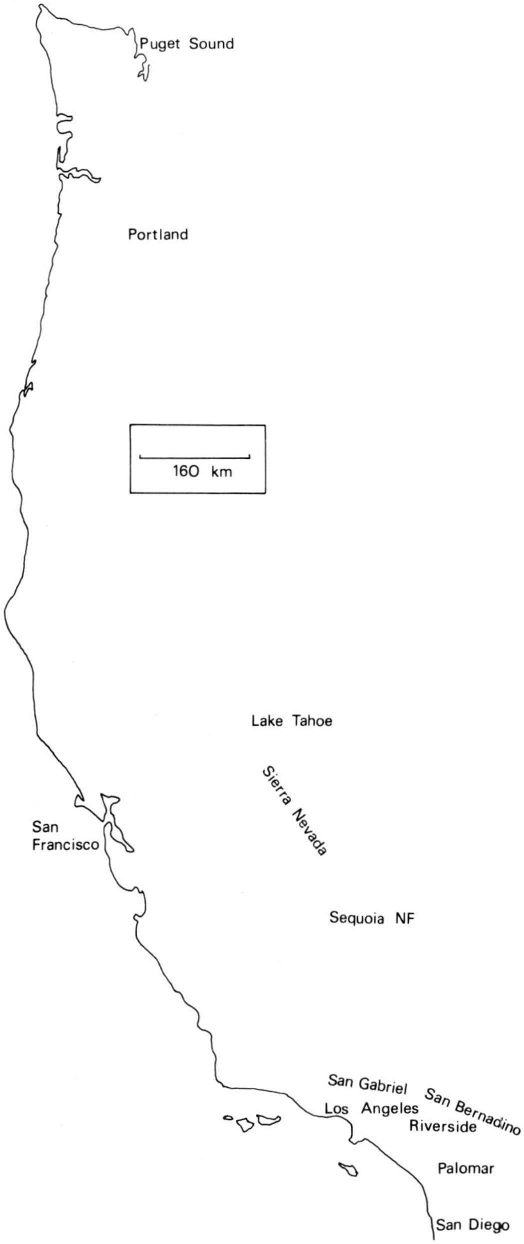

Figure 2.3 Map of the western seaboard of the United States, showing some of the locations mentioned in the text.

have also been reported in the Puget Sound region (Edmonds, 1988; Böhm and Vandetta, 1989), although this can be attributed to local sources of pollution. In pristine areas, such as the Olympic National Park in Washington State, ozone concentrations above 60 ppb have not been recorded (Böhm, 1989).

There are considerable spatial and temporal variations in the concentrations of ozone and other pollutants in forest areas in the west. Ozone concentrations in the San Bernardino National Forest are greater in the west and north than in the south and east (Miller *et al.*, 1986). During daylight hours, there is little variation in ozone concentration with altitude in the San Bernardinos, but night-time concentrations are considerably higher at mountain locations than at low-lying sites. Maximum hourly average concentrations in the period 1974–78 in the San Bernardinos were 200–40 ppb, with a maximum 24-hour average ranging between 90–140 ppb (Miller *et al.*, 1986). Similar variations in ozone concentrations have been reported by Böhm and Vandetta (1989). In the Sierra Nevada range, ozone concentrations are highest in the southern portion of the range. There is a gradual decrease in concentrations northwards (Reams *et al.*, 1990), although this pattern is disrupted by local incursions up valleys (Vogler, 1982).

Symptoms

Damage to several species, notably Jeffrey pine (*Pinus jeffreyi* Murr.) and ponderosa pine (*Pinus ponderosa* Dougl.) has occurred. It was first documented in the late 1950s and early 1960s in the coniferous forest zone of the San Bernardino Mountains, inland from Los Angeles (Parmeter *et al.*, 1962; Miller *et al.*, 1963; Miller and Parmeter, 1967; Evans and Miller, 1972), although pollution damage to ponderosa pine had also been recognized elsewhere (Adams *et al.*, 1952). Damage is characterized by chlorotic mottling, tip necrosis, reduced needle retention, a tufted appearance to the crown, reduced needle length and needle density, increased branch mortality in the lower crown and reduced growth (Fig. 2.4). Damage has since been reported further to the north in central California along a 250 km portion of the southern Sierra Nevada (Miller and Millecan, 1971; Pronos *et al.*, 1978; Pronos and Vogler, 1981; Allison, 1982, 1984a; Warner *et al.*, 1982; Duriscoe and Stolte, 1989; Mattson *et al.*, 1990) and around Lake Tahoe (Allison, 1984b; Pedersen, 1989). The damage at Lake Tahoe requires confirmation, as the area is not one in which ozone damage would be expected. The furthest inland that damage has been reported is on pines in the Rincon Mountains of the Saguaro National Monument, near Tucson, Arizona (Miller and McBride, 1989).

Affected trees in the San Bernardino Mountains were killed by the western pine bark beetle (*Dendroctonus brevicomis* LeCont.) (Stark *et al.*, 1968; Miller, 1983) or fungal problems, such as root rot and butt rot

Figure 2.4 Damaged ponderosa pine in the San Bernadino Mountains. The trees have a tufted appearance caused by only one year of needles being present.

(James *et al.*, 1980). In western forests, the damage to ponderosa pine was greatest at elevations of 1200–1800 m along west-facing ridges. This is believed to represent the altitudinal range at which the pollution trapped in the inversion layer reaches the mountains. Duriscoe and Stolte (1989) found a trend for decreasing chlorotic mottling with altitude (over a range of 1700–2700 m) in a survey of Jeffrey and ponderosa pines in Sequoia, Kings Canyon and Yosemite National Parks and a similar trend was reported further north, at Lake Tahoe (Pedersen, 1989).

The damage has been attributed to injury by photochemical oxidants. Histologically, the damage can be distinguished by experienced observers from the effects of a variety of other stresses, including winter desiccation, winter freezing (which can cause mottling of the upper surfaces of needles), drought and salt (Carlson and Gilligan, 1983; Kärenlampi, 1986). Other symptoms, such as basal needle necrosis, have sometimes been attributed to air pollution, but satisfactory cause and effect sequences have yet to be shown (c.f. Edmunds and Alstad, 1985).

Experimental studies

The most sensitive tree species in the region appears to be ponderosa pine and it is with this species that the greatest number of experiments has been undertaken. Chlorotic mottling is believed to be the symptom most representative of ozone damage, and it has been reproduced by exposing seedlings to above-ambient and, more recently, ambient ozone concentrations (Miller *et al.*, 1963; Richards *et al.*, 1968). Evans and Miller (1972) studied the histological and histochemical effects of acute exposures of ozone on ponderosa pine. They fumigated sensitive clones with 450 ppb ozone for twelve hours per day (representing an exposure much greater than ambient). Within five days, chloroplasts and carbohydrate stain accumulated in the peripheral portions of mesophyll cells. Simultaneously, the normally even distribution of proteins and nucleic acids was disrupted. Cell wall destruction occurred in mesophyll cells after significant intercellular damage. Visible symptoms were not apparent in the seedlings until two to three weeks after the start of the fumigation. Carlson and Gilligan (1983) found that hypertrophy and hyperplasma of the vascular parenchyma, endodermis collapse and intense vascular staining were all characteristic of fluoride injury in both ponderosa pine and Douglas fir.

Detailed histological investigations were also made by Kärenlampi (1986) and Kärenlampi and Houpis (1986), following fumigation with ozone and sulphur dioxide, respectively. In the ozone study, Kärenlampi (1986) studied needles from sensitive and tolerant genotypes of ponderosa pine in the Sequoia National Park and San Bernardino Mountains. In the mesophyll cells of apparently healthy needles, the central vacuole was filled with a fine granular material, the chloroplasts had large starch grains, the nuclei with nucleoli were clearly discernible and only a few liquid droplets could be seen. The mesophyll of chlorotic two-year-old needles had degenerated cells with strong accumulations of electron-dense droplets on the tonoplast and poor resolution of the chloroplasts, other organelles and the cytoplasm. Under normal conditions, needle senescence is characterized by the contents of the central vacuole becoming very coarse and accumulating near the tonoplast, and large bodies of lipid-like material appear. Chloroplasts of senescent needles may still have starch grains present. The cells of sensitive and strongly chlorotic needles did not have large lipid bodies, in contrast to healthy needles. Because of these observations, Kärenlampi (1986) suggested that some features of the chlorotic needles were attributable to the slow death of malfunctioning tissue rather than the direct effects of ozone.

High concentrations of ozone can cause changes in the root biomass of ponderosa pine seedlings (Andersen and Hogsett, 1991). Ozone significantly reduced the starch reserves available in roots immediately before budburst the following spring and reduced spring root growth. In the experiment by Andersen and Hogsett (1991), effects occurred in treatments where the exposure regimes reflected the air quality found in the western USA. Effects started after a dose of 122 ppm-hours, calculated by summing hourly means over the exposure period.

Ponderosa pine seedlings fumigated during the day with sulphur dioxide (hourly average concentrations of 75 or 150 ppb) for 26 months showed very different patterns (Kärenlampi and Houpis, 1986). Visible injury took the form of brown tips to the needles. As this was also recorded on the controls, the injury may have been due to the very warm growing conditions during the experiment. Some chloroplasts of the control needles (0 ppb SO_2 exposure) had thick grana and some had starch grains. No large droplets of lipid-like material were present and the condition of the central vacuole was very variable. At 75 ppb SO_2, the brown tips to the needles were longer and a bleached area was present behind the browning. The bleached areas frequently showed large droplets of lipid-like material. However, the amount of starch and the proportion of depressed cells were similar to the control needles. At 150 ppb SO_2, the brown tips were even longer and the remainder of the needle was yellowish. The chloroplasts of the yellow tissue were rounded and had no starch present. The thylakoids were commonly swollen, particularly the stromal thylakoids in the marginal part of the chloroplast. Vacuoles contained dark deposits and large droplets of lipid-like material.

In contrast to the above experiments based on fumigation, Bytnerowicz *et al.* (1990) exposed two-year-old ponderosa pine seedlings to ambient and charcoal-filtered air in the San Gabriel Mountains over a single growing season. There was no significant effect of air pollution on the development of injury, stem growth or the concentrations of pigments in the foliage. These results contrast with exposure studies which have induced injury at higher-than-ambient ozone concentrations (e.g. Miller *et al.*, 1969; Taylor *et al.*, 1986), but agree with earlier studies with ambient air made by Bytnerowicz *et al.* (1989).

Foliar injury and growth may be related through the effects of ozone on specific leaf weight (Coyne and Bingham, 1982; Ewell *et al.*, 1990). The study by Ewell *et al.* (1990) was based on samples of ponderosa pine foliage obtained from the Sequoia National Park. Trees with the greatest amount of foliar injury (mottling present on third year or younger needles) had the lowest specific leaf weights and the lowest number of needle whorls per branch. However, average needle weight and average needle area showed no relationship with the degree of foliar injury. Similar results were obtained by Coyne and Bingham (1982). The relationship between specific leaf weight (SLW) and foliar injury suggested that the photosynthetic capacity of injured trees was reduced, since SLW and photosynthesis are related, at least in Norway spruce (Oren *et al.*, 1986; Oren and Zimmermann, 1989).

Most work has concentrated on ponderosa pine, which is generally considered the Californian pine species that is most sensitive to ozone. Foliar injury, however, also occurs on Jeffrey pine (Pronos *et al.*, 1978; Williams, 1980; Wallner and Fong, 1982; Warner *et al.*, 1982; Duriscoe, 1987; Patterson and Rundel, 1989). Experimental fumigation studies have suggested that Jeffrey pine is sensitive to ozone (Evans and Miller, 1972), but not quite to the same extent as ponderosa pine, although it is commonly found in the field with symptoms of ozone injury. Field measurements of photosynthetic activity of trees growing in the Sequoia National Park showed that activity was greatest in the mornings and in the early summer months (June and July) (Patterson and Rundel, 1989). Photosynthetic rates varied between trees showing high levels and low levels of ozone injury, with tolerant trees showing lower stomatal conductances and lower maximum photosynthetic rates in one year compared with more sensitive trees. The reverse occurred in the second year of measurements as the foliage of the sensitive tree became mottled and photosynthesis decreased. Sensitive trees had a 25% slower initial increase in photosynthesis in response to increasing intercellular CO_2 concentrations (Fig. 2.5), which Patterson and Rundel (1989) attributed to a lower carboxylation efficiency. Carbon dioxide saturated photosynthetic rates were also lower on sensitive trees. Water potentials did not differ between resistant and sensitive trees. All ages of needles had a positive carbon budget, although sensitive trees with fewer needle whorls produced less carbon than more tolerant trees.

The interactive effects of ozone and other sources of stress have been investigated in some detail in California. Experimental work (Fenn *et al.*, 1990) with ponderosa pine and ozone found that fumigation with realistic (near ambient) concentrations of ozone predisposed trees to increased infection by black stain root disease (caused by the fungus *Leptographium wageneri* var. *ponderosum* Harrington and Cobb), although this pathogen is absent in the main areas affected by elevated ozone concentrations. Ozone also predisposed pines (Jeffrey and ponderosa) to other fungal pathogens, including *Heterobasidion annosum* (James *et al.*, 1980). Similar effects are assumed to occur in eastern white pine, as increased levels of infection by *Leptographium procerum* have been seen in pollution-

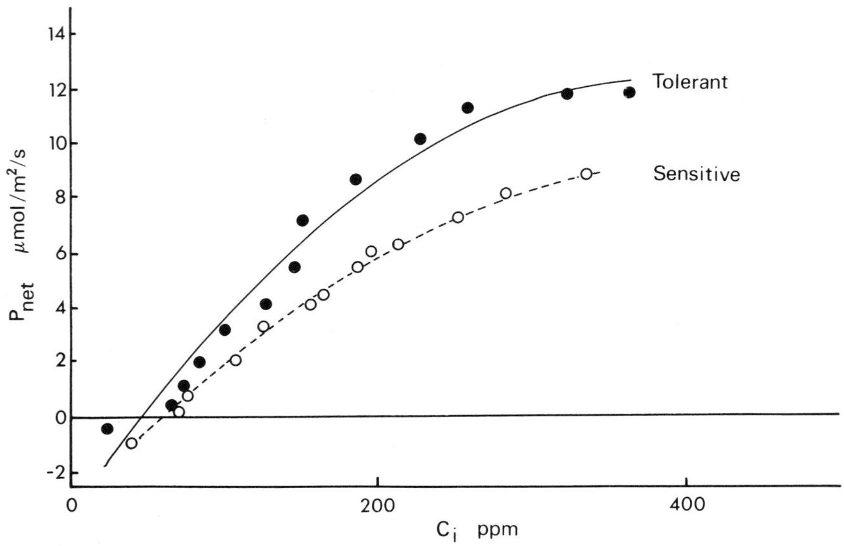

Figure 2.5 Photosynthetic assimilation (P_{net}) responses of Jeffrey pine to internal carbon dioxide levels (c_i). Sensitive individuals had approximately 20% chlorotic mottle over their needle surface area. From: Patterson and Rundel (1989).

sensitive trees (Lackner and Alexander, 1983), although Leininger *et al.* (1990) found no association between presumed ozone injury and the incidence of *Heterobasidion annosum.* In contrast to studies elsewhere (e.g. Keller and Matyssek, 1990), nutrient stress and exposure to ambient ozone do not appear to interact with ponderosa and Jeffrey pine (Bytnerowicz and Takemoto, 1990; Bytnerowicz *et al.*, 1990).

There has been relatively little interest in the possible effects of acidic deposition in the western United States. McColl and Firestone (1987) conducted experiments with simulated acidic rain on seedlings of Douglas fir (*Pseudotsuga menziesii* (Mirb.) Franco), sugar pine (*Pinus lambertiana* Dougl.) and ponderosa pine growing in two different forest soils. The effects differed between the two soil types used and were usually only apparent in the most acidic treatment (pH 3.0). Few or no visual symptoms were noted, nor were there any apparent effects on seedling growth or nitrogen concentration in the shoots. Changes in soil chemistry were seen, with decreases in exchangeable cations and soil pH, and increases in soil aluminium, manganese, iron and zinc. The results of this study are, however, difficult to apply to forests. The mean acidity of wet deposition in the Sierra Nevada and coastal areas of northern California is about pH 5.0 and minimum values rarely fall below pH 4.5 (McColl *et al.*, 1982; Bicknell, 1989). However, these figures do not consider cloud-water deposition, which may form a very important source of moisture for trees during growing season droughts. For example, moisture input under forest canopy in Oregon may be 25–40% greater than in the open (Isaac, 1965; Harr, 1982) and may equal rainfall in the coastal redwood (*Sequoia sempervirens* (D.Don) End.) and Douglas fir forests of California (Oberlander, 1956; Azevedo and Morgan, 1974). The latter difference reflects the low rainfall in California rather than an increased frequency of cloud immersion (Muir and Böhm, 1989). Basabe *et al.* (1989) recorded cloudwater pHs of less than 4.0 in western Washington and pHs as low as 2.5 have been reported from the Los Angeles Basin (Brewer *et al.*, 1983; Richards, 1989) and elsewhere in southern California (Muir and Böhm, 1989), suggesting that on occasion, trees are exposed to highly acidic moisture.

Hogsett *et al.* (1990) used a series of different exposure treatments to look at the interactions of sulphur dioxide, acidic fog and ozone on seedlings of five western species, namely Douglas fir, ponderosa pine, lodgepole pine, western hemlock and western red cedar (*Thuja plicata* D.Don). They had mixed results, with significant differences usually consisting of growth reductions when compared to controls. However, growth stimulations also occurred, particularly in sulphur dioxide (variable concentrations) and acidic fog (pH 2.1 and 3.1) treatments. Ponderosa pine and western hemlock were the most responsive to ozone-acidic fog treatments, and lodgepole pine showed the least response. Western red cedar showed increased growth in response to this combination. Winter-time sulphur dioxide fumigation sometimes caused an increase in spring bud elongation, whereas the ozone-acidic fog treatment reduced it. Their study was particularly valuable because of their attempts to simulate natural exposure regimes.

Increment studies

Peterson and co-workers (Peterson and Arbaugh, 1988a, 1988b; Peterson *et al.*, 1987, 1989, 1991) have investigated the relationships between visible injury and radial growth in Jeffrey and ponderosa pine. One of the first studies by this group, and the first study to show a clear reduction in the growth of forest trees associated with ozone injury, involved the relationship between growth and ozone-symptom occurrence in Jeffrey pine growing in the Sierra Nevada (Peterson *et al.*, 1987). Growth was investigated at five symptomatic and three asymptomatic sites. Trees were rated for ozone injury by counting the mean number of needle years retained on branches and assessing the amount of chlorosis. Growth reductions occurred in symptomatic trees, with the greatest reductions being in larger trees and those more than 100-years-old. The growth reductions were widespread, with mean annual increment being about 11% lower in symptomatic trees than in asymptomatic trees.

In addition, the response of symptomatic and asymptomatic trees to annual climatic fluctuations differed from 1956 onward, with symptomatic trees becoming more responsive to climate. Despite compelling evidence (Peterson *et al.*, 1987) it was difficult to attribute the growth differences directly to ozone injury.

The work on Jeffrey pine was subsequently extended to ponderosa pine (Peterson and Arbaugh, 1988). Changes in ponderosa pine growth rates were assessed using the same techniques as for Jeffrey pine. In marked contrast to the Jeffrey pines, symptomatic and asymptomatic ponderosa pines showed no differences in growth, and there was no evidence for an increase in climatic sensitivity. Given that ponderosa pines are generally considered to be more sensitive to ozone than Jeffrey pines, these findings are of considerable interest. They suggest that the use of individual foliar symptoms (in this case, chlorotic mottle) to assess air pollution impact on forests cannot be extended to infer growth changes, something that has been widely done in Europe (see Chapter 10). However, the sample size was relatively limited and a larger sample, described below, revealed rather different results. One possible mechanism for the apparent discrepancy is the ecology of the two species. Jeffrey pine occurs at sites experiencing higher levels of environmental

Figure 2.6 Jeffrey pine at a typical site in the Sierra Nevada. This species occurs at higher altitudes than ponderosa pine and is also more prevalent on unfavourable sites.

stress than does ponderosa pine (Fig. 2.6). Consequently, where it grows on extreme sites, it may be less able to repair ozone injury than trees growing at more favourable sites.

Peterson *et al.* (1989, 1991) subsequently extended the work on Jeffrey and ponderosa pine in the Sierra Nevada to a much wider range of sites. Ponderosa pines from asymptomatic (Fig. 2.7) and symptomatic (Fig. 2.8) sites were sampled in a transect extending from Tahoe National Forest in the north to Sequoia National Forest in the south. Tree ring series from these sites revealed a highly dynamic situation, with both growth increases and growth decreases occurring. Growth decreases were widely recorded in the 1920s and increases in the 1930s. Since the 1950s, several stands at sites in the southern portion of the transect have shown growth decreases. The greatest reductions occurred at sites with the highest levels of needle injury and lowest levels of needle retention and the highest exposure to ozone. The authors suggested that these spatially limited growth decreases were probably due to ozone exposure. The results from the wider area were consistent with the earlier study by Peterson *et al.* (1988). Ozone injury was present throughout the western portion of the forests occurring in the Sierra Nevada, although its severity increased markedly towards the south. The work of Peterson and co-workers suggested that foliar injury was a more sensitive response to ozone than growth and that, as a result, single measures of foliar injury could not be used to infer growth effects. However, this conclusion may be reversed if better methods of characterizing foliar injury are used when examining growth losses.

Graybill and Rose (1989) also adopted a dendroecological approach. They looked at recent growth rates of ponderosa pine (*Pinus ponderosa* vars. *scopalorum* and *Arizonica*), pinyon pine (*Pinus edulis* Engelm.), south-western white pine (*Pinus strobiformis* Engelm., syn. *Pinus ayacahuite* Ehrenb. var. *brachyptera* Shaw), white fir (*Abies concolor* (Gord.) Hildebr.) and Douglas fir in Arizona and New Mexico. Their approach was to sample 15–25 dominant and codominant trees at each site. Each growth series was individually standardized, involving the supposed removal of any growth trend attributable to the ageing of the tree. Several models were used to do this, ranging from straight line fits to negative exponentials to cubic splines. While this process is standard in dendroclimatological applications, its application to the growth decline problem has been questioned by Innes and Cook (1989) as some growth trends of interest may be removed. Standardized growth series were subsequently averaged for a site, a process that can also lead to the loss of potentially valuable information (Van Deusen, 1991). A further manipulation adopted by Graybill and Rose (1989) was to remove any persistent autocorrelative structure in the series, using best-fit ARMA models. This also is a questionable procedure in view of the nature of the problem. The resulting series were then used to compare actual and predicted growth (Fig. 2.9) of the ponderosa pine and Douglas fir.

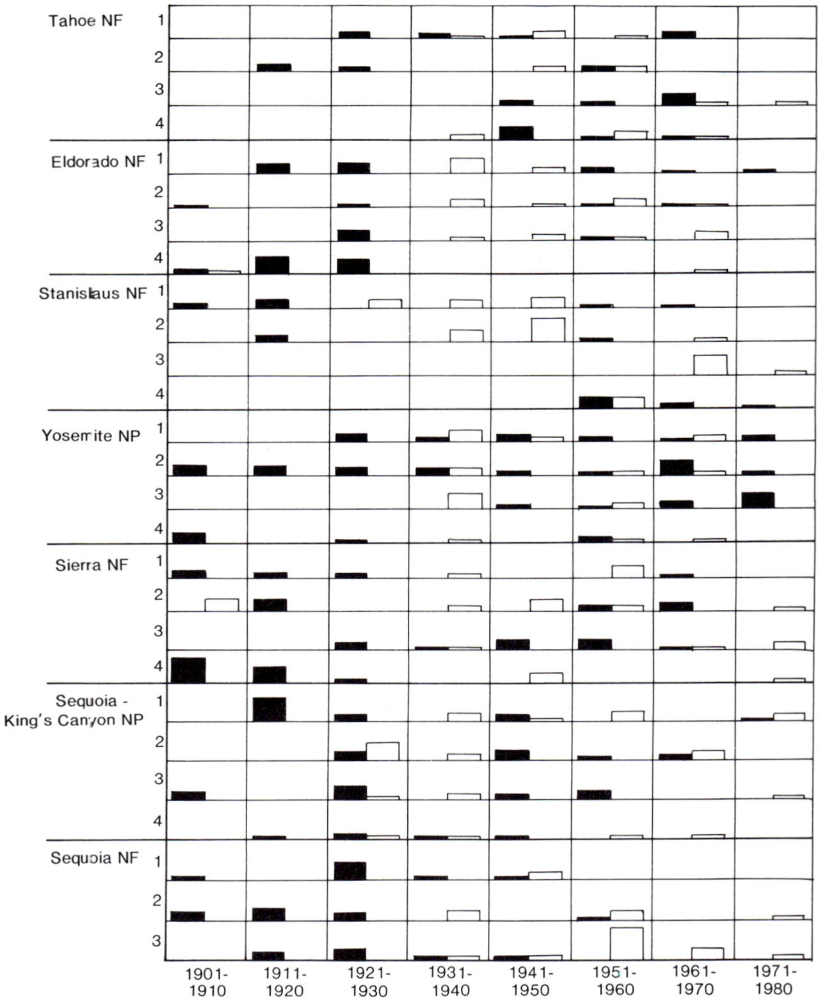

Figure 2.7 Growth decreases (solid bar) and increases (open bar) of ponderosa pine as determined by Kalman filter analysis, summarized by decade for asymptomatic sites. The percentage of trees with growth changes at each site is indicated, with the column height ranging from 0 to 60%. From: Peterson *et al.* (1991).

One of the most striking findings of Graybill and Rose (1989) was the large number of missing rings in the ponderosa pine series, particularly in southern Arizona. Up to 40 rings were missing from the period 1920–86, with the greatest frequency of missing rings occurring after 1950. An increase in the frequency of missing rings in the middle of the century was expected because of the general prevalence of drought conditions then.

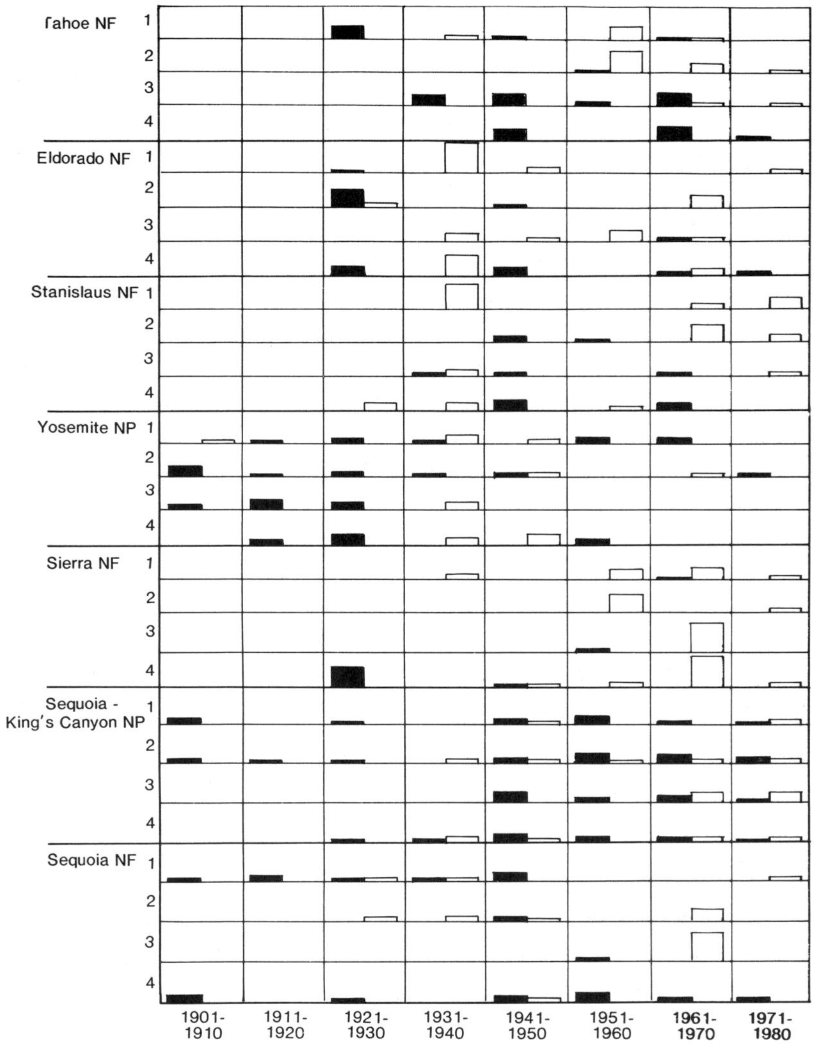

Figure 2.8 Growth decreases (solid bar) and increases (open bar) of Jeffrey pine as determined by Kalman filter analysis, summarized by decade for symptomatic sites. The percentage of trees with growth changes at each site is indicated, with the column height ranging from 0 to 60%. From: Peterson *et al.* (1991).

However the trees failed to show the growth increases seen in other species at the same site associated with the wetter conditions in the 1970s and 1980s (Rose *et al.*, 1982; Graybill, 1986). Growth anomalies were identified at several sites. Twenty-one of the 41 series had less than predicted growth in the 1951–86 period, with greater growth decreases

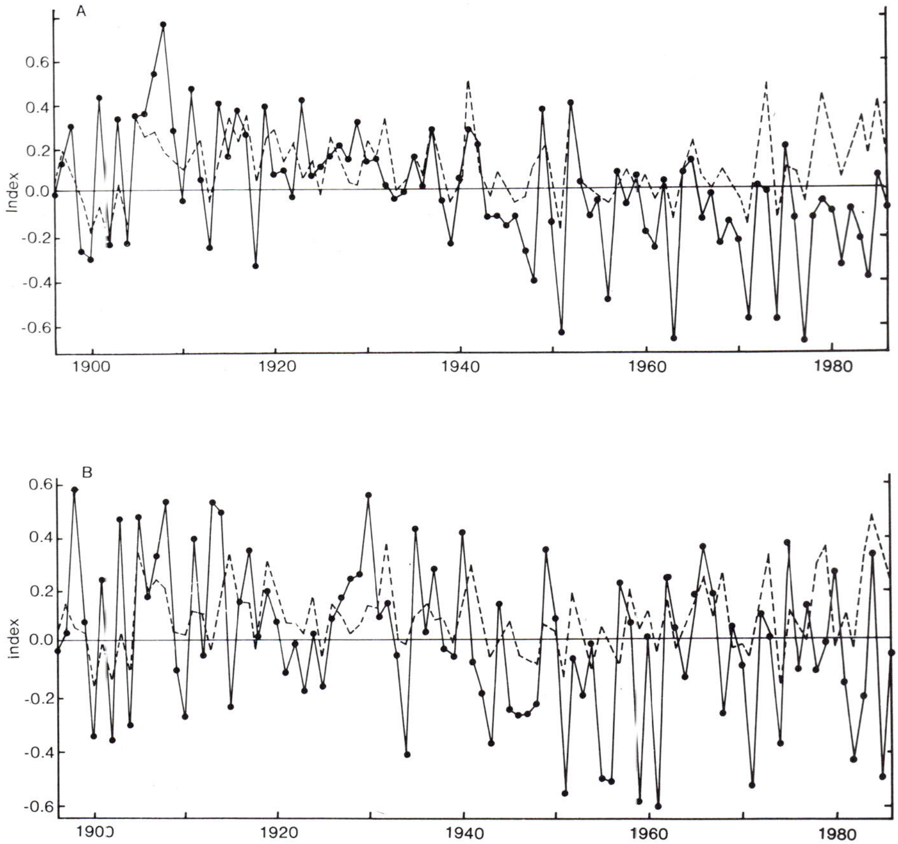

Figure 2.9 Actual and expected growth of ponderosa pine from Slate Mountain, central Arizona (a) and Douglas fir from the Santa Rita Mountains, southern Arizona (b). Actual growth is shown by the solid lines and predicted growth by the dashed lines. From: Graybill and Rose (1989).

being reported from the south of the study area. The authors went as far as to say that some ponderosa pine series showed growth patterns which suggested that mortality was imminent. Patterns of growth are however difficult to interpret because of climate-growth interactions. For example, Armstrong *et al.* (1988) found marked annual differences in the needle length of white pine (*Pinus albicaulis*) and lodgepole pine (*Pinus contorta* var. *murrayana* (Grev. and Balf.) Engelm.) which they attributed to climatic fluctuations associated with El Niño.

D'Arrigo and Jacoby (1991) derived a 1000-year record of tree-growth fluctuations for Douglas fir, ponderosa pine and pinyon pine from western New Mexico. As with the study by Armstrong *et al.* (1988), the primary aim of this work was to relate variations in tree dynamics to climatic fluctuations associated with the El Niño. However, the results may have considerable significance for air pollution studies in the American west. In their study area, D'Arrigo and Jacoby (1991) found that November-May precipitation was the most important climatic factor explaining annual fluctuations in growth. Their study showed that two of the most prolonged severe droughts in the 1000-year record occurred in the last 100 years, at the turn of the century and in the 1950s to early 1960s. Given that the latter period was when a marked increase in the number of missing rings occurred in the series studied by Graybill and Rose (1989), the possibility that the missing ring sequences were a function of drought effects rather than air pollution should be considered. Graybill and Rose (1989) recognized the effects of this drought, but contrasted the near negligible growth of the ponderosa pine and Douglas fir since the 1960s with the recovery of other species at the same sites after the drought. The cause of the growth suppression remains unknown.

Status and damage to other species

The extent of forest damage in the western part of the United States is relatively stable in some areas at present. In the San Bernardino Mountains, crown condition, as measured by a combined oxidant injury score, improved between 1974 and 1988 at nine out of 11 sites (Miller and McBride, 1989; Miller *et al.*, 1989). This may reflect a trend in hourly and monthly maximum ozone concentrations. These declined in the 1970s and have remained relatively constant since 1980, currently being about 38% below their 1974 values (Miller and McBride, 1989). Conversely, the incidence of pines with chlorotic mottling in the areas of the Sierra and Sequoia National Forests with the highest levels of injury increased from 47% in 1980–2 to 79% in 1984–5 (Duriscoe and Stolte, 1989). The studies in the San Bernardino Mountains suggested that an ozone concentration of 0.07–0.08 ppm, 24-hour average, May to October, was the minimum exposure required to maintain the decline in tree health (Miller *et al.*, 1989).

There has been increasing interest in the possibility that other western species have been adversely affected by air pollution. Douglas fir in the Seattle area is believed to experience ozone concentrations that could cause foliar injury. However, no visible injury has been reported to date (Edmonds, 1988) and experimental studies suggest that this species, together with Engelmann spruce (*Picea engelmannii* (Parry) Engelm.) is tolerant of exposures of up to 71 ppb (Miller et al., 1988). In contrast, white fir and subalpine fir (*Abies lasiocarpa* (Hook.) Nutt.) both showed increasing injury with increasing ozone levels (Miller *et al.*, 1988). In Canada, experimental studies with jack pine (*Pinus banksiana* Lamb.) and white spruce (*Picea glauca* (Moench) Voss) (Abouguendia and Baschak, 1987) have suggested that acidic rainfall could damage white spruce needles, although no significant effect on either growth or chlorophyll concentrations was noted. However, other stresses, particularly those associated with winter conditions, appear to have a much greater impact on forests than air pollution, with major canopy damage occurring in some areas every 9–16 years (e.g. Grier, 1988).

Studies have also been conducted with giant sequoia (*Sequoiadendron giganteum* (Lindl.) Buchholz) (Miller, 1988; Grulke *et al.*, 1990), as chlorotic mottling has been recorded in shaded seedlings in the field (Fig. 2.10). Visible injury has not been reported on mature individuals of this species, but this may be partly related to the difficulties of examining the foliage of 100 m tall trees. Experiments were conducted over several years. In one year, seedlings and rooted cuttings were exposed in open-top chambers for a single growing season to charcoal-filtered air, ambient air or ambient air with added ozone (+50% of ambient concentration). Considerable variation was found within treatments, complicating the interpretation of the results. Significant differences in the light compensation point, net assimilation at light saturation and dark respiration occurred, with higher light compensation points and reduced net assimilation in ozone-exposed treatments. Differences in response were found between seedlings grown in full sun and seedlings grown in shade, with more ozone injury occurring on seedlings grown in shade. The differences

Figure 2.10 Regeneration site for giant sequoia in the southern Sierra Nevada. Seedlings at sites like this are the most likely to show chlorotic mottling.

appeared to be related to the moisture of the soil, with seedlings growing in drier soils developing less injury.

California black oak (*Quercus kelloggii* Newb.) has also been examined following field studies that revealed foliar injury (Miller *et al.*, 1979). It was less sensitive to ozone than either ponderosa or Jeffrey pine. When injury occurred, it took the form of interveinal chlorosis, chlorotic mottle and necrosis and depressed radial increment. Injury has been seen in the San Bernardino Mountains and in the Sequoia-King's Canyon National Parks of the Sierra Nevada. Experimental studies with seedlings produced similar results to the work on giant sequoia, with seedlings that experienced less direct sunshine being more susceptible to ozone. Whether ozone injury is partly to blame for the poor regeneration of black oak and other species (white oak [*Quercus lobata* Née], live oak [*Quercus wislizenii* A. DC.], maul oak [*Quercus chrysolepis* Liebm.], and blue oak [*Quercus douglasii* Hook. & Arn.]) in the southern Sierra Nevada (White, 1966; Griffin, 1971, 1976, 1980; Standiford *et al.*, 1991) is unknown. Given the acknowledged failure to regenerate (e.g. Jepson, 1910; White, 1966), it is likely that other factors are at least involved. Low ground water levels have probably been a factor (Barnhardt *et al.*, 1991; Brown and Davis, 1991), as has exposure to direct sunlight (in the case of leather oak, *Quercus durata* Jeps.) (Williams *et al.*, 1991) and browsing damage (McBride, 1974; Griffin, 1980; Davis *et al.*, 1991).

While emphasis has been placed on montane forests, there is also the possibility that desert communities are being affected, particularly in the Mojave Desert, inland from the southern central Californian valley. However, Bytnerowicz *et al.* (1988) found no effect of near-ambient ozone on a range of desert annuals and only three species, *Camissonia claviformis* (Torr. & Frem.) Raven, *Camissonia hirtella* (Greene) Raven and *Erodium cicutarium*, showed injury at the highest levels of fumigation used (133 ppb). The levels of total leaf injury (percentage of leaves injured multiplied by the average injury to affected leaves) were relatively low, being about 1–2%. Similar results (injury at hourly average ozone concentrations of 100 ppb) for *Camissonia claviformis* and *Camissonia hirtella* were obtained by Thompson *et al.* (1984). Temple (1989) examined the effects of ozone on plants in the part of the Mojave Desert known as the Joshua Tree National Monument, another inland desert area. The greatest ozone concentrations occurred at night, when most plant species would be less sensitive to ozone. Although no damage was recorded in monitoring plots, injury occurred during fumigation experiments on *Rhus trilobata* Nutt, with very little damage seen on the Joshua tree (*Yucca brevifolia* Engelm.).

Conclusions

Damage to ponderosa pine, Jeffrey pine and other species in the San Bernardino Mountains seems to be a clear case of air pollution injury to a forest ecosystem. The damage has coincided both spatially and temporally with the occurrence of a pollutant that is known to cause the type of symptoms that have been observed in the forests. Trees stressed by foliar loss undergo a reduction in vigour, particularly as expressed by sapwood area. This, in turn, makes the trees more susceptible to secondary biotic agents, particularly the western pine bark beetle and various fungal diseases. However, other factors may also have been involved in some areas. The role of drought needs to be evaluated in greater detail, particularly in some declines reported from outside the San Bernardino Mountains. In addition, the apparent lack of effects reported by Bytnerowicz and co-workers in some recent exclusion experiments needs to be examined in detail, particularly in relation to the known reduction in ambient ozone concentrations in some areas.

The growth studies undertaken in the area appear to show that the foliar symptoms *per se* are not necessarily associated with growth declines. The techniques used to identify the growth declines were rather unusual and the data on pine growth collected from the Sierra Nevada by Peterson and co-workers warrant further examination. The work on the growth of red spruce in the east, described below, suggests that the adequate interpretation of tree-core data is difficult and requires several different approaches. This is clearly an area where further work could be done in the West. By looking at growth, it should be possible to spatially delimit the extent of any decline attributable to photochemical pollution and to assess its severity.

Eastern white pine

There has been considerable debate over the possibility of a regional decline in the health of eastern white pine (Fig. 2.11). The species is sensitive to sulphur dioxide, hydrogen fluoride and various oxidants (Berry, 1961, 1973; Dochinger and Seliskar, 1970; Stone and Skelly, 1974; Gerhold, 1977), although current laboratory studies no longer show it as extremely sensitive, a trend that may be due to natural selection (Bennett *et al.*, 1989; Berrang *et al.*, 1991). As with other species, several studies, such as those made by Dochinger and Seliskar (1965), Houston and Stairs (1973), Berry (1973), Genys and Heggestad (1978, 1983) and Anderson *et al.* (1988) have shown a high degree of variation in the susceptibility of particular individuals to different forms of pollution. It is therefore possible that the most sensitive genotypes have already been lost from the population (Houston, 1974; Genys and Heggestad, 1978, 1983; Thor and Gall, 1978; Kriebel and Leben, 1981; Mullins, 1983; Armentano and Menges, 1987; Karnosky, 1989; Karnosky *et al.*, 1989). Surprisingly, this line of argument does not appear to have been developed in the same way for other instances of purported decline in the United States.

Figure 2.11 Healthy eastern white pine growing near Lake Placid, New York.

The potential decline is important as eastern white pine is a valuable timber species, with a growing stock of 250 million cubic metres in the eastern United States (Wharton and Powell, 1985). However, past exploitation has depleted the resource, with current levels of timber production being about 10% of the peak at the turn of the century (NAPAP, 1990). Any potential decline should be seen against this background.

Symptoms

Several symptoms have been documented. Berry and Hepting (1964) described premature shedding of older needles, chlorosis followed by necrosis of the distal parts of current needles (known as tip-burn), stippling or mottling of the foliage, gradual onset of discoloration any time of the year and a gradual reduction in shoot and radial growth, culminating in the premature death of the trees. Thin crowns, short needle length and reduced needle retention have also been reported (Hayes and Skelly, 1977; Blanchard *et al.*, 1979; Usher and Williams, 1982).

Concern exists over the symptomatological aspects of foliar injury, since tip-burn is not unique to air pollution. Chlorotic mottling is generally believed to be due to air pollution, but ice and insect damage can cause very similar symptoms. Bennett *et al.* (1989) reviewed 61 surveys of foliar injury and found only five that had sample sizes sufficiently large to be reliable. Two patterns of injury were found: low numbers of trees with high severity of injury and high numbers of trees with low severity of injury. The absence of records combining high severity with high incidence suggested that no widespread decline was occurring and there have been no observations of mortality associated with mottling. In addition, there are records of recovery of mottled trees. Consequently, Bennett *et al.* (1989) considered that eastern white pine was not a proven case of forest decline. However, as already suggested, this may have been because susceptible individuals have already been removed from the population, resulting in selection for air pollution tolerance.

Increment studies

Growth analyses have shown that the annual radial increment of sensitive individuals has declined since about 1955 (Benoit *et al.*, 1982), although McLaughlin *et al.* (1982) dated the onset of reduced growth to much later, about 1965. However, this latter study only involved nine trees, so the significance of the results is very questionable. Growth losses were usually higher in trees showing symptoms of damage than in unaffected trees (Benoit *et al.*, 1982) and Anderson *et al.* (1988) reported that symptomatic trees had 49% less volume than healthy trees. Around point sources of sulphur dioxide, reductions in radial growth were clearly associated with levels of pollution (Stone and Skelly, 1974; Phillips *et al.*, 1977), but regional trends were more difficult to identify. Although most work has concentrated on visible injury to foliage, the premature shedding of older needles may also be significant. Kuser *et al.* (1989) reported that trees that retained their older needles grew faster, possibly because early season photosynthesis was enhanced and early experimental work by Mann *et al.* (1980) also suggested such a link. This is a phenomenon that requires further investigation.

Field studies have not supported the work comparing the growth rates of symptomatic and asymptomatic trees. For example, Hornbeck *et al.* (1988) reported continuous increases in average annual basal area growth of eastern white pines in New England over the period 1950 to 1980 (Fig. 2.12). One possibility is that the effects on susceptible trees may have been masked by a much larger population of insensitive trees (Hornbeck *et al.*, 1988), and the relatively young age of the trees used by Hornbeck and his co-workers should also be considered. Another possibility is that much of the New England area does not experience the concentrations of ozone necessary to induce damage.

Experimental studies

Under controlled laboratory conditions, ozone fumigation of eastern white pine led to the development of tip necrosis, foliar chlorosis, pigmented

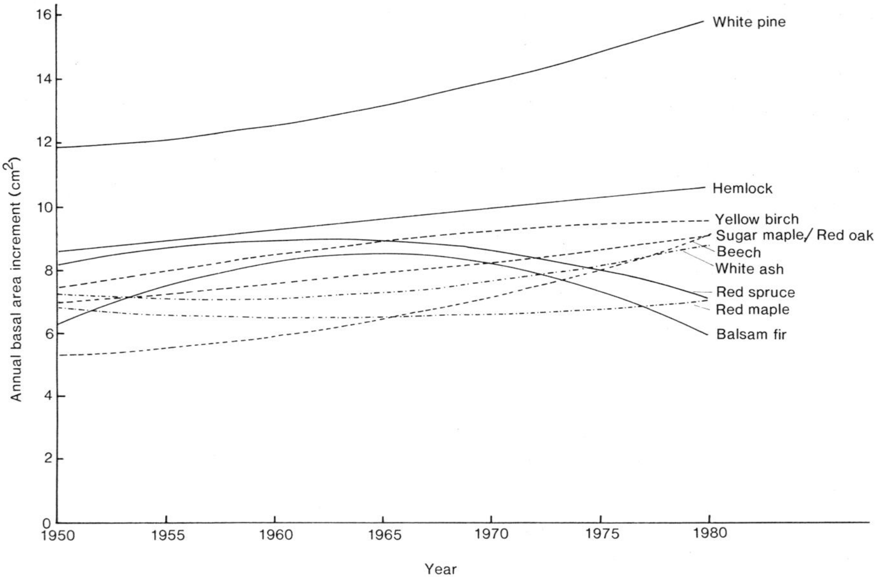

Figure 2.12 Regional growth curves for a number of forest species present in New England. From: Hornbeck *et al.* (1988).

mottling and necrotic banding (Houston, 1974; Yang *et al.*, 1982), with the most consistent response to fumigation being a reduction in net photosynthesis (Barnes, 1972; Botkin *et al.*, 1972; Yang *et al.*, 1983; Boyer *et al.*, 1986). Over short-term (four hours) exposures, the threshold ozone concentration for inhibition of photosynthesis occurred at about hourly average concentrations of 0.5 $\mu l\ l^{-1}$, whereas longer-term exposures (7–136 days) of only 0.05–0.10 $\mu l\ l^{-1}$ were found to depress the uptake of carbon dioxide. However, the resulting doses were generally well above the exposures that would occur in the field, and experiments using realistic doses have not produced visible injury (Reich *et al.*, 1987; Eberhardt *et al.*, 1988). In the field, needle injury has been associated with repeated exposure to ozone concentrations equal to or exceeding 0.8 μl/l (Skelly *et al.*, 1983).

Although most studies have attributed the foliar damage and growth reductions to ozone, other pollutants may also be involved. In parts of its United States range, eastern white pine is exposed to acidic deposition, particularly as acidic fog. Maurice and Crang (1986) looked at the impact of near-ambient misting with solutions of sulphuric and nitric acids (2 : 1 ratios, respectively) on the needles of eastern white pine seedlings. Although no changes in the surface or internal structural characteristics were seen, increases in needle widths were noted in the more acidic treatments. This trend was attributed to the nutritional impact of the mists.

Nightly misting for six weeks produced no foliar injury, even at pHs as low as 2.6. Longer term (24 weeks) misting produced foliar necrosis in all treatments below pH 5.6 (Maurice and Crang, 1989). Over the longer period, growth and needle morphology were unaffected. These results contrast with those obtained by Wood and Bormann (1977) who found increased growth of seedlings treated with acid solutions, but no foliar damage at pHs above 2.3.

The results of many of these experiments are difficult to extend to field situations in view of the strictly controlled environmental conditions. However, in a field-based study, Lovett and Hubbell (1991) exposed the upper canopy branches of three mature trees to various combinations of ozone (25, 70 and 140 ppb) and acidic mist (pH 5 or 3.8). The ozone treatments did not affect foliar leaching, nor were visible symptoms recorded, but acidic mist almost doubled the rate of leaching of calcium and magnesium ions compared with the pH 5.0 treatment. Hydrogen ions were taken up by the foliage in the pH 3.8 treatment, but released at pH 5.0. Visible injury was not recorded in the mist treatments.

Conclusions

The recognition and explanation of the decline of eastern white pine in the north-east is extremely complex. Some earlier studies purporting to show a decline were clearly based on inadequate or inappropriate sampling designs. Consequently, more recent information has to be used, but this is biased by the possibility that significant losses of more sensitive genotypes have occurred. Currently, studies of the basal area dynamics of the whole New England population show that the basal area of eastern white pine is increasing. This is not consistent with the hypothesis of a widespread decline. Detailed symptomatological studies have revealed that high incidence of symptoms is not associated with high severity, again suggesting that a widespread decline is not currently occurring.

Experimental studies, together with investigations of the genetic composition of the remaining population suggest that a decline in more sensitive individuals may have occurred in the recent past. Whether these genotypes have been totally eliminated is unclear; they may survive in relatively pristine areas. The loss of individuals sensitive to a particular type of stress from one area is part of the normal process of plant population dynamics. In eastern white pine, it appears that air pollution may have contributed to this process. However, there is no evidence of a current widespread decline of the total population of eastern white pine that can be attributed to air pollution.

The Forest Response Programme

As already indicated, several different types of forest decline are believed to be present in North America. To some, the presence of widespread effects of air pollution has been clear for some time. For example, the American Forestry Association felt in a position to state categorically that further controls of sulphur, oxides of nitrogen, hydrocarbons, ozone and heavy metals were required to protect forest ecosystems (American Forestry Association, 1987), a position that was justified by demonstrable effects of atmospheric pollution in some forest areas. Concern about the effects of air pollution led to the creation of the Forest Response Programme (FRP) under the auspices of the National Acid Precipitation Assessment Programme (NAPAP). The FRP recognized four major forest regions or types and investigated problems in each. The four groups were: spruce-fir forests, western conifers, eastern hardwoods and southern commercial pines. A large amount of work has been conducted on the declines and some results are described here. As a documented case of decline attributable to air pollution, the results from the western conifers region have been described earlier in this chapter.

Eastern hardwoods

Of all the purported recent declines described from North America, the decline of hardwoods in the north-east had the greatest similarity to the reported declines of forest health in Europe. The same level of concern was expressed in both Europe and North America, yet the underlying data were not as concrete as some might suppose. There was clearly a decline in the crown condition of some species, particularly sugar maple (*Acer saccharum* Marsh.), in some areas in the 1980s. However, similar declines have occurred in the past. Connections with management practices (particularly the use of sugar tapping in North American sugar maples) were vigorously denied, as in Europe. Again, as in Europe, there was a strong political element to the controversy. Canadian maple producers were quick to place the blame on long-range transboundary air pollution, particularly from the United States. The following section reviews some of the controversy surrounding the purported decline of sugar maple and other species.

Past declines

There are several reviews of the historical occurrence of declines in north-eastern trees (e.g. Millers *et al.*, 1989; Auclair, 1989). In Canada, three major dieback episodes have been recognized (Auclair, 1989), involving black ash (*Fraxinus nigra* Marsh.) and white ash (*Fraxinus americana* L.) in the 1920s, birch in 1937–57 and sugar maple since the early 1970s. The decline of ash was widespread in southern Quebec and was particularly severe in trees growing on wet soils and in isolated trees growing on thin soils (Pomerleau, 1944). Although the problem more or less disappeared in the 1940s and 1950s, it reappeared in 1960 and again in the late 1970s to mid 1980s. No clear cause has been attributed to the syndrome.

The dieback of birch occurred in the 1930s, with the onset of the phenomenon varying between regions (Auclair, 1989). It was particularly widespread and severe in the period 1940–9, with heavy mortality being reported from the Province of New Brunswick in 1940–1 (Clark and Barter, 1958). Recovery was virtually complete by 1952. Since then, there appears to have been a resurgence of problems with white birch (*Betula papyrifera* Marsh.) in some areas in recent years (1979 to 1986), particularly along the Bay of Fundy (Magasi, 1989). Dieback has been recorded there since 1979 and, in the north-eastern United States, dieback has been reported since 1984 (Weiss and Rizzo, 1987). The dieback is preceded by the browning of leaves and premature leaf fall. It is not clear whether the resurgence is the same problem, and it seems likely that different factors may be involved.

Earlier declines were linked to drought stress, the presence of overmature stands and insect attacks in some areas and to high water tables causing root mortality in others (Kessler, 1965). The more recent dieback was initially restricted in the Bay of Fundy area to a narrow coastal strip and coastal islands such as Campobello, Deer and Grand Manan Islands (Magasi, 1989). It has since been recorded in a number of counties in New Brunswick, Cape Breton Island and Nova Scotia (Magasi, 1990). The dieback has been tentatively associated with direct damage by

Table 2.1 Condition of white birch along the Bay of Fundy, New Brunswick, 1982-1990 (assessments of 11 plots, 50 trees per plot)

	% of trees in each class			
Year	No dieback	Twig dieback only	Twig and branch dieback	Dead
1982	92.9	1.5	4.7	0.9
1983	83.7	8.6	6.0	1.7
1984	64.0	24.9	7.8	3.3
1985	45.3	34.9	14.4	5.4
1986	14.5	47.3	31.3	6.9
1987	0.0	42.6	49.8	7.6
1988	0.0	38.0	54.0	8.0
1989	0.0	43.0	47.0	10.0
1990	0.0	47.0	42.0	11.0

From Hall (1991)

acidic fogs (Cox *et al.*, 1989), an explanation which is supported by the extension of the problem to several other species, including mountain paper birch (*Betula papyrifera* var. *cordifolia* (Reg.) Fern.), speckled alder (*Alnus rugosa* (DuRoi) Spreng.), largetooth (*Populus grandidentata* Michx.) and trembling aspen, mountain ash (*Sorbus americana* Marsh.) and mountain maple (*Acer spicatum* Lam.) (Addison, 1989). Moderate to severe leaf browning has not been observed in the area since 1986 (Hall, 1991), although trees continued to deteriorate in crown condition until 1988. Since then, the condition of trees has improved (Table 2.1). The results of forest pest surveys in the Maritimes showed that in 1991, a substantial recovery occurred, with the proportion of trees with combined twig and branch dieback more than halving (Magasi, 1992).

Lake Michigan and other specific areas

Within the eastern hardwood area, Armentano and Menges (1987) documented widespread needle injury in natural populations of jack and white pines growing on the sand dunes along the south shore of Lake Michigan. Air pollution was believed to be the cause of the injury. Chlorotic mottle and tip necrosis occurred in all the populations of jack pine and eastern white pine that were studied, with one or both symptoms occurring in most trees. The levels of foliar injury were low in view of the proximity of pollution sources. Juvenile trees showed the greatest injury levels and, with jack pine, there appeared to be a connection between the degree of effect and soil moisture conditions. Levels of air pollution were not particularly high in comparison to some areas around point sources, with mean 7-hour summer ozone concentrations of 40–50 ppb (maximum values of 124–155 ppb) and maximum sulphur dioxide concentrations of 50–360 ppb. However, pollutants may have been incorporated into lake fogs, leading to much higher concentrations.

Several pollution gradients exist in the north-east (Fig. 2.13), providing an opportunity for gradient studies (Glass and Loucks, 1986; David *et al.*, 1988; Frelich *et al.*, 1989). There have been several attempts to exploit these. Brooks (1989) looked at 200 mixed species plots in Pennsylvania extending from the west (high levels of pollution) towards the east (low levels). A weak, negative association between stand growth and wet sulphate and nitrate deposition was found, concurrent with significant negative influences of stocking density, drought and insect defoliation.

Holdaway (1988) examined a gradient extending from Minnesota, through Wisconsin to Michigan. He assessed a total of 2408 plots, with the majority having been first measured in the 1960s and subsequently being remeasured in the early 1980s. A marked correlation existed between sulphate deposition and winter temperature, confounding any attempt to isolate the impact of sulphate disposition. Holdaway concluded that sulphate was not a major factor influencing forests but that it might

Figure 2.13 Sulphate ion deposition (kg ha^{-1} yr^{-1}) in the eastern United States. Source: National Acid Precipitation Assessment Program.

represent an additional problem to climatically-stressed trees. Using the same gradient, but only 169 stands, Ohmann and Grigal (1990) found a trend in sulphur concentrations in the wood of forest trees which corresponded to the trend in deposition rates. Burton *et al.* (1991) also worked with the Minnesota-Michigan gradient, using it to look at variations in leaf area index of hardwood forests dominated by sugar maple. Substantial year-to-year variations in leaf area were identified, related to seed years and to defoliation by insects. The primary controls on canopy transmittance, maximum leaf area index, allometric relationships and patterns of specific leaf area were more related to soil characteristics and moisture availability than to any pattern in the levels of acidic deposition. However, with only five sites located along the 800 km gradient, the conclusions of this study must be treated with extreme caution.

Bockheim *et al.* (1989) examined the growth and the chemical characteristics of red pine (*Pinus resinosa* Ait.) foliage and the forest floor composition along a precipitation-chemistry gradient in Wisconsin. Although sites were carefully selected in an attempt to remove unwanted 'noise', there was still greater within-site variation than between-site variation for the measured growth parameters. Annual radial increment for the period 1982–6 was more strongly influenced by the timing of the most recent thinning than by precipitation chemistry. Similarly, there were few correlations between foliar chemistry and precipitation chemistry, suggesting that the original soil chemistry probably had more importance than precipitation chemistry in determining the foliar composition of the needles.

Mielke *et al.* (undated) looked at the health of sugar maple and other northern hardwoods across the rainfall pH gradient in northern Wisconsin. They found no evidence of differences in the occurrence of dieback or decline of dominant and codominant sugar maple across the gradient. However, the differences in pH were not very large (ranging from means of 4.5–4.6 in the east to 4.8–5.0 in the west), with the possibility that a gradient in dieback might have been found if areas with a wider range of rainfall acidities had been examined.

These studies suggest that there may be regional variations in the impact of air pollution on forest ecosystems in the eastern hardwood zone. However, they do not show any regional effects on trees. This presents a problem. If there is no evidence of a gradient in forest health from areas known to have relatively high levels of air pollution to areas with relatively low levels of air pollution, then the establishment of air pollution as a major cause of forest ill-health in the north-east is likely to be extremely difficult. However, it is not clear if the indices used to characterize pollution were particularly useful as indices of pollution impact (particularly if ozone was involved). Further gradient studies, using more detailed descriptions of symptoms, would therefore be of value. This is supported by the success of gradient studies in Californian forests (e.g. Miller *et al.*, 1979).

Sugar maple decline

Sugar maple decline has been a driving factor behind the concern about the state of North American forests. Worries were first expressed by maple syrup producers in the autumn of 1978. The prominence given to the purported decline can be directly attributed to the economic importance of maple syrup production and maple timber and to the supposed transboundary nature of the pollution affecting the trees. The declines received widespread media coverage, both in North America and in Europe, although much of the information presented by the media was inaccurate. The assertions contrasted with field observations,

Figure 2.14 Decline of sugar maples, Vermont. The upper photograph shows typical dieback symptoms whereas the defoliation of younger trees in the lower photograph is atypical.

which have suggested that the health of sugar maple in Vermont (Kelley and Eav, 1987), Ontario (Moody, 1988), northern Wisconsin (Mielke *et al.*, undated) and the north-eastern United States generally (Gansner *et al.*, 1987) was quite good in the mid to late 1980s. However, the widespread distribution of sugar maple and its economic importance both for sugar production and as a timber species means that any potential threat to this species is likely to receive much attention and to be investigated in detail.

Past declines

There is a considerable literature documenting declines in sugar maple health in the past (c.f. Hartley and Merrill, 1915; Aikman and Smelser, 1938; Anderson, 1959; Sinclair, 1964; Westing, 1966; Lachance, 1985; Auclair, 1989; Millers *et al.*, 1989). For example, Borie (1987) described a decline affecting 4000 ha of forest in Wisconsin, Michigan and Minnesota in 1957, which was attributed by Giese *et al.* (1964) to a combination of insect defoliation (by *Sparganthus* sp., *Achleris* sp. and the maple webworm *Tetralopha asperatella*), fungal attacks, bacterial problems and injury to buds occurring after an early autumn frost. The reports in the literature suggest that sugar maple has been regularly involved in declines, and that drought is one of the most frequent triggers. Millers *et al.* (1989) have given a full account of past declines which should be referred to for further information.

Symptoms

The symptoms of sugar maple decline have included abnormally small leaves, loss of foliage from the ends of branches, chlorosis and marginal leaf necrosis, early leaf discoloration, branch dieback, tap holes in the trunk taking longer to clear, exfoliated bark, lower starch levels in storage tissues, reduced increment and mortality of trees (Bernier and Brazeau, 1986; Gagnon *et al.*, 1986; McIlveen *et al.*, 1986; Bauce and Allen, 1991; Renaud and Mauffette, 1991) (Fig. 2.14). Root damage has also been reported (McLaughlin *et al.*, 1985), with trees with dieback having lower fine-root biomass and fewer rootlet tips than apparently healthy trees (Bauce and Allen, 1992). However, the work of Bauce and Allen (1992) did not show any evidence for increased root mortality in defoliated trees. In the most recent period of decline, an increase in the numbers of trees showing symptoms of decline was documented for the period 1983–87. The initial dieback in some areas may well have been largely attributable to defoliation in two or three consecutive years by the forest tent caterpillar (*Malacosoma disstria*) and Bruce spanworm (*Operophtera bruceata*) (Bernier *et al.*, 1989), although some stands that were not defoliated by these insect outbreaks had decline symptoms.

Development of the decline

The problem was first reported in the Beauce region in 1978 (Bordeleau, 1986) and, by 1982, it was present in various parts of Quebec province, including Beauce, Mégantic, Frontenac and Arthabaska counties. A decline was also reported from Ontario (Muskoka County) in 1978, shortly after severe defoliation by the forest tent caterpillar. In 1985, an aerial survey of 5336 km^2 found that 60% of the surface area covered by sugar maple could be classed as having less than 10% of the crown missing. Light dieback was recorded in 35% of the trees and moderate mortality in 4% (Bordeleau, 1986). The amount of damage in 1985 was greater than in 1983 (Gagnon *et al.*, 1986), but there were problems over the comparability of data collected in different years (Bordeleau, 1986). A survey of the Quebec region showed that about 83% of the maple stands had some evidence of dieback, but the extent was classed as moderate or severe in only 6% of stands (Millers *et al.*, 1989). In 1987, an investigation of 42 sugarbushes located in Wisconsin indicated that 95% of the trees had healthy crowns (Drillias *et al.*, 1990).

Since the early surveys, the classification of crown damage has been changed, and in a recent survey (Brooks *et al.*, 1991), 97% of trees in New England were considered to have no or only light (<20%) dieback and 99% had normal (<30% reduction) crown transparency. Similar results were obtained by Millers *et al.* (1992) who found a substantial improvement in the condition of sugar maple stands throughout the north-eastern area between 1989 and 1990. There have been few long-term investigations of sugar maple health, but it is worth pointing out the study by Burns (1990) which found no change in health in 54 Vermont sugarbushes over the period 1977 to 1985.

In the United States, a decline in the annual radial increment of sugar maple starting in 1979–82 was reported from Vermont and New Hampshire (Hornbeck, 1987). However, this was only temporary and, in New Hampshire, growth strongly recovered after 1983; no comparable data were available for Vermont. The growth declines coincided with severe outbreaks of several insect pests, notably the gypsy moth (*Lymantria dispar* L.), forest tent caterpillar and saddled prominent (*Heterocampa guttivata*), a phenomenon also noted by Lachance *et al.* (1984) and Bauce and Allen (1991). No long-term decline in the growth of sugar maple has been identified in New England (Hornbeck *et al.*, 1987; Leak, 1988) and nothing on the scale of that seen for red spruce (see below) has been documented anywhere.

Results from the Canadian ARNEWS monitoring plots are presented in Fig. 2.15. There has been a gradual increase in the number of dead trees over the period 1986–91, but the rate of mortality (1.1% per annum) is within that expected from natural processes. The proportion of 'healthy' trees gradually increased over the period 1987–91. The data do not provide any suggestion of a decline.

An independent source of information is the volume of syrup production. This showed a marked increase between 1979 and 1981 (Fig. 2.16), then fell dramatically before recovering to record levels in 1989 (NAPAP, 1990). The sharp fall in production in 1981–2 is not consistent with a long-term decline in tree health. Instead, it suggests that a major stress occurred in 1981/2. This period was one when major insect defoliations were occurring in some areas, particularly the region south of the St. Lawrence area (Lachance *et al.*, 1984). The subsequent recovery is also inconsistent with the idea of a large-scale, irreversible decline.

Possible causes of the sugar maple decline

Natural abiotic factors

Bernier *et al.* (1989) described a severe frost that occurred in early June 1980, killing 75–100% of the sugar maple foliage in the Appalachians. A freeze-thaw cycle in February 1981 and a summer drought in 1983 were also very important. Lachance (1985) suggested that root damage by frost probably occurred in many areas during the winters of 1981 and 1982 due to unusually shallow snow depths. The 1983 drought was followed by

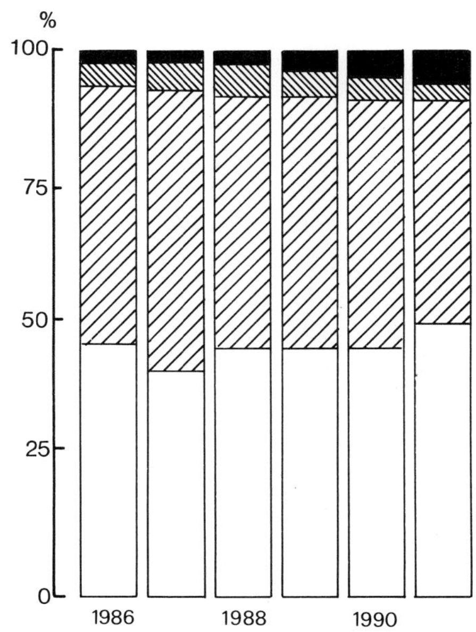

Figure 2.15 Trend in the crown condition of sugar maple assessed in the Canadian ARNEWS programme between 1986 and 1991. Black: dead or dying trees; dense shading: declining trees; light shading: intermediate trees; open: healthy trees. The cumulative percentage of the sample trees is given. Source: ARNEWS (1992).

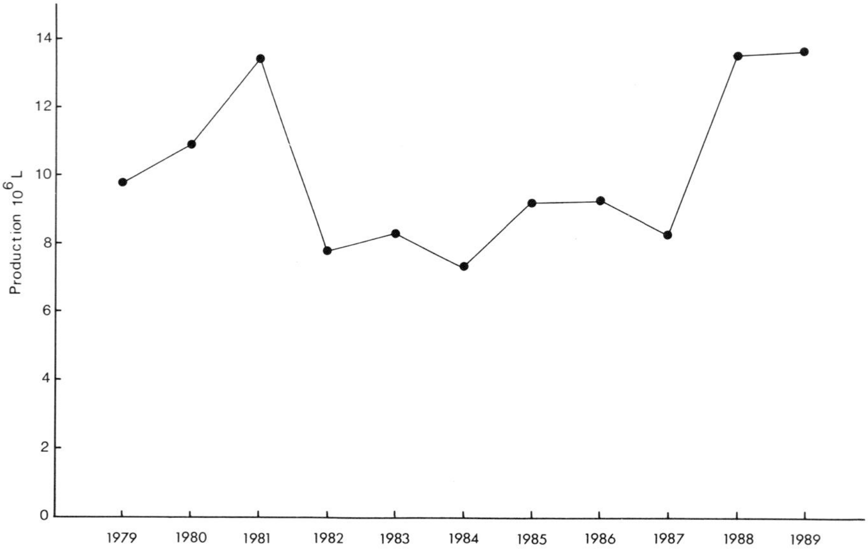

Figure 2.16 Annual production of sugar maple syrup between 1979 and 1989. Source: NAPAP (1990).

heavy fruiting of trees in the summer of 1984, which may well have upset the nutritional balance of many trees (Bernier *et al.*, 1989). In the spring of 1985, significant shoot mortality was seen, a phenomenon consistent with reported cases of crown deterioration during or immediately after years of heavy seeding (Gross and Harden, 1968; Gross, 1975; Skelly *et al.*, 1987). However, McIlveen *et al.* (1986) considered the decline to be a slow-developing, non-specific response to one or more inciting agents, primarily insect defoliation and drought.

Studies of the nutrient contents of the foliage revealed complex patterns of inter-relationships between nutrition and the decline of trees (Bernier and Brazeau, 1988a). The level of stand decline was negatively correlated with foliar potassium concentrations (Fig. 2.17), suggesting that potassium-deficiency might be involved in the decline. The deficiency was evident as abnormally dark green leaves and upward curling of leaf margins (Bernier and Brazeau, 1988b). In cases with particularly severe deficiency, marginal and interveinal necrosis occurred. Potassium deficiencies have been seen in symptomatic trees by Ouimet and Fortin (1992). Phosphorus deficiencies were reported, taking the visible form of small purple leaves, particularly those exposed to the sun (Bernier and Brazeau, 1988b). Leaf curling is thought to represent water stress, caused by increased water loss by transpiration because of suboptimal potassium levels. Magnesium deficiency was also involved in the declines, although it was only seen in the Lower Laurentians, apparently being absent in the

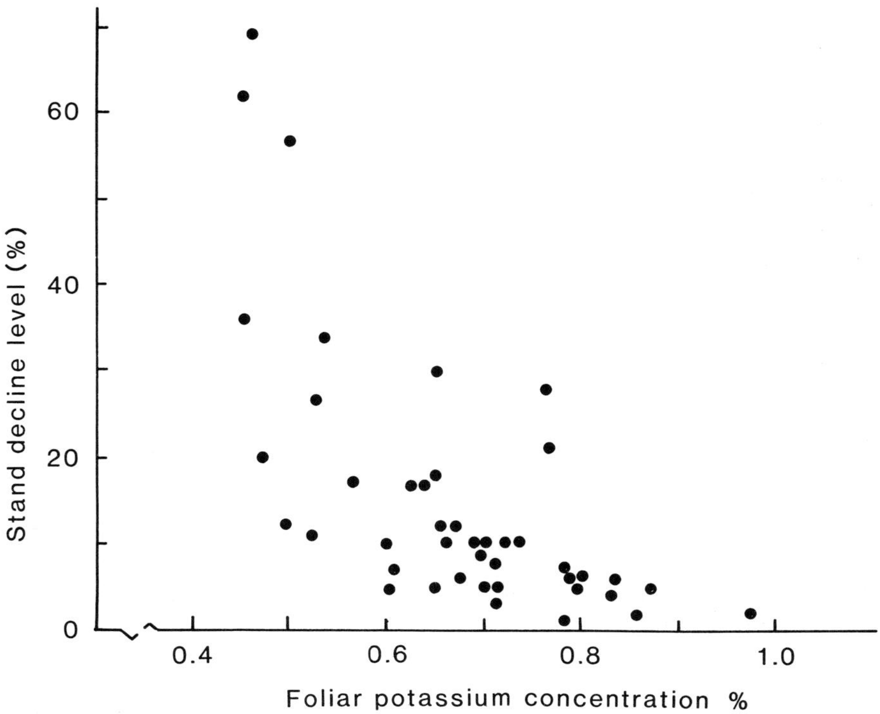

Figure 2.17 Foliar potassium concentrations and level of sugar maple stand decline in the Quebec region, 1985. From: Bernier and Brazeau (1988a).

Quebec Appalachians (Bernier and Brazeau, 1988c). The magnesium deficiency took the form of vivid orange-yellow interveinal chlorosis in extreme cases, being preceded by a light-green stage and mottling. All three nutrient deficiencies are believed to be a recent phenomenon as there are few reports of similar problems in the past.

Paré and Brazeau (1989) proposed an interesting explanation for the phosphorus deficiency. They found that phosphorus concentrations in sugar maple foliage were lower on less acidic soils characterized by mull humus, whereas phosphorus concentrations were adequate on mor-dominated sites. They proposed that the phosphorus cycling on mull sites had been severely disturbed. As a result of the biological activity in mull soils, iron and aluminium hydrous oxides were reaching the surface where they acted as binding sites for phosphorus, locking it up and preventing it from being reabsorbed by the tree roots. Various factors could have resulted in the initial perturbation to the cycle which was then maintained by positive feedback loops.

As in Europe, experimental fertilization of stands showing nutrient deficiencies has proved promising. Hendershot (1991), working with sugar maple stands in the Beauce-Mégantic region south of Quebec, found that application of fertilizers containing calcium, magnesium and potassium reduced the incidence of increased foliar transparency, leaf discoloration and abnormally small leaves. Fertilization also increased the rate of closure of old tap holes. The changes in health were accompanied by increases in the foliar and soil concentrations of base cations, but did not affect either nitrogen or phosphorus concentrations. These results point strongly to a nutritional problem in the stands that he investigated. Burke *et al.* (1992) found that increased soil nitrogen significantly increased the production of foliage, stems, coarse roots and fine roots, suggesting that there may be annual changes associated with nitrogen availability.

Air pollution

The role of air pollution was, and still is, uncertain. Bernier and Brazeau (1986) reported that trees were in decline in stands with a calcium : aluminium ratio of less than 1 in the soil solution at 20–30 cm depth. However, some declining stands had calcium : aluminium ratios of up to 2.5, suggesting that the ratios were not a particularly useful index in relation to sugar maple (similar results have been found for other species in Europe – Chapter 8). Sugar maple is generally quite tolerant of ozone, both to short-term high concentrations and to longer-term lower concentrations (Hibben, 1969; Lacasse and Treshow, 1978; Kress and Skelly, 1982; Reich *et al.*, 1986; Jensen and Dochinger, 1989), and experimental work with mature trees using ozone concentrations of up to 140 ppb has failed to produce visible injury (Lovett and Hubbell, 1991). However, sugar maple is sensitive to other forms of pollution and its frequency in forests declines markedly close to point sources of sulphur dioxide and fluoride pollution (McClenahen, 1978).

Sugar maple is also quite tolerant of acidic precipitation. Dustin and Raynal (1988) investigated the effect of artificial acidic precipitation on root growth. They showed significant reductions in root hair growth over short time scales but growth of primary and secondary roots was unaffected by pH 3.0, well below ambient levels of snowmelt. Similarly, Raynal *et al.* (1982a) found no effect of substrate acidity (down to pH 3.0) on the germination rate of sugar maple, although seedlings exposed to pH 3.0 precipitation showed an increased tendency for bacterial infection (Raynal *et al.*, 1982b). It is also unlikely that foliar leaching induced by acidic precipitation is having a major effect on the nutritional status of the trees (Muir, 1992). Consequently, there is little evidence that air pollution is or has been involved in the declines.

Increment studies

Several groups have looked for evidence of growth declines associated with visible symptoms caused by air pollution injury. Bauce and Allen (1991) undertook stem analyses of trees at a site in Dubuar Forest in

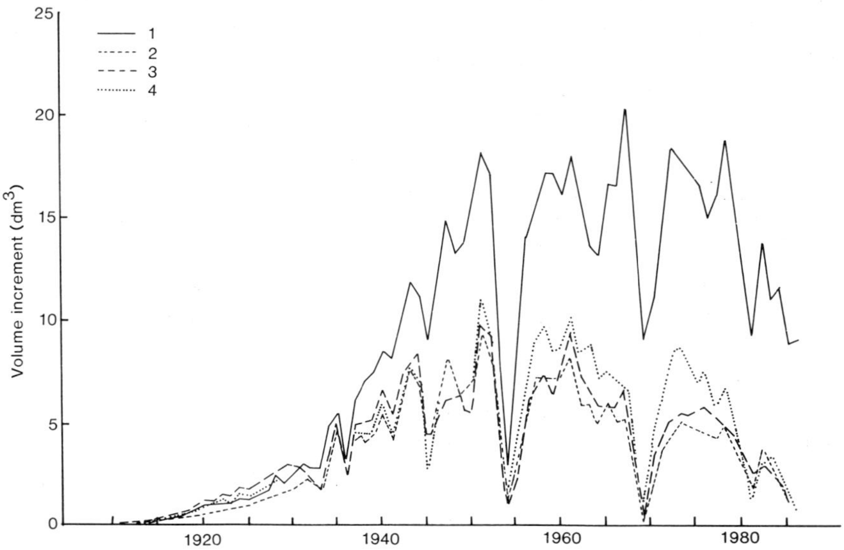

Figure 2.18 Annual volume increment of dominant sugar maples in Dubuar Forest, New York, with different stages of crown dieback (n = 12). 1: healthy trees; 2: early dieback (15–33%); 3: mid-stage dieback (34–50%); 4: late dieback (>50%). From: Bauce and Allen (1991).

New York State. Past insect defoliations (by the forest tent caterpillar in 1953–4 and the saddled prominent in 1967–8) had a major influence on growth. A decline in growth, starting in 1959, was apparent and was accompanied by increasing climatic sensitivity, particularly to warm winters with periods of thawing and to summer droughts. The annual volume increment (Fig. 2.18) and annual radial increment (Fig. 2.19) were related to the degree of dieback, although the sample size (12) on which this conclusion was based was small. Bauce and Allen concluded that stand-level competition played an important role. This conclusion is supported by the work of Roberge (1988), who found a recovery in the growth of trees following thinning, and by records of rapid recovery of remaining trees following mortality in sugar maple stands (Griffin, 1965; Ohman, 1969). Reports by Gagnon *et al.* (1986) that dieback in Quebec sugar maple stands was lower in stands with higher basal areas do not contradict this idea as basal area of an individual species and stand-level competition are not necessarily related.

There appears to be a relationship between the annual radial increment and the degree of dieback in individual trees (Bauce and Allen, 1991; Renaud and Mauffette, 1991), but the significance of this is uncertain. Stand level problems may well have played an important part, as many stands have developed on old fields abandoned in the early and mid-

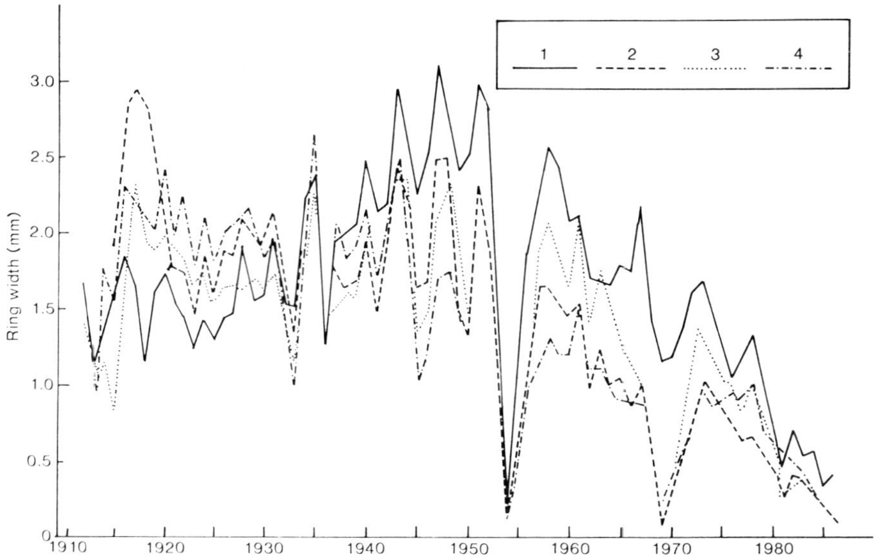

Figure 2.19 Mean annual radial increment of dominant sugar maples in Dubuar Forest, New York, with different stages of crown dieback (n = 12). 1: healthy trees; 2: early dieback (15–33%); 3: mid-stage dieback (34–50%); 4: late dieback (>50%). From: Bauce and Allen (1991).

nineteenth century (Borie, 1987). This is a feature of forests throughout the north-eastern area (Brooks *et al.*, 1991).

Gansner *et al.* (1987) reported increases in stocking levels, growth and total volume of sugar maple in Vermont and New Hampshire as occurring over the period 1973–83. The reproduction strategy of sugar maple is to carpet the forest floor with seedlings which then take advantage of any canopy gaps. Several periods of release and suppression may occur before individual trees reach the canopy (Fayle and MacDonald, 1977; Lorimer, 1980; Oliver, 1981; Hibbs, 1982; Canham, 1985; Houle, 1990). As sugar maple may be competitively dominant in many forests (Eickmeier, 1988), the role of stand dynamics requires careful evaluation.

Detailed investigations of the growth of sugar maple and other hardwood species have been reported by Hornbeck (1987), Hornbeck *et al.* (1987, 1988) and Smith *et al.* (1989). Hornbeck (1987) described growth analyses of 521 sugar maples from three New England states (Fig. 2.20). Annual basal area increment has increased from 1–2 $cm^2\ yr^{-1}$ in 1900 to 9–10 $cm^2\ yr^{-1}$ in 1980. Declines in growth were recorded in Vermont and New Hampshire starting in 1979–82, although a strong recovery was present in New Hampshire from 1983 onwards (there were no data for Vermont so it is unclear whether recovery also occurred there). The declines can be attributed to severe outbreaks by insect pests, particularly

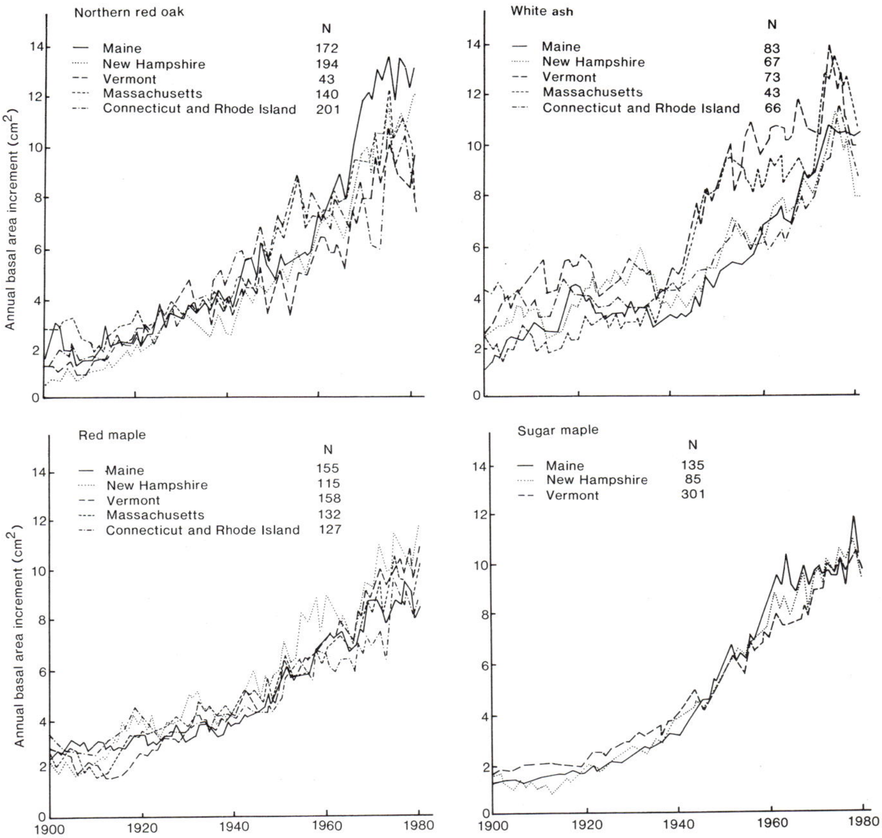

Figure 2.20 Regional mean annual basal area increment for sugar maple and three other eastern hardwoods in New England.

gypsy moth, forest tent caterpillar and the saddled prominent. Similar conclusions were reached by Leak (1988), who found no evidence of a decline in the growth of a very small sample (9 trees) of sugar maples in New Hampshire. One possibility for the absence of a marked growth decline in sugar maple associated with dieback is that most of the cores used in the growth studies by Hornbeck and co-workers stop in the early 1980s, which is when the recent dieback problems were first reported (Hornbeck *et al.*, 1988). This remains to be evaluated.

Hornbeck (1987) also looked at growth patterns of red oak (750 trees) and red maple (*Acer rubrum* L.) (657 trees) from all six New England states. Annual basal area increment of red oak has increased markedly since 1900, although declines coinciding with outbreaks of the gypsy moth have occurred. Red maple has increased its annual basal area increment, and again showed growth declines associated with gypsy moth outbreaks in recent years. A different pattern was present in white ash (Hornbeck *et al.*, 1987), which had uniform growth until about 1940, then a steady increase until the early 1970s followed by a consistent downward trend in annual basal increments. White oak (*Quercus alba* L.) also showed variable growth patterns (Phipps and Whiton, 1988), with declines being recorded at some sites in the 1950s and 1960s, but not subsequently.

Houle (1990) examined the growth patterns of trees responding to gaps created by dieback. Working near Trois-Rivières, Quebec, Houle found a marked decrease in the growth of apparently healthy trees in the early 1980s, followed by a growth increase at declining sites in the period 1985–7. Houle attributed the growth decrease in the early 1980s to the unusually mild winter conditions of 1979–80 and 1982–3 and the cold winters of 1980–1 and 1981–2. There was no universal response in the growth of healthy trees to canopy openings. If these results can be extrapolated to other sites, which is always a problem, then it seems possible that the cores of Hornbeck and others were too short to record the 1980s growth trends.

The role of management

There has been some speculation that the sugar maple decline may partly lie in the excessive use of tapping. The study by Drillias *et al.* (1990) in Wisconsin found that 33 out of 42 sugarbushes had been overtapped, encouraging decay. In a historical survey of sugar maple decline in Ontario (McIlveen *et al.*, 1986), bad management was reported to be the cause of decline in four cases. McLaughlin *et al.* (1985) found that tapped trees in Ontario showed more signs of decline than untapped trees. These results were partly confirmed by the results of the North American Sugar Maple Decline Project. It found higher crown transparency and dieback in managed sugar maple stands than in unmanaged ones (Fig. 2.21) (Millers *et al.*, 1992), but the differences were marginal. Clearly, tapping, and particularly over-tapping, represents a stress to sugar maples. If the trees are already weakened by other factors, the use of tapping may push the trees into decline. Another important factor affecting sugar bushes is the excessive soil compaction caused by vehicular operations that may occur because of vehicular operations associated with the tapping (Houston *et al.*, 1990). This would increase the sensitivity of the roots to stress.

Conclusion

There is no doubt that a widespread decline in the health of sugar maples occurred in some areas in the late 1970s and early 1980s. This decline was not unprecedented, with several similar declines occurring in the past. The decline appears to have been genuine, in that it was characterized by crown dieback and, in severe cases, mortality. However, its causes remain very uncertain. Natural factors were important, particularly insect defoliation and severe climatic conditions. As elsewhere, stand dynamics may

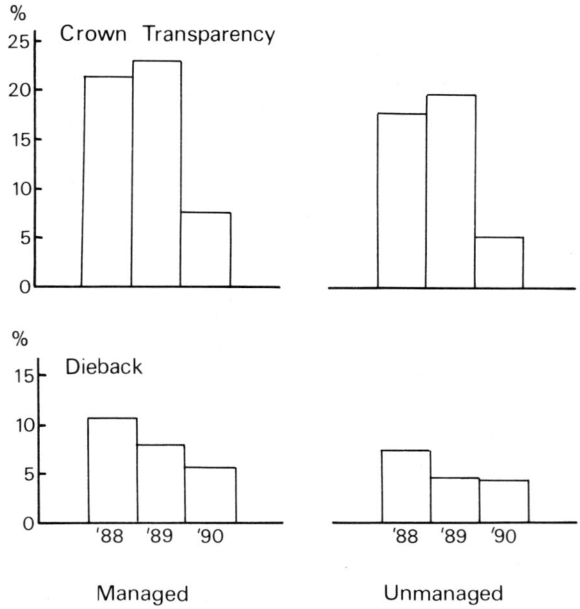

Figure 2.21 Average stand crown transparency and dieback in managed and unmanaged sugar maple stands assessed by the North American Maple Project between 1988 and 1991. From: Millers *et al.* (1992).

have contributed to the general stress levels in the trees. Nutrient depletion, both through the sequestration of soil nutrients in the increasing standing volume and the removal of nutrients in syrup, may have reduced the ability of trees to recover. However, during the mid to late 1980s, the health of sugar maple appears to have recovered in most areas, suggesting that the decline was only temporary.

The role of air pollution in the decline is controversial. Direct effects of gaseous pollutants appear unlikely. If there are any effects, then the most probable mechanism is through nutritional disturbances via changes in the forest soils. In this respect, the hypothesis that phosphorus cycling on some mull soils has been disturbed by acidification put forward by Paré and Brazeau (1989) is particularly interesting and may be applicable in other areas. However, it remains to be tested over a wider region.

Other hardwood species

While attention in the north-east has focused on sugar maple, damage to other species is also of concern, although the field evidence to justify such worries is very limited. Results from the 1990 survey of forest health in New England are presented in Table 2.2 (Brooks *et al.*, 1991) and, with

Table 2.2 Results of the 1990 survey of forest health in New England

Species	Percent crown dieback class				Foliage transparency class		
	None (0–5%)	Light (6–20%)	Moderate (21–50%)	Severe (>51%)	Normal (0–30%)	Moderate (31–50%)	Severe (>51%)
Balsam fir	91.4	7.2	1.1	0.3	99.7	0.3	0.0
Red spruce	92.7	6.0	1.1	0.2	99.8	0.2	0.0
Eastern white pine	92.4	6.6	0.8	0.2	95.5	4.5	0.0
Northern white-cedar*	82.8	12.1	4.0	1.0	91.9	7.6	0.5
Eastern hemlock*	93.0	3.5	2.9	0.7	97.9	2.1	0.0
Red maple*	67.2	26.5	4.4	1.9	95.6	3.5	0.9
Sugar maple	87.0	10.1	2.4	0.5	98.9	0.8	0.3
Yellow birch*	77.9	18.8	1.4	1.9	96.2	1.9	1.9
Paper birch*	68.6	27.0	3.1	1.4	92.8	6.5	0.7
American beech	54.7	32.1	7.5	5.7	86.8	6.9	6.3
White ash	71.3	25.0	1.5	2.2	94.9	5.1	0.0
Northern red oak	50.0	49.4	0.0	0.6	90.4	4.8	4.8

From Brooks *et al.* (1991)

*Northern white cedar: *Thuja occidentalis* L.; Eastern hemlock: *Tsuga canadensis* (L.) Carr.; Red maple: *Acer rubrum* L.; Yellow birch: *Betula lutea* Michx.; Paper birch: *Betula papyrifera* Marsh.

the exception of American beech (*Fagus grandifolia* Ehrh.) (damage by the beech bark disease complex) and northern red oak (*Quercus borealis* Michx. f.) (gypsy moth and other defoliators), very little damage was present when the forests of the region as a whole were considered.

Yellow poplar

Various experimental studies have been conducted with species commonly found in the north-east. Yellow poplar (*Liriodendron tulipifera* L.) has been widely used, reflecting its value as biomonitor, although some field studies in the north-east have not shown visible injury in this species (Winner *et al.*, 1989). Chappelka *et al.* (1985) looked at growth responses of nine-week old seedlings of yellow poplar when exposed to a variety of pollutants. A combination of ozone (100 ppb O_3, 4 hours per day, 5 days per week) and sulphur dioxide (80 ppb SO_2, 4 hours per day, 5 days per week) reduced height growth and dry matter production, whereas a combination of gases and acidic precipitation (pH 3.0 and 4.3, 1 hour per day, 2 days per week at 0.75 cm per hour) caused a range of effects. As acidity increased, ozone fumigation resulted in a linear decrease in root dry weight and an increase in leaf area and mean relative growth rate. Exposure to sulphur dioxide and acidic precipitation resulted in an increase in root dry weight.

Similar results were obtained by Dochinger and Jensen (1985), who misted one-year-old yellow poplar seedlings with artificial mist at pH

2.5, 3.5, 4.5 and 5.5, besides fumigating with various combinations of ozone, sulphur dioxide and nitrogen dioxide. Height growth of seedlings exposed to acidic mist alone was significantly decreased after eight weeks, with growth reductions increasing with higher levels of acidity. Leaf weight and the weight of new stems were only affected by the pH 2.5 treatment. Interestingly, the addition of one or more gases did not have any effect on the basic relationship established with acidic mist alone. This work is difficult to interpret in the light of later studies. For example, Jensen and Patton (1990) found that acidic precipitation down to pH 3.0 stimulated height growth. In contrast, Roberts (1990) found decreases in net photosynthesis, leaf stomatal conductance and xylem water potential at pH 3.0. These conflicting results suggest differential responses between genotypes, a widespread problem in experimental studies with trees and other plants.

Green and white ash

Nine-week-old green (*Fraxinus pennsylvanica* Marsh. var. *lanceolata* (Borkh.) Sarg.) and white ash seedlings exposed to the same treatments as used by Chappelka *et al.* (1988) for yellow poplar, also showed differential responses (Chappelka *et al.*, 1988). White ash biomass was suppressed by exposure to ozone and the cumulative elongation of shoots of green ash was also reduced. Crang and McQuattie (1986) looked at combinations of acidic rain and ozone on yellow poplar, but nitrogen dioxide was their third experimental pollutant and they used older (one year) seedlings. Differences among the three treatments were found and significant interactions were identified. The results agreed with earlier work that showed an effect of acidic misting (pH 4.5, 3.5 and 2.7) on the leaves of yellow poplar (Crang *et al.*, 1983). In this earlier study, marked reductions in leaf size were noted at the most acidic level of treatment.

Elliott *et al.* (1987) exposed two- and three-year-old green and white ash to ambient levels of acidic precipitation and ozone. During the three-year study, potted seedlings showed no signs of ozone toxicity, despite up to 78 hours exposure in excess of 1.2 ppm. Acidic precipitation with a pH down to 3.6 also caused no visible injury. Neither pollutant had any effect on shoot growth when applied together.

Red oak

Red oak seedlings can be affected by soil acidification under experimental conditions (Joslin and Wolfe, 1989), although as red oak is particularly sensitive to aluminium (Kelly *et al.*, 1990), the results should not be extrapolated to other species. Acidification treatments led to significant reductions in fine root and foliar biomass production and in fine root branching. However, the treatments were considerably greater than ambient and the authors concluded that current levels of aluminium in the soil and current acidic deposition rates did not pose a threat to red oak in the southern portion of its range. On particularly acidic soils in the northern part of its range, acidic deposition is greater and further investigations are required before any conclusions can be reached. In the field, red oaks on more acidic soils in the northern part of the red oak range (Pennsylvania) had lower root tissue concentrations of potassium, calcium, iron and aluminium than on less acidic soils, but no visible damage was recorded (Majumdar *et al.*, 1989). The implications of this have still to be assessed.

General

A problem affecting all eastern forests is the possibility of base cation depletion. As trees grow, they take up nutrients, some of which are stored permanently in the wood. These are then removed during harvesting, resulting in a net loss to the system. Whole tree harvesting severely aggravates this loss, and it may lead to substantial losses of nutrients from the system (Mann *et al.*, 1988; Johnson and Taylor, 1989; Johnson and Todd, 1990). The problem is exacerbated by acidic deposition which, under certain circumstances, will lead to nutrient losses from the soil and may alter nutrient cycles through increased foliar leaching (e.g. Kelly and Strickland, 1986; Parker, 1990). Some studies (e.g. Wood and Bormann, 1975; Scherbatskoy and Klein, 1983; Schier, 1987; Lovett and Hubbell, 1991) have suggested that the level of acidity required to obtain such an effect is greater than that which currently occurs in the north-east.

Input-output balances for a variety of sites were examined by Federer *et al.* (1989a). Their greatest concern was for total soil and biomass calcium, which they estimated could be reduced by 20–60% within 120 years by current rates of leaching and removal in harvesting operations. In Minnesota, changes in the calcium status at some sites have occurred over the last 40 years (Alban, 1982) and net losses of calcium and magnesium have been reported from much further south, in forest ecosystems in Tennessee (Johnson and Todd, 1990).

Conclusions

The evidence concerning the widespread decline of hardwood forests in the north-east is at best controversial and is dependent on how decline is defined. Growth studies have not revealed any widespread decline (Figs 2.20 and 2.22); instead, they show that basal areas have been increasing for some time. This is inconsistent with the idea of a universal decline in forests. Gradient studies have failed to reveal any significant gradients in forest health, and monitoring has not, to date, shown any evidence of widespread poor tree health. Given that concern was first expressed almost 15 years ago, the frequent occurrence of poor tree health would now be expected.

Sugar maple and other species have undergone periodic declines in health for as long as they have been studied. Several insect defoliators regularly reach epidemic proportions on a scale that is rarely, if ever, encountered in Europe. In addition, the climatic conditions of the north-eastern hardwood zone are much more severe than those

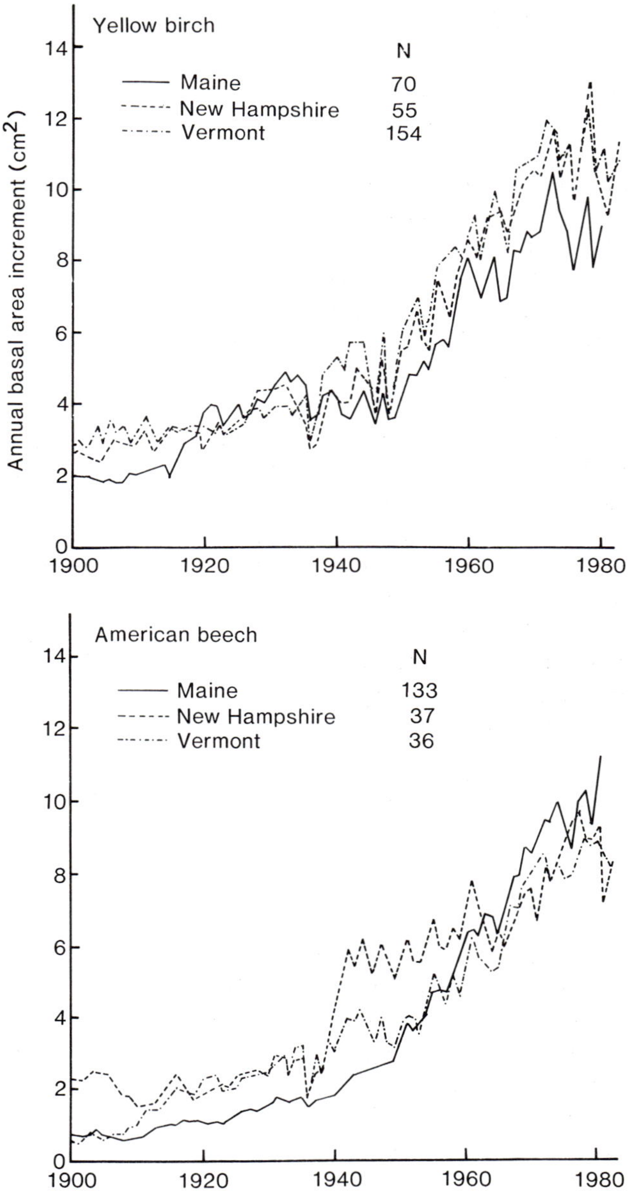

Figure 2.22 Regional mean basal annual area increment for yellow birch and American beech in New England. From: Smith *et al.* (1990).

Figure 2.23 Decline of high-altitude spruce-fir forest at Whiteface Mountain, New York.

encountered in the European hardwood zone. Although trees can withstand both occasional defoliation and the severe winter conditions, the occurrence of unusual conditions, such as defoliation in two or more successive years or thaw periods in mid-winter, appears to severely stress them. The early 1980s were characterized by several unusual stresses for forests, and these alone may have been sufficient to cause the temporary decline in the condition of some species. Air pollution represents an additional stress, but its role has yet to be convincingly demonstrated.

Spruce-fir forests

Red spruce is widely distributed at low altitudes in Maine, where spruce-fir forests cover about 3.2 million hectares (Powell and Dickson, 1984; Reams *et al.*, 1990) and the total area of spruce-fir forest in the United States is estimated to be more than 10 m ha, about 0.3% of which occurs in the south. Further large areas of spruce-fir forest occur in eastern Canada.

High-altitude northern red spruce

In the northern Appalachians, red spruce forms an important component of several high-altitude forest ecosystems, although in montane mixed spruce-fir-birch stands it rarely exceeds 25% of the density by species or 40% of the basal area (Friedland, 1989a). High-altitude red spruce occurs in association with balsam fir with spruce-fir forests covering about 132 000 ha in the Adirondacks and northern Appalachians (Barnard *et al.*, 1989). The spruce-fir forests over much of the area have been subjected to major disturbances. Logging of the virgin forests began in the mid–1880s, with trees on the lower slopes being cut first, followed by higher altitude stands (Korstian, 1937; Westveld, 1951). Usually, stands were clear-cut. Occasionally, the logging resulted in the replacement of spruce forest with mixed hardwood forest, particularly at lower altitudes. At higher elevations, balsam fir expanded into areas formerly occupied by red spruce (Oosting and Reed, 1944; Weiss and Millers, 1988). As a result of these past disturbances, much of the forest is relatively young. For example, Seymour (1985) reported that over 50% of the red spruce forest in Maine was 50–80 years old, although much of this consisted of low-altitude stands.

Considerable concern has been expressed over the apparent decline of balsam fir and red spruce at several locations in the north-east United States. The decline is most pronounced at high-altitude locations (Reams *et al.*, 1990) (Fig. 2.23). The decline was first reported in the early 1980s when a reduction in growth rates of red spruce and balsam fir was noted, together with mortality of red spruce (Siccama *et al.*, 1982; Johnson and Siccama, 1983a, 1983b), although the mortality had probably been present since the early 1960s. In examining the decline, it is important to distinguish between the northern and southern Appalachians, as the decline symptoms in the two areas are very different. In addition, there are differences between the high- and low-altitude populations of red spruce.

Red spruce has been affected by major declines in the past. The first documented major decline occurred in the period 1871–85, over an area extending from New York to Maine. Up to 50% mortality of red spruce in the Adirondack Mountains of New York was reported (Hopkins, 1901). Since then, occasional periods of damage have been reported, but it is the most recent phase that has been of greatest concern.

Symptoms

The spruce-fir forests at many sites in the northern Appalachians give the impression of having been heavily disturbed in the recent past. Numerous dead trees are present at some locations. However, the vigorous regeneration, particularly of balsam fir, suggests that the disturbance has ended and that the forest is quickly regenerating. Visible symptoms of declining red spruce in the northern Appalachians include needle discoloration (although pronounced chlorosis is rare) and needle death, crown thinning, altered branching patterns (bunching of branches because of secondary shoots developing after terminal branch dieback) and the eventual death of branches and trees (Friedland *et al.*, 1984; Bruck, 1985; Johnson and McLaughlin, 1986). Typically, loss of needles occurs from the youngest to the oldest (from the outside of the tree inwards) and from the top of the crown downwards (Johnson and Siccama, 1983b; Friedland *et al.*, 1984). Death of trees occurs in all size classes and is frequently associated with attack by the spruce beetle (*Dendroctonus rufipennis* Kirby) (Souto and McCreery, 1988), which is probably the main agent responsible for the final deaths of many of the trees. Other insect pests that may be common include the ghost moth (*Hepialus gracilis* Grote), the grey spruce looper moth (*Caripeta divisata* Walker), the conifer swift moth (*Korscheltellus gracilis* Grote) and the four-eyed spruce beetle (*Polygraphus rufipennis* Kirby) (Grossman, 1988). Damage by the conifer swift moth in particular appears to have been underestimated in many studies (Leonard *et al.*, 1991; Wagner *et al.*, 1991; Bergdahl *et al.*, 1992). Other problems commonly associated with the decline are stem resinosis (often due to *Fomes pini*) and Eastern dwarf mistletoe (*Arceuthobium pusillum* Peck). Mortality in the subalpine forest is patchy, with affected areas commonly covering about one hectare.

A detailed description of the decline of northern high-altitude red spruce has been made by Miller-Weeks *et al.* (1988a) and Miller-Weeks and Cooke (1989). Eighty plots were established in the north-east of the United States. A variety of symptoms were assessed, including defoliation and discoloration, branch abundance, proportion of dead, broken or contorted branches and the proportion of branches with brooms, bunched or tufted foliage, or resinosis. Branch samples were collected for more detailed analysis. Very little tree discoloration was present, but about 65% of the red spruce had 11–50% foliage loss. About 10% of the trees had more than 50% foliage loss. Needle retention was generally 7–9 years, with 15 years observed at one site in New York.

In marked contrast to Norway spruce in Europe, the defoliation mainly took the form of dieback or mortality of individual branches, with over 75% of the trees having individual branch mortality. In West Virginia, some of the branch dieback was attributed to Cytospora canker (*Valsa kunzei* Fr.) (the pycnidial stage of *Cytospora kunzei* Sacc.), but the causes of the dieback elsewhere were uncertain.

The incidence of discoloration is unclear. Miller-Weeks *et al.* (1988a) recorded frequent instances of mottling of red spruce foliage. Over 50% of the trees in New Hampshire, Vermont and Tug Hill (New York) were affected, with fewer being affected in the Adirondacks and West Virginia. Chlorosis, particularly tip chlorosis, was also common. These results contrast with those from elsewhere, possibly because of differences in what is deemed to be discoloration. In balsam fir, chlorosis of older foliage was quite widespread at Camel's Hump and Whiteface Mountain when I visited these sites in summer 1992. There was no evidence of discoloration of the 1992 needles.

Status of the forests

Several studies have been made of the status of individual northern red spruce forests. Although the spruce-fir forests are thought to have been under climatic stress for some time (Griggs, 1946), most surveys revealed an increase in mortality in the late 1970s and early 1980s.

Johnson *et al.* (1989) reported extensive mortality in the Adirondack, Green and White Mountains between the mid–1960s and mid–1980s, with up to 60% of the standing spruce being dead at higher altitudes. Similar quantitative studies have been conducted elsewhere, such as at Whiteface Mountain, New York (Scott *et al.*, 1984; Friedland and Battles, 1987; Battles *et al.*, 1988, 1989; Friedland, 1989a), where 57% of the standing spruce was dead at an altitude of 1000 m. At Camel's Hump, Vermont, the basal area of spruce declined by 60–70% between 1965 and 1983 (Vogelmann *et al.*, 1985; Friedland and Battles, 1987; Perkins, 1988). At high-altitude sites in New Hampshire, New York and Vermont, Silver *et al.* (1988) reported increases in the numbers of standing dead trees between 1982–87. The highest percentages of standing dead were in the Adirondack and Green Mountains, with lower percentages recorded on mountains in Maine (Fig. 2.24) (Friedland, 1989b; Craig and Friedland, 1990). Although dead trees were present on the more eastern mountains, many of these consisted of old trees that died before the most recent decline episode (Craig and Friedland, 1990). In the Adirondacks and Green Mountains, the proportion of standing dead trees increased with altitude, but no such trend has been recorded on mountains to the east. Above 950 m in the Adirondacks, over 50% of the canopy spruce were dead. In the Green Mountains, mortality was higher, with over 75% of the canopy spruce dead above 950 m at Camel's Hump and Mount Abraham.

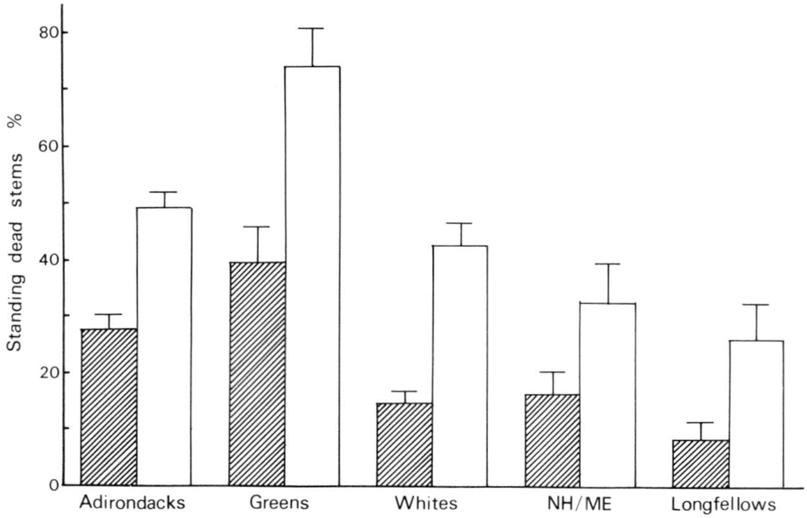

Figure 2.24 Mean and standard error of recent standing dead (shaded) and total standing dead (open) red spruce on 19 mountains in the north-east. All data were collected between 950 and 1100 m above sea-level. From: Craig and Friedland (1990).

Detailed surveys have been undertaken at Whiteface Mountain. Stand basal area at higher altitudes decreased by 20–30% between the mid 1960s and the early 1980s. This was mainly due to a reduction of over 50% in the live basal area of red spruce (Scott *et al.*, 1984). Decline in the basal area of balsam fir was restricted to wind-disturbed sites and, as this is a site affected by fir waves (see Chapter 1), an exposure-related explanation may be appropriate. Trees with severe foliage loss in the early 1980s have since died, whereas those then classed as healthy have remained so (Battles *et al.*, 1988).

Battles *et al.* (1988) established 331 sample plots on Whiteface Mountain in 1986–87. Below 1000 m approximately 25% of the standing red spruce trees were dead (Fig. 2.25) and at above 1000 m, approximately 55% were dead. Interestingly, there was no evidence of a gradient of increasing mortality with altitude when plotted in 100 m elevation bands. Rather, there was a marked step in mortality occurring at about 1000 m, a feature that was particularly noticeable on the north-west-facing side of the mountain (Fig. 2.26). The number of larger standing dead trees was greater than the number of smaller ones, although this may partly reflect the decay rate of trees. The issue is complex, as there were also more larger red spruce at higher altitudes, suggesting an altitude-mortality-age structure interaction.

At Mount Moosilauke, in the White Mountains of New Hampshire, a

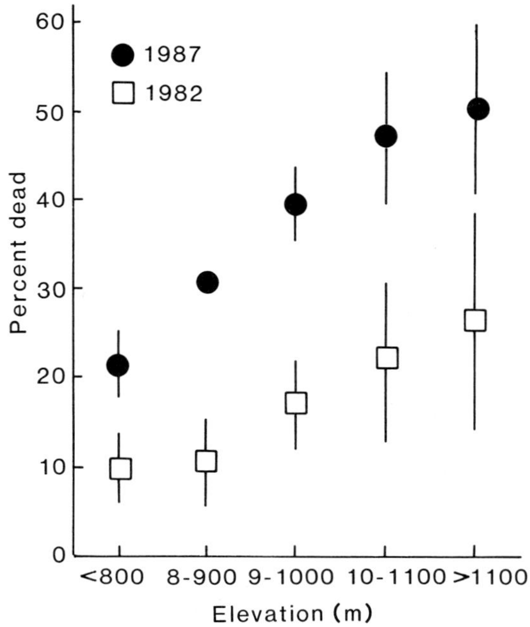

Figure 2.25 Increase in the percentage of standing dead red spruce at various altitudes on Whiteface Mountain, New York. The proportion of standing dead increased between 1982 and 1987, but the marked increase in mortality with altitude was maintained. From: Battles *et al.* (1988).

different pattern was evident (Peart *et al.*, 1988). As with many other areas in the north-east, the spruce-fir zone at Mount Moosilauke has a complex history of disturbance from logging and also experienced a major hurricane in 1938 (Peart *et al.*, 1988). The percentage of standing dead spruce and fir in 1987 was 29–34% and 16–17%, respectively. With spruce, there was a very slight trend of increasing mortality with altitude on east-facing slopes, whereas west-facing slopes showed a marked peak in mortality at middle elevations (ca. 1000 m). Fir also showed the latter trend on west-facing slopes, but mortality on east-facing slopes declined with altitude. There was an interaction with soil type, with greater mortality on cryofolists than spodosols, but this relationship was not apparent on both aspects. The occurrence of standing dead trees appeared to form a mosaic, but the causes of this were unknown.

The decline of red spruce at Camel's Hump, Vermont, was not reported as being accompanied by a decline in balsam fir (Fig. 2.27). Perkins (1988) failed to detect any consistent changes in the vigour of balsam fir on the mountain during the period 1979–86, yet a marked decline of red spruce occurred during the same period. However,

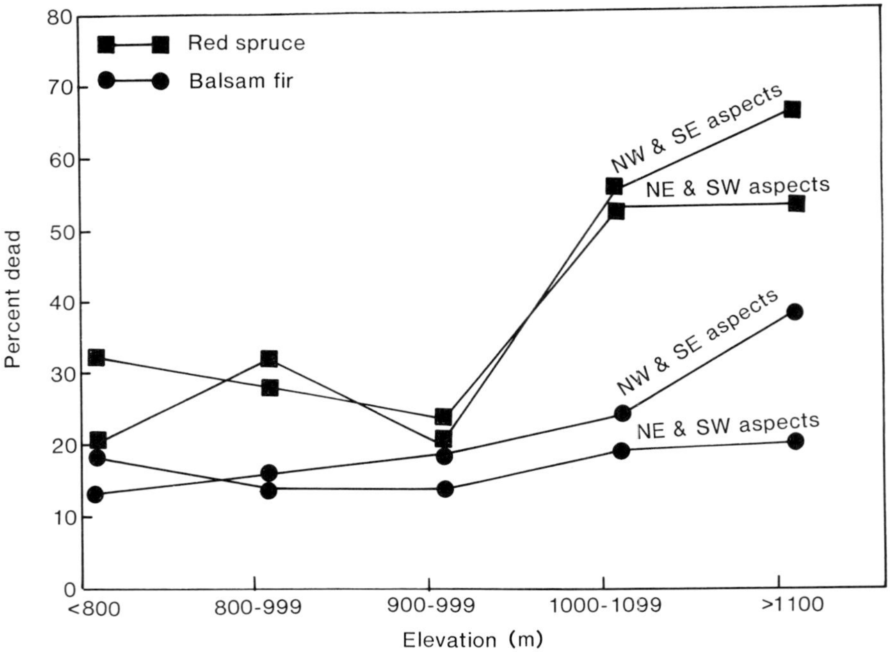

Figure 2.26 Percentage of dead red spruce and balsam fir by aspect and altitude at Whiteface Mountain, New York. From: Battles *et al.* (1988).

Vogelmann *et al.* (1988) found a decline in the basal area of balsam spruce on the mountain (but no decline in the density). The red spruce showed several foliar symptoms, including the development of secondary shoots, yellowing, defoliation and occasional short shoots and needles, whereas no symptoms were apparent on the balsam fir. Despite the absence of symptoms, the balsam fir in the mixed spruce-fir forest had not responded to the decline and mortality of the red spruce through competitive release, suggesting that some factor had prevented the improved growth of the trees. However, it is possible that the mixed species forest represents a near-optimal growth environment; in such circumstances, competitive release might not occur.

There is some uncertainty over the reports of an absence of decline in the balsam fir at Camel's Hump. Perkins *et al.* (1992) refer to 'asymptomatic declines' of balsam fir and white birch at the site. Standing dead trees were present during a survey of canopy gaps on the mountain. Above the mixed forest, wave mortality of balsam fir can clearly be seen, indicating that mortality is definitely present. Over much of the mountain, there is a vigorous understorey of balsam fir saplings, and these may have compensated for any decline in basal area caused by the mortality of mature trees.

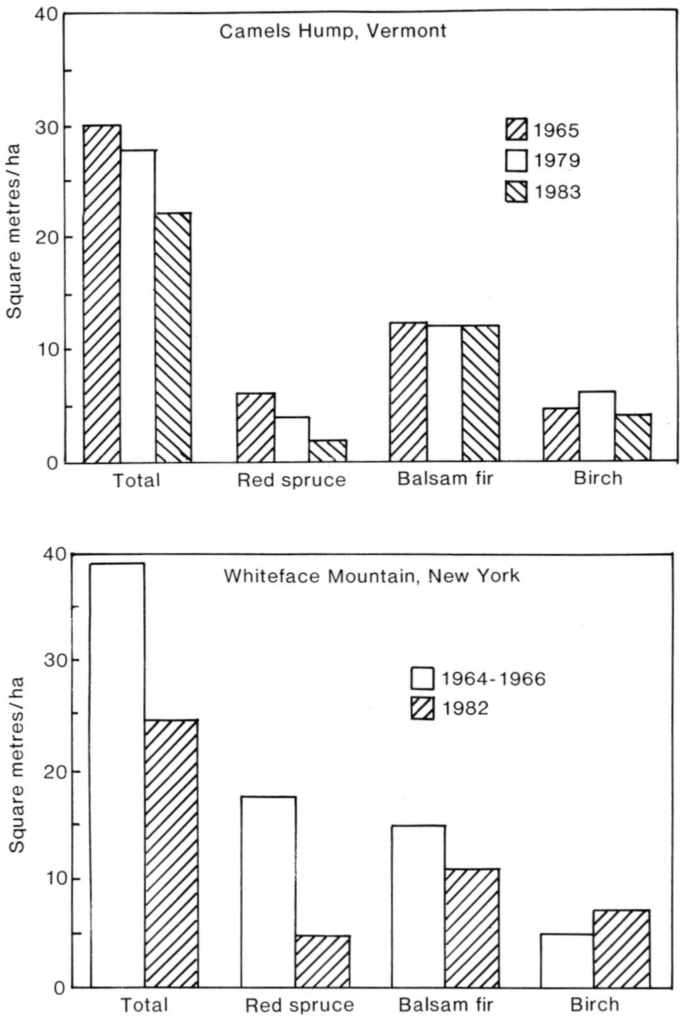

Figure 2.27 Changes in the basal area of three species at Camel's Hump, Vermont (live trees < 2 cm dbh) and Whiteface Mountain, New York (live trees < 10 cm dbh). Data drawn from several sources and compiled by Friedland (1989a).

The occurrence of dense regeneration seems to be a characteristic of the sites affected by decline in the 1970s and early 1980s. The regeneration appears to consist primarily of balsam fir, which may be a successional process or may be related to high mortality of regenerating red spruce as a result of damage by the conifer swift moth (*Korscheltellus gracilis* (Grote)) (Grehan *et al.*, 1992). The effects of this insect appear to have been considerably underestimated in many studies of red spruce mortality. It will be interesting to see how the even-aged regeneration develops in the future, as it may be particularly susceptible to wave-type declines.

Increment studies

The decline in crown condition in the north-east of the United States appears to have been paralleled by a decline in growth rates. Johnson and Siccama (1983a) found growth decreases at Whiteface Mountain, New York, Mount Mansfield, Vermont and Mount Washington, New Hampshire, starting in the mid–1960s. Conkey (1988) recorded major changes in the growth of a sample of 26 red spruce at Elephant Mountain, Maine, starting in the early 1960s, and noted that whereas ring width declines of a similar magnitude occurred in 1862–76 and 1907–20, the decline in ring density appeared to be unprecedented in magnitude. Similarly, Cook (1987) and Johnson *et al.* (1988a,b) recorded growth changes and a lower correlation between growth and climate variables in red spruce from the Adirondack Mountains of New York and elsewhere starting in the 1960s (Fig. 2.28). In addition, Hornbeck and Smith (1985) reported an onset of decline in red spruce growth primarily in New Hampshire and New York, starting about 1960 (Fig. 2.29). Many of these purported declines are difficult to interpret because of the techniques that were used to derive the time series. Dendroclimatic analyses have been similarly complicated by averaging, detrending and delagging (e.g. Cook, 1986, 1987; Federer *et al.*, 1988, 1989b).

A major problem with the interpretation of the reported growth declines has been the role of stand dynamics. For example, Hornbeck *et al.* (1986) reported that about 80% of their sample of over 4000 trees

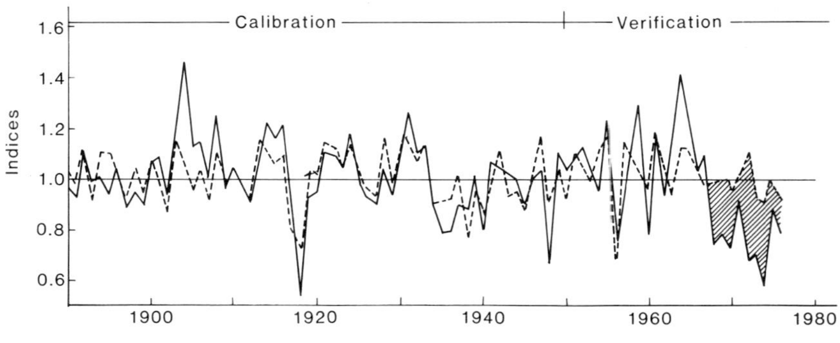

Figure 2.28 Observed and predicted growth of red spruce in the Appalachian Mountains. Predictions were based on the derivation of a climate–growth relationship for a control period (1890–50) and then forecasting growth for the recorded annual climatic conditions in the period 1951–76. The recent breakdown in the old climate-growth relationship is emphasized by shading.

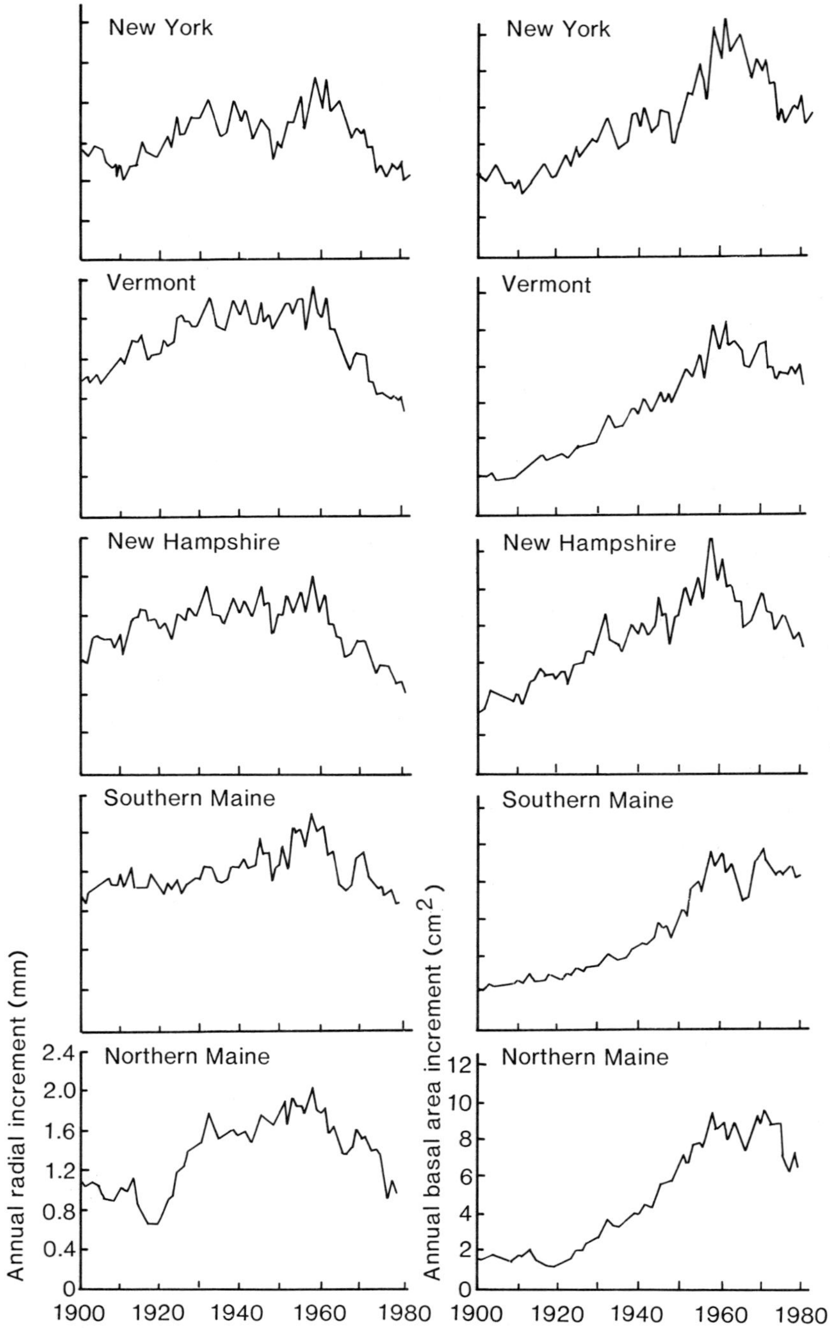

Figure 2.29 Mean annual radial increment (left) and mean basal area increment (right) for red spruce in New England. Sample sizes were New York: 224, Vermont: 303, New Hampshire: 391, Northern Maine: 1681 and southern Maine: 402. From: Hornbeck and Smith (1985).

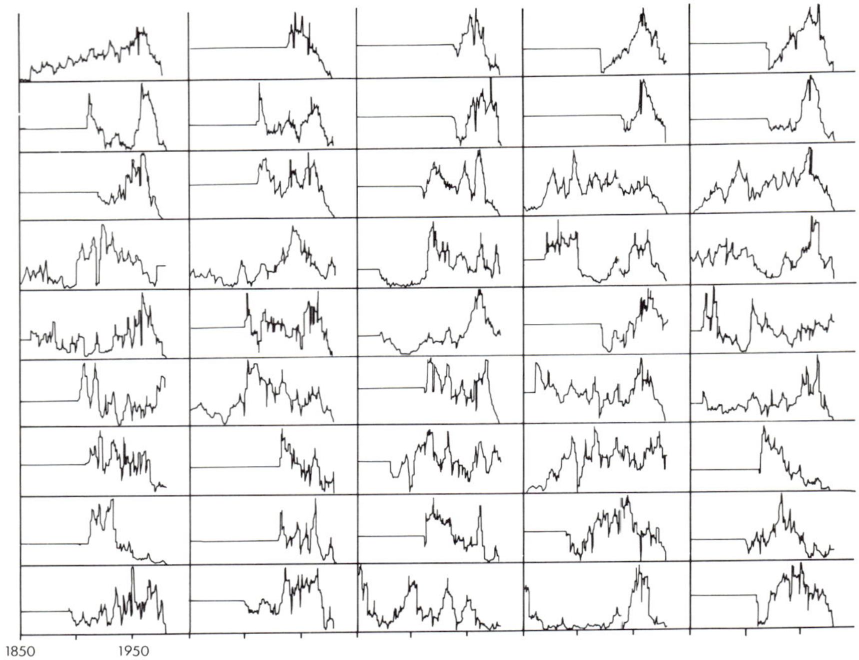

Figure 2.30 Raw ring widths from Whiteface Mountain, New York. The upper left hand corner is the average chronology and the remaining plots are of individual cores. The years are 1850–1981. The *y*-axis is scaled so each plot-window is fully utilized, so the scales are different among plots. Missing values are replaced with a straight line at the mean for the series. From: Van Deusen (1990).

germinated either in the period 1900–20 or in the period 1850–1900. The trees were severely suppressed until 1900–20 when selective logging resulted in their release and much faster growth rates. Selective logging is known to have been widely used in many mid-Appalachian forests (Adams *et al.*, 1985), and the importance of its effects on stand dynamics should not be underestimated (e.g. Federer and Hornbeck, 1987a, b).

Van Deusen (1987, 1990) and Van Deusen *et al.* (1991) argued that many observed growth declines in individual trees occurred in trees which show post-suppression releases. His work, together with studies by Cook (1990), demonstrated the dangers of using mean chronologies derived from large numbers of trees, such as the one produced by Hornbeck and Smith (1985). When Van Deusen split the data into individual tree series, two distinct populations were found. One showed release from the understorey less than 100 years ago, and the other showed release more than 100 years ago (Fig. 2.30). The younger trees showed a decline in radial increment starting in about 1955, whereas older trees showed both

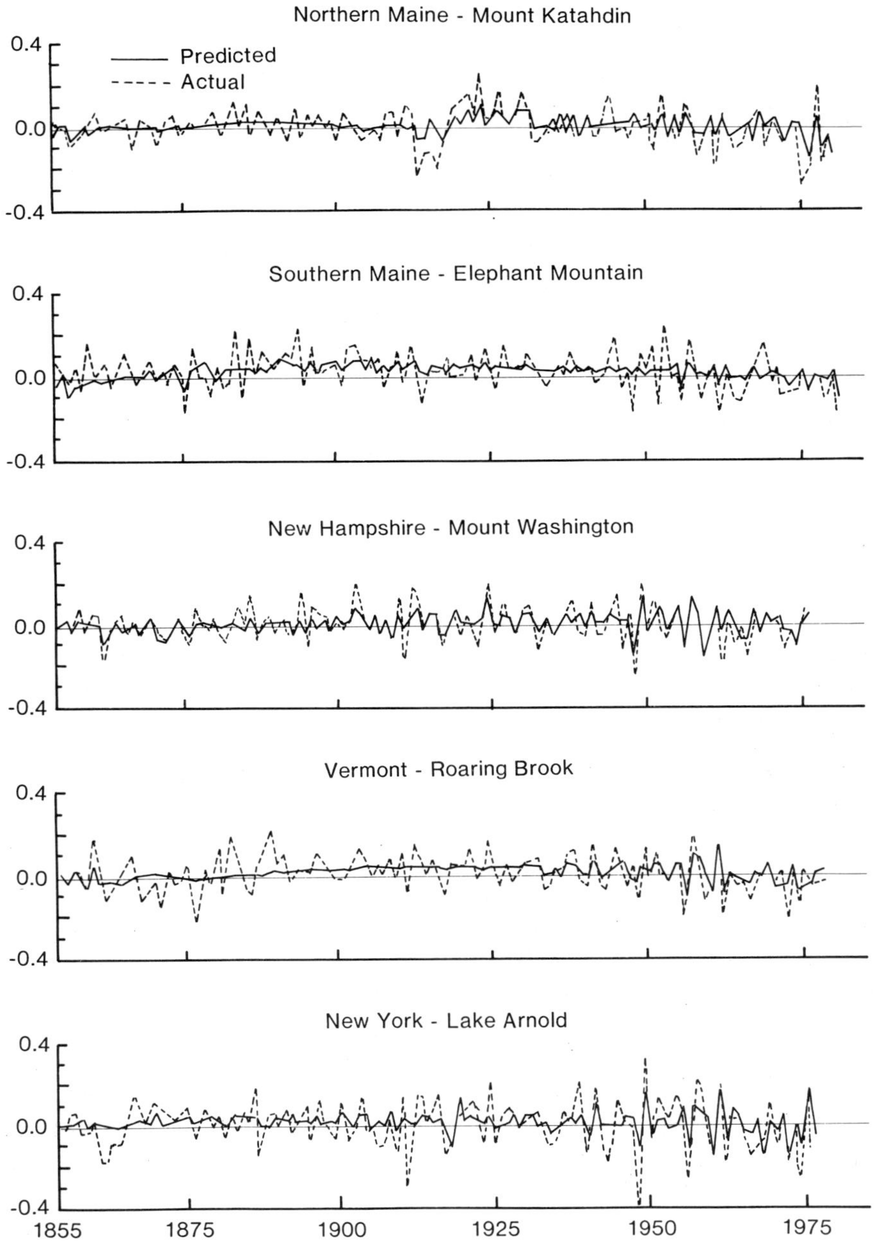

Figure 2.31 Kalman filter predictions of second-growth increment compared to old-growth increment of red spruce at a number of sites in the north-east. Predictions are shown by a solid line, actual increment by the dashed line. From: Van Deusen (1987).

increases and decreases since 1960. The increment of old-growth stands showed a high level of agreement with the predicted increment of second-growth stands (Fig. 2.31). Similar trends were apparent from low-altitude sites (Van Deusen, 1987), and increased growth of red spruce in northern Maine in the 1950s and elsewhere may be related to competitive release following the documented birch decline in the 1940s (Reams and Huso, 1990; Cook, 1990; Reams and Peterson, 1992). Stand dynamics have also been proposed as a reason for the increased mortality (Hyink and Zedaker, 1987), particularly in relation to nitrogen cycling (Pastor *et al.*, 1987). However, other factors are also involved as growth declines and mortality have been reported from stands reported to be undisturbed by fire or logging activities (Johnson *et al.*, 1988a, b; Cook, 1990) and, sometimes, declines apparently bear no relation to competition (LeBlanc, 1990a).

LeBlanc (1990a) sampled 46 dominant and codominant red spruce from sites at 900 m and 1100 m at Whiteface Mountain, New York. Both sites were believed to be virgin spruce-fir forest, and canopy red spruce had more than 70% mortality. Using a different technique from most other authors to define a decline (LeBlanc recognized a decline as occurring when there was a significant decrease in mean annual volume increment between two 10-year periods), LeBlanc argued that decline in growth was more severe at 1100 m and in larger trees. He found no evidence that the growth declines between 1964–84 were due to stand maturation, with trees aged between 50–200 years having reduced growth. However, the occurrence of growth declines in trees of a variety of ages does not rule out the possible effects of stand dynamics as, despite the apparent lack of evidence, the area may have been selectively logged or otherwise disturbed at some stage in the past. The remaining trees might then show a growth increase due to competitive release, followed by a decline at some stage following the attainment of canopy closure.

One of the problems with the study by LeBlanc was the use of annual volume. It is not always clear how this relates to radial increment, although LeBlanc (1990b) derived a relationship to link the two which was quite consistent. Detailed stem analyses of the trees provided further information on the history of the decline. LeBlanc and Raynal (1990) examined the relationships between apical and radial growth in the trees. After 1970, there was a high incidence of terminal leader injury, with the injury being greater at the higher site. However, throughout the last 100 years, there were more terminal leader injuries at the higher site than at the lower site. LeBlanc and Raynal (1990) found that the decline of leader growth and radial increment coincided on emergent red spruce. This suggested the presence of some external forcing factor rather than, for example, a decline caused by an internal mechanism such as calcium deficiency. The calcium deficiency hypothesis (see page 115) implies a decline in radial increment before a decline in other indices, which was not apparent in

LeBlanc and Raynal's study. They suggested that the growth decline may have been related to the loss of photosynthetic tissues, pointing out that the post–1960 declines coincided with a widely reported period of anomalous weather in the northeast United States (Namias, 1970; Karl *et al.*, 1984; Hamburg and Cogbill, 1988).

The role of air pollution

Pollution, particularly ozone, has been considered as a major factor involved in the decline. This, however, runs contrary to the evidence that red spruce and balsam fir are both tolerant of ozone (Fincher *et al.*, 1989; Johnson and Taylor, 1989; Laurence *et al.*, 1989; Kohut *et al.*, 1990; Lee *et al.*, 1990; Thornton *et al.*, 1990; DeHayes *et al.*, 1991; Patton *et al.*, 1991; Fincher and Alscher, 1992; Rebbeck *et al.*, 1992). In addition, Taylor *et al.* (1986), Jensen *et al.* (1988) and Lee *et al.* (1990) have reported the absence of any interactive effects between ozone and acidic precipitation, although the co-occurence of these two forms of air pollution, particularly in the southern part of the red spruce range, has caused concern (Vong and Guttorp, 1991). Despite the decline in red spruce having been reported from several different environments (Hornbeck *et al.*, 1986; Jagels, 1986; Hamburg and Cogbill, 1988), it does not appear to extend to other species (except for balsam/Fraser fir). For example, most broad-leaved species in the region (mainly present at lower altitudes) had a significant increase in basal area between 1965 and 1983 (Barnard and Scott, 1988; Hornbeck *et al.*, 1988).

The forests at the highest altitudes are exposed to acidic cloudwater and precipitation, with the mean pH for these being 3.5 and 4.2, respectively (Mohnen *et al.*, 1990). Consequently, considerable attention has been given to the possible role of atmospheric acidity.

In a detailed investigation of the effects of ambient ozone and cloudwater concentrations on indigenous red spruce seedlings (and also seedlings from the Great Smoky Mountains) at Whitetop Mountain, Thornton *et al.* (1990) used open-top chambers to evaluate the effects of removing either ozone or cloudwater or both. Net photosynthesis and respiration were unaffected by exposure to either type of pollutant and, after one growing season, there were no significant differences in the chlorophyll-a and b concentrations or the concentrations of carotenoids. However, needles of trees in treatments involving exposure to cloudwater had greater quantities of nitrogen. There was also a significant increase in the biomass of needles on trees grown in chambers where pollutants had been excluded. They attributed these apparently contradictory results to an alteration in the use or the translocation of assimilates or to the increased use of assimilates by pollution-exposed trees for repairs. Despite high levels of sulphate in the cloudwater, there was no evidence of increased sulphur levels in foliage exposed to the cloudwater. This agrees with the observations made by Robarge *et al.* (1989) in the southern

Appalachians that foliar sulphur contents were not unusually high. It is not clear from the work of Thornton *et al.* (1990) whether the nitrogen uptake was from nitrate or ammonium in the cloudwater (or both), since both were present in roughly equal quantities.

The study by Thornton *et al.* (1990) was extended to cover three years of exposure (Thornton *et al.*, 1992). There was little evidence of any effects of the reduced exposure to pollution after three years. Seedling diameters were unaffected and the only height growth differences suggested greater growth in the ambient ozone and cloudwater treatment. Removal of both ozone and cloudwater did not affect photosynthetic rates. However, there was some evidence, in the form of higher nitrogen concentrations, of a nitrogen fertilization effect in the ambient treatments.

Other impacts of cloudwater on red spruce are also uncertain. Deans *et al.* (1990) looked at the effects of artificial acidic mists on two-year-old red spruce seedlings. Mists of pH 2.5, 3.0 and 5.0 were applied twice-weekly from late July to early October in a single growing season. Misting did not affect biomass accumulation or height growth of the seedlings. However, marked differences in root morphology were noted, with significantly larger amounts of coarse roots produced at pH 5.0 and significantly more fine roots at pH 3.0. Misting increased the foliar content of nitrogen and sulphur, and stimulated the uptake of potassium, phosphorus and calcium. The uptake of nitrogen by the trees examined in this study and that of Thornton *et al.* (1990) contrasts with the results of Bowden *et al.* (1989). The discrepancies may be due to deficient levels of nitrogen which, in the study by Deans *et al.* (1990) were below sufficiency levels in all treatments except pH 2.5. In addition, the numbers of mycorrhizal fruiting bodies of *Thelephora terrestris* Ehren.:Fr. were significantly greater at pH 5.0, suggesting that acidic mist had an adverse effect of mycorrhizae.

Using material from the same experiment as Deans *et al.* (1990), Eamus and Fowler (1990) examined the photosynthetic and stomatal conductance responses of red spruce seedlings exposed to acidic mist treatments and Eamus *et al.* (1989) studied the water relations. The pH 2.5 treatment increased both the rates of light-saturated photosynthesis and the CO_2-saturated rate, when expressed per unit leaf area. The apparent quantum yield and the chlorophyll-a and b contents also increased. Acidic mist did not affect the cuticular resistance of the needles or the rates of day or night-time transpiration. This suggested that neither stomatal nor cuticular regulation of water loss were affected by the treatments. Although there has been some suggestion that the late start of the misting may have prevented some effects from occurring, red spruce buds generally flush between mid-June and the end of July (Evans and Biesemeyer, 1988), suggesting that the misting period coincided approximately with the active vegetation period.

One difficulty with the misting technique is that it is usually difficult to tell whether any uptake has been directly through the foliage or indirectly via the rooting medium. The two processes may have different effects, with foliar uptake resulting in growth reductions (Jacobsen *et al.*, 1989; Leith *et al.*, 1989b) and growth increases occurring if the uptake is via the rooting medium (Schier, 1986; Taylor *et al.*, 1986; Temple, 1988). In addition, the nature and frequency of the misting may have significant effects, depending on whether drying occurs between events (Jacobsen *et al.*, 1989, 1990). Usually, it is assumed that foliar uptake has occurred (e.g. Jacobsen *et al.*, 1989), but the work with radioactively labelled nitrogen described above suggests that uptake via the rooting system may be the more important process.

In one of the longer misting studies (two years) reported in the literature, Geballe *et al.* (1990) exposed red spruce seedlings to combinations of acidic mist and heavy metals (cadmium, copper, lead, manganese, nickel and zinc). As in other studies, height growth was unaffected by the pH 3.0 misting, even after two years. Lead only had effects at 2000 ppm, well above any exposure level of natural red spruce stands.

Other studies have looked at the effects of acidic precipitation on the reproductive characteristics of red spruce. Generally, these have not found any effect on cone scale weight, seed weight, number of seeds per cone, percentage germination of filled seed or germination (Scherbatskoy *et al.*, 1987; Feret *et al.*, 1990), although Feret *et al.* (1990) recorded a decrease in the percentage of filled seed per cone as the pH of the simulated acidic precipitation (pH 5.6 to 2.2) declined. These results suggest that the reduced cone size, seed size and germination capacity of red spruce at Camel's Hump, Vermont, which Klein and Perkins (1988) reported, should be attributed to factors other than acidic deposition.

Winter damage

Some damage to northern high-altitude red spruce is related to frost and/or winter desiccation, the two symptoms being difficult to differentiate (Wardle, 1981; Sheppard *et al.* 1989). However, although winter desiccation does occur (Herrick and Friedland, 1991), it seems to be quite rare in montane red spruce (Kincaid and Lyons, 1981), suggesting that most of the reported damage is related to freezing injury. When desiccation does occur, it is associated with periods of low relative humidity, high radiation levels, calm conditions and temperatures below –4°C (Boyce *et al.*, 1991).

Winter damage to red spruce is a well-known phenomenon and has been widely reported in the literature (e.g. Curry and Church, 1952; Tegethoff, 1964; Wheeler, 1965; Hadfield, 1968; NAPAP, 1990). Major events occurred in the Adirondacks in 1947–8 and 1959–64, in Vermont in 1959, 1961 and 1962 and in New Hampshire in 1962–5 (Curry and Church, 1952; Pomerleau, 1962; Morgenstern, 1969; Friedland *et al.*, 1984;

Weiss, 1985). Besides these specific cases, persistent reports of winter injury have occurred in most years in the north-east (Curry and Church, 1952; Perkins, 1988).

The normal pattern of bud development in red spruce is for three buds to form at the end of each twig, although occasionally only two may be present (Evans and Biesemayer, 1988). In addition to the three terminal buds, one or more lateral buds may be produced along a young twig. Evans (1986) described damage to one or two of the three buds forming at branch apices. Damage occurred in late winter and early spring 1985. Injury usually occurred to all needles on a twig, with no injury on adjacent twigs. Evans and Biesemeyer (1988) argued that this was caused by variations in the timing of periderm formation in autumn; twigs that broke late in spring had later periderm formation and were therefore more susceptible to winter injury.

Friedland *et al.* (1984) described needle loss from branch tips and from the apex of the crown without any appreciable chlorosis occurring and with only slight macroscopic damage to needles. Browning was only observed between February and April in spring 1984 in New York and Vermont. In Vermont in 1984, the year classes most frequently recorded as necrotic or missing were from 1980 and 1983. They attributed the damage to freezing injury occurring after de-hardening caused by warm spells in February 1981 and February 1984.

Damage to red spruce at Mount Moosilauke, New Hampshire, during the winter of 1988–9 was reported by Peart *et al.* (1991). During the winter, an average of 29% of the current-year foliage was killed. In addition, the foliar injury appeared to be related to the failure of buds to break in 1989. The damage increased with crown height at higher elevations but decreased with crown height at lower elevations (Fig. 2.32). The higher levels of injury in the lower parts of the crown at lower altitudes suggested that desiccation was not the cause of the injury. In addition, injury was recorded on all aspects of the crown, whereas desiccation injury would be expected to be greatest on the south sides. Consequently, freezing injury was the more likely explanation.

An association between winter injury and growth decline was shown as early as 1952 (Curry and Church, 1952). Johnson *et al.* (1988a, b) proposed that the declines in the northern Appalachians (Catskill, Adirondack, Green and White Mountains) could be triggered by severe winter damage to foliage and buds, and they provided dendroecological evidence to support this hypothesis. At higher elevations, the July and August temperatures in the year preceding the year of ring formation were negatively correlated with ring width. December and January temperatures before ring formation were positively correlated with ring width. Cook *et al.* (1987) provided further evidence from high altitude sites in New York and Vermont. They argued that periods of significant temperature stress occurred in the 1870s, the late 1930s, the 1940s and the late

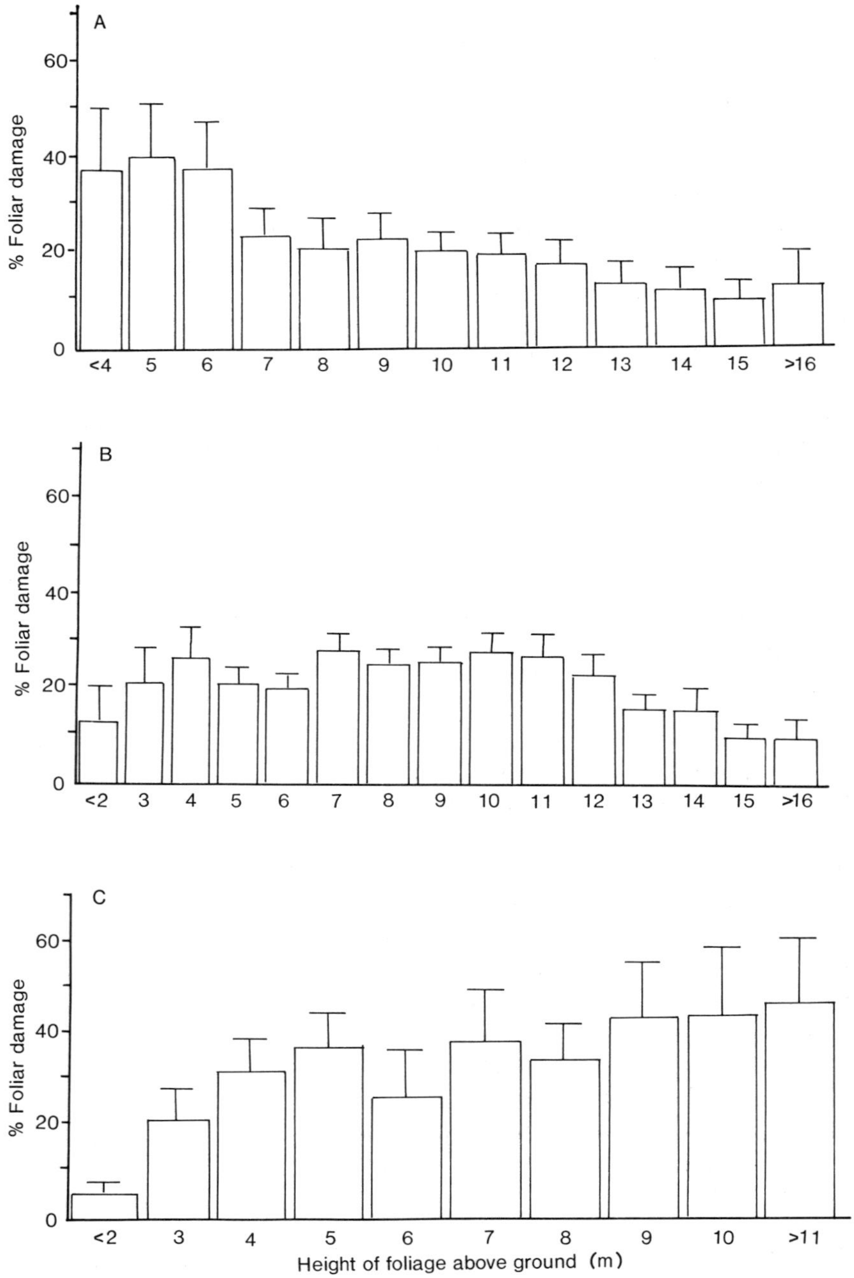

Figure 2.32 Effect of height above ground on injury to current-year foliage of red spruce in the spruce-fir zone at Mount Moosilauke, New Hampshire. A: low elevation (880 m); B: mid-elevation (990 m); C: high elevation (1140 m). Values are means for each 1 m height plus 1 standard deviation. 157 trees were sampled in 1402 crown sections. From: Peart *et al.* (1991).

1950s–early 1960s. These periods all coincide with documented periods of above-average red spruce mortality. The period 1959–64 was one of above-average incidence of winter injury in Vermont, New Hampshire and western Maine (Johnson *et al.*, 1988b; Reams *et al.*, 1990).

In contrast, Peart *et al.* (1992) found no relationship between the severity of winter injury and growth rates in the three years preceding the injury, suggesting that the severity of the injury was unrelated to the vigour of the tree. The severity of the injury did not affect growth rates in the two subsequent years. However, they acknowledged that repeated injury might trigger growth declines.

The damage reported in the field is different to the damage usually associated with winter freezing injury. Under normal conditions, this occurs on the current-year needles, and the buds and twigs are generally unaffected (DeHayes *et al.*, 1990). Current-year needles of red spruce are much less tolerant to cold than older needles, with older needles being able to withstand temperatures 8–12°C below the temperature that will induce damage in current-year needles. This differentiation is absent in balsam fir. Discoloration first appears as a yellow-green mottling on current-year needles early in spring, after the onset of consistently above-freezing weather. These needles then turn red-brown and are dropped before the onset of the new season growth (DeHayes *et al.*, 1990; Wilkinson, 1990). The damage reported in New York and Vermont by Friedland *et al.* (1984) in the spring of 1984 closely follows this pattern. Damaged shoots usually flush successfully, but Friedland *et al.* (1984) and Evans (1986) recorded damage to buds.

Examination of winter injury in provenance trials (DeHayes *et al.*, 1990; Wilkinson, 1988, 1990) revealed marked variations in the extent of winter injury (in trees from low altitude sites), with trees from Massachusetts and New Hampshire being consistently more injured than trees from either further north or further south. There have been several studies that have shown that introgression of black spruce results in increased frost resistance (Morgenstern, 1969; Eckert and O'Malley, 1988; Wilkinson, 1988; Eckert, 1989) and one of the reasons for the susceptibility of some populations may be reduced genetic variability (Eckert, 1989; DeHayes and Hawley, 1992; Hawley and DeHayes, 1992). Wilkinson (1990) examined the growth of the damaged trees in provenance trials and found that winter injury was closely related with growth decreases. Repeated injury in three successive years caused up to 59% reductions in basal area increment and up to 30% reductions in height increment. Although provenances with significant introgression by black spruce were as equally injured as the pure provenances, recovery of growth was faster in the provenances with significant black spruce components. The genetic affinities of the high-altitude red spruce populations are unknown.

Soil effects

A possible cause of the decline is calcium deficiency. In the Adirondacks, there is evidence of a progressive loss of calcium from the upper horizons of some soils over the last 50 years (Anderson, 1988). What remains uncertain is the cause of this loss. It may be due to leaching of calcium from the upper horizons or to uptake into biomass, with the latter being more likely.

However, there is little to suggest that the calcium deficiency hypothesis is applicable to the decline of red spruce. At Whiteface Mountain, New York, Johnson *et al.* (1989b) found few indications of calcium deficiency in the foliage of red spruce. There was no relationship between exchangeable base cations or base cations : aluminium ratios in the soils and base cation concentrations in foliage. Furthermore, there was no relationship between the base cation content of foliage and the amount of crown dieback.

Red spruce is quite sensitive to aluminium, with concentrations as low as 50 mg l^{-1} having effects on the roots of seedlings grown in nutrient cultures (McQuattie and Schier, 1990). However, it is unclear how applicable nutrient solution studies are to the field. Raynal *et al.* (1990) reported increasing reductions in fine root biomass with increasing aluminium concentrations from 0.1 to 2.5 mmol l^{-1}. However, they did not identify any consistent relationship between root aluminium concentrations and root biomass growth. Problems arise when relating the effects of aluminium to the calcium concentrations of the solution. Although significant reductions in biomass production were recorded at low calcium : aluminium ratios (0.03–1.0 in solution), reductions were seen at a wide range of calcium concentrations (up to a calcium : aluminium ratio of 37.8 in the soil). High aluminium concentrations and low calcium : aluminium ratios have been recorded in areas exposed to high levels of cloud deposition, and root development of red spruce in such areas may have been inhibited (Joslin and Wolfe, 1992). However, Miller *et al.* (1992) did not record high concentrations and it seems that some systematic studies of soil chemistry will be required to resolve the apparent discrepancies in the results of the various studies.

Generally, effects on root elongation have only been identified at aluminium concentrations much greater than those encountered in the field, with Schier (1985) and Thornton *et al.* (1987) reporting effects at 3.0 mmol l^{-1} and 0.25 mmol l^{-1}, respectively. Concentrations in the soils of spruce–fir stands are usually 0.03–0.08 mmol l^{-1} (Raynal *et al.*, 1990). At Mount Moosilauke, Huntington and Ryan (1988) found that shallow organic soils (cryofolists) predominate at higher elevations (1200 m). These generally had lower reserves of base cations and higher concentrations of aluminium than soils at lower altitudes (840 m). In addition to the above studies, Johnson (1983) found reduced aluminium concentrations and reduced aluminium : calcium ratios in fine roots with increasing elevation at Camel's Hump in the northern Appalachians. These results

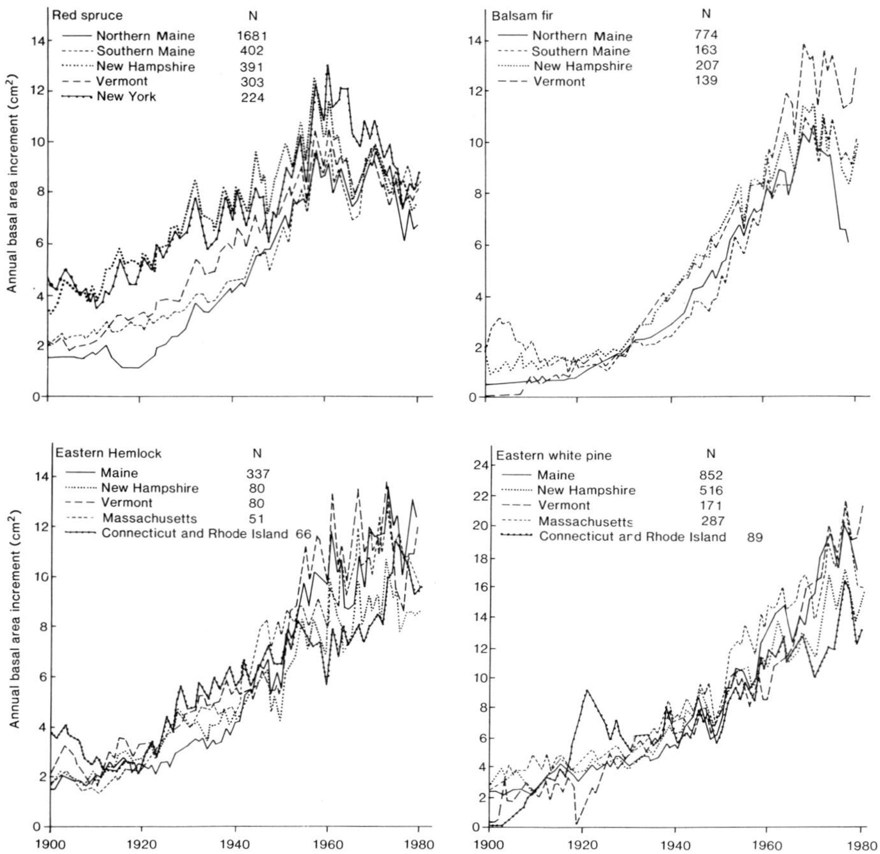

Figure 2.33 Annual basal area increment of red spruce and three other coniferous species in the north-east over the last 80 years. From: Smith *et al.* (1989).

suggest that the direct effects of aluminium toxicity can be ruled out as a cause of the decline of red spruce.

Northern low-altitude spruce decline

Friedland *et al.* (1988) defined low-altitude red spruce stands as those lying at elevations below 650 m ASL. Using annual basal area increment, Smith *et al.* (1989) documented a decline in the growth of predominantly low-altitude red spruce starting in the early 1960s, following a long period of increasing increments between 1900 and 1960 (Fig. 2.33). Hornbeck *et al.* (1988) and Federer *et al.* (1988) argued that the decreasing basal area growth rates recorded in balsam fir and red spruce represented the normal ageing pattern of low-elevation, even-aged spruce-fir forest. Solomon *et al.* (1989) reached a similar conclusion after examining growth data from 774 plots in the north-east United States and in Canada. The stand ageing argument is supported by the expected growth trends following extensive outbreaks of the spruce budworm (*Choristoneura fumiferana* Clem.) in Maine and northern New Hampshire between 1910 and 1920. These resulted in the mortality of between 33% and 66% of the low-elevation spruce-fir forests (Seymour, 1985). The forests now consist largely of even-aged stands of spruce and fir in which the mean annual basal area increment should have peaked in about the early 1960s (Seymour, 1985; Reams *et al.*, 1990). Mortality in the low-altitude red spruce forests is within the range that would be expected from natural factors (Reams *et al.*, 1990; ARNEWS, 1992), throwing further doubt on the suggestion that an abnormal decline in condition has occurred.

In Canada, there is no evidence of a decline in the condition of red spruce in the Maritimes and Quebec regions (Hall, 1991). Mortality is present, but only at very low levels (>1% per annum), and has been attributed to windthrow, porcupine damage, beetle attack and root rot fungi (*Armillaria mellea* Vahl: Fr.).

There is a possibility that the same 'fog-related' problem that affected birch on the shores of the Bay of Fundy also affected red spruce on the Maine coast (Jagels, 1986; Jagels *et al.*, 1988). A gradient study in the area (Jagels *et al.*, 1988) revealed different symptoms to high-altitude stands, including chlorosis, necrosis, defoliation, reduced terminal growth and reduced radial growth. The chlorosis was not related to magnesium deficiency, and soil acidification was also inapplicable as a causal mechanism. The most severe symptoms were seen at sites with thin soils and south or south-easterly exposures. A decline in growth rates accompanied the symptoms. Unfortunately, the regional extent of the problem was unknown and the description of symptoms was restricted to a single location (Isle au Haut) where unique site factors may have played an important part.

Few studies have looked at the possible effects of air pollution on low-lying spruce-fir forests in the north-east. Lawrence and Fernandez (1991) studied the nutrient budgets of a site in the Howland Forest in east-central Maine. Acidic deposition was present, but the soil chemistry and soil solution buffering were largely controlled by organic matter and mineral weathering. Mobilization of aluminium was minimal and ratios of calcium : aluminium were well below the thresholds where effects were likely to occur (Meiwes *et al.*, 1986). Consequently, acidic deposition does not seem a major factor affecting the condition of the trees.

Southern high-altitude red spruce and Fraser fir

In the southern Appalachians, spruce-fir forests form relict populations near the summits of the higher mountains. The current area covered by forest is only about 10% of its former extent, with commercial logging and forest fires during the early 1900s having almost eliminated the ecosystem type (Korstian, 1937). Since the logging, large periodic wind blows in forests near clearcuts and roads have further reduced the area (Pyle and Schafale, 1988). About 26 600 ha remain, much of which (74%) is located in the Great Smoky Mountains National Park (Dull *et al.*, 1988; Groton and Eagar, 1988; Reams *et al.* 1990). Although the forests cover a very small area, they form an important component of the landscape.

The uppermost parts (above ca. 1890 m) of the mountains usually consist of pure Fraser fir (*Abies fraseri* (Pursh.) Poiret) stands, which grade downwards into mixed spruce-fir forest and then forest dominated by spruce. The fir extends down to about 1670 m and the spruce to 1370 m before grading into hardwood forests. These mountains are much more humid than the mountains carrying spruce-fir forest in the north-east, but temperatures in the equivalent vegetation zones are similar. The symptoms in the southern Appalachians are different to those further north, with loss of needles and mortality being the main phenomena (Bruck, 1985).

Status of the forests

The southern spruce fir forests have probably provided more dramatic media images than any other forest type in North America. At several sites, the death of all canopy Fraser fir has occurred (Fig. 2.34). This has been accompanied by the decline of red spruce lower down in the mixed spruce fir forests.

Figure 2.34 Death of Fraser fir at Mount Mitchell, North Carolina, in 1987.

Some of the most spectacular mortality has been associated with the pure Fraser fir stands at higher elevations. This can be directly attributed to the balsam woolly adelgid (*Adelges piceae* Ratz.), a species introduced from Europe around 1900. The insect started to affect the southern spruce-fir forests in the mid–20th century, with the first record at Mount Mitchell, North Carolina, being in 1957 (Speers, 1958). A few years later, the insect was reported as occurring throughout the Black Mountains of North Carolina and the Great Smoky Mountains of Tennessee (Nagel, 1959; Ciesla *et al.*, 1963). Spraying with insecticides to control this pest was stopped in the early 1970s (Zedaker *et al.*, 1989), since when mortality of fir has been substantial. Infestation by the adelgid usually kills mature Fraser fir within nine years, and mortality can occur as quickly as three years after the initial infestation (Amman and Speers, 1965).

Dull *et al.* (1988) analysed an aerial survey of a variety of southern spruce-fir forests conducted in 1985. Of the sites sampled, over 24% of the area had more than 70% of the dominant and codominant standing trees dead. However, there were marked variations in the condition of trees at different sites (Fig. 2.35). The highest mortality was present in Fraser fir stands severely affected by the balsam woolly adelgid. However, it is important to remember that standing dead trees do not necessarily

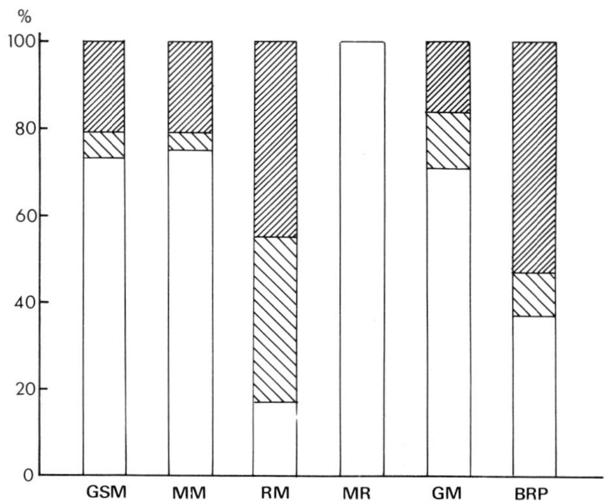

Figure 2.35 Extent of mortality in spruce fir forests in the south-east. Areas with heavy shading had severe mortality (>70% standing dominant and codominant trees dead, heavy mortality (30–70% trees dead) is shown by light shading and light (<30%) mortality by open bars. GSM: Great Smoky Mountains National Park; MM: Mount Mitchell; RM: Roan Mountain; MR: Mount Rogers/White Top Mountain; GM: Grandfather Mountain; BRP: Blue Ridge Parkway. Data from Dull *et al.* (1988).

represent recent mortality: Zedaker *et al.* (1989) considered that standing dead stems may last up to 25 years before falling or rotting away.

A network (129 plots) of sample sites was established by Zedaker *et al.* (1988) in three separate areas: Mount Rogers (Virginia), the Black Mountains of North Carolina and the Great Smoky Mountains of Tennessee/North Carolina. Survey data suggested that the lower elevation stands had basal areas consistent with those predicted by Meyer (1929). Site quality decreased with altitude but increased southwards and current basal area was closely related to site quality. The basal areas reported by Zedaker *et al.* (1988) were greater than those reported by McLaughlin *et al.* (1987) but were consistent with those of Oosting and Billings (1951). McLaughlin *et al.* (1987) found lower stand densities and, because of this, argued that stand dynamics were unlikely to be an important factor governing mortality in the stands. However, the 25–50% higher figures found in the more representative survey by Zedaker *et al.* (1988) strongly suggest that stand dynamics may be critical in some stands. Zedaker *et al.* (1988) drew an important distinction in the response of mixed stands of red spruce to mortality of Fraser fir. They suggested that in stands with low densities of fir, competitive release may occur, enabling the remaining red spruce to grow better. In stands with initially high densities of Fraser fir, the microclimatic changes occurring with the opening of the canopy may have led to mortality of the spruce.

At Clingman's Dome, Fraser fir was still the dominant species at higher altitudes in the mid–1980s, despite a basal area decrease of 47% (Busing and Clebsch, 1988). This can be attributed to the abundant regeneration of fir in damaged areas. Elsewhere, regeneration has been less successful and, at Mount Collins in the Great Smoky Mountains, successful establishment did not occur in the second half of the 1980s, despite good seed production (Nicholas *et al.*, 1989a; Pauley and Clebsch, 1990). A possible cause for this was competition with the thornless blackberry (*Rubus canadensis* L.). There has also been concern that the mortality occurring in some of these high-altitude stands may cause severe problems in the future. The problems are related to the microclimatic changes occurring at these and other sites, which inhibit adequate regeneration and stand development (Perkins *et al.*, 1987). However, modelling studies suggest that the continued persistence of Fraser fir is likely (Dale *et al.*, 1991).

At Mount Mitchell, North Carolina, Bruck and Robarge (1988) reported a mosaic of decline. The development of decline has been monitored in 16 stands 25–100-years-old, with a decline index being based on estimates of the percentage of the crown length affected and the number of needle years in the affected part of the crown. Normal needle retention was assumed to be six years in red spruce and five years in Fraser fir. The red spruce figure was considerably lower than the observed needle retention (up to 15 years) of red spruce at sites in the northern Appalachians (Miller-Weeks *et al.*, 1988), but this may reflect genetic

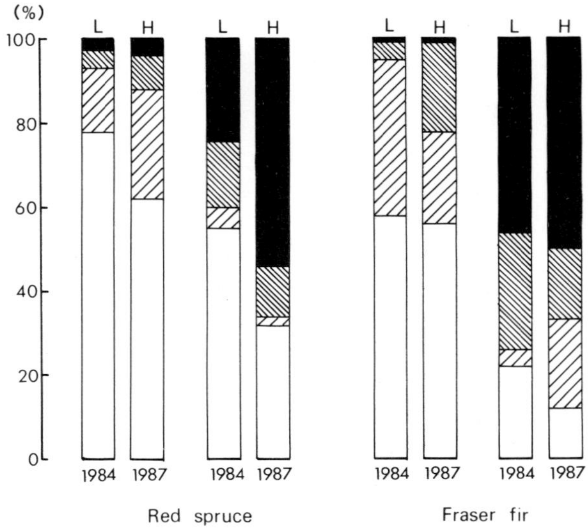

Figure 2.36 Changes in high-altitude (H: >1800 m) and low altitude (L: <1800 m) red spruce and Fraser fir at Mount Mitchell between 1984 and 1987. Black: standing dead; heavy shading: 51–99% foliar loss; light shading: 11–50% foliar loss; open: 0–10% foliar loss. Redrawn from Bruck and Robarge (1988).

differences. Since 1985, forest health has deteriorated and, by 1987, 41% of the red spruce and 49% of the Fraser fir were dead (Fig. 2.36). This was thought to be related to the severe regional drought in the spring and summer of 1986 and an intense rime-ice event on 11 December 1986. Trees on west-facing slopes were consistently worse than trees on east-facing slopes (Fig. 2.37).

A different pattern was apparent at Mount Rogers, Virginia, where an increase in basal area of red spruce has been documented by Friedland and Battles (1987) between 1954–61 and 1982–3. However, they pointed out that there were problems with the degree to which the two assessments were comparable. These results contrast with the results of an air photo survey of stands in West Virginia (Mielke *et al.*, 1986; Mielke, 1988), which showed that about a third of the red spruce volume was dead or severely declining (see below). The results from Mount Rogers revealed an inconsistency with the purported decline in the south. While decline at some sites has been severe, other sites that are apparently similar have shown no decline. The reasons for the spatial differences in the nature, extent and severity of the decline remain unclear, although the continued negligible effect of the balsam woolly adelgid on the Fraser fir at Mount Rogers is undoubtedly significant.

Comparisons between the different sites are complicated by the

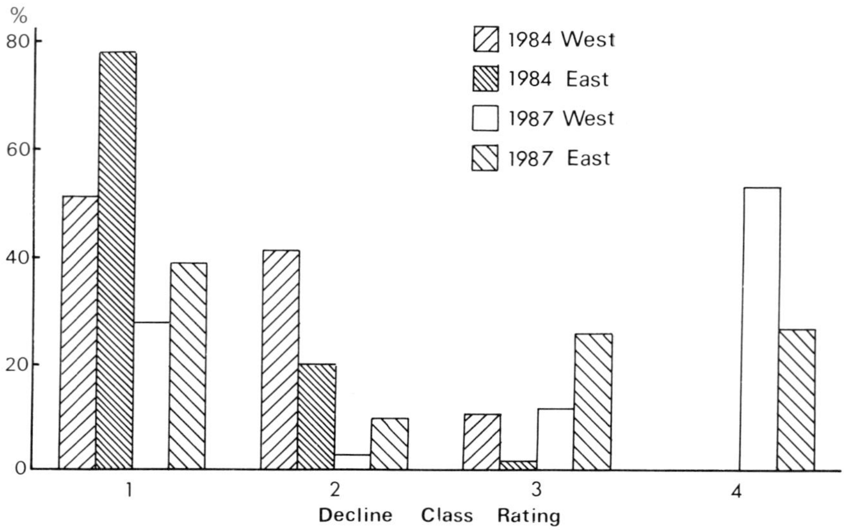

Figure 2.37 Percentage of red spruce at Mount Mitchell in each decline class as a function of aspect for the 1984 and 1987 growing seasons. 1: 0–10% foliar loss; 2: 11–50% foliar loss; 3: 51–99% foliar loss; 4: standing dead. From: Bruck *et al.* (1989).

apparent inability of the different research teams to standardize their methods of assessment. However, Nicholas *et al.* (1989b) reported the results of surveys conducted between 1985 and 1988 at three different sites in the south (Fig. 2.38). At Mount Rogers, there was little change in the condition of red spruce between 1985 and 1988. In contrast, there were marked increases in the numbers of defoliated trees over the same period in the Black Mountains and the Great Smoky Mountains. In the Black Mountains, the increase was particularly noticeable at higher elevations.

Mielke (1988) used aerial photography to assess the condition of red spruce in West Virginia where the stands are intermediate between the southern and northern forest types. As in other areas, the surviving red spruce forests are only a fragment of their former size. Most areas were logged by 1920 (Clarkson, 1978). Attacks by bark beetles (*Polygraphus rufipennis* Kirkby and *Dendroctonus frontalis* Zimm.) in the late 19th century also resulted in heavy losses (Hopkins, 1899). Mortality of red spruce was limited, with only about 7% of the total area of spruce-type forest having more than 10% mortality. However, expressed as a percentage of the number of red spruce trees, the situation was considerably worse. Thirty-five percent of the basal area, 40% of the number of trees and 33% of the volume was classified either as dead or as declining (more than 40% foliage loss). As with the study by Miller-Weeks *et al.* (1988a)

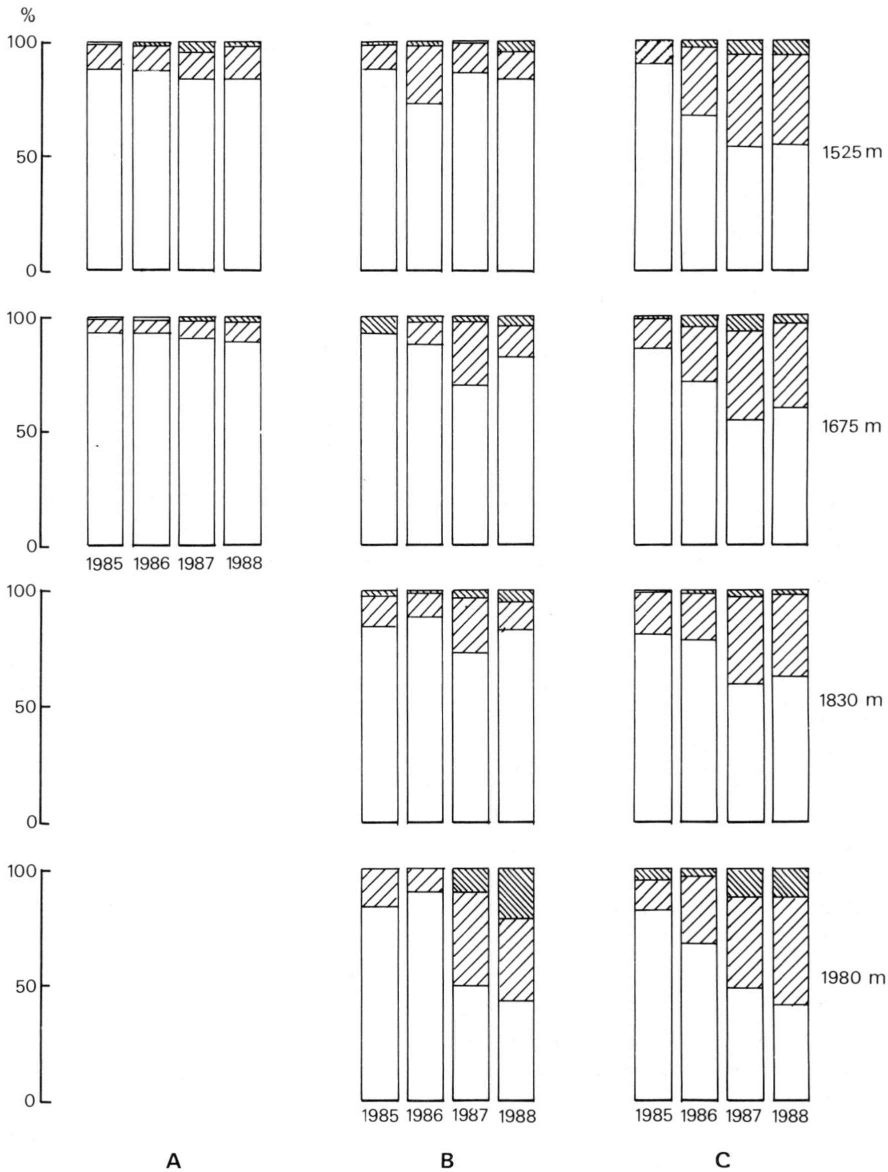

Figure 2.38 Comparison of crown condition at three sites in the southern Appalachians. Heavy shading: 51–99% foliar loss; light shading: 11–50% foliar loss; open: 0–10% foliar loss. A: Mount Rogers National Recreation Area; B: Black Mountains; C: Great Smoky Mountains. The altitudinal bands of the plots are given on the right-hand side. Data from Nicholas *et al.* (1989b).

in the north-east, the primary symptom associated with declining red spruce was branch mortality. This was commonly associated with the fungus *Valsa kunzei*. Although this fungus is known to kill trees, it normally only invades as a secondary pathogen; a primary stress, such as winter injury, is required to predispose the tree to infection (Waterman, 1955; Lavalle and Bard, 1978).

In contrast to the northern Appalachians, some studies have suggested that there is not a marked decline in the health of red spruce with increasing altitude in the south (Johnson and Siccama, 1983), although the elevational range over which red spruce occurs is limited. The impression of increased mortality at higher altitudes is partly the result of increased proportions of Fraser fir on higher ground, and it is this species that has suffered the greatest mortality. However, Zedaker *et al.* (1988) reported increasing mortality of both red spruce and Fraser fir at higher altitudes at some sites and higher mortality at higher altitude was also found by Bruck and Robarge (1988). Consequently, it seems that suggestions that mortality was evenly distributed across the altitudinal range of red spruce in the south were premature.

Increment studies

Increment analysis of red spruce and balsam fir in the Appalachians of Virginia suggested that the onset of the growth decline in individual trees was variable at a given site (Adams *et al.*, 1985). Many trees went into decline in the mid 1960s, a time of increased drought stress in many areas. A drought in about 1930 was also associated with growth declines, but the trees recovered quickly, whereas there was no indication of recovery in the cores studied by Adams *et al.* (1985).

As with the northern red spruce, there is good evidence that stand dynamics have been involved in the decline of basal area of red spruce at some sites (Hyink and Zedaker, 1987; Zedaker *et al.*, 1987). At lower altitudes, the basal area increment of red spruce in the relatively even-aged stands is consistent with the predictions of Meyer (1929). Disturbance in stands, particularly those at lower altitudes, has been considerable (Wentworth *et al.*, 1988), with Pyle *et al.* (1985) reporting that over 60% of the lower elevation red spruce stands in the Black Mountains and at Mount Rogers show signs either of logging or of burning. Disturbance in the Great Smoky Mountains seems to have been much less, with only 20% of the stands showing evidence of burning or logging.

Pollution

The strongest proponents of a pollution-related causal complex for the southern spruce-fir forests have been Bruck and co-workers (Bruck, 1985; Bruck *et al.*, 1989). Bruck (1985) argued that nitrogen-induced frost susceptibility might be involved, since frost degree, nitrogen deposition and injury all increase with altitude. In a later paper, Bruck *et al.* (1989)

argued that acidic cloud water might accelerate the occlusion of wax plugs in the stomata, leading to a decrease in photosynthetic efficiency and diffusion resistance of water on the needles. This could cause a reduction in tree growth and early senescence of the needles. Given that Mount Mitchell has the most acidic cloudwater episodes in the Appalachians (Li and Aneja, 1992), Bruck's hypothesis requires further study.

Johnson *et al.* (1991a) examined the nutrient budgets of spruce-fir forest sites near Clingman's Dome in the Great Smoky Mountains National Park. They were particularly interested in the hypothesis of Shortle and Smith (1988) that inhibition of calcium ion uptake by elevated soil solution aluminium concentrations might result in calcium deficiency. This in turn would lead to reduced cambial growth, as the demand for calcium per unit cambium is constant, leading to reduced functioning of the sapwood and, ultimately, reduced leaf area. Johnson *et al.* (1991) suggested that existing records did not support this hypothesis. Most of the high elevation soils in the spruce-fir range are Histosols, in which the build-up of potentially toxic concentrations of aluminium is unlikely. Furthermore, it is widely believed that it is foliar area that normally determines the extent of sapwood area, rather than the reverse, with Tyree *et al.* (1987) showing that stems have a water-conducting capacity 50–1000 times more than the crown foliage requirements. Elsewhere, there is limited evidence of calcium deficiency, with Robarge *et al.* (1989) reporting low foliar levels of both calcium and magnesium and high levels of aluminium from a variety of southern spruce-fir sites. However, McLaughlin and Tjoelker (1992) have emphasised the possible importance of acidic deposition on the nutritional status of the soils.

Johnson *et al.* (1991b) found no evidence to support the contention that increased aluminium concentrations were inhibiting calcium uptake. There was no indication of calcium deficiency, nor were potassium or magnesium concentrations deficient. Based on the nutrient fluxes, Johnson *et al.* (1991) considered that calcium limitations were the most likely potential nutritional problem. However, there was no net loss of calcium from the system, suggesting that it is being tightly cycled. The soils were found to be well within the aluminium buffering range and were therefore unlikely to acidify much further through base cation loss, despite an apparent net loss in magnesium from the systems. A major factor influencing the calcium budget at the sites was the atmospheric input of calcium ions, a source that is likely to be important throughout the eastern USA, and the budgets showed that calcium was accumulating in the forest floor.

Further evidence that the aluminium hypothesis is not applicable to the study sites comes from the root distributions. Despite red spruce being classified as a shallow-rooted species (Fowells, 1965), roots extended down to 60 cm or more, suggesting that they were not being inhibited by toxicity. Similar rooting depths were recorded by Kelly and Mays (1989), although they found that the rooting depths varied significantly between soil types, being greatest in Pachic Haplumbrepts and least in Lithic Borofolists. They also found that exchangeable aluminium concentrations were generally low (< 2.0 cmol kg^{-1}).

Other factors

Warm periods during the winter are a further possible factor in the decline (Friedland *et al.*, 1985). In mixed stands of Fraser fir and red spruce, mortality of Fraser fir caused by the balsam woolly adelgid has resulted in the rapid exposure of red spruce to climatic stress. This appeared to have adverse effects on the spruce, triggering decline and mortality (Zedaker *et al.*, 1989). The exposure has been combined with other stresses, particularly three severe droughts between 1985 and 1987 and two major ice storms in the winter of 1986–7. These processes do not appear to occur in the northern spruce-fir forests nearly to the same extent.

The role of extreme climatic events may have been underestimated in the area. Nicholas and Zedaker (1989) reported extensive injury caused by two severe ice storms during the winter of 1986–7 in the Black Mountains of North Carolina. The storms were of sufficient intensity to cause mortality, with 14.1% of the dominant and co-dominant trees dying between 1986 and 1987. Overall, the mortality was greater among red spruce (8.1%) than Fraser fir (2.4%). Among the surviving trees, ice damage was apparent on 12.4% of the stems, causing, in red spruce, an average height reduction of 2.3 m (17%). The damage was greatest on dominant and co-dominant trees, with 28% of trees losing their tops or having another type of obvious ice damage. As might be expected, the extent of the damage increased with altitude. Since storms of a similar magnitude occur every 2–4 years in the southern Appalachians (Nelson and Zillgitt, 1969), damage from them is likely to occur frequently.

In situ physiological studies

There has been increasing interest in examining the physiology of trees growing at high altitudes. Andersen and McLaughlin (1991) looked at the water relations of current-year shoots at two elevations (1720 m and 1935 m) at Clingman's Dome, Tennessee. Here, data from increment cores have suggested that the degree of growth reduction increases with increasing altitude. The reduced growth at higher altitudes was confirmed by measurements of relative height growth. These showed that the relative height growth of trees at the lower site was 68% greater than at the higher site. Andersen and McLaughlin (1991) concluded that although water stress occurred during the growing season, it was not the dominant factor affecting growth, although there was some evidence that during dry periods the trees at the higher site were under greater moisture stress. Several factors could have caused this, including the shallower soils at higher altitudes and the more open canopy, which had 50–75% closure at the upper site and more than 75% at the lower site. These field-based results appear to corroborate the conclusion of Roberts and Cannon (1989) that major disruptions to plant-water relationships in red spruce were unlikely to be caused by air pollution and that drought was a more important factor. Andersen and McLaughlin (1991) found no evidence of winter desiccation stress.

Andersen *et al.* (1991) looked at the foliar pigment composition of red spruce at three sites in the Great Smoky Mountains National Park in the southern Appalachians. Foliar flecking occurred at all sites and, on two mountains, it increased with altitude. The degree of foliar flecking increased during the winter, suggesting a winter-related cause. The extent of the flecking was generally very low (<3% and only 39 of 290 current-year needles affected), except at Clingman's Dome, where up to 10.6% of the upper surface areas of needles was affected. Pigment concentrations in the needles were unrelated to the flecking frequency. Andersen *et al.* (1991) considered that the flecks were distinctly different to those associated with photo-oxidant injury, 'winter damage' and nutrient deficiencies (c.f. Fincher *et al.*, 1989; Liedecker and Klein, 1989) and attributed them to localized freezing injury, similar to that on Californian pines in the Sierra Nevada of California and elsewhere.

In a further study at the same sites, McLaughlin *et al.* (1990, 1991) looked at patterns of photosynthesis and respiration of red spruce in an attempt to explain the slower growth of saplings at the higher site. The difference in the growth rates of the saplings was associated with increases in dark respiration and less favourable net photosynthesis to dark respiration ratios. The higher site had lower foliar calcium and magnesium contents, reduced foliar chlorophyll and higher foliar aluminium concentrations. These trends resulted in an association between calcium concentrations in the foliage and dark respiration rates. The results suggested that ozone could be ruled as a factor affecting the physiology of the saplings and pointed to the possible involvement of reduced calcium availability. Aluminium toxicity was unimportant, the low levels of calcium being the result of reduced availability rather than calcium–aluminium antagonisms.

Studies throughout the spruce-fir range

Several studies have attempted to compare red spruce throughout or across large parts of its range. Hanson *et al.* (1982), in one of the earliest studies into the possible effects of pollution in the region, looked at changes in the chemical composition of the forest floor in subalpine balsam fir stands on 14 mountains extending from New England to eastern Canada. As the mean precipitation acidity increased, the concentrations of calcium and magnesium in the forest floor decreased, and lead and the concentration of hydrogen ions increased. However, Johnson *et al.* (1989b) found no systematic change in pH in extremely acidic soil horizons sampled in the Adirondack Mountains between 1930–2 and 1984. There was no systematic change in the amount of extractable calcium in the B horizons, although there was evidence of acidification of the E horizons. Soils that initially had higher pH showed a pH decrease and a reduction in extractable calcium. This trend was consistent with the normal soil changes associated with the development of forest over 50 years.

Bondietti *et al.* (1990) looked at the divalent cation contents of red spruce wood from a variety of sites in New England, Tennessee and North Carolina. They found an apparent anomaly in the cation contents, with the concentration of calcium and other divalent cations increasing in wood formed in the mid 1900s and then abruptly decreasing. At Camel's Hump, the peak in divalent cation concentrations occurred in 1960. The trends were not universal, with trees from Beddington, Maine, showing no trend. Bondietti *et al.* (1990) attributed the changes to changes in calcium availability. Linking this to acidic precipitation, they argued that calcium availability had been reduced by acidity-induced leaching. However, as with all studies of elements in tree rings (e.g. Donnelly *et al.*, 1990; Momoshima and Bondietti, 1990), there were major problems associated with determining the extent of translocation within rings once formed. In addition, studies relating foliar concentrations to crown condition (e.g. Huntington *et al.*, 1990) have not found a good relationship between the two, nor has marked calcium deficiency been observed in the foliage. However, there has been a suggestion that phosphorus deficiencies may be linked with the condition of trees (Huntington *et al.*, 1990), and this warrants further investigation.

In other cases, the reasons for looking at trees from more than one location are unclear. For example, Meier *et al.* (1989) took trees from Maine and potted them in a Typic Haplumbrept soil from Mount Mitchell, North Carolina, in a study of the effects of acidic deposition on ectomycorrhizae. After this rather curious procedure, the potted trees received either 20 or 50 applications of simulated acidified rainfall, with treatments at pH 5.5, pH 3.5 and pH 2.5. This short-term exposure had no apparent effect on the total numbers of ectomycorrhizal root tips, but the number of tips formed by *Cenococcum geophilum* showed an increase after 50 applications of the more acidic solutions.

Winter damage

An increase in frost sensitivity (particularly to autumn frosts), if such a trend has occurred, may have been associated with decreased hardiness caused by acidic mists (Fowler *et al.*, 1989a; Cape *et al.*, 1991; DeHayes *et al.*, 1991; L'Hirondelle *et al.*, 1992; Vann *et al.*, 1991, 1992), although Sheppard *et al.* (1989) found little evidence that insufficient hardening had occurred on trees at Whiteface Mountain, New York, and Clingman's

Dome, Tennessee. In addition, Jacobson and Lassoie (1988) found no indication that acidic mist affected the cold tolerance of red spruce. During the last 22 years, there have only been two occasions (1980 and 1981) when temperatures at Whiteface Mountain fell below the level at which significant injury would be expected (Sheppard *et al.*, 1989). In contrast, DeHayes *et al.* (1991) found a decrease of frost hardiness in the October to January period of 3–5°C associated with exposure to ambient cloudwater at Whitetop Mountain, Virginia. However, Sheppard *et al.* (1989) questioned whether predisposition to frost injury by pollutants alone would account for the observed decline.

The role of cloudwater has also been thrown into question by studies of nitrogen uptake. For example, Bowden *et al.* (1989) conducted experiments with ^{15}N in mist. Accumulation of ^{15}N in foliage, stems and roots was very low, suggesting that nitrogen in clouds supplies only a small (<1.5%) fraction of the trees' nitrogen requirements. This work supported the findings of Friedland *et al.* (1988) that increased nitrogen uptake by red spruce at higher altitudes in comparison to uptake at lower altitudes did not occur. Further support comes from the work of Norby *et al.* (1989), who found that exposure of red spruce to acidic mist containing nitrate did not result in the induction of the enzyme nitrate reductase, a likely prerequisite for the assimilation of foliar-absorbed nitrogen. In contrast, nitrogen dioxide and nitric acid vapour (at concentrations higher than those normally encountered in the field) induced a response. Lindberg *et al.* (1986) found that 80% of the total NO_x deposition at the Walker Branch watershed in Tennessee was accounted for by the dry deposition of particles and vapours, with 75% of the dry-deposited nitrate being in the form of nitric acid vapour. They argued that the longer retention time in the canopy of dry-deposited material, as opposed to wet-deposited material, resulted in higher uptake.

The estimated deposition rates of nitrogen at high-altitude sites in the north-east (37.5–44.1 kg N ha^{-1} yr^{-1}) are well below the nitrogen fertilization rates that have been applied in native spruce-fir forests without inducing injury (DeHayes *et al.*, 1989) and well below the estimated uptake requirements of 100–130 kg N ha^{-1} yr^{-1} (Strader *et al.*, 1989). However, the continuous inputs provided by atmospheric deposition may generate a very different response to the episodic inputs of fertilizers. Several reports in the literature have suggested that late-season applications of nitrogen either have no effect on subsequent frost damage or decrease it (Benzian, 1966; Pellett and White, 1969a, 1969b), although reports from Finland have suggested that nitrogen application to Norway spruce can result in increased microscopic winter injury (Soikkeli and Kärenlampi, 1984). DeHayes *et al.* (1989) fertilized red spruce seedlings with rates equivalent to up to 3000 kg N ha^{-1} yr^{-1} without any effect on the rates of cold acclimation, depth of midwinter hardiness or rate of de-acclimation of red spruce seedlings. Timing of acclimation affected the response, with late fertilization resulting in increased hardiness. Similarly, Klein *et al.* (1989), using ion leakage as a measure of freezing injury, found no evidence to support the hypothesis that anthropogenic nitrogen significantly reduced the winter hardiness of red spruce foliage. However, the relationship between nitrogen fertilization and winter hardiness requires further study as the response of trees to late-summer fertilization is very variable.

Leith *et al.* (1989a) reported the effects of exposing two-year-old red spruce seedlings to mists of varying pH in open-top chambers (located in Great Britain). Needle necrosis was recorded in the pH 3.0 treatments and in more acidic treatments. On 80% of the seedlings, necrosis only occurred on the current needles. It took the form of light brown or light orange discoloration later turning a deep red. There was no evidence that acidic mists caused bud mortality. In marked contrast to the reports from forests, damage under controlled experimental conditions was reported in the autumn and developed very rapidly. The development of the visible injury coincided with the application of a pyrethrum insecticide and with the first low night temperatures ($< 5°C$) over a six-day period. The insecticide can be ruled out as a causal factor as trees removed from the open-top chambers before the insecticide treatment had the same level of necrosis. Although these results revealed that repeated exposure to below-average pH mists might cause visible injury, the absence of reports from the field suggest that this process is either absent or rare under natural conditions.

Wind

The importance of wind has been given surprisingly little attention in the debate on the causes of the spruce decline. Wind, which increases with altitude, is likely to be important in both the northern and southern cases. Wind damage to shoots occurs from the outer shoots inwards, as seen in declining red spruce stands. Sudden exposure to wind is also known to cause severe reductions in red spruce growth. This has been shown by studies of the impact of felling associated with the opening up of pistes for skiing (Harrington, 1986; Worrall and Harrington, 1988). The damage to the crowns is probably caused by a combination of direct mechanical damage and damage to the root systems, particularly the important water-absorbing roots. Marked root breakage occurred after the canopy of red spruce in the White Mountains was opened up (Harrington *et al.*, 1984), and this process may well act as a positive feedback loop in some declining red spruce stands (see Fig. 1.10).

The importance of wind was, however, discounted by Wargo *et al.* (1989), who studied the root systems of declining red spruce at Mount Abraham in Vermont. They found no evidence of damage to the root systems that might be associated with wind stress, although the canopies of the trees also showed no signs of wind stress (such as flagging).

Consequently, their results may not be appropriate to areas where trees are exposed to wind.

Wind damage may have increased following mortality caused by another factor. Perkins *et al.* (1992) refer to the 'pulse of tree mortality' that occurred between 1965 and 1983 at many sites. This created gaps in the canopy that have since continued to expand, primarily as a result of wind action. Severe flagging is present at all the high-altitude sites, indicating that wind stress plays an important part in controlling the amount of foliage on individual trees and the increased mortality on wind-exposed slopes as compared to sheltered slopes is a striking feature of most of the mountain sites.

Dendroclimatic studies of spruce-fir forests

In both the northern and the southern spruce-fir stands, there is good evidence that climate is involved in the decline. A number of techniques are available to test this supposition, mostly using dendroclimatology. Federer *et al.* (1989b) used the data set reported by Hornbeck and Smith (1985) to look at climate-growth relationships of predominantly low-altitude red spruce. Regionally averaged basal-area increments were related to regionally averaged climatic data. The mean series were detrended and delagged before analysis. In line with many similar studies (e.g. Conkey, 1979; Cook *et al.*, 1987; McLaughlin *et al.*, 1987; Kimball and Keifer, 1988), diameter growth was negatively correlated with July and August temperatures of the preceding year and positively correlated with early winter temperature. However, there was evidence that extreme values had distorted the relationships, particularly the September 1961 temperature, which coincided with a very wide ring.

In the southern Appalachians, although there has been a change in the climatic response of red spruce since the mid-1960s, only the highest-elevation sites show evidence of a growth decline (Cook, 1988). This may be related to the increased opening up of high-altitude stands due to the adelgid-induced mortality of Fraser fir. Cook (1988) argued that red spruce in the south has been more sensitive to drought since 1966 and to summer temperature since 1977. Similarly, Van Deusen (1988) suggested that trees were more sensitive to climate in the 1980s than previously. The conclusions of Cook and Van Deusen were based on the analysis of climatic response functions, which showed changing relationships between growth and specific climatic variables. The reasons for this were unclear, but in mixed Fraser fir–red spruce stands, the red spruce are now less sheltered than in the past. Drought and other climatic extremes affect isolated trees more than trees sheltered by others nearby, and this may have been involved in the change. These conclusions however are not supported by the work of LeBlanc *et al.* (1992) who argued that the incidence on individual tree growth declines in the period 1965–85 did not exceed the incidence of such declines in the past.

Cook (1987) and Cook *et al.* (1987) looked at pre- and post-decline climatic relationships. Samples were obtained from a variety of sites in the Appalachians. Before 1960, the most important climatic predictors of growth were the July and August temperatures of the preceding growing season, November and December temperatures of the year before growth and May and September temperatures of the current growing season, with above-average August and below-average December temperatures having a particularly negative effect on growth. However these relationships broke down in the post–1960 period. Cook and Johnson (1989) re-analysed some of the data and found that November temperature in the year before ring formation and July temperature in the same year as ring formation explained much of the post–1960 annual variations in tree growth. Dividing the data by altitude, Cook and Johnson (1989) found that some low-altitude sites did not show the post–1960 breakdown in climatic responses. They used techniques developed by Visser (1986) and Visser and Molenaar (1986) to model changing responses of trees to specific climatic conditions, and found that the peak response to prior-August temperatures occurred at all elevations in the period 1950–60.

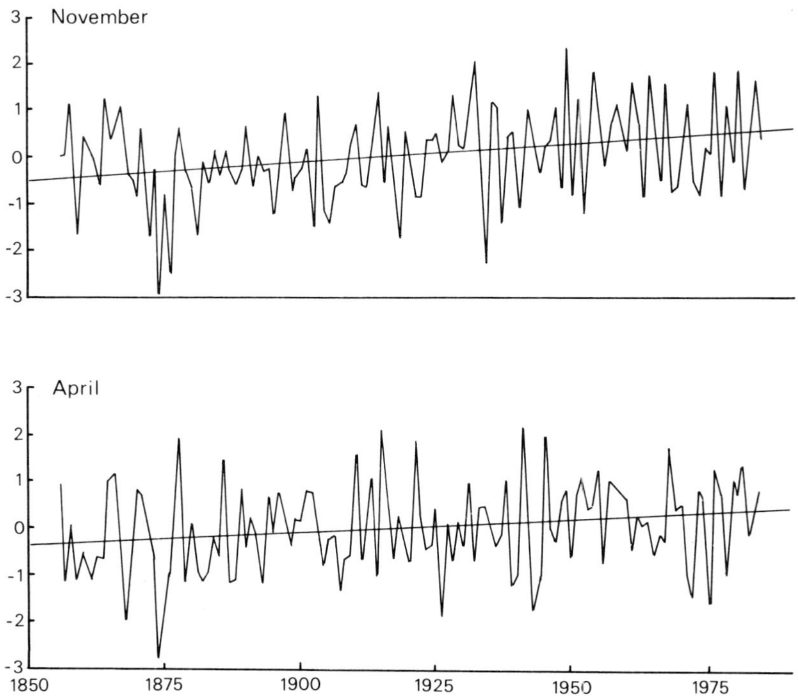

Figure 2.39 Plots of November and April normalized temperature departures (°C) from a variety of sites in the north-east. From: Cook and Johnson (1988).

However, as abnormally warm prior-August temperatures became increasingly time-dependent with decreasing altitude, contrary to the elevational gradient in the degree of decline, adverse August temperatures were unlikely to be the cause of the decline. Prior-December temperatures were influential at all altitudes, but were only time-dependent at high altitudes. Prior-November temperatures were strongly time-dependent (Fig. 2.39) and the temperature response could be associated clearly with a trend in the temperature record. Consequently, Cook and Johnson (1989) proposed that the red spruce trees were responding to climatic change.

Johnson *et al.* (1988a) looked at red spruce cores from the Catskill, Adirondack, Green and White Mountains. At all sites, ring widths began to decrease after about 1960, apparently regardless of site disturbance history or tree age. Climate-growth relationships were derived for the period 1885–1940 and were then used to predict growth in the periods 1856–84, 1941–60 and 1961–81. Usually, prior-July and prior-August temperatures were important variables. There were good agreements between predicted and actual growth for the period 1941–60, but very poor agreement for the period 1961–81. The reasons for this were uncertain: the addition of a factor such as pollution was only one possibility. Other possibilities included changes in climate or changes in the responses of trees to climate. Examining historic evidence, they found that periods of extensive and severe red spruce mortality occurred after periods of warm late summers and/or early cold winters. The most pronounced periods of climatic stress were in 1871–80, 1935–50 and 1956–80, which coincided with documented increases in mortality starting in 1871, 1938 and 1962. It is possible that the occurrence of severe climatic stresses led to the breakdown of the normal recovery process because of high recovery costs in trees that were already in a marginal situation.

The climate-growth relationships of red spruce in the White Mountains of New Hampshire were examined by Kimball and Keifer (1988). An increase in growth in the 1940s and 1950s was followed by a decline starting in the late 1950s. An inverse relationship occurred between ring width and the prior-July and prior-August temperatures. Similarly, Conkey (1979), working with high-elevation red spruce in western Maine, found a negative relationship between the temperature of the preceding July and August and a positive relationship with December and January temperature during the winter preceding the ring formation.

McLaughlin *et al.* (1987) analysed two data sets first reported by Johnson and Siccama (1983b) and McLaughlin *et al.* (1983) from the Appalachian Mountains. Intervention analysis was used to identify changes in growth, with two types of intervention being recognized. The first involved pulses and was caused by abnormal single annual observations. The second type comprised step interventions, which consisted of sustained and significant changes. Of the 2013 cores used in the analysis, 75% had steps and 80% had pulses. A strong upswing in the

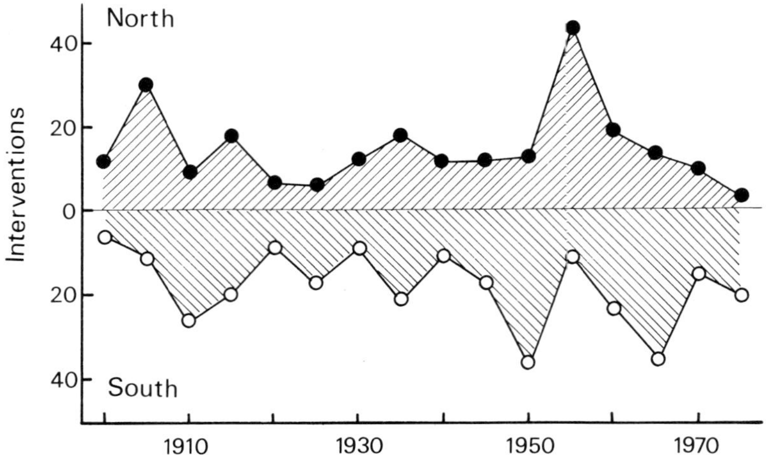

Figure 2.40 Frequency of negative step interventions in 5- year intervals, normalized to a population size of 500 cores for both North and South. From: McLaughlin *et al.* (1987).

number of positive steps was identified as starting in about 1885 and peaking between 1900 and 1904. In contrast, the biggest increase in the number of negative steps occurred between 1955 and 1960 (Fig. 2.40). They attributed the increase in the growth of trees at the turn of the century to competitive release following logging and mortality of trees in the period immediately before 1900. The sharp increase in the number of negative steps in 1955–60 occurred in the north and was not identified in southern stands. However, an increase in the number of negative steps in the south in the period 1950-5 was probably a response to the 1953–4 drought. A second increase, in the period 1965–9, was more difficult to explain. The results of this study strongly suggest that the regionally synchronous increase in mortality of red spruce in predominantly high-elevation uneven-aged stands in the north-east was not unprecedented, although the most recent decline appeared to be the most severe in the tree ring record.

Working with data from the southern Appalachians, Cook (1988) suggested that red spruce in the area has shown irregular behaviour since the mid–1960s. At high elevations only, there has been a step-like decline in growth since 1966. He did not identify any apparent climatic correlation, but found altitudinal effects, suggesting a role for climate and/or air pollution. In marked contrast to northern high-altitude red spruce, there was no indication of a breakdown in the dendroclimatic relationships after 1960.

The dendroclimatic studies all suggest changing relationships between tree growth and climate. In addition, there is evidence of changing climatic conditions. The problem remains, however, of evaluating the cause of the change in the climatic response of trees. The response may change because a critical climatic variable has become less important compared with others. Alternatively, an external factor may alter the response of a tree to a specific climatic factor. This problem has yet to be resolved.

Conclusions

The decline of red spruce and balsam and Fraser fir is, like many other declines, complex. In Fraser fir, the decline and most tree mortality can be assigned to the adverse impact of the introduced balsam woolly adelgid. In the north, balsam fir mortality is sometimes related to wave phenomena (such as on parts of Whiteface Mountain). In the southern forests, the sudden exposure of red spruce growing in the mixed spruce-fir zone following the mortality of the Fraser fir is likely to have exerted a major stress on the remaining trees. Sudden exposure is an inappropriate explanation for the decline of red spruce in the north as the waves are primarily restricted to pure balsam fir stands, and balsam fir mortality in the mixed stands is negligible.

There is convincing evidence that factors other than air pollution play a major role in the dynamics of the spruce-fir forests. Most have been severely disturbed, either by logging or by severe climatic events. Frequently, the declines in the growth of individual trees can be related to a reduction in growth rates following a release from suppression. Although this process has been discounted at some sites, the evidence for past disturbance does not appear to have been evaluated as carefully as it should have been. The role of stand dynamics in the occurrence of mortality is also uncertain. Decline and mortality would be expected because of self-thinning. Although many assumptions have been made about the survival of standing dead trees, the actual rate of decomposition/destruction of the stems does not appear to have been properly established. Consequently, doubt must be placed on those studies that have calculated decline rates based on single assessments of the numbers of standing dead trees.

Winter injury is clearly of major importance and is responsible for triggering the decline of trees at some locations. The precise nature of the injury is still disputed, with early or late frost, direct freezing injury during mid-winter and drought-induced injury caused by transpiration during mild periods when the soil is still frozen all being cited. In addition, ice storms have a major effect where they occur. The injury caused during these events may be sufficient to disrupt carbon allocation patterns for several years, thereby causing a temporary or even permanent breakdown of the normal physiological processes in the trees.

The role of air pollution is uncertain. Direct effects of gaseous pollutants have been eliminated as a cause, but acidic deposition, particularly in cloudwater, has been vigorously advocated by some groups.

The evidence for a major effect of acidic cloudwater is both controversial and inconsistent and requires further study before it can be included in explanations or ruled out as a factor. At present, the available evidence only suggests that it might be involved. Aluminium toxicity does not appear to be important in the northern sites, but it could become significant at some southern sites with soils that are already acidic. One question that does not appear to have been answered is whether the red spruce in the southern Appalachians shows the same lack of sensitivity to ozone as trees from the northern sites. Most gaseous exposure studies have concentrated on northern red spruce, but the southern population may consist of very different genotypes. Consequently, results obtained for northern genotypes should not be extrapolated to southern genotypes.

A major lesson to be learned from the study of the red spruce decline is that although a superficially similar decline occurred simultaneously over a large area, prompting speculation concerning a single cause, more careful analysis revealed that there were major differences between the declines reported from different parts of the red spruce range. Not only were the symptoms different, but growth analyses revealed differences in the nature and the timing of the growth decline. Consequently, a single explanation for the phenomenon can be discounted.

Southern commercial pines

The pine forests of the south and south-east of the United States form an important timber resource for the country. About two thirds (16.6 m ha) consists of natural stands, whereas a further 8.5 m ha have been planted (NAPAP, 1990). Recently there has been some suggestion that the rate of growth of these forests has declined during the last 20 years (Sheffield *et al.*, 1985; Knight, 1987; Sheffield and Cost, 1987; Zahner, 1988; Zahner *et al.*, 1989; Bechtold *et al.*, 1991; Van Deusen, 1992). This growth decline is highly controversial and is still being vigorously debated (e.g. Hyink and Zedaker, 1987; Lucier, 1988; Warren, 1990). Several species are involved, including loblolly pine (*Pinus taeda* L.), yellow or shortleaf pine (*Pinus echinata* Mill.) and slash pine (*Pinus elliottii* var. *elliottii* Engelm.). Several other pine species occur in the area, including longleaf pine (*Pinus palustris* Mill.), pond pine (*Pinus serotina* Michx.), sand pine (*Pinus clausa* (Engelm.) Vasey ex Sarg.), Virginia pine (*Pinus virginiana* Mill.) and eastern white pine, but declines in these have not been described so far.

Symptoms

Heavy mortality occurs as a normal process in the southern pine region, with annual mortality attributable to insects and pathogens being estimated at 31 million cubic metres (USDA Forest Service, 1978). The main agents for this mortality are fusiform rust (*Cronartium fusiforme*) and the southern pine beetle (*Dendroctonus frontalis*). However, unlike many other studies described in this book, the decline in southern pines is based solely on growth measurements. Although reliable systems for visually assessing the crowns of loblolly and shortleaf pines have been developed (Anderson and Belanger, 1987; Belanger and Anderson, 1988), visible injury associated with the growth decline has not been reported. This does not mean that effects of air pollution are absent as, except for a few sensitive genotypes, the southern pine species show fewer visual symptoms in response to given concentrations of ozone than the western pines.

The belief that a decline was occurring was based on the measurements made by the Forest Inventory and Analysis (FIA) programme. This involved the assessment of approximately 27 000 permanent inventory plots located across the region (Florida, Georgia, North Carolina, South Carolina and Virginia). The measurements suggesting a growth decline were based on natural stands that have mostly developed in abandoned agricultural areas (mainly formerly used for cotton production). The inventories were undertaken over three periods: 1956–61, 1961–72 and 1972–82. Loblolly pine in the Piedmont and Mountain regions of Georgia showed reduced growth in both 1961–72 and 1972–82 when compared to 1956–61, with the difference in average annual radial increment between 1961–72 and 1972–82 being about 15% (Fig. 2.41). Shortleaf pine in the same area showed an even larger reduction in growth (28%) between 1961–72 and 1972–82. On the coastal plain of North Carolina and Virginia, loblolly pine showed substantial growth reductions during the 1960s but, since then, growth has mostly increased. Slash pines in the Georgia coastal

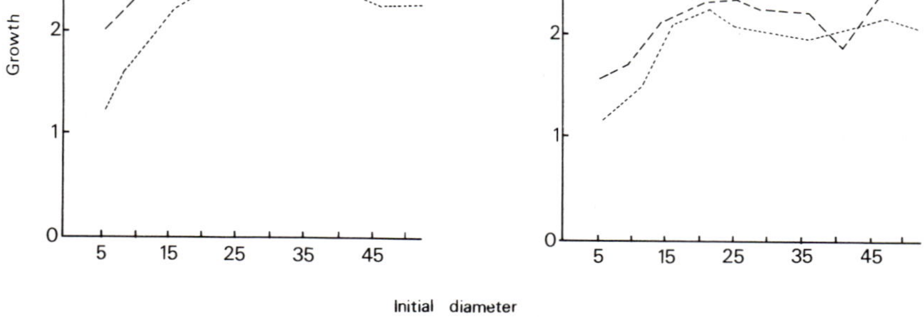

Figure 2.41 Changes in average annual radial increment (cm), by initial diameter class (cm), for loblolly (left) and slash (right) pines in natural stands in Georgia. From: Sheffield and Cost (1987).

plain showed large reductions in annual radial increment between 1956–61 and 1961–72, with much smaller reductions in the latest inventory period. Over the whole region, Sheffield *et al.* (1985) proposed that a reduction of 30–50% in the growth of natural pine stands had occurred between 1957–66 and 1977–85.

Bechtold *et al.* (1991) reported mean annual growth rate reductions of the pine component in natural stands of Piedmont and Mountain loblolly pines of 19%, 28% reductions in Piedmont and Mountain shortleaf pine stands and 28% reductions in Coastal Plain slash pine stands, although Gertner (1991) questioned whether the decreases were really as large as Bechtold *et al.* (1991) estimated. The growth decreases were not just restricted to pines, with overall mean annual growth rates for all trees on the plots also showing substantial reductions. Data from plantations did not show significant differences in growth rates among the three periods, a point that is of considerable significance. The decline apparently extended outside Georgia and North Carolina, with Ruark *et al.* (1991) reporting growth declines in Alabama and Zahner *et al.* (1989) reporting a growth decline in South Carolina.

Unfortunately, interpretation of the growth data is not as straightforward as it might seem. The 1961 inventory was based on a basal area factor (BAF) ten prism which was used to tally all live trees with a diameter at breast height (DBH) of 12.7 cm or greater in each plot. This was supplemented by a 35.2 m^2 circular fixed-radius plot to record all trees with a DBH of between 2.5 cm and 12.4 cm. In 1972, the inventory used three BAF 37.5 prism plots, spaced 21.3 m apart to tally larger trees, supplemented by three 13.5 m^2 plots for smaller trees. However, neither Hafley (1985) and Hyink (1991) considered this discrepancy to represent a problem, if the sampling was undertaken exactly as described. This is disputed by Lappi and Bailey (1987), who suggested that the bias introduced by using prism data could account for some of the growth reduction. In addition, there are several other difficulties with data that rely on comparisons of radial growth rate data (Thomas and Parresol, 1989). Other differences included the use of increment cores to assess average annual radial increment in the 1957–66 assessments and diameter measurements in subsequent assessments.

Partially independent evidence comes from a study by Zahner *et al.* (1989), who used FIA based data to look at radial growth increment. Their study involved increment cores from 131 even-aged (25–80 years), naturally stocked loblolly pine stands in the Piedmont region of Georgia, North Carolina and South Carolina. The results revealed a decline of 38% in radial growth between 1949 and 1984, with no apparent association between the extent of the decline and stand age or site productivity. They proposed that drought might have triggered the decline. A similar decline in the growth of natural loblolly pine in the Piedmont region of Georgia was documented between 1960 and 1984 by Zahner and Saucier (1989).

Possible causes of the decline

Stand dynamics

Several possible reasons for the decline have been investigated. Some southern pine species, particularly loblolly pine, have very strong relationships between growth and leaf area and canopy structure (Vose and Allen, 1988). However, changes in stand density alone would not account for the reductions in growth as the growth reductions have occurred over a wide range of stand densities. Stand age similarly appeared to have very little effect (Bechtold *et al.*, 1991). However, as basal area and the numbers of stems have increased over time (Zahner *et al.*, 1989; Ruark *et al.*, 1991; Bechtold *et al.*, 1991), it was difficult to completely rule out the effects of stand dynamics. Van Deusen (1992) looked at tree ring data from natural loblolly pine stands from the Piedmont regions of Georgia and South Carolina. He confirmed that a growth decrease occurred after the mid 1970s for all tree age classes. However, he also showed that the inventory data for 1961–72 spanned a period of rapid growth whereas the inventory data for 1972–82 covered a period when growth was decreasing. He attributed this, at least in part, to increased intraspecific competition and to increased competition from hardwoods. Warren (1990) went further and argued that the growth decline could be almost entirely explained by competition with hardwoods.

Knight (1987) listed several factors that might be involved in the purported decline. The area of timberland in the south-east has decreased recently, resulting in lower total volume increments for particular regions. Regeneration following harvesting is inadequate, particularly on privately-owned land. Annual mortality has increased sharply, with insects accounting for 35–45%; suppression and disease are also important. Sheffield and Cost (1987) added drought, problems arising from previous land use and competition from hardwoods to this list.

Reams *et al.* (1990) reported a study by Christensen and Peet (1988) that may well have significance for the work on the decline of pines. They found that the growth of loblolly pine showed cyclical variations, with periods of increasing growth coinciding with periods of decreasing stand density. What is important is that they found that silvicultural treatments, particularly thinning, greatly affected growth, which may provide a possible explanation for the absence of any growth decline in managed plantations. Such findings can hardly be considered as new since they form the basis for much plantation management, but their implications do not appear to have been fully considered in the southern pine debate.

Stand dynamics may therefore play an important part in the observed decline of pines in natural stands in the south. However, there is still a major gap in our knowledge of the detailed growth of young natural stands, as is clearly shown by the level of debate about the pine decline.

Hopefully, the current controversy will stimulate some high-quality dendroecological research on the subject.

Nutrient availability

Nutrient availability may play an important part in the growth decline. Newly established stands developing on old fields are likely to have the benefit of any residual levels of fertilizer in the soil (Coile and Schumacher, 1964). Most of the soils in the southern pine region are Ultisols, with low levels of cation supply once the initial supply is exhausted. In addition, past agricultural practices led to severe nitrogen depletion. Consequently, any residual fertilizer is likely to be important. As this becomes exhausted, growth may slow.

Binkley *et al.* (1989) compared soils under loblolly pine five and 25 years after the establishment of a plantation on an old-field site. A 20–80% decline in extractable calcium, magnesium and potassium was found, together with a decline in soil pH. The pH reduction was largely the result of the reduction in the base saturation of the exchange complex. These effects are important as there is an interaction between substrate quality and the response of loblolly pine to ozone fumigation, with plants in more fertile soils being less sensitive (Chevone *et al.*, 1984; Horton *et al.*, 1990). However, this altered response does not seem to be a function of the magnesium levels in the soil (Edwards *et al.*, 1992). In addition, changes in the acidity of the soil may lead to increased availability of potentially damaging elements, such as aluminium. However, comparisons through time are difficult: Adams *et al.* (1987) examined seasonal variations in the nitrogen and phosphorus contents of root and foliage tissues of loblolly pine, finding marked but irregular seasonal trends in the concentrations, which they attributed to climatic effects.

Nutrient depletion because of tree growth may be an important factor on some soils, as has been found with subtropical pine plantations (see previous chapter). Some evidence exists for this from repeated measurements (Wells, 1980; Binkley *et al.*, 1989b), but this work was based on a single soil in South Carolina and it is unclear how representative either the soil or the site was. The process is likely to be accelerated in the southern soils by leaching induced by acidic deposition. Some soils with very acidic A and E horizons may be particularly susceptible. Turner *et al.* (1986) estimated that these underlaid about 10% of the forest area, mostly on the Atlantic coastal plain of North and South Carolina and Georgia.

There has been a considerable amount of work done on the toxicity of aluminium to loblolly pine and other southern pine species. Raynal *et al.* (1990) reported unpublished work by Thornton, Schaedle and Raynal that did not record any significant effect on either root or shoot biomass of loblolly pine at aluminium concentrations of less than 1.5 mmol l^{-1} and Williams (1982) considered the threshold for root elongation effects to be 1.5 mmol l^{-1}. However, Raynal *et al.* (1990) reported 10% reductions in root elongation at 1.2 mmol l^{-1} and Paganelli *et al.* (1987) found a reduction in the numbers of new roots formed at a concentration of 0.19 mmol l^{-1} aluminium. Optimal root elongation has been reported to occur at concentrations of 0.13 to 0.16 mmol l^{-1} (Lyle and Adams, 1971). Tepper *et al.* (1989) described reductions in the elongation of primary roots of loblolly pine due to aluminium exposure. At 0.6 mmol l^{-1} Al, root growth after seven days was reduced by 26%, but it was not clear whether this reduction would be maintained over longer periods. Given the concentrations of aluminium recorded in southern soils, toxic effects seem unlikely.

Pollution

The role of pollution in the decline is also very uncertain. Ozone concentrations in the south-eastern United States have been summarized by Lefohn *et al.* (1988) and the gas appeared to be the most important of the major types of pollution present in the area (Pinkerton and Lefohn, 1987).

Many studies have looked at the effects of ozone on various aspects of southern pines. Shafer *et al.* (1987), Shafer and Heagle (1989), Kress *et al.* (1982, 1989), Horton *et al.* (1990), Manderscheid *et al.* (1992) and Mudano *et al.* (1992) all found reduced growth of loblolly pine fumigated with above-ambient concentrations of ozone, and root growth may also be affected (Edwards *et al.*, 1992). In contrast, Lee and Chevone (1990) found no significant effect of ozone on any biomass variables or gas exchange rates in loblolly pines fumigated with near-ambient levels of ozone. The absence of effects in this study may be related to the substantial between-family differences in tolerance to ozone (Kress *et al.*, 1982; Qiu *et al.*, 1992). Shafer and Heagle (1989), working over three growing seasons, found reductions in the amount of growth suppression by ozone after three years, compared with after two years, suggesting acclimatization. Stem height and branch length were inconsistent measures of ozone response. However, ozone fumigation in one study resulted in major changes in the carbon allocation patterns of loblolly pine (as with most other conifers), with increased carbon accumulation in primary branches and reduced translocation to the roots (Spence *et al.*, 1990), and reductions in photosynthesis on exposure to ozone have also been documented (at 1.5 x ambient) (Sasek *et al.*, 1991).

Exposures of 0.15 μl l^{-1} ozone for 28 days (six hours a day) did not affect Virginia pine seedlings (Kress and Skelly, 1982), and this species seems to be relatively tolerant of ozone. In contrast, the growth of slash pine responded adversely to ozone exposures over 112 days of a seven hour average of 0.104 μl l^{-1} (Hogsett *et al.*, 1985). Similarly, ozone (1.7 $\times$ ambient) caused growth reductions in shortleaf pine (Boulton and Flager, 1990).

Although no visible injury was observed, Stow *et al.* (1992) found reduced leaf areas in young trees exposed to ambient and higher levels of

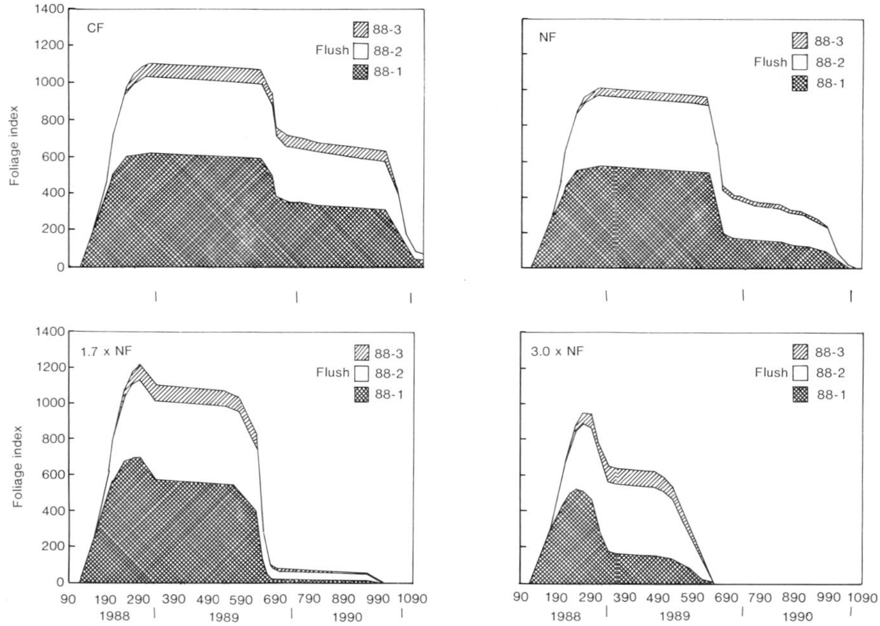

Figure 2.42 Leaf area duration of the 1988 cohort of loblolly pine during 1988, 1989 and 1990 at the branch position for charcoal-filtered, non-filtered, 1.7 × non-filtered and 3.0 × non-filtered treatments. From: Stow *et al.* (1992). Reprinted from *Forest Science* published by the Society of American Foresters, 5400 Grosvenor Lane, Bethesda, MD 20814–2198. Not for further reproduction.

ozone in North Carolina. At higher (1.7 × and 3.0 × ambient) than ambient ozone concentrations, accelerated leaf abscission was recorded (Fig. 2.42). Ozone exposure had little effect on needle production.

Edwards and Kelly (1992) fumigated loblolly pine seedlings with sub-ambient (charcoal-filtered), ambient and above-ambient (× 2) ozone concentrations and simultaneously exposed them to differing levels of rainfall acidity (pH 3.8 or 5.2). Over a three-year period, seedlings exposed to twice ambient concentrations of ozone had smaller percentages of total ectomycorrhizal short roots, whereas the acidic treatments had no discernible effect. As ozone does not penetrate the soil, the differences must have been caused by above-ground tree changes, such as the pattern of carbon allocation. Similar adverse effects of ozone on ectomycorrhizae have been found by Meier *et al.* (1990) and Adams and O'Neill (1991), but over shorter fumigations.

Acidic rainfall (pH 3.3) had no effect on the photosynthesis or growth of loblolly pine (e.g. McLaughlin *et al.*, 1988) and no interaction between acidic rainfall and ozone has been found (Chappelka *et al.*, 1990; Sasek *et al.*, 1991; Qiu *et al.*, 1992). In slash pine, Dean and Johnson (1992) found interactions between higher than ambient acidic deposition and ozone treatments, but there was evidence of reduced effects through time. Volume increment was the only growth variable that showed a response to the treatments. However, Boulton and Flager (1990) reported decreases in the photosynthetic rates of shortleaf pine at 2.5 × ambient ozone concentrations as simulated rainfall acidity increased. Acidic treatments alone did not affect the growth or the measured physiological variables. Consequently, acidic deposition can be discounted as a direct cause of the decline in the growth.

Light- and CO_2-response curves have shown significant decreases in light-saturated and CO_2-saturated photosynthetic capacities at 2 × ambient ozone compared to ambient (Sasek and Richardson, 1989). Various biochemical changes were identified following ozone fumigation, including changes in the fatty acid composition of needles (Fangmeier *et al.*, 1990), protoplast composition (Sigal *et al.*, 1988) and phloem photosynthate concentration (Spence *et al.*, 1990). Chappelka *et al.* (1990) investigated interactions between ozone and low temperature in loblolly pine. They found that seedlings fumigated with above-ambient (1.7 × and 2.5 × ambient) concentrations of ozone showed greater foliar injury following an early growing season frost than controls. This work was supported by the results of Edwards *et al.* (1990), who found an interaction between above-ambient (2 ×) ozone fumigation, magnesium nutrition and cold sensitivity. Interestingly, this study found that seedlings grown in soils with high magnesium concentrations were more susceptible to cold stress, possibly because of their greater physiological activity. However, as mentioned above, the growth decline of pine in the southeastern United States has not been accompanied by visible injury and the role of ozone-low temperature interactions is therefore uncertain. Further problems include the differential responses between field- and laboratory-exposed loblolly pine seedlings (e.g. Hanson *et al.*, 1988) and variations in the responses according to the length of the study and the family used (Kress *et al.*, 1982; Shafer and Heagle, 1989). These make the interpretation of the results of some studies difficult.

The links between ozone and growth in the field are extremely difficult to evaluate. However, Van Deusen (1989) described the occurrence of ozone injury on sensitive tree species (yellow poplar and black cherry) at many sites throughout North and South Carolina and Georgia. Radial increment of loblolly pine tended to be lower at sites where ozone injury to these two sensitive species was recorded and there was also an indication of increased needle loss.

Climatic factors and climate-pollution interactions

Climate is an important factor influencing the growth rates of southern pines (Chang and Aguilar, 1980; Friend and Hafley, 1989). Growth is sensitive to climatic conditions in the autumn of the previous year and to the spring and late summer conditions of the current year. However, there does not appear to have been any attempt to determine whether there have been any changes in any of these parameters over the last 30 years that might explain the changes in growth rates.

Loblolly pine is known to be responsive to drought, with the amount of needle fall occurring in a given year being related to the rainfall and temperature conditions during the year of needle formation (Hennessey *et al.*, 1992). Consequently, an association between climate and growth seems likely but, so far, this has been difficult to substantiate in any detail, possibly because of the young age of the plantations and their fast growth rates.

As elsewhere, there is a possibility of pollution-climate interactions. Ozone and cold stress interactions have already been described but, in contrast to many studies with other species, ozone does not appear to affect the stomatal control of loblolly pine (Sasek and Richardson, 1989). However, plants fumigated with near-ambient ozone concentrations had a more rapid decline in net photosynthesis rates as drought intensity increased (Lee and Chevone, 1990). The causes of the effects may be direct or indirect, occurring for example because of changes in the root-shoot ratios (Horton *et al.*, 1990). Other results (Taylor *et al.*, 1986) have suggested that exposure to high levels of ozone increases the efficiency of water use by the foliage, with ozone-fumigated trees being less sensitive to drought than either trees grown in charcoal-filtered air or ambient air.

Conclusion

It now appears indisputable that some individual trees and some stands have had reduced growth rates in recent years. The reduction is limited to natural stands, strongly suggesting that stand dynamics and natural successional processes have played an important part in the decline. The role of other factors remains very uncertain. Ozone may have reduced productivity, but is difficult to say why this has not occurred in managed stands (if it hasn't). Possibilities include the inadvertent selection of tolerant genotypes in plantations and the better growth conditions available to individual trees in plantations, enabling more resources to be devoted to repair mechanisms. Soil acidification may have placed the trees under additional stress, but the concentrations of aluminium that have been reported in the field are below those thought to inhibit root development. Climate is also likely to be involved, either as short-term extreme stresses, such as drought, or in longer-term, more subtle changes in seasonal climate.

Summary

Although air pollution has been widely cited as a factor affecting the health of forests in North America, the evidence for anything other than localized declines around specific point sources is limited. The well documented cases on the west coast of the United Sates and around Mexico City appear to provide good examples of regional scale effects of air pollution on forests. Even in these cases, the effects seem to be limited to a few species. In such areas, there is an urgent need to reduce the levels of specific pollutants if the long-term sustainability of the forests is to be guaranteed. However, other cases are much more controversial. Tree health clearly fluctuates on a range of timescales and, in an area the size of North America, it is hardly surprising that declines of particular species have occurred in particular areas in recent years. These declines appear to be the norm, as the work by Millers *et al.* (1989) has shown.

Despite the good historical evidence for periodic declines of trees, the declines that have been reported in recent years, often identified from very unsatisfactory evidence, have been quickly attributed to air pollution. The enthusiasm with which air pollution has been adopted as a cause of purported declines appears to have resulted in the importance of many other factors being badly underestimated. In many North American cases where a genuine decline has been identified, natural factors provide an adequate explanation. Furthermore, there is historical evidence of similar occurrences in the past, when regional pollutants were not a problem. The inclusion of air pollution as a likely contributing factor in some purported declines appears to be unjustified, but it is always possible that it plays a minor contributory role. Ozone is strongly implicated in forest effects in the West and is probably having an effect on the growth of some species elsewhere. However, qualification of these effects remain elusive and its importance needs to be determined through further field studies.

3 Forest Declines in Europe

> Through the necessary and anticipated reduction of pollutant emissions even for other than silvicultural reasons, it will become apparent that the predisposition of the Middle European woodlands to forest disorders is at heart a silvicultural problem.
> *Cramer, 1984*

Introduction

Europe, with its long history of forest management, has numerous records of forest and tree declines. Individual instances have been caused by many different factors, ranging from natural phenomena, such as severe insect defoliations and droughts, to the effects of atmospheric pollution. The most persistent reports have come from the heavily polluted triangle comprising Bohemia, southern Poland and the south of the former German Democratic Republic. Problems exist over a much larger area in these countries, extending northwards to the Baltic States, but the decline of high-altitude Norway spruce stands has been particularly emphasized.

Northern Czechoslovakia, Southern Poland and Eastern Germany

The local effects of air pollution in eastern Europe have been recognized for over 100 years. Sulphur dioxide and heavy metals are implicated in the decline and mortality of several species. An indication of the levels of pollution is provided by what is considered to be relatively clean in Czechoslovakia: an annual mean sulphur dioxide concentration of 20 $\mu g\ m^{-3}$ (Kubiková, 1991). In polluted areas, annual means in excess of 120 $\mu g\ m^{-3}$ have been recorded, and peak concentrations of 3000 $\mu g\ m^{-3}$ are not unknown. Pollution is widespread in Poland, with particularly high levels of pollution occurring in Szczecinek, Pulawy, Gdansk, Wroclaw, Torún and Katowice (Smykala, 1985; Mazurski, 1990).

The most important source of sulphur dioxide in the area is the soft coal region of north-west Bohemia. Major sources of sulphur dioxide

Figure 3.1 Damaged Norway spruce in the Krusné Hory area of northern Czechoslovakia. The upper photograph was taken at 750 m ASL, the lower one at 700 m. Photographs by J. Jirgle.

include the Most power plant in Czechoslovakia and the Hirschfelde and Bergsdorf plants in the east of Germany (Mazurski, 1986). In 1982, 700 000 ha of the forested area in Bohemia and Moravia and 263 000 ha of the area in Slovakia were damaged (Cejchan, 1985). However, the criteria for recognizing the damage were questionable. The damage was primarily identified at higher altitudes and was often localized. For example, in the Ore Mountains of northern Czechoslovakia, otherwise known as Krusné Hory, the main pollution sources were located in the valley bottoms. The damage, which included the mortality of substantial areas of Norway spruce forest, was located on the surrounding mountains at or above 800–900 m ASL. Average annual concentrations of sulphur dioxide in the damaged forests ranged from about 40 to about 80 $\mu g\ m^{-3}$ (Fig. 3.1). This suggested that gaseous air pollution was not the only cause of the damage, as pollution concentrations were greater in the valleys than at higher altitudes. The most likely explanation was a combination of gaseous pollution, occult and wet deposition and climatic stress (Materna, 1988; Kubiková, 1991). An additional factor may have been the poor nutritional status of some mountain soils, as the tolerance of trees to sulphur dioxide is partly controlled by their nutritional status (Rohmeder, 1960; Bjorkman, 1970; Smith, 1990; Dittrich *et al.*, 1991), although some experiments in the area have been inconclusive (e.g. Materna, 1962). Studies of the foliar composition of needles from trees in affected stands showed that magnesium contents were low and that they had declined since the 1960s (Materna, 1989). Nitrogen contents were also low. Soil acidification was not a major cause of the defoliation in the area (Mejstrik, 1987), although it is occurring (Materna, 1988).

Fabijanowski (1986) documented 654 000 ha of forests with visible damage in the Upper Silesian and Sudeten regions of south-west Poland, with a further 4 m ha believed to have latent damage. Again, the criteria for recognising the damage were questionable, throwing uncertainty over the figures. Greszta and Maczynski (1989) claimed that in 1988, 2.8 m ha showed signs of damage, based on the number of needles per shoot and the annual shoot growth. Sanitation felling of unhealthy forests, amounting to 16 million m^3 annually, was the sole source of timber in the mid-80s, healthy forests being left untouched (Mazurski, 1986). In parts of south-west Poland, selective sanitation felling has resulted in the almost complete removal of species such as Norway spruce, and only species tolerant of sulphur dioxide, such as beech, remain (Fig. 3.2). Beech forests are, however, changing, with a variety of herbaceous plants such as small balsam (*Impatiens parviflora* DC.) and herb bennet (*Geum urbanum* L.) increasing in frequency, suggesting changing environmental conditions.

High sulphur contents have been found in Norway spruce foliage in this area. Greszta *et al.* (1989) reported that in the Sudetes, the sulphur content of second-year needles ranged from 0.173% to 0.202%. In the Silesian Beskedes, they ranged from 0.152% to 0.183% and on the Silesian and

Figure 3.2 Beech forest near Katowice, Poland, in good condition, despite high levels of pollution in the area and severe subsidence at the site.

Wieliczka Plateaux from 0.168% to 0.235%. Mazurski (1986) found sulphur contents of one-year-old Norway spruce needles in the Sudetes between 0.143% and 0.194%. The highest sulphur contents were reported at high-altitude sites, where damage has also been reported as greatest, but the data for Norway spruce are unconvincing (Fig. 3.3).

The trend for increased sulphur concentrations with altitude is interesting in view of the well-known relationships between exposure to sulphur and its concentration in foliage. Currently, there is insufficient information to explain the increase with altitude (if it exists), although it seems unlikely that gaseous concentrations of sulphur should increase away from the source. A possibility is that the high levels of sulphur and potentially associated damage may be related to higher levels of deposition at altitude.

The larch bud moth (*Zeiraphera diniana* (Guenee)) has killed the Norway spruce in some parts of this area (Baltensweiler, 1985), and the tussock moth (*Lymantria monacha* L.) has been a major problem in parts of northern Poland (Mazurski, 1986). In the Gorce Mountains of southern Poland, the death of Norway spruce at higher altitudes was attributed to

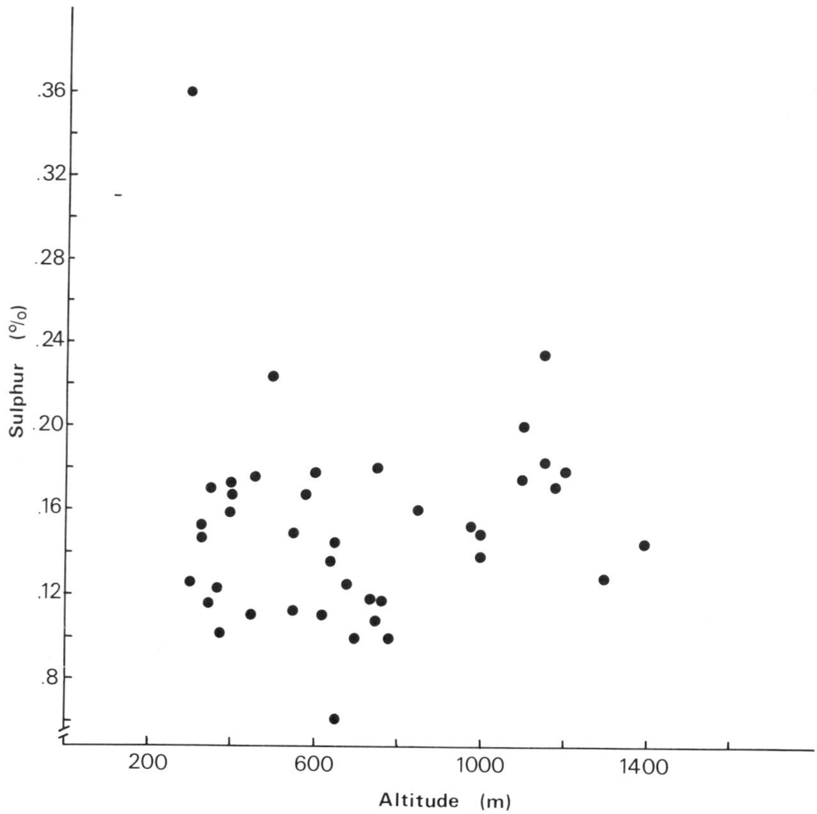

Figure 3.3 Sulphur content (%) of Norway spruce needles in southern Poland plotted against altitude (m). From: Greszta *et al.* (1989).

the sawfly *Cephalcia falleni* Dalm. (Medwecka-Kornas and Gawronski, 1990). Some pest outbreaks can be attributed to the rise in temperatures in 1890–1900 and 1940–50. Drought is particularly important in predisposing trees to attacks by the larch bud moth, and subsequent outbreaks generally last five to seven years. The tussock moth is able to take advantage of damage caused by severe winters, and insect damage following the winter of 1979–80 was particularly marked in northern Poland (Mazurski, 1986, 1990). Within the region, there is evidence of increasing levels of insect activity, with more frequent outbreaks of insects and rare insects becoming more common. However, although weakened trees may be less resistant to attack, they are poor substrates for breeding, suggesting a self-limiting process.

The decline of silver fir in Czechoslovakia has been documented by Cerny (1989). It has remained at a constant level since the early 1980s. Naturally regenerated silver fir growing at 800–1150 m ASL in areas affected by air pollution had no symptoms of decline, suggesting a more complex problem than at first sight. The main pest involved in the silver fir decline was the silver fir woolly adelgid (*Adelges nordmannianae* Eckst.). However, direct injury by pollution is definitely a factor near Krakow. Silver fir there has been badly affected by high levels of sulphur dioxide, with heavy mortality occurring, particularly among the youngest and oldest trees (Medecka-Kornas and Gawronski, 1990).

At Bíly Kríz in the Moravskoslezké Beskydy Mountains of north-central Czechoslovakia, the main source of pollution is the Ostrava-Karviná conurbation. Damage to trees was first noticed in the 1960s and the stress caused by a combination of air pollution and low temperatures was particularly important in 1978–9 (Kratochvílová *et al.*, 1989). Since then, damage by the spruce bark beetle (*Ips typographus* L.) has been important as has drought and continued air pollution. Heavy mortality occurred in the period up to 1984–5.

In some of the most polluted areas, very few trees can survive. Zareba (1984) recommended the use of yew (*Taxus baccata* L.) in heavily polluted areas of Poland, as it is tolerant of sulphur dioxide. However, this suggestion does not appear to have been widely adopted. Elsewhere, remedial measures have been necessary, and pollution has had a marked influence on management practices.

While the decline of forests in the region has been attributed to air pollution, it is important to consider other factors. For example, in southern Poland, many stands are quite young and have low diversity (Dzialuk, 1987). This makes them susceptible to a range of stresses and creates an opportunity for decline and mortality associated with stand dynamics. However, the scale of the mortality, with 60–80% of the growing stock in the Snieznik Massif of the Sudety Mountains being dead or nearly so, suggests that the conditions there are very unusual. The very high concentrations of sulphur dioxide, combined with massive attacks by secondary pests, appear to be the main causes of the mortality. Similarly, poor provenance or even species selection has undoubtedly played a part in the breakdown of forest ecosystems in northern Czechoslovakia, southern Poland and south-east Germany.

The available evidence all points to the damage in the north of Czechoslovakia, south Poland and south-east Germany being the same as the 'smoke damage' reported from heavily polluted areas over the last 150 years. The damage is relatively localized, but the scale of the pollution means that quite a large area is affected. The direct and indirect effects of pollution have clearly interacted with a number of other stresses, cold stress and drought being particularly important. In addition, there have been very serious outbreaks of insect pests, which may well relate to the weakened condition of many trees. Other factors that may have been involved include poor soil conditions, poor provenance selection and inappropriate management practices.

Other European tree and forest declines

Many studies have documented past declines of individual species or forests, similar to those described in the 1970s and 1980s. Cramer (1984) reviewed the literature on forest declines in central Europe, one of the areas of most concern today. All the species affected in the 1980s have shown similar declines previously. Despite this, it was frequently argued that the scale of the recent degeneration surpassed anything seen earlier (Forschungsbeirat Waldschäden, 1986). Such arguments failed to take into account the greatly increased awareness of foresters of the possibility of decline and their much more critical attitude towards the health of individual trees. The change in attitude undoubtedly contributed to the high levels of perception of 'forest decline'.

In Europe, declines have occurred in most species, particularly Norway spruce, silver fir, Scots pine, oak, beech, elm, ash and alder. Occasionally, the declines may have been caused by the use of fast-growing tree provenances on unsuitable sites (Manion, 1985; van Goor, 1985) and there is some evidence that faster-growing trees may be more susceptible to stress (e.g. Oleksyn, 1988, 1989; Innes and Neumann, 1991). Oleksyn (1988) showed this by a provenance experiment undertaken near the Poznan phosphate fertiliser plant in Poland. Provenances from throughout mainland Europe were used. The degree of damage was expressed as the magnitude of needle necrosis (Fig. 3.4). The number of necroses increased with plant height confirming earlier work that needle sensitivity was greater on more productive provenances (Oleksyn and Biabalok, 1986). Similar results were obtained by Havranek *et al.* (1990) who found that trees with higher gas exchange rates were more sensitive to extreme ozone concentrations (120 ppb for 12 weeks). Other

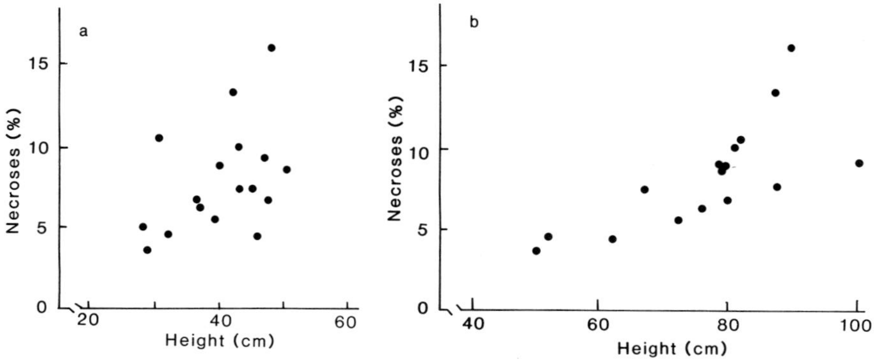

Figure 3.4 Proportion of Scots pine trees of differing provenance with needle necrosis grown (a) near a pollution source and (b) in an 'unpolluted' environment in Poland. From: Oleksyn (1988).

provenance trials (e.g. Tzschacksch and Weiss, 1972) have suggested that northerly and high-altitude provenances of several species are less sensitive to sulphur dioxide than other origins, again suggesting that speed of growth affects sensitivity.

The second Forschungsbeirat Waldschäden report (FBW, 1986) acknowledged that some unsuitable sites had been planted with conifers in Germany. Nevertheless, it discounted inappropriate management as a major factor in forest decline because silver fir allegedly first showed decline symptoms in natural stands and provenance selection did not explain damage to broadleaves. The logic of the latter point is debatable, since provenance selection in broadleaves has been widespread. Other accounts also discounted inappropriate management (e.g. Burschel, 1986; Flückiger *et al.*, 1986), although results from Germany clearly show that the nature of thinning can affect the extent of the damage (e.g. Decker and Gürth, 1985). In addition, Denstorf *et al.* (1984) identified greater damage in more open stands than stands with closed canopies. As discussed below, there is a substantial body of evidence that suggests that the decline of silver fir was a very specific phenomenon which should not be linked to the decline of other species.

In examining past declines, it is most appropriate to take each major species in turn. Schütt (1989) argued that more than 20 species were involved in the recent 'forest declines' in Europe, but the involvement of several of these was very questionable. In practice, concern focused on a few major species. Kandler 1992 concisely summarized much of the problem of European 'forest decline' research: 'All kinds of symptoms have been lumped together, and the resulting mixture has been declared as typical of a new complex disease of forest ecosystems caused by an undefined mixture of pollutants. This careless habit is totally contrary to the detailed symptom analysis procedure that has been practised successfully for more than a century of classical phytopathology'.

Norway spruce

Individual stands of Norway spruce have frequently been reported as declining. This phenomenon extends outside central Europe, although the most detailed accounts are those from southern Germany and nearby areas. Several distinct forms of decline have been described.

Past declines

Detailed accounts of early declines of Norway spruce can be found in a variety of publications, including those by Schwappach (1889), Von Tubeuf (1906), Wiedemann (1925), Rubner (1934), Hitschold (1934), Mülder (1934) and Priehäusser (1943). Many past periods of decline were directly linked to droughts, particularly on sites where Norway spruce did not form the natural forest (Wiedemann 1925, Rubner 1934, Cramer and Cramer-Middendorf 1984; Haupolter, 1987). As non-indigenous Norway spruce provenances have been planted over much of Europe, few stands now exist where the trees are natural. Norway spruce is very responsive to drought stress, with needle fall occurring during and shortly after drought conditions. The losses are followed by a relatively rapid recovery of the needle biomass (Gruber, 1990). Consequently, where drought-sensitive genotypes have been planted on drought-prone sites, adverse effects are entirely predictable.

Some declines of Norway spruce reported in the literature may be related to the rather misleadingly named phenomenon of 'top-dying'. This involves the rapid decline and often the subsequent mortality of trees. It is characterized by reddening of needles in the upper half of the crown, usually followed by progressive dieback from the top to the base of the crown (Murray, 1954). The development of symptoms may be extremely rapid, occurring over a few days in mid- to late summer (Innes, 1989a), although reddening may also occur in late winter. It is not clear what triggers the onset of the discoloration. Innes (1990a) examined the meteorological conditions leading up to the onset of discoloration of trees in a stand in Hampshire in England, but no obvious triggering factor was present (Fig. 3.5). Occasionally, needles may first turn yellow before turning reddish-brown (Innes, 1990a). It is relatively widespread, having been reported in Great Britain, The Netherlands, Germany, Sweden, Norway and Denmark (Diamandis, 1978). The causes of the problem are still unknown, although the available evidence strongly suggests a climate-site-management interaction (Murray, 1957; Diamandis, 1979).

In 1990 and 1991, widespread reports of reddening occurred in Great Britain and Denmark. Some of these reports clearly referred to top-dying,

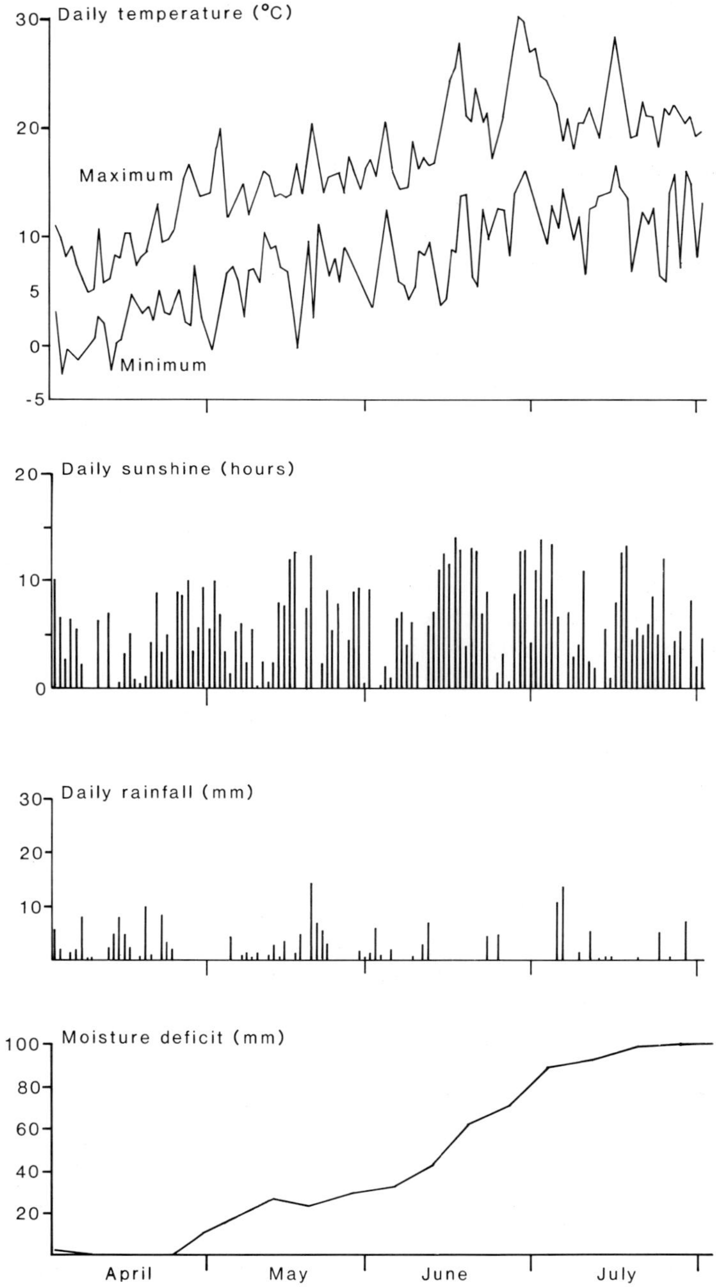

Figure 3.5 Meteorological conditions leading up to the onset of rapid discoloration in Norway spruce at Alice Holt Forest, England. Yellowing began on 25 July. From: Innes (1990a).

but others were more ambiguous. Often, the reddening developed irregularly on individual branches, and the clear progression from the top of the tree to the crown base was not apparent. Reddening was sometimes restricted to current-year needles whereas in others all needle years were affected. Sometimes, the discoloration was associated with medium-term (< 10 years) disturbances to shoot extension patterns. At other sites, the discoloration appeared without any obvious growth disturbances. It was unclear whether these reddening phenomena were all part of the same problem or whether different processes were responsible.

The top-dying syndrome has been associated with mild winters combined with stand disturbance. However, this combination does not always trigger the syndrome, and it is uncertain just how closely the syndrome is linked to mild winters. The absence of a consistent relationship may be related to variations in the susceptibility of different provenances to the problem. In addition to mild winters, the occurrence of spring and/or summer drought may exacerbate or even be a cause of the problem.

Lees (1991) looked at the growth rates of trees suffering from top-dying and compared them with the growth rates of asymptomatic trees. She found that affected trees generally had faster growth rates in the early stages of growth, which suggested that a factor related to stand dynamics might be involved in the decline. No clear relationship was found between winter temperatures and the onset of growth declines, with the implication that the climatic factors involved in the decline may be more complex than previously thought.

Symptoms of Norway spruce decline

The main symptoms of decline in Norway spruce are (from Verein Deutsche Ingenieure, 1987):

1. severe chlorosis, particularly of exposed needles, usually starting with the oldest needles and affecting either the entire needles or only the tips;
2. shedding of older needles;
3. loss of secondary shoots in the upper crown;
4. increased development of adventitious buds;
5. death of fine roots;
6. correlation between chlorosis and nutrient deficiency, particularly magnesium, calcium and potassium; and
7. symptoms first seen on dominant and exposed trees.

These are not ubiquitous, and the particular combination of symptoms, with the region affected, often defines the type of decline. In Germany, five main types of decline have been distinguished by symptomatology and the spatial distributions of the declines, although others undoubtedly exist.

It is important to recognize that some of these declines overlap spatially and that within any given area, more than one type may be present.

Type 1

Distribution and characteristics

By far the greatest amount of research has been concerned with Type 1 Norway spruce decline. This is probably because of all the different types, it is Type 1 decline that has been alleged to result in the greatest amount of tree mortality. The decline involves the yellowing of older needles on trees at higher (usually $>$ 900 m) elevations in the German 'Mittelgebirge', such as the Black Forest and Bavarian Forest, although it has also been reported outside Germany. Yellowing of Norway spruce was seen in the Donon Pass in the Vosges Mountains, France, as early as 1968 (Becker, 1985; Bonneau and Landmann, 1988) and even earlier reports exist for the area (Fourchy, 1951). The decline in France has been studied in detail, and much the same symptoms as seen in areas such as the Black Forest have been reported. Respiration of needles from defoliated trees in the Vosges was significantly higher than that of fully-foliaged trees (Fig. 3.6). This could be linked to the observed higher levels of NAD malic enzyme and cytochrome oxidase in the needles (Garrec, 1991). Garrec *et al.* (1988, 1989) examined the pH of current year and 2nd year needles of Norway spruce and silver fir in the Vosges Mountains, France. The pH of the foliar tissue showed no difference between asymptomatic and symptomatic trees. Similarly, the alkalinic buffering capacities of the needles were either similar (Norway spruce) or showed only slight differences (silver fir). The buffer capacity of Norway spruce foliage at a polluted site in Czechoslovakia was found to be much lower than at an unpolluted site and experimental fumigation with sulphur dioxide produced a similar result (Pasuthová, 1981).

Yellowing is followed by the loss of needles and, sometimes, by the death of the trees. Discoloration starts at the tips of the needles and is usually much greater on the upper surfaces. However, yellow-tipping does not necessarily indicate that those needles will subsequently become chlorotic. Chlorosis is usually associated with magnesium deficiency (Fig. 3.7). It often, but not always, appears at magnesium concentrations lower than about 300 $\mu g\ g^{-1}$ dry weight (Bosch *et al.*, 1983; Zöttl and Mies, 1983; Hüttl, 1985a, 1985b; Zech *et al.*, 1985; Evers *et al.*, 1986; Zöttl and Hüttl, 1986; Glatzel *et al.*, 1987; Kandler *et al.*, 1987; Forschner and Wild, 1988; Roberts *et al.*, 1989). The magnesium deficiency is typical of forests growing on very old soils. Liu and Hüttl (1991) described the chlorosis as mainly taking the form of spots, flecking or horizontally striped discolorations. Tip-yellowing occurs mainly on the upper surface of older needles, resulting in it frequently being missed in ground-based surveys. Chlorosis usually occurs on the upper surfaces of older needles

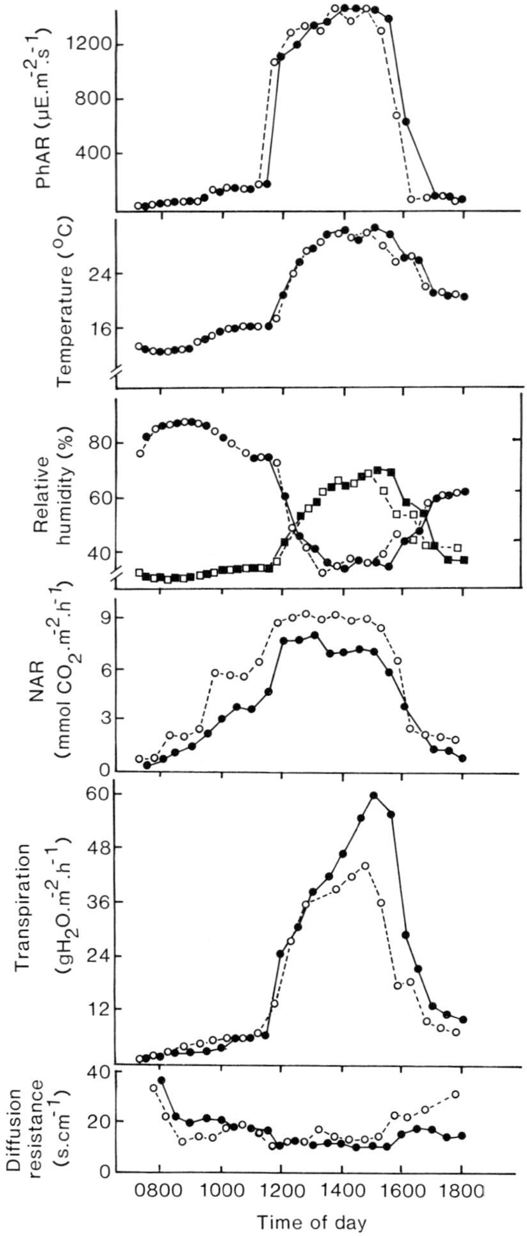

Figure 3.6 Daily time course of incident light intensity (PhAR), air temperature, relative humidity in the assimilation cuvette, vapour pressure deficit (VPD), net assimilation rate (NAR), transpiration and diffusion resistance (water vapour). Asymptomatic trees: dashed lines; symptomatic trees: solid lines. From: Benner *et al.* (1988).

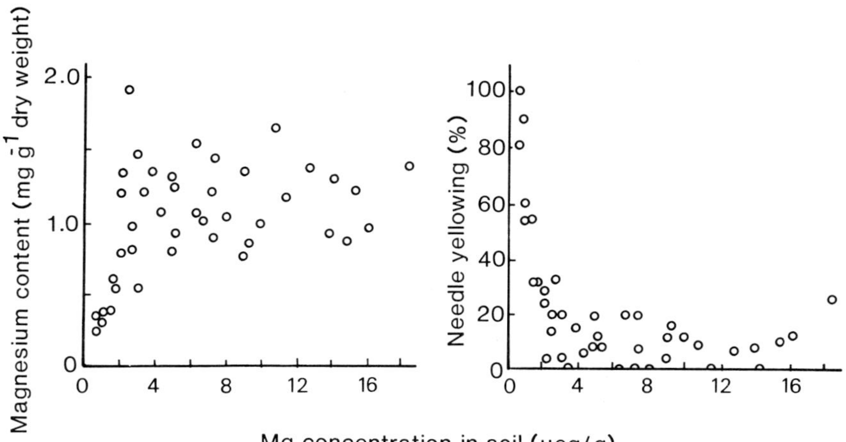

Figure 3.7 Relationship between (left) magnesium content of first-year needles and (right) the amount of needle yellowing, and the magnesium content of the soil. From: Liu and Trüby (1989).

soon after the development of new needles. It can be attributed to the greater-than-normal removal of magnesium from older needles so as to supply the new needles with adequate amounts of this essential element (Mies and Zöttl, 1985; Zech *et al.*, 1985; Lange *et al.*, 1987). This process was clearly shown by removing terminal buds from branches which then failed to develop the yellowing symptoms (Fig. 3.8) (Lange *et al.*, 1989a; Weikert *et al.*, 1989).

The chlorosis is reversible (Beyschlag *et al.*, 1987; Kandler *et al.*, 1987), and a decrease of chlorosis in Germany was reported in the second half of the 1980s. In addition, fertilization trials showed that chlorotic needles can revert to green (Hüttl, 1985b). The reversal of chlorosis in German forests may be related to improved nitrogen : magnesium ratios following the end of a series of dry summers, which would have increased mineralization and nitrification (Zöttl *et al.*, 1989).

Similar chlorosis has been reported in the past and some stands near Baden-Baden in the Black Forest were affected in the 1960s and the mid-1980s. Records also exist from elsewhere in Germany, particularly on magnesium-deficient soils (Wehrmann, 1960; Moll, 1961; Baule and Fricker, 1970), where the chlorosis was successfully treated by adding magnesium fertilizers. In such cases the development of sporadic chlorosis was related to the planting of provenances of varying sensitivity to magnesium deficiency (Münch *et al.*, 1990).

There have been several attempts to relate the extent of Type 1 damage to altitude. The general trend for increasing damage with increasing altitude described by, for example, Decker and Gürth (1985) and

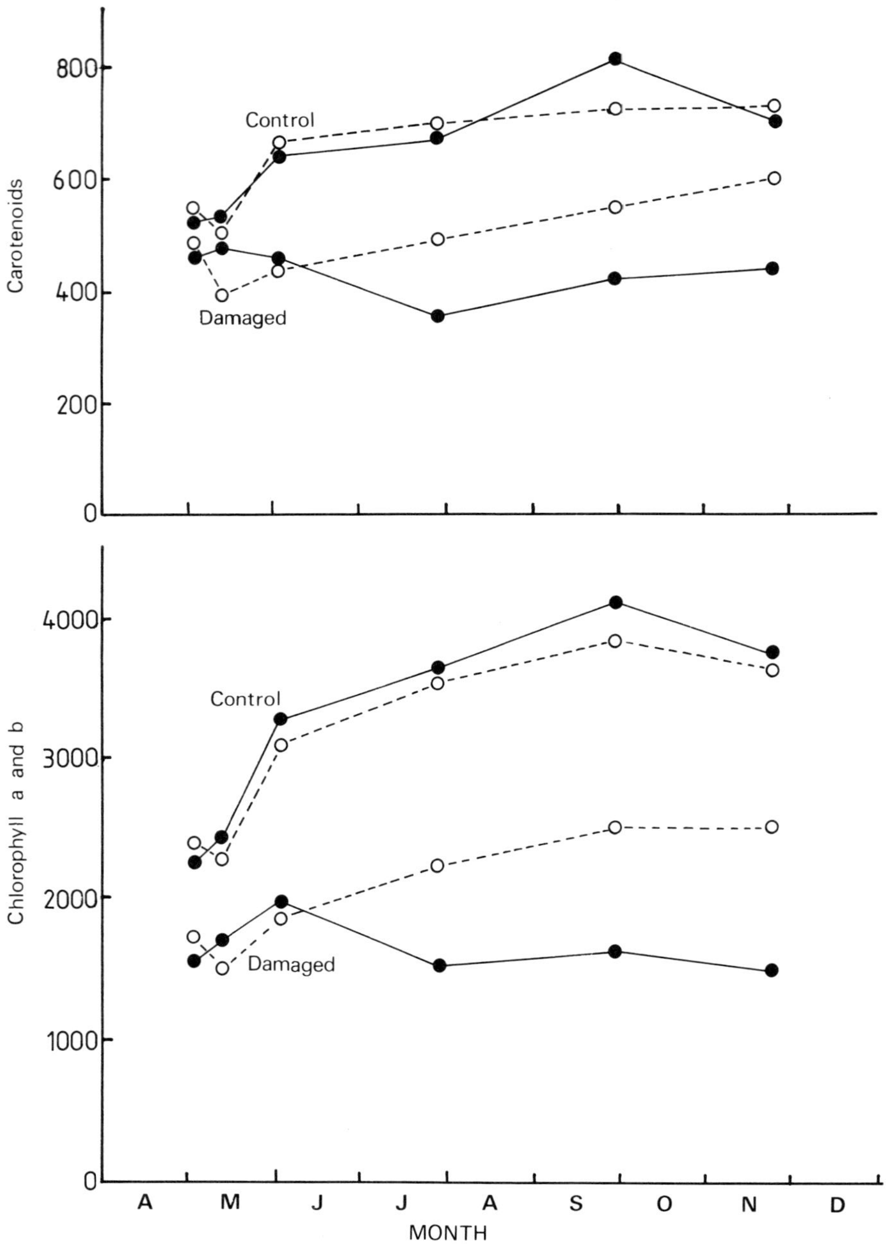

Figure 3.8 Seasonal time courses in 1986 of dry-weight related chloroplast pigments of the previous year's spruce needles for an asymptomatic control Norway spruce tree and for a symptomatic ('damaged') tree. Branches with intact flush of the current year needles (solid circles) and experimental branches without the new flush (open circles). From Lange *et al.* (1987).

Heinsdorf *et al.* (1988) has not been supported by subsequent studies. Earlier studies (e.g. Denstorf *et al.*, 1984) also failed to find a relationship between the degree of damage and altitude. However, Nebe and Rossbach (1990) found that the severity of nutritional disturbance in the Harz Mountains of Germany increased with altitude. Reichelt (1983) reported that damage in south-west Germany increased with altitude, but only on granites. It seems likely that within certain areas, damage increases with altitude. However, in larger scale studies such as that conducted by the Commission of the European Communities, this trend is obscured by using data from a variety of areas (Fig. 3.9). For example, the upper tree-line in Scotland is much lower than in central Europe and a better measure of altitudinal effects might be to relate defoliation to altitude as measured by height below the treeline. Even this could not account for regional variations in altitudinal environmental gradients.

The micromorphology of Type 1 defoliated Norway spruce has been examined by Fink (1986a, b). There was no significant difference in the differentiating tracheids undergoing lignification in symptomatic and asymptomatic trees. No change in the lignin content was detected by UV-photometry in the walls of the tracheids of affected trees. However, affected trees showed much lower starch reserves in the parenchymatous elements in autumn and spring. In addition, the activities of enzymes involved in the mobilization and transportation of carbohydrates was lower in the spring.

Fink (1991) also examined the structural changes in needles that occur because of magnesium deficiency. In cross-section, healthy, green needles had an intact epidermis, turgescent mesophyll cells and a vascular bundle

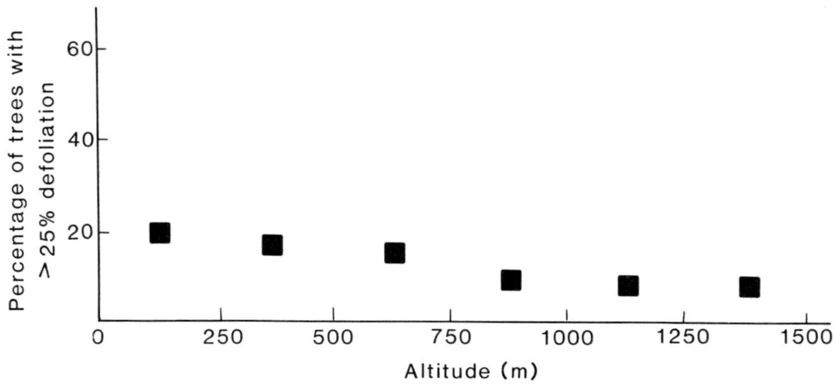

Figure 3.9 Relationship between the proportion of trees with more than 25% transparency and altitude in the European Community countries. A slight decrease with increasing altitude is evident. Source: *European Community Forest Health Report* 1990.

with open xylem, cambium and phloem cells. These were surrounded by transfusion cells and a closed circular bundle sheath. The chloroplasts in the mesophyll cells revealed a well-developed thylakoid system, few plastoglobuli and no or only small starch grains, a phenomenon that was dependent on the time of sampling. Needles with visible symptoms of magnesium deficiency showed several distinctive features. The sieve cells in the phloem had collapsed and the adjoining parenchyma and cambial cells were partially enlarged. The mesophyll cells were turgescent, as in the mesophyll cells of normal needles, but they were filled with starch grains. The thylakoid system of the chloroplasts was reduced and disorganized and the thylakoid membranes were partially separated, resulting in the formation of electron-translucent gaps. Similar findings were reported by Schmitt *et al.* (1986) who concluded that the damage could not be primarily attributed to air pollutants invading the needle tissue via the cuticle and/or the stomata. Wild (1987) also found collapsed sieve cells and damage to the thylakoids and associated loss of chlorophyll and of electron transport activity.

Possible causes

The chlorosis is associated with light-dependent pigment bleaching (Elstner and Osswald, 1988). The bleaching occurs as an oxidative process, initiated by an oxygen activating reaction followed by cofactor activation, such as lipid peroxidation. Elstner and Osswald (1988) documented several changes associated chlorotic needles. These included increases in the concentrations of antioxidative metabolites, ascorbate and glutathione, enhanced catalase activity, increases of tocopherol, superoxide dismutase(s) and glutathione reductase, reduced levels of unsaturated fatty acids and the formation of stress ethylene. In addition, chlorosis is normally accompanied by the loss of structural integrity of the needles (Hasemann and Wild, 1990; Hasemann *et al.*, 1990).

The ultrastructural changes described for 'new' decline areas are different from those found in areas affected by classic sulphur dioxide pollution (Zellnig *et al.*, 1989a). Norway spruce needles exposed to sulphur dioxide had lightened plastoglobuli and deformed chloroplasts. In contrast, needles from areas affected by 'new forest decline' had changes to the thylakoid system and an increase in the number of plastoglobuli. However, needles from an area experiencing relatively low levels of pollution revealed changes to the chloroplasts, including thylakoid swelling, accumulation of plastoglobuli and the deformation of chloroplasts. Unfortunately, it is not clear which type of spruce decline was being assessed. The annual mean sulphur dioxide concentrations in areas affected by Type 1 decline are generally low (less than 10 to 15 $\mu g\ m^{-3}$), so direct injury by sulphur dioxide to the trees can be discounted.

Although the most frequent cause of Type 1 decline is magnesium deficiency, calcium and zinc deficiencies have also been noted at some sites

(Hüttl, 1985b; Evers *et al.*, 1986; Evers and Schöpfer, 1988; Fiedler, 1988; Forschner and Wild, 1988). Fiedler (1988) found a strong geological control on the incidence of zinc and calcium deficiency, with zinc contents being lowest on siliceous rocks and increasing on base-rich and alkaline rocks. As the solubility of zinc increases with acidity, there is a risk of leaching losses in acidified soils. Zinc availability to plants is also reduced when the soil pH rises above about 5.0 to 5.5. Latent to acute magnesium, zinc and calcium deficiencies have also been found on moderate to poor nutrient status soils in eastern Germany (Nebe and Lucas, 1990).

According to Liu and Hüttl (1991), stands with needle nitrogen contents above 17 mg g^{-1} were prone to chlorosis, as were stands with a phosphorus content of less than 1.5 mg g^{-1}. Chlorosis also occurred in stands with less than 0.7 mg g^{-1} magnesium, less than 2 mg g^{-1} calcium and less than 25 μg g^{-1} zinc. Consequently factors other than magnesium may sometimes be applicable.

There is some evidence that leaching of elements from the soils has occurred. Grünhage and Jäger (1988) found reduced levels of calcium and magnesium in spruce needles in the Taunus Forest, Germany, between 1972 and 1983 (Fig. 3.10). However, it is not clear from this study whether the reduction was due to reduced uptake or reduced availability.

Hauhs (1989) reported net losses of magnesium from the Lange Bramke catchment in Germany. Magnesium deficiency in Norway spruce needles had been reported there since 1982 (Hauhs and Wright, 1986), and Reemtsma (1986) and Nebe and Rossbach (1990) related low foliar magnesium levels in the area to soils with poor nutritional status. In addition, Reemtsma (1986) noted that magnesium contents of Norway spruce needles had declined since observations were started in 1964. Hüttl (1990) described reductions in the magnesium contents of Norway spruce in the Black Forest between 1975 and 1983 and Zöttl (1987) described losses of calcium, potassium and magnesium in streams draining the Black Forest.

Studies by Ke and Skelly (1989) in North America found that the magnesium contents of Norway spruce foliage were positively correlated with mean soil pH, exchangeable soil magnesium levels and the percentage base saturation of magnesium. They therefore suggested that reduced magnesium levels were more likely to be associated with magnesium availability than reduced uptake. Murach (1988) argued that low magnesium levels in the foliage of Norway spruce were related to low soil magnesium levels, reporting close correlations between foliar magnesium and the magnesium contents of the fine roots. Liu and Trüby (1989) found that the needle contents of potassium, magnesium, calcium and aluminium were positively correlated with the exchangeable amounts of these elements in the soil (Fig. 3.11). Stock (1988) and Matzner *et al.* (1989) reported that chlorosis in the Harz Mountains primarily occurred in acidic soils with little available magnesium. Where magnesium-rich seepage

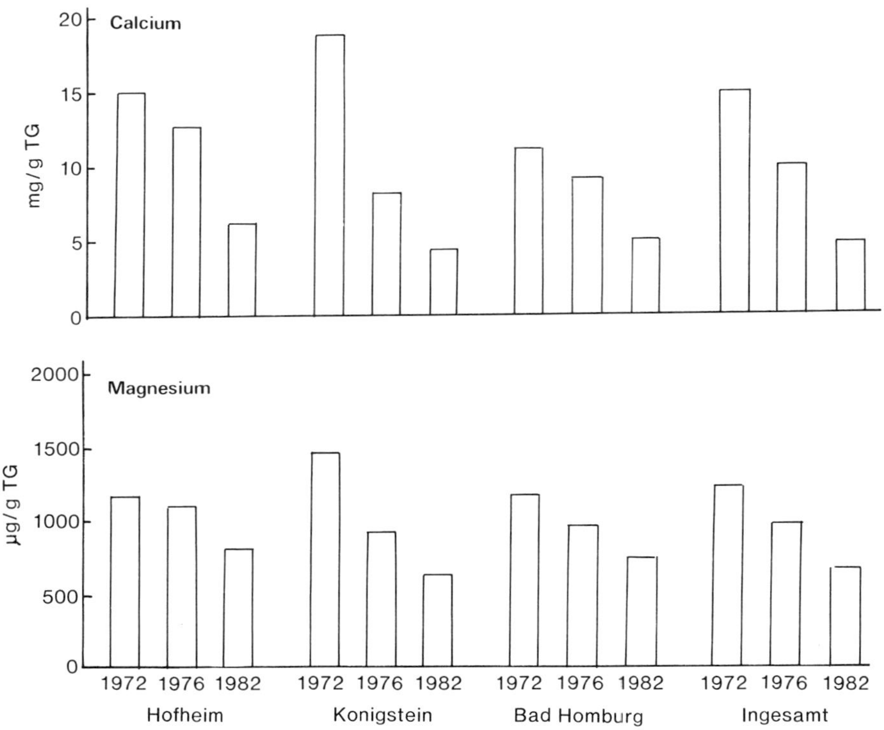

Figure 3.10 Changes in the calcium (upper) and magnesium (lower) contents of Norway spruce needles in the Taunus Forest, Germany, between 1972 and 1982. From: Grünhage and Jäger (1988).

water occurred, trees were healthy. Soil magnesium contents were particularly closely related to the incidence of yellowing of older needles. In the Black Forest, the same pattern occurred, with chlorosis occurring on soils that were deficient in magnesium (Feger and Raspe, 1992). Much the same has been reported by Estivalet *et al.* (1990) and Becker (1991) for Norway spruce and silver fir in the Vosges Mountains of France. The depletion of soil cations is discussed in detail in Chapter 8.

Rehfuess (1989) suggested that there was no evidence of a reduction in exchangeable magnesium in unfertilized soils in Germany, but this argument did not take into account the studies mentioned above. He has pointed out that magnesium deficiency also occurred on rocks such as metagabbro where weathering rates should be sufficient to replenish magnesium losses through leaching (Rehfuess, 1987). This argument favours inhibition of uptake as the main mechanism, possibly because of damage to the fine root system caused by acidification (Blaschke, 1986; Vogelei and Rothe, 1988) or because of inhibition of uptake by the blocking action of aluminium ions (e.g. Stienen and Bauch, 1988).

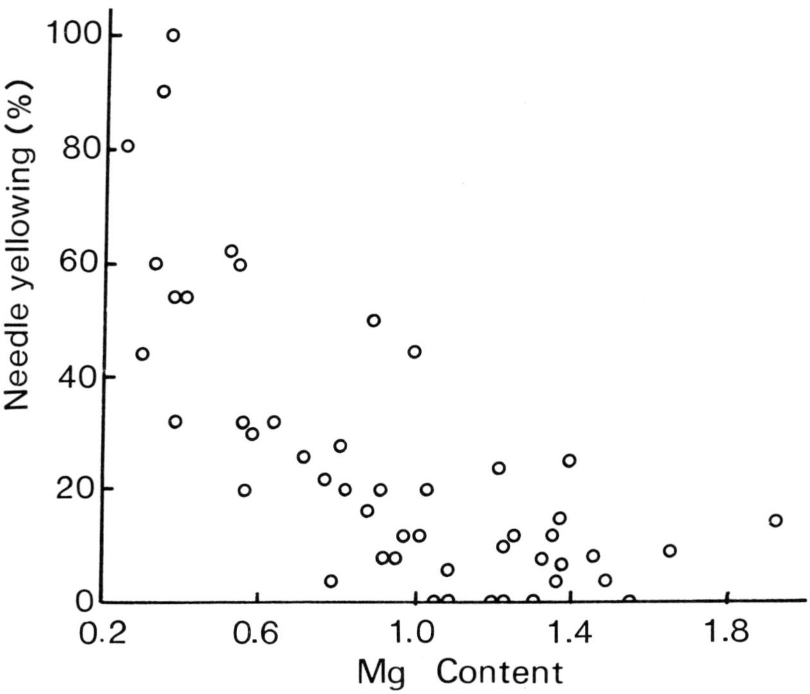

Figure 3.11 Relationship between the percentage of yellowing in fourth-year needles and the magnesium content (mg g^{-1} dry weight) of first-year needles. From: Liu and Hüttl (1991).

The extent of root damage in declining stands is unclear, with some workers (e.g. Matzner *et al.*, 1986) having alleged that damage has been recorded in all investigated stands showing decline symptoms and others reporting no differences in fine root biomass (e.g. Zöttl and Hüttl, 1986; Rehfuess, 1988a; Schneider, 1989; Schneider *et al.*, 1989). Root damage has certainly been recorded in some situations (e.g. Meyer, 1987; Schulze *et al.*, 1987) and it seems likely that the total fine root biomass used by some workers is a poor measure of the condition of the roots. Important factors include the distribution of the roots within the soil profile, their turnover rates, the rate of formation of active root tips and the mycorrhizal associations. As in other situations, the difficulty of adequately defining a 'damaged' tree may be confusing the issue. Trees with thin crowns caused by site factors (particularly by poor soil conditions) would be expected to have poor fine-root development. In such cases, a causal relationship exists between root and crown symptoms but it has little, if anything, to do with forest decline.

Roberts *et al.* (1989) suggested that the reduction in magnesium uptake was associated with drought effects. This argument was supported by the widespread evidence of improved tree condition in the mid to late 1980s, followed by a deterioration in some areas at the end of the decade. These trends parallel the widespread occurrence of droughts in Europe. The main development of chlorosis occurred in the period up to 1984, which was marked by a sequence of particularly dry years. Drought conditions could have impaired root development, leading to a possible reduction in magnesium uptake. The position is complicated by the role of nitrogen mineralization and its relation to soil moisture and temperature. Increased soil nitrogen associated with drought conditions could result in increased magnesium demands. In soils already low in magnesium, deficiency might then arise, particularly in areas where absorption of magnesium was interfered with by ammonium. Evidence for this comes from the association between the improvement in crown condition and the reduction in foliar nitrogen concentrations that has been reported in south Germany (Rehfuess, 1989). In addition, it is uncertain how important the volume of the fine roots are in relation to magnesium uptake. For example, trees may compensate for the loss of fine roots by increasing uptake per unit volume of roots. This needs to be examined experimentally.

The purported increase in the degree of decline at higher elevations led Bosch *et al.* (1986) and Bosch and Rehfuess (1988) to argue that winter cold was a trigger for Type 1 decline. Frosts were unusually frequent in the period 1979–87, but this period covered both the decline and recovery of trees. In addition, frosts should have been more prevalent in valley bottoms, with cold temperatures being the critical factor on the valley sides.

Gaseous air pollution was initially believed to be involved in the decline. However, studies of forest lichen floras have not revealed a clear relationship between the lichen vegetation and the pattern of forest damage in areas such as the Black Forest (Fig. 3.12) (Bartholmess, 1989; Grabherr and Mucina, 1989). The use of lichens is probably only appropriate for sulphur dioxide pollution. Lichens seem to be tolerant of ozone and, with daily mean summer ozone concentrations reaching 50–80 ppb and some hourly means above 100 ppb in some of the areas affected by Type 1 decline (Arndt *et al.*, 1982; Lefohn and Mohnen, 1986; Eiden *et al.*, 1989; Ulrich, 1991), ozone could be problem. However, studies in the Bavarian Forest of the gas exchange and growth of shoots of a mature, 36 m tall tree in filtered and ambient over a three-year period provided no indication of adverse effects of gaseous air pollution (Koch and Lautenschlager, 1988; Koch, 1989). In addition, Garrec *et al.* (1990) concluded from a range of physiological tests that, if pollution played a part in the decline, it was very minor.

Experimental studies involving the fumigation of trees with ozone (up to 306 $\mu g\ m^{-3}$ for 60 days) have failed to reveal any effect of ozone on magnesium nutrition or the development of chlorotic symptoms on older needles (Brown *et al.*, 1987). Instead, acute injury by ozone on Norway spruce and silver fir grown under experimental conditions has

Figure 3.12 Lichens sensitive to sulphur dioxide growing abundantly on the branches of Norway spruce showing symptoms of Type 2 decline. Glaubenbielental, Switzerland.

usually taken the form of a fine chlorotic mottle affecting needles of all ages (Guderian *et al.*, 1985a; Prinz *et al.*, 1985; Senser *et al.*, 1987). Necrosis of older needles which, in experimental studies, has been induced by high-concentration ozone doses combined with cold stress (Brown *et al.*, 1987; Barnes and Davison, 1988), has not been reported from forest areas. Some biochemical changes seen in fumigated plants are also absent in forest trees (Senser *et al.*, 1987; Kandler *et al.*, 1987). Grafting experiments conducted by Hüttl and Mehne (1988) showed conclusively that the magnesium deficiency in the needles was not the result of ambient air pollution levels; grafts from chlorotic to healthy trees indicated that chlorotic shoots recovered when attached to healthy trees, a trend that would not have occurred if gaseous air pollution was the primary cause of the problem. In marked contrast, in similar grafting experiments with symptomatic shoots on asymptomatic stocks of ponderosa pine in California, the shoots failed to show a similar recovery (Parmeter and Miller, 1968).

Studies of carbon dioxide and water-gas exchange in 20–25 year old trees in the Hunsrück Mountains of Rheinland-Pfalz showed that trees with needle chlorosis and needle loss had reduced photosynthesis and significantly lower water use efficiency and apparent photorespiration

(Benner and Wild, 1987; Benner *et al.*, 1988). Diurnal patterns of transpiration strongly suggested that stomatal regulation was impaired. The results were inconsistent from year to year, apparently depending on ambient weather conditions. In a similar study, Schulze *et al.* (1987) found that photosynthetic rates declined below a magnesium concentration of 0.3 mg g^{-1}.

Gas exchange measurements have been made on a fertilized tree and a nearby tree showing marked magnesium deficiency symptoms (Beyschlag *et al.*, 1987). Carbon dioxide uptake was lower in the unfertilized tree, but transpiration rates were similar. Stomatal behaviour was identical, contrasting with the study in the Hunsrück Mountains. Needles of the unfertilized tree had a lower maximum photosynthetic capacity and a lower carboxylation efficiency. Consequently, the magnesium deficiency was attributed to damage in the rooting zone. This study, together with those by Schulze *et al.* (1987) and Benner *et al.* (1988), all show a link between magnesium deficiency and physiological processes, but also suggest that site-specific factors may be critical.

Another potential explanation for the problem was the leaching of magnesium and other nutrients from the foliage (Prinz *et al.*, 1982). The proposition was that exposure to acidic mists could enhance foliar leaching. However, experiments involving the experimental fumigation of young trees with acidic mists failed to produce significant leaching losses (Seufert and Arndt, 1986; Bosch *et al.*, 1986; Mengel *et al.*, 1987). The work in this area is an example of how inadequate characterisation of the chemistry of the fumigations can result in faulty conclusions. Although several studies suggested that magnesium leaching could be enhanced by combinations of ozone and acidic mist, Brown and Roberts (1988) showed that this was due in some cases to the deposition of anhydrous nitric acid rather than the two pollutants *per se.* Foliar leaching occurs as a normal process and its exacerbation by acidic deposition is a complex issue, outside the scope of this review. A good summary of the phenomenon has been provided by Klemm (1989).

The most detailed study of Type 1 Norway spruce decline has been conducted by E.-D. Schulze and co-workers. Work was conducted in the Fichtelgebirge, Germany, and has been reported in a variety of publications. The Fichtelgebirge is a mountain area in the north of Bavaria, close to the Erzegebirge of Czechoslovakia. It experiences higher levels of sulphur dioxide (some mean monthly concentrations above 0.12 mg m^{-3}, and winter peak concentrations of 740 $\mu g\ m^{-3}$) than most other remote areas in Germany. Much of the area was formerly covered by beech forest which was replaced in the nineteenth century by Norway spruce (Lenz and Haber, 1990).

Detailed comparisons were made of the physiology of two stands, one showing signs of Type 1 decline, the other apparently being unaffected. The height, stem density and leaf areas of the two stands were similar

(Oren *et al.*, 1988d). However, stemwood production was 35% lower in the affected stand, suggesting major differences in carbon dynamics. The trees in the affected stand maintained their carbohydrate levels in the foliage, and Oren *et al.* (1988d) argued that this was at the expense of stemwood production.

Green needles from trees in both stands showed the same carbon dioxide assimilation rates (Fig. 3.13) (Zimmermann *et al.*, 1988), although one-year-old needles had lower uptake rates at the unaffected site. Green needles at both sites showed a general reduction in photosynthetic rates and stomatal conductance with age (Fig. 3.14). There was no evidence of a dose-response relationship that might explain the difference in tree condition between the two sites (Zimmermann *et al.*, 1988), nor was there any evidence that green needles in the affected stand were photosynthesizing at a different rate. Furthermore, green and partly yellow needles in both stands responded similarly to changes in air humidity and photon flux density, showing that there was no difference between the sites in the stomatal responses.

Lange *et al.* (1989a, b) found that the photosynthetic performance of the trees was correlated with the mineral nutrition of the needles.

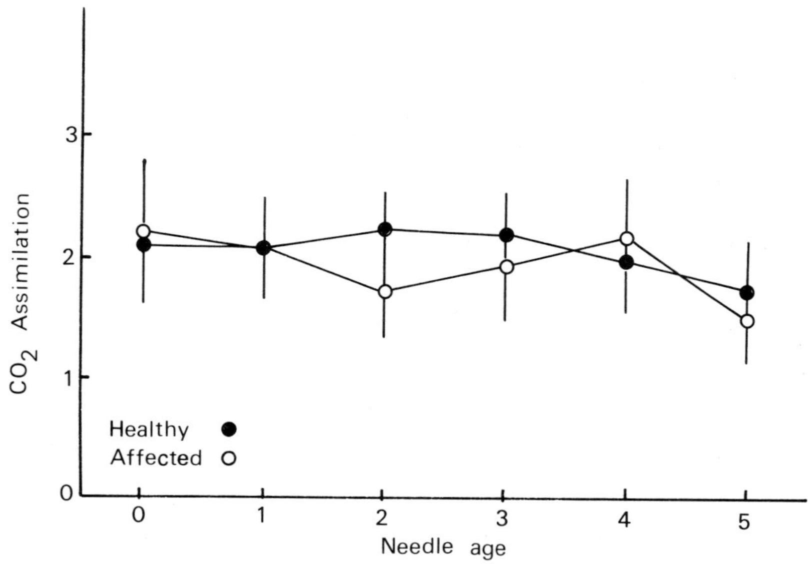

Figure 3.13 Carbon dioxide assimilation rates (μmol m^{-2} s^{-1}) of Norway spruce needles of different age classes in asymptomatic (Wülfersreuth) and symptomatic (Oberwarmensteinach) stands in Germany. Rates were calculated for a leaf conductance (*g*) of 30 mmol m^{-2} s^{-1}. Values are on all-sided needle area basis. Data from the asymptomatic site are indicated by filled circles; open circles represent the symptomatic site. From: Zimmermann *et al.* (1988).

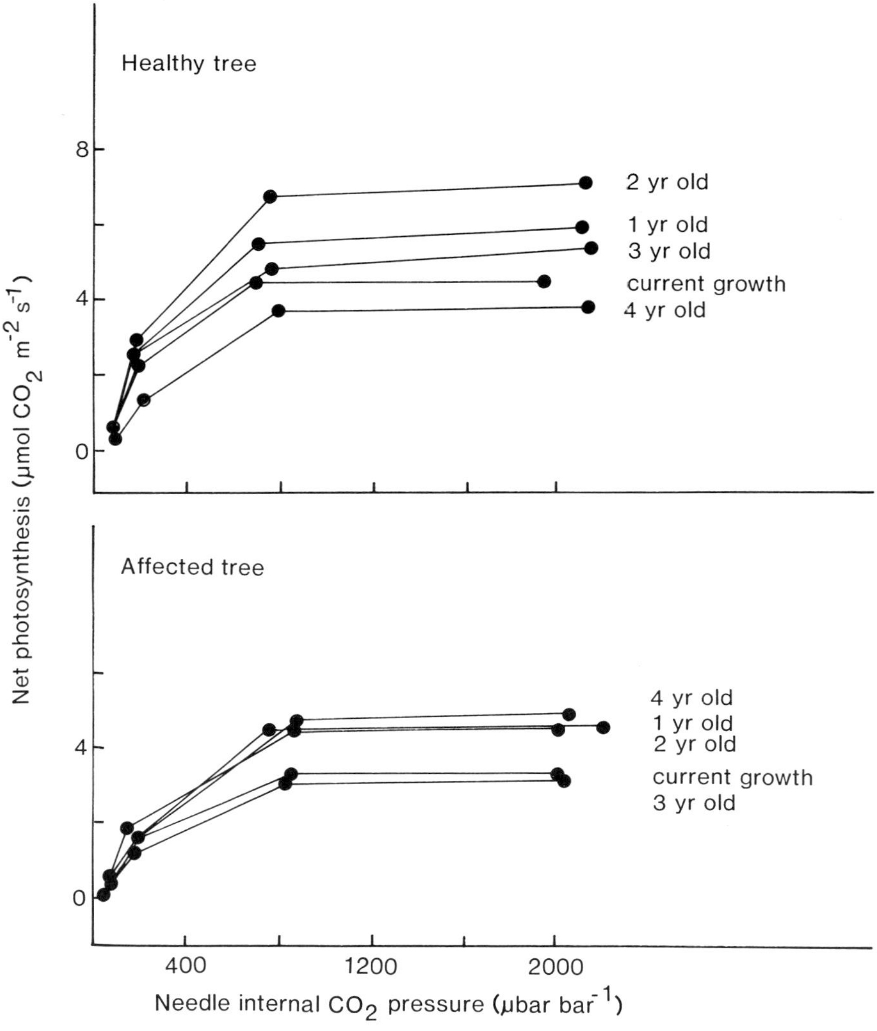

Figure 3.14 Carbon dioxide response of net photosynthesis under saturating photosynthetically active radiation flux, standard temperature and air humidity for different needle age classes of a healthy and an affected spruce tree. Oberwarmensteinach, Germany. From: Lange *et al.* (1989), based on Weikert (1986).

Damaged needles showed reduced rates of carbon dioxide assimilation, which seemed to be the result of a decrease in the biochemical-biophysical capacity of the photosynthetic apparatus rather than reduced stomatal conductance. Chlorotic needles showed a strong depression of the photosynthetic capacity and carboxylation efficiency, but there was no progression in the relative amount of needle damage with increasing needle age. This is consistent with the results of Weidner and Kraus (1987) who found no significant difference in the ribulose-1,5-bisphosphate carboxylase activity in needles from 'healthy' and defoliated trees.

A comparison between the photosynthesis and nutrition of Norway spruce stands in the Fichtelgebirge and New Zealand (Schulze *et al.*, 1990) has been made. Absolute rates of carbon dioxide uptake, the slope of the carbon dioxide response curve at 240 μl l^{-1} internal carbon dioxide concentration and the change in photosynthetic rates with needle age at saturated and ambient carbon dioxide concentrations were virtually identical in the Fichtelgebirge and in New Zealand. Needles with high photosynthetic capacity in New Zealand had lower nitrogen and higher calcium concentrations per unit dry weight, but had higher concentrations of nitrogen, phosphorus, potassium, magnesium and calcium per unit leaf area. There was no evidence of a gradual reduction in potassium concentrations with age at the polluted site, nor was there any evidence of damage to the photosynthetic process in the trees from the Fichtelgebirge.

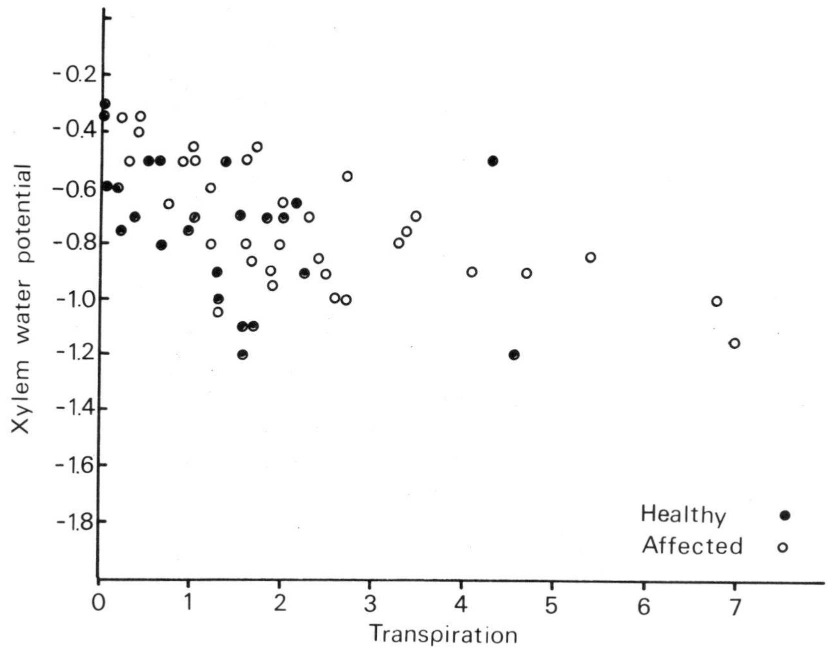

Figure 3.15 Relationship between xylem water potential (MPa) measured on bagged branches at the base of Norway spruce trees and the corresponding transpiration (mol m^{-2} sapwood s^{-1}) rate. Transpiration is expressed per unit area of hydroactive xylem so that the relationship reflects a measurement of the xylem hydraulic conductance. From: Werk *et al.* (1988).

The diurnal pattern of transpiration was largely controlled by the meteorological conditions at the sites, with vapour pressure deficit being the most important factor. There were no differences in the rates of transpiration of trees at the two sites, nor were there any differences in the water relations of the trees (Fig. 3.15). Werk *et al.* (1988) concluded from the absence of any differences that neither loss of stomatal control nor disruption of the water conducting system explained the differences in condition between the two stands.

Chlorotic needles had reduced levels of magnesium, calcium, manganese and zinc, whereas potassium and aluminium concentrations were higher. Sulphur contents in healthy and damaged trees were similar. In contrast, Liu and Hüttl (1991), working in south-west Germany, found no clear relationship between the aluminium content of foliage and the amount of yellowing. Fertilization with magnesium resulted in a reduction in chlorosis. The magnesium content of the foliage was correlated with the magnesium content of the xylem (Fig. 3.16), suggesting that reduced supply of magnesium to the canopy rather than foliar leaching was the cause of the low magnesium concentrations (Osonubi *et al.*, 1988).

The nutrient dynamics between the two sites were different, besides

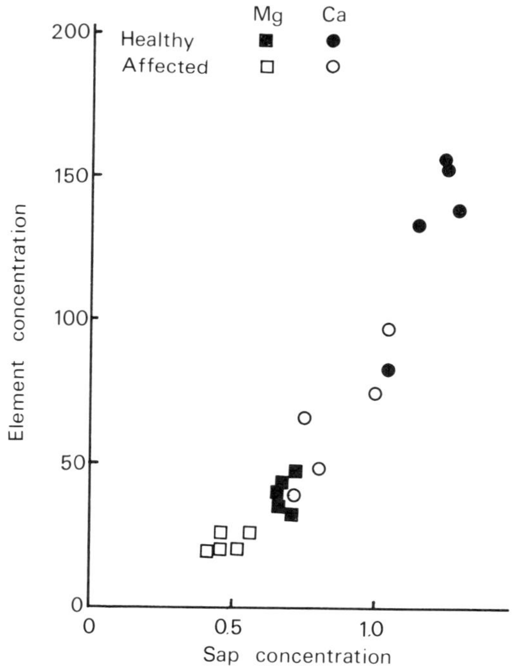

Figure 3.16 Relations between element concentrations of one-year-old needles (μmol g^{-1}) and the average xylem concentration (mmol l^{-1}) for the measuring period in October 1985. From: Osonubi *et al.* (1988).

the differences in absolute levels of several nutrients that have already been described. During canopy growth, the needle and twig production and nitrogen concentrations were similar in both stands, although calcium and magnesium concentrations were lower in the affected stand. This suggests that nutritional imbalances rather than external stresses were present at this stage. Translocation of magnesium from old to new needles occurred at both sites but, whereas it accounted for 75% of the magnesium in trees at the unaffected site, it only accounted for 50% at the affected site. Consequently, trees at the affected site were more reliant on soil magnesium. However, uptake was probably restricted by limited root expansion (because of unfavourable soil conditions) and lower soil concentrations of magnesium (compared to the unaffected site) (Oren *et al.*, 1988b, c). The restriction of the rooting to shallow soil layers would also increase the susceptibility of the affected trees to drought.

An independent study was made by Manderscheid and Jäger (1992), working in the northern part of the Black Forest, Germany. They undertook an exclusion experiment, using open-top chambers. Despite high summer concentrations of ozone, no evidence of any gaseous pollutants on a variety of biochemical markers was detected. At a forest site, magnesium deficiency was evident, but there was no evidence that trees suffering from this were subject to greater oxidative stress than trees with no or slight magnesium deficiency symptoms.

Type 1

Norway spruce decline was clearly a real phenomenon, resulting in the mortality of trees in a number of separate areas. It was caused by severe magnesium deficiency, which reduced the vigour of the trees such that they were more susceptible to other stresses. The causes of the magnesium deficiency were complex and have not been fully evaluated. Using the results from the Fichtelgebirge, Schulze *et al.* (1989) produced a model which attempted to account for all the different processes (Fig. 3.17).This illustrates the complexity of the problem.

Type 2

The second type of spruce decline is similar to Type 1 in that it occurs in the 'Mittelgebirge' (Fig. 3.18). However, yellowing of needles does not normally occur, the trees only showing thinning. The spatial distributions of Type 1 and Type 2 decline may overlap at a regional scale. Within the Lange Bramke catchment in Germany, both types of decline were present, depending on the local geology (Hauhs, 1989). Trees with symptoms of Type 1 decline occurred on granites and sandstones and tended to be restricted to 30–60-year-old stands. Older (> 60 years) stands on quartzite and silicate-poor greywackes tended to have symptoms of Type 2 decline.

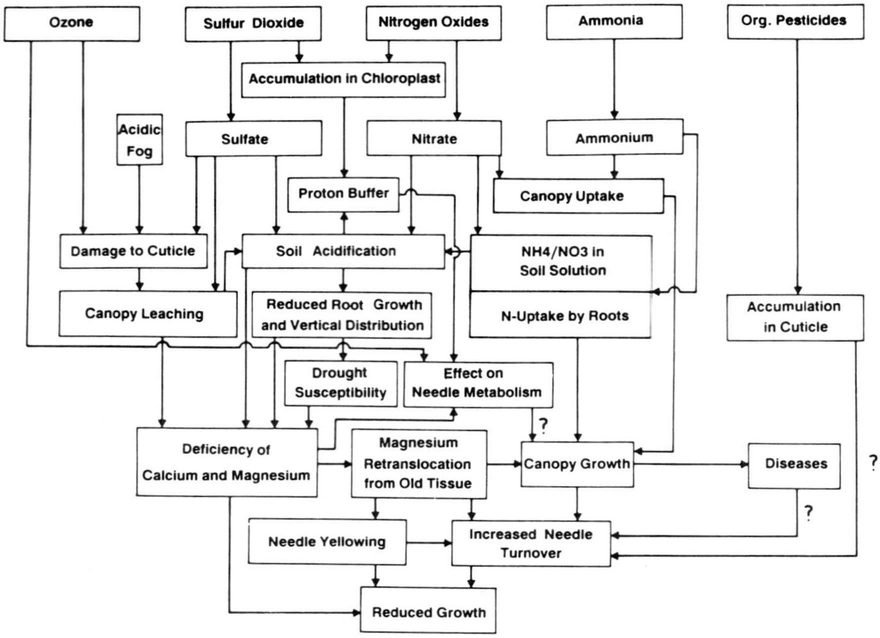

Figure 3.17 Possible events and processes leading to the decline of Norway spruce in the Fichtelgebirge, Germany. From: Schulze *et al.* (1989).

Similar results have been obtained at a larger scale with soil type being found to have a major influence on the occurrence of all types of forest decline in Germany (Schröder and Vetter, 1989).

Thinning of trees is apparent from several older studies, with thin-crowned Norway spruce at St. Moritz, Switzerland, showing reduced increment in a study conducted in 1921 (Keller and Imhof, 1987; Schweingruber, 1989). Bucher (1988) argued that trees classed 70 years ago as having an 'ideal crown form' would currently be scored as having 35–40% needle loss. Consequently, the extent of Type 2 decline is very unclear. Much of the 'defoliation' reported in the European assessments of Norway spruce falls into this category on the basis of the described symptoms. However, there are few reports of mortality associated with it and little evidence that the trees actually deteriorated into their current state. Furthermore, there is little evidence for a decline in transparency of Norway spruce across Europe (Chapter 6). Similarly, the causes of the decline are very unclear. Drought stress is probably involved, and this may have been aggravated by pollution (e.g. Cornic and Pierre, 1987; Maier-Maercker, 1991). The relationships between drought stress and pollution are complex and have still to be properly investigated, as the results that are currently available are very conflicting.

Type 2 decline has been associated with mineral deficiency. Although

Figure 3.18 Norway spruce showing the transparent crowns typical of trees classed as having Type 2 decline.

yellowing is not apparent, nutrient analyses have sometimes revealed low levels of potassium, magnesium, calcium and zinc (Kenk *et al.*, 1985; Lutz and Breininger, 1986; Mengel *et al.*, 1986; Zech *et al.*, 1983, 1985; Hüttl, 1989).

The crown transparency that is a supposed characteristic of Type 2 decline is the most common symptom on Norway spruce in Great Britain. Consequently, it has been examined in detail in Chapters 6 and 7.

Type 3

In southern Germany, the rapid reddening of needles was described from several low-altitude areas in the early 1980s. Reddening started in the autumn of 1982 and needles were shed in November and December (Rehfuess, 1987, 1988b; Rehfuess and Rodenkirchen, 1984, 1985). Several age classes of needles were involved and the reddening developed simultaneously across these and very rapidly (within two weeks). It occurred both on the edges of stands and within them and in both vigorous and slow-growing stands. The syndrome was only recorded in older (> 60 years) stands. The symptoms were specific to Norway spruce, with adjacent silver fir apparently being unaffected.

Needle loss occurred among the older foliage towards the bases of shoots and involved both primary and secondary shoots. Secondary shoots forming after the first year were unaffected during subsequent recurrences of the symptoms. They were characterized by 'tinselling', with shoots hanging vertically from the main branch. The outer parts of the crowns remained green, although where the symptoms occurred in successive years, the crowns became progressively thinner. Growth losses occurred when the needle loss exceeded 40–50% (Rehfuess, 1988b), although no long-term losses have been documented. Most affected trees have shown signs of vigorous recovery.

According to Liu and Hüttl (1991), the foliar contents of phosphorus, potassium, calcium and magnesium were closely correlated with the amount of needle reddening. Reddening could be intense at phosphorus contents of less than 1.2 mg g^{-1} and was very rare when the phosphorus contents exceeded 1.5 mg g^{-1}. Similarly, when potassium contents exceeded 5 mg g^{-1}, reddening was rare, as it was when calcium levels were below 5 mg g^{-1}. However, Rehfuess (1987) argued that there was no association between the incidence of needle reddening and the trees' nutrition. Such diametrically opposing conclusions clearly need to be resolved.

It is not certain whether this type of forest decline is related to fungal problems. In the late autumn of 1982 and 1983, older stands in southern Germany were heavily attacked by several needle-cast fungi, including *Lophodermium piceae* (Fckl.) Höhn., *Lophodermium macrosporum, Lophodermium filiforme* and *Rhizosphaera kalkhoffii* Bubak (Rehfuess, 1985a, 1987; Rehfuess and Rodenkirchen, 1984, 1985) and the red needles showed clear signs of infection by these needle pathogens. The problem was much less apparent in 1984 than in the two previous years (Rehfuess, 1985b).

This type of decline is a special case. It is probably unrelated to the reddening seen in Great Britain and Denmark as no mortality was reported in south Germany, whereas mortality was a frequent feature associated with the reddening in Great Britain and Denmark. Rehfuess (1987, 1988b) postulated that frost shocks in the late 1970s and early 1980s might have caused sub-lethal damage to the needles with subsequent fungal infection being favoured by high rainfall and high air humidities in the summers of 1982 and 1983. Schütt (1985) has argued that needle-inhabiting fungi are secondary pathogens, and the association of the infestations with a preliminary stress therefore seems likely.

Type 4

Needle-yellowing has been reported from trees growing at higher elevations in the calcareous Alps, particularly in southern Bavaria, although it has also been seen in the French Alps (Landmann, 1992). The decline is believed to have started in about 1981 on south-facing slopes and was still in progress in the late 1980s (Rehfuess, 1987, 1988b). Needle loss followed the chlorosis, resulting in progressively thinner crowns.

Ammer *et al.* (1988), working at Wank in the Bavarian Alps, found a bimodal pattern of damage, with peaks at 900–1200 m and at 1600 m and a similar pattern has been described by Paffrath and Peters (1988). However, Mößmer *et al.* (1990), working at Garmisch-Partenkirchen, did not find a second peak in crown transparency at 1600 m, the bimodal (1987, 1989) or irregular distributions (1986) of transparency peaking at different heights in different years.

Table 3.1 Annual mean concentrations of ozone in relation to the number of hours above 120 $\mu g\ m^{-3}$ in Switzerland 1987

Site	Altitude (m)	Annual mean ($\mu g\ m^{-3}$)	Number of hours $>120\ \mu g\ m^{-3}$
Zürich	410	20	96
Dübendorf	431	37	293
Payerne	490	47	238
Lägeren	685	52	381
Davos	1660	64	19
Jungfraujoch	3580	71	16

From Gehrig (1992)

Ammer *et al.* (1988) associated the peaks in damage with altitudinal peaks in ozone concentrations. This failed to consider factors such as seasonal differences in ozone concentrations and the relationship between peak concentrations and the annual means. In Table 3.1, data from several Swiss stations are presented. Although the highest annual means were found at the highest site, this also had the lowest number of hours above 120 μg m^{-3} (Gehrig, 1992). A further problem lies in the temporal pattern of ozone concentrations, which in the northern Alps have been steadily rising over the period 1977–91 (Seiler, 1991). This is not consistent with the reduction in the defoliation of Norway spruce seen in the same area between 1987 and 1989 (Mößmer *et al.*, 1991). In addition, field measurements on mature trees revealed no direct effects of ozone or photo-oxidants on gas exchange measurements (Dunkl and Knoppik, 1991). Yet more evidence against the direct involvement of ozone comes from studies of superoxide dismustase activity of trees in an altitudinal transect at Wank, Bavaria (Polle and Rennenberg, 1989, 1992). The activities showed no evidence of a rise with altitude, suggesting that the trees higher up the slope were not utilizing this antioxidative system. In contrast, the antioxidative system used to scavenge hydrogen peroxide (ascorbate, ascorbate peroxidase, glutathione and glutathione reductase) showed altitudinal patterns, suggesting that this pollutant might be affecting trees. However, it is not clear how specific these antioxidants are to hydrogen peroxide.

Occasionally, the yellowing has been associated with potassium deficiency (Rehfuess, 1983, 1987; Fink, 1991), with yellow-tipping to the needles being widespread. The potassium deficiency had several effects on the structure of the needles (Fink, 1991). Changes to the vascular tissue were similar to the effects of magnesium deficiency (see above). However, the collapse of the sieve cells and starch accumulation in the mesophyll cells occurred less rapidly. In addition, the chloroplasts showed a strong reduction of the thylakoids, which were reduced to a few isolated membranes without any grana framework.

Manganese deficiency has also been implicated in Type 4 decline (Zöttl, 1985; Zöttl and Hüttl, 1986; Rehfuess, 1987, 1988) and is believed to be associated either with low reserves in the soil or biological fixation within the rooting zone, making it unavailable for uptake. The deficiency causes a yellow to pale discoloration of the youngest needles. The soils on which this occurs are derived from dolomite and are rich in calcium and magnesium but poor in available potassium and manganese. At Wank, potassium, calcium and magnesium were all present in the soil in adequate quantities. Instead, nitrogen and phosphorus were limiting (Rehfuess, 1998b; Liu *et al.*, 1991). The prevalence of the decline on shallow soils on steep, south-facing slopes suggests that desiccation may also be important. Soils at many alpine sites are very poor and susceptible to drought (Liu *et al.*, 1991), and this is a likely cause for the nutritional stresses that have been recorded. The differences between sites reflect differences in the composition of the source materials. Soils derived from dolomite tend to be low in manganese whereas those on limestone have low potassium and iron contents.

In the Jura, France, trees with Type 4 decline have also been seen. Defoliation has decreased since 1984, but yellowing was particularly high between 1986 and 1988 (Landmann, 1992). Badot *et al.* (1988) reported that the water content of needles of affected trees was lower than that of apparently healthy trees, and that there was no evidence of potassium, calcium or magnesium deficiency. However, Bonneau (1989) suggested that the chlorosis was associated with potassium deficiency in particular and also phosphorus and nitrogen deficiency. In the area, drought and air pollution interactions are believed to be responsible for the decline of Norway spruce (Badot, 1991). However, soil depth and the relative crown length also play an important part (Landmann, 1992), and drought appears to be the dominant factor causing the decline.

As with Type 3 decline, the occurrence of this form of decline appears to be relatively limited, being restricted to the calcareous Alps and neighbouring mountains. Little attention has been paid to it and its spatial extent and causes remain to be fully evaluated.

Type 5

The final type of spruce decline involves the thinning of crowns in coastal areas in northern and eastern Germany, without the development of marked discoloration. This type of damage is associated with the deposition of excessive amounts of nitrogen and affects an area extending from Belgium to Denmark and probably also the Baltic states. A variety of other species are affected, including Douglas fir, Scots pine and Corsican pine (*Pinus nigra* var. *maritima* (Ait.) Melville). It has been associated with magnesium, potassium and phosphorus deficiency (Reemtsma, 1989). The damage is relatively local, extending about 1000 m from the source of the ammonia (De Temmerman and Coosemans, 1989). However, in areas such as the Peel in The Netherlands, so much nitrogen is spread on the fields that regional concentrations build-up, with consequent effects on forests throughout the area.

Meyberg *et al.* (1988) examined the ultra-structural condition of mesophyll cells of green needles from Type 5 defoliated and undefoliated trees from northern Germany. In healthy trees, the mesophyll cells were characterized by large numbers of chloroplasts lining the peripheries of the cells. The central vacuoles contained tannins in the form of small or large droplets. The chloroplasts had well-developed internal membrane systems, with moderate stacking of the thylakoids. Occasionally, a slight swelling of the thylakoids was recorded. In older needles, the number of plastoglobuli increased. Starch grains were small and scarce. In defoliated trees, there was no difference in the nucleus, mitochondria, microbodies or endomembrane system. However, dilatations of the thylakoids were recorded, ranging from limited swelling to extreme extensions of the thylakoid lumina.

A detailed study of a declining Norway spruce stand has been made at Wingst in the coastal area of northern Germany (Krzak *et al.*, 1988). This forest receives 82 kg N ha^{-1} yr^{-1} deposition, with 46 kg as NH_4-N. Ozone concentrations had an annual mean of 0.061 mg m^{-3}, with peak hourly means of 0.309 mg m^{-3} (Krzak *et al.*, 1988). Higher rates of net apparent photosynthesis in the one-year-old and two-year-old needles were found on trees with thinner crowns (Fig. 3.19), although current-year needles had similar rates in the two tree types. As the increment of defoliated trees was lower than that of 'healthy' trees, Krzak *et al.* (1988) suggested that the assimilate was being used to support the fine-root system. The defoliated trees had less total chlorophyll in all the needle years that were sampled (current, one-year and two-year-old needles). The stomatal conductance to water vapour ratio was higher in all three defoliated trees (Fig. 3.20). Magnesium levels were lower in the defoliated trees, although the calcium levels were similar and the defoliated trees had lower potassium contents in their older needles. Nitrogen levels in the current-year needles of symptomatic trees were low, bordering on deficient. It was concluded that the problems were occurring because of a combination of high soil acidity and low nutrient reserves, these being caused by high levels of ammonia and acidic deposition.

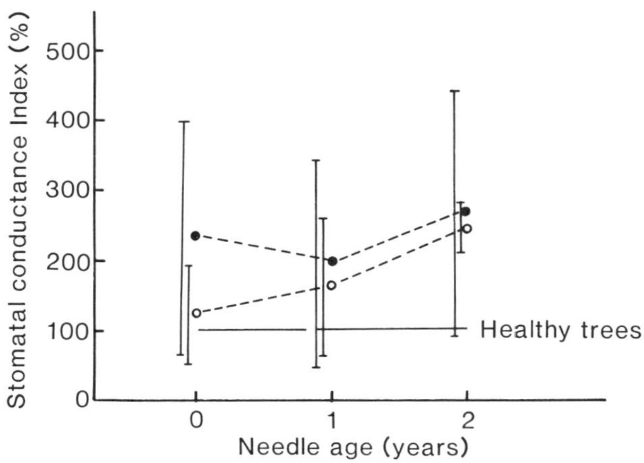

Figure 3.19 Mean rates of net apparent photosynthesis of 68-year-old Norway spruce in the Wingst forest, northern Germany. Solid lines: healthy to slightly transparent crown; dashed line: very transparent (>60%) crown; open circles: west aspect: closed circles: east aspects. From: Krzak *et al.* (1988).

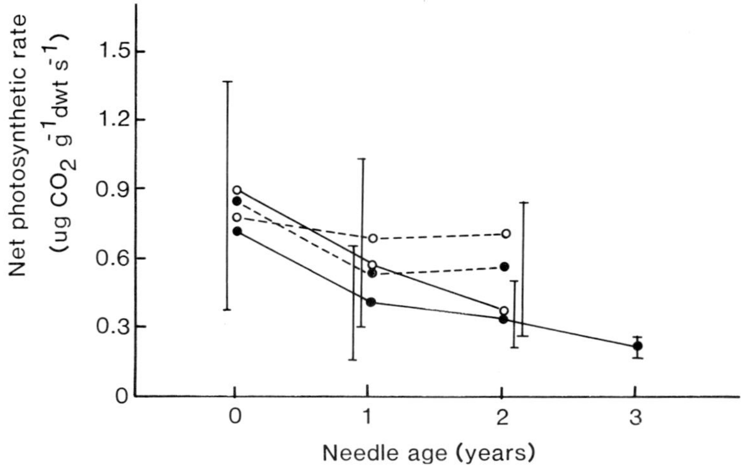

Figure 3.20 Stomatal conductance to water vapour ratio (G_s). The index shows one value for symptomatic trees as a percentage of the conductance of healthy Norway spruce trees. Open circles: west; filled circles: east. From: Krzak *et al.* (1988).

In the Belgian and French Ardennes, symptoms were characterized by the total loss of needles four or more years old and complete yellowing or tip-yellowing of younger needles, particularly those exposed to the sun (von Praag and Weissen 1986; Nys, 1989; Weissen *et al.*, 1990). The problem was related to nutritional disturbance. It was not apparent until after the winter of 1982–3 (Weissen *et al.*, 1988). In the Ardennes, fertilization with calcium has resulted in a reduction in the needle loss (Nys, 1989; Belkacem *et al.*, 1992). The soils in the area generally consist of calcium-deficient acid brown earths with moder humus, with acidic pseudogleys on flat plateau areas (von Praag and Weissen, 1986; Weissen *et al.*, 1990), explaining the foliar response to calcium fertilization. Fertilization with nitrogen led to increases in chlorosis and loss of needles, together with a reduction in growth. Consequently, it has been argued that nitrogen deposition was the cause of the nutritional disturbances (Hambuckers and Remacle, 1991, Belkacem *et al.*, 1992). Magnesium, in contrast to elsewhere, does not appear to be a critical factor controlling the condition of the trees. As elsewhere, the problem may be partly related to past management with current coniferous plantations frequently dating from the end of the nineteenth century, when extensive afforestation of degraded coppice and heathland occurred. These areas would have had low initial reserves of base cations, a problem aggravated by leaching and biomass uptake. In the area, nitrogen contents of needles were generally high (1.5–1.7%), and an interaction between increased demand associated with increased growth and low levels of magnesium and calcium in the soil is likely to be involved.

The occurrence of this type of decline again appears to be a special situation, closely related to the spreading of the waste products of high-intensity stock rearing on fields. However, as agricultural slurry applications are widespread throughout Europe, it seems that some additional factor is involved in those areas where the nitrogen inputs are not especially high.

Other types of Norway spruce decline

Sutinen (1990) described damage to Norway spruce and Scots pine that has occurred in southern Finland since 1984. His study site was Jokioinen, where the trees were native to the site and were about 50–90 years old. Discoloration of the needles was noted from the winter of 1983–4 onwards. Needles turned from green to dirty greyish-green, with or without yellowing. The discoloration was more pronounced in the upper crowns, on the southern sides of trees and on the upper sides of branches. Damage progressed from edge trees towards the centres of the stand, with the tallest trees being affected first. In October 1986, the current-year needles on many trees were dead.

A structural investigation of needles from the damaged trees and from healthy trees at a site in central Finland was undertaken. The needles of the healthy trees had intact cell structures in the mesophyll cells and in the vascular tissue. Summer needles had chloroplasts filled with large starch grains, usually one per plastid. There was an increased amount of plastoglobuli in the chloroplasts. In winter, this was only apparent in one cell layer under the upper epidermis, but sometimes extended to one or two cell layers under both the upper and lower epidermis. Green needles from visibly undamaged trees at the study site had shorter mesophyll chloroplasts. The stroma were electron-dense and rich in ribosome-like granules. Starch grains were present, but not in all chloroplasts. The amount of plastoglobuli increased in the same fashion as needles from the control area. The damage was attributed by Sutinen (1990) to ozone, because the nature of the damage did not appear to be the same as that caused by natural factors such as drought and frost.

Elsewhere in Europe, it is difficult to determine whether trees are in decline. In Sweden, defoliation of both Norway spruce and Scots pine is reported to be higher at northern and high-altitude sites (Andersson, 1990). Concern existed that the 'decline problem' may have spread from Germany, but the increase of decline-like symptoms in southern Sweden in 1983 is believed to be related to the very dry summers of 1982 and 1983 (Bengtsson, 1985). Consequently, the evidence for a widespread decline in Norway spruce in the northern Scandinavian countries is very limited.

Within the European Community, the proportion of trees with less than 10% defoliation decreased from 64% to 56% between 1987 and

1990 (Anon., 1991b) with no further changes in 1991. The proportion of trees with more than 25% loss of density has increased slightly from 12.4% to 17.2% between 1987 and 1991. These figures suggest that crown transparency has increased, but they should be treated with considerable caution due to problems over the standardization of assessments (see Chapter 5).

Summary

Of the five different types of decline that have been identified in Germany, the one characterized by needle yellowing, foliage loss and mortality, which occurs at higher elevations in several medium-altitude mountain ranges, has received the most attention. Many now consider that the basic mechanisms underlying the decline are understood. The chlorosis in older needles develops soon after flushing of the new year's growth. It is caused by magnesium deficiency, the magnesium being withdrawn from older needles and transferred to the new needles (Mies and Zöttl, 1985; Zech *et al.*, 1985; Lange *et al.*, 1987, 1989a). This is a normal procedure: the deficiency symptoms occur when there is insufficient magnesium in the older needles and uptake has to occur primarily from soil magnesium (Oren *et al.*, 1988a, b, c).

The causes of the reduced uptake of soil magnesium have yet to be fully determined. It may represent a reduction in the ability of trees to take up magnesium, a reduction in the available magnesium in the soil or nutritional imbalances created by increased nitrogen availability. The first mechanism seems unlikely except where soil magnesium is already very low, as trees fertilized with magnesium normally recover (Bosch *et al.*, 1983; Zöttl and Hüttl, 1986; Hüttl and Wisniewski, 1987; Kaupenjohann *et al.*, 1987). Soil magnesium reductions have been reported by various authors (e.g. Ulrich *et al.*, 1979; Hauhs, 1985; Hildebrand, 1986), and this is clearly a factor in some cases. Increased nitrogen uptake is likely to increase the demand for magnesium (Schulze *et al.*, 1987; Roberts *et al.*, 1989; Schulze, 1989) and this, combined with a reduction in the available soil magnesium, seems to be the most likely cause of the deficiency. It is not clear whether the reduced magnesium availability is due to losses through soil leaching (possibly accelerated by soil acidification) or whether it is the cumulative result of various forest management practices that have depleted the levels of magnesium over the last 100 years. A combination of the two seems likely. In the two most detailed field studies undertaken to date, the direct effects of pollution were ruled out as a cause of the decline in the Fichtelgebirge (Zimmermann *et al.*, 1988; Lange *et al.*, 1989; Schulze, 1989) and at Davos and Lägeren in Switzerland (Häsler, 1991).

Although many experimental studies have been undertaken to evaluate the effect of air pollution on Norway spruce, there are major differences in the responses of different provenances to specific levels of individual pollutants (Bergmann and Scholz, 1987). This makes any interpretation of the results extremely difficult. In addition, experimental studies have frequently failed to address the specific set of pollutants that may be present in an area, or have used very artificial conditions. This further complicates any analysis.

The role of drought appears to be complex. Many of the earlier declines were associated with drought conditions, and drought can cause the death of trees in the absence of pollution. The role of different types of pollution in aggravating drought effects is unclear. Drought conditions seem to reduce the impact of pollution by reducing the stomatal conductance, thereby reducing pollutant uptake (Dobson *et al.*, 1990). However, the extent to which exposure to particular pollutants subsequently makes a tree more sensitive (or tolerant) to drought is unknown.

Scots pine

Past studies

There are fewer historical records of decline for Scots pine than there are for Norway spruce, despite Scots pine being the most common forest species in western and central Europe. However, Wiedemann (1925) described declines occurring in association with drought and, in some areas, Scots pine appears to be very sensitive to this stress.

In some parts of southern Scotland, decline of Scots pine stands has been observed (Redfern *et al.*, 1987). The decline was characterized by infection by *Lophodermium seditiosum* Minter, Staley & Millar and *Brunchorstia pinea* and by attacks by the pine shoot beetle, *Tomicus piniperda*. The cause of the severity of the damage associated with this complex was unclear. Its apparent restriction to sites at higher altitudes suggested that climatic stress was probably involved, possibly in combination with incorrect provenance selection.

In another example, problems were experienced with Scots pine in the Pennine Hills of northern England in the 1960s and 1970s (Farrar *et al.*, 1977; Lines, 1984), although difficulties in the area had been recognized much earlier (Fisher, 1908). A survey conducted in the 1970s revealed an unusually low occurrence of Scots pine in a corridor extending to the north-east of Liverpool and Manchester. A negative correlation was found between the occurrence of Scots pine and the winter mean level of sulphur dioxide, which ranged from 100 to 150 $\mu g\ m^{-3}$ in rural areas (Fig. 3.21). Those Scots pine present in the corridor had very high levels of sulphur in the needles, ranging from 3 to 4 $mg\ g^{-1}$. While sulphur dioxide was an important factor limiting the presence of Scots pine, poor soils and adverse climatic conditions must also have played a part.

Damage to Scots pine from sulphur dioxide has also been reported

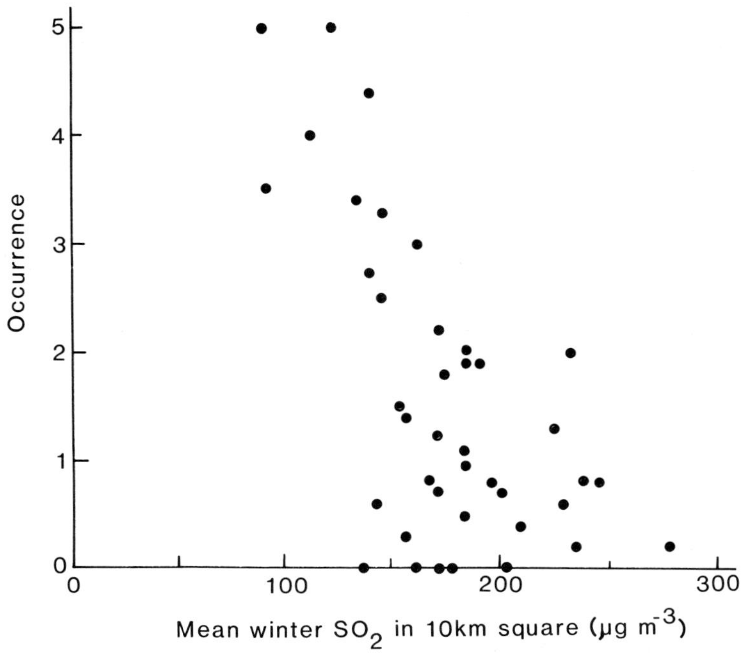

Figure 3.21 Relationship between occurrence (number of 2 km squares per 10 km with Scots pine present) of Scots pine and mean winter sulphur dioxide concentrations in the English Pennines in the mid-1970s. From: Farrar *et al.* (1977).

from the former German Democratic Republic (Fiedler *et al.*, 1990). Damage from excessive nitrogen deposition has been seen in eastern Germany, usually when deposition of nitrogen exceeded 35 kg ha^{-1} yr^{-1} (Hoffmann *et al.*, 1990). Scots pine trees are extremely variable and substantial differences in susceptibility to pollution injury have been reported between provenances (e.g. Huttunen and Törmälehto, 1982; Oleksyn and Bialobok, 1986; Oleksyn, 1988). Provenance trials (Oleksyn and Bialobok, 1986) showed that provenances from northerly and southerly were the least sensitive to sulphur dioxide. This again suggests that fast-growing provenances may be less tolerant to certain pollutants than slower-growing ones.

In the 1980s, trees from the Düben Heath area of Germany showed symptoms that differed from classic sulphur dioxide damage. Needle chlorosis occurred mainly in autumn and winter and was restricted to the upper sides of branches. The degree of chlorosis was dependent on the climatic conditions in the preceding summer and varied significantly on individual trees within stands. Examination of the needles of affected trees revealed reduced potassium levels and reduced water contents in first- and second-year needles, with no apparent effect on magnesium, calcium or zinc contents. Needle surfaces were contaminated with particulate silica, aluminium and iron, suggesting that air pollution was involved in the problem.

Status of Scots pine

Despite the above studies, the attention that has been given to Scots pine is sparse. This is surprising in view of its known sensitivity to some types of pollution. One reason may be the difficulty of assessing this species. Scots pine is one of the most difficult species to assess as features such as male flowering and crown form have such a great influence on crown transparency. However, these problems are not insurmountable, and considerably more attention could have been given to the species. The Verein Deutsche Ingenieure (1987) identified several symptoms occurring in Scots pine:

1. chlorosis of older needles;
2. shedding of older needles;
3. reduction of height increment, resulting in a flattening of the crown; and
4. symptoms prevalent on dominant and exposed trees.

Besides these, several other phenomena have been observed, including increased rates of flowering, changes in branch structure and shortened needles. As with other species, these are not universal.

Within the European Community, a small increase in the proportion of Scots pine trees with more than 25% loss of foliage has been recorded between 1987 and 1991. The proportion of trees with less than 10% loss in crown density has fallen from 67.3% in 1987 to 53.0% in 1991. The data derived by the UN-ECE could have been relevant as they include results from northern Europe, where Scots pine is particularly abundant. However, the figures that are presented in their annual reports are aggregated for the eleven pine species assessed each year. As these species vary significantly in their responses to pollution (Innes, 1990a), it is almost impossible to interpret the aggregated figures.

Causes of decline

Chlorosis in Scots pine has been associated with magnesium deficiency (Küppers *et al.*, 1985). Severe needle tip chlorosis was recorded at magnesium concentrations below 0.3 mg g^{-1} and magnesium deficiency is known to reduce the stomatal response to light. Schulze and Küppers (1985) recorded magnesium deficiency in Scots pine growing on acidic soils at altitudes between 700 m and 1000 m in Germany. The deficiency was

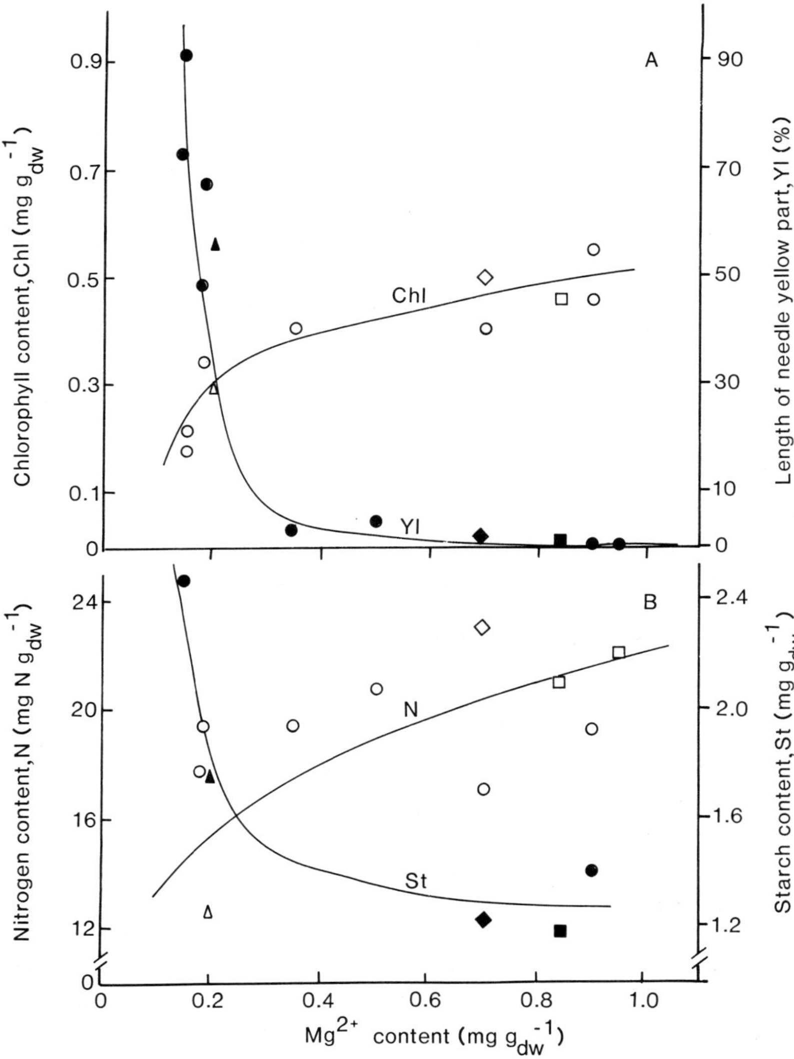

Figure 3.22 The relationship between leaf parameters and magnesium contents of Scots pine. Data are from samples taken from the upper whorl of the stem (open) or from the distal whorls of first order lateral branches (filled). Untreated trees indicated by triangles, magnesium-fertilized trees by diamonds and Mg-N-P-K fertilized trees by squares. From: Küppers *et al.* (1985).

linked to reduced carboxylase activity and decreased photosynthetic capacity. Although plant-water relations were unaffected, the decreased photosynthetic capacity resulted in a deterioration of the water use efficiency. The magnesium-deficient needles were shorter, thinner and lighter in weight, all of which would contribute to reduced estimates of crown density (Fig. 3.22).

Crossley and Fowler (1986) examined the weathering rates of epicuticular wax from Scots pine needles taken from a polluted and a relatively unpolluted site in Great Britain. Large variations in the needle waxes between different provenances occurred at the same sites. Using the Altyre provenance alone, clear differences in needle longevity were present between the two sites. Fifty per cent of the third-year needles were retained at the unpolluted site and only 10% at the polluted site, where soot deposition resulted in increasing occlusion of the stomata as the needles aged. The rate of weathering at the polluted site was greater, but a much wider range of sites would be required before any reliable inferences could be drawn from this finding.

Summary

Scots pine was susceptible to the high concentrations of sulphur dioxide that were present in industrial areas before the introduction of emission controls. Substantial, unambiguous damage was reported in such areas. It is also one of the main species affected by high levels of ammonia deposition, a problem that is most apparent in The Netherlands. This particular type of decline seems to represent an extension of the Type 5 Norway spruce decline to other species.

In unpolluted areas, Scots pine is subject to a number of difficulties. In particular, it seems that climatic stress may trigger a number of problems, which together, can cause the decline and death of individual trees. The decline may be the result of incorrect provenance selection in some areas. However, there does not seem to be any evidence of widespread decline in remote parts of Europe.

Silver fir

Past declines

The information related to historical declines of silver fir is even more detailed than that for Norway spruce. Early descriptions, such as those by Neger (1908), Scheidter (1919) and Wiedemann (1927), described several years of stunted growth followed by tree mortality. The crown usually deteriorated from the base upwards, followed by loss of needles from the remaining part of the crown and a reduction in the vigour of the leading shoot. This led to the characteristic 'stork's nest' appearance of the tree. Usually, the decline was associated with the development of a wet core in the lower trunk. An important historical account of silver fir decline in the Limpurg region of Swabia in 1784 describes the problem as being worst on dry, stony soils, providing a definite link between the decline and drought stress (Rommel, 1989).

Cramer (1984) and Brandl (1985) surveyed the literature related to these early descriptions of decline in silver fir, finding records extending back as far as ca. 1500 AD (Krauss 1957), with many records throughout the 19th and early 20th centuries. Often, these early declines occurred towards the eastern and north-western limits of the range of silver fir (Seitschek, 1982). Krehan (1989) identified three main areas where declines of silver fir had been reported and he associated the declines with drought, severe frosts, competition with Norway spruce, root pathogenic fungi and air pollution. The main area affected by decline in the early 1980s was also the area where the greatest numbers of past reports of decline have occurred, again suggesting that the phenomenon was not unique.

Symptoms and status

The detailed symptomatology of the recent silver fir decline is poorly described in the literature. However, as Cramer (1984) pointed out, there are good descriptions of past declines and the symptoms associated with these are the same as those seen recently. The full set of symptoms associated with the decline in Germany (from Verein Deutsche Ingenieure, 1987) are:

1. chlorotic lesions, especially on the upper sides of needles, leading to chlorosis of entire needles, beginning with the oldest needles;
2. premature reduction in height growth and formation of the 'stork's nest' crown;
3. shedding of needles from the trunk outwards and from the bottom to the top of the tree;
4. development of epicormic shoots on the main stem in closed stands;
5. pathological extension of wet heartwood into sapwood at the tree base;
6. death of fine roots; and
7. symptoms first seen on dominant and exposed trees.

Most reports describe the gradual shortening of the crown from the base upwards. At the same time, height growth slows, resulting in the development of a platform ('stork's nest') crown. Often, epicormic branches occur in large numbers on the stem exposed by the primary branch dieback.

Barbu (1987, 1991) recognized four types of silver decline in Romania. These included the slow death of the crown, particularly in older trees (120–160 years) in uneven-aged mixed stands on pseudogleys, and the rapid death of trees, usually in even-aged pure and mixed stands, 60–100 years old. In the other two types, needles fell during a period of six to eight months and this was followed by tree mortality. Alternatively, tree crowns lost their foliage due to abrasion, a phenomenon that was restricted to very dense stands 60–80 years old.

Needle retention in silver fir is usually about eight to ten years. This decreases under stress, and affected trees often have only five or six years of needles. In extreme cases, even fewer years of needles may be retained (Scheidter, 1919).

A considerable amount of work has been done on declining silver fir stands in the Vosges Mountains of eastern France. Here, there have been some interesting trends in the development of defoliation and discoloration (Landmann, 1992). Defoliation peaked in 1984, since when it has been decreasing. However, this concealed a difference between stands below and above 750 m ASL. Lower stands improved between 1984 and 1988, whereas older stands continued to deteriorate until 1986 and then remained roughly constant. In contrast, discoloration peaked in 1985, then dropped. A shift occurred from chlorosis on mineral-rich sites to chlorosis restricted to mineral-poor sites (Landmann and Fabbri, 1989).

In some parts of the French Pyrenees, silver fir shows intense chlorosis (Bonneau and Landmann, 1988), but the cause of this has yet to be fully determined. The decline of silver fir in the Pyrenees accelerated in the late 1980s (Cheret *et al.*, 1987; Fromard *et al.*, 1991), in contrast to elsewhere. The decline was characterized by marked chlorosis, sometimes followed by needle loss. In some cases, needle loss occurred without discoloration (Fromard *et al.*, 1991). Loss of apical dominance was common, resulting in the development of 'stork's nest' crowns. The symptoms were most apparent in mature, dominant trees, but were also evident in younger ones. Potassium deficiency occurred in chlorotic trees whereas magnesium deficiency was absent. In addition, excessive amino acid levels were present in chlorotic trees, which Fromard *et al.* (1991) attributed to disturbances associated with unusually high levels of ammonium.

Defoliation and yellowing of silver fir have also been seen in the French Alps. The occurrence of symptoms was strongly related to the local geology, with healthy trees occurring on deep, acidic brown to brown podzolic soils. Conversely, chlorotic and thin-crowned trees occurred on calcareous, karren-type soils (Becker, 1991).

Timing of the declines

Observations have suggested that periods of decline have occurred, rather than specific years when trees have come under severe stress (e.g. Wiedemann, 1927; Wachter, 1978; Seitschek, 1982). However, increment studies have shown that while periods of decline may occur over relatively long periods, the start of each decline can frequently be traced back to a particular drought year. For example, Wiedemann (1927) associated the decline of silver fir in some parts of Germany between 1890 and 1905 with drought years in 1858, 1886, 1894 and 1904. Increment studies of silver fir seen to be declining in the 1980s have shown a good relationship between radial growth and summer rainfall. Further work has found that the divergence of growth of vigorous and declining trees is associated with drought years (Lévy and Becker, 1987; Becker, 1987, 1989; Becker *et al.*, 1989), with 1976 being particularly important. However, some authors (e.g. Brandl, 1985) have maintained that past incidents of silver fir decline were much less severe in terms of both the degree of damage encountered and the extent of the area affected. This latter view does not seem to be supported by the available evidence.

Concern about silver fir decline peaked in 1983–5. This was precisely when studies of forest productivity in the Vosges Mountains suggested that silver fir growth was at a peak (Bonneau and Landmann, 1988). Similarly, increment studies revealed no sign of a decrease in productivity at this time in the Jura region of Switzerland (Bräker and Z'Graggen, 1987; Schmid-Haas, 1989a; Bräker, 1990) or in the Black Forest, Germany (Kenk and Spiecker, 1988). Silver fir is generally thought to be one of the most damaged species, although European Community surveys have not supported this (Anon., 1989b, 1990a, 1991b). For example, in 1990, *Abies* sp. generally had lower transparency scores than the average for all species.

The decline was believed to have started in the mid to late 1970s or early 1980s, following the severe droughts of 1975 and 1976. Studies on permanent plots showed that the decline did not develop further after autumn 1986 (Mettendorf *et al.* 1988) and the change may have occurred even earlier (Schröter, 1985; Schröter *et al.* 1985). The recovery was most marked in stands less than 60 years old (Prinz and Krause 1989). It has occurred in most areas (Gerecke 1990) with both crown condition and growth rates improving. The marked increase in growth seen in silver fir in the Vosges Mountains and other areas has not been repeated in the Jura. However, growth in the Jura has recovered from a period of low growth in the 1970s (Bert and Becker, 1990). Some silver fir stands in the Jura were still declining at the end of the 1980s, although most had recovered (Bert, 1992).

Silver fir in the Vosges Massif of France has been the subject of extensive dendroecological investigations. No significant trend in the radial growth variation was evident between the start of the 20th century and the mid-1980s, suggesting that the trees were not in immediate danger of dying out (Becker 1987a, b). Stands suffered a period of severe stress beginning in 1973, peaking in 1976 and almost ending by 1983. Comparable stress periods have occurred in the past, in 1943–51 and 1917–25. All three periods were correlated with periods of severe drought.

Eckstein and Saß (1989) documented the period of decline in Bavaria as extending from the mid–1960s to 1982, with marked growth increases in 1983 and 1984. A similar period of decline was noted by Gerecke (1986) who dated the strongest decrease in growth to 1976. However, Kenk (1983), Lenz *et al.* (1988), Torelli *et al.* (1986) and Kenk *et al.* (1984b) all dated the onset of increment reductions to the early to mid–1950s, and Mitscherlich (1983) described major growth anomalies in the Black Forest in the period 1920–60. Some of these differences may be due to the ages of the trees involved, with the study by Mitscherlich involving particularly old trees.

The available evidence strongly points to a series of declines, with different timing in different areas. The decline of individual trees appears to have been triggered by drought stress. Many of the trees involved in the declines have subsequently recovered. Consequently, it seems that the declines of silver fir in the 1980s were a repeat of previous, well-documented events.

Causes of the declines

A detailed analysis of the climatic conditions during and before periods of silver fir decline was made by Cramer and Cramer-Middendorf (1984). Their work clearly showed that declines in particular areas followed specific climatic conditions, namely below-average rainfall during the growing season. This strongly points to a link between climate and decline in silver fir.

One possibility is that silver fir is not sufficiently adaptable to stress. The areas experiencing decline were also those areas where genetic variability was least (Larsen, 1986; Ukleja-Dobrowolska, 1989) and areas with a high degree of variation generally did not show any decline. Drought is known to have had an impact on the decline of silver fir and different populations show differing resistance to drought (Schütz, 1985). Kral and Mayer (1985) found marked differences in the drought resistance of silver fir from the subatlantic region of France and from high-altitude sites in Switzerland. In addition, silver fir is known to show marked between-provenance variation in its tolerance to acidic deposition (Bottacci, 1991), although the relationship between rate of growth and susceptibility to pollution is unknown.

Lautenschlager (1984), in an early study of the recent decline of silver fir, examined the osmotic potential and gas exchange of two trees in the Bavarian Forest, Germany. One was transparent, the other had no symptoms of decline. The transpiration measurements were conducted in tandem, limiting the interpretation of the results. Transpiration in the symptomatic tree was much lower than in the ‘healthy' tree (Fig. 3.23), although secondary shoots on the defoliated tree had higher transpiration values than primary shoots. Photosynthesis by the shoots of the defoliated tree was also lower. The osmotic potential in the two trees was very similar, although primary shoots on

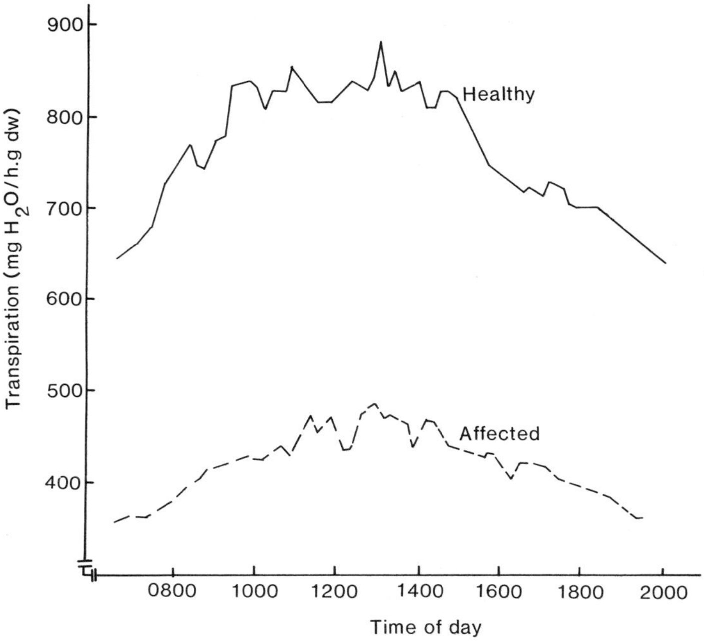

Figure 3.23 Transpiration of a healthy and a symptomatic shoot from two silver fir trees in the Bavarian Forest, Germany. From: Lautenschlager (1984).

the defoliated tree had higher osmotic potential than other shoots. Lautenschlager (1984) concluded from these results that there was no evidence that pollution was resulting in stomatal interference which in turn might have caused an uncontrolled increase in transpiration.

Becquer *et al.* (1990) examined the nitrogen production in a 110-year-old stand in eastern France consisting of 90% silver fir and 10% beech. The stand was located on a typic dystrochrept, and had a mull-type humus. Inputs of ammonia in throughfall were low (1.4 kg N ha^{-1} yr^{-1}) and inputs of nitrate in throughfall amounted to 19 kg N ha^{-1} yr^{-1}. Sulphur inputs were less than in many areas (27 kg S ha^{-1} yr^{-1}). Nitrification occurred throughout the year, producing about 65 kg N ha^{-1} yr^{-1}. While most of the nitrate was produced in summer, 36% was produced outside the growing season. The authors felt that there was abnormal temporal discoupling of nitrate production, suggesting that excess nitrification was not the cause of the decline of the trees.

Subsequent work has shown the importance of silvicultural treatments in predisposing silver fir to decline in the Vosges (Becker and Lévy, 1988; Becker *et al.*, 1989; Gerecke, 1990; Becker, 1991; Landmann, 1992) and in Italy (Hippoliti, 1989). Decline was most severe in over-crowded stands with poorly developed crowns, particularly those with high height/diameter ratios and low relative crown lengths. Although stands affected by the decline had faster basal area growth rates up to 1962, over-stocking was considered to have increased the demand for water per unit area, making the trees more susceptible to drought. Similar results were obtained in Tuscany, Italy, where summer droughts had a much greater effect in dense stands than in open ones (Paganucci, 1989). Silver fir in lower Normandy, right on the edge of its range, was severely and unambiguously affected by the 1976 and 1989 droughts.

Summary

Silver fir is the one of the most westerly species of a large, well-differentiated family that includes a number of European and Middle East species such as King Boris's fir (*Abies borisii-regis* Mattf.), Bornmüller's fir (*Abies bornmuellerana* Mattf.), Cilician fir (*Abies cilicica* (Ant. & Kotschy) Carr.) and Grecian fir (*Abies cephalonica* Loud.). Only the rare hedgehog fir (*Abies pinsapo* Boiss.) occurs naturally further west in Europe. Over its range, the silver fir is very variable in its genetic diversity, probably reflecting the occurrence of severe natural selection of genotypes in the past. Over the last 400 years, there has been a steady reduction in the forest area covered by silver fir in western Europe (Cramer, 1984), strongly suggesting a major, long-term problem. The decline has occurred both gradually and in a step-like fashion, with particular stress events (mainly droughts) causing a sharp increase in the incidence of declines. It seems likely that the recent phase of decline was only a part of this long-term trend, although if it is true that the genotypic variation in silver fir is decreasing, particular stress events may be having an increasingly significant impact.

Oak

Past declines

Periodic reports of oak decline have occurred in the literature, frequently associated with frost damage and attacks by the oak leaf roller moth (*Tortrix viridana* L.) and the winter moth (*Operophthera brumata* L.). Widespread, regional declines of oak have been reported in Europe in the past (Delatour, 1983; Hartmann and Blank, 1992) and similar declines have occurred in North America (Millers *et al.*, 1989). Severe declines in oak in Germany was reported following droughts in 1904 and 1911 (Wiedemann, 1925) and it appears that several species of oak present in Europe may be susceptible to drought stress. Other periods of oak decline include 1739–48, 1929–34 and 1939–44 (Hartmann and Blank, 1992). In

France, four periods of decline have been identified: in the early 1920s, the late 1940s to early 1950s, the late 1970s to early 1980s and the late 1980s. All these periods corresponded with periods of marked drought stress (Turc, 1927; Rol, 1951; Becker and Lévy, 1982; Nageleisen *et al.*, 1991; Landmann, 1992). In Germany, Hartmann *et al.* (1989) associated the onset of oak decline in the mid-1980s with a series of dry years, together with acute stress caused by a series of severe insect defoliations. Secondary damage was caused by *Agrilus biguttatus* Fabr., *Cytospora intermedia* Sacc. and *Pezicula cinnamomea* (DC:) Sacc.

Young (1965) established a connection between summer drought and both the growth and mortality of pedunculate oaks in Great Britain. However, Führer (1992) found no relationship between outbreaks of oak decline over the last 15 years in Hungary and unfavourable weather conditions, suggesting that drought is not always responsible for the onset of oak declines.

Symptoms and timing

Symptoms of the most recent oak decline are:

1. discoloration of foliage;
2. premature shedding of leaves and death of fine shoots;
3. extensive development of epicormic shoots in the crown; and
4. dieback of larger shoots and eventually the entire crown.

The decline is characterized in the first year by a small amount of discoloration followed in subsequent years by severe discoloration and increases in transparency. Foliage may be restricted entirely to epicormic shoots and mortality may follow. In Great Britain, discoloration of oak leaves is not particularly significant, and there has been no evidence of a trend through time. There was a marked increase between 1989 and 1990 in the number of trees with a marginal amount of dieback (see Chapter 7). However, it is not clear whether this was a temporary response to the two droughts of 1989 and 1990 or whether it was part of a longer-term trend.

The syndrome was first reported from continental Europe (e.g. Eichholz, 1985; Balder and Lakenberg, 1987; Donaubauer, 1987; Jakucs, 1988; Hartmann *et al.*, 1989), although it was later found in Great Britain (Innes and Boswell, 1991; Greig, 1992). The extent of the decline and its progression through Europe is difficult to establish because of inadequate symptomatological descriptions in many reports. It has been documented for some time in Europe; Donaubauer (1987) attributed the sudden and widespread appearance of decline in Austria to the delayed and cumulative effect of many biotic and abiotic stress factors during the last 20 years. He and Eisenhauer (1989) attached particular significance to drought stress, suggesting that outbreaks of defoliating insects such as the oak leaf roller moth and winter moth may have aggravated the effects of the drought. Frost is also believed to have been a problem.

In The Netherlands, the decline of pedunculate oak first became evident in 1985 (Oosterbaan and Nabuurs, 1991), probably having started in 1982–3. Both 1982 and 1983 in The Netherlands were characterized by hot dry summers, and there was also severe frost damage in 1984–5 (Oosterbaan and Borgesius, 1986). Severe defoliation by the winter moth and oak leaf roller moth may also have been involved (Leffef, 1988). Studies of the distribution of the mortality suggested that it was greatest on soils with strongly fluctuating groundwater tables in the rooting zone (Oosterbaan and Nabuurs, 1991). Trees in such areas were characterized by shallower rooting systems and a higher proportion of dead roots.

Within the European Community region, the condition of oak has varied according to species (and, by implication, region). English oak (*Quercus robur* L.) has shown marked variations between 1987 and 1991, but the percentage of healthy trees increased in 1989 and 1990. No trends in discolouration have been seen at a European scale. Sessile oak (*Quercus petraea* Mattuschka) has shown a different trend, with the percentage of healthy trees decreasing in 1989, 1990 and 1991. The amount of discoloration increased during the period 1987–90 but dropped in 1991. Holm oak (*Quercus ilex* L.), plots of which are located in the Mediterranean region, steadily improved between 1987 and 1990, with both defoliation and discolouration decreasing, but a marked increase in defoliation occurred in 1991. Conversely, another southern species, the cork oak (*Quercus suber* L.), deteriorated markedly between 1989 and 1991. The reason for this is unknown, but drought and management practices are likely to have been involved (Anon., 1991b).

Causes of the declines

A detailed study has been made of the decline of oak in the Forest of Tronçais in France. Symptoms included crown thinning, stunted leaves and chlorosis, leading to the eventual death of trees. It was first seen in 1978 and accelerated in 1979 and 1980, although tree ring analyses showed that the increment of pedunculate oaks dropped significantly in 1976. Only pedunculate oak was affected, despite sessile oak being widely distributed in the forest (Landmann, 1992). The work on the decline suggested that the soil conditions were a very important factor, with unfavourable soil moisture regimes creating unsuitable rooting conditions for pedunculate oak (Guillaumin *et al.*, 1983). With the rooting systems sensitive to damage, the 1976 drought provided the trigger for the decline in health of the trees (Bonneau and Landmann, 1988). A decline in the condition of oak in the Pyrenees following the 1976 drought appears to have a similar explanation (Durand *et al.*, 1983).

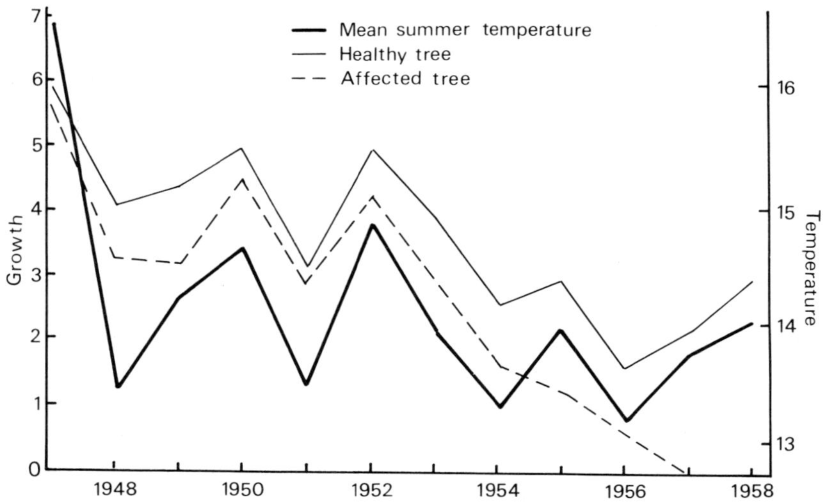

Figure 3.24 Annual mean increment of healthy and affected oaks in relation to summer temperature (May-August, inclusive). Note the increasing differentiation in growth rates the year after temperature peaks. From: Young (1965).

Similar interactions between pedunculate and sessile oaks and soil moisture conditions have been described by Ackermann and Hartmann (1992) in northern Germany. A colour infrared survey revealed that pedunculate oak tended to have more transparent crowns than sessile oak, probably because it tended to occur on more extreme sites than sessile oak. Both species had more damage on sites where soil moisture was stagnant or intermittent.

The hypothesis that drought is a major factor in the decline is supported by early studies by Young (1965). He drew attention to the association between drought and oak decline in Great Britain in east England in 1958 (Fig. 3.24). He suggested that drought had weakened the trees which were subsequently adversely affected by insect and fungal problems.

Experiments involving the fertilization of English oak seedlings (Harmer, 1989) have indicated that nutrient supply can influence the numbers of buds and the numbers of branches on interwhorl sections of the shoot. Trees supplied with high levels of nitrogen flushed more frequently and produced more branches per bud. NPK fertilization increased the number of branches in the lammas flush. NPK fertilization of 10-year-old trees resulted in a more branched crown, with reduced apical dominance of the terminal bud. These are all indices that could probably be recorded in mature crowns. Consequently, some of the structural characteristics of oaks may be related to nutrition.

Summary

The declines of oak seem to be rather different from most of the other declines recently reported in Europe. They did not develop at the same time as the coniferous declines and, although they seem to be related to drought, the available evidence suggests that secondary effects (e.g. insect defoliation) are extremely important. Defoliation by insects can be particularly severe in oak, and the incidence seems to be determined by the synchronicity of the leaf flush and the hatching of defoliators. The severity of the impact depends on the vigour of the trees. When they have already been weakened, the defoliation may be particularly significant. However, much more research is required on the ecological aspects of the declines before any more substantive conclusions can be reached.

Beech

Past declines

Beech has also been the subject of decline reports, with Wiedemann (1925) and Bergmann (1927) describing major declines in radial increment associated with droughts in 1892, 1893 and 1904. These early declines were accompanied by other symptoms, particularly green leaf fall (Wiesner, 1904), a phenomenon also associated with the 1989 and 1990 droughts in Great Britain (Innes and Boswell, 1991). The available literature strongly suggests a link between the occurrence of declines in beech and unusually dry conditions (Cramer, 1984), with more recent droughts also being associated with growth declines (Rehfuess *et al.*, 1983).

Symptoms and timing

Beech is very variable in its response to stress (Müller-Starck, 1985). Consequently, it is often difficult to ascribe symptoms to particular causes. The Verein Deutsche Ingenieure (1987) listed the main symptoms associated with beech decline as:

1. chlorosis and necrosis of leaves;
2. shedding of green leaves and premature autumnal discoloration;
3. formation of epicormic shoots in closed stands;
4. reduced shoot growth and dieback of lateral branches; and
5. disruption of natural rejuvenation of stands.

However, the symptoms that are apparent at any particular site vary considerably, making it difficult to draw up any general indices of stand condition.

As with oak, reports of a recent decline in beech started later than in conifers. Heavy depressions in annual radial increment associated with the

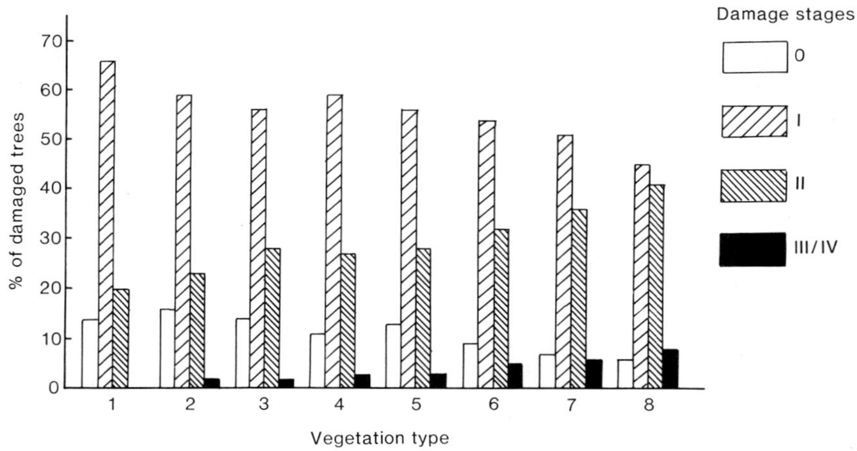

Figure 3.25 Proportion of symptomatic beeches in different forest communities in Germany. 'Damage stages': 0: 0–10% transparency; 1: 11–25% transparency; 2: 26–60% transparency; 3+4: >60% transparency. Vegetation types: 1: Elmyo-Fagetum corydaletosum; 2: Elymo-Fagetum typicum; 3: Elymo-Fagetum (Geum urbanum-type); 4: Asperulo-Fagetum typicum; 5: Asperulo-Fagetum festucetosum; 6: Violo-Quercetum; 7: Luzulo pilosae-Fagetum Matusz.; 8: Elymo-Fagetum (Lathyrus vernus-type). From: Härdtle (1990).

droughts of 1976 and 1983 occurred in beech growing in upland areas in Germany (Abetz, 1988), with recovery being slower on drier soils than moister ones. The 1983 drought appears to have been particularly important in causing the differentiation of trees in older stands throughout Germany. The link with site quality is important; Härdtle (1990) found marked differences in the condition of beech in different forest communities (Fig. 3.25). The 1976 drought also had a major influence on increment elsewhere in Germany (Eckstein, 1985; Eckstein *et al.*, 1984). Abetz (1988) recorded that younger trees were currently growing better than older trees when they were young; Eckstein (1985) attributed this to reduced stand density following mortality after the 1976 drought. Wahlmann *et al.* (1986) documented a growth depression beginning after 1970, with healthy trees today (as defined by the Roloff system – see Chapter 4) recovering from 1977 onwards and damaged trees failing to recover. They considered the low growth in the early 1970s to be due to low annual precipitation.

Several trends are apparent from the more detailed analyses that have been conducted on beech condition. Defoliation increases with tree size (Hunziker *et al.*, 1988) and is greater in heavily thinned or very open plots than in lightly thinned plots or plots with a closed canopy (Runkel and Kenneweg, 1986; Keller and Imhof, 1987), a phenomenon also recorded in Great Britain (Innes and Boswell, 1990a, 1991). The amount of masting is higher in thinner-crowned trees (Gärtner, 1988; Innes, 1992a). Beech can be subject to marked discoloration although the significance of this remains unclear. Chlorosis is particularly common and, in most years, develops on trees growing on chalky soils by mid-July.

Within the European Community, the proportion of beech trees without defoliation decreased between 1987 and 1991 (Anon., 1991b). The proportion of trees with more than 25% defoliation has increased (from 12.8% in 1987 to 18.2% in 1991). In Great Britain, the sample of trees is too small to draw any reliable conclusions. However, an improvement in health occurred between 1987 and 1989, followed by a sharp decline in 1990, particularly in the south of England (Innes, 1992a). The condition of trees in 1991 was worse than in 1990, but the major decline in health seen between 1989 and 1990 was not repeated.

Possible causes

A study of 37 beech forests in north-west Switzerland (Flückiger *et al.*, 1986a, b; Flückiger and Braun, 1992) revealed increases in defoliation between 1984 and 1986 and increased formation of short shoots and decreased growth of terminal shoots over the period 1980–4. However, in 1987 and 1988, the average condition of the trees improved, probably because of favourable (wetter) climatic conditions. Nutrient analyses suggested that potassium deficiency might be involved as this element was insufficient or scarce at half of the investigated sites. This deficiency may have been related to acidic deposition, as a good correlation was found between precipitation acidity and the concentration of potassium in stemflow. This, in turn, suggested that leaching of potassium in the canopy was occurring. The low soil potassium levels were recorded at two-thirds of the sites, suggesting that the potassium was being lost from the system. Nitrogen levels were high (1.9–2.5%) to excessive (>2.5%) at all sites except one, and increased relative to other nutrients between 1984 and 1987. As a result, an unfavourable nitrogen : potassium ratio was present at most sites.

Lonsdale *et al.* (1989) examined the condition of beech in a transect across southern England. The past extension growth of five trees was assessed at each of 15 plots. The effects of drought and ageing were 'removed' from the time series. There was a significant growth loss and a significant deviation from the expected performance of the trees from 1975 onwards. Growth was less than expected at eight of the plots, close to the expected rate at five, and above the predicted growth rate at two. The greatest reductions were found in the south-west of the country. However, the methods used in this study were questionable and the authors did not take sufficient account of the recovery periods required following drought stress. This is evident in the work by Flückiger and Braun (1992) who found marked variations in the rates of recovery of shoot growth of beech in Switzerland following the 1976 and 1983 droughts.

Comparative studies of the physiology of defoliated and healthy trees have not thrown any light on the problem. For example, Hotz *et al.* (1990), working near Würzberg in Germany, found no differences in the chlorophyll fluorescence, chlorophyll contents, chlorophyll-a to chlorophyll-b ratios, or the water soluble ion contents of leaves taken from defoliated and healthy trees. The results strongly suggested that the photosynthetic apparatus of the defoliated trees was no different from the healthy trees.

There has been speculation that pollution may be inhibiting regeneration. However, Dohrenbusch (1990) recorded adequate regeneration of beech at Solling following the excellent mast year of 1976 and good regeneration has occurred in England following the 1990 mast year (Innes, unpublished). The main factor limiting regeneration of beech and other species in many continental stands is the artificially high deer populations, which cause excessive browsing damage (Motta and Quaglino, 1989; Schäffner and Gürth, 1992).

The decline of German forests

The greatest concern during the most recent phase of reports on forest decline has been in Germany and it is worth examining their situation from a national rather than species-by-species perspective. Reports from Germany in the mid–1980s suggested that a major environmental catastrophe was occurring. The national 1984 damage report (Anon., 1984) claimed that 3.7 m ha (50% of the forest area) was subject to visible damage. However, simultaneously, it was pointed out that with one or two exceptions, dying forest stands extending over large areas had not been found. Particular emphasis was placed on silver fir, with 150 000 ha (87% of the total area of this species) allegedly being damaged. Damage was also reported for 1 500 000 ha of Norway spruce (50%), 630 000 ha of beech (50%) and 270 000 ha of oak (43%). Areas with particularly severe damage included the border mountains in eastern Bavaria, the foothills of the Alps, the Black Forest, the Palatinate Forest, the Weser Mountains and the Harz Mountains, with lesser damage in the Neckar region, the Rhine-Main lowlands, the coastal regions of Saxony, the Westphalian lowlands and the lower Rhine region. Although several claims were made about increasing damage between 1983 and 1984, in retrospect it has become clear that the 1983 survey was not comparable with the 1984 survey. The comparisons were therefore suspect.

In 1985, the classification of what formed 'damaged' forests in Germany was revised, resulting in a drop in the proportion of damaged forest from 50% to 19% (Anon., 1985). If the change in the assessment classification is taken into account, it is clear that the large-scale decline in forest health, reported from the incomplete surveys of previous years, did not continue into 1985, when the condition of some tree species (e.g. pine) even improved. These national trends were not necessarily reflected at a local level. Within individual growth regions or even within states, trends away from the average were observed. For example, the forest condition in Saarland deteriorated by 6.8% whereas the condition of forests in North Rhine-Westphalia improved by 5.1% (Anon., 1985). Subsequent surveys revealed a similar pattern, with the condition of trees fluctuating both by species and by area (Anon., 1989c).

A major problem with the German data is that results are extrapolated from sample plots to forest areas. For example, if 40% of the trees assessed in a region are Norway spruce, it is assumed that 40% of the forest comprises Norway spruce. This technique has been shown to be liable to error in view of the sampling procedures being used (Schmidtke, 1987; Neumann, 1989b) and this should be considered when examining the results.

Despite an independent board of experts (Forschungsbeirat Waldschäden, 1986) suggesting that there was no such thing as a single forest damage, the idea is firmly entrenched in the German literature and has been adopted elsewhere. The different types of Norway spruce decline described earlier in this chapter have not been distinguished in the surveys of forest condition. This has resulted in the assumption that a similar phenomenon is affecting all areas. However, it is now clear that most forests in the western parts of Germany are not in a state of decline. On the other hand, some of those areas experiencing Type 1 spruce decline (< 5% of the total forest area) are seriously affected and have continued to decline in the mid- to late–1980s.

Boehmer-Christiansen (1988) examined the political context of forest decline in Germany. She argued that the 'decline' coincided with the wish of the right-wing Länder in southern Germany to push through a nuclear energy programme, together with a perceived need to be 'green' to win votes. The Bundesministerium für Ernährung, Landwirtschaft und Forsten in Bonn was, for several reasons, quick to take action. Consequently, it was placed in the international arena whereas in the past it had been fairly weak in relation to the Länder. The continued role of Germany as lead country in the UN-ECE International Co-operative Programme on Assessment and Monitoring of Air Pollution Effects on Forests is entirely consistent with this. As with many environmental issues, media interest peaked and then rapidly declined. In Germany, the number of reports concerning forest decline was declining as early as 1985 (Krämer and Suda, 1987).

It is probable that some triggering factor(s) acted to start the near-simultaneous decline in several different areas. The most likely candidates are climatic extremes. Although most authors discounted extremes of weather as a factor in forest decline, there is compelling evidence that they were involved as a major factor. This was pointed out early on during the debate in Germany (Cramer, 1984; Schmidt, 1985; Schütz, 1985). The frequency of late frosts and winter-desiccation events was higher than usual in the early 1980s (Mayer *et al.*, 1988; Schulze, 1989) and the 1976 and subsequent droughts have clearly played a role. The way in which climatic factors have continued to be discounted in Germany does not reflect well on the quality of much German forestry science.

Forest decline in The Netherlands

Another area of major concern within Europe is The Netherlands. Here, a monitoring programme with one of the densest sampling networks in Europe has been achieved (den Boer and van den Tweel, 1985). Although a preliminary survey was undertaken in 1983, their base-line was 1984. Surveying was conducted on a 1 km × 1 km grid, with trees at each grid intersection being assessed. The sample trees were assessed along a line transect extending inwards from the southern-most edge of each stand. Five trees were examined at four 20 m intervals. A non-standard classification was used: 'less vital', 'hardly vital' and 'non-vital', although these appeared to be based on the UN-ECE Task Force classification of moderately damaged, severely damaged and dead categories. Most problems were identified in medium-aged stands (40–100 years), with damage being lower in very young and older stands. The worst trees were Scots pine, although the substantial difficulties of assessing this species were recognized. In addition, Douglas fir was in poor condition.

One problem in The Netherlands is the juxtaposition of forests on poor, sandy soils with intensive stock-rearing units (van Breemen and van Dijk, 1988; Roelofs *et al.*, 1985). This means that many forests are exposed to particularly high levels of ammonia. Annual means in intensive livestock-farming regions are 10–25 $\mu g\ m^{-3}$, and peak local concentrations of >500 $\mu g\ m^{-3}$ occur (Van der Eerden and Perez-Soba, 1992). The average level of nitrogen deposition in The Netherlands is 51 kg N $ha^{-1}\ yr^{-1}$ (Heij and Schneider, 1991) and deposition in some forest areas exceeds 100 kg N $ha^{-1}\ yr^{-1}$ (Roelofs *et al.*, 1985). The condition of forests does not match the distribution of sulphur dioxide, the oxides of nitrogen or ozone, with the greatest incidence of symptoms occurring in remote stock-rearing areas.

Within forests, changes towards a more nitrogen-loving ground flora have been seen, with species such as broad buckler fern (*Dryopteris dilatata* (Hoffm.) A.Gray), climbing corydalis (*Corydalis claviculata* (L.) DC.), common hemp-nettle (*Galeopsis tetrahit* L.), black nightshade (*Solanum nigrum* L.), groundsel (*Senecio vulgaris* L.) and common chickweed (*Stellaria media* L.) becoming more abundant. Damage was

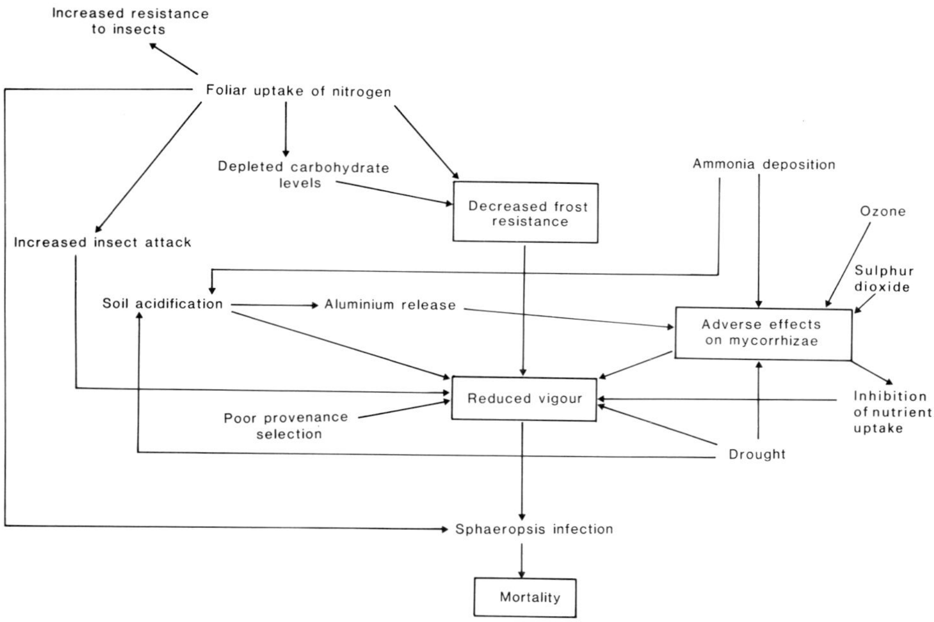

Figure 3.26 Possible events and processes leading to the decline of trees in intensive livestock-rearing areas in The Netherlands. From: Innes (1992).

particularly concentrated in the Peel area (van Dijk and Roelofs, 1988), with chlorosis of pine needles being widespread. The damage has been attributed to ammonia emitted from intensive livestock units, a problem that has also been identified in Belgium (de Temmerman and Coosemans, 1989), Germany (Krzak *et al.*, 1988; Hunger, 1989; Kaupenjohann *et al.*, 1989; Heinsdorf, 1991) and near fur farms in Finland (Ferm *et al.*, 1988). The ammonia causes major imbalances in the nutritional status of forests, making them more susceptible to other stresses (Mohren *et al.*, 1986).

The damage caused by ammonia occurs primarily through changes in the soils and the subsequent effects on tree nutrition. Ambient levels of pollution were not found to have any long-term significant effects on the wax morphology of Douglas fir, either in the field or in fumigation experiments (Thijsse and Baas, 1990). Although short-term effects were seen in the form of an acceleration of wax degradation over the first year, by the second year, the degree of degradation was the same as control trees. Similarly, Van der Eerden (1982) reported that direct foliar injury by ammonia only occurred at very high concentrations (>600 μg NH_3 m^{-3}) and Van der Eerden *et al.* (1992) found no visible injury to Douglas fir at 180 μg m^{-3} ammonia for 14 weeks. Scots pine exposed to up to 240 μg m^{-3} for three months also showed no visible injury (Van der Eerden and Perez-Soba, 1992).

The problem in The Netherlands appears to be an extension of the Norway spruce Type 5 decline described earlier in this chapter. The applications of very large amounts of slurry result in the release of high volumes of ammonia. This has an adverse effect on nearby trees, although direct injury does not usually occur, even at the very high ammonia concentrations found in The Netherlands. Instead, the damage seems to occur through the nutritional imbalances created by the abundance of nitrogen and shortages of some other critical nutrients. In addition, adverse effects associated with soil acidification are likely to have occurred. The resulting loss of vigour of the trees makes them more susceptible to insect and fungal problems. Some of the processes are shown in Fig. 3.26, which reveals the complexity of the situation.

Mediterranean forests

Most declines studied in Europe have been concerned with the northern European species. Little attention has been given to the Mediterranean forests, but these form an important part of the European forest resource. Decline has been reported in these forests (e.g. Bottacci *et al.*, 1988; Clauser *et al.*, 1989; Bussottii *et al.*, 1991). The symptoms generally take the form of loss of foliage, but other symptoms, such as crown deformation, unusually high levels of flowering and chlorosis have also been reported (Clauser *et al.*, 1989).

Bussotti *et al.* (1991a) studied the decline of trees in the Forest of Pixinamanna in southern Sardinia. The forest consists of a mixture of species at lower altitudes, with holm oak and strawberry tree (*Arbutus unedo* L.) prevalent at higher altitudes. Approximately 700 ha has been planted with stone pine (*Pinus pinea* L.) at higher latitudes. Problems have been experienced with the stone pine and with some species in the low-altitude maquis vegetation, particularly Mediterranean buckthorn (*Rhamnus alaternus* L.) and Phoenician juniper (*Juniperus phoenicea* L.). The decline appeared to be caused by a combination of drought, poor soil conditions and pollution from nearby industrial sources, although it is not clear how important the pollution was, as the conclusion was based on evidence of increased levels of sulphur, phosphorus and chlorine in annual rings of the stone pine (Ferretti *et al.*, 1992).

The causes of the decline elsewhere are uncertain, with local pollution playing an important part (Barbolani *et al.*, 1986; Heliotis *et al.*, 1988; Ferretti *et al.*, 1990; Bussotti *et al.*, 1992). Decline has been seen in stone pine, Corsican pine (*Pinus nigra* Arnold), maritime pine (*Pinus pinaster* Ait.), Aleppo pine (*Pinus halepensis* Mill.), European hop-hornbeam (*Ostrya carpinifolia* Scop.) and holm oak and has been reported from several different areas in Italy (Clauser *et al.*, 1989; Bussotti *et al.*, 1991b) and elsewhere.

Discussion

Prinz and Krause (1989) summarized the state of knowledge about the causes of forest deterioration at that time. They drew attention to several points. Firstly, the lack of differentiation between decline symptoms was a major drawback to the study of the problem. Many factors other than air pollution were clearly involved. They drew attention to the late 1980s improvement in forest health, which threw into question the role of air pollution as a stress factor and, lastly, they emphasised the poor quality of many studies.

Many of these reservations still hold, despite a series of conferences since 1988 in which recent work has been summarized. Several species have not been examined in sufficient detail to draw up detailed descriptions of their symptoms. This represents a major drawback to the evaluation of the health of individual tree species in Europe. The improvement in forest health identified in 1988 extended into 1989 but, in some countries, it did not continue into 1990 or 1991. This can be attributed to the major effects on tree health of the 1989 and 1990 droughts.

The last point made by Prinz and Krause (1989), namely the poor quality of many studies, deserves further mention. While the Forest Response Programme in the United States adopted rigorous quality-control assurance procedures, a similar system was not adopted in Europe. This is reflected in the quality of many studies currently appearing in the literature. Too often, results are accepted uncritically and there is an outstanding need for a thorough and critical evaluation of the experimental work that has been done to date.

However, several conclusions can be reached. The extent and magnitude of forest decline in Europe is much less than initially believed. The use of crown density as an index of tree health has resulted in very inflated figures for forest 'damage' which cannot now be justified. Most countries took several years to establish consistent and internally reliable surveys of forest health. If early surveys are discounted because of inconsistent methodology, as Bennett *et al.* (1989) has done with surveys of eastern white pine in North America, then there is little evidence for a large-scale decline of tree health in Europe. However, there is good evidence for many relatively small-scale, localized declines in tree health.

This situation is not helped by the international reports on forest health produced by the United Nations Economic Commission for Europe. Many countries have attempted to make these reports more scientifically rigorous. However, the reports continue to emphasize the absolute differences in forest health between countries, despite clear evidence that such comparisons fail to account for the differences in assessment standards between countries (see Chapter 5). This makes any interpretation of the data extremely difficult. The reports produced by the

Commission of the European Communities were more helpful in that they broke down forest health by broad climatic regions. This was a useful approach, although the regions that were used were too broad to be of much value.

To obtain an accurate picture of the development of forest decline in Europe, it would be necessary to break down the data not only by species but also by decline type. This has yet to be done, even for the well-known Norway spruce. Currently, the level of detail of the field assessments undertaken in most countries precludes any such division, but future monitoring may require a more detailed assessment of tree decline types. Furthermore, there are clear problems in the determination of the evolution of forest decline through time. The results of early surveys are not comparable with later ones, even in countries where such claims have been made (see, for example, the criticisms of the German results by Moehring, 1987). Until longer time series become available, a degree of uncertainty will surround the issue.

There is overwhelming evidence for the role of adverse climatic conditions in triggering the decline of certain species. Historical evidence has shown that such declines have been associated with climatic stress for as long as records have been maintained. This suggests that the argument that pollution has made trees more susceptible to climatic stress may have to be reconsidered. There appears to have been an increase in the incidence of tree declines in recent years. However, several possible factors that might influence this conclusion remain to be evaluated, including the role of more critical observation of tree health, the increased use of faster-growing provenances (which also appear to be more susceptible to stress) and the unusual climatic conditions experienced over the last few decades. While many uncertainties remain, there is a clear need for continued monitoring of forest health. It is only through careful field observations that any statements will be possible about the status of individual tree species and forest ecosystems today and in the future.

4 Methods of Assessment

> In human health management, physicians have developed several routine procedures to monitor general wellness. These include body temperature, blood presuure and chemistry, heart rate, and weight. We must monitor comparable tree parameters to monitor forest health.
> *Smith, 1990*

Introduction

The reports of declining forests in several parts of Europe described in the previous chapter led the Economic Commission for Europe of the United Nations (UN-ECE) to establish an unprecedented international co-operative programme on forest monitoring in 1985 under the auspices of the Convention on Long-Range Transboundary Air Pollution. This was one of a variety of such programmes (ICPs) established under the supervision of a Working Group on Effects. The ICP concerned with forests was termed the 'International Co-operative Programme on Assessment and Monitoring of Air Pollution Effects on Forests'. Its members comprised representatives from all signatory countries of the Convention on Long-Range Transboundary Air Pollution.

The ICP-Forests quickly produced a manual describing monitoring methods. This was published in 1985 but, since then, it has been subject to severai major modifications as knowledge has developed. By 1990, three levels of monitoring had been identified in the manual, each more detailed than the last. Most countries had adopted Level 1 monitoring by the late 1980s, which is based on the systematic survey of the transparency and discoloration of tree crowns in forest plots. Plots were established wherever forest coincided with a 16 km × 16 km grid, with the centre of the forest sample plot being located on the grid intersection. This density was not chosen at random. Many countries used greater plot densities and the 16 km × 16 km grid represented a compromise between those countries with a very large forest area to cover (e.g. Finland) and those with small areas (e.g. Luxembourg). Several countries have used a 4 km × 4 km network (e.g. Germany) and the 16 km network enables these more detailed grids to be sub-sampled. Level 2 entailed the more detailed assessment of trees at fewer sample plots. Level 3 involved intensive monitoring at two or three sites in each country. In practice, the distinction between Levels 2 and 3 monitoring is uncertain and there is some overlap between Level 3 and the ICP on integrated monitoring. A review (Thomsen, 1991) showed that very few countries had set up plots for Level 3 monitoring that adhered to the recommendations of the Manual.

The UN-ECE group was followed by the Commission of the European Communities (CEC), which established a compulsory forest health monitoring programme for Member States in 1987 (Regulation No. 1696/87). This programme was also based on a 16 km × 16 km grid, but was not necessarily the same as used by the United Nations programme. The CEC was more specific about the sizes of trees to be sampled, stating that all trees above 40 cm height were to be included. In countries such as Great Britain, with vigorous afforestation programmes, this resulted in a sample heavily weighted towards younger trees.

I have criticized the systematic sampling design used in both the CEC and the UN-ECE programmes on the basis that while it may be useful for some countries, it is inappropriate to apply it rigidly across the whole of Europe when the nature and distribution of forests differs markedly between countries (Innes, 1987, 1988a, 1990b). For example, in Great Britain, the distribution of forests is such that there are large areas of central and eastern England with few forests. Pollution levels there are high compared to the rest of the country. Consequently, this is precisely the area of greatest interest for studies of the effects of gaseous air pollution on forests. However, the small forest area meant that very few plots were located in the region. Many different sampling schemes exist and, in Great Britain, a stratified random design was clearly more useful when establishing a monitoring programme (Innes, 1988a). A further problem with the systematic design was that, in many cases, it was unable to cope with unforeseen circumstances. This is a major requirement of long-term monitoring programmes (Skalski, 1990). For example, there were very few plots located in forest areas in Great Britain receiving high levels of cloud immersion. The pollutant deposition associated with cloud immersion may be very high, but its possible effects on trees would have been largely missed using a 16 km × 16 km grid.

There has been considerable debate over the exact nature of the surveys. The European surveys aim to assess the condition of trees in any given year. However, Kimmins (1990) provided a convincing argument against the instantaneous assessment of condition, proposing that forest degradation, from whatever cause, should be assessed in relation to the deviations from the expected pattern. This enables recovery processes to be considered. However, the UN-ECE and CEC programmes (which were combined in 1991) are well-established and instantaneous assessments of condition are likely to continue for some time.

Plot and tree selection

The standard method of plot assessment is to locate four sub-plots on each plot and then to assess between five and eight trees on each sub-plot. Subplots are at the cardinal points of the compass, with their centres usually being 25 m from a central point (Fig. 4.1). At the end of each 'arm', the five or more nearest trees are assessed. In some countries, suppressed trees are included, but in the UN-ECE and CEC programmes, only dominant, co-dominant and intermediate trees are examined. The inclusion of suppressed trees means that a false impression of the condition of the stand canopy may be gained, but important information on the condition of regeneration would otherwise be lost. In plantations, suppressed trees are frequently removed by selective thinning and, for example, no suppressed trees are included in the plots used in the British monitoring programme.

Innes and Boswell (1990b) investigated the reliability of the sample sizes generally used in the monitoring programmes. To simplify the

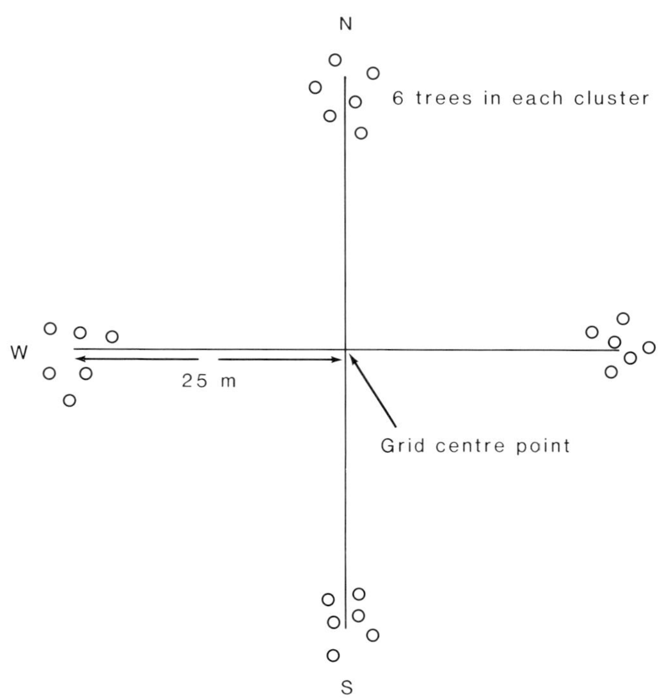

Figure 4.1 Plot design for the United Nations and Commission of the European Communities' forest health surveys. Trees are located at the end of each 25 m arm.

Table 4.1 Results of a hierarchical analysis of variance test on the British monitoring programme data (1988 data)

	df	MS	*F*-ratio	Variance component
Sitka spruce				
Between stand	64	17.24	6.35	0.61
Between edge	195	2.71	2.55	0.28
Between tree	1300	1.06		1.06
Norway spruce				
Between stand	78	23.10	9.68	0.86
Between edge	237	2.39	2.34	0.23
Between tree	1580	1.02		1.02
Scots pine				
Between stand	79	28.52	16.39	1.12
Between edge	240	1.74	1.5	0.1
Between tree	1600	1.16		1.16
Oak				
Between stand	49	16.90	7.79	0.61
Between edge	150	2.17	1.58	0.13
Between tree	999	1.38		1.38
Beech				
Between stand	30	12.83	6.29	0.45
Between edge	93	2.04	2.09	0.18
Between tree	620	0.98		0.98

analysis, we examined the even-aged, single-species stands assessed in the British monitoring programme. The trees in these stands were believed to be less variable in condition than those in mixed-age, mixed-species stands. We used a hierarchical analysis of variance to examine the variation between and within sub-plots and plots (Table 4.1). All sub-plots were located on the edges of stands and are described as such in Table 4.1. All species had significant between-tree variation, as might be expected from other studies (e.g. Müller-Starck, 1985). At a higher level, there was significant variation between edges within stands, although this source of variation was considerably lower than that attributable to between-stand variation. Generally, the amount of variation between stands was greater than the amount of variation within stands.

The results are important as although variation between stands was higher than within, variation within stands was still significant, even in these relatively homogeneous plots. This suggests that a sample size of 24 trees is generally inadequate to characterize the condition of trees in the stand. However, the optimum sample size has yet to be found.

In assessing the current extent of forest declines, it is important to examine the methods used to generate the data that are so widely reported. Most surveys are based on the assessment of a sample of trees within a stand. While some surveys examine a variable number of trees within a fixed area, the majority are based on the assessment of a fixed number of dominant, co-dominant or intermediate trees within the stand. Even here there is potential for error, since the classification of tree crown position is subjective and liable to error (Nicholas *et al.*, 1991). Generally, sample sizes used in the UN-ECE and CEC programmes have been less than 30.

Many indicators can be assessed to develop a picture of the condition of the stand (Table 4.2). Inevitably, what is assessed in the forest represents a compromise between what is desirable and what is practical. For example, the national forest health monitoring programme and several

Table 4.2 Effects or symptoms of stress and damage in forests and trees

Effects on individual trees

- Visible symptoms of injury:
  - Change in the shape, size, surface properties, colour and number of leaves
  - Change in the normal patterns of growth and development
  - Change in normal patterns of foliage flushing or senescence
  - Decreased annual height growth or radial increment
- Alteration of physiological processes:
  - Photosynthesis
  - Respiration and metabolism
  - Transpiration
  - Mineral nutrition
  - pH and buffer capacity of foliar tissues
  - Transport and allocation of photosynthate
  - Hormonal control of growth
  - Symbiotic relationships with other organisms
- Changes in susceptibility to other stress factors
- Life-history changes
  - Decreased longevity; early onset of dieback
  - Changes in reproductive behaviour

Effects on forests

- Decreased productivity of stands
- Changes in age-class distributions
- Changes in normal patterns of competition and mortality
- Changes in normal patterns of community succession
- Changes in species composition
- Changes in nutrient cycling
- Changes in hydrologic behaviour, watershed functions, or wildlife habitat
- Changes in genetic structure of populations

Based on Committee on Biologic Markers of Air Pollution Damage in Trees (1989) other monitoring programmes in the United States are based on the intensive investigation of a small number of stands (e.g. Dwire *et al.*, 1990), and some authors, such as Muir and McCune (1987), have stressed the need to collect information on a range of different symptoms in any assessment study. However, in many European countries, only two indicators are assessed: 'defoliation' and crown discoloration. The method of assessment of 'defoliation' varies from country to country. Observers in some countries base it on an estimate of the amount of light passing through the crown (crown density or transparency) whereas others base it on the average number of years for which conifers retain their needles (needle retention), with the score being given to the oldest needle-year class with more than 25% of the needles present.

In the following account, indices have been broadly divided into those which are visible and those which are not readily visible. The latter category frequently involves destructive sampling of the foliage or other parts of the tree. There is some overlap between the two groups. For example, foliage discoloration can be estimated visually, or the pigment concentrations can be measured. Remote sensing does not fall readily into either category and has been considered separately.

Visual assessments

There have been several major developments in the assessment of tree vigour by visual methods over the past five years. The work of Lesinski and co-workers and Gruber on Norway spruce is of particular note (Lesinski and Westman, 1987; Lesinski and Landmann, 1988; Lesinski, 1989; Gruber, 1990). Regrettably, most of these improvements have still to be incorporated into standard assessments of tree vigour.

Foliage discoloration

Foliar discoloration is one of the two main indices estimated in the European forest health monitoring programmes. As used in the forest, it is an estimate of the extent of discoloured foliage in the crown. Even this simple index is interpreted differently, with some countries only including chlorosis (yellowing) and others including all forms of discoloration.

Foliage discoloration is a key factor in the diagnosis of several types of pollution injury. It is particularly useful for oxidant injury and has been used extensively as an index of damage in the western USA. For example, Pronos *et al.* (1978) used a relatively simple index based on the youngest needle whorl to show evidence of chlorotic mottle (yellow spots on the foliage). More complex indices involve combinations of the amount of damage to different year classes and to different positions within the crown. However, as these indices become more detailed, the potential for error increases. Work in California has shown that assessments made by binoculars do not correlate well with assessments made by hands-on inspection of the foliage (Muir and Armentano, 1988), particularly for upper crown assessments. This represents a considerable problem.

At least two of the types of Norway spruce decline are characterized by yellowing and a third is characterized by browning. However, European-wide surveys have revealed that discoloration is rare and, in many areas, its incidence has declined since about 1986. For example, within the European Community area, 89% of the 48 014 trees that were assessed in 1991 had less than 10% of the foliage discoloured. Only 2.6% of the trees had more than 25% discoloration (Anon., 1992). Many discoloured trees were located in the drought-affected Mediterranean region, with discoloration being less common in the Atlantic, sub-Atlantic and alpine regions.

Problems exist over the definition of discoloration. In Europe, Gruber (1989) emphasised the need to distinguish between chlorosis (yellowing) and necrosis (browning) on the needles of Scots pine. Much the same can be said about all species, as the two different types of discoloration can be caused by many different factors. However, experience in Great Britain has suggested that it can be very difficult to distinguish between the two as there is a tendency for conifer needles to follow a sequence from green to yellow to brown, with various intermediate stages. One possibility is to specify a threshold level for discoloration and Brooks *et al.* (1991) recommended that at least 50% of a needle should be discoloured before it is scored. Such a threshold would mean that phenomena such as tip-burn and tip-chlorosis would be excluded unless they were recorded separately. There is some justification for such separation in that tip-chlorosis in conifers may well occur much more regularly than is generally recorded.

It is also important to distinguish between discoloration of current-year foliage and older foliage. The former has not been exposed to any forms of winter damage and the presence of discoloration on older foliage may reveal a past problem. In addition, some forms of mineral deficiency may first become apparent through discoloration of the older foliage as minerals are withdrawn to supply the new foliage (Stone, 1953; Will, 1985; Lange *et al.*, 1989a). On some pine species, it is possible to specify the levels of damage to particular needle cohorts. With spruce and fir, this may be more difficult, but it should still be possible for hand-held specimens.

A method used by Ewell *et al.* (1990) to characterize the amount of visible injury on ponderosa pines in California is particularly appropriate. They based their assessments on the extent of visible injury (chlorotic mottling) on different year classes of foliage. The rationale behind this approach was that the dose received by needles increases through time and, consequently, the amount of visible injury should also increase. Sensitive individuals, or individuals exposed to higher doses, would show injury earlier. Consequently, by assessing the age of foliage on which injury first appeared, it was possible to develop an index of injury to pollution.

Foliar pigment concentrations

While not a visible symptom as such, a quantitative measure of the colour of foliage can be obtained by measuring the pigment concentrations. One of the most difficult changes to assess visually is the shift from 'healthy' green foliage to chlorotic foliage. The processes involved in this change are well-known. Chlorophyll is easily transformed to pheophytin through the loss of the central magnesium ion (Heath, 1989), and the resulting yellow-green pigmentation is characteristic of magnesium deficiency. The different types of chlorophyll change at different rates and this can be assessed. Chlorophylls have distinct spectral characteristics and although the blue band of the spectra overlaps with the carotenoid spectra, the red band can be used to distinguish chlorophyll-a from chlorophyll-b.

Field measurements

Hoshizaki *et al.* (1988) examined the pigment concentrations of red spruce at Camels Hump, Vermont, and Norway spruce in the Black Forest, Germany. Chlorophyll-a levels and chlorophyll–carotenoid ratios were lower in needles from sites with high levels of damage, whereas chlorophyll-a to chlorophyll-b ratios were higher (at Camels Hump only). At the high-damage sites, the total levels of chlorophyll were lower. In contrast, Kandler *et al.* (1987) found no change in the chlorophyll to carotene ratio in Norway spruce during pigment reduction associated with yellowing. Oßwald *et al.* (1987) recorded similar chlorophyll concentrations in green Norway spruce needles from trees with and without discoloration, and chlorophyll only decreased in yellow needles. Similarly, Hotz *et al.* (1990) found no differences in the chlorophyll contents or the chlorophyll-a to chlorophyll-b ratios in defoliated and healthy beech trees growing near Würzberg in Germany, and Cornic (1987) concluded that sulphur dioxide did not affect chlorophyll-a to chlorophyll-b ratios. Mehlhorn *et al.* (1988) and Wolfenden *et al.* (1988) argued that the ratio of two carotenoid pigments, violaxanthin to antheraxanthin, might produce a useful index of stress, but there were inconsistencies in the results that suggested that the index was of limited value (Fig. 4.2).

Ashenden and Williams (1988) looked at the spectral characteristics of silver birch (*Betula pendula* Roth.) seedlings grown under experimental conditions and subjected to various degrees of acidity. After 75 weeks of treatment, the dip in reflection between the green and red wavelengths (at about 0.65 μm) disappeared. As this is a chlorophyll-adsorption band, they suggested that pigment damage may have occurred. Chang and Yong

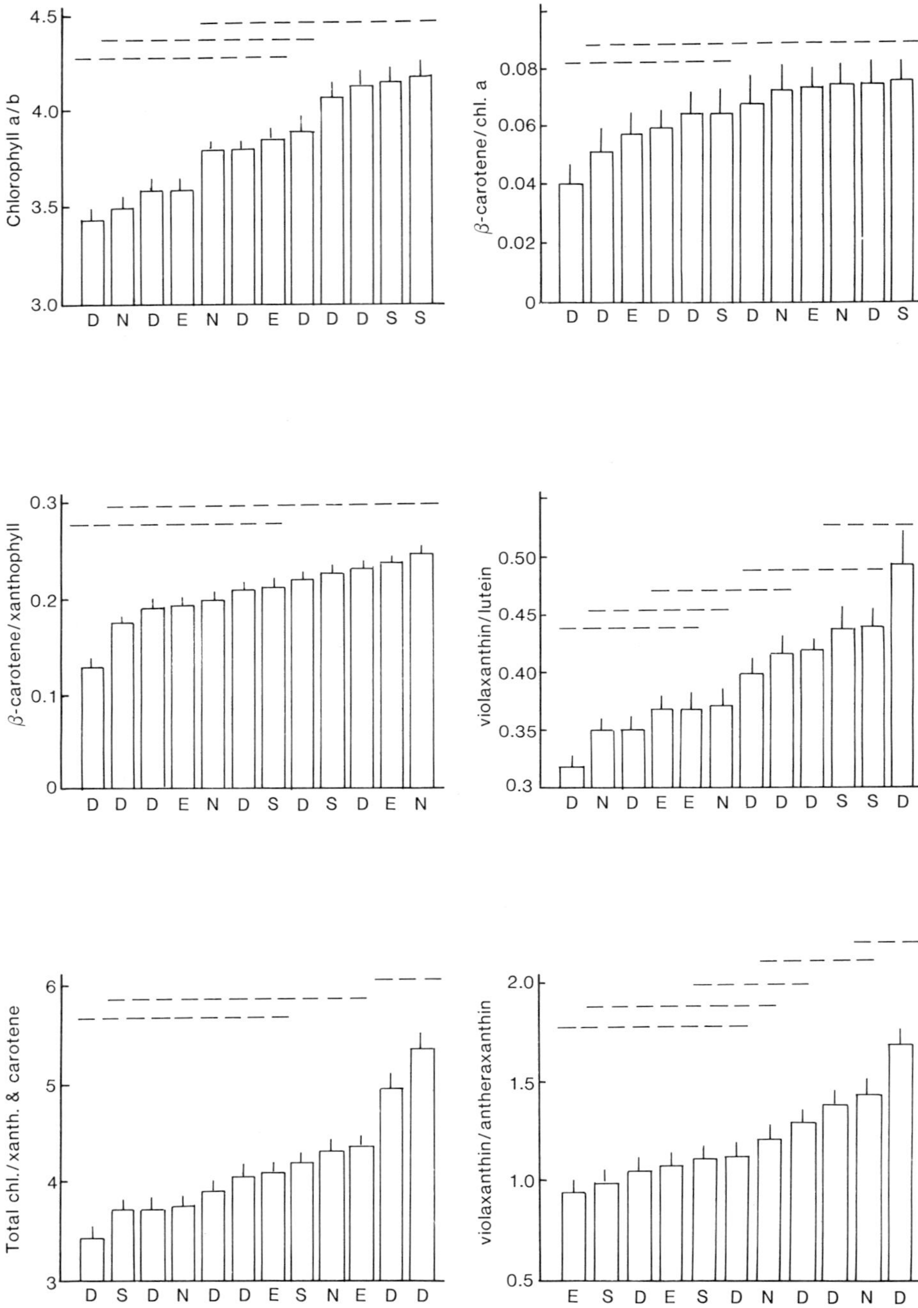

Figure 4.2 Pigment ratios from 2-year-old Norway spruce needles showing various pigment ratios at different sites in Germany (D), the Netherlands (N), England (E) and Scotland (S). Sites under the same horizontal bar are not significantly different from each other.

(1987) also reported changes (reductions) in chlorophyll-a and –b contents following acidic treatments (in this case, with the Korean pine, *Pinus koraiensis* Sieb. and Zucc., and privet, *Ligustrum obtusifolium*).

There are several difficulties with the interpretation of chlorophyll in needles. Köstner *et al.* (1990) examined the seasonal changes in the chlorophyll contents of Norway spruce needles in an area affected by Type 1 decline in the Fichtelgebirge, Germany. In green trees, as elsewhere, the chlorophyll content increased from May to October, remained high throughout the winter and then decreased markedly between March and early May. The annual maximum chlorophyll content increased with needle age during the first four years of the needles' life and then decreased. This relationship is not universal. Penka and Cervená (1985) reported decreases in chlorophyll content with increasing needle age in Norway spruce and a similar trend has been reported for slash pine and Caribbean pine (*Pinus caribaea* Morelet) by Doley (1988). The dynamics of the pigments of chlorotic needles corresponded with those of green needles. Together with the decrease in chlorophyll contents, a reduction in the chlorophyll to carotenoid ratio was observed and a simultaneous increase in the xanthophyll to carotene ratio occurred. In the study by Penka and Cervená (1985), the periods of needle chlorophyll reduction did not correspond with periods of pollution by sulphur dioxide. Instead, chlorophyll formation in trees affected by yellowing occurred when the new flush was sprouting, suggesting nutritional deficiencies as the cause. Interestingly, the chlorophyll contents of chlorotic needles stabilized after more than two years, suggesting the lack of a dose-response relationship.

A further difficulty is the lack of specificity in pigment changes. These are caused by a wide range of different stresses, together with natural ageing processes, making their usefulness as a bioindicator of pollution damage limited. These problems suggest that the value of pigment content analysis is limited. It is likely to be of most value in cases where discoloration is the main symptom. When the decline is characterized by other symptoms, pigment analysis is likely to be less useful.

Chlorophyll fluorescence

A related technique is chlorophyll fluorescence. Baillon *et al.* (1988a, b) described the process in detail. As light is absorbed by the different pigments in the foliage, it is transferred to the reaction centres where it is transformed in energetic substances (ATP and NADPH) during the photochemical reactions of photosynthesis. During this process, some energy is dissipated into heat, but some is also reemitted as light from the photosystem 2. This re-emitted light is the fluorescence. The intensity of the fluorescence depends on the efficiency with which the light energy is transformed within the photosynthetic apparatus, and it therefore provides a means of assessing the apparatus. The fluorescence displays an induction kinetics that has been termed the Kautsky effect. It rises quickly (over microseconds) to a peak, F_m and then decreases slowly (over minutes) to a steady state, F_s. The coefficient of fluorescence, R_{fd}, is expressed by:

$$R_{fd} = \frac{F_m - F_s}{F_s}$$

The coefficient tends to be higher in current-year needles than in older ones. Baillon *et al.* (1988a) recorded changes to chlorophyll fluorescence early in a treatment inducing magnesium deficiency, but the effect was limited to older needles. This is consistent with the transfer of magnesium from older to younger needles when it is scarce and also reflects the importance of magnesium in the photosynthetic process.

Van Kooten and van Hove (1988) found changes in the fluorescence of poplar (*Populus euroamericana* L.) leaves exposed to ammonia and Schreiber *et al.* (1978) and Schneckenburger and Frenz (1986) have seen fluorescence changes in various species following exposure to ozone. However, when Hotz *et al.* (1990) compared the chlorophyll fluorescence of leaves from defoliated and healthy beech trees in Germany, they found no difference between the two.

A problem with the interpretation of fluorescence data is that it is affected by various environmental stresses. For example, Strand and Lundmark (1987) and Strand and Öquist (1988) found that frost affected the fluorescence of Scots pine seedlings, the degree depending on the amount of shade that the seedlings were subjected to. Weather patterns were also important, with recovery of photoinhibitory damage occurring on cloudy days. They suggested that natural frosts might increase the susceptibility of young trees to photoinhibition during the following day(s). These problems place considerable limitations on the use of chlorophyll fluorescence in standard monitoring programmes.

Foliage size and shape

The size and shape of the foliage of trees provide an important indication of their vigour.

Foliage size

Trees under stress frequently produce smaller leaves or needles because of the re-allocation of their available resources before or during foliage development. Schütt (1989) described abnormally small foliage in beech, ash, hornbeam and pine as a common symptom of 'forest decline'. However, research in Great Britain and elsewhere has shown that leaf size in beech (Fig. 4.3) can be directly related to climatic conditions, particularly drought (Roloff, 1988; Innes, 1992a). Nihlgård *et al.* (1985) and Glavac (1988) also argued that the leaf size of beech was an important indicator of stress and Klumpp *et al.* (1988) noted leaf deformation in

Figure 4.3 Abnormally small leaves in beech. The leaves are the same size as the mast. The sample was collected from the upper crown on a mature beech at Selborne, England, following the severe summer drought of 1990.

three-year old beech trees following fumigation with a mixture of pollutants. In pines, needle size varies annually (Innes, 1990c), and can again be directly related to climatic conditions. This can be clearly seen on pines in northern locations, where the long retention (seven or more years) enables detailed *in situ* comparisons to be made. Consequently, unusually small foliage can be interpreted as showing that a particular tree has been under stress, but the stress may not have originated in the same year as the foliage developed.

Riederer *et al.* (1988) studied the dimensions of 38 000 needles collected from 48 130-year old Norway spruce trees growing in the Fichtelgebirge, Germany. There was a large amount of variation in needle size, with no consistent effects of site being found (Figs 4.4 and 4.5). Needle lengths and surface areas did not correlate with either total chlorophyll content or the degree of crown transparency, suggesting that needle dimensions are not a good indicator of defoliation. A bimodal distribution of needle lengths was found, with peaks at 12–15 mm and 17–20 mm. The effects of provenance and/or tree age are clearly apparent in this study when it is compared with other work. For example, Masuch *et al.* (1989) found that the average length of three-year-old Norway spruce needles was 10 mm, dropping to 8 mm in

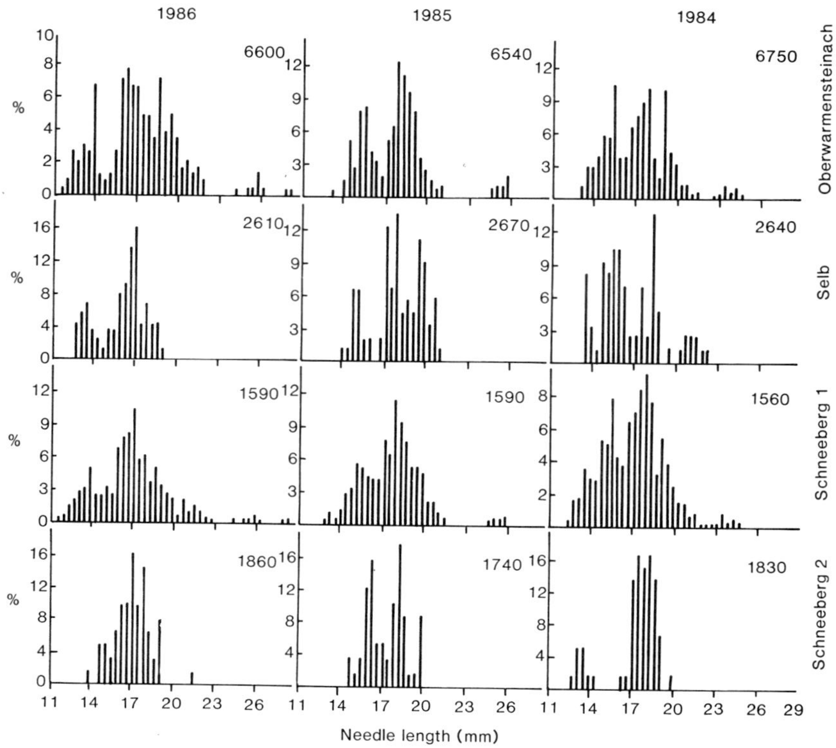

Figure 4.4 Frequency distribution of the lengths of needles of Norway spruce. Samples (with sample sizes indicated on the top right of each histogram) were taken from the three youngest needle age classes of 25, 10, 6 and 7 trees (approximately 130 years old) at Oberwarmensteinach, Selb and two locations at Schneeberg (all in Germany). From: Riederer *et al.* (1988).

trees exposed to acidic fogs. Consequently, the results of Riederer *et al.* (1988) are difficult to interpret, and more information on the factors affecting needle dimensions are required. Similar studies in other species are also needed.

Leaf shape

Leaf-rolling can occur in several species and is believed to be an indication of water stress (Skelly *et al.*, 1987). It normally starts in the upper crown, with only a few shoots showing light levels of rolling. However, as the stress increases, the rolling becomes more severe and gradually extends throughout the crown. In Germany, leaf-rolling was higher in thinner crowned trees (Gärtner, 1988), suggesting a relationship of some sort. The

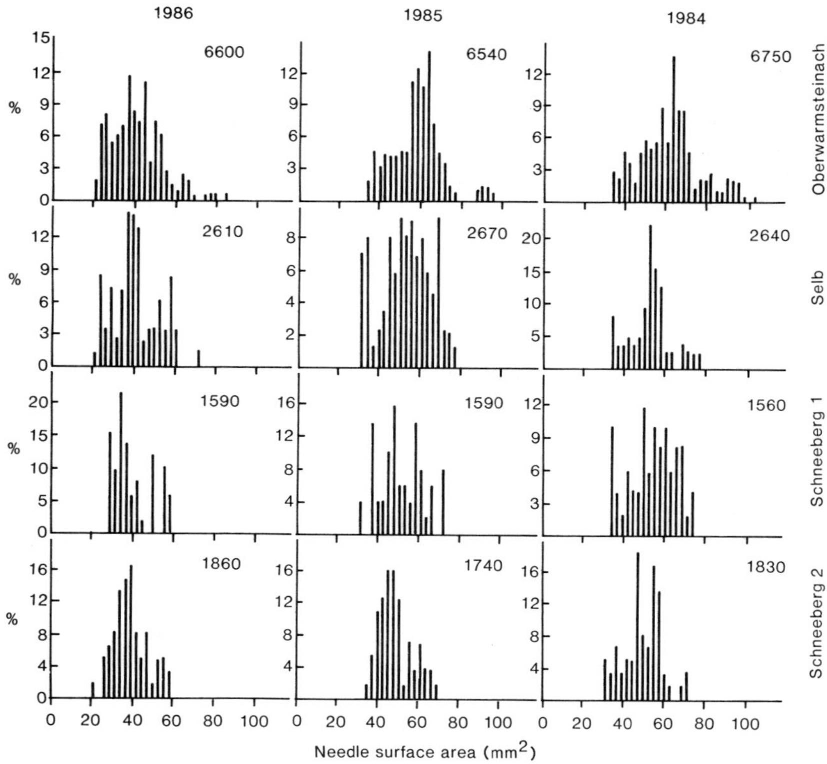

Figure 4.5 Frequency distribution of the surface area of needles of Norway spruce. Details as in Fig. 4.4. Surface areas were estimated from projected needle areas using a conversion factor of 2.74. From: Riederer *et al.* (1988).

significance of the phenomenon is unknown, although it may be related to water stress.

Deformation of foliage may also occur, but this seems to be largely undocumented within the context of monitoring. Insects commonly cause deformation but, on occasion, deformation occurs without any insect or pathogen being involved. With modern image-analysis techniques greatly speeding up the quantitative assessment of shape, rapid progress in this area is likely.

Foliage surface properties

The surface of the foliage is the first point of impact for gaseous pollutants, acidic precipitation and other external substances. Consequently, foliage surfaces might be expected to show some of the earliest signs of alteration following a change in the environment (Cape, 1988).

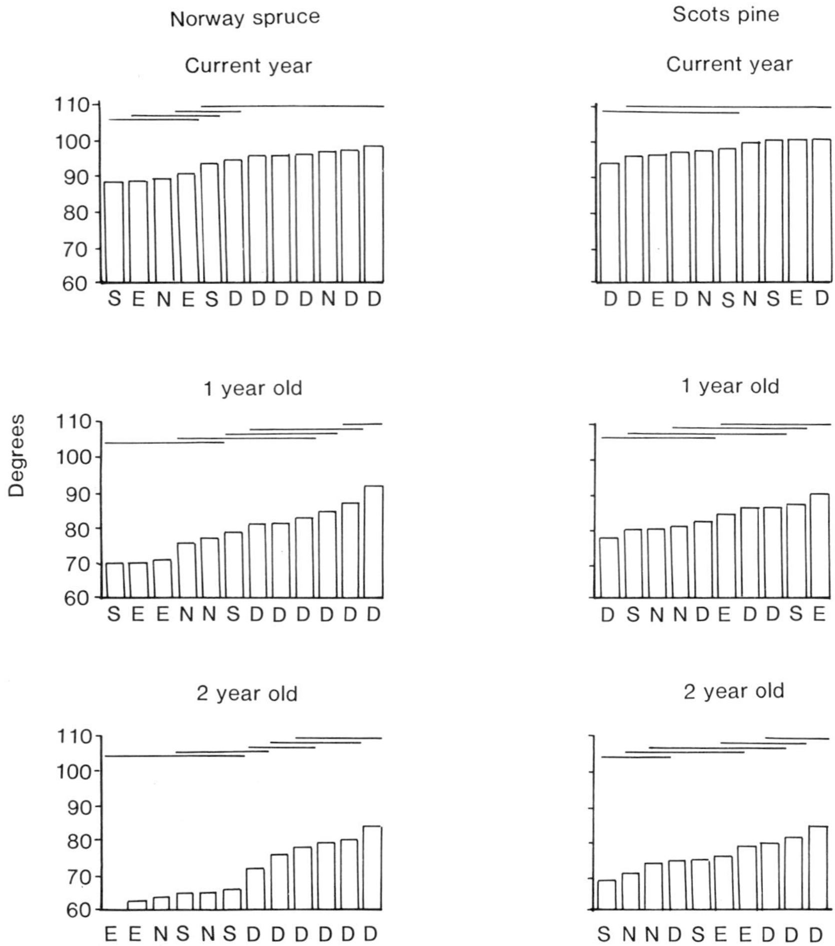

Figure 4.6 Contact angle of deionized water droplets on needle surfaces of Norway spruce and Scots pine. Year 0 = 1986 needles; year 1= 1985 needles; year 2 = 1984 needles. Location of sites are given on the *x*-axis: E: England; S: Scotland; N: the Netherlands; D: Germany. Values under the same horizontal bar are not significantly different (p = 0.05, Duncan's Multiple Range Test). From: Cape *et al.* (1989).

Cape *et al.* (1989) reported the results of a study of the surface properties of Norway spruce and Scots pine needles from several sites in Europe. As a surrogate for more detailed examination, they examined the leaf wettability of needles. There were no differences in the contact angles between asymptomatic and transparent trees within sites, but significant differences between sites existed, particularly for Norway spruce (Fig.

4.6). Contact angles decreased with increasing needle age. No differences in the amounts of wax were observed between symptomatic and asymptomatic sites, nor were there any patterns in the amounts of dust on the needles. Consequently, they concluded that variations in leaf surface properties did not give a clear correlation with variations in visible damage symptoms. Fink (1988) reached a similar conclusion. He argued that although wax degradation on Norway spruce needles occurred faster in polluted environments, there were too many other factors involved to use it as a diagnostic tool. The observed patterns suggested that the needle surface properties reflected environmental exposure rather than plant response, a conclusion also reached by Riederer *et al.* (1988) and Günthardt-Georg (1989).

Several other studies have looked at surface waxes. Euteneuer *et al.* (1988a) found that epicuticular waxes on Norway spruce melted and eroded with increasing needle age and Euteneuer *et al.* (1988b) concluded that the effects of air pollution on epicuticular waxes could not be separated from the effects of microclimate. Increasing deterioration of the wax with age has also been reported for a variety of other species, including Colorado spruce (*Picea pungens* Engelm.) (Reicosky and Hannover, 1976) and maritime pine (Franich *et al.*, 1977). Sauter and Voß (1986) found progressive deterioration of the surface waxes on Norway spruce needles through time, but needles of a given age could show various stages of degradation. They interpreted this as revealing the degree of environmental stress that the needles had experienced.

A similar approach was adopted in an experimental study by Barnes

Table 4.3 Wax structures on the needles of Norway spruce

Category	Condition of wax
1	Undamaged crystalline structure consisting of a dense array of fine wax crystalloids, up to 2.5 μm long and 200 nm diameter
2	Semi-crystalline structure, but wax crystalloids beginning to fuse, tubes 200–700 nm diameter
3	Increased fusion of wax tubes, giving the wax a melted appearance, approximately half of the epistomatal chamber occluded by structurally degraded wax, rodlets 700–1200 nm diameter
4	Almost total agglomeration of the wax crystalloids retaining very little structural integrity, epistomatal chamber filled with degraded wax, tubes of indeterminate diameter
5	Total occlusion of the epistomatal chamber by a flat plate of completely degraded wax, with no structure and forming an amorphous layer

From Barnes *et al.* (1988)

et al. (1988), who distinguished five different categories of wax structure. Their classification is given in Table 4.3. All five of these categories were seen on trees grown in carbon-filtered air, with the incidence of types 3–5 increasing with needle age. These results, together with those of Sauter and Voß (1986), showed a high level of variation in the degradation of surface waxes.

Grill *et al.* (1987) examined the waxes on 90-year-old Norway spruce in the Aichfeld, Germany. Young needles of defoliated and asymptomatic trees showed no differences in the epicuticular waxes. However, after the first winter, the waxes of needles exposed to higher levels of sulphur dioxide had altered more than the waxes of trees in cleaner conditions. Waxes tended to be eroded and fused at needle sulphur contents of about 0.12% on first-year needles and 0.15–0.16% on second year needles. The changes were not accompanied by differences in defoliation. Similarly, Günthardt-Georg (1989), working at Lägeren and Davos in Switzerland, found that the crystalline wax fibrils covering the youngest needles of Norway spruce became progressively fused, flattened and eroded with age. The rate of erosion was higher at the more climatically extreme site. This suggested that the needles of more transparent crowns were less protected and might result in faster wax erosion than in dense crowns.

Raddi and Rinallo (1989) found that wax degradation on 70-year-old silver fir increased with age. The rate of degradation was faster in a polluted environment than in a relatively clean area. In addition, there were marked differences between the rates of degradation on two different provenances (Vallambrosa and Serra San Bruno). However, more work is required on silver fir before any generalisations can be made.

Further evidence of a strong climatic effect on wax degradation comes from North America. Hadley and Smith (1989) looked at the effects of wind on the wax erosion of Engelmann spruce. As the winter progressed, needles remaining exposed above the snowpack showed rapid erosion of the waxes, until there was 60% less wax on exposed needles than on protected needles. Wax amounts declined with increasing altitude, a trend also observed on subalpine fir. Wax erosion was greatest close to the surface of the snowpack, a feature that was attributed to blowing snow. Wind may also cause leaves to rub against each other and the abrasion may affect the surface waxes (Wilson, 1984). Baker and Hunt (1986) have even found that water droplets 250–400 μm in diameter travelling at 0.25–5 m s^{-1} can fracture the wax tubes on the leaves of blue gum (*Eucalyptus globulus* Labillardère) and other species. Besides wind and raindrop impact, wax structures may be physically damaged by extreme cold (Huttunen *et al.*, 1990).

Experimental studies of wax degradation by pollutants

In one experimental study, ozone (8 hour daily mean of 156 μg m^{-3} over three consecutive summers) did not affect wax amounts but significantly affected the wettability, suggesting changes to the wax structure of Norway spruce needles (Barnes *et al.*, 1990). However, Günthardt-Goerg and Keller (1987) found no effect of ozone (300 μg m^{-3} for 1215 hours) on the wax structure of Norway spruce needles. Instead, decreases in the wax quantity occurred. Similarly, holly seedlings exposed to high levels of several pollutants showed no epicuticular wax structural changes after treatment (Garrec and Kerfourn, 1989). When compared with other studies, these results suggest that long-term fumigations at low concentrations may produce greater effects than short-term treatments at relatively high concentrations. Further studies have suggested that ozone might be retarding the formation of wax rather than eroding it (Günthardt-Goerg and Keller, 1988), although this explanation is inconsistent with the findings of Barnes *et al.*, (1990). Norway spruce seedlings exposed to simulated acidic precipitation at pH 3.5 also resulted in degradation of wax structures (Rinallo *et al.*, 1986).

Despite the difficulties encountered in the interpretation of field studies of surface waxes, there has been a great deal of interest in the possibility of using wax degradation as an early warning indicator (e.g. Gullvåg, 1986; Karhu and Huttunen, 1986; Tuomisto, 1988). Tuomisto (1988), in particular, produced some very convincing arguments in favour of wax morphology as a diagnostic tool. He reasoned that it was important to look at the abundance of preserved tubular wax and intermediate erosion stages, with the amount of amorphous wax and the general condition of the wax in the epistomatal chamber being sufficiently specific to suggest pollution damage. However, these studies failed to consider the important variation that occurs in the nature and rates of natural degradation of wax between genotypes (Barnes *et al.*, 1988; Percy and Baker, 1990).

Wax composition

The above studies are nearly all concerned with the visible identification of wax degradation. Consequently, they run the risk of subjective interpretation. The chemical analysis of the wax provides an objective approach to changes in the cuticle which is yet to be fully explored. For example, Simmleit and Schulten (1987) found that yellow needles from Norway spruce at Kleiner Feldberg in the Taunus, Germany, had changes in the 10-nonacosonal and high-mass signals from estolides. The higher percentages of estolides in the yellow needles revealed polymerization of the wax, the same result that would have arisen from natural ageing. In necrotic needles, the estolides were completely absent. Despite these differences, the total amounts of wax on green, chlorotic and necrotic needles were the same.

Conclusion

Wax quality, quantity and composition appears to offer a non-specific measure of the stress that a needle has been subject to. As such, it is of potential value as a general indicator of needle condition. However, its importance as a specific diagnostic tool is much more limited, and any use of wax in a monitoring programme would have to be with extreme care.

Patterns of foliage flushing and senescence

Patterns of foliage flushing and senescence provide an important means of establishing the nature, timing and extent of stress. The patterns are related to branch structure, and some indices overlap between the two. While most monitoring studies are concerned with the senescence of foliage, the date and rate of flushing may be very important.

Flushing

Phenological studies (e.g. Crawley and Akhteruzzaman, 1988) have shown that trees that flush early tend to do so consistently from one year to the next. The date of budburst of a sample of 36 oak trees varied within any given year by almost one month. The date of budburst did not affect defoliation, with early flushing trees having the same level of defoliation as late-flushing trees. To establish the timing of budburst requires repeated visits early in the season, adding to the cost of any monitoring programme. However, it appears to be an important part of a monitoring project and at least one early-season visit should be made to monitoring plots.

Longevity of needles

It is not clear how long Norway spruce should keep its needles. Kramer (1986a) argued that healthy Norway spruce in Germany had a maximum of seven full needle years. However, many Norway spruce in Great Britain and elsewhere hold their needles for much longer than this (Innes and Boswell, 1991). As with many other species, needle retention increases with altitude (Ewers and Skied, 1981) and also increases with latitude. It appears that both environmental and genetic factors control retention, but some controlled experiments are required to confirm this.

Any defoliation that occurs in spruce (and other species) without involving loss of woody biomass can be attributed either to needle fall or needle loss (Gruber, 1990). The former involves the use of the abscission apparatus and is an internal physiological process. The latter is a mechanical process induced by external stresses. In spruce, needle fall can be related to needle desiccation and, according to Gruber (1990), it occurs if:

- the whole tree or a part of it suffers from water deficiency;
- the needle surface is irreparably damaged by some agent;
- the internal water transport system is disrupted; or
- an interruption occurs between the water vessels of the needle trace and the shoot axis.

The last of these processes is generally insignificant. The process of needle fall in Norway spruce occurs throughout the year but, in most areas, there are maxima in late winter and early spring (Gruber, 1990). Needle fall may however occur any time following drought conditions.

The techniques used to assess needle retention in Norway spruce are applicable to other spruce species, including Sitka spruce. In Scotland, needle and shoot extension in Sitka spruce occurs over a four to nine week period between May and July, usually ending by the second week of July (Chandler and Dale, 1990). The exact timing of the apical development depends on provenance (Burley, 1966; Cannell and Willett, 1975; Cannell, 1978) and environmental conditions preceding and during the period of extension (Ford *et al.*, 1987a, b). Needle initiation begins at the apices in March and April, with about 15% of the primordial forming by mid-July and 85% by the end of September (Cannell, 1987). Most of the upper canopy of Sitka spruce consists of current year and one-year old foliage (Cannell, 1987), although foliage is generally retained for six to eight years or longer. Photosynthesis continues throughout the winter. Needles develop acropetally along the shoot, with the basal needles completing their development first. The largest needles occur on the middle of the shoot (Chandler and Dale, 1990). Needle number of Sitka spruce was not correlated with shoot length (Chandler and Dale, 1990).

Needle retention is difficult to assess on spruce, particularly on trees with high retention or with large numbers of secondary shoots. This has sometimes resulted in underestimates of needle retention and, in Great Britain, trees with more than seven years of needle retention are aggregated into a single class. There is also a problem of differential retention throughout the crown, a source of variation that has still to be studied in detail.

Pines have a different pattern and, generally, needle retention times are shorter. Miller and Van Doren (1981) made repeated counts of the surviving needles in single annual whorls of various species of young pines subjected to ozone in California. The counts were made in three successive years, enabling the construction of needle life tables. The rate of needle loss was proportional to ozone exposure, suggesting that the technique was of value. However, it requires non-destructive, hands-on examination of the whorls, which may only be practical for young trees.

Needle fall in Scots pine primarily occurs in the autumn (Alvera, 1973; Elagin and Cherkashin, 1987) but, as with spruce, moisture conditions also exert a strong control on the fall of older needles (Orlov, 1980). High July temperatures led to enhanced litter fall in the same and the next year (Kouki and Hokkanen, 1992). Retention periods vary from less than two years to eight or more years, with the period being closely related to provenance. For example, northern provenances of Scots pine tend to retain their needles much longer than southern provenances. There is some evidence that retention times have dropped in some areas. Skuhravy

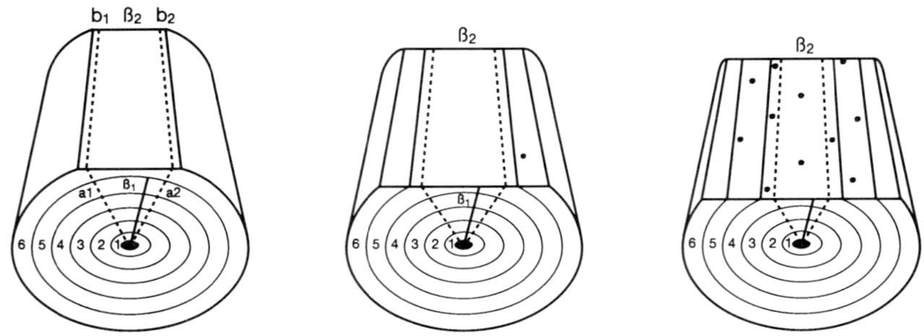

Figure 4.7 Sections through the stem of a Scots pine, showing the appearance of vascular bundles as the core of the stem is approached. From: Jalkanen and Kurkela (1990).

(1990) found that mean retention periods had dropped from 3.5 years in 1950 to 2.6 years in the late 1980s in central Bohemia and southern Moravia, Czechoslovakia. In southern Bohemia, retention had dropped from 2.6 years to 1.8 years over the same period.

A new technique for examining past needle retention in pines has been developed in Finland (Jalkanen and Kurkela, 1990; Kurkela and Jalkanen, 1990). It involves the examination of the vascular bundles in cross section. The vascular bundle supporting each needle pair continues to elongate until the needle pair is shed, after which the cambium grows over it (Fig. 4.7). This can clearly be seen in cross-section, enabling past needle retention periods to be determined. The technique is particularly useful in northern Scandinavia where needles on Scots pine may be retained for up to 10 years. However, its applicability further south, where needle retention is much shorter, is yet to be established. The technique enables the construction of needle life tables through time, as recommended by Miller and Van Doren (1981) and McKay (1988), and therefore has considerable potential. However, a problem with the vascular bundle method is that the relationship between needle retention on the stem and on individual branches is uncertain. This is an important limitation. For example, needle retention on the stems of loblolly pine is more sensitive to ozone exposure than it is on branches (Stow *et al.*, 1992).

Although variations in needle retention occur, the importance of such fluctuations remains uncertain. In species with very long retention periods, such as the single-leaf pinyon pine (*Pinus monophylla* Torr. and Frem.), the main use of older needles is as a long-term nutrient reserve rather than as a mobile nutrient pool (Everett and Thran, 1992). With species having shorter-lived needles, the situation is much less clear (Raitio, 1987) and urgently needs to be investigated. One possibility is that trees vary their retention times in response to short-term fluctuations in nutrient availability. When certain nutrients are in short supply, retention times may increase, resulting in higher nutrient use efficiency (Escudero *et al.*, 1992).

Needle density

The needles themselves are also of importance since they can vary in density, length and angle. Niehaus (1989) recognized five needle characteristics that influenced the state of a Scots pine crown: frequency of needles of the same year's growth, the proportion of the shoot with needles, the density of needles, the needle length and the needle angle.

There is some evidence that needle density in Scots pine decreases with increasing tree age (Jukola-Sulonen *et al.*, 1987; Levin, 1985), but it is not clear how age affects other aspects of crown condition. Needle density may also change following attacks by certain pathogens, such as *Lophodermium seditiosum*. Needle length varies from about 2 cm to more than 10 cm, although a variety of factors influence this. For example, in Great Britain, native Scots pine generally has short needles whereas the introduced continental provenances have much longer needles.

An important feature of Scots pine is the degree to which it differs across its natural range (Wright and Bull, 1963). Shoot extension varies markedly between provenances and several other variables are also dependent on provenance. Needle length is longest in trees originating from central Europe and decreases to the north and the south. As needle length also varies markedly from year to year at a site due to climatic factors, the difficulties of cause–effect attribution are clear.

Another important feature of Scots pine is that a relatively healthy tree, with long needles and shoots, may be more transparent than a stagnating tree with short needles and shoots (Gruber, 1989). This is also apparent to a lesser extent in other species and it is a major limitation for the use of crown density as the only index of a tree's vigour.

Premature shedding of leaves

The premature shedding of green leaves in beech was described as a symptom of the new form of forest decline by Schütt and Cowling (1985). However, as with other species, the cause of the loss is important. Green leaf fall may occur because of mechanical damage (particularly by wind) and a certain amount of leaf shedding is caused by insect damage. Consequently, it is difficult to attribute this symptom wholly to injury by air pollution.

Flowering and fruiting

In pines, flowering affects needle density and therefore crown density. In Scots pine, male flowering occurs temporarily on shoots as they expand. At the end of the growth season, a shoot that has had male flowers on it will show an area without needles towards its base (Fig 4.8). In cases where flowering is particularly abundant, the bare area may extend for 75% or more of the length of the shoot. To a certain extent, the amount of flowering can be related to crown density (Table 4.4). In Great Britain, heavy flowering in both the top and bottom of the crown has been associated with lower crown densities. Philipson (1990) drew attention to the influence of environmental conditions (such as drought) in the establishment of flower buds, suggesting that stress might trigger both flowering and an increase in transparency.

It is uncertain whether the heavier flowering in Scots pine occurs as a response to the high transparencies or whether the thinness associated with the flowering is the reason the trees were scored as having higher transparencies. There was no indication that flowering was more common on trees with very thin crowns, suggesting that the latter explanation may be more likely, but this requires further investigation. There is the added complication that flowering will directly affect the crown density score unless it is considered in the scoring. Some countries, such as Sweden, have avoided this problem by excluding the effects of flowering on the defoliation estimates (Andersson, 1990), but it is not clear how this can be

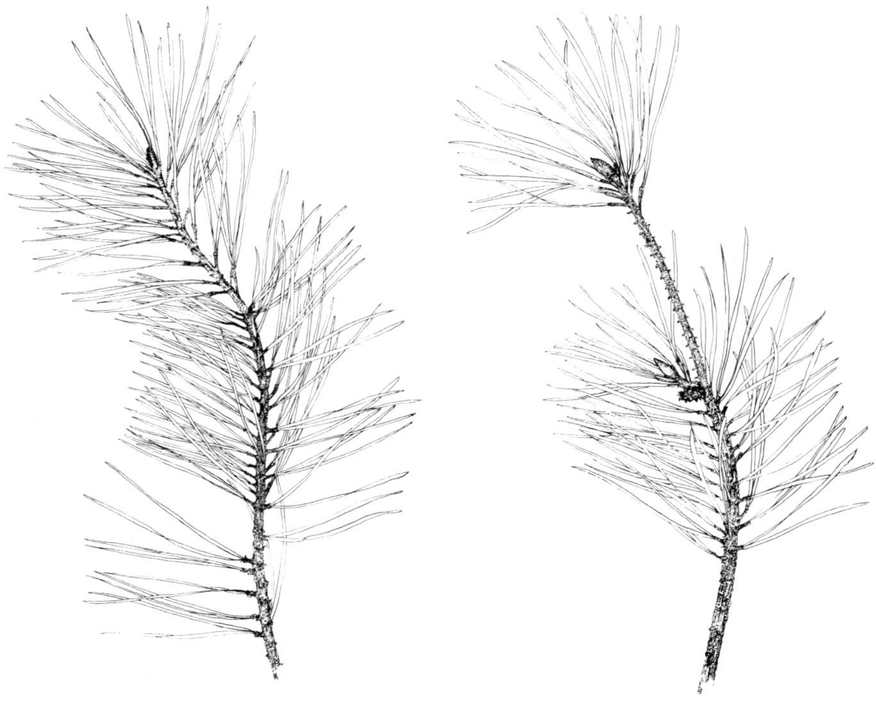

Figure 4.8 Effects of male flowering on the needles of Scots pine. The left-hand shoot has not flowered, whereas the right-hand one has.

Table 4.4 Relationship between flowering and transparency in Scots pine in Great Britain in 1991. Figures are the percentage of trees in each transparency category with a given level of flowering (i.e. columns sum to 100%).

	Flowering top of crown					Flowering base of crown				
Crown density score	0	1	2	3	4	0	1	2	3	4
0	0.4	0	0	0	0	0.9	0.6	0	0	0
1	0.5	0.2	0.9	0.5	0	0.5	0.9	0.5	0	0.8
2	2.8	2.1	1.1	1.0	3.0	1.9	4.0	1.8	1.7	1.8
3	7.7	6.2	6.8	6.3	0	6.9	10.0	6.9	6.3	5.7
4	15.0	12.8	12.0	17.3	3.0	13.4	17.1	13.5	14.7	12.2
5	17.9	20.0	22.2	21.6	9.1	11.6	17.1	20.4	20.3	22.4
6	16.4	22.1	20.8	19.7	21.2	11.6	15.6	20.0	19.0	23.0
7	11.9	16.2	16.0	13.0	18.2	9.3	10.6	12.8	17.5	15.2
8	9.9	9.0	10.0	10.1	24.2	7.9	8.4	12.4	10.2	9.6
9	6.0	5.0	6.3	6.7	9.1	9.3	8.1	4.8	5.4	4.7
10	3.4	2.6	1.7	1.9	3.0	8.3	2.8	2.3	1.9	1.6
11	1.5	1.9	0.9	0.5	0	1.4	1.9	1.4	1.5	0.8
12	1.0	1.0	1.1	1.0	3.0	1.9	0.9	1.4	0.6	0.8
13	1.0	0	0	0	0	1.4	0.3	0.7	0.2	0.2
14	0.9	0.2	0	0	0	1.4	0.9	0.5	0.2	0
15	0.8	0	0.3	0	0	2.8	0	0.2	0.2	0
16	0.3	0.5	0	0	6.1	0.9	0	0.2	0	0.8
17	0.4	0.2	0	0	0	0.9	0	0.2	0.2	0.2
18	0.1	0	0	0.5	0	0	0.3	0	0	0.2
19	0.2	0	0	0	0	0.5	0.3	0	0	0
20	1.7	0	0	0	0	7.4	0	0	0	0

done objectively. The extent of the problem varies regionally, as the incidence of male flowering in Scots pine is affected by provenance (Wright and Bull, 1963).

In lodgepole pine, the effect of flowering and coning on needle density and other parameters has been quantified (Dick *et al.*, 1990). Branches with male cones had on average 35 fewer needle pairs than the equivalent vegetative branches. Formation of male cones was associated with a 27–50% reduction in the number of needles per cone-bearing branch. No similar study appears to have been made in European pines.

The situation with many broadleaved species is much less clear. The production of acorns by oaks in Great Britain is frequently prevented by the flowers being killed by late spring frosts and a similar situation exists for beech. The amount of masting on beech is strongly related to previous summer temperatures (Matthews, 1955) although the effects on masting of two drought years in succession have yet to be described. Masting is often higher on thinner-crowned trees (Gärtner, 1988; Innes and Boswell, 1991) and on older trees (Gärtner, 1988), suggesting a relationship with other forms of stress.

Patterns of growth and development

Some of the most detailed methodological work on the assessment of crown vigour has been related to crown architecture. This has received particular emphasis as the architectural characteristics of trees potentially provide long-term indications of crown vigour. Crown architecture varies markedly between species and, in the following sections, spruce, pine, fir and broadleaves are dealt with separately. Further studies will no doubt enable the elaboration of architectural changes in several other species groups.

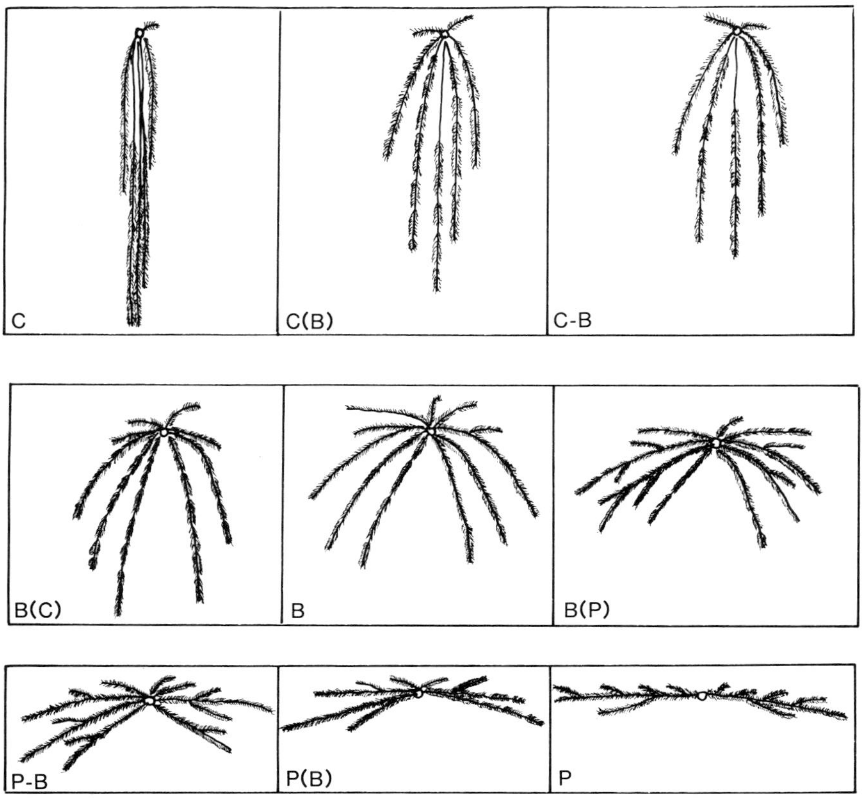

Figure 4.9 Cross-sections of different branching types in Norway spruce. C: comb-type; C(B): comb-type with brush-type tendencies; C-B: mixed comb and brush type; B(C): brush-type with comb-type tendencies; B: brush-type; B(P): brush-type with plate-type tendencies; P-B: mixed brush and plate type; P(B): plate type with brush-type tendencies; P: plate-type. From: Gruber (1989).

Spruce

In Norway spruce, several different types of branching occur that can be related to specific phenotypes (Gruber, 1986, 1989). Various branching types are shown in Fig. 4.9. Some of these are easily distinguished in the field, but more than one type may occur on the same tree, making assessment difficult. In particular, juvenile branches of some forms tend to be brush-like, and it may be difficult to decide which category an individual tree lies in. Gruber (1990) described an ideal Norway spruce crown as consisting of juvenile-type branches in the upper crown, comb- and brush-type branches in the middle crown and plate-type branches in the lower crown. However, individual trees may be dominated by one or more of these types. In addition, the different types form a continuum, making it difficult to place some trees in a particular category. Genuine comb-type trees (Figs 4.9, 4.10 and 4.11) should be distinguished from brush-type trees that have undergone stress, resulting in drooping branches. It is still not entirely clear whether the drooping or hanging brush-type also represents another phenotype or whether it is purely the result of stress.

The nature of the branching pattern also affects the nature of secondary branching (secondary shoots being defined here as shoots that develop several years after the parent shoot has formed) and the tree's ability to replace lost foliage (Gruber, 1987; Lesinski and Westman, 1987). Trees with comb-type branching structures have the most vigorous

Figure 4.10 A typical comb-type branch from a Norway spruce.

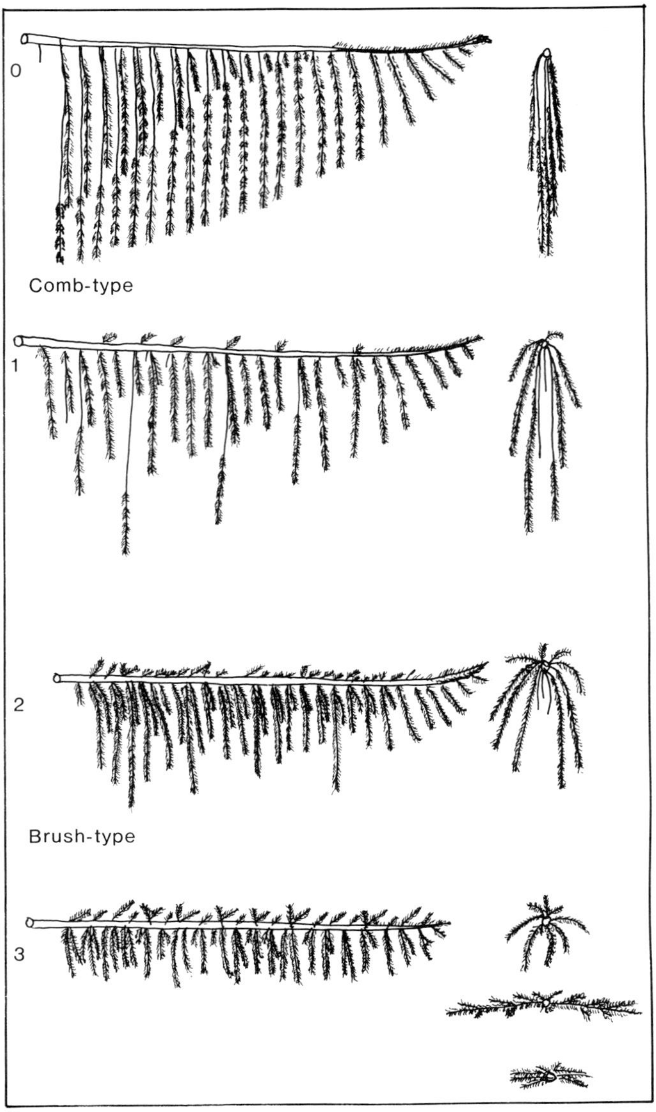

Figure 4.11 Visual gradation between genuine comb-type branch and 'drooping' brush-type branches of Norway spruce. From: Gruber (1989).

secondary shoots (Fig. 4.12) whereas trees with a brush-like branch structure have the least vigorous ones. Secondary shoots are normally produced continuously in Norway spruce (Rehfuess and Rehfuess, 1988), although the rate may increase following severe defoliation.

Gruber (1990) recognized three phases of shoot development during the lifetime of a Norway spruce and he termed these the proleptic,

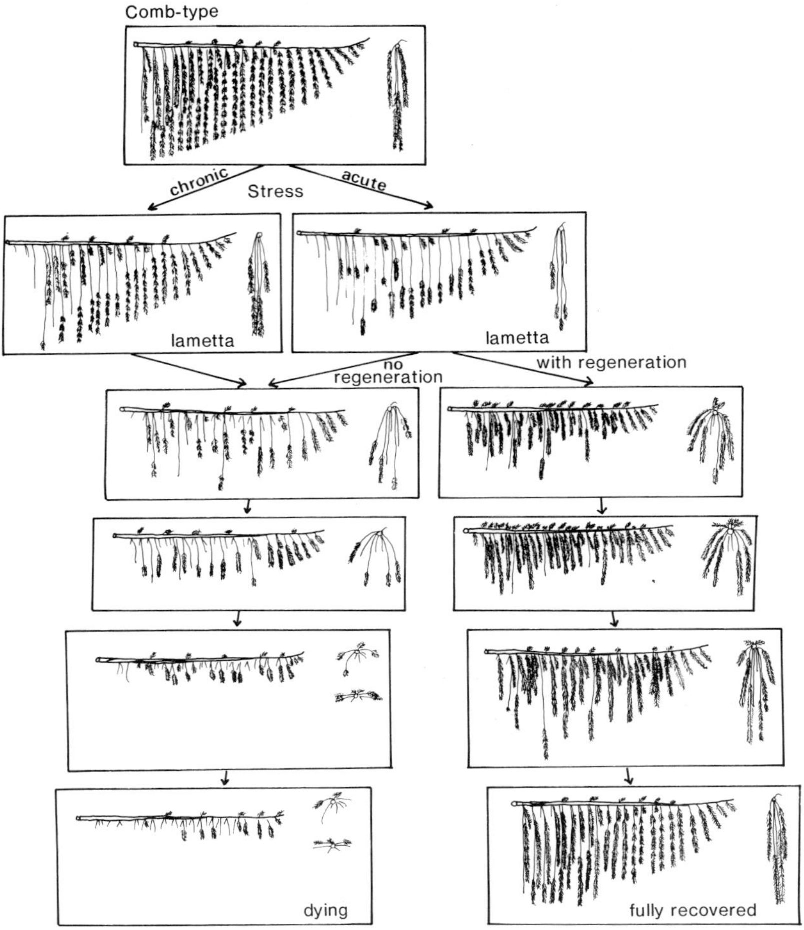

Figure 4.12 Recovery (or deterioration) of a comb-type branch of Norway spruce following different types of stress. The recovery takes place through the development of secondary shoots. From: Gruber (1989).

regular and proventitious (secondary) shoot formation types. Proleptic shoots develop from embryonic buds and only occur on young trees growing under good conditions. Regular shoots develop each year from one-year-old winter buds. They are present throughout the tree's life, and form the dominant mass of branching within the crown. The regular shoots are responsible for the expansion of the crown and formation therefore takes place in the crown periphery. As the tree ages, secondary shoots become increasingly important. The shoots develop from dormant buds at the bases of annual shoots and on the upper side of near-horizontal branches. They enable bare shoots to be replaced by photosynthetically active shoots, improving the efficiency of light utilization within the crown. The presence of large numbers of secondary shoots within a crown reveals a strong recuperative capacity in the tree. This, in turn, implies an unimpaired uptake and transport system for water and nutrients (Westman and Lesinski, 1985; Rehfuess and Rehfuess, 1988; Gruber, 1990). They may however suggest that an external stress has affected the tree in the past.

The formation of secondary shoots is not an abnormal condition, as suggested by Schütt (1989). Lesinski and Landmann (1988) recognized three types of branch injury related to the pattern of secondary branching, although a further category has since been defined. The four groups are:

- Group D, with obvious loss of biomass;
- Group DR, with loss of biomass evident, but some secondary shoots present;
- Group RD, with the loss of biomass being more-or-less balanced by secondary shoots
- Group R, where no loss of biomass is evident because of complete refoliation with secondary shoots.

The actual nature of the defoliation varies according to the overall pattern of defoliation in the tree and the particular branching types present (Fig. 4.13). Depending on the nature of the stress, defoliation may develop in different ways. Six types are generally recognized, shown in Fig. 4.14.

Lesinski and Landmann (1988) showed that the number of secondary branches changes with increasing age of the tree. Normal crown thinning involves the loss of the oldest needles and the eventual drying out

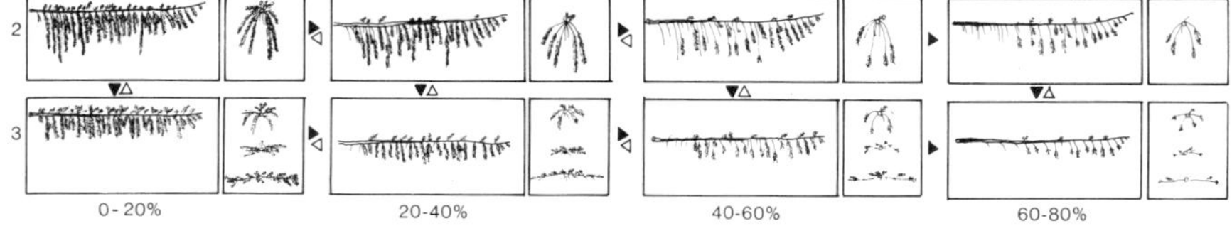

Figure 4.13 Defoliation development in different branch-types of Norway spruce. The estimated loss of needles is given as a percentage of the total amount of foliage that should be present. Movement from one branch-type to another is possible, as indicated by the arrows. From: Gruber (1989).

and loss of the oldest lateral shoots. These are gradually replaced by secondary shoots, so that the proportion of secondary shoots in the crown increases through time. These shoots should therefore be included in defoliation assessments.

Pine

Niehaus (1986, 1989) has made a detailed study of the visible indicators of crown vigour in Scots pine. She drew attention to the need to examine

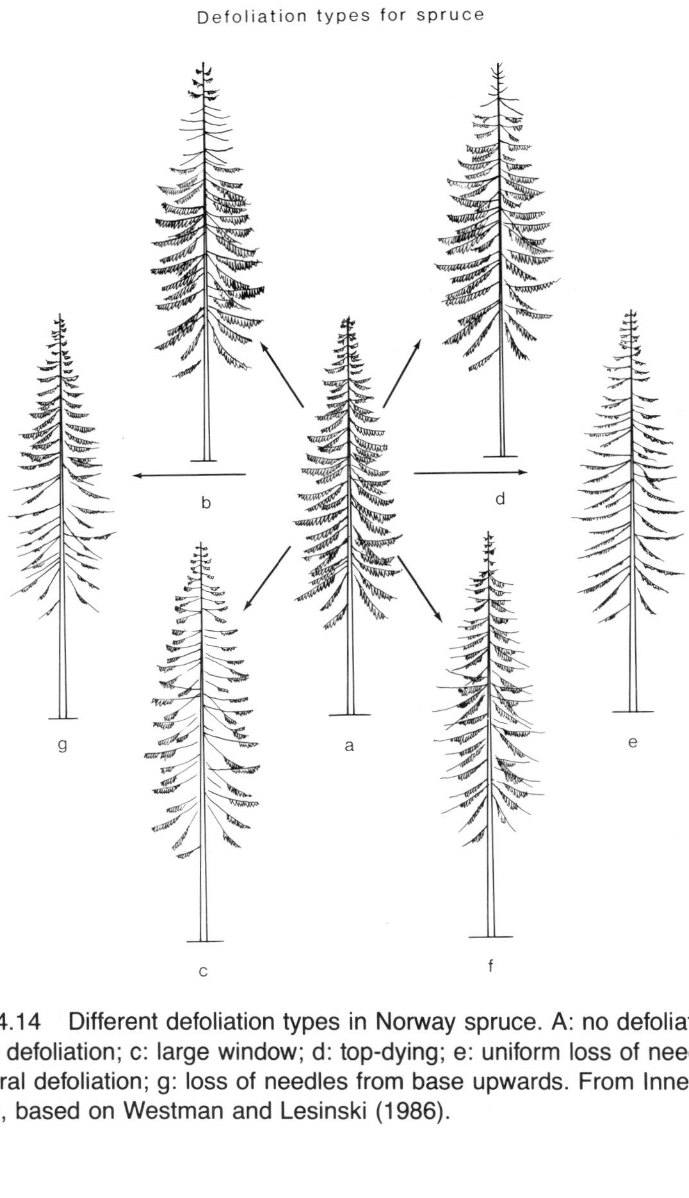

Figure 4.14 Different defoliation types in Norway spruce. A: no defoliation; b: sub-top defoliation; c: large window; d: top-dying; e: uniform loss of needles; f: peripheral defoliation; g: loss of needles from base upwards. From Innes (1990c), based on Westman and Lesinski (1986).

Figure 4.15 Different shoot forms in beech. Upper: good extension growth of primary shoots and laterals. Lower: development of 'claw' shoots caused by poor development of laterals and upward growth of primary shoots.

Upper: disturbance of normal extension growth evident. Lower: Severe disturbance of growth with very poor growth of laterals. All pictures taken at Selborne, England.

a range of different factors, including the density of the primary branching system, the density of branching (number of shoots per branch), the branch arrangement, the length of needled shoots and the branching angle. Niehaus was particularly concerned about the use of standard reference trees when assessing Scots pine, in view of the effects of genetic, local and climatic variations on needle growth. Changes in branching structure associated with needle loss have been reported in Scots pine (Pretzsch, 1989), but an adequate symptom description is still lacking.

The density of the primary branching system was defined by the distance between different branch systems, with the density of branching within each primary branch system defining the mass of needles in each system. The length of the annual shoots may vary from a few centimetres to more than 30 cm and provides an indication of the general vigour of the tree.

Changes to branch architecture through time are particularly marked in Scots pine (Niehaus, 1989). The changes reflect the gradual loss of dominance of the apical shoots, resulting in the flattening of the crown in older trees. This process can be accelerated by stress, but there is also an interaction with provenance which may complicate some assessments.

Branch structure classification systems also exist for other pine species (e.g. Shelbourne and Low, 1980; Raymond and Cotterill, 1990), but have yet to be applied within the context of forest health. The method used by Raymond and Cotterill (1990) involved a subjective six-point score based on branch diameter and angle. Other indices examined by Raymond and Cotterill included the number of visible ramicorn branches, the number of forks and the presence of a uninodal or multinodal branch habit. The six-point score was adopted because it resulted in large additive genetic variance and a higher heritability, leading to good discrimination between genotypes. However, it is likely that another combination of symptoms would be more useful in monitoring and classification studies.

Silver fir

In silver fir, reduced height increment results in the development of 'stork's nest' crowns, whereby the top of the tree becomes almost flat. This occurs naturally with increasing age, but it has frequently been associated with reduced vigour of younger trees. It is reversible, and recovery of the leader growth has been noted in many cases. No reliable classification system appears to have been developed for this phenomenon. Sometimes, the development of secondary shoots near the base of the stem has been associated with severe defoliation. The long-term implications of this major change have yet to be assessed.

Broadleaves

The use of crown architecture as an index of vigour is quite common in beech and other broadleaved species (Roloff, 1985a, 1985b; Dobler *et al.*, 1988; Härdtle, 1990). The technique is based on the nature of the upper shoots in the crown, which may show a range of different forms (Fig. 4.15), depending on the vigour of the tree (Nihlgård *et al.*, 1985). However, as the vigour of a tree is dependent on its age, some changes in crown architecture can be directly related to the age of the tree. Thiébaut (1988) also questioned the use of crown architecture in surveys of forest health because many variations are likely to be genetically controlled and therefore do not provide a good indication of tree response to external stress.

The architectural studies are mostly concerned with the relationship between long and short shoots. The functions of these differ. Short shoots carry 79% of the leaves in the crown (Thiébaut, 1988). Long shoots extend the height and spread of the crown and account for more than half the annual dry weight production. Dobler *et al.* (1988) examined branch structure in beech using Roloff's methods. Both young and old stands showed evidence of decline over the last 12 years, with the least growth occurring in 1978. Lonsdale *et al.* (1989) and Hotz *et al.* (1990) also used the extension growth of beech to look at growth dynamics, although the paper by Lonsdale *et al.* (1989) nicely illustrates the problems associated with accounting for long-term ageing trends in such data. In the study by Hotz *et al.* (1990), a clear difference in the extension growth of an asymptomatic and a defoliated tree was identified (Fig. 4.16). However, as only two trees were examined, this study should be treated with caution.

Athari and Kramer (1989a, 1989b) reported a major study of the crown architecture of beech in Germany. Trees in five stands were classified according to the Roloff system. Stem analyses were then done for each of the 77 trees. The rate of growth was inversely proportional to the degree of damage. The environment and social class of the trees were important factors influencing the impact of the degree of damage on growth. Stand density was generally less in stands with damaged trees and this may have had a bearing on increment. In trees with equal levels of competition, the rate of growth decreased with increasing levels of damage. Roloff scores seriously conflicted with degree of foliage loss, although Roloff (1991) suggested that this was predictable. The branch structure of beech, as classified by the Roloff method, was also poorly correlated with radial growth at breast height (Mahler *et al.*, 1988a; Perpet, 1988; Athari and Kramer 1989a, 1989b; Fischer and Rommel, 1989), but stand structure may have significantly affected the relationship in individual cases.

These inconsistencies represent a problem for any proposed monitoring programme or survey. Although there is a broad relationship between the architecture of beech crowns and their health, a variety of other factors also appear to play a part, particularly in the middle part of the range of scores. Consequently, it would be unwise to rely on branch architecture alone when examining beech crowns.

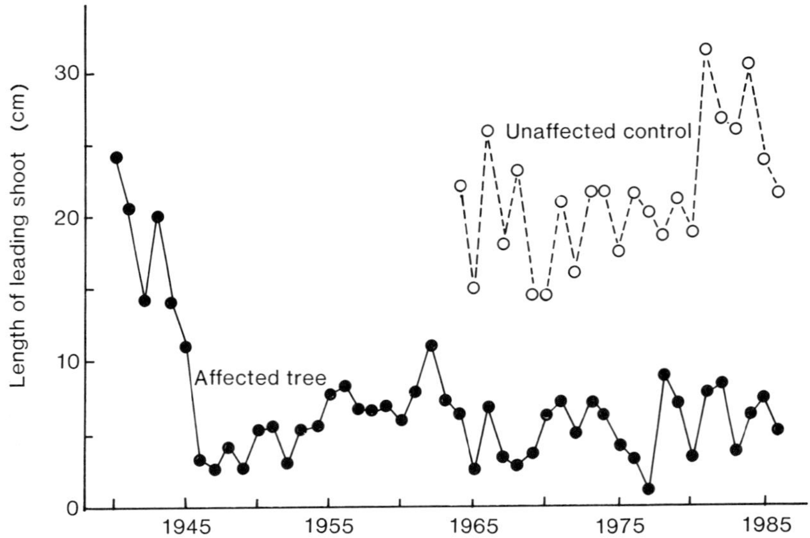

Figure 4.16 Growth of the leading shoot of a symptomatic and asymptomatic beech tree in Würzberg, Germany. From: Hotz *et al.* (1990).

Roloff (1989) extended his earlier work on the branch structure of beech to oak and other species but, on some species, the symptom is difficult to record and is of uncertain value as a diagnostic of stress. It has also been extended to oak by Clauser and Gellini (1986) and these authors recommended assessment in winter, when the branch structure is likely to be much clearer. However, winter assessments involve an extra visit to the monitoring plot and consequently they may be avoided because of financial constraints.

Much of the work of Westman (1989) on birch can be applied to the assessment of other broadleaves. I adapted Westman's system to oak and beech, drawing on the main characteristics of the two species and eliminating some systems that were only apparent in birch (Innes, 1990c). The nature and extent of leaf loss in the crown, dieback and adventitious shoots were the most useful indices. The nature of leaf loss is variable and several different patterns can be recognized, ranging from small gaps caused by sparse foliage to large gaps with some branches without leaves. Dieback, which is defined as branch mortality beginning at the tip of the branch and extending inwards towards the trunk (Brooks *et al.*, 1991), can take various forms and may occur in different parts of the crown. Its nature and extent is a useful indication of the vigour of the tree and may also provide an indication of both short-term and long-term stresses. A particularly useful way of assessing this is to estimate the proportion of primary branches that have died back (in 5% or 10% steps) (Podger *et al.*,

1980; Kile *et al.*, 1981). Secondary shoots can provide an indication of the amount of light penetrating the crown, but their frequency and extent may also be genetically controlled, complicating any interpretation.

Defoliation levels

Crown transparency (also termed crown density, thinning, transparency, defoliation and needle/leaf loss) is an almost universal index of tree health, but it is influenced by many different factors. This lack of specificity represents a serious shortcoming to its use in assessments of tree condition. Sometimes, transparency may be directly attributable to the shedding of foliage. Alternatively, the transparency may be related to changes in branch architecture. As assessments are made at a single point in time, it is frequently impossible to state whether the trees have become thinner; the crowns may always have been thin. Schweingruber (1989) neatly showed this by comparing trees shown in old postcards with their crowns in the 1980s. The definition of crown transparency is particularly difficult for faster-growing trees, where the apical shoot extension may dominate the overall form of the tree.

There are several different ways of assessing crown transparency, although it is frequently (and incorrectly) assumed that these produce similar results. The simplest approach is based on the comparison of the crown density of individual trees with photographic standards. Various standards have been published, including ones for Great Britain (Innes, 1990c), Switzerland (Sanasilva, 1986; Müller and Stierlin, 1990) and the Mediterranean region (Cadahia *et al.*, 1991). A more complex approach involves the comparison of trees with theoretical standards for the area. This approach is heavily dependent on the knowledge of the observer. Yet another approach is to calculate the defoliation based on the amount of missing foliage. This again assumes that the normal amount of foliage is known. In conifers, it is very difficult to determine the normal amount of foliage as needle retention is influenced by so many factors.

Establishing the nature of the defoliation is often difficult. For example, distinguishing between a crown that has lost branches and one in which the branch development is very poor may be almost impossible. However, such a distinction is critical in cause–effect studies.

The relationship between crown thinning and damage is equally complex. The term damage implies a causal agent, yet at least some of the variations in crown density are genetically controlled. This is illustrated by the debate over what level of defoliation should be considered as damage. In early German studies, 10% was taken as a critical threshold. Subsequently, this was raised to 25%. In Sweden and Finland, a loss of 20% of the foliage is taken as the threshold (Bengtsson, 1985; Jukola-Sulonen *et al.*, 1987) and this value has also been adopted in some other studies (e.g. Mandallaz *et al.*, 1986). In the north-eastern United States, trees with less than 30% foliar loss are considered to be within the normal range (Brooks *et al.*, 1991). However, Becker (1987a) failed to find any significant increment loss in silver fir at crown defoliation values of <40% and similar figures were cited by Rehfuess (1988b) for Norway spruce. Consequently, it seems that the current threshold of 25% may be too low in some situations. This issue is discussed further in Chapter 10.

Factors affecting crown transparency

The degree of canopy closure plays a part in determining crown condition, with increased crown transparency occurring in more open stands of most species (Denstorf *et al.*, 1984; Prinz, 1985; Hunziker *et al.*, 1988; Levin, 1985; Keller and Imhof, 1987; Neumann, 1989a; Innes and Boswell, 1990a). Crown transparency of Norway spruce is known to increase with increasing age (Jukola-Sulonen *et al.*, 1987; Innes and Boswell, 1988, 1989a, 1990a; Neumann, 1989a) although Reichelt (1983) has suggested that this increase does not extend in Norway spruce and silver fir beyond trees aged over about 70 years. It is difficult to determine whether this trend is a result of a natural ageing process or whether older trees are more susceptible to those stresses that promote needle loss. There is good evidence that crown transparency of Scots pine increases with increasing tree age (Jukola-Sulonen *et al.*, 1987; Levin, 1985), but it is not clear how age affects other aspects of crown condition. In broadleaves, clear relationships are not readily apparent, probably because so many other mitigating factors are present.

Crown transparency of Norway spruce

As with other species, crown transparency in Norway spruce can be caused by several processes (Westman and Lesinski, 1985). It may be due to needle loss, through a reduction in needle longevity, a reduction in needle density or a reduction in needle biomass. Alternatively, it may be the result of branch destruction, underdeveloped shoots, malformed branches or the absence of secondary shoots. The nature of the defoliation plays an important part in evaluating the cause of the injury.

In Norway spruce, natural needle loss starts from the stem and proceeds outwards and, in dense stands, from the base of the tree upwards (Westman and Lesinski, 1986). This means that the oldest needles are shed first. Other types of thinning tend to start elsewhere in the crown. For example, top-dying results in the death of the upper crown of the tree whereas peripheral thinning involves the loss of the youngest needles. Crown transparency tends to be underestimated on co-dominants in lower damage classes and on trees with no distinct boundaries between damaged and undamaged parts of the crown (Westman and Lesinski, 1986). Coning can result in a very irregular crown pattern.

Assessment systems

In most studies, only one or two visible indices have been used to classify the condition of trees. Consequently, it has been difficult to draw up multivariate classification systems. There are few long-term studies which can be used to assess the efficiency of different classification systems using a number of different variables. The detailed classification developed by Lesinski has been used to monitor changes in permanent plots of Norway spruce in Sweden from 1984–91 (Lesinski, 1991a, b). Nine types of change were identified:

- absence of any change in injury symptoms;
- length of the most damaged part of the crown increased;
- needle loss in the upper part of the crown increased;
- uninjured crown (<20% needle loss in the upper part of the crown) changed to an injured category;
- crown transparency type moved from one class to another;
- contribution of secondary shoots within the existing green biomass occurring in the most injured part of the crown increased;
- length of the most injured part of the crown decreased;
- needle loss in the upper part of the crown decreased; and
- an injured crown classification changed to an uninjured classification.

Although changes were more evident over eight years than over four, the changes were more apparent in the 1988–91 period than in the period 1984–88. The most common change was an increase in the length of the injured parts of the crown. In the sample, the most frequent type of needle loss involved the uniform loss of needles throughout the crown, with loss of needles from the base being the second most important type. Quite marked changes occurred in the defoliation type, with changes between window-type and uniform-type being the most frequent. Healthy trees in 1984 which subsequently deteriorated mainly moved into the uniform-type and base-type defoliation categories.

Lesinski *et al.* (1991) compared the different types of assessments of Norway spruce undertaken in Sweden and Poland. The Polish system was based on that described by Dmyterko (1991), a combination of defoliation, height growth reductions and the decline of lateral shoots. Trees were assessed on a four-point scale of no injury, slight injury, distinct injury and severe symptoms. The Swedish system was based on the degree of defoliation, as determined by crown transparency, the condition of the upper part of the crown and the branch malformation pattern. The condition and malformation assessments were based on a number of different indices.

The defoliation estimates obtained using the two systems were very similar, not varying by ± 5%. This suggests that defoliation systems for Norway spruce, as estimated by needle loss and crown transparency, do not necessarily differ significantly. However, the Polish and Swedish systems revealed some differences when the results were compared in greater detail. These were mainly related to one or other system containing more information and Lesinski *et al.* (1991) therefore recommended combining the best features of the two when making assessments of Norway spruce.

Roger Boswell and I attempted to develop a composite index of crown condition based on the various indices used in the British monitoring programme (Innes and Boswell, 1988). We used principal components analysis to derive mathematical relationships between the different variables and the resulting principal components were then mapped or used in regression analyses. In oak, 81% of the variation in the first component was explained by a range of estimated variables (hydrogen ion concentration in rainfall, tree age, mean October-December temperature, site slope and the ruggedness of the local topography). However, the regression analyses were complicated by the presence of multicollinearity among the explanatory variables, making interpretation difficult.

Roger Whittaker and I used a rather different approach to classify the crown condition variables (Innes and Whittaker, 1993). Two-way indicator species analysis distinguished a number of groups of sites by crown condition (as determined by a number of different variables). We then assessed the relationships between these groups and several environmental variables using multiple discriminant analysis. The technique appeared to offer considerable potential and we are exploring the method further.

Miller (1973, 1989) used a combined index to assess ozone-injured pines in the San Bernadino National Forest of California. It was based on assessments of the average number of needle whorls in the upper and lower sections of the crown, the condition and length of the needles present and lower branch mortality. The scores were combined into a health index with seven categories and was successful for describing both Jeffrey and ponderosa pine. Such combined assessments have generally not been used elsewhere.

Salemaa and Jukola-Sulonen (1990) used factor analysis to find out whether transparency was related to a combination of other condition variables. The first two factors were interpreted as 'vitality' indices but the coefficients of determination were relatively low (16.9% and 15.2%, respectively). However, this study and the others described above suggest that it should be possible to construct indices of crown condition based on multiple observations of individual trees.

Summary

Visual estimates of crown condition and vigour provide an important tool in forest health surveys and in forest monitoring programmes. A variety of different indices are available, some more useful than others. Crown thinning, or defoliation as it is frequently termed, is one of the most widespread indices. It is an integrative index, combining a number of different parameters. As such, it provides a reasonably good indication of the overall condition of the crown but, on its own, it has little value in diagnostic studies. In such situations, it is important to make more detailed records. In addition, it is important to obtain data from different periods. A single survey will provide very little information of the dynamic aspects of forest condition, and such data will only be obtained after a number of years of survey. The use of combined indices of crown condition appears to offer considerable potential, but has yet to be fully exploited.

Remote sensing

Remote sensing lies somewhere between visual assessments and non-visual assessments. The most successful applications have involved invisible wavelengths and this section has therefore been separated from the visual assessments section. A selection of early applications of remote sensing to vegetation damage is provided by Anon. (1978). More recently, there has been a great deal of interest in the use of remote sensing to quantify forest damage. Much of this stems from the relatively encouraging results obtained by studies of the pigment contents of foliage and chlorophyll fluorescence tests already described. However, many early studies were conducted using black and white or colour infrared imagery alone; the latter has been particularly useful (e.g. Murtha, 1983; Hartmann and Voretzsch, 1985; Hartmann *et al.*, 1985, 1986).

Two different types of remote sensing imagery can be examined, one using satellites as a platform, the other based on imagery derived by aircraft. The data obtained from such studies are particularly amenable to statistical treatment and a number of techniques have been developed for looking at the distribution of asymptomatic and symptomatic trees (e.g. Pielou, 1963; Vithayasai, 1971, 1972; Ratkowsky *et al.*, 1980). In this section, only canopy reflectance in the visible and invisible wavelengths are considered. Techniques such as side-ways looking radar have not been included.

Air photographs

At its most basic level, standard aerial photography can provide good information on the incidence of chlorosis, severe needle loss and mortality within a forest. The technique has been successfully applied in areas such as Lower Saxony in Germany, where chlorosis has been widely documented (e.g. Niehaus, 1986, 1990) and elsewhere in Germany (e.g. Kenneweg, 1973; Hildebrandt, 1980; Ammer *et al.*, 1983). Generally, colour infra-red (CIR) imagery provides more useful information than black-and-white imagery, despite the denser shadows on CIR imagery. CIR film is sensitive in the 0.4 to 0.9 μm wavelengths and reflection from vegetation in this range appears to be sensitive to the type of vegetation and its condition (Hildebrandt and Kadro, 1984; Ciesla and Hildebrandt, 1986). The scale of photography has varied. In individual German states, scales ranging from 1 : 3000 to 1 : 16 000 have been used (Kenneweg, 1989), whereas 1 : 8000 has been used in the United States (Kelley and Eav, 1988) and 1 : 9000 in Switzerland. In Sweden, 1 : 20 000 photographs were more difficult to interpret than 1 : 10 000 ones (Holmgren and Wastenson, 1985) and the optimum scale appears to be between 1 : 5000 and 1 : 10 000. For the interpretation of individual tree crowns, Murtha (1983) found that a scale of 1 : 4000 was adequate. Similarly, Wardlaw (1989) found that a scale of 1 : 4000 was sufficient to distinguish between dead and healthy trees, and that larger scales provided no increase in health information.

A number of problems have been encountered with CIR photography. The photography, like ground assessments, should only be done during a relatively narrow time window, after the end of new growth and before the onset of autumnal senescence. Bad weather during this period may severely restrict flying time in some years. The sensitivity of CIR film varies between batches and this has been an important obstacle when comparing different films (Kleman, 1985; Voretzsch *et al.*, 1986), although it is now resolved. Interpretation of the imagery can be difficult due to systematic differences in the colour across the images (in addition to the colour differences caused by variable film colour balance). The problems with the colour interpretation have led to recommendations to use the structural aspects of the crowns as the main source of information on the different damage classes (Kenneweg, 1989). The structural aspects primarily concern defoliation; discoloration data are obtained from the distribution, saturation, brightness and tone of the images. Training and practice in the interpretation of the photographs is also required as significant differences may occur between different interpreters (Mauser, 1991).

Miller-Weeks *et al.* (1988b) used CIR aerial photographs to investigate the incidence of red spruce and balsam fir mortality in the northeastern United States. The survey covered about 1.8m ha and photographs were obtained at a scale of 1 : 24 000 or 1 : 20 000. The method provided a relatively quick method of broadly assessing the extent of mortality within stands. Three mortality categories were used: light (less than 10% standing dead), moderate (10–30% standing dead) and heavy (greater than 30% standing dead). So far, only data from Vermont and the White Mountains of New Hampshire have been presented, but these illustrate the potential value of the method.

Multispectral imagery

Much of the work involving remote sensing relates to the spectral characteristics of green vegetation in the visible and near-infrared (NIR) wavelengths (e.g. Senoo and Honjyo, 1988). Green leaves have marked pigment absorption in the 0.4–0.7 μm range. Two peaks occur within this range, one in the violet/blue band at 0.36–0.4 μm and the other in the red band of the visible spectrum at about 0.67 μm. Leaf water absorption occurs in the 1.3–2.5 μm region. Individual pigments, such as chlorophyll-a, chlorophyll-b, xanthophyll and carotene all have different absorption characteristics and changes in the relative of abundance of these can affect the reflection characteristics of the foliage. As leaves age, changes in the spectra in the 0.4–0.8 μm region occur. Pigment absorption is maintained in the blue whereas green and red reflection increase. The chlorophyll-a absorption near 0.68 μm is maintained while the red edge of the chlorophyll absorption (0.7–0.8 μm) moves to shorter wavelengths (Elvidge *et al.*, 1989). These changes (Fig. 4.17) can be measured using a variety of spectrometers, ranging from hand-held ground and boom-mounted instruments to airborne and satellite-borne instruments.

Rock *et al.* (1988b) argued that the position and slope of the red edge of the reflection curve is related to leaf chlorophyll concentrations and Horler *et al.* (1983) documented a linear relationship between the wavelength of the maximum slope of the red edge and leaf chlorophyll concentrations. The blue shift appears to be associated with a reduction in chlorophyll-b content. Minor changes of the red edge to shorter wavelengths are known as the blue shift. A blue shift of the red edge of about 5 nm has been found in declining Norway spruce from the Black Forest and red spruce from Vermont (Fig. 4.18). A blue shift of the red edge has been induced experimentally by exposing seedlings to ozone (Westman and Price, 1988b), but the results from Vermont and Germany should not be interpreted as necessarily indicating ozone damage. The chlorophyll content of the canopies can be monitored by the wavelength position of the red edge (Collins, 1978; Rock *et al.*, 1986, 1988a). The position of the red edge is a very sensitive indicator, but its specificity remains to be fully assessed. While the technique appears to be very promising, most airborne multispectral scanning and thematic mapper data do not provide the necessary resolution to identify the blue shift (Hermann *et al.*, 1988).

As with visible symptoms, there are no unique spectral features of vegetation that can be used to identify damage by air pollution exclusively. Elvidge *et al.* (1989) described a number of spectral signatures which may, under certain circumstances, be associated with damage by air pollution. Ustin and Curtiss (1990) described in detail the effects of ozone fumigation on the spectral reflection of the seedlings of several species (Figs 4.19 and 4.20). Ozone affected the reflectance after a single growing season. The

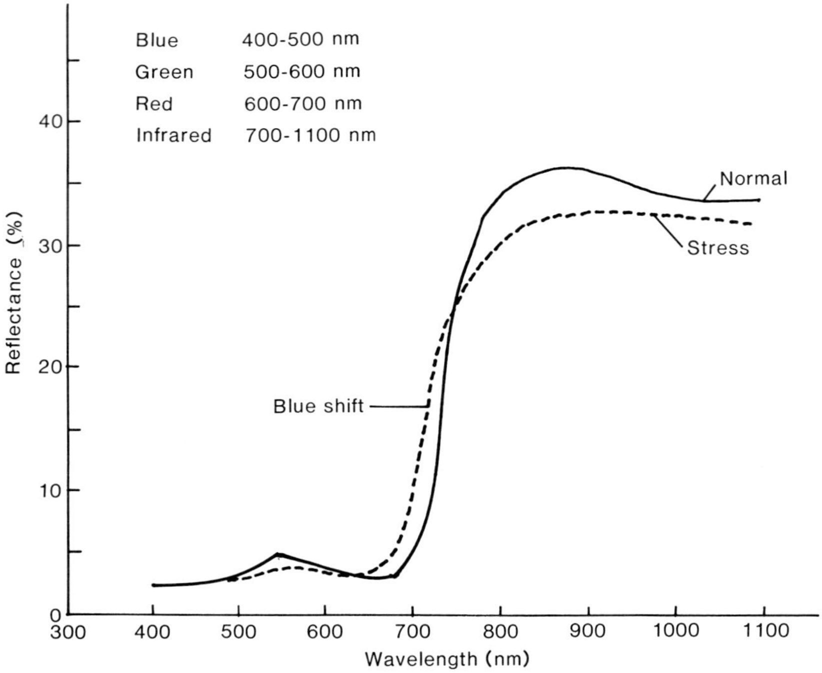

Figure 4.17 Stress induced spectral changes in the reflectance of foliage. From: Reid (1987).

greatest changes occurred in ponderosa pine, reflecting the sensitivity of this species to ozone (see Chapter 2). Douglas fir showed no spectral responses to the treatments. The greatest changes were found in the chlorophyll absorbance spectrum (600–725 nm), as shown in Fig. 4.20.

Other aspects of the canopy can be assessed by multispectral scanning. Changes in leaf area indices were identified by Peterson *et al.* (1987) by contrasting the reflectance in the red and near infra-red wavelengths. Leaf water content of the canopy can be monitored by combining bands from the near infra-red plateau (0.8–1.3 μm) with bands in the water absorption areas (1.3–2.5 μm) or by using radar imagery (Hunt *et al.*, 1987). Changes in the species composition of the canopy can be assessed using multiple bands and multiple dates of imagery. Dieback and mortality can be characterized by an increase in the intensity of the lignin/cellulose absorption bands (Elvidge, 1988).

In the following account, a very brief summary of some of the available techniques is provided. However, this is a rapidly developing field in which new instruments are constantly being developed. For example, the potential of the visible infra-red reflection absorbance fluorescence spectrometer described by Buschmann *et al.* (1991) is still to be fully

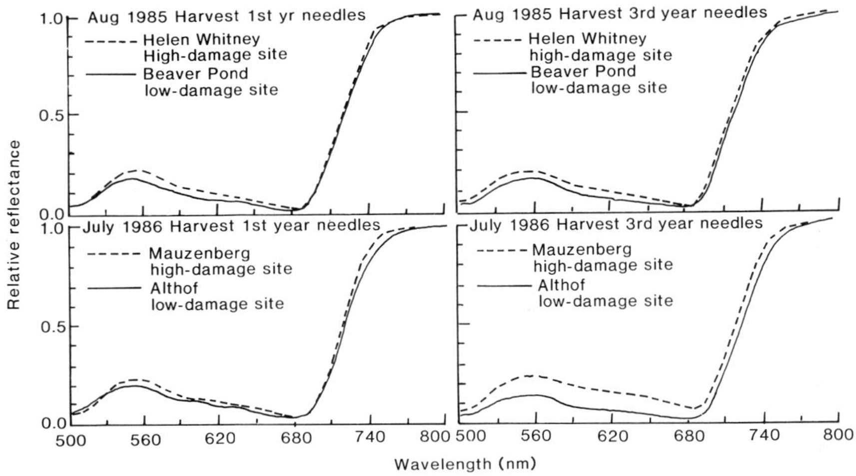

Figure 4.18 Normalized VIRIS (visible infra-red intelligent spectrometer) reflectance spectra of needles taken from red spruce in Vermont (upper two diagrams) and Norway spruce in Germany (lower two diagrams). From: Hoshizaki *et al.*, 1988).

determined. Consequently, by the time of publication, this section more than most others may have been overtaken by recent developments.

Satellite imagery

The use of satellite imagery was initially less successful than aerial photography, primarily because of the relatively poor spatial resolution of the early LANDSAT satellites (Kenneweg *et al.*, 1989). For example, a single pixel in an early LANDSAT image covered 6241 m^2 on the ground. A single pixel from the visible/near infra-red bands of LANDSAT TM, however, covers 900 m^2. The SPOT–2 satellite images have a higher resolution, with each pixel covering 100 m^2, but the increased accuracy is at the cost of fewer channels and a smaller range of wavelengths. Despite these scale problems, Kenneweg (1989) reported that a comparison of damage inventories made from LANDSAT imagery and CIR imagery had a 70% agreement (Fig. 4.21). Kenneweg (1989) identified a number of trends in the misclassified stands. Older stands tended to be classed as more severely damaged in the LANDSAT interpretation than in the CIR interpretation. Conversely, the degree of damage in young stands was underestimated by LANDSAT. Discoloration could not be detected in the LANDSAT imagery and interpretation became increasingly inaccurate in mixed-species stands.

There has been considerable interest in the use of satellite imagery (both multispectral scanner [four bands on LANDSAT–5 and three bands

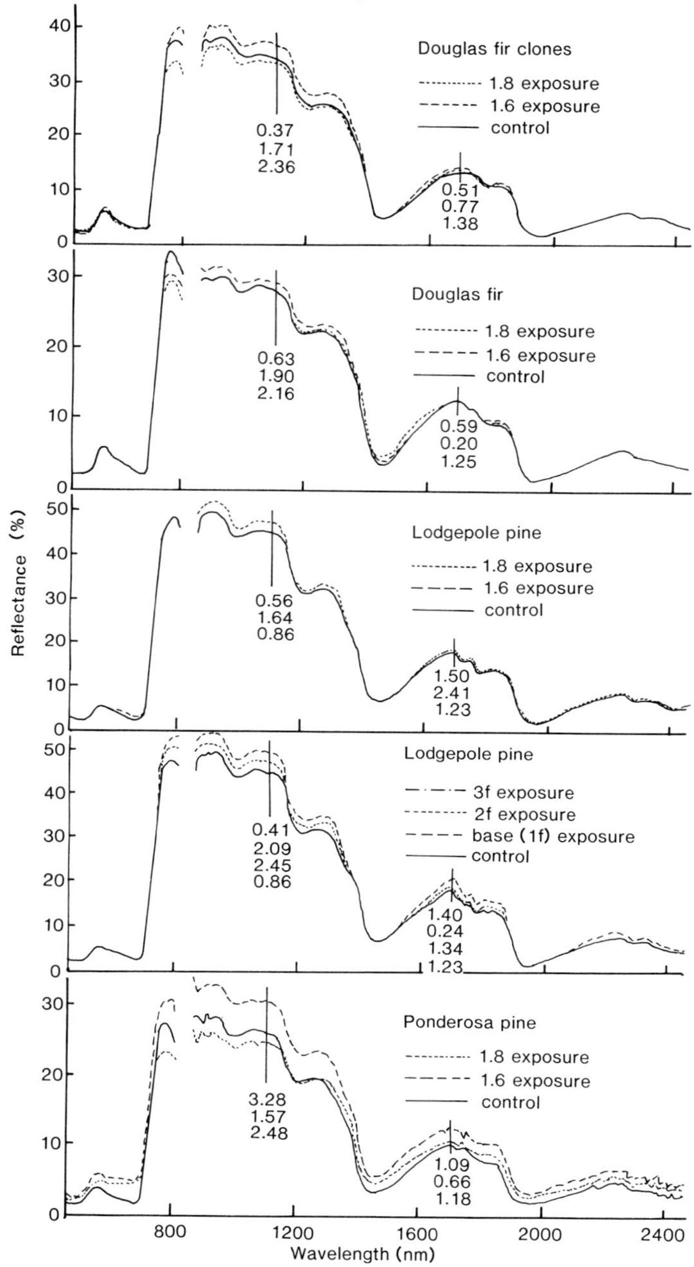

Figure 4.19 Mean spectra for three species for the 450-2450 nm wavelength interval. One standard deviation of the mean for each ozone treatment is shown at 1100 and 1700 nm. The number of live seedlings per tray varied between 25 and 30. Treatment means are from the open-top chamber means. Reprinted with permission from Ustin and Curtiss (1990), Copyright 1990, Pergamon Press Ltd.

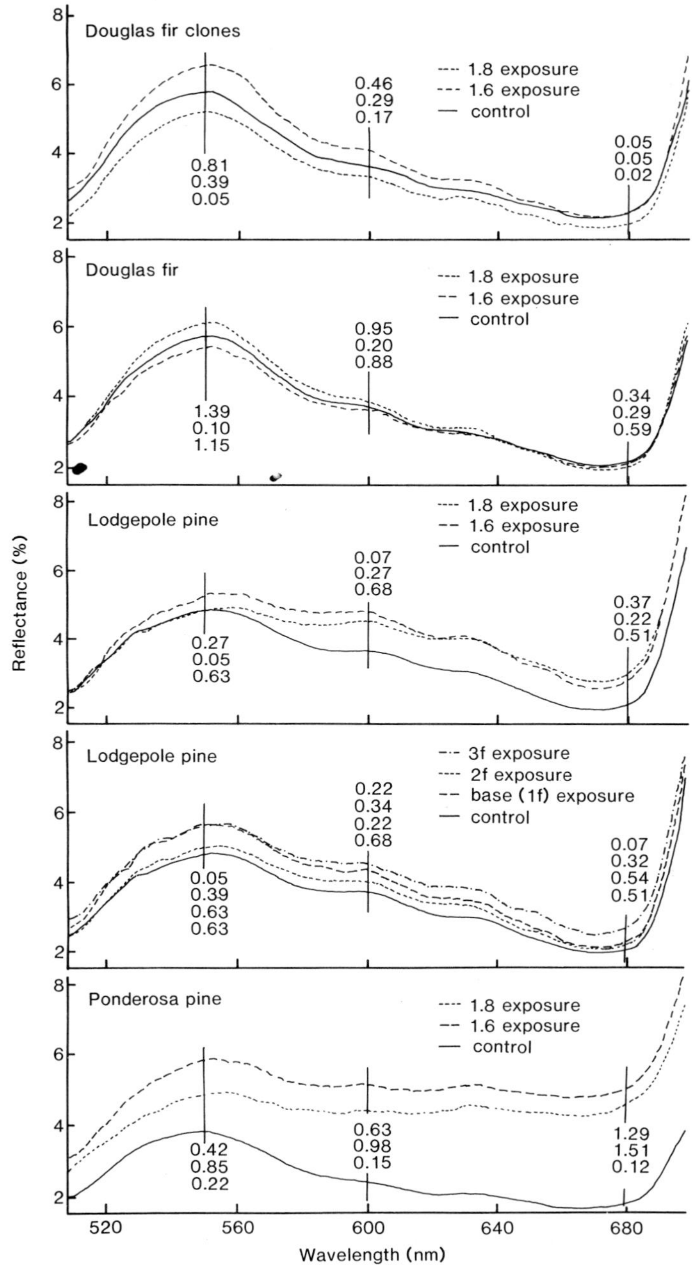

Figure 4.20 Mean spectra for three species for the 505-705 nm wavelength interval. One standard deviation of the mean for each ozone treatment is shown at 550, 600 and 680 nm. Other details as in Fig. 4.19. Reprinted with permission from Ustin and Curtiss (1990), Copyright 1990 Pergamon Press Ltd.

on SPOT and thematic mapper data [seven bands on LANDSAT TM) to assess damage to forests. Much of the early work concerned insect defoliation (e.g. Beaubien and Jobin, 1974; Dottavio and Williams, 1983, Hall *et al.*, 1983; Nelson, 1983) or fungal problems and these applications continue to produce useful results (e.g. Vogelmann and Rock, 1989; Joria *et al.*, 1991; Nakane and Kimura, 1992). Multispectral scanning data have been used to assess the extent of the impact of the Sudbury smelters (see Chapter 1) (Pitblado and Amiro, 1982; Allum and Dreisinger, 1987), albeit on a very coarse scale, and to look at the decline of maritime pine and stone pine in Italy (Conese *et al.,* 1989). Thematic mapping has also provided an indication of leaf area index in some species (Herwitz *et al.*, 1989).

Elvidge *et al.* (1989) made a general comparison between satellite and aircraft imagery. They emphasized that the most useful data were multispectral, with data being available from 1972 to the present. The French SPOT satellite and the Thematic Mapper (a successor to LANDSAT) provided more useful data than LANDSAT, although SPOT was limited by the absence of a mid-infra-red band on the HRV sensor. The Japanese Maritime Observation Satellite, although only having a ground resolution of 50 m, had similar spectral bands to the LANDSAT MSS, but has not been used much in forest studies. The altitude of satellite-borne sensors (LANDSAT–4 onwards at 707 km, SPOT at ca. 830 km) presents a problem as there is an increased amount of atmospheric absorption. This reaches maximum levels at 0.95, 1.1, 1.4, 1.8, 2.6, 2.8 and 6.2 μm for water vapour, 4.3 and 15 μm outwards for carbon dioxide and 2.9 and 9.6 μm for ozone. These would not affect measurements of pigment changes, but could affect other assessments.

Satellite imagery applications

Ciesla and Hildebrandt (1986) found that areas characterized by trees with less than 25% defoliation could be separated from areas with greater levels of defoliation, but shadowing presented a major problem. Kadro and Kuntz (1986) also distinguished asymptomatic and symptomatic trees using MSS data, but the results were only clear in areas where the extent of damage was relatively uniform. In addition, problems were encountered in the critical 25–60% damage category, with trees in this category frequently being misclassified.

The potential of multispectral scanning in the assessment of forest decline has been examined in detail by Rock and co-workers (e.g. Rock *et al.*, 1986, 1987, 1988a, b, 1989; Rock and Vogelmann, 1986; Vogelmann and Rock, 1986, 1988). There are several sources of such data although Rock and co-workers used LANDSAT Thematic Mapper data. The bands used in the studies were centred at 0.56 μm (Band 2; 0.52–0.60 μm), 0.66 μm (Band 3; 0.63–0.69 μm), 0.83 μm (Band 4; 0.76–0.90 μm) and

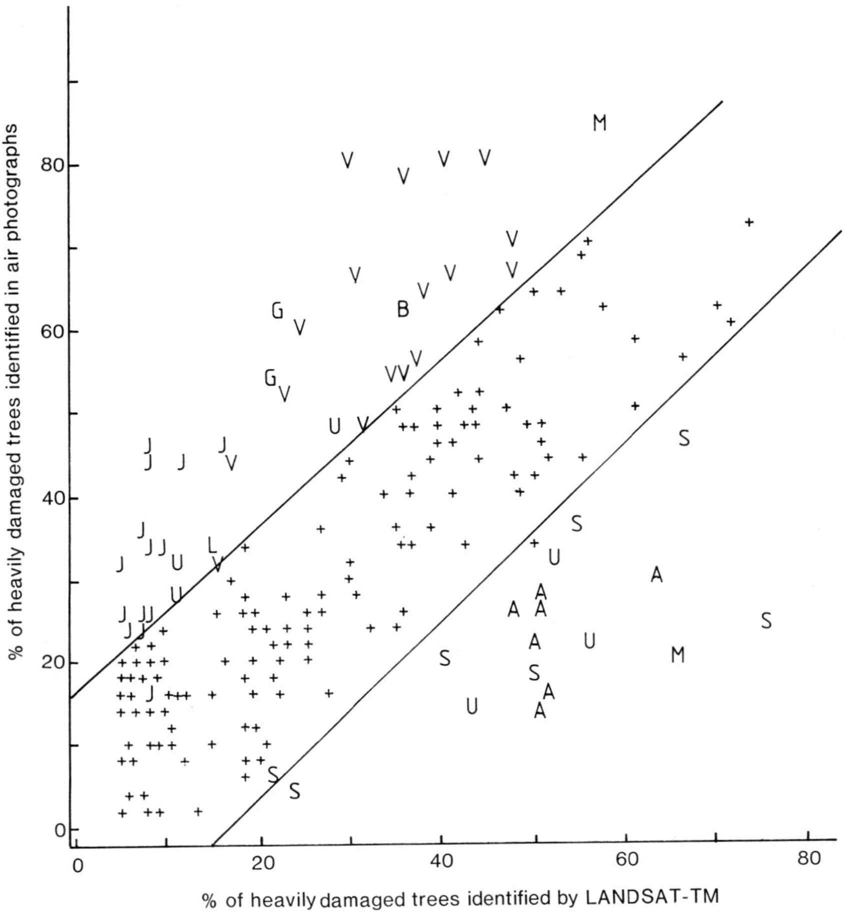

Figure 4.21 Comparison of the results of air photo interpretation and LANDSAT TM classification for 188 Norway spruce in the Harz Mountains, Germany. From: Kenneweg (1989). Points with a maximum of ±15% variation are indicated by a +. Stands with greater differences are classified as J: young stands; V: chlorotic stands; G: stands with vigorous grass growth; L: stands containing some hardwoods; S: open canopy with no grass; A: old stands; M: mixed pixels dominant and U: unexplained.

1.65 μm (Band 5; 1.55–1.75 μm). Data were calibrated using ground-based reflection spectra, obtained from a hand-held ratioing radiometer.

Bands 4, 3 and 2 in red, green and blue colour order were used to produce a false-colour mimic image. In this combination, hardwood forests were represented by red areas whereas fields and populated areas were represented by white and bluish areas. Coniferous forests were represented by dark reddish-brown areas. This particular band combination was found to be useful for separating coniferous and hardwood forests, but the combination gave no information on the health of the coniferous forests.

Health was assessed using the TM band 5 (1.65 μm) to TM band 4 (0.83 μm) ratio assigned to red, band 5 to green and band 4 to blue. Using this combination, areas of damaged forest appeared orange or red, whereas undamaged forest appeared as blue or bluish-green. Interestingly, it seemed that damage associated with fir waves could be distinguished from other damage types using colour density slice images. The damage to red spruce was associated with three components of the spectral signature (Rock *et al.*, 1989):

- a blue shift of the chlorophyll well/red edge;
- a drop in the reflection of the near infra-red (NIR) plateau; and
- a relative increase in the shortwave infra-red (SWIR) reflection values.

Koch and co-workers (Koch, 1987; Koch and Kritikos, 1984; Koch *et al.*, 1984, 1989), in a particularly useful set of studies, also looked at the ratio of TM bands 5 to 4. After some manipulation, it was possible to define stand ages, density and species composition, but they were less successful with the definition of defoliation levels. However, with increasing levels of defoliation, the degree of canopy closure decreased, resulting in increased reflection from the ground vegetation and increased shadows, both of which could be discerned from the imagery. Similar results have been reported from the United States, where reflection in TM bands 3, 4 and 5 have been related to the leaf area index of the forest (Spanner *et al.*, 1990). Of particular importance, the angle of incidence had a major effect, suggesting that the area that can be reliably assessed on a single image is strictly limited.

Airborne multispectral scanning

Airborne multispectral scanning data are now available and these provide a much better spatial resolution than satellite data. This has the advantage of enabling the disruptive effects of shadows to be more easily controlled. In addition, the amount of atmospheric noise tends to be less and can be more easily accounted for (Spanner *et al.*, 1990). Further advantages include control over the time of the over-passes and the greater number of available wavelength bands (Leckie and Ostaff, 1988). Airborne fluorescence line imagery has provided particularly useful data (Reid, 1987; Hoshizaki *et al.*, 1988). It works in 288 narrow bands, allowing detection of the blue shift of the chlorophyll and the red edge in the canopy. The programmable multispectral imager which collects the data appears to be extremely flexible and is likely to be of considerable value.

Herrmann *et al.* (1988) examined the potential of an airborne imaging spectrometer (AIS–2) for assessing forest decline in the Haunstetter forest of Bavaria. They were particularly interested in the near infrared/short wave infrared range (0.70–2.40 μm), which past studies had shown to have considerable potential in decline studies (Rock and Vogelmann, 1986; Rock *et al.*, 1986, 1987; Vogelmann and Rock, 1986, 1988). The forest in question was composed of 60% Scots pine, 20% Norway spruce and 20% broadleaves. All species showed signs of defoliation, with the Norway spruce being the most affected, and the Scots pine showed severe chlorosis. Several other instruments were also used. Herrmann *et al.* (1988) reported data from the airborne imaging spectrometer (AIS–2), which collects data in 128 channels with a band width of 10.6 nm. The wavelength range used in the study was 0.8–2.2 μm. The ground resolution of the AIS–2 scanner was 9.4 m with a flying height of 4500 m. The Daedalus airborne thematic mapper scanner (ATM scanner) collected data in 11 channels, from 0.42–13.00 μm.

The study suggested that the reflection properties associated with increasing levels of damage may differ between Norway spruce and Scots pine. This was probably related to the nature of the damage, with chlorosis being more obvious in the pine than in the spruce. Both visible and near infrared and short-wave infrared reflection features increased with increasing levels of damage in Scots pine. This contrasts with studies made of Norway spruce and red spruce, where near infrared reflection tends to drop with increasing levels of damage. Problems were encountered as a result of the general thinness of Scots pine canopies, which meant that branches contributed significantly to the reflectance.

Several studies have looked at Bendix 11-channel scatter data, the scanner having been modified to simulate the data collected by the LANDSAT Thematic Mapper (Koch and Kritikos, 1984; Kritikos *et al.*, 1985; Kadro and Kuntz, 1985). The sizes of the ground areas covered by each pixel varied, ranging from 0.75 m at a flying height of 300 m to 10 m on the ground at 4000 m. Only the 0.75 m resolution enabled individual crowns to be assessed and, with data obtained from higher flights, atmospheric influences caused problems with the interpretation of the data.

A study by Leckie and Ostaff (1988) was primarily concerned with evaluating the extent of defoliation in spruce fir forests caused by the spruce budworm. Eleven channel multispectral scanning data were obtained using a Daedalus 1260 multispectral scanner mounted on DC–3 aircraft. The scanning was undertaken at 3650 m altitude, giving an image with a spatial resolution of approximately 9 m. This scale was suitable for their particular investigation as although each pixel consisted of several trees, damage tended to be grouped. Bands with wavelengths of 0.39–0.42 μm and 0.42–0.45 μm were not used owing to their low signal-to-noise ratios. The near infra-red bands at 0.77–0.90 μm (Band 9), 0.68–0.78 μm (Band 8) and 0.87–1.04 μm (Band 10) were the best for distinguishing current defoliation and the green bands at 0.50–0.55 μm (Band 4) and 0.55–0.60 μm (Band 5) also provided useful information. The three infra-red bands were highly correlated and the inclusion of data from more than one of these bands did little to improve the quality of the data. They concluded that only three levels of current defoliation could be reliably distinguished (heavy [>35%], light [0–35%] and healthy), with confusion between moderate (35–70%) and severe (>75%) defoliation being particularly marked.

The Daedalus 1240/1260 scanner was also used by Felten *et al.* (1985). Various flying heights were tested, ranging from 330 m to about 2500 m, with ca. 830 m being the most effective. Differential reflectances according to the amount of defoliation were identified, with reflectance in the infrared ranges being proportional to the amount of damage. The ratio of orange (600–650 nm) to infra-red (700–1100 nm) reflectance was found to be useful, although shadowed trees were misclassified as damaged, making a masking procedure necessary.

Comparison of satellite and airborne multispectral scanning

Westman and Price (1988a) compared LANDSAT Thematic Mapper data with airborne Thematic Mapper simulator data in relation to oxidant-induced vegetation injury in southern California. A Daedalus 1268 Thematic Mapper Simulator was used for the airborne studies. The data revealed that although visible symptoms were present, these were not sufficiently marked to register as spectral changes above the background noise. These results are important because the forests of southern California have some of the most obvious symptoms of air pollution to be found anywhere in the world, outside those areas affected by point sources. The failure of the Thematic Mappers to identify any clear-cut spectral changes in this area suggests that they are unlikely to be of great use in uncontrolled studies of chronic pollution impacts. However, the potential for more general studies of stress is clear.

Summary

Despite the successful studies described above, multispectral scanning has not been widely used. Smith (1988) suggested the following reasons for the lack of enthusiasm over the technique:

- stand classifications vary with stand age;
- stand classifications vary with tree density within the stands;

- the presence of more than one species within a stand complicates the interpretation;
- shadow may affect the scanning data;
- undergrowth and soil visible from the air may affect the data;
- dead trees and branches show up well, but dead trees may be removed during sanitation operations;
- the degree and distribution of the damage to the trees in the stands affects the reflection characteristics;
- good ground reference data are not always available; and
- complications arise from the directional reflection properties of coniferous stands and the geometric properties of the imagery.

The high cost of data collection, particularly from airborne multi-spectral scanners, is an additional factor that must be considered. If these problems can be overcome, then the role of remote sensing in surveys and monitoring is likely to increase dramatically.

Non-visual assessments of individual tree condition

Emphasis in the above sections has been placed on the use of visible symptoms of injury. However, there has been a great deal of interest in the use of other indices, some of which have been considered to be more 'quantitative' than visual assessments. Interest has also been expressed in the use of 'early warning' indicators, on the logical premise that biochemical and physiological changes should occur before the development of visible symptoms. Some of these changes will be ephemeral and of no direct relevance to tree health. However, others may be more important. In particular, biochemical and/or physiological changes may occur that affect the susceptibility of the tree to visible injury caused by other stresses. A well-known example of this is the increased winter injury in otherwise asymptomatic trees exposed to high concentrations of sulphur dioxide.

There are two main problems with the majority of studies that have been undertaken using invisible markers. Either the studies have been made with young trees under strictly controlled environmental conditions without any attempt to apply the results to the environmental conditions found in forests or field studies have been based on the assumption that a tree with a transparent crown is affected whereas one with dense foliage is not.

Mehlhorn *et al.* (1988) reported that only three out of 30 variables investigated showed significant relationships with the crown condition of Norway spruce exclusive of site effects. These were the difference between dry weight to fresh weight ratios of current and two-year-old needles, the rates of ethylene formation in two-year-old needles and the ratio of violaxanthin to antheraxanthin pigments in two-year-old needles. It is possible that further rigorous studies will provide some potential indicators in the future. What is more important, crown condition (as assessed by defoliation and discoloration) does not provide a good index of tree health and the results of Mehlhorn *et al.* (1988) may reflect this. A range of possible biomarkers is discussed below.

Cellular structure

The nature of cellular damage provides an important diagnostic tool in investigations of forest condition (Palta and Li, 1978; Soikkeli and Tuovinen, 1979; Carlson and Gilligan, 1983; Fink, 1988; Zöttl *et al.*, 1989). Different forms of air pollution and other types of stress may affect foliage in different ways. Acidic deposition directly affects foliage from the epidermis through the mesophyll to the vascular bundle. Gases affect the mesophyll first, especially close to the stomata. Mineral deficiency, on the other hand, first causes changes to the vascular bundle followed by changes to tissues closer to the surface of the foliage.

Some of these changes have been successfully used as indicators of stress. For example, Kärenlampi and Houpis (1986) described the cellular alterations in ponderosa pine following fumigations with sulphur dioxide. However, different types of stress may produce similar symptoms (e.g. Zwiazek and Shay, 1987), complicating any interpretation and statements of the uniqueness of a particular observation are rarely made.

Fink (1988) produced a classification of the different types of ultrastructural changes seen in needles in response to stress (Table 4.5). He concluded that most alterations were the result of pathological alterations caused by changes in, for example, membrane permeability or osmotic conditions. As the responses at a cellular level vary considerably between species and between situations, it is almost impossible to draw up any generalizations. Where relevant, cellular changes have been described in Chapters 2 and 3.

Photosynthesis

Portable gas exchange systems are being used increasingly to measure photosynthesis in the field. Alternatively, foliage samples can be collected and the rate of photosynthesis measured at intervals after the initial detachment. Both methods enable field investigations to be undertaken, but applications so far have only been in research studies, rather than in monitoring or surveys. The approach involving the detachment of shoots has the major disadvantage that the detachment causes stress and the results are therefore difficult to interpret in relation to intact shoots.

Respiration

A number of important physiological changes occur in response to specific types of stress. For example, several pollutants are known

Table 4.5 Different types of pathological damage to the chloroplasts

1. A normal, young chloroplast has a healthy mesophyll cell, showing a well-developed thylakoid system with definite grana.
2. Chloroplasts with large numbers of grana, typical of cells from shade leaves. Certain herbicides may have a similar effect.
3. Granulation of the stroma, which is a widely reported response to certain air pollutants.
4. Chloroplasts becoming crooked, typically towards the cell wall, with the development of thylakoid-free parts of the stroma caused by separation of the membrane phase from the stroma. Sometimes seen in response to ozone, nitrogen dioxide or acidic rain exposure.
5. Curling of the internal membranes, leading to an undulated appearance of the thylakoids. Seen in plants suspected as having been damaged by sulphur dioxide or ozone or by cold-treatments.
6. Shrinkage of the chloroplast leading to a stretching of the chloroplast envelope. Recorded following exposure to gaseous pollutants.
7. Swelling of the thylakoids by separation and dilatation of their double membrances. Can be caused by pollutants or mineral deficiency.
8. Formation of vesicles and vacuoles within the chloroplast or between the double membrane of the chloroplast envelope. Seen following exposure to sulphur dioxide, but is more commonly associated with biotic damage.
9. Dilatation of the parallel alignment of the thylakoid systems and the formation of electron-transparent gaps in the swollen stroma. Can be caused by frost, artificial dehardening, potassium deficiency or air pollutants.
10. Similar to 9, but accompanied by the development of large starch grains. Associated with magnesium deficiency, damage by photochemical oxidants and drought stress.
11. Underdeveloped chloroplast typically caused by nitrogen deficiency.
12. Plastids with reduced membrane systems lacking or having poorly-developed grana. Associated with air pollutants, etiolated leaves or phosphorus or iron deficiency.
13. Pronounced shortening of the thylakoids, often being reduced to single isolated grana. Associated with manganese deficiency.
14. Accumulation of electron-dense plastoglobuli. Normally, this is a sign of senescence and it has been seen in normally senescing foliage and also in foliage senescing prematurely.
15. Dissolution and disruption of the chloroplast envelope and the inner membranes, leaving only the plastoglobuli at the sites of the former chloroplasts. Seen in cases of acute frost damage and following intensive UV radiation.

As listed by Fink (1988)

It should be noted that many of the responses listed above were identified in plants other than trees.

to affect respiration (Black, 1984, 1985; Ballantyne, 1984; Dizengremel and Citerne, 1988; Keller, 1991), often through effects on stomatal closure. Consequently, respiration and metabolism can provide important information on the condition of the tree (Black, 1985; Brenninger and Tranquillini, 1983; Keller and Häsler, 1986). However, respiration is extremely difficult to measure as it is so small on a leaf/needle or shoot basis.

As with other processes, measuring the respiration of a mature tree presents rather more problems that measuring the respiration of a seedling or tissue under controlled laboratory conditions. However, it is possible, as Cropper and Gholz (1991) and Gholz and Cropper (1991) have shown. They used a controlled temperature respiration chamber system, with respiration being measured by an infra-red gas analyser. Measurements were possible at a variety of positions in the canopy through the use of 20 m high towers. This again was a research application, and it will be some time before the techniques can be applied to monitoring programmes.

Biochemical substances

Many biochemical substances could be assessed in surveys or monitoring programmes. These range from substances involved in basic metabolism to substances specifically generated to detoxify materials introduced to or generated within the plant. Some examples are given below, although a deliberate attempt has been made to restrict the review to studies involving trees. Further information can be obtained in a number of specific reviews (e.g. Jäger, 1982; Jäger *et al.*, 1986; Schulte-Hostede *et al.*, 1988).

Myo-inositol

Myo-inositol is an important precursor, along with other molecules, of cell wall compounds and membrane lipids. In addition, ascorbic acid is formed from *myo*-inositol. Landolt *et al.* (1988, 1989a) found that this substance decreased significantly in first-year Scots pine needles fumigated with ozone, although a similar response was not evident in Norway spruce or silver fir.

Detoxification systems

Plants contain a number of substances that can be used to detoxify phytotoxic substances. Some of the most toxic substances are the free radicals, which can be generated by pollutants such as sulphur dioxide, the oxides of nitrogen and ozone. At the cellular level, several enzymes exist which detoxify free radicals, including superoxide dismutase, peroxidase and catalase (Richardson *et al.*, 1989). In addition, a number of non-enzymatic defensive compounds exist, including glutathione, α-tocopherol, ß-carotenes and ascorbic acid (Di Giulio and Richardson, 1987).

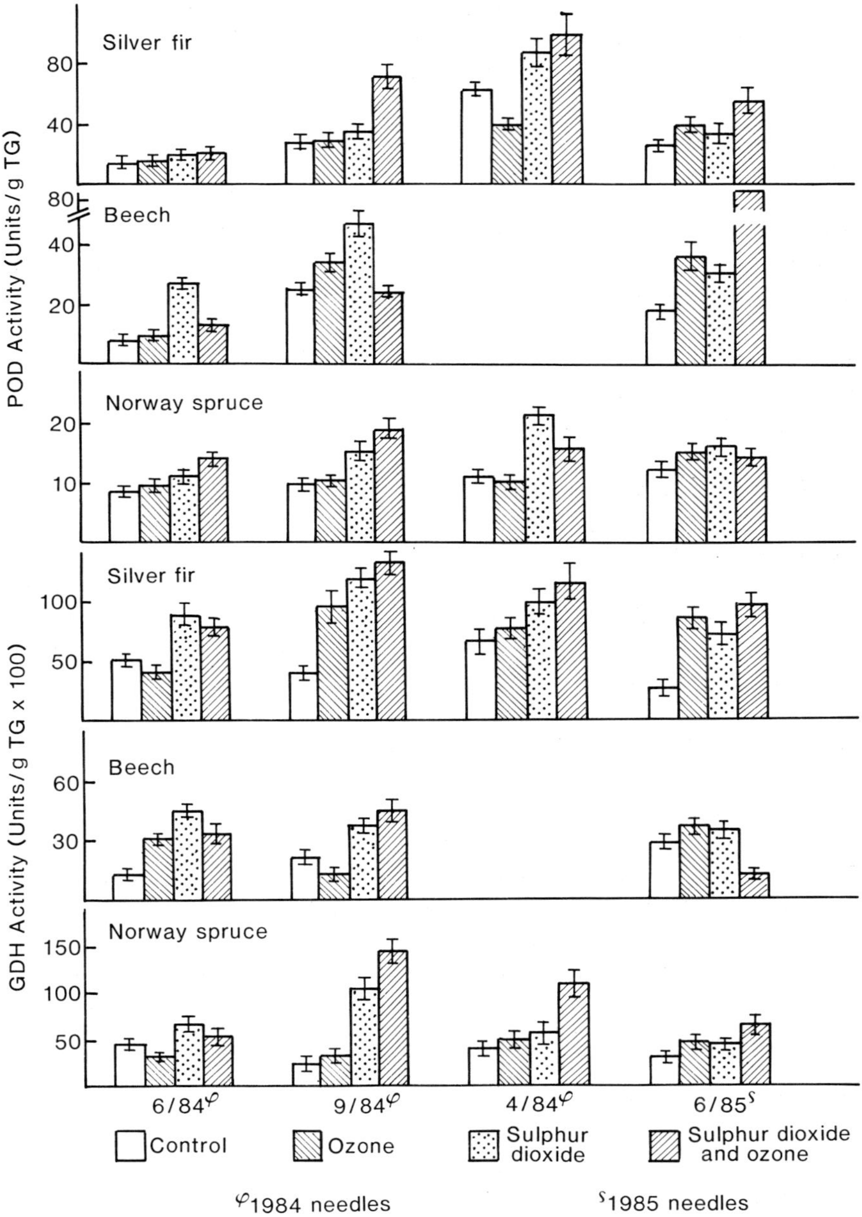

Figure 4.22 Peroxidase (POD) and glutamate dehydrogenase (GDH) contents of the foliage of several species exposed to ozone, sulphur dioxide and ozone and sulphur dioxide combined. From: Bender *et al.* (1986).

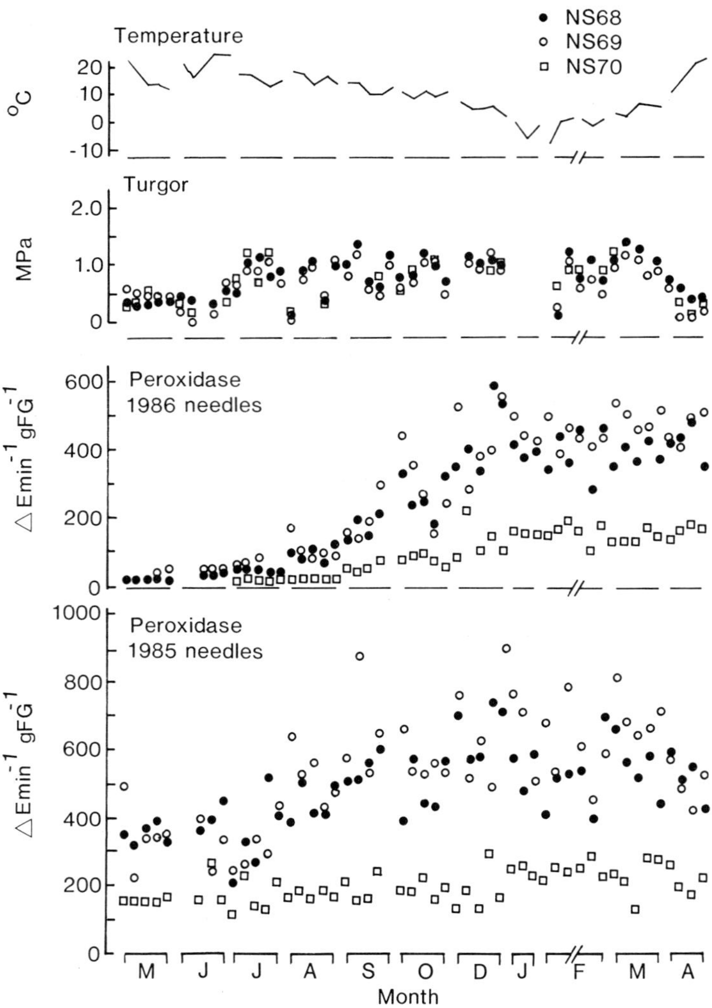

Figure 4.23 Seasonal changes in 1986 of total peroxidase activity and turgor in needles of three adjacent trees near Ratzeburg, Germany. Peroxidase activity was measured in 1986 in the current-growth needles (centre) as well as in the needles formed in 1985 (below). Each value is the median of three separate measurements, and each measurement was repeated three times. Turgor: each value is the median of three measurements for each tree. From: Ludewig *et al.* (1989).

Peroxidase activity

There has been considerable interest in the use of peroxidase activity as an indicator of stress although the interpretation of the specificity of the results has been a recurring problem. Peroxidases are enzymes required to oxidize peroxides that occur in cells as a result of ageing or stress. Under experimental conditions, ozone and sulphur dioxide have been shown to increase the peroxidase activity in the needles of Norway spruce and several other species (Fig. 4.22) (Bender *et al.*, 1986, 1989; Schulz, 1986). Manes *et al.* (1988) found increasing levels of peroxidase in maritime pine associated with simulated acidic rain treatment, and Patton and Garraway (1986) found increased peroxidase associated with ozone-induced foliar necrosis on the leaves of hybrid poplars.

From the above findings, peroxidase activity might be considered to be a useful index of pollution stress. However, in a series of detailed experiments on peroxidase activity, Keller (1981a, b, 1984a), Mejnartowicz (1984) and others found that peroxidase activity was a poor diagnostic tool because of its lack of specificity. A likely reason for this was the wide range of stresses that can stimulate peroxidase activity, together with substantial diurnal, seasonal (Fig. 4.23) and plant-to-plant variations in the response of peroxidase levels to particular stresses (Huttunen, 1980; Castillo, 1986; Patton and Garraway, 1986; Ludewig *et al.*, 1989). In addition, there are problems with the measurement of the activity as phenolics, organic acids and various ions can interfere with the determinations. Some of these problems can be overcome by the use of extracellular peroxidase activity as an index (as opposed to internal activity). Separation of different types of peroxidase may also be helpful, since phenylenediamine peroxidase has been found to respond differently to ascorbate peroxidase (Manderscheid and Jäger, 1990).

The results of Manderscheid and Jäger (1990) further complicate any interpretation. They reported the results of an open-top chamber experiment involving 16-year-old Norway spruce trees at Edelmannshof in the Swabian-Frankonian Forest of Germany. Trees were grown in either ambient or charcoal-filtered air for 17 months (at the time of reporting). Peroxidase activity was dependent on the position in the crown and the time of sampling. There appeared to be no influence of pollution on ascorbate peroxidase and there was only a weak effect on phenylenediamine peroxidase.

An area still requiring further investigation is peroxidase activity in the roots of trees under stress (Markkola *et al.*, 1990). Positive responses of peroxidase activity to several potential indicators of root stress were found by Markkola *et al.* (1990), but the amount of scatter was very high. Penka and Cervená (1985) recorded increased peroxidase activity in needles from mature Norway spruce in Czechoslovakia, which they interpreted as indicating latent damage. Consequently, while it might be used as a general indicator of root stress, its specificity to pollutants is questionable.

Castillo *et al.* (1989) looked at the effects of ozone and sulphur dioxide on the composition of the apoplast of current-year needles of four-year-old Norway spruce, under experimental conditions. Exposure to ozone (200 μg m^{-3} for 12 weeks) inhibited ascorbate and guiacol peroxidase activities in the intercellular washing fluid, but increased the contents of ascorbic acid and protein. Sulphur dioxide (70 μg m^{-3} for 10 weeks) increased ascorbate peroxidase activity and the contents of ascorbic acid and protein.

These problems all suggest that the value of peroxidase activity to a forest health monitoring programme is limited. It provides a useful index of stress, but its lack of specificity severely reduces its potential value.

Superoxide dismutase

Experimental fumigation of Norway spruce with high levels of ozone caused an increase in superoxide dismutase in the intercellular fluid (Castillo *et al.*, 1987). A similar effect has been documented in Scots pine needles exposed to sulphur dioxide (Schulz, 1986). However, Hausladen *et al.* (1990) recorded reduced superoxide dismutase activity in the needles of red spruce fumigated with ozone, suggesting that responses are inconsistent between species.

α-tocopherol

α-tocopherol increased in Norway spruce and silver fir fumigated with ozone, sulphur dioxide or a combination of the two (Mehlhorn *et al.*, 1986). However, its value as a bioindicator is unknown.

Ascorbic acid

The foliage of plants contains substantial amounts of ascorbic acid, with the crowns of conifers in well-lighted positions having particularly high contents (Keller and Schwager, 1977). The acid is a reductant, responsible for the photoreduction of protochlorophyllide, and is also an important electron donor in the photosynthetic process. It is mainly concentrated in the chloroplasts and changes to these will be reflected in the ascorbic acid content.

Ascorbic acid can be higher in chlorotic Norway spruce needles than in green needles (Oßwald and Elstner, 1987; Oßwald *et al.*, 1987, 1990). However, as chlorotic needles are generally located on shoots exposed to the sun and ascorbic acid contents are higher in sun-exposed needles, the relationship between chlorosis and ascorbic acid content may be coincidental. The largest increases were found in two- to three-year old needles. Keller and Schwager (1977) found that ascorbic acid contents of needles in laboratory-grown trees peaked in the third or fourth season of the needles' life. Ascorbic acid contents dropped as a result of fumigation with sulphur dioxide and Keller (1981a) recorded lower contents of ascorbic acid in the needles of Norway spruce fumigated with sulphur dioxide during winter. Similarly, Castillo *et al.* (1987) found decreases in ascorbic acid contents of Norway spruce fumigated with ozone. However, Kunert and Hofer (1987) and Castillo *et al.* (1989) reported increased ascorbic acid contents in Norway spruce in response to ozone or sulphur dioxide fumigation. These inconsistencies suggest that some factor other than air pollution has a strong influence on the levels of ascorbic acid in trees.

Glutathione

In the field, glutathione behaves similarly to ascorbic acid, being higher in chlorotic needles than in green needles (Oßwald and Elstner, 1987; Oßwald *et al.*, 1987, 1990). Kunert and Hofer (1987) recorded increases in glutathione in Norway spruce, silver fir and beech foliage following exposure to relatively low doses of sulphur dioxide and ozone and advocated the use of glutathione reductase as a diagnostic tool. Increased glutathione in silver fir and Norway spruce exposed to ozone, sulphur dioxide or a combination of the two was reported by Mehlhorn *et al.* (1986), as shown in Fig. 4.24. A similar trend occurred in red spruce exposed to ozone (Hausladen *et al.*, 1990); they also showed the

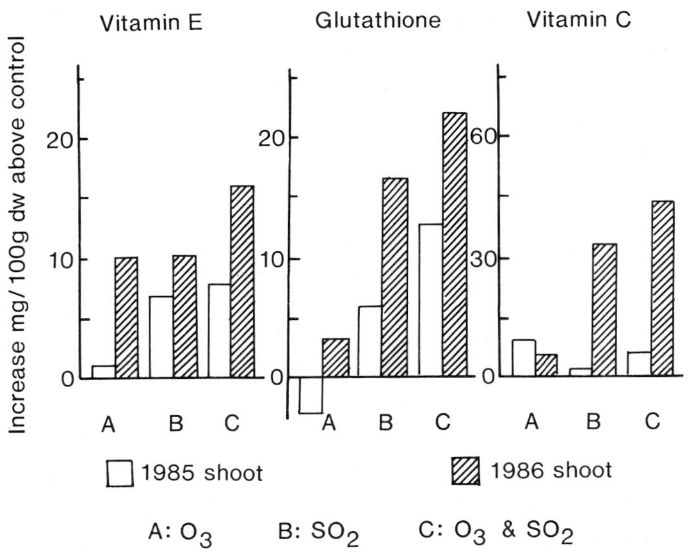

Figure 4.24 Increase of antioxidants above control in 1-year-old (1985) and 2-year-old (1984) needles of Norway spruce exposed in open-top chambers to different air pollutants. One-year-old control needles that were treated with charcoal-filtered air had 21.5 ± 1.0 mg/100 g dry wt of total glutathione. Two-year-old control needles had 22.4 ± 1.7 mg/100 g dry wt of vitamin E, 143 ± 7.2 mg/100 g dry wt of vitamin C, and 18.9 ± 0.9 mg/100 g dry wt of total glutathione. Needles were collected in October after exposure to ozone. The data represent the means ± SE of five samples of four different trees. From: Mehlhorn *et al.* (1986). Reproduced with the permission of the copyright owner.

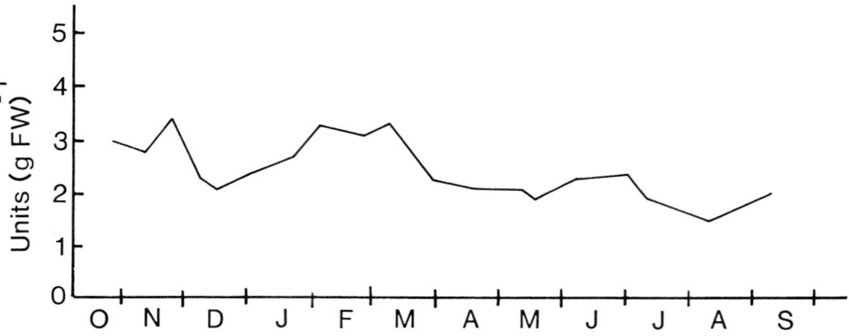

Figure 4.25 Seasonal (October-September) variations in the glutathione reductase activity in current needles of Norway spruce growing at Stubalpe, Austria (1700 m ASL). From: Grill *et al.* (1988).

importance of distinguishing between total glutathione and oxidized glutathione contents. In contrast, Manderschied and Jäger (1990) found that glutathione contents were the same in trees exposed to ambient and charcoal-filtered air in open-top chambers.

Grill *et al.* (1988) examined the water-soluble thiol contents of Norway spruce needles. Water-soluble thiol contents consisted mainly of glutathione, with only a little cysteine being present. The content increased with altitude, and young trees transplanted from 400 m ASL to 1700 m ASL showed substantial increases in the thiol contents. The results showed that water-soluble thiol contents were strongly affected by climatic conditions and were therefore of little diagnostic value in pollution-related studies. In addition, marked seasonal variations in the thiol contents of the needles were found (Fig. 4.25). However, repeated sampling at individual sites may provide information on climatic stress.

Other enzyme and amino acid contents

If particular types of air pollution have any distinctive effects on tree species, then these should be apparent in the enzyme composition. Rothe *et al.* (1988) looked at a variety of enzymes in Norway spruce from the Kleiner Feldberg, Germany. Trees with more than 60% defoliation had reduced activities of a number of enzymes, including α-galactosidase, α-mannosidase, glucose6-phosphate DH, glucose6-phosphate isomerase, isocitrate DH, malate DH, 6-phosphogluconate DH, shikimate DH and peroxidase. A similar multivariate approach to establishing variations in enzyme content has been used by Schulz (1989).

Looking at ten 80-year-old Norway spruce from the Taunus Mountain Range, Germany, Zedler *et al.* (1986) found that amino acid contents of trees without discoloration increased with increasing levels of needle loss. The increases in amino acid contents were attributed to elevated uptake of ammonia by the trees. Schmeink and Wild (1990) found increased levels of amino acids, including glutamine, asparagine and proline, in trees with 25–60% needle loss compared to trees with less than 10% needle loss.

Arginine, histidine, tryptophan and putrescine

Villanueva *et al.* (1986) examined the arginine contents of Norway spruce needles from the Massif du Taenchel and Massif du Donon areas of the Vosges Mountains, France. They found that the contents of arginine were higher in healthy trees than defoliated ones and speculated that this might be used as a diagnostic tool. However, arginine contents in Norway spruce are very variable and do not reliably reflect the state of health of the tree (Schmeink and Wild, 1990). In a later study, Villanueva and Santerre (1989) found differences in the histidine and tryptophan contents, with evidence that the levels of metabolites were higher in defoliated trees than in healthy ones. Putrescine was also much higher in the needles of defoliated trees (Villanueva and Santerre, 1991).

Glutamic acid, aspartic acid, glutamine and asparagine

Various amino acids are known to increase (in beech) in response to nitrogen fertilization, including glutamic acid, aspartic acid, glutamine and asparagine (Balsberg Påhlsson, 1992). However, there is insufficient information about the natural dynamics and variability of these particular substances for them to be of much use as biomarkers at the present time. Their dynamics in other species are unknown.

Adenine nucleotides and pyridine nucleotides

Working at Kälbelescheuer in the southern Black Forest (an area affected by Type 1 Norway spruce decline), Keil *et al.* (1988) looked at the contents of adenine nucleotides (ATP, ADP:AdN) and reduced (NADH, NADPH) and oxidized (NAD, NADP) pyridine nucleotides in the needles of healthy and defoliated Norway spruce. As the age of the needles increased, there was a significant decrease in the total AdN-pool and a simultaneous increase in the redox levels of both pyridine nucleotide systems (NADH/NAD; NADPH/NADP). This trend was evident for both classes of needles. However, the ratios of ATP:ADP, NADH:NAD and NADPH:NADP were higher in needles from symptomatic trees in comparison to needles from asymptomatic trees. They interpreted these patterns as showing accelerated senescence, and found a good correlation with the degree of damage, although they did not state their sample size. Seasonal and annual changes in the ratios suggest that they might not be particularly useful in early diagnosis studies, nor is there sufficient information to assess their specificity to particular causes of damage.

Nitrate reductase

If elevated nitrogen levels are part of the explanation of forest decline, then the occurrence of nitrate reductase might be a useful biomarker. Guerrero *et al.* (1981) have pointed out that the reduction of nitrate to nitrite in foliage is controlled by the enzyme nitrate reductase, with levels of activity being controlled by the supply of nitrate (Beevers and Hageman, 1969). Because of this, Rowland *et al.* (1985) and Norby (1989) proposed that the activity of the enzyme might be used as an indicator of the extent to which foliar-absorbed nitrogen oxides are assimilated by trees. However, as nitrate reductase activity can be induced by other forms of nitrogen, notably ammonium, and can also be induced by soil-borne nitrogen (Klumpp *et al.*, 1989; Tjoelker *et al.*, 1992), its use as an indicator of foliar-absorbed nitrogen oxides is questionable. In addition, nitrate reductase shows marked seasonal variations and is strongly affected by ambient temperatures (Pietiläinen *et al.*, 1991). There are several uncertainties surrounding the differences in nitrate reduction mechanisms and locations in different species, and these need to be resolved before the technique can be widely applied.

pH of foliar tissue

The pH of foliar tissue has been examined in Norway spruce and silver fir in the field (Garrec *et al.*, 1988, 1989). No differences were noted between asymptomatic and symptomatic trees. However, sulphur dioxide and other pollutants are known to alter the buffering capacity of needles (Grill, 1971; Nieboer *et al.*, 1984) and the phenomenon may provide a non-specific indicator of stress in some situations.

Fatty acid composition

Fangmeier *et al.* (1990) looked at the effects of ozone fumigation on the fatty acid composition of loblolly pine foliage. They found a significant effect, with reduced $C_{18:3}$ and increased $C_{18:2}$ and $C_{18:1}$ contents following fumigation. Fatty acid contents varied with both needle age and tree age and there was, in addition, some indication that the genetic composition of the trees had an effect. This severely limits the use of fatty acids as a biomarker.

Protoplast composition

Sigal *et al.* (1988) investigated the effects of ozone and acidic deposition on the protoplasts of loblolly pine. They recorded decreased fluorescences for esterase activity, protein, neutral lipids and RNA following treatment with above-ambient concentrations of ozone. However, their value as bio-indicators is unknown.

Foliar terpene composition

The concentrations of individual terpenes may provide an index of stress. However, Merk *et al.* (1988), Schönwitz *et al.* (1990a) and Chang and Hanover (1991), among others, have shown that these are extremely variable in tree populations, severely limiting their use as bioindicators. Even in clonal material, terpene composition varies with needle age and the position of the needle within the crown. In addition, the terpene contents of needles change dramatically during the course of the growing season (Schönwitz *et al.*, 1990b), making the time of sampling critical. Consequently, their potential use in surveys of forest condition and forest monitoring is very limited.

Electrical conductance

Injury to plant tissues often results in the leakage of electrolytes from cells, and the conductivity of tissue diffusates is a widely used index of, for example, frost injury. Unfortunately, as electrolyte leakage is a general response to stress, it cannot be used to identify specific problems and it is therefore of rather limited use outside experimental studies (e.g. Feiler, 1985; Keller, 1986).

Experiments with a variety of species have revealed that there was no relationship between the sensitivity of individual trees to sulphur dioxide and the electrical impedance of the shoots (Bialobok, 1989). The relationships with other pollutants are unknown.

Transpiration

Transpiration measurements are now conducted routinely in some physiological studies (e.g. Benner *et al.*, 1988; Ellsworth and Reich, 1992). Typically, they reveal disturbances to the stomatal regulation of needles in affected trees (see Chapter 3). Such disturbances have been shown in fumigation experiments (e.g. Keller and Häsler, 1984, 1986; Selinger *et al.*, 1986) and the technique therefore offers potential in diagnostic studies. However, the stomata respond differently to different pollutants (Black, 1985; Winner *et al.*, 1988) and, as pollutants rarely occur alone, the identification of the cause of stomatal disruption under field conditions may be impossible.

Mineral nutrition

Mineral nutrition represents one of the most important diagnostic tools available to foresters. It has been widely used and a large volume of results is available. Consequently, this subject has been dealt with separately (see Chapter 9).

Transport and allocation of photosynthate

Assimilate levels

The dynamics of carbohydrates in plants are known to respond to pollutants (e.g. Koziol, 1984; Koziol *et al.*, 1988). As a result, carbohydrate levels have frequently been cited as a useful bioindicator in air pollution studies. The theoretical basis for this is the possibility that the mobilization of photosynthate in the needles of affected trees may be delayed due to a reduction in the assimilate transport capacity of the phloem, resulting in an autumnal increase in starch concentrations. Pollutants have been shown to affect carbon dynamics more through decreased translocation than decreased photosynthesis (McLaughlin *et al.*, 1982; Lorenc-Plucinska, 1986). An autumnal increase in starch concentrations has been documented in one (but not two other) areas in Germany affected by decline (Forschner *et al.*, 1989). However, until recently, little information was available on their daily or seasonal dynamics within individual trees growing in forest environments. Gholz and Cropper (1991) looked at the carbohydrate dynamics in mature slash pine. They found marked seasonality in the starch levels in all tissues, with the highest levels occurring in late winter. Sugars did not show this pattern and exhibited little seasonality. Conversely, Villanueva and Santerre (1991) recorded marked seasonality in the total acid-insoluble sugars in Norway spruce in France and Fromard *et al.* (1991) reported seasonal variations in total sugar contents in silver fir in the Pyrenees.

Fengel (1987) examined the sugar contents of the sap of freshly-felled Norway spruce, Scots pine and beech in a number of different defoliation categories. Trees were obtained from several sites in Bavaria, Germany. Considerable variation was present, irrespective of the degree of defoliation. However, defoliated Norway spruce and Scots pine tended to have reduced levels of fructose, glucose and sucrose in the sap. In beech, the sugar content of the sap of defoliated trees was higher than that of undefoliated trees. The nutrient contents of the sap are known to show major seasonal variations (Glavac *et al.*, 1990a, b) and similar variations in the sugar content are likely, throwing the index into further doubt as a diagnostic tool. The results of Fengel (1987) contrast quite markedly with those of Feiler (1985), Debus (1988) and Villanueva *et al.* (1988), who all found an increase in the soluble sugar content of needles from defoliated Norway spruce, possibly because of the seasonal variations described above. An alternative explanation is that the sugar contents of the sap and of needles are unrelated. Fengel (1987, 1988) found no relationship between the chemical composition of sap or lignin solubility and the degree of visible damage. As with many other indices, there are appears to marked seasonal variations in the carbohydrate composition of tree tissues (Fig. 4.26), presenting a major problem in any analysis.

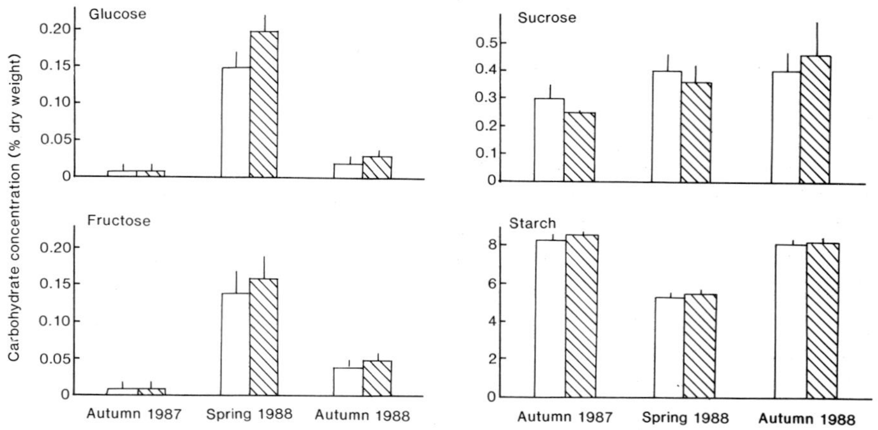

Figure 4.26 Concentration of glucose, fructose, sucrose and starch in root tissues as related to the crown condition. Means and SE are given. Open columns: trees with no crown deterioration (n = 27); hatched columns: trees with dieback (n = 28). From: Renaud and Mauffette (1991).

Assimilate transport

Many experimental studies have suggested that the transport of photosynthate and other substances may be affected by pollution. For example, Lorenc-Plucinska (1986) found that the inhibition of translocation in Scots pine seedlings exposed to sulphur dioxide was much greater than the inhibition of photosynthesis.

The production, movement and storage of assimilates represents an area of intensive research. It is likely that techniques will be developed which can be used in monitoring studies. However, at present, no reliable methods exist.

Relationships with other organisms

Trees under stress undergo altered relations with their fungal and bacterial symbionts (e.g. Jansen *et al.*, 1988; McCool, 1988; Marx and Shafer, 1989), pathogens (Treshow, 1975; Dowding, 1988; Smith, 1990) and insect pests (Flückiger *et al.*, 1988). The mycorrhizal associations appear to be particularly important and although much is known about them, there has been surprisingly little work done in this area in relation to forest health (Jansen, 1988). One of the greatest difficulties is the identification of the mycorrhizae, with over 5000 ectomycorrhizal species having been described. There are very few people who can reliably make the identifications, and this has severely restricted any large-scale study as part of a monitoring programme. Within the present context, the most important group is the vesicular-arbuscular mycorrhizae (VAM). The fungi forming VAM belong to the family Endogonaceae and the individual species within this family show very little host specificity. They play an important part in the nutrient uptake of trees, but may also be important in enhancing water uptake and increasing resistance to heavy metal toxicity and drought (Marx and Shafer, 1989). These latter factors may be very significant within the context of forest health.

As many microbial species can occur on a single length of root, there is potential for the use of indices of 'mycorrhizal health'. However, the ecology of individual species and the ecology of the symbiotic relationships are poorly known and much more information will be required before the incidence of particular species or species combinations can be used as a diagnostic tool. In the meantime, monitoring of the species composition and abundance of the symbionts seems to have considerable potential.

Root condition

A fundamental part of a tree is the root system. Damage to the root system may well affect the rest of the tree and damage to the canopy may be reflected in the roots. Evidence from arboricultural studies suggests that a substantial proportion of the root system may be damaged without any immediate effect on the canopy. However, during subsequent periods of stress, the reduced ability of the tree to withstand such stress may become evident. Although the condition of the root system is an important index of tree health, it has not been used widely in the field owing to difficulties of assessment. This is unfortunate, as a common response of young trees to air pollution is a change in the root-shoot ratio (e.g. Cooley and Manning, 1987; Lechowicz, 1987).

Fine root turnover can be determined by periodic sampling of the live and dead roots and the soil litter fractions (Persson, 1983; McClaugherty *et al.*, 1982, Van Praag *et al.*, 1988; Richards, 1989). The technique suffers from many of the difficulties associated with the soil analyses that are described in Chapter 8, including a very high spatial variability in the process. Calculation of the rates of fine root production from repeated sampling data is difficult and liable to error (Singh *et al.*, 1984; Vogt *et al.*, 1986a, b; Kurz and Kimmins, 1987). In addition, the method is time-consuming and laborious, making it impractical for routine monitoring studies. However, with studies such as that by Norby *et al.* (1992) revealing a link between carbon dioxide enrichment and fine root turnover, a reliable method of routine assessment is urgently required.

An alternative is to use root observation chambers or tubes. This technique has been widely used in experimental studies, but its application to field studies is limited by practical considerations. However, a number of studies have succeeded in examining root growth in trees growing under natural conditions (e.g. Reich *et al.*, 1980; Teskey and Hinckley, 1981; Kuhns *et al.*, 1985) and it seems that the technique could be used in the more specialized monitoring systems used at Levels 2 and 3 within the UN-ECE classification.

Summary

Intuitively, the use of measured values of crown condition should be more valuable than visually assessed variables. However, in many cases, the measurements are costly and/or time-consuming and frequently involve destructive sampling. This means that the variation in a particular measure has often been inadequately assessed. Within a tree, there are usually substantial variations depending on the location of the sampling. Variations often occur in the same tree through time, both on a seasonal basis and over the longer term as the tree ages. On top of these internal sources of variation, there may be substantial differences between trees and between sites. This severely limits the interpretation of many measures at the present time. However, as more data become available, it is likely that the potential value of some of the measures will increase.

One outstanding problem that is yet to be resolved is the value of experimental studies with young trees in the development of diagnostic tests. Problems arise because of the differences between young trees and mature ones. For example, Steele *et al.* (1989) documented many physiological differences between 3- and 35-year-old Sitka spruce, particularly in relation to needle morphology. Edwards and Kelly (1993) found a number of differences in the response of seedlings and mature individuals of northern red oak to ozone, further indicating that extrapolation of results from seedlings and young trees to mature trees cannot be readily undertaken with our current level of knowledge.

Another major problem is the interpretation of data from comparative studies (‘affected’ versus ‘unaffected’ trees). In most cases, the comparison is being made between trees with and without visual symptoms. However, in forest situations, it is extremely difficult to establish whether a tree without visual symptoms has been affected. This is an area where laboratory experimental work could offer considerable assistance.

Conclusions

A variety of methods exist for the assessment of the vigour of trees growing in forests. However, many that have been used to date have been inadequately researched, and their interpretation is difficult. Nowhere is this more true than with crown transparency. Most scientists now accept that crown transparency is the result of a variety of different stresses, although air pollution is still invoked by some as the dominant factor. It is clear that crown transparency can occur without any impact of pollution (trees were not immortal before the advent of atmospheric pollution), and assessing the contribution of air pollution to crown transparency at any particular (polluted) location may be impossible.

Modern surveys increasingly use a number of different indices to assess forest condition. This chapter has focused on the various indices used to assess individual trees within forests, but it is important that long-term monitoring programmes include soil parameters and details of the status of other vegetation types (e.g. ground flora, lichens, soil flora). Monitoring of animal populations, both in terms of their absolute numbers and their physiological and biochemical condition, will also provide useful data. Collating, analysing and interpreting such data presents a major challenge for the future.

5 Quality Assurance

> Many components of measurement variation are controllable through quality control procedures such as calibration, analysis of reference materials, and use of defined operating procedures and control charts.
> *Cline et al.*, 1989

Introduction

Quality assurance has been one of the most neglected areas of study in European forestry research. This criticism applies to both health-related research and more standard measurements such as growth analyses. The latter were frequently assumed to be reliable, but recent work by Päivinen *et al.* (1992) suggests otherwise. While the principles of quality assurance have been widely adopted in forestry research in the United States, very little has been done to ensure similar standards in Europe. This lack of concern is particularly applicable to the reports of the assessments of forest condition that are conducted in Europe. Initially, great emphasis was placed on quality assurance (e.g. Schöpfer, 1985; Schmid-Haas, 1987) but, since then, the lack of consistency of observations made within different countries has been a recurring problem (Lick and Krapfenbauer, 1986; Neumann and Stowasser, 1987; Innes, 1988b, 1990b, d; Mahrer, 1989; Köhl, 1991; Schadauer, 1991; Innes *et al.*, 1992). The problem has been less apparent in North America, where quality assurance has been built into air pollution research programmes from the outset (e.g. Mickler and Medlarz, 1987; Medlarz, 1988; Cline and Burkman, 1989; Burkman, 1990) but has occurred in health studies elsewhere (e.g. Ellis *et al.*, 1980).

Table 5.1 shows the various activities that are brought under the heading of 'Quality assurance' from which four main activities can be identified (Cline and Burkman, 1989): quality management, quality assurance, quality control and quality evaluation. Quality management involves the proper design and documentation of the project, ensuring that appropriate data are collected in an appropriate way. Quality assurance is the first part of the evaluation of the reliabilty of the data collection, including the use of standards and the documentation of

Table 5.1 Basic aims of 'Quality assurance'

Concern	Benefit
Quality management	
Proper design	Considers if the right questions are being asked
Planning	Allows comparability considerations
Models	Identifies critical measurement variables
Definition of data quality requirements	Considers data users' needs
Relevant samples	Evaluates numbers and representativeness
Error structure	Partitions sampling and measurement error
Peer evaluation and review	Judges 'correctness'
Documentation	Records 'design' process
Implementation of QA programme	Achieves data quality requirements
Quality assurance	
Data quality objectives	Aids methods selection
Use of standards	Allows quality control and evaluation
Use of standard operating procedures (SOPs)	Provides consistent, efficient use of method with known and verified data quality
Verification and validation	Documents sample integrity and data consistency
Documentation	Provides evidence of activities and quality
Quality control	
Evaluation samples	Basis for statistical control
Training and use of SOPs	Promotes statistical control
Precision determinations	Defines random variation and allows accuracy assessment
Calibration	Reduces or eliminates bias
Control charting	Documents statistical control
Quality evaluation	
Use of standards	Allows precision and accuracy determinations
Replication	Provides ongoing evaluation
Blanks	Monitors contamination
Inspections and audits	Provides objective evaluation and basis for comparability

As specified by Cline and Burkman (1989)

standard operating procedures. Quality control ensures that data collection is being adequately undertaken and that the quality assurance is being carried out. Quality evaluation normally involves the statistical evaluation of the data quality.

The lack of quality assurance in the European forest health assessment programme has been a principal cause of uncertainty in the data. For example, it is not clear whether the differences in defoliation reported from different countries are due to real variations in forest health or are attributable to inconsistencies in the standards used between countries. Preliminary work suggests that the latter explanation is the more likely (Bouhot *et al.*, 1992; Innes *et al.*, 1992), but there has been considerable resistance to exploring this area further in view of the 'damage' that it might cause to international monitoring programmes. These have always been keen to emphasise the consistency of their observations and any doubts about this consistency have been unwelcome. Another problem is that the evolution of damage during the mid-1980s is very unclear as early assessments were inconsistent with later ones (e.g. Möhring, 1987).

In this chapter, the problem of quality assurance is examined principally using information gained in the British quality assurance programme. Emphasis has been placed on the evolution of the assurance programme, as it has become apparent that early surveys were inconsistent with later surveys. This chapter provides the basis for the data examined in the next two chapters.

The British programme in 1984

The British monitoring programme was established in 1984 (Binns *et al.*, 1985). The first survey involved 99 plots and was restricted to three species: Sitka spruce, Norway spruce and Scots pine. Assessments were made using the German categorical system, with crown defoliation (transparency) being estimated in five classes: 0 (0–10% needle loss), 1 (11–25% loss), 2 (26–60%), 3 (61–99% loss) and 4 (dead). Assessments were made in the late autumn and early winter. Several teams were used, and the authors acknowledged that 'The use of a number of different teams has almost certainly produced some variation, particularly in the observation of more subjective features such as crown density' (Binns *et al.*, 1985, p. 11). Needle retention (number of years) was also recorded and was considered objective and unlikely to be biased by any particular team. However, no attempt was made to assess the degree of variation between surveyors. In each case, a particular team worked within a specific region, providing potential for regional bias.

The British programme in 1985

The 1985 monitoring programme in Great Britain introduced a basic attempt to assess the extent of variation between the different surveyors (Binns *et al.*, 1986). Only four observation teams were used. Again, particular teams worked in specific regions. To assess the possibility of regional bias, observers from adjacent regions independently assessed a series of 'connectivity plots'. In the 1985 report, Binns *et al.* (1986) stated that significant spatial bias was present but that this was removed using the results obtained from the connectivity plots. The authors again acknowledged that the problem required attention: 'The use of connectivity plots has demonstrated considerable observer bias. It has been possible to allow for this to some extent in the analyses, but clearly a more rigorous system is required and will be instituted in the 1986 survey' (Binns *et al.*, 1986, p.14). A major difference between the 1984 and 1985 surveys was that crown transparency in 1985 was assessed in 10% classes instead of the five-category German system.

The use of connectivity requires further evaluation since it represents a potentially useful technique to assess observer bias. As already indicated, the British teams were not assigned randomly to the plots, rather they worked in specific regions. Connectivity plots were located on the boundaries of each region. As there were four regions, three boundaries existed and therefore data potentially included three sets of plots (one each for Sitka spruce, Norway spruce and Scots pine). In the field, the sampling was incomplete, with three complete plots of Norway spruce being assessed, two and three quarters plots of Scots pine and two plots of Sitka spruce. The results of the crown transparency assessments are given in Table 5.2. All scores were obtained by averaging the relevant scores for each index. The scores for crown transparency and needle retention represent true averages. However, for needle browning, scores were based on the continental classification system (0 = 0–10%, 1 = 11–25%, 2 = 26–60%, 3≥60%), and the averages therefore need to be interpreted with care.

The results presented in Table 5.2 illustrate the variations between the different surveyors. Crown transparency scores showed a consistent trend passing from observer 1 to observer 4. In all three species, observer 2 had a lower average score than observer 1. Observer 3 was also consistently lower in his estimates of defoliation than observer 2. In the two assessments that were made, observer 4 was lower in his assessments of crown defoliation than observer 3. The greatest differences occurred between observers 2 and 3 with Sitka spruce, with only seven of the 24 trees being recorded in the same 10% category and 10 trees differing by more than 10%.

Needle retention, the score that was considered in 1984 to be objective and therefore free of bias, showed less variation than the crown transparency estimates. In Scots pine, there was good agreement between assessors, probably because all trees in the plots had either two or three years needle retention. There was less agreement over the assessments of needle retention for the two spruce species. In particular, observers 2 and 3 differed markedly in their assessments, with differences of up to three years in the retention estimates.

Table 5.2 Comparison between surveyors' assessments of connectivity plots in 1985

Species	Assessor	Plot score	Assessor	Plot score	Assessor	Plot score
Average crown transparency						
Sitka spruce	1	14.6	2	21.7		
	2	9.2	3	8.3		
Norway spruce	1	11.3	2	7.1	3	10.4
	2	7.1	3	5.0	4	9.6
Scots pine	1	10.8	2	10.0	3	7.8
	2	8.3	3	5.0	4	7.2
Average needle retention						
Sitka spruce	1	7.3	2	6.2		
	2	7.4	3	5.3		
Norway spruce	1	7.0	2	7.8	3	6.5
	2	7.0	3	6.3	4	6.2
Scots pine	1	2.5	2	2.5	3	2.4
	2	2.5	3	2.6	4	2.2
Average amount of needle browning						
Sitka spruce	1	1.3	2	1.7		
	2	1.9	3	0.0		
Norway spruce	1	1.4	2	2.3	3	0.4
	2	2.0	3	0.0	4	0.1
Scots pine	1	1.0	2	2.8	3	2.2
	2	2.7	3	0.4	4	0.8

The estimates for needle browning showed the highest levels of variation. In two cases, observer 3 recorded no browning within the plot whereas observer 2 recorded a substantial amount. This revealed a major problem in the 1985 assessments of needle browning, throwing into serious doubt the value of the results for this index obtained in the main assessment programme.

Two other indices were assessed in the connectivity exercise. These were branch habit and defoliation type. Branch habit was divided into comb, brush, brush/plate, plate and 'other' whereas defoliation type was divided into larch-, subtop dying-, top dying- and peripheral-types. In spruce, there was agreement over branch habit in only 46 of the 120 trees (38%). In the trees with defoliation (67 trees), agreement over defoliation type was reached in only six cases (9%). This level of agreement could not under any circumstances be considered acceptable and the data from the main assessment programme must therefore be discounted.

The figures in Table 5.2 are based on the mean differences per plot. Consequently, they provide information on the potential bias within the data set. However, they do not indicate the scatter of the estimates, which were sometimes considerable. The results for crown transparency and needle retention are given in Table 5.3. Sitka spruce showed the greatest variation for crown transparency estimates, with estimates for five trees varying by 30% and the mean difference being more than 10%. Needle retention was also very variable for Sitka spruce, with the mean difference being more than one year, underlying the lack of precision in the estimates for this index. Although Scots pine showed less variation in the needle retention estimates, there was still an important amount of variation, and two standard deviations about the mean gave an expected variation of more than one year.

The British programme in 1986

Because of the problems identified in 1985, efforts were made to improve the reproducibility of the assessments in 1986. A training programme was introduced based on techniques used in Germany (Schöpfer, 1985). However, there was insufficient time to organise this properly, and only half a day of training was given. A course this short would have been insufficient to complete the learning curve for crown transparency assessments (Figs 5.1 and 5.2) (Innes, 1988b). Nonetheless, it was sufficient to significantly change the assessments made by some observers (Innes *et al.*, 1986). In addition, the use of two observers per team was introduced to reduce the degree of subjectivity involved when using only one observer.

Two different forms of quality assessment were adopted. First, single plots of Sitka spruce, Norway spruce and Scots pine were assessed both before and after the main survey period by all four teams of observers. Although the assessments were made independently, they were undertaken at the same times on the same days to reduce variation caused by environmental conditions. Besides these tests, I visited all four teams in

Table 5.3 Average differences (and standard deviations) in crown transparency and needle retention scores obtained on the connectivity plots in 1985

Species	Crown transparency		Needle retention	
	Mean	Standard deviation	Mean	Standard deviation
Scots pine	4.39	6.82	0.27	0.45
Sitka spruce	10.21	9.1	1.14	0.77
Norway spruce	5.21	6.18	1.07	0.89

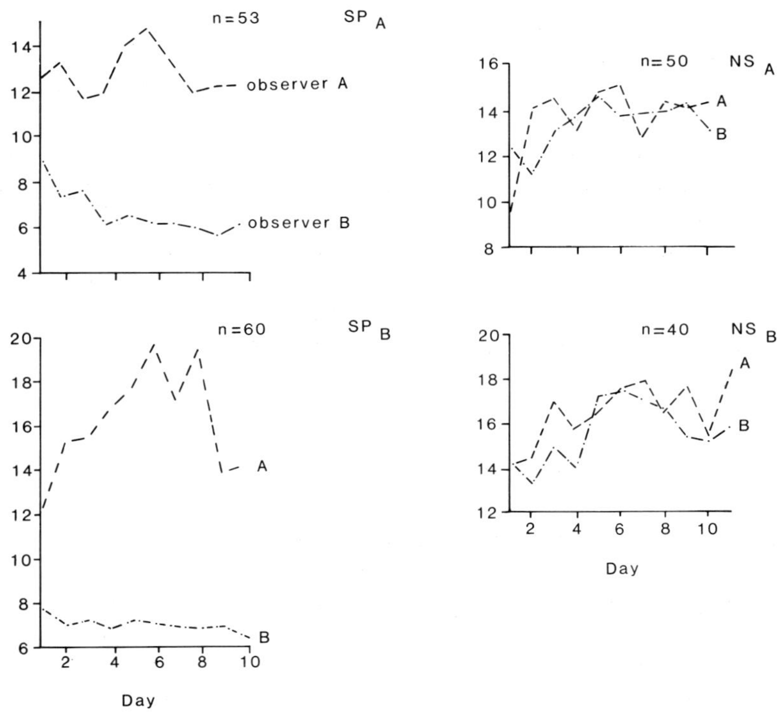

Figure 5.1 Learning curves for two (A and B) inexperienced observers repeatedly assessing crown transparency in two plots of Scots pine (SP) and two plots of Norway spruce (NS) over a ten-day (two week) period. The assessments for Scots pine remained very different, but the Norway spruce assessments showed some stabilization through time and a degree of convergence between the two observers. From: Innes (1988b).

their respective field areas, conducting simultaneous, but independent, assessments in 10 (7%) of the plots.

The results of the comparisons made before the start of the survey (and before the training) revealed differences between surveyors. In this study, each observer made his estimates independently, rather than in collaboration with his co-worker. The results are presented in Table 5.4. Crown transparency estimates differed significantly between surveyors, with some recording no trees with any defoliation and others reporting substantial amounts of defoliation. Needle retention was equally variable, with up to three years difference in the mean needle retention. There was insufficient needle discoloration at the sites to assess the extent of the variation in this index.

The exercise undertaken at the end of the survey was different from previous ones. For the first time, a 'control observer' was introduced,

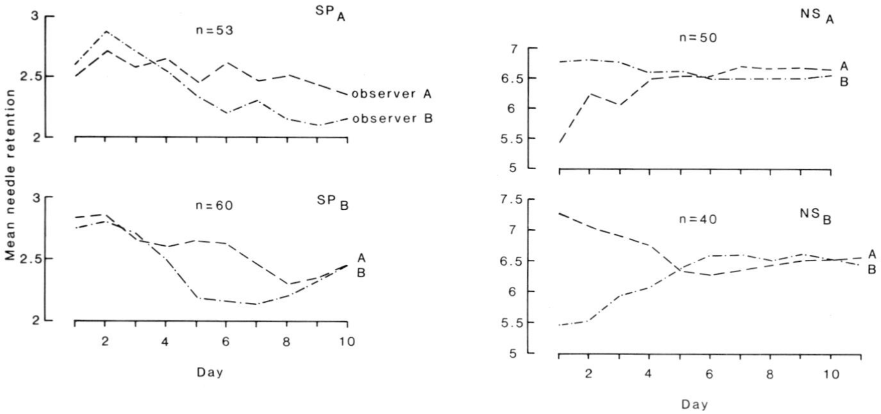

Figure 5.2 Learning curves for two (A and B) inexperienced observers repeatedly assessing needle retention in two plots of Scots pine (SP) and two plots of Norway spruce (NS) over a ten-day (two week) period. Stabilization and convergence is apparent. From: Innes (1988b).

whose sole function was to assess the degree of variation between observers. The control (me) had received limited training in Freiburg, Germany, earlier in the year and was also responsible for checking the observations made in the field by the different observation teams. The observers made their assessments in the teams used in the field. The results are presented in Table 5.5. Unfortunately, as different sites were used for the exercises at the start and end of the survey, it is impossible to gauge the extent to which individual observers or observation teams changed their estimates during 1986. The results in Table 5.5 suggest that some

Table 5.4 Results of the pre-training exercise in 1986, using eight observers

	Observer							
Pre-training	1	2	3	4	5	6	7	8
Crown transparency								
Scots pine	6.3	7.1	0	2.5	0	1.7	1.3	1.7
Sitka spruce	11.7	9.2	1.3	1.7	0	0.8	0.4	0.4
Norway spruce	11.7	14.2	1.7	5.8		1.7	9.6	9.6
Needle retention								
Scots pine	2.4	2.3	2.6	2.3	2.1	2.7	2.3	2.3
Sitka spruce	8.0	7.7	7.5	7.4	5.0	5.6	6.8	7.1
Norway spurce	6.4	6.0	6.6	6.7		6.6	5.8	6.1

The observers are arranged according to their teams: observers 1 and 2 formed team A, observers 3 and 4 formed team B and so on.
Mean crown transparency and needle retention estimates are given.

Table 5.5 Results of the post-survey exercise in 1986, using eight observers

	Team				
Post-survey	1	2	3	4	C
Crown transparency					
Scots pine	–	25.8	6.6	9.6	14.6
Sitka spruce	–	29.6	19.2	14.2	22.9
Norway spruce	–	22.5	5.4	9.6	12.5
Needle retention					
Scots pine	–	2.2	2.0	2.2	2.0
Sitka spruce	–	7.7	7.3	6.7	7.3
Needle retention	–	7.0	8.0	6.2	6.3

The observers are arranged according to their teams: observers 1 and 2 formed team A, observers 3 and 4 formed team B and so on.
'C' in the post-survey section of the table refers to the control observer.
Team 1 was not present for the exercise.

improvement had occurred, but there were still significant differences between teams. The control observer made estimates roughly in the middle of the range of the observation teams.

Quality assurance in 1987

The problems encountered in 1986 resulted in a major reassessment of the manner in which quality assessments were undertaken. Several changes to the survey programme were introduced at this time. These included an increase in the number of conifer plots assessed and the introduction of a few oak and beech plots. The control observer attended a full training course in the assessment of tree crowns conducted in Freiburg and also obtained copies of the photographic standards prepared as part of the Swiss Sanasilva programme (Sanasilva, 1986). A week-long training course was introduced for the observers.

It is not possible to assess the impact of the introduction of Swiss photographic standards and the training in defoliation assessment in Germany. Although some of the same trees were assessed in both 1986 and 1987 by the control observer, it is not known whether the condition of those trees changed in the interim. No assessments were made immediately before and after the training which could then have been used to assess the effects of the training. However, it is strongly suspected that crown defoliation was marked much more heavily following the training. Several observers recorded in their field notes that although they were marking the trees more heavily, they did not believe that the trees had changed in condition. Regrettably, there is insufficient data to resolve this issue.

Quality assurance 1988 and 1989

Quality assurance checks were maintained in 1988 and 1989 (Innes and Boswell, 1989a, 1990a). In 1988, the number of observation teams involved in the programme was reduced from ten to eight, despite an increase in the number of sample plots. With one exception, the observers were the same as in 1987. I undertook the quality assurance, visiting 13% of the sample stands at the same time as the stands were being assessed by the observation teams. Overall, the values for crown transparency obtained by the observation teams were within 0.4% of the values obtained by the control. However, the use of a mean figure is misleading as it gives no information on the degree of scatter. Most observations of crown transparency made by the individual teams (95%) were within ±14% of the observations made by the control. This represented a 1% improvement over the 1987 value.

In 1989, three observers were new to programme. Each was accompanied in the field by an observer with at least two years previous experience of work on the health survey. A control was again used to assess the extent of inter-team variation. Only 7% of stands were examined by the control, although this figure was still higher than the 5% recommended by the Commission of the European Communities' Regulation. There were significant differences in the accuracy of the estimates for crown transparency between tree species. With Scots pine and beech, 95% of the assessments made by the observation teams fell within ±12% of those made by the control. The figures for Sitka spruce, Norway spruce and oak were ±13%, ±14% and ±17%, respectively. The values for oak represented a deterioration in the quality of the data as compared to 1988.

Quality assurance 1990 and 1991

In view of the apparent deterioration in the quality of some crown transparency observations made in 1989, a renewed effort was made in 1990 to improve and then assess all aspects of the data quality. The need for this was accentuated by a change from two-observer teams to single observers, a move induced by financial constraints on the monitoring programme. All surveyors involved in the assessments in 1990 received training for one week. Of the nine surveyors involved, seven had worked on the 1989 programme and one of the remaining two had been involved between 1984 and 1987. Consequently, only one person was new to the programme. In 1991, the same level of training was used, but three surveyors were new to the programme, and the plots were distributed between 10 observers.

Training was helped by the development of a set of standard operating

Table 5.6 Percentages of observations where the control team agreed with the field observers in 1990 and 1991

Variable	Sitka spruce 90	Sitka spruce 91	Norway spruce 90	Norway spruce 91	Scots pine 90	Scots pine 91	Oak 90	Oak 91	Beech 90	Beech 91
Sample size	*71*	*216*	*168*	*192*	*198*	*187*	*167*	*120*	*88*	*144*
Branch density	83	51	76	55	53	51				
Branch pattern	89	63	83	64						
Browning of current needles	100	99	99	91	94	79				
Browning of leaves							98	95	91	65
Browning of older needles	100	93	97	89	96	88				
Butt damage – type	100	85		73	96	93		93		
Butt damage – extent	100	98				92		93		
Canopy closure around tree	76	71	63	52	68	53	53	62	72	60
Crown transparency	92	34	68	30	72	40	70	45	86	58
Crown dieback extent							74		55	
Crown dieback location							59	38	78	40
Crown dieback type							81	52	84	63
Crown form					85	73	77	67	94	67
Crown pattern							72		86	
Defoliation type	86	50	79	33	72	52		59		94
Degree of leaf-rolling									84	76
Epicormics on branches							60	50		
Epicormics on the stem							58	62		
Flowering in lower crown					56	49				
Flowering in upper crown					65	51				
Frequency of leaf-rolling									85	49
Fruiting	99	92	88	96	73	64	92	78	77	99
Fungal presence	100	100	97	96	100	100	100	100	100	100
Insect damage	85	44	92	53	90	65	48	57	60	35
Leader condition	97	81	85	70	70	76				
Leaf size									94	90
Mechanical damage to crown – type	93	72	97	66	79	72	80	68	73	73
Mechanical damage to crown – extent	92	65	97	55				66		71
Needle retention	99	46	84	34	88	63				
Overall discoloration	94	90	99	84	91	87	99	89	99	79
Premature leaf loss									91	60
Secondary shoot frequency	87	42	77	39						
Secondary shoot location	83	19	75	17	58					
Shoot death extent	82	40	75	40	59	43				
Shoot death location	78	42	75	36		37				
Stem damage – type	99	92	94	85	94	90	99	96	91	
Stem damage – extent	100	91		85			99	96		
Tree dominance	80	70	80	79	83	75	94	79	98	79
Type of leaf browning							90	57	90	60
Type of leaf yellowing							99	93	95	69
Type of needle yellowing	100	66	100	83	95	93				
Yellowing of current needles	97	100	100	93	99	98				
Yellowing of leaves							99	93	92	76
Yellowing of older needles	96	64	100	74	95	100				

procedures (Innes, 1990c). These replaced the Swiss Sanasilva guide (Sanasilva, 1986) used in previous years, although great care was taken to ensure standardization between the two guides.

Field checking involved visits by myself to each surveyor in his respective field area. In 1990, 71 (5.3%) of the 1344 Sitka spruce trees assessed in the programme were checked and the corresponding figures for the other species were 168 (9.6%) of the 1752 Norway spruce, 198 (10.2%) of the 1944 Scots pine, 167 (9.6%) of the 1741 oak and 88 (10.2%) of the 864 beech. These figures were an increase on 1989 but, in percentage terms, were less than in 1988, because of an increase in the number of plots over the three years (from 309 in 1988 to 319 in 1990). In 1991, the numbers of trees checked were 216 (12.4%) Sitka spruce, 192 (10.4%) Norway spruce, 187 (9.6%) Scots pine, 120 (6.5%) oak and 144 (9.8%) beech.

The results for each species and index are presented in Table 5.6. The percentages of observations for which there was agreement are given. The results are not comparable between species or between years, as the sample sizes varied between years and not every observer was checked on all species (although every observer was checked on at least two species). The degree of agreement for some indices was occasionally very poor, falling below 50%. This suggested that the precision of some observations was low and still required improvement, despite the training that had been given. No observation quality objectives exist for any of the indices and the setting of these will depend on further work to determine what is and what is not feasible.

In cases where there was disagreement over the scoring, the differences were frequently significant (Table 5.7). In each case, a Wilcoxon matched pairs test was used to identify significant differences between the observations made by the control observer and the surveyors. An exception was for crown transparency, for which t-tests were used. The Wilcoxon test shows whether two sets of data have been drawn from different populations, but does not provide an indication of overall or individual observer bias. The t-tests provide an indication of any general bias in the data but they do not reflect any bias apparent in individual observers.

The results presented in Table 5.6 reveal that in most cases, significant differences were recorded between the scores obtained by the surveyors and the control. One possibility is that the scores of the control observer were out of phase with those made by the surveyors and that there was little variation among surveyors. This was checked by examining the signs of the differences between the surveyors and the control. In all cases, the observations made by the control lay roughly in the middle of the range of differences, indicating that the maximum differences among the surveyors were greater than the maximum differences between the surveyors and the control.

Table 5.7 Significant ($p < 0.05$) differences recorded between the observation teams and the quality assurance team in 1990 and 1991

Variable	Sitka spruce		Norway spruce		Scots pine		Oak		Beech	
	90	91	90	91	90	91	90	91	90	91
Branch density	x	x	x	x	x	x				
Branch pattern		x	x	x						
Browning of current needles				x	x	x				
Browning of leaves							x		x	
Browning of older needles		x	x	x	x	x				
Butt damage – type				x		x		x		
Butt damage – extent		x		x		x		x		
Canopy closure around tree		x	x			x	x	x	x	x
Crown transparency		x	x	x		x	x	x		
Crown dieback extent								x		x
Crown dieback location							x	x		x
Crown dieback type								x		x
Crown form									x	
Diameter at breast height										
Defoliation type		x	x	x	x	x	x	x		
Degree of leaf-rolling										x
Dominance		x	x		x					
Epicormics on branches							x			
Epicormics on the stem							x			
Flowering in lower crown					x	x				
Flowering in upper crown					x					
Frequency of leaf-rolling										
Fruiting			x				x	x		
Fungal presence				x						
Insect damage	x	x	x	x			x	x	x	
Leader condition		x	x	x	x	x				
Leaf size										
Mechanical damage to crown – type	x	x	x	x	x		x	x	x	
Mechanical damage to crown – extent		x		x	x	x		x	x	x
Needle retention		x	x	x		x				
Overall discoloration				x	x	x		x		
Premature leaf loss									x	x
Secondary shoot frequency			x	x						
Secondary shoot location		x	x							
Shoot death extent	x			x	x	x				
Shoot death location	x	x	x	x	x	x				
Stem damage – type				x				x		
Stem damage – extent				x	x	x		x		
Tree dominance										
Type of leaf browning										
Type of leaf yellowing								x	x	
Type of needle yellowing										
Yellowing of current needles										
Yellowing of leaves								x		
Yellowing of older needles		x	x	x						

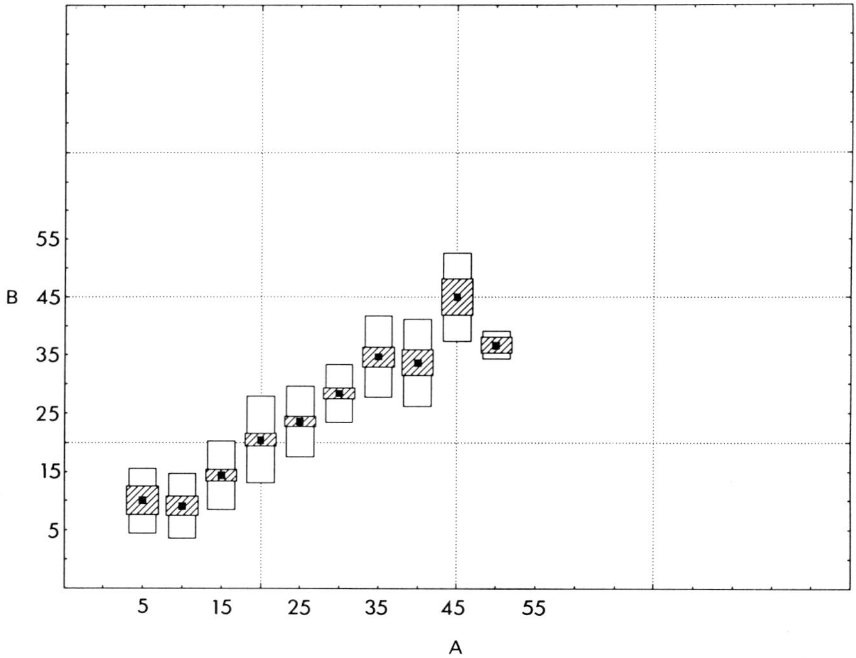

Figure 5.3 Comparison between crown transparency estimates made by the observation teams (A) and the control observer (B) for Sitka spruce in Great Britain in 1991. Means, standard errors (shaded) and standard deviations (unshaded) for each 5% transparency category recorded by the observations are given.

Crown transparency comparisons for 1991 between the control observer and the surveyors are given in Figs 5.3–5.7. A non-linear pattern was found for Sitka spruce (Fig. 5.3). The control scored trees at the 'better' end of the scale as having greater transparency levels than the surveyors. The pattern for Norway spruce was different (Fig. 5.4). The surveyors generally scored transparency lower than the control on healthier trees, but the position was reversed in the middle of the range. Complete agreement was obtained for severely defoliated trees. The discrepancies between the surveyors and the control were larger for Norway spruce than for Sitka spruce.

In Scots pine (Fig. 5.5), agreement between the control and the surveyors was good for trees with dense crowns. However, agreement progressively deteriorated as the crowns became thinner, with the control scoring the trees lower than the surveyors. In contrast, agreement for oak was good (Fig. 5.6). There was little evidence of any drift in the differences, which remained evenly distributed across the range of trees assessed. Beech showed a similar pattern to oak (Fig. 5.7),

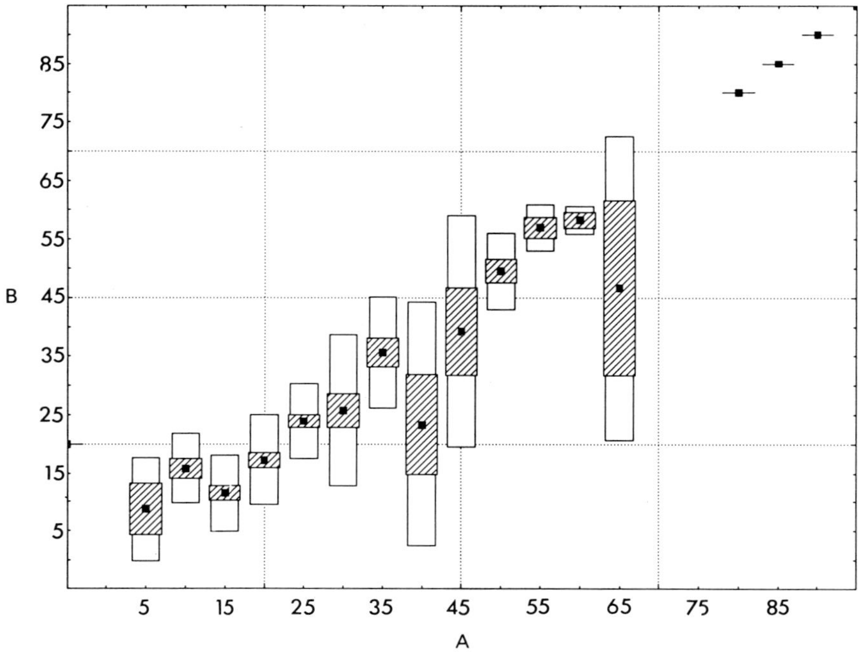

Figure 5.4 Comparison between crown transparency estimates made by the observation teams (A) and the control observer (B) for Norway spruce in Great Britain in 1991. Other details as in Fig. 5.3.

although there were major differences in the scores given to some heavily defoliated trees.

Evidently, there were several problems associated with the consistency of the observations in both 1990 and 1991. Ideally, all assessors should be brought to the same locations and asked to assess the same trees simultaneously. It would then be possible to determine the true degree of scatter among them. If such an excercise were conducted, it might be possible to set observation quality objectives.

Quality assurance in other countries

Several other countries have begun to look formally at the internal consistency of their data. Mahrer (1989) found that 70% of trees assessed by a check team in Switzerland were within ± one 5% assessment interval of the crown transparency assessment of the original team. There was no indication of bias being present in the results. However, the standard deviation of the differences was about 8%, suggesting that problems might be encountered when interpreting the original data. In Austria, Schadauer

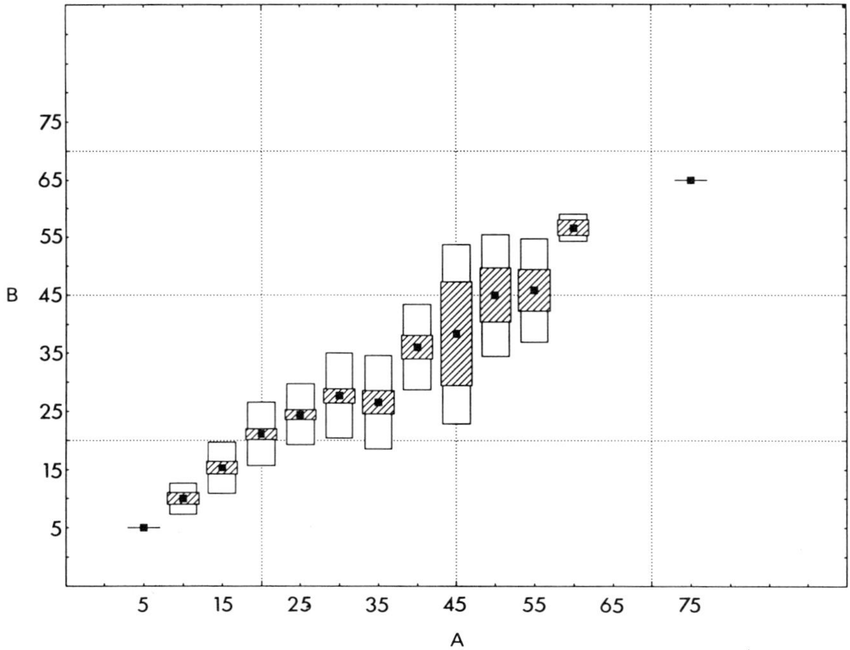

Figure 5.5 Comparison between crown transparency estimates made by the observation teams (A) and the control observer (B) for Scots pine in Great Britain in 1991. Other details as in Fig. 5.3.

(1991) reported that 75% of the assessments made between observation teams and the control team were identical, with 95% of observations lying within ±0.8 5% classes. Consequently, he concluded that the assessment error was less than or equal to the actual variation from year to year. These values were aggregated over the entire survey; individual species may well have shown greater levels of variation (Schlaepfer, 1985).

Lick and Krapfenbauer (1986) examined the variation in the classification of needle retention in Norway spruce between several Austrian survey teams. The variation was significant, despite the teams having been trained in the assessment of this index. The actual classification caused problems, as individual teams produced different scores when repeating assessments on individual trees. Similar results were obtained by Neumann and Stowasser (1987).

In the United States, great emphasis has been placed on the role of quality assurance programmes, although the results of such studies are not always readily available. Burkman and Alerich (1991) reported the quality of the data collected in the New England forest monitoring programme. They found that there was 67% agreement for crown dominance class, with most (96%) observations falling within the ±1 class.

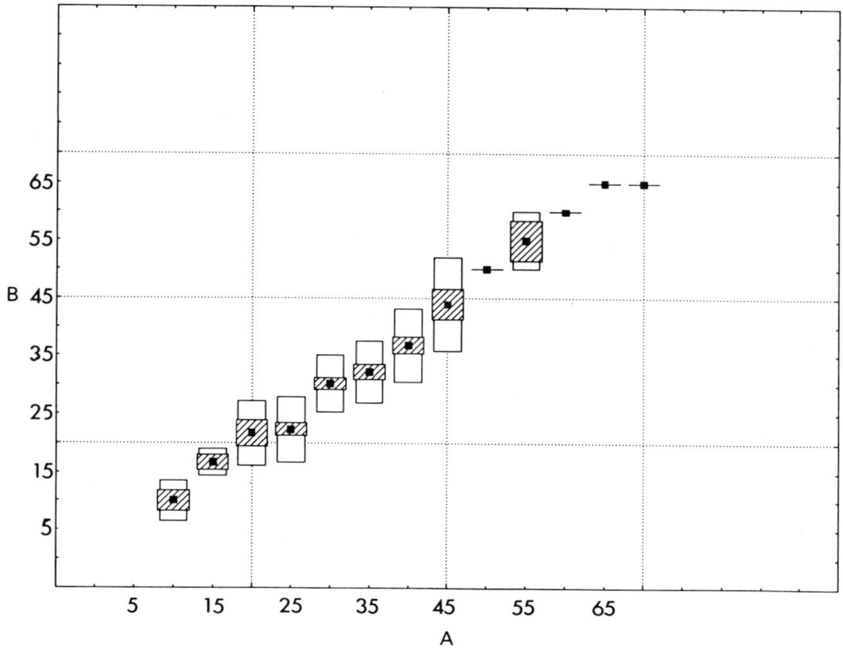

Figure 5.6 Comparison between crown transparency estimates made by the observation teams (A) and the control observer (B) for oak in Great Britain in 1991. Other details as in Fig. 5.3.

In another study, Nicholas *et al.* (1991) found that only 38% of the dominants in a spruce-fir forest received the same classification on a second visit, although 81% of the trees classed as co-dominant received the same classification on the second visit. Altogether, only 77% of the trees were placed in the same category on the second visit. The results obtained in Great Britain were consistently better than this, with agreement varying between 70 and 90% for particular species in any given year. This would be expected, as the British stands were exclusively even-aged, which the Kraft classification system is better adapted to. Crown dieback remeasurements revealed that 92% of the observations in the New England forest health monitoring programme were within 10% of the original value, although the percentage of agreements was not given. Rather higher figures were obtained for the North American Maple Project (97–99%). As dieback extent is recorded differently in Great Britain, no direct comparison has been possible. As in Great Britain, the low incidence of foliage discoloration meant that the precision of assessments of this index was generally good.

Foliar transparency re-assessments showed that 87% of the observations in the New England programme were within 10% of the original

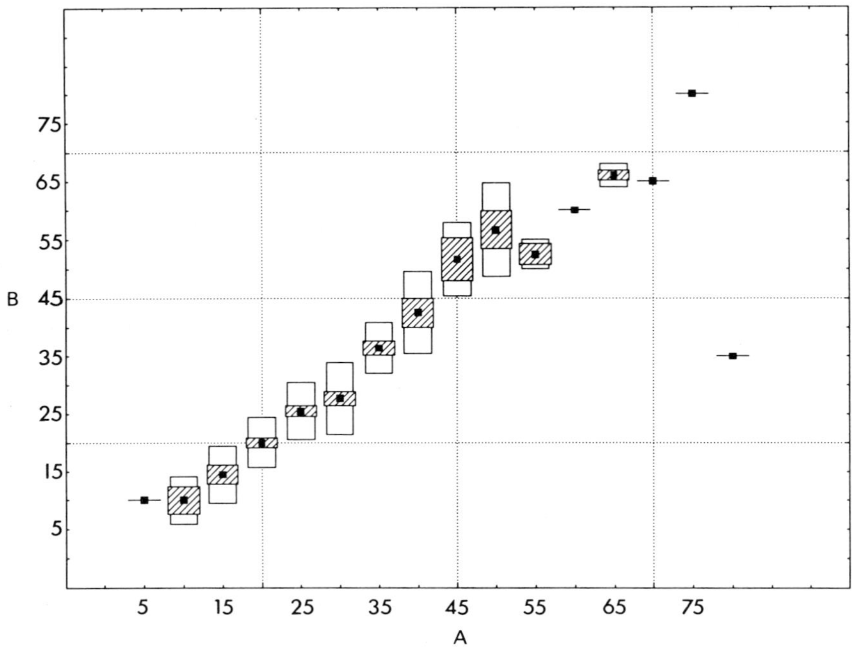

Figure 5.7 Comparison between crown transparency estimates made by the observation teams (A) and the control observer (B) for beech in Great Britain in 1991. Other details as in Fig. 5.3.

value. The assessments varied between species, ranging from 71% for eastern white pine to 94% for red spruce. The corresponding figures for Great Britain for 1990 and 1991 respectively were 100% and 92% for Sitka spruce, 97% and 90% for Norway spruce, 96% and 88% for Scots pine, 98% and 94% for oak and 100% and 94% for beech. The precision of the British estimates therefore seems to be higher than that of the American estimates.

Changes during the assessment period

A different problem, but still related to quality assurance, is the changes in crown condition that occur during the monitoring period each year. The changes have important implications for any spatial analyses of the data. Since 1987, the annual forest monitoring programme has been undertaken in Great Britain in July and August each year. As indicated above, earlier surveys were undertaken later in the year, providing another reason for viewing their results with caution. In 1990, prompted by the observation of marked summer drought effects during the previous year

(Innes *et al.*, 1989), assessments of several plots surveyed at the start of the monitoring programme were repeated at the end of the programme. This procedure was partially repeated in 1991, when no major drought was experienced.

In 1990, the assessments were made between 10 July and 31 August. During this time there was little, if any, rainfall over much of England. For example, July and August rainfall figures for Bournemouth were 14.6 mm and 27.3 mm and for Sheffield were 21.7 mm and 38.2 mm, respectively. In Scotland, rainfall was considerably higher, with the July and August figures for Aviemore being 30.7 mm and 88.6 mm and 86.5 mm and 87.4 mm for Rothesay (on the Isle of Bute). Consequently, in the south of the country, there was progressive evidence of drought stress in trees during the survey period. This represented a problem as trees assessed at the beginning of the period were unlikely to be affected by the drought whereas those assessed at the end of the period might well be expected to show symptoms associated with drought.

To assess the effects of the 1990 drought on the repeatability of assessments, seven plots each of Norway spruce, Scots pine and beech were re-assessed at the end of the survey period. The interval between sampling was approximately seven weeks. The results are difficult to interpret as any difference in score is likely to be due to a real change combined with differences associated with observer variation (attributable to, for example, differences in the light conditions between the two

Table 5.8 Changes in scores for various indices in 1990

	Norway spruce	Scots pine	Beech
Crown transparency	**−6.06**	−1.65	**−4.93**
Shoot death extent	**−2.95**	−1.51	
Needle retention	**5.27**	**2.90**	
Current needle browning	−1.74	**−3.67**	
Older needle browning	1.00	**−4.17**	
Current needle yellowing	**−2.39**	−1.42	
Older needle yellowing	−1.00	−1.91	
Leaf browning			**−2.61**
Leaf yellowing			**−6.58**
Overall discoloration	**−2.36**	**−5.39**	**−7.25**
Dieback			**−2.23**
Leaf size			0.82
Leaf rolling frequency			**−4.98**
Leaf rolling degree			**−4.55**
Premature leaf loss			**−2.58**

t-values are given.
A negative value indicates that the scores for a particular variable tended to be higher on the repeat assessment.
Significant ($p < 0.05$) changes are shown in bold type.

Table 5.9 Changes in scores for various indices in 1991

	Sitka spruce	Norway spruce	Scots pine	Oak	Beech
Crown transparency	**−2.93**	**−4.17**	**3.73**	**−2.49**	1.45
Shoot death extent	1.16	**−3.81**	**−3.17**		
Needle retention	1.88	**4.41**	−1.93		
Current needle browning	—	**2.7**	**−4.8**		
Older needle browning	1.43	**4.03**	**−4.97**		
Current needle yellowing	—	**−2.07**	**−3.84**		
Older needle yellowing	—	1.00	−1.14		
Leaf browning				**3.14**	**−2.88**
Leaf yellowing				0.53	**−6.37**
Overall discoloration	—	**−0.85**	**−6.3**	−0.45	**−7.82**
Dieback				**2.82**	**2.47**
Leaf size					**4.16**
Leaf rolling frequency					**−9.25**
Leaf rolling degree					**−5.06**
Premature leaf loss					
Insect damage			**−8.14**	**2.2**	0.56
Acorns/mast				**−5.48**	−0.45

t-values are given a negative value indicates that the scores for a particular variable tended to be higher on the repeat assessment.
Significant ($p < 0.05$) changes are shown in bold type.

assessments). Although the latter were minimized by using the same surveyor for each pair of plots, small changes in the scores for particular indices between the two visits are likely to be of little relevance unless there is definite evidence of a trend. The differences associated with the observations should be randomly distributed and, consequently, any directional change, indicated by a significant difference between the pairs of observations, should reflect a real change in condition.

In 1991, a few plots (2 Sitka spruce, 1 Norway spruce, 6 Scots pine, 4 oak and 5 beech) were re-assessed. The small sample sizes for some species means that the data should be interpreted with care. These data provide an interesting comparison with those from 1990, as there was no marked drought during the growing season.

The changes for particular indices were examined using t-tests for paired samples. The results are given in Tables 5.8 and 5.9. Many indices were significantly different, suggesting that changes had occurred. The changes were always in the expected direction.

Sitka spruce

Sitka spruce is generally planted in the north and west of Great Britain, in areas with relatively high rainfall and low soil moisture

Table 5.10 Changes in the crown condition of Sitka spruce between the beginning and end of the 1991 survey

	Change in crown transparency (%)						
Assessment at start	−15	−10	−5	None	+5	+10	+15
0	0	0	0	**0**	0	0	0
5	0	0	0	**0**	0	0	0
10	0	0	0	**2**	0	0	0
15	0	0	0	**5**	0	0	0
20	0	0	0	**2**	2	0	0
25	0	0	1	**4**	2	1	0
30	0	0	0	**5**	1	0	0
35	0	0	2	**4**	0	0	0
40	0	0	2	**1**	0	0	0
45	0	2	4	**0**	0	0	0
50	0	3	1	**0**	0	0	0
55	1	1	1	**0**	0	0	0
60	0	0	1	**0**	0	0	0
65	0	0	0	**0**	0	0	0
70	0	0	0	**0**	0	0	0
75	0	0	0	**0**	0	0	0
80	0	0	0	**0**	0	0	0
85	0	0	0	**0**	0	0	0
90	0	0	0	**0**	0	0	0
95	0	0	0	**0**	0	0	0

Values are the number of trees in each category.

deficits. Consequently, significant changes in crown condition of the trees were not expected, and this species was not examined in 1990. Two plots were re-assessed in 1991. Of the parameters tested, only crown transparency was significantly different on the second visit, with needle loss (transparency) being scored as greater on the second visit. There were many differences between the two visits, with only 23 of 48 trees receiving the same score on both visits (Table 5.10).

The pattern of differences is interesting. Trees with low levels of defoliation generally changed little or slightly improved between the two visits (Fig. 5.8). In contrast, trees with thin crowns at the start of the survey period generally deteriorated by the time of the second visit. This pattern is unlikely to be related to observer bias and is probably real. Needle retention did not change significantly over the period, suggesting that the crown transparency changes were caused by the loss of needles from within needled shoots. The cause of the loss was unknown, nor was any information available as to whether this was a normal event.

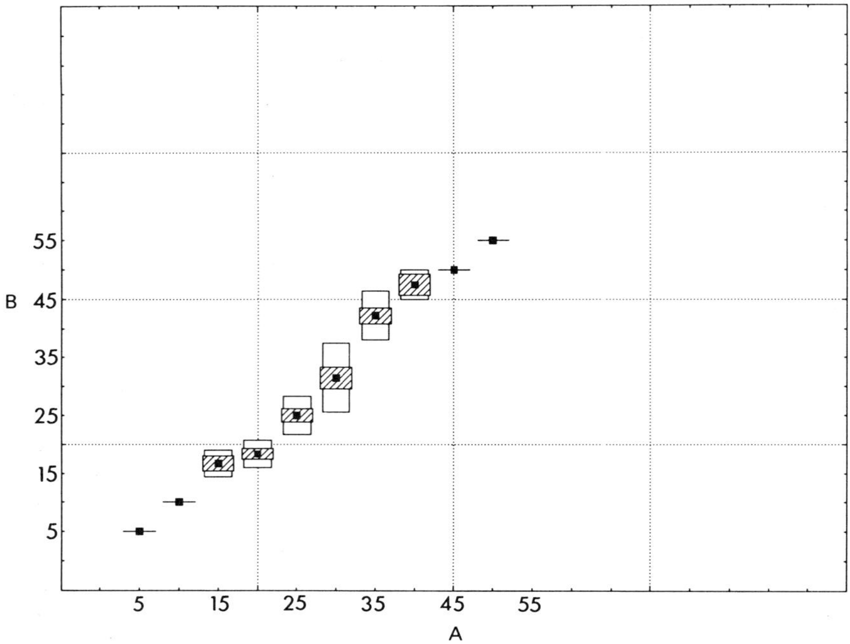

Figure 5.8 Changes in the crown transparency assessments of Sitka spruce assessed at the start (A) and end (B) of the survey period in 1991. Means of the transparency estimates (%), standard errors (shaded) and standard deviations (unshaded) for each 5% transparency category recorded by the observations are given.

Norway spruce

The crown condition of Norway spruce trees changed during the survey period in both 1990 and 1991, although the sample size in 1991 (24 trees) was based on only one site. In 1990, crown transparency scores in the repeat assessments were generally lower than on the first assessment, revealing that crown thinning had occurred (Table 5.11). Sometimes, the changes were substantial, amounting to between 20% and 25% of the foliage. Twenty-three trees showed an improvement during the 1990 survey period, with the majority involving an improvement of 5% (one class).

The extent of shoot death in the crown also increased during the 1990 assessment programme. As expected, very few trees (1) showed less shoot death on the repeat visit. Eleven (7%) of trees had higher scores for shoot death. Of these, nine were trees where shoot death was first classified as scattered or common. Only one of the 69 trees

Table 5.11 Changes in the crown condition of Norway spruce between the beginning and end of the 1990 survey

	Change in crown transparency (%)								
Assessment at start	−15	−10	−5	None	+5	+10	+15	+20	+25
0	0	0	0	**4**	2	0	0	0	0
5	0	0	1	**6**	1	1	0	1	0
10	0	0	1	**3**	8	1	1	0	0
15	0	0	2	**4**	4	3	1	0	1
20	0	0	2	**12**	9	2	1	0	0
25	0	0	1	**12**	7	1	1	1	0
30	0	0	4	**11**	5	2	0	0	0
35	0	0	1	**7**	2	7	0	0	0
40	0	0	4	**4**	1	0	0	1	0
45	0	1	2	**1**	4	1	0	0	0
50	0	1	1	**1**	2	0	0	1	0
55	0	0	0	**0**	1	0	0	0	0
60	0	0	0	**1**	0	0	0	0	0
65	0	0	0	**2**	0	1	0	0	0
70	0	1	1	**0**	0	0	0	0	0
75	0	0	0	**0**	0	0	0	0	0
80	0	0	0	**1**	0	0	0	0	0
85	0	0	0	**0**	1	0	0	0	0
90	0	0	0	**0**	0	0	0	0	0
95	0	0	0	**1**	0	0	0	0	0

Values are the number of trees in each category.

without shoot death on the first visit had shoot death recorded during the repeat assessment.

Needle retention was significantly lower on the second visit. However, this is an index that is subject to error (Innes and Boswell, 1989a), as illustrated by the nine trees which had an apparent increase in needle retention (from seven years to more than seven years). One tree was scored as having five years of needles on the first visit and only one year on the repeat assessment. Eleven trees scored as having seven years of needles on the first visit, had five or less on the second. The directional changes were consistent with the changes in crown transparency. However, a comparison of the trees that changed transparency and those that changed needle retention, revealed some inconsistencies. A direct relationship between crown transparency and needle retention was not always apparent. This was probably because the most common type of crown thinning experienced in Great Britain involves the uniform loss of needles throughout the crown rather than losses from the oldest needles to the youngest.

Needle discoloration showed little change, although there was a significant increase in the yellowing of current needles and in overall discoloration. Eleven trees with less than 10% of current needles showing yellowing at the beginning of the programme had more yellowing at the end, with the amount of yellowing increasing to the 25-60% category (two cases) in five of the trees.

In 1991, significant increases in transparency (Fig. 5.9), the extent of shoot death, the amount of yellowing of current-year needles and in overall tree discoloration were recorded. Conversely, needle retention and the amount of browning of current-year and older needles decreased over the period. However, these data should be treated with caution in view of the small sample size. This is apparent in the very irregular nature of the changes shown in Fig. 5.9. The presence of browning and the very low needle retention scores (2–5 years) strongly suggest that the site had trees suffering from the needle-reddening syndrome described in Chapter 3 and the changes should therefore be interpreted with caution. On the first visit, necrotic needles were present. By the time of the second visit, these needles had fallen, resulting in lower needle retention scores and lower crown densities. In addition, four trees had developed a noticeable amount of current-year needle chlorosis at the time of the second visit.

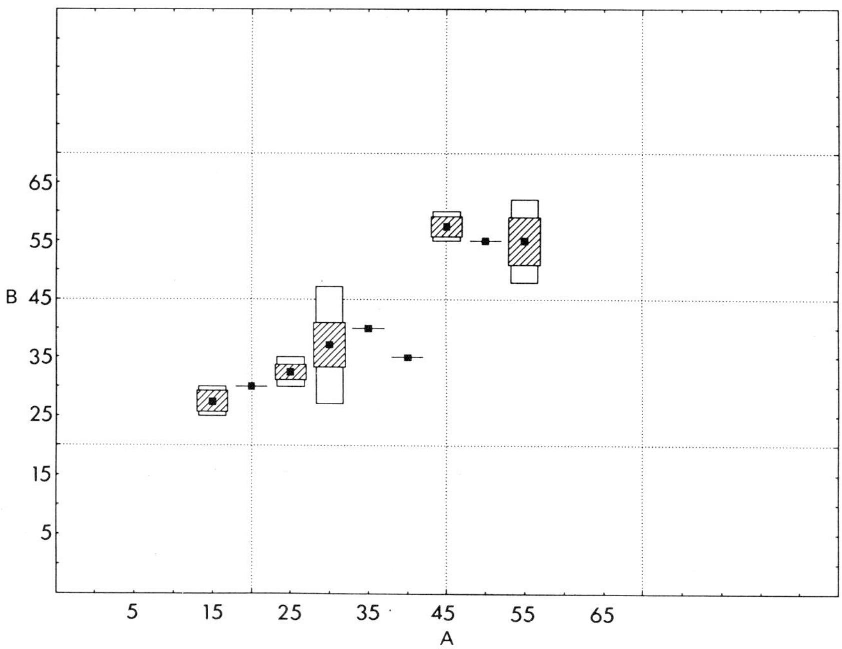

Figure 5.9 Changes in the crown transparency assessments of Norway spruce assessed at the start (A) and end (B) of the survey period in 1991. Other details as in Fig. 5.8.

These results suggest the progression of top-dying during the late summer rather than any 'normal' changes in crown condition.

Scots pine

Generally, there was little change in crown transparency during the 1990 survey period. Only one tree had an assessment differing by more than 10% and, as this was in a positive (improving) direction, it was unlikely to be real. The extent of shoot death was also unaffected. Needle retention was significantly lower on the second visit. The change mainly involved the trees scored as having three years of needles on the first occasion, with 15 (out of 88) of these having two years retention on the second visit. This was consistent with normal autumnal needle shedding. There was an increase in the browning of both current and older needles. The increase primarily involved trees that had been scored as having less than 10% browning on the first visit. The changes resulted in an increase in the number of trees scored for overall discoloration, and can probably be attributed to progressive damage by the pine shoot beetle (1st year needles) and autumnal senescence (older needles).

In 1991, several significant differences were identified between the two visits. The extents of shoot death, browning of current and older needles, yellowing of current needles and overall discoloration were all higher, whereas defoliation was lower (Fig. 5.10). The extent of shoot death can be directly related to attacks by the pine shoot beetle during the late summer. These would also have resulted in the increases in the amount of current-year needle necrosis. The increase in older-needle necrosis is probably related to normal autumnal senescence.

The changes in crown transparency (Table 5.12) are much more difficult to explain. In both 1990 and 1991, there was a tendency for trees to improve in both years, although the trend was not significant in 1990. Innes and Boswell (1991) dismissed the apparent trend in 1990, arguing that it was unlikely to be real as there were no ostensible reasons why trees should have denser crowns over the summer. However, the trend, which is clearly apparent in Fig. 5.10, was much more marked in 1991, suggesting that real changes do occur during the survey period.

The results for 1990 and 1991 reveal that substantial changes in the crown condition of Scots pine can occur during July and August, and that these changes occur despite the climatic conditions. If plots are assessed at the same time each year, annual time series for individual plots are unlikely to be unduly affected. However, analyses involving aggregated data from all plots will be severely constrained.

Oak

No repeat plots of oak were assessed in 1990, however in 1991 four plots (totalling 96 trees) were re-assessed. As with the other species, significant

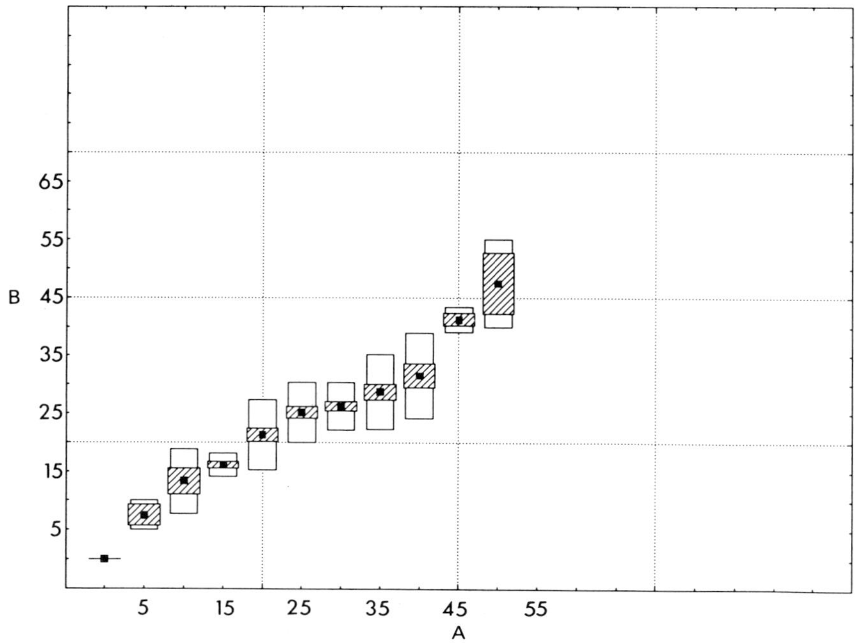

Figure 5.10 Changes in the crown transparency assessments of Scots pine assessed at the start (A) and end (B) of the survey period in 1991. Other details as in Fig. 5.8.

differences were apparent. Crown transparency generally increased between the first and the second visits. Sometimes, the reductions in foliage were substantial, amounting to 40% of the foliage in three trees (Table 5.13 and Fig. 5.11). The extreme change weighted the mean results and it is apparent from Fig. 5.11 that generally, transparency was lower on the second visit than on the first. One possible explanation is that Lammas growth may have occurred on some trees between the two visits, resulting in an increase in the total leaf area of the crown. The reasons for the severe cases of defoliation during the survey period are unknown, but are likely to be related to damage by insects.

A significant decrease in browning can probably be explained by damaged leaves being shed during the summer. The reductions in the amount of dieback and the extent of insect damage are more difficult to explain, particularly in view of the results for crown transparency. With insect damage, it is again possible that some affected leaves are quickly shed. Only the extent of dieback decreased; there was no change in the severity. In both cases, the development of Lammas growth might have affected the scoring.

On the second visit, two sites had a significantly greater occurrence

Table 5.12 Changes in the crown transparency scores of Scots pine during the 1991 programme

	Change in crown transparency (%)									
Assessment at start	−25	−20	−15	−10	−5	None	+5	+10	+15	+20
0	0	0	0	0	0	**1**	0	0	0	0
5	0	0	0	0	1	**1**	0	0	0	0
10	0	0	0	0	0	**2**	1	0	0	0
15	1	2	1	0	7	**11**	1	0	0	0
20	0	1	2	3	8	**12**	3	2	0	0
25	0	0	1	4	10	**14**	5	0	0	0
30	0	0	0	3	8	**11**	3	1	0	0
35	0	0	0	0	4	**6**	0	0	1	0
40	0	0	0	1	3	**3**	1	0	2	1
45	0	0	0	0	0	**1**	0	0	0	0
50	0	0	0	0	0	**0**	0	0	0	0
55	0	0	0	0	0	**0**	1	0	0	0
60	0	0	0	0	0	**0**	0	0	0	0
65	0	0	0	0	0	**0**	0	0	0	0
70	0	0	0	0	0	**0**	0	0	0	0
75	0	0	0	0	0	**0**	0	0	0	0
80	0	0	0	0	0	**0**	0	0	0	0
85	0	0	0	0	0	**0**	0	0	0	0
90	0	0	0	0	0	**0**	0	0	0	0
95	0	0	0	0	0	**0**	0	0	0	0
Dead	0	0	0	0	0	**0**	0	0	0	0

Values are the numbers of trees in each category.

of acorns than on the first visit. Problems with this index were suspected earlier in the season when stands known to have shown good flowering failed to receive any scores for acorns. The size of acorns, and therefore their visibility, increased markedly during the survey period. The occurrence of acorns at some sites was badly underestimated early in the survey, throwing into question the use of this variable. Consequently, the assessment of acorn frequency was dropped from the survey programme and has not been included in the section dealing with fruiting in Chapter 7.

Beech

Of the three species investigated in 1990, beech showed the greatest changes in crown condition during the period of the programme. All the main indices changed significantly between the two assessments. Crown thinning occurred, with one tree increasing its transparency score by 45%

Table 5.13 Changes in the crown transparency scores of oak during the 1991 survey programme

Assessment at start	Change in crown transparency (%)										
	−10	−5	None	+5	+10	+15	+20	+25	+30	+35	+40
0	0	0	**0**	0	0	0	0	0	0	0	0
5	0	0	**0**	0	0	0	0	0	0	0	0
10	0	0	**0**	0	0	0	0	0	0	0	0
15	0	0	**1**	0	0	0	0	0	0	0	0
20	1	0	**1**	3	0	0	0	0	0	0	0
25	1	2	**6**	6	1	0	0	0	0	0	0
30	1	3	**5**	4	3	1	0	0	1	0	0
35	1	3	**7**	4	0	0	0	0	0	0	0
40	1	2	**2**	3	1	2	0	0	0	0	3
45	1	3	**6**	2	0	0	0	0	0	0	0
50	2	2	**4**	2	1	0	0	0	0	0	0
55	0	2	**0**	0	1	0	0	0	0	0	0
60	0	0	**0**	0	0	0	0	0	0	0	0
65	0	0	**0**	0	0	0	0	0	0	0	0
70	0	0	**0**	0	0	0	0	0	0	0	0
75	0	0	**0**	0	1	0	0	0	0	0	0
80	0	0	**0**	0	0	0	0	0	0	0	0
85	0	0	**0**	0	0	0	0	0	0	0	0
90	0	0	**0**	0	0	0	0	0	0	0	0
95	0	0	**0**	0	0	0	0	0	0	0	0
99	0	0	**0**	0	0	0	0	0	0	0	0

Values are the numbers of trees in each category.

between assessments (Table 5.14). Several trees showed minor improvements, but it was not clear whether this represented the error surrounding the assessment of beech or a genuine improvement caused by a late flush of leaves. Of the two, the former was more likely as the drought would have reduced or prevented any late flushes. This is supported by the absence of any change in the frequency of small leaves.

Discoloration was significantly worse on the second visit, with both browning and yellowing increasing. The increase in yellowing (Table 5.15) was much more apparent than the increase in browning, which mainly involved 12 (out of 144) trees scored with less than 10% browning on the first visit. With yellowing, the changes again primarily involved trees scored with less than 10% yellowing on the first visit, with 43 (out of 141) showing an increase. Overall discoloration was similarly affected, with 35% of the trees scored as having less than 10% discoloration on the first visit showing discoloration on the second visit.

Increased dieback was also recorded. This is difficult to explain as

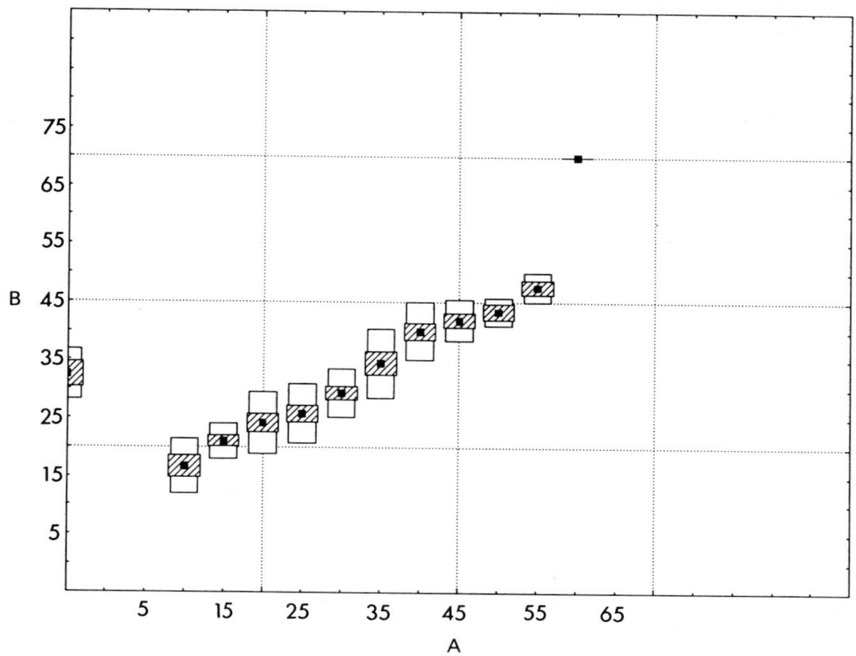

Figure 5.11 Changes in the crown transparency assessments of oak assessed at the start (A) and end (B) of the survey period in 1991. Other details as in Fig. 5.8.

dieback is normally regarded as a long-term process. However, observations suggest that the loss of shoots and small branches may occur quite quickly (Innes and Boswell, 1991). The main change was from 'leaf loss only' to 'dieback restricted to small branches', suggesting the loss of first-order, leafless shoots during the period. There is also the possibility that the premature dropping of leaves led to the upper crown being more visible, resulting in minor dieback missed on the first occasion being seen on the second visit.

Premature leaf loss, indicated by green leaves on the ground, was not recorded on the first visit. However, on the second visit, it was recorded as infrequent under 29 trees, common under seven and abundant under one. Little importance can be attached to this, as some leaf loss occurs throughout the summer because of insect activity and mechanical damage.

The degree and extent of leaf-rolling both significantly increased during the programme. This is as expected, as leaf-rolling in broadleaves is normally interpreted as a symptom of drought (Bernier *et al.*, 1989; Skelly *et al.*, 1987). The increase primarily involved trees scored as without

Table 5.14 Changes in the crown transparency scores for beech during the 1990 survey programme

	Change in crown transparency (%)											
Assessment at start	−10	−5	None	+5	+10	+15	+20	+25	+30	+35	+40	+45
0	0	0	**2**	0	0	0	0	0	0	0	0	0
5	0	0	**4**	2	0	0	0	0	0	0	0	0
10	0	0	**6**	3	2	1	0	0	0	0	0	0
15	0	2	**5**	1	1	0	0	0	0	0	0	0
20	0	1	**10**	7	4	0	0	1	0	0	0	0
25	0	5	**6**	7	4	1	0	0	0	0	0	0
30	0	1	**13**	6	4	0	0	0	0	0	0	0
35	2	11	**14**	4	2	0	0	0	0	0	0	1
40	0	2	**5**	3	4	0	0	0	0	0	0	0
45	0	5	**2**	2	1	0	0	0	0	0	0	0
50	0	1	**2**	0	0	0	0	0	0	1	0	0
55	0	1	**0**	1	0	0	0	0	0	0	0	0
60	0	1	**0**	0	0	0	0	0	0	0	0	0
65	0	0	**1**	0	1	0	0	0	0	0	0	0

Values are the numbers of trees in each category.

rolling on the first visit. On trees where rolling was recorded on the first visit, the degree of rolling showed little change. However, the extent of rolling generally increased, whatever the extent on the first visit.

In 1991, changes also occurred during the survey period, but they were less marked. There was no change in the crown transparency of trees, although there is a suggestion in Fig. 5.12 that an increased sample of more defoliated trees might have revealed a pattern of improvement during the summer. The amount of discoloration increased, yellowing more so than browning. The amount and degree of leaf-rolling also increased whereas there was a reduction in the proportion of trees with

Table 5.15 Changes in the extent of crown chlorosis (yellowing) in beech during the 1990 programme

	First visit			
Second visit	0–10%	11–25%	26–60%	>60%
0–10%	**98**	0	0	0
11–25%	29	**8**	0	0
26–60%	14	0	**2**	0
>60%	0	2	0	**5**

Values are the numbers of trees in each category.

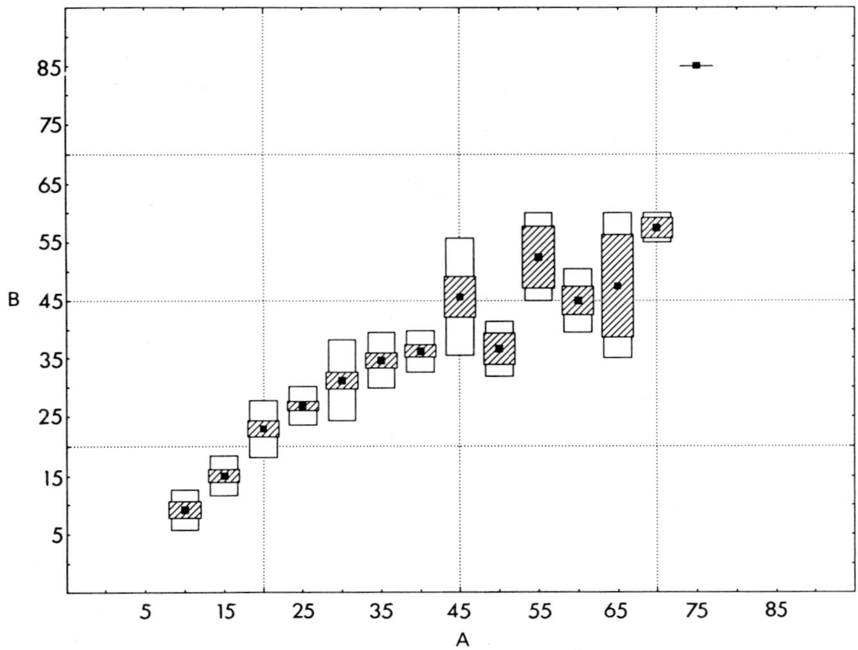

Figure 5.12 Changes in the crown transparency assessments of beech assessed at the start (A) and end (B) of the survey period in 1991. Other details as in Fig. 5.8.

unusually small leaves (Fig. 5.13). As with oak, the extent of dieback in the crowns was scored lower on the second visit.

Summary

The results of the repeat observations in 1990 and 1991 suggest that changes in crown condition occur over a short period in mid- to late-summer. These changes can be accentuated by drought and other environmental factors, but also occur as a normal process. Consequently, when conducting a monitoring programme, it is important to ensure that plots are examined at the same time of year each year. When prolonged survey periods are used, there is a real possibility of significant differences occurring between trees assessed at the start of the period and those assessed at the end of the period. This places a major constraint on spatial analyses of the data. Areas visited late in the survey season may have more symptoms of stress than those visited early in the season, purely because of the time of observation. This whole area needs to be investigated in more detail.

Figure 5.13 Rolling in beech leaves. The index is very subjective and liable to bias.

Quality assurance on a European scale

In the annual reports of both the United Nations Economic Commission for Europe and the Commission of the European Communities, great emphasis is placed on the degree of harmonization that has been achieved between countries. As only two indices are assessed, crown defoliation and crown discoloration, it might be hoped that this reported degree of compatability between the different surveys was real. However, examination of the data presented in the following chapter reveals some interesting trends. Why, for example, does Austria have so little defoliation in comparison with its neighbours? Why has Ireland the least 'damage' and the United Kingdom the most?

Some attempts have been made to look at the degree of standardization achieved in different countries. This has been primarily on international training courses, when assessments of a small sample of trees (usually less than 20) of a few species have been made. Data obtained from the 1987 training course conducted in Germany are presented in Fig. 5.14 (data supplied by B. Mettendorf). At the time, it was felt that these data indicated a degree of compatability between the different teams. However, it is obvious that major differences existed, not only between the different teams and the control team from Germany but also among the different

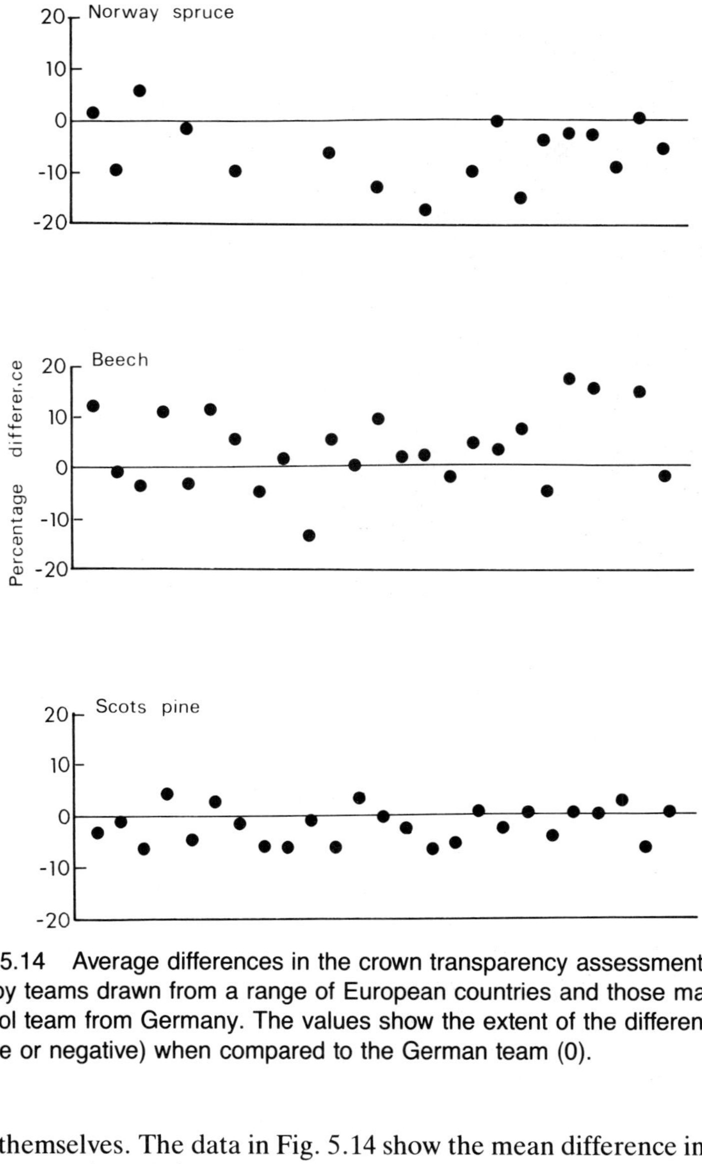

Figure 5.14 Average differences in the crown transparency assessments made by teams drawn from a range of European countries and those made by a control team from Germany. The values show the extent of the difference (positive or negative) when compared to the German team (0).

teams themselves. The data in Fig. 5.14 show the mean difference in crown transparency (density) estimates for each species. Most teams underestimated the amount of defoliation on Norway spruce and Scots pine in comparison to the German control team. In beech, there was a scatter in both directions. In both Norway spruce and beech, the extents of the mean divergences were up to 15% or more. The transparency of Norway spruce and Scots pine was consistently underestimated relative to the German assessment whereas beech was overestimated. Comparisons made during the 1990 international training course suggested that the differences identified in 1987 have not been resolved.

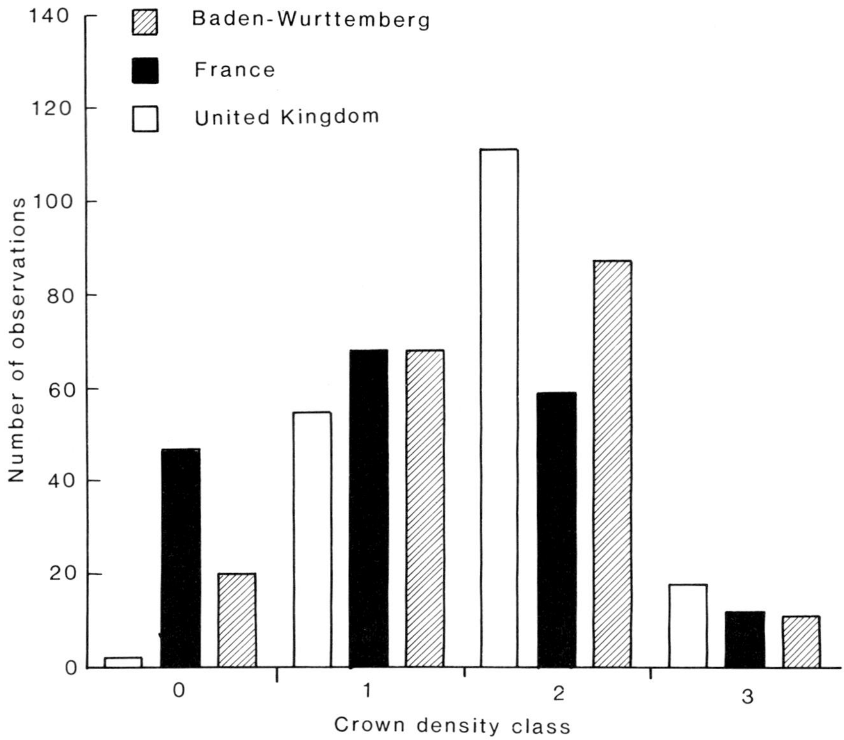

Figure 5.15 Comparison of crown transparency classes for Great Britain, France and the State of Baden-Württemberg, Germany. The teams assessed the same trees. The crown transparency classes are in the European classification, whereby 0 represents 0–10% transparency, 1: 11–25% transparency, 2: 26–60% transparency and 3: >60% transparency.

Detailed comparisons

In one of the few attempts to quantify the degree of variation between the assessments made in different countries, Innes *et al.* (1992) compared observations made simultaneously by assessors from Great Britain, France and Germany. Each team had at least one member who was responsible for the training of observation teams in his region or country. In the following account, the German team is referred to as being from Baden-Württemberg rather than from Germany, as differences in assessment techniques exist between regions in Germany, each of which conducts its own survey.

Normal assessment procedures were followed, with crowns being examined from the ground and foliar assessments (needle/leaf loss or crown transparency) made in 5% classes. Assessments were made using binoculars and with reference to any photographic or documentary aids normally used by the respective teams.

General differences among the three countries

Major differences were found in the overall assessments made by teams from the different countries. Scores by the French team were generally lower (more optimistic) by about 6% than the team from Baden-Württemberg which, in turn, scored trees lower (again by about 6%) than the team from Great Britain.

The results obtained for each species showed substantial differences. Between France and Great Britain, the greatest mean difference (22.5%) was for estimates of the crown transparency of oak. This species also had the highest difference for the British and Baden-Württemberg teams and was above average between the French and Baden-Württemberg teams. Scots pine also revealed substantial differences in the scoring, although the differences were, on average, considerably less than with oak. The smallest differences were for silver fir, with the British team's scores lying between those made by France and Baden-Württemberg.

Using correlation analysis to compare percentage scores, the lowest level of agreement was with Scots pine, with a mean correlation coefficient of 0.62. The other four species had mean correlation coefficients of between 0.81 and 0.91. This suggests that estimates of crown transparency for four of the species were approximately in accord, with all teams recognising increasing levels of defoliation. However, as shown by the means, there were considerable differences in the absolute values.

Only a few stands were assessed and it is therefore difficult to identify the extent of variations attributable to age differences of the trees. Substantial differences existed between stands, suggesting that stand age may have an influence on the assessment of crown defoliation. No clear trend was apparent in the relationship between the mean defoliation difference and stand age, nor between the standard error of the differences and stand age.

In every species, there was a significant ($p < 0.05$) increase in the differences between teams with increasing levels of defoliation. The most obvious trends were for Norway spruce and oak ($p < 0.01$).

Dutch–British comparisons

In a similar study, very large differences emerged when observations taken by British and Dutch observation teams in 1990 were compared. Although the British team was usually more pessimistic than the team from the Netherlands, there were exceptions which made the use of a standard weighting factor impossible. The differences can be attributed to several factors, including the method used to estimate defoliation and the area of the crown included in the estimates. The British team primarily assessed defoliation by crown transparency whereas, for the conifers, the Dutch team primarily assessed it by needle retention.

Mediterranean countries

Another study involved the comparison of teams from Mediterranean countries. This was in part undertaken because of the criticism (actually incorrect) of the study by Innes *et al.* (1992) that surveyors were working in very different environments to their normal conditions. Forty trees each of downy oak (*Quercus pubescens* Willd.), cork oak, Holm oak, stone pine, maritime pine and Aleppo pine were assessed. Significant differences in the defoliation assessments were found, with teams from Spain, Greece and France producing lower transparency scores than teams from Turkey, Italy, Portugal and Germany (Fig. 5.16). Differences of up to 95% between the maximum and minimum estimates were found on individual trees. The results confirm the findings of Innes *et al.* (1992) and have been further corroborated in an unpublished comparison made in 1992 between countries assessing montane forests in the Alps.

The differences reported by Bouhot *et al.* (1992) and Innes *et al.* (1992) throw into question the use of tables comparing the absolute levels of

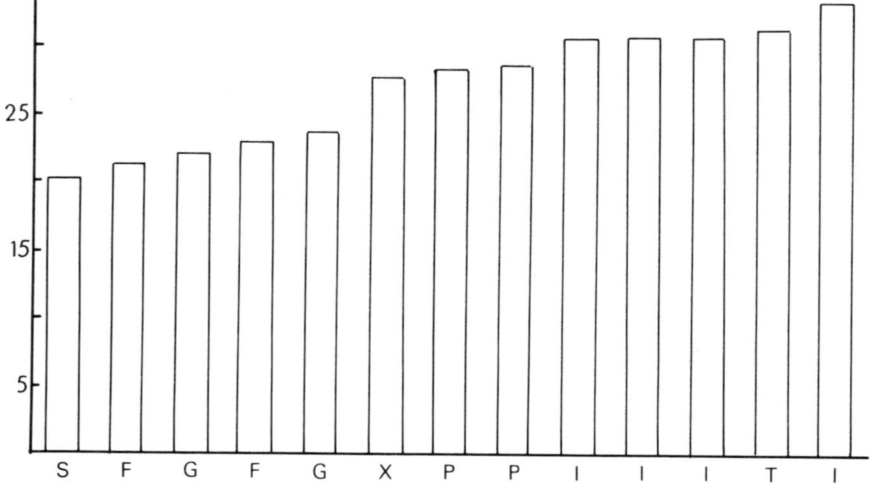

Figure 5.16 Average transparency estimates from several teams from southern Europe when assessing the same trees. S: Spain; F: France; G: Greece; X: Average; I: Italy; T: Turkey. Some countries were represented by more than one team. From Bouhot *et al.* (1992).

crown defoliation in different countries. This has important implications for those studies that have used comparative tables of crown defoliation as a basis for forecasting damage development and for assessing the economic impact of forest decline. As it is highly unlikely that individual countries will change their methods of assessment, greater recognition of the dangers of making international comparisons is required.

Summary

Although crown transparency is favoured by some as an index of stress (e.g. Anderson and Belanger, 1987), there are many problems associated with its estimation. Essentially the estimates of crown transparency and other symptoms are based on visual assessments made from the ground by one or two skilled observers, a process that may lead to the underestimation of symptom frequency (Muir and Armentano, 1988). The transparency estimates are made in relation to a reference tree, which may either be an ideal crown or a crown that is considered good for that particular environment. Point intercept methods, such as that described by Cooper *et al.* (1987), have generally not been used. Given sufficient training, the estimates made by a particular observer will become consistent and repeatable, although the area with biggest errors may be in the crucial 25–75% range (Belanger and Anderson, 1988). Problems arise when estimates made by different observers are compared or when estimates made in different years are compared (Lick and Krapfenbauer, 1986). Both difficulties usually can be overcome by using a rigorous training programme (e.g. Schöpfer, 1985; Mickler and Medlarz, 1987; Bonneau and Landmann, 1988; Mahrer, 1989; Burkman, 1990; Schlaepfer and Haemmerli, 1990; Innes and Boswell, 1991) and by adopting a standardised set of operating procedures (e.g. Sanasilva, 1986; Innes, 1990c). However, obviously the operating procedures adopted by different countries vary, sometimes very significantly (Köhl, 1987). The training programme and specific operating procedures must be backed up by a rigorous checking programme supported, if necessary, by regular refresher courses for the observation teams. Few countries appear to have installed such a programme, although Switzerland is a notable exception (Köhl, 1991).

The proportion of the crown assessed is an important factor leading to observer variation. This was identified as a problem early in the British programme and resolved. Elsewhere, the problem remains and, for example, Andersson (1990) has reported that the portion of Scots pine crowns assessed in Sweden changed from the whole crown to the upper two-thirds during the period 1984–87. Other sources of variation include the way in which mechanical damage to the crown is accounted for and whether dieback is included.

A criticism that is frequently levied against visual assessments of tree condition is that they are subjective. However, this is only a problem if the assessments are not repeatable. While it has been assumed that they are, it is evident that some assessments are made with greater precision than others. In cases where the precision is low, the assessment should not necessarily be discarded. The quality assessment of forest health surveys is at an early stage and some of the problems are only just being recognised. It is likely that better training and more specific operating procedures will improve the precision of some estimates, and particular observations should only be discarded if attempts at improvement fail.

While observation quality objectives are already being considered, the importance of changes in tree condition has not been addressed. The results presented here suggest that this is a major omission which could significantly affect the results of a long-term monitoring programme. Several options are available and there are undoubtedly others:

- More surveyors could be used to reduce the length of the observation period. This could, however, result in a decreased precision of the observations.
- Particular plots could be monitored on the same day each year. This would help the development of time series for individual plots, but between-plot comparisons could be invalidated.
- Data could be presented accompanied by the necessary error ranges. Such an approach involves several assumptions, some of which may not be valid (e.g. that different plots will change in the same way).

It seems likely that the eventual choice will depend on the practicalities involved. However, it is important that in future, the changes in tree condition that occurs during any particular survey season are documented. Without such information, interpretation of the data will have an added layer of uncertainty.

6 Assessment of Crown Transparency in Great Britain and Elsewhere

> Interestingly it was not the symptoms [of forest decline] that were 'new'; most of the obvious symptoms like needle thinning, needle yellowing, and branching anomalies had been observed for a long time. The really 'new' facts rather were (1) the rapid yearly increases in visible damage, (2) the inability of establishing cause-effect relationships with available knowledge, and (3) the vast areas of affected forests.
>
> *Orthofer and Kienzl, 1990*

> In a few years with hindsight, one may question the reality of the 'rapid generalized damage on the European scale in the early 1980s', that has been presented as a specific trait of the current decline.
>
> *Landmann, 1991*

Introduction

Most forest health surveys concentrate on only two indices of crown condition: crown transparency and crown discoloration. In practice, crown discoloration is rare so by far the greatest amount of emphasis has been placed on crown defoliation. This is unfortunate, since the shedding of foliage is a common response to stress and it is usually very difficult to assign causes to it. An alternative approach is to assess a variety of other indices of tree condition and to use these to determine the seriousness of the defoliation estimates. This approach has been adopted in Great Britain and, in this chapter, the detailed results from Great Britain for the past five years are presented. In view of the problems of standardization associated with the data obtained from the period 1984–86, only data for 1987–91 have been presented.

Definition of crown transparency

Considerable confusion has arisen over the terminology of what is actually assessed in surveys of crown condition. Initially, it was thought that needle/leaf loss was being assessed, and this was frequently called damage if it exceeded 10% or 25% of the foliage of the tree. However, the photographic guides that have been published (Sanasilva, 1986; Innes, 1990; Müller and Stierlin, 1990; Cadahia *et al.*, 1991) are based on crown transparency (which, in Europe, is the same as crown density). Countries that do not use photographic guides (e.g. The Netherlands) base their estimates on indices such as needle retention, and their figures are therefore more closely related to needle loss.

Crown transparency seems to be the optimum term for visual assessments made by comparison with photographic standards. It is preferable to crown density, as increasing 'density' scores suggest increasing density, making the scoring the inverse of defoliation (where increasing scores suggest increasing defoliation). In both Europe and North America, increasing scores indicate increasing levels of defoliation and the use of an inverse score in Great Britain would only be confusing. Consequently, the term crown transparency has mostly been used in preference to crown density. When transparency scores have increased, the tree has been deemed to be deteriorating whereas reductions in transparency scores indicate improvement. An important point to note is that crown transparency does not imply a loss of foliage; it is merely a statement of the transparency of the crown at a particular point in time.

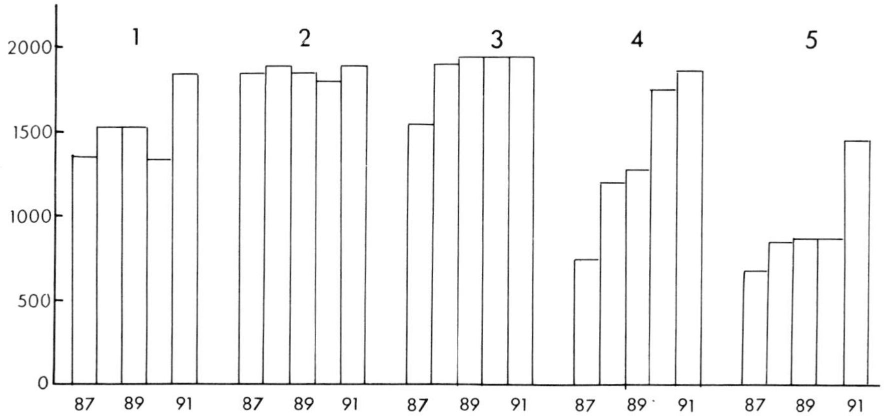

Figure 6.1 Changes in the numbers of plots in Great Britain over the period 1987-91. 1: Sitka spruce; 2: Norway spruce; 3: Scots pine; 4: Oak; 5: Beech.

Crown transparency in Great Britain

Sample sizes for crown transparency gradually improved in Great Britain between 1987 and 1991. The numbers of oak and beech plots steadily increased and sample sizes for Norway spruce and Sitka spruce also increased (Fig. 6.1). The project had a target of 80 plots for each of the five main species. Although several new plots were added each year, plot losses also occurred. In 1987, a number of plots were lost from the southeast of the country due to storm damage. Storm damage also occurred in 1990, resulting in the loss of a further nine plots. In addition, endemic windthrow and routine felling operations led to the removal of individual trees and several plots during the programme. Losses were greatest for the two spruce species (Table 6.1), with 14% of trees lost during the five years. It should be stressed that no losses were attributable to mortality, as dead trees are retained within the sample.

The distribution of plots of each species in 1991 is given in Figs 6.2 and 6.3. The lack of plots in the south-east is primarily the result of the storms of 1987 and 1990. This is revealed in the reductions in sample sizes shown for Sitka spruce and Norway spruce in 1990 (Fig. 6.1). Sitka spruce is rare in the south and east of the country and this is also reflected in the sampling design. Similarly, beech is increasingly rare towards the north of the country and the sampling density expresses this trend.

There are many ways in which the results can be presented, each having its relative advantages and disadvantages. Throughout this and the

Table 6.1 Sample sizes of each species

	Sitka spruce	Norway spruce	Scots pine	Oak	Beech
Total number of trees assessed 1987–91	2034	2143	2139	1916	1516
Number of trees assessed in 1991	1746 86%	1840 76%	1944 91%	1847 96%	1464 97%
Number of common sample trees 1987–91	1040 51%	1377 64%	1318 62%	724 38%	631 42%
Number of trees lost 1987–91	288 14%	303 14%	195 9%	69 4%	52 3%

The percentage of trees assessed relative to the total number of trees of each species involved in the programme is given.

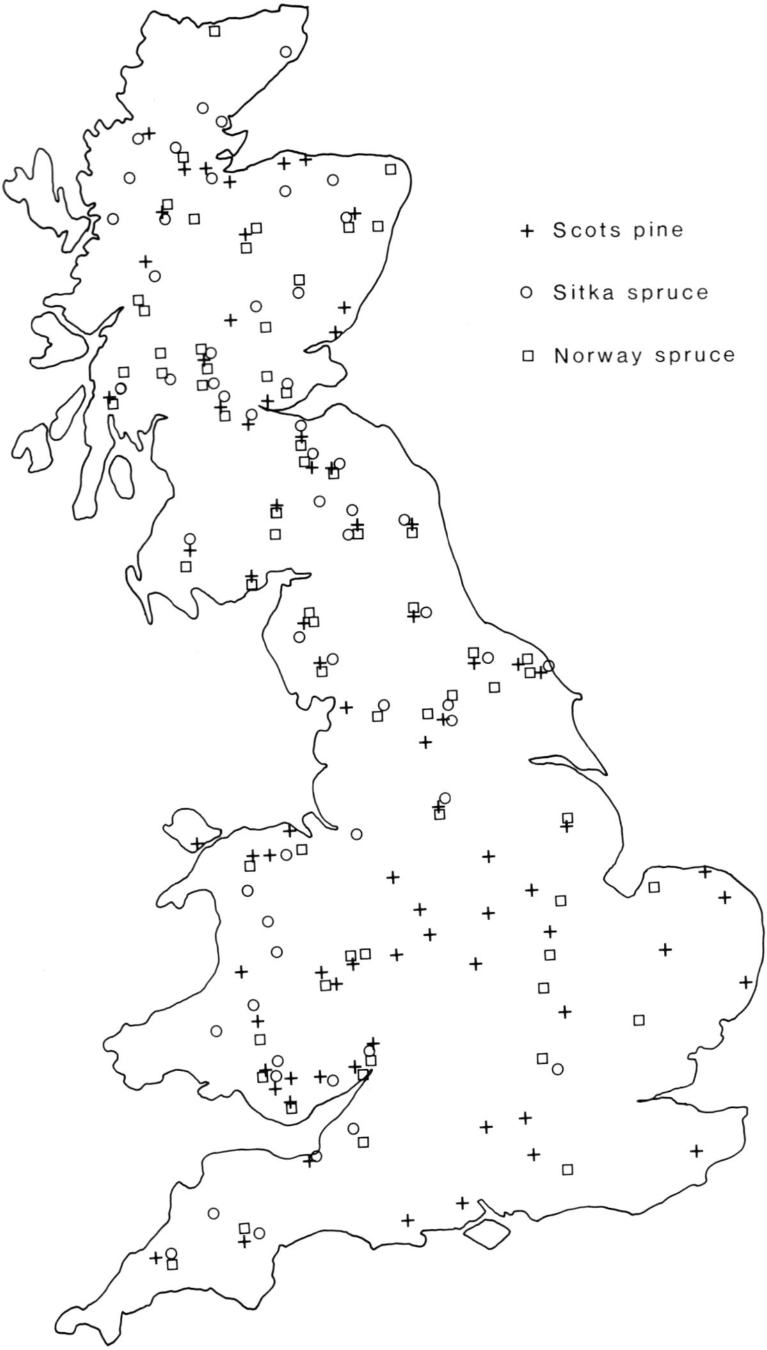

Figure 6.2 Location of monitoring plots for coniferous trees in Great Britain, 1991.

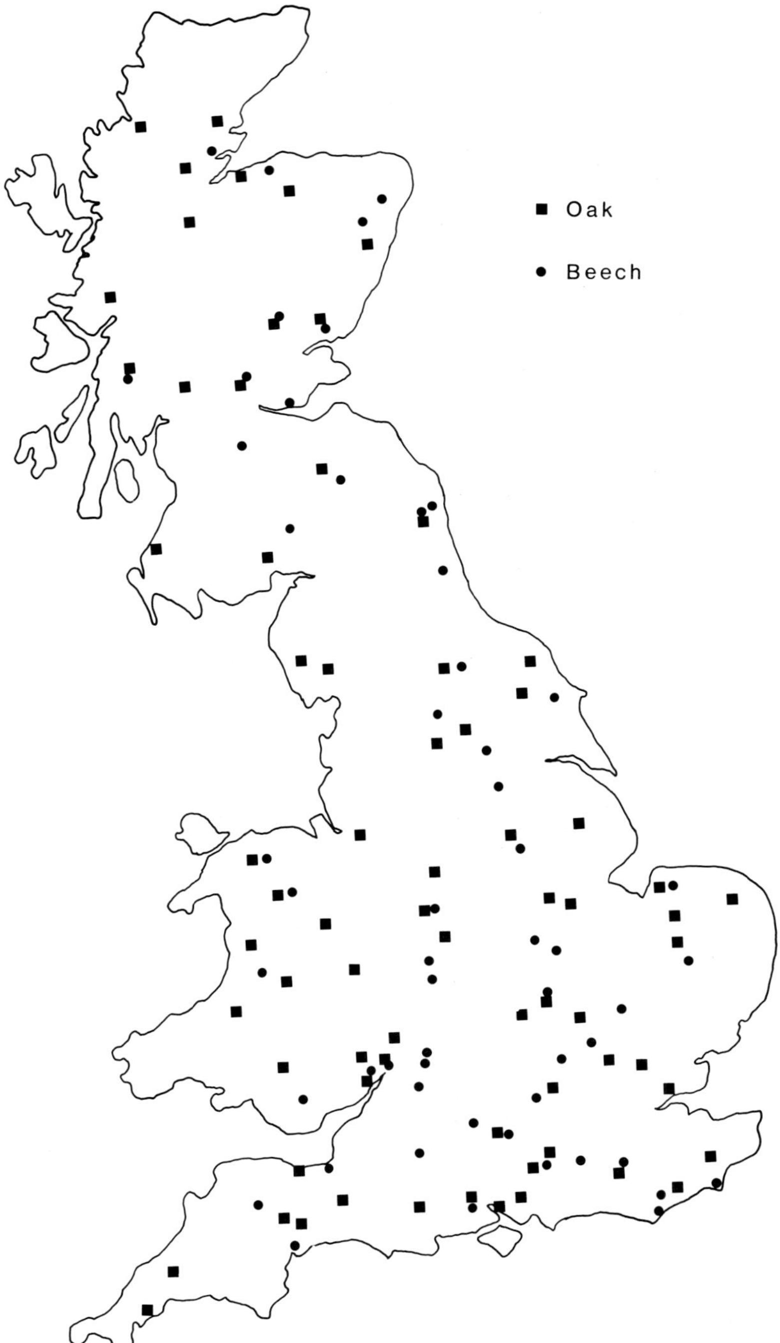

Figure 6.3 Location of monitoring plots for broadleaved trees in Great Britain, 1991.

Table 6.2 The percentage of trees of Sitka spruce in each crown transparency category

Score	1987	1988	1989	1990	1991
0			2.8	1.9	0.5
5			3.7	3.2	2.2
10	16.0	9.0	6.5	6.0	5.3
15			10.1	13.5	8.8
20	23.0	23.0	11.9	12.9	15.9
25			12.0	13.8	18.2
30	25.0	27.0	12.0	12.1	16.2
35			9.6	10.9	13.4
40	18.0	22.0	9.4	8.0	7.2
45			5.3	6.2	4.9
50	11.0	13.0	4.7	3.1	2.9
55			2.4	2.1	1.3
60	5.0	5.0	2.6	2.1	1.1
65			1.6	1.9	0.7
70	2.0	1.0	1.4	0.7	0.8
75			1.0	0.6	0.2
80	0	0	1.0	0.9	0.2
85			1.3	0.1	0.1
90	0	0	0.7	0	0.1
95			0	0	0
Dead	0	0	0	0	0

A score of zero indicates no transparency.
A score of 5% represents 1–5% transparency and so on.
Data collected in 1987 and 1988 were grouped in 10% classes, with a score of 10 indicating 0–10% transparency, a score of 20 indicating 11–20% transparency and so on.

following chapter, two types of results are described. The first concerns the results of assessments made in 1991 and refers to all trees. The 1991 results have also been used to draw comparison with results from previous years. However, in the latter case, only trees common to the comparison period have been used (unless otherwise stated). Consequently, the sample sizes for these common sample trees (the CST sample) are considerably smaller than the sample size for 1991 alone (Table 6.1). For example, only 38% of the oak trees assessed in the period 1987–91 were assessed in all five years, whereas 96% of the sample was assessed in 1991. In all the tables in this chapter, figures have been rounded to whole numbers or to one decimal place. Consequently, percentages may not always exactly total to 100.

In 1987 and 1988, crown transparency was assessed in 10% reduction classes relative to a fully foliaged tree, using the photographic standards for Switzerland (Sanasilva, 1986; Magasi, 1988). In the period 1989–91, the level of detail was increased and crown transparency was assessed in 5% classes. From 1990, photographic standards for Great Britain

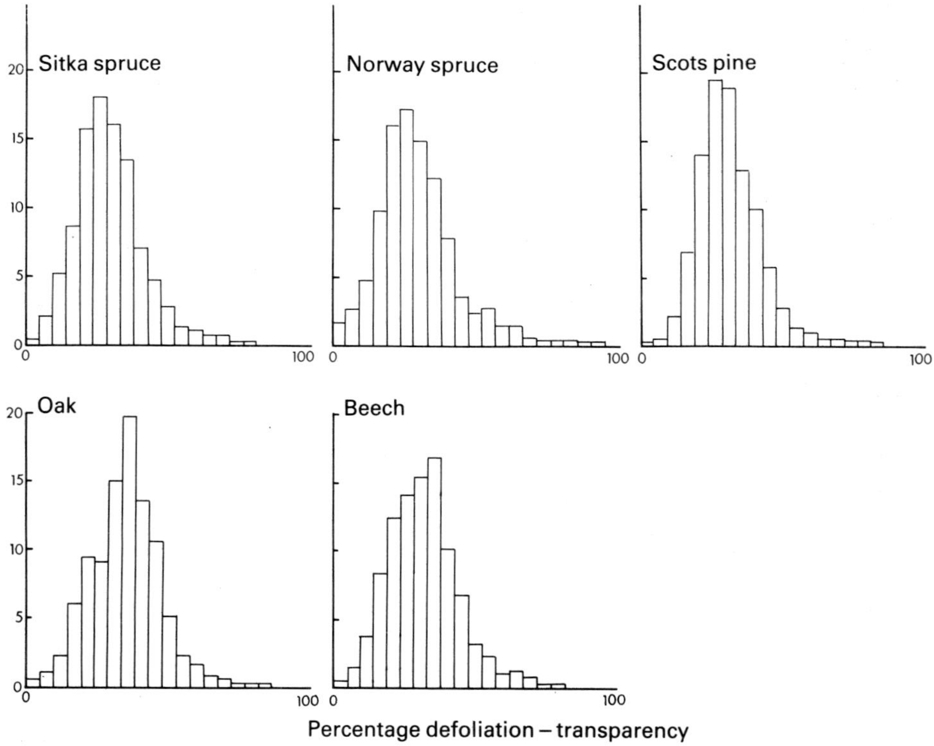

Figure 6.4 Frequency distributions, in 5% classes, of the crown transparency of trees assessed in Great Britain, 1991.

were adopted (Innes, 1990c), although great care was taken to ensure that the standards were consistent with earlier assessments. The 5% assessment classes adopted in 1989 followed the recommendations of the Commission of the European Communities and the International Co-operative Programme on Monitoring and Assessment of Air Pollution Effects on Forests of the United Nations Economic Commission for Europe. The complete data for crown transparency for each species are presented in Tables 6.2 to 6.6. In each case, the total sample assessed has been given. As the sample sizes vary between years, the results are not directly comparable.

The distributions of the transparency classes of the different species were similar (Fig. 6.4). The modal class varied from 25% for the three conifers to 35% for the two broadleaved species. The proportion of trees scored with no transparency was always less than 2%. The histograms strongly suggest that the general pattern of transparency is similar between species.

Comparison between years was made using the CST sample. The

Table 6.3 The percentage of trees of Norway spruce in each crown transparency category

Score	1987	1988	1989	1990	1991
0			7.1	5.8	1.7
5			8.6	6.7	2.6
10	20.0	21.0	11.4	9.6	4.7
15			13.4	13.9	9.9
20	24.0	27.0	14.6	14.2	16.0
25			11.6	13.5	17.1
30	24.0	24.0	10.4	11.9	14.8
35			8.2	9.8	12.1
40	16.0	17.0	5.8	5.4	7.9
45			3.2	3.0	3.5
50	9.0	7.0	2.8	2.3	2.4
55			0.9	1.1	2.7
60	4.0	3.0	1.1	0.5	1.4
65			0.7	1.0	1.3
70	2.0	1.0	0.3	0.4	0.5
75			0.1	0.1	0.3
80	1.0	0	0.1	0.2	0.3
85			0	0.3	0.3
90	0	0	0	0.1	0.2
95			0	0.1	0.2
Dead	0	0	0	0.1	0.1

A score of zero indicates no transparency.
A score of 5% represents 1–5% transparency and so on.
Data collected in 1987 and 1988 were grouped in 10% classes, with a score of 10 indicating 0–10% transparency, a score of 20 indicating 11–20% transparency and so on.

comparisons are given in Table 6.7. In this instance, only the data for those trees assessed in all five years have been presented. The use of the CST sample represents a problem, as the sample size is dependent on the size of the initial sample and will gradually decrease through time as trees are lost from the sample. Consequently, a third method of presenting the changes through time has been used in Fig. 6.5. Here, the sample trees common to two consecutive years have been used. For example, the comparison between 1987 and 1988 involved only those trees sampled in both years and the comparison for 1988 and 1989 involved the trees sampled in both of those years. This method allows the increased amount of data for oak and beech associated with the increasing sample size to be used in the analysis.

Mostly, a complex pattern of changes is evident. The mean values give a very approximate indication of the changes encountered from one year to the next, but do not indicate the extent of changes of individual trees. In any given year, the fluctuation in the mean is caused by the net changes that have occurred. Reference to Fig. 6.5 clearly suggests that in any given

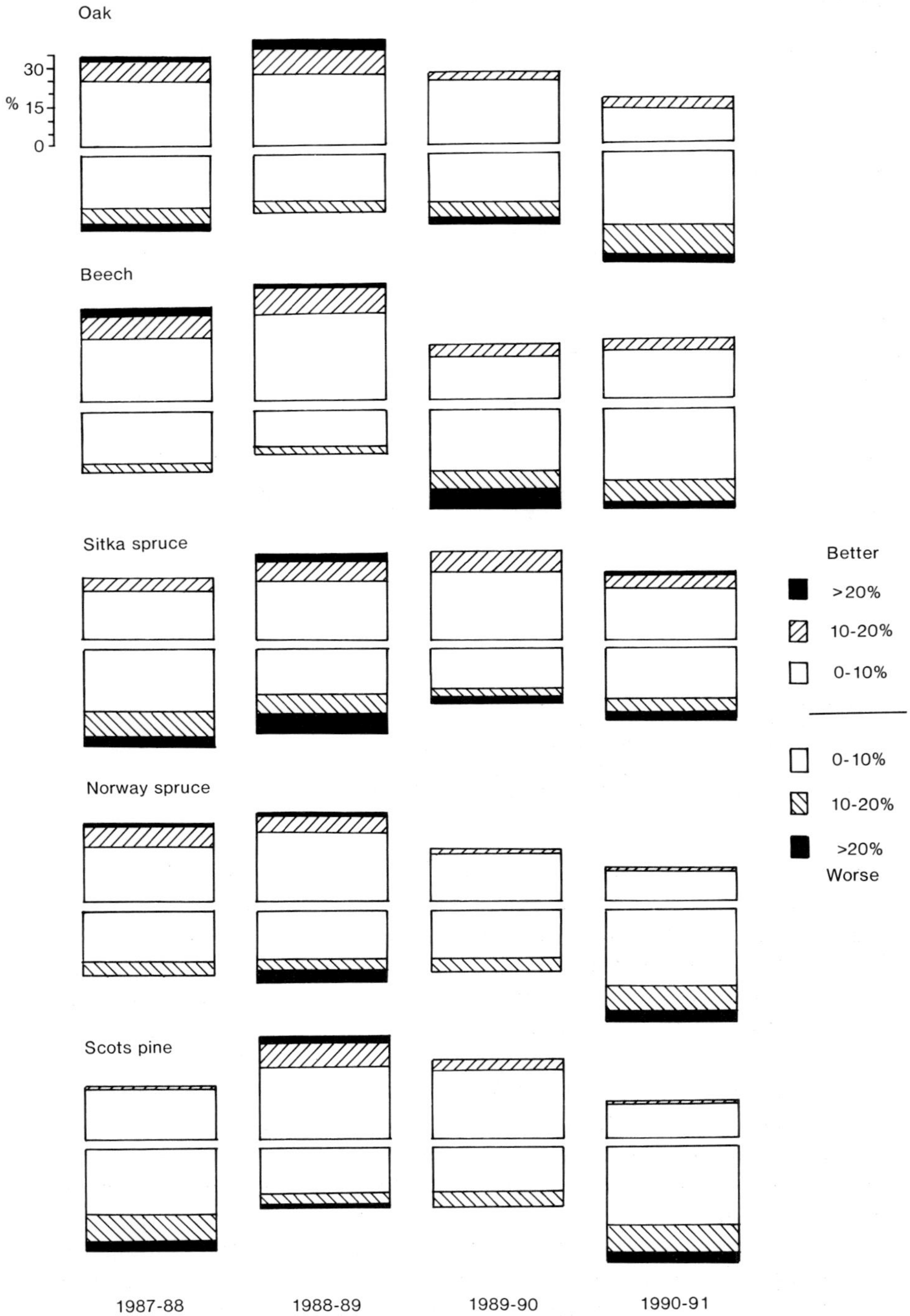

Figure 6.5 Changes in crown transparency of trees in Great Britain over the period 1987-91, taking the trees common to two adjacent years.

Table 6.4 The percentage of trees of Scots pine in each crown transparency category

Score	1987	1988	1989	1990	1991
0			1.7	2.2	0.2
5			3.5	3.4	0.5
10	19.0	8.0	7.8	7.4	2.2
15			14.4	11.8	6.9
20	24.0	23.0	16.6	17.2	14.0
25			14.5	17.2	19.4
30	25.0	32.0	12.5	14.9	18.9
35			8.9	10.3	13.8
40	18.0	21.0	6.1	6.6	10.0
45			4.9	2.9	5.9
50	8.0	9.0	2.6	1.6	2.8
55			1.2	0.8	1.3
60	3.0	3.0	1.7	1.0	1.0
65			1.1	0.2	0.5
70	1.0	1.0	0.9	0.3	0.5
75			0.4	0.3	0.4
80	1.0	1.0	0.6	0.7	0.4
85			0.5	0.2	0.3
90	1.0	1.0	0.5	0.1	0.1
95			0.3	0.3	0.1
Dead	0	1.0	0.7	0.7	0.8

A score of zero indicates no transparency.
A score of 5% represents 1–5% transparency and so on.
Data collected in 1987 and 1988 were grouped in 10% classes, with a score of 10 indicating 0–10% transparency, a score of 20 indicating 11–20% transparency and so on.

year there may be many trees both improving and deteriorating. However, representation of this is extremely difficult when 20 classes and several years are involved. Below, the data have been examined by species, which makes interpretation slightly easier.

Sitka spruce

Over the five-year period, the crown transparency scores for Sitka spruce fluctuated considerably. Transparencies mostly increased between 1987 and 1988 and, in 1989, some trees improved while others deteriorated because of heavy defoliation by the green spruce aphid.

The transparency of trees generally improved in 1990. This was mainly the result of increased needle mass following the severe defoliation caused by the green spruce aphid in 1988–89. Some trees assessed in 1989 had only one year of needles in the crowns and, by 1990, these mostly had two years of needles (see below). Trees with two or three years' needles in 1989 also showed a tendency for increased needle longevity.

Table 6.5 The percentage of oak trees in each crown transparency category

Score	1987	1988	1989	1990	1991
0			0.8	1.7	0.6
5			1.6	4.0	1.1
10	8.0	4.0	4.6	7.5	2.3
15			9.2	9.5	6.2
20	15.0	16.0	12.8	11.6	9.6
25			15.6	11.4	9.1
30	23.0	28.0	15.4	13.3	15.1
35			16.3	14.9	19.7
40	29.0	29.0	9.7	9.9	13.6
45			6.2	5.4	10.7
50	14.0	13.0	3.8	3.5	5.3
55			2.1	2.2	2.2
60	5.0	6.0	2.2	1.6	1.7
65			0.6	1.2	0.9
70	4.0	2.0	0.4	1.0	0.6
75			0.1	0.5	0.3
80	2.0	1.0	0.1	0.5	0.4
85			0.1	0.4	0.3
90	0	1.0	0	0	0.1
95			0	0.1	0.1
Dead	0	0	0	0	0

A score of zero indicates no transparency.
A score of 5% represents 1–5% transparency and so on.
Data collected in 1987 and 1988 were grouped in 10% classes, with a score of 10 indicating 0–10% transparency, a score of 20 indicating 11–20% transparency and so on.

The improvements were particularly apparent among the more severely defoliated trees, confounding views that severely affected conifers are unlikely to reçover (e.g. Jukola-Sulonen *et al.*, 1987). In 1991, there was a slight deterioration in condition. While many trees improved, a greater number showed higher levels of transparency. The reasons for this are unknown.

Following individual trees through time can produce important information on changes in forest condition (Innes and Boswell, 1990b), although some uncertainties are involved (c.f. Mahrer, 1989). Since 1987, the transparency of many trees has changed (Table 6.8). Both deteriorations and substantial improvements were apparent, with individual trees losing up to 50% of their transparency. Most trees changed by at least one class between 1987 and 1991. However, the figures shown in Table 6.8 underestimate the numbers of trees that have changed. No account is taken of those trees that changed in 1988, 1989 or 1990 and returned to their 1987 score in 1991. As percentages have been used, the values in the higher transparency should be interpreted with caution as the sample sizes are small.

Table 6.6 The percentage of beech trees in each crown transparency category

Score	1987	1988	1989	1990	1991
0			2.8	3.6	0.5
5			6.8	4.4	1.5
10	8.0	9.0	9.4	8.6	3.8
15			15.3	8.6	8.4
20	22.0	24.0	15.7	12.2	12.5
25			15.7	13.3	14.1
30	26.0	33.0	13.3	13.0	15.5
35			10.0	12.3	16.9
40	28.0	23.0	6.0	7.8	10.2
45			2.5	6.3	6.8
50	12.0	9.0	1.5	4.5	3.3
55			0.5	2.9	2.4
60	3.0	2.0	0.5	0.6	1.1
65			0.1	1.0	1.3
70	1.0	0	0	0.2	0.8
75			0	0.1	0.3
80	0	0	0	0.2	0.2
85			0	0.3	0.1
90	0	0	0	0.1	0.1
95			0	0.1	0
Dead	0	0	0	0	0

A score of zero indicates no transparency.
A score of 5% represents 1–5% transparency and so on.
Data collected in 1987 and 1988 were grouped in 10% classes, with a score of 10 indicating 0–10% transparency, a score of 20 indicating 11–20% transparency and so on.

A particularly important feature of Table 6.8 is the proportion of trees assessed as having between 21–30% in 1987 which were in that class in 1991. A threshold of 25% is often assumed to reflect damage, but the results suggest that trees in this range may be relatively stable. If this is the case, then it implies that the trees normally have this level of defoliation and that damage has not occurred. Another important feature is the very small proportion of trees that initially had higher transparency scores that remained in the same transparency category. Mostly, trees with high initial transparency values improved, including one initially scored with 81–90% transparency. The results suggest that trees at the higher transparency levels are unstable (they may either improve or deteriorate), but this may be an artifact of the small sample sizes. The most likely explanation for the change is the recovery of trees following severe defoliation by the green spruce aphid. In most cases, defoliation by the insect is transient, and the vigorous nature of Sitka spruce means that new growth quickly fills in the transparency created by the defoliation.

Table 6.7 Comparison of tree transparency results for the period 1987–91

	Crown transparency class									
	0 0-10%	1 11-20%	2 21-30%	3 31-40%	4 41-50%	5 51-60%	6 61-70%	7 71-80%	8 81-90%	9 91-100%
Sitka spruce										
1987	15.4	22.8	24.1	18.8	11.3	5.2	2.2	0.2	0.1	0
1988	8.0	23.5	26.0	22.3	13.6	5.2	1.1	0.3	0	0
1989	11.2	22.8	24.3	19.7	10.2	4.7	2.9	2.2	1.6	0.5
1990	17.3	24.7	23.0	17.8	7.2	4.0	2.3	2.1	1.3	0.3
1991	7.5	22.9	35.2	18.8	8.7	3.4	2.4	0.8	0.2	0
Norway spruce										
1987	21.2	23.7	24.1	14.8	8.9	4.4	2.0	0.6	0.3	0
1988	20.2	26.2	24.2	16.6	7.3	3.6	1.2	0.3	0.3	0
1989	25.2	27.6	22.2	14.4	6.2	2.4	1.3	0.6	0.3	0
1990	20.9	28.6	25.2	15.4	5.8	1.7	1.4	0.3	0.5	0.1
1991	8.3	26.2	32.0	19.4	6.4	4.6	1.9	0.4	0.6	0.4
Scots pine										
1987	17.8	23.1	24.9	19.1	8.2	3.2	1.0	1.3	0.9	0.6
1988	6.6	22.2	32.1	20.8	10.1	3.6	1.4	1.3	0.9	1.0
1989	10.9	29.7	28.0	16.6	7.5	3.1	1.9	1.2	0.6	1.0
1990	9.9	28.4	33.5	17.2	5.1	2.2	0.7	1.3	0.4	1.3
1991	2.3	19.7	37.9	24.3	9.5	2.7	1.0	0.9	0.5	1.4
Oak										
1987	8.1	14.8	22.0	29.6	14.2	5.7	3.9	1.5	0.3	0
1988	4.1	16.2	29.7	28.7	12.6	5.7	1.9	0.6	0.6	0
1989	6.4	20.8	30.4	26.6	10.4	3.2	1.3	0.4	0.5	0
1990	9.5	24.3	32.2	21.7	8.0	2.6	1.0	0.3	0.4	0
1991	3.9	16.0	23.4	32.1	15.6	5.5	2.1	0.8	0.4	0.1
Beech										
1987	8.3	22.6	26.2	27.0	12.5	2.7	0.8	0	0	0
1988	9.7	24.2	32.4	22.5	8.9	1.6	0.5	0.3	0	0
1989	18.6	30.5	28.9	16.1	4.2	1.1	0.6	0	0	0
1990	17.0	20.9	26.0	19.7	10.8	3.3	1.3	0.4	0.5	0.1
1991	8.2	21.2	30.4	22.7	10.1	3.5	2.7	0.6	0.5	0

Ten per cent categories have been used as the 5% class interval was only introduced in 1989.
Class 0 represents a 0–10% transparency, class 1 represents 11–20% transparency, class 2 represents 21–30% transparency and so on.
All figures are given in percentages of trees in each class in each year (i.e. rows sum to 100).

Norway spruce

The crown transparency of Norway spruce remained similar during the five-year period. However, there was a marked deterioration in 1991, reflected by the high proportion (45%) of trees that had higher transparency scores. Changes in the common sample trees between 1987 and 1991 are shown in Table 6.9. Percentages of trees with the same score in 1987 and 1991 were higher than for Sitka spruce, indicating greater stability. Again, there was evidence that individual trees could improve by

Table 6.8 Changes in the crown transparency of individual Sitka spruce trees between 1987 and 1991

Crown transparency class in 1991 n=1040	Crown transparency class in 1987									
	0 0-10%	1 11-20%	2 21-30%	3 31-40%	4 41-50%	5 51-60%	6 61-70%	7 71-80%	8 81-90%	9 91-100%
0	**21.9**	11.4	7.0	0.0	0.8	1.8	0	0	0	0
1	29.0	**34.2**	30.2	17.3	10.9	3.6	0	0	0	0
2	29.0	33.8	**47.4**	43.4	33.6	27.3	17.4	0	0	0
3	15.5	11.0	19.1	**27.0**	27.7	27.3	30.4	0	0	0
4	1.9	6.3	9.3	8.1	**12.6**	25.5	30.4	50.0	100.0	0
5	1.3	1.2	0.9	6.5	7.6	**5.5**	13.0	50.0	0	0
6	1.3	1.7	0.9	3.2	5.9	5.5	**4.3**	0	0	0
7	0	0	1.9	0	0.8	3.6	4.3	**0**	0	0
8	0	0.4	0.5	0	0	0	0	0	**0**	0
9	0	0	0	0	0	0	0	0	0	**0**

Values represent the percentage of trees in a given class in 1987 that were in the specified class in 1991 (i.e. columns sum to 100).

Table 6.9 Changes in the crown transparency of Norway spruce between 1987 and 1991

Crown transparency class in 1991 n=1377	Crown transparency class in 1987									
	0 0-10%	1 11-20%	2 21-30%	3 31-40%	4 41-50%	5 51-60%	6 61-70%	7 71-80%	8 81-90%	9 91-100%
0	**24.6**	9.5	3.6	0.5	1.6	0	0	0	0	0
1	39.9	**41.6**	22.3	13.5	7.0	4.8	0	0	0	0
2	24.6	34.6	**39.9**	35.6	28.9	19.4	7.4	11.1	0	0
3	8.0	8.9	23.2	**34.1**	37.5	17.7	25.9	11.1	0	0
4	1.4	3.4	6.3	9.6	**10.2**	25.8	7.4	11.1	0	0
5	1.4	0.6	3.9	4.8	7.8	**24.2**	33.3	0	0	0
6	0	1.5	0.6	1.4	4.7	3.2	**18.5**	33.3	0	0
7	0	0	0	0	1.6	1.6	7.4	**0**	0	0
8	0	0	0	0	0.8	0	0	33.3	**100.0**	0
9	0	0	0.3	0.5	0	3.2	0	0	0	**0**

Values represent the percentage of trees in a given class in 1987 that were in the specified class in 1991 (i.e. columns sum to 100).

up to 50%. Unlike Sitka spruce, some trees deteriorated markedly. These trees were mostly affected by the needle-reddening problem. The occurrence of this syndrome in 1990 and 1991 was independent of the transparency of the trees in 1987. Trees scored as having 91–100% defoliation in 1991 were initially scored in several different classes, ranging from 21–30% to 51–60%.

Changes in the crown transparency of Norway spruce have been examined by several authors. For example, Maier (1985) studied the dynamics of 4100 Norway spruce from 43 plots in Bavaria, Germany. Between 1983 and 1984, trees classed as severely damaged in 1983 changed little whereas previously healthy trees deteriorated markedly. This is the same pattern that was identified in Great Britain between 1990 and 1991. The loss of needles in Norway spruce was related to drought stress (Maier-Maercker, 1986; McKay, 1988), among other factors.

Scots pine

The condition of Scots pine, as revealed by the transparency scores, fluctuated markedly from year to year. Trees deteriorated in 1988, improved in 1989 and 1990 and then deteriorated again in 1991. Many individual trees have shown considerable changes during the period (Table 6.10), changing by as much as 90%. Trees with higher initial levels of transparency generally improved and there was no evidence of a progressive deterioration in the transparencies of large number of trees.

Table 6.10 Changes in the crown transparency of individual Scots pine between 1987 and 1991

Crown transparency class in 1991 *n*=1318	Crown transparency class in 1987									
	0 0-10%	1 11-20%	2 21-30%	3 31-40%	4 41-50%	5 51-60%	6 61-70%	7 71-80%	8 81-90%	9 91-100%
0	**7.8**	2.6	0.9	0.4	0	0	0	0	0	0
1	32.9	**25.5**	22.2	11.2	1.8	4.7	7.7	0	0	0
2	37.7	35.6	**42.9**	47.2	33.9	9.3	15.4	5.9	0	0
3	15.2	26.8	22.2	**28.8**	36.7	34.9	15.4	0	8.3	0
4	3.5	6.9	8.2	8.4	**19.3**	34.9	38.5	41.2	0	0
5	1.7	0.9	2.4	2.0	5.5	**9.3**	7.7	5.9	25.0	0
6	0.9	0.3	0.6	1.2	0.9	4.7	**7.7**	5.9	0	0
7	0	0.6	0	0.8	0.9	0	0	**17.6**	25.0	12.5
8	0	0.3	0.3	0	0	2.3	7.7	5.9	**8.3**	0
9	0.4	0.3	0.3	0	0.9	0	0	17.6	33.3	**87.5**

Values represent the percentage of trees in a given class in 1987 that were in the specified class in 1991 (i.e. columns sum to 100).

Oak

Oak has also shown variable patterns of transparency development. Between 1987 and 1988, more trees improved than deteriorated. This trend continued into 1989 when the number of improving trees considerably exceeded those deteriorating. The improvements continued into 1990, although the balance between trees deteriorating and improving between 1989 and 1990 was approximately level. In 1991, a substantial deterioration in the transparency of the trees occurred, with almost three times as many trees having higher transparency scores as lower ones.

The changes of individual trees between 1987 and 1991 reveal an interesting pattern (Table 6.11). Trees initially scored as having high transparencies nearly all improved. The deterioration in crown transparency was among those trees initially scored with low transparency. Over the entire period, the threshold for change appears to be about 30% transparency. Above this level, trees improved, whereas below it, they tended to deteriorate. However, examination of the 1990–91 data revealed that the marked deterioration in 1991 occurred in trees with up to 40% transparency. The differential response of trees may be related to drought stress (Innes and Boswell, 1991).

Beech

Beech has also shown substantial variations in crown condition. Between 1987 and 1988, many trees changed transparency class, but the proportions improving slightly outnumbered those deteriorating. This trend was accentuated between 1988 and 1989, with substantial improvements in transparency levels. However, in 1990, a marked deterioration occurred, and this pattern continued into 1991.

The patterns of crown transparency change shown by individual trees were different to the patterns shown by oak (Table 6.12). Trees initially scored with high (>60%) transparency have all subsequently recovered. There was much less evidence of the differential changes depending on initial crown transparency, although many trees with 40–60% initial transparency improved.

Interpretation of the British results

Several trends can be seen in the transparency data. Some studies have suggested that severely defoliated trees (those with more than 60% foliage loss) are incapable of recovery. This is not supported by the British data or by data from Switzerland (Mahrer, 1989). The recovery of trees following severe defoliation is of interest, as it suggests that damage occurred, as opposed to those trees that have a constantly

Table 6.11 Changes in crown transparency of individual oak trees between 1987 and 1991

Crown transparency class in 1991 *n*=725	Crown transparency class in 1987									
	0 0-10%	1 11-20%	2 21-30%	3 31-40%	4 41-50%	5 51-60%	6 61-70%	7 71-80%	8 81-90%	9 91-100%
0	**20.3**	8.4	3.8	0	0	2.4	0	0	0	0
1	40.7	**30.0**	21.9	9.8	1.9	4.9	0	9.1	0	0
2	22.0	28.0	**29.4**	23.8	20.4	7.3	14.3	9.1	0	0
3	13.6	23.4	31.3	**38.8**	35.0	34.1	39.3	45.5	50.0	0
4	3.4	6.5	10.0	17.8	**31.1**	24.4	21.4	9.1	50.0	0
5	0	2.8	1.9	7.5	7.8	**14.6**	7.1	18.2	0	0
6	0	1.9	1.3	0.9	2.9	7.3	**10.7**	0	0	0
7	0	0	0.6	0.5	0	4.9	7.1	**0**	0	0.5
8	0	0	0	0.5	0	0	0	9.1	**0**	0
9	0	0	0	0.5	1.0	0	0	0	0	**0**

Values represent the percentage of trees in a given crown transparency class in 1987 that were in the specified class in 1991 (i.e. columns sum to 100).

high level of transparency. The latter trees may include those with up to 40% transparency.

An analogy can be drawn from North America. The effects of defoliation on balsam fir have been intensively studied in relation to spruce

Table 6.12 Changes in crown transparency of individual beech trees between 1987 and 1991

Crown transparency class in 1991 *n*=631	Crown transparency class in 1987									
	0 0-10%	1 11-20%	2 21-30%	3 31-40%	4 41-50%	5 51-60%	6 61-70%	7 71-80%	8 81-90%	9 91-100%
0	**28.8**	16.4	4.2	2.3	1.3	5.9	20.0	0	0	0
1	34.6	**36.4**	24.6	10.5	6.3	5.9	0	0	0	0
2	23.1	29.3	**34.1**	31.6	30.4	17.6	20.0	0	0	0
3	11.5	13.6	25.7	**27.5**	30.4	17.6	20.0	0	0	0
4	1.9	2.9	6.6	17.5	**17.7**	11.8	40.0	0	0	0
5	0	0.7	3.0	5.8	5.1	**11.8**	0	0	0	0
6	0	0	1.2	4.7	3.8	23.5	**0**	0	0	0
7	0	0.7	0.6	0	2.5	0	0	**0**	0	0
8	0	0	0	0	2.5	5.9	0	0	**0**	0
9	0	0	0	0	0	0	0	0	0	**0**

Values represent the percentage of trees in a given class in 1987 that were in the specified class in 1991 (i.e. columns sum to 100).

budworm outbreaks. This is a serious pest in eastern North America, requiring major efforts to control outbreaks. Two important factors were found contributing to the recovery of trees following defoliation (Piene, 1989). First, the ability of the trees to produce secondary shoots, combined with increased retention of older needles, considerably aided recovery. Second, the development of new foliage at the expense of volume growth was important. Two years of severe defoliation did not impair the recovery of volume increment, but recovery did not occur following four years of defoliation. Partial defoliation of the current-year foliage led to increased shoot production for three years. Severe defoliation, on the other hand, triggered the production of both secondary shoots and increased shoot production.

The results from the spruce budworm studies are not universally applicable, although they seem to be relevant for most spruce species. The means of recovery vary among species, with Scots pine rarely producing secondary shoots and consequently taking longer to recover from severe defoliation (Nilsson, 1974; Austarå *et al.*, 1986; Ericsson *et al.*, 1985). Each species should therefore be examined separately.

Westman and Lesínski (1985) examined various factors affecting crown transparency. They argued that there were many reasons why tree crowns may thin, including changes in needles loss (overall biomass, needle longevity and needle cover on branches), branch destruction, undeveloped shoots, malformed branches and a lack of secondary shoots. Crown thinning was believed to be a normal process associated with tree ageing. Many of these have been examined in the following chapter.

Most trees vary substantially in their condition from year to year, oscillating around a steady state (Kandler 1988, 1989; Hägi, 1989; Innes and Boswell, 1991). This creates problems when examining the general forest health statistics for a region, as subtle trends may be masked. In Switzerland, the dynamics of forest damage have been examined by Hägi (1989). She drew attention to the marked variation that can occur in individual trees (Fig. 6.6), a trend that has also been seen in Great Britain. The implication of this work is that a long time series will be required before any long-term trends can be reliably identified. The length of the time series that is required is unknown, but will depend on the species in question.

International results

Many countries started surveying the condition of their forests as early as 1982. However, the first surveys were frequently done on a rather *ad hoc* basis, and the results are not comparable with later, more rigorous, investigations. While five or more years of continuous data may be available in individual countries, considerable care must be taken in their

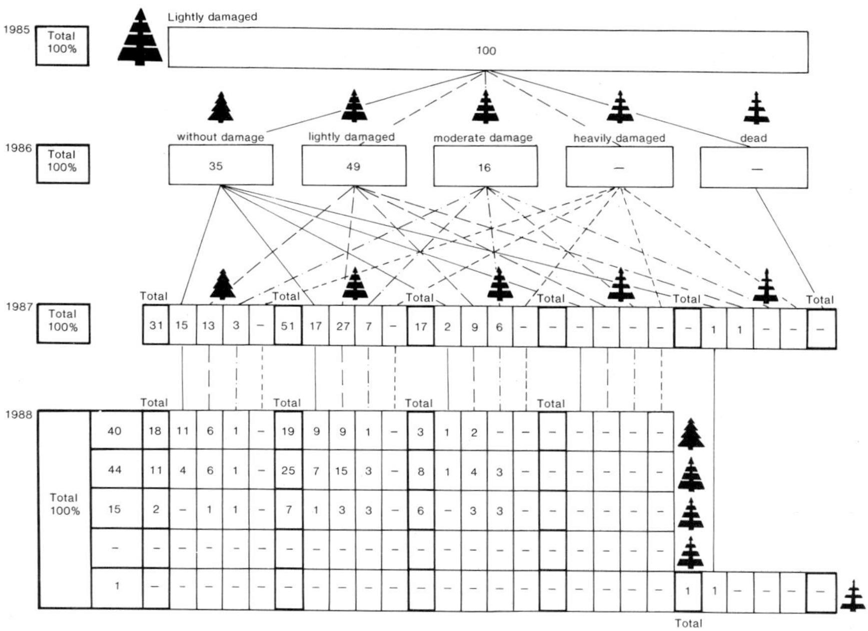

Figure 6.6 Changes in the crown transparency of individual trees in Switzerland. The changes of trees with 10–25% transparency in 1985 are shown. From: Hägi (1989).

interpretation. The results for many countries are not as internally consistent as might be believed, and the studies by Innes *et al.* (1992) and Bouhot *et al.* (1992), described in the previous chapter, have revealed major differences in the defoliation estimates between countries. Consequently, any attempt to compare the absolute amounts of defoliation between countries must be made with extreme care (if at all).

The results for the 1991 European survey are presented in Table 6.13. These were the most recent data available at the time of writing. There were substantial differences in the reported results between countries. However, nothing else can be deduced from the data in Table 6.13 because of differences in assessment standards. For example, the difference between France and the United Kingdom can be wholly explained by differences in standards (Innes *et al.*, 1992).

A more useful way to examine the results is to look at the changes within countries from one year to the next (Table 6.14 and 6.15). The differences in standards still affect these results, but not to the same extent as in Table 6.13. Sometimes, some marked changes have occurred. Changes in excess of 10% of the proportion of either conifers or broadleaves with more than 25% transparency were quite frequent.

However, no synchronous pattern was apparent when all the countries were examined. Countries with deteriorating trees usually outnumbered those with improvements and, in most countries, transparency has increased over the period that trees have been assessed.

The Commission of the European Communities has published a detailed analysis of the forest health data collected by Member States (Anon., 1991b). Unfortunately, the data suffer from the same problem as that collected by the United Nations programme (the data are actually the same, only the Commission of the European Communities programme involves fewer countries). Less emphasis is placed on comparisons between countries, but those analyses that are undertaken are based on the assumption that the differences are randomly distributed. This may be

Table 6.13 Defoliation estimates of all trees for a selection of European countries and regions in 1991

Country	n	Defoliation amount		
		0–10%	11–25%	>25%
Austria	6890	54.6	37.9	7.5
Belgium	602	43.4	38.7	17.9
Bulgaria	4724	44.7	33.5	21.8
Czechoslovakia	14369	24.1	34.6	41.3
Denmark	1626	41.5	28.6	29.9
Finland	3896	64.5	19.5	16.0
France	10255	76.4	16.5	7.1
Germany	219876	35.8	39.0	25.2
Greece	1960	51.8	31.3	16.9
Hungary	20858	48.3	32.1	19.6
Italy	5818	58.4	25.2	16.4
Liechtenstein	480	32.0	49.0	19.0
Lithuania	22836	24.6	51.5	23.9
Luxembourg	1151	55.8	23.4	20.8
The Netherlands	34100	52.5	30.3	17.2
Norway	8147	49.4	30.9	19.7
Poland	29862	9.2	45.8	45.0
Portugal	4588	47.5	22.9	29.6
Romania	243473	61.6	28.7	9.7
Slovenia	13176	62.9	21.2	15.9
Spain	10554	64.3	28.4	7.3
Sweden	11733	54.7	33.3	12.0
Switzerland	8244	32.0	49.0	19.0
United Kingdom	8843	6.0	37.3	56.7

Source: Report of the 1991 forest damage survey in Europe, published by the United Nations Global Environment Monitoring System, 1991.
Direct comparisons between countries are not possible because of differences in assessment standards.
Geographical terminology is that applicable at the time of the survey.

roughly true for the programme as a whole, but when the data are broken down by climatic region, age or otherwise, the potential for significant bias becomes considerable. Consequently, these analyses have not been described here.

A problem that affects the interpretation of the results of the monitoring programme is the removal of dead trees from the sample population. No general agreement has been reached whether or how this should be done and it remains a substantial problem. The surveys undertaken by most countries are of the current state of the forest, and no account is taken of dead or dying trees removed from the forest. It has

Table 6.14 Changes in the percentage of conifers assessed as having a defoliation score greater than 25%

	1986–87	1987–88	1988–89	1989–90	1990–91
Austria	—	—	−1.9	−1.8	−1.3
Belgium	—	+6.1	+4.2	−4.3	+12.7
Bulgaria	−0.9	+3.8	+25.3	+4.5	−10.9
Byelorussia	—	—	—	−19.0	—
Czechoslovakia	−0.8	+11.4	+5.0	+18.3	−4.3
Denmark	—	−3.0	+3.0	−5.2	+12.6
Estonia	—	—	+19.5	−8.5	+8.0
Finland	—	+3.5	+1.7	−0.7	−1.6
France	—	—	−1.9	−0.6	−0.1
Germany – east	+1.1	+5.2	+2.0	+14.0	
Germany – west	−3.6	−1.9	−0.8	—	
Germany – combined	−3.6	−1.9	−0.8	+1.8	+9.8
Greece	—	—	−1.0	+3.3	−2.8
Hungary	—	—	+3.9	+10.0	−5.5
Ireland	—	—	+8.4	−7.8	+9.6
Italy – Bolzano	—	+2.1	−0.9	+2.6	
Kaliningrad	—	—	—	−10.3	
Lithuania	—	−11.8	+21.0	−1.1	+4.9
Liechtenstein	+5.0	−4.0	−10.6		
Luxembourg	−0.4	+7.3	−1.6		
The Netherlands	−10.2	−4.2	+3.2	+3.7	0
Norway	—	—	−6.0	+2.3	+1.9
Poland	—	—	+10.3	+6.2	+6.2
Portugal	—	—	+8.1	+15.9	−5.9
Slovenia	—	—	—	—	−3.3
Spain	−7.5	−3.4	−3.8	−0.4	+4.2
Sweden	−5.5	+5.7	+0.6	+3.2	−3.8
Switzerland	−2.0	+1.0	−1.0	+5.0	+2.0
Ukraine	—	—	—	+1.6	
United Kingdom	—	+4.0	+7.0	+11.0	+6.5

Positive values indicate increases in defoliation.
Source: Report of the 1990 forest damage survey in Europe, published by the United Nations Global Environment Monitoring System, 1991.

been argued that sanitation felling will lead to improved forest health statistics, but there is little evidence of this. In Great Britain, if a tree dies, it is retained in the survey figures in future years, even if it is subsequently removed. However, trees that have been lost because of mechanical damage or normal thinning operations are replaced. In Switzerland, approximately 2% of the sample trees are removed annually (Schlaepfer and Haemmerli, 1990). Extraction of dead and damaged trees has had no significant effect on the general health statistics for the country (Hägi, 1989). The surveys have not revealed any unusual patterns of mortality in Switzerland, the observed patterns being within the range expected to occur naturally (Hägi, 1989). However, the removal of dead trees and their replacement by live trees is likely to have accelerated any apparent recovery shown by the data.

Table 6.15 Changes in the percentage of broadleaved trees assessed as having a defoliation score greater than 25%

	1986–87	1987–88	1988–89	1989–90	1990–91
Austria	—	—	−0.9	−0.8	−3.8
Belgium	—	−6.0	−1.9	−2.9	+8.3
Bulgaria	−0.9	+5.7	+7.4	+1.1	−2.0
Byelorussia	—	—	—	+11.6	
Czechoslovakia	—	—	+7.9	−3.1	−10.2
Denmark	—	−6.0	+16.0	−4.6	+1.9
Finland	—	+3.2	+4.7	−1.0	−3.9
France	—	—	−1.5	−2.9	−0.3
Germany – east	+1.2	+6.2	+3.9	+36.1	
Germany – west	+2.4	−2.7	+3.9	—	
Germany – combined	+2.4	−2.7	+3.9	+3.4	+2.7
Greece	—	—	−10.1	+8.1	+2.0
Hungary	—	—	+5.5	+9.0	−1.6
Italy	—	−0.7	+6.6	+7.2	+0.4
Liechtenstein	−3.0	−2.0	+4.0		
Lithuania	—	—	+15.0	−0.2	−0.9
Luxembourg	+4.5	+2.2	+0.8	—	
The Netherlands	+13.3	−0.9	−12.3	−1.6	
Norway	—	—	—	—	+6.9
Poland	—	—	+10.6	+7.9	+9.2
Portugal	—	—	+7.8	+25.5	+2.5
Slovenia	—	—	—	—	+1.4
Spain	0	−6.9	−3.6	+1.2	+3.0
Sweden	—	—	—	—	−13.0
Switzerland	+7.0	−8.0	−1.0	+6.0	+1.0
Ukraine	—	—	—	+1.3	
United Kingdom	—	0	+1.0	+7.8	+36.8

Source: Report of the 1990 forest damage survey in Europe, published by the United Nations Global Environment Monitoring System, 1991.

The consistency that has been achieved within the countries means that internal trends can be assessed. This has been done below for a selection of countries.

France

Results for France have been summarized by Landmann (1988a, b, c, 1991) and a series of national reports (Anon., 1990c, 1991c, 1992b). The monitoring network for France was established over several years, with the longest series of data being restricted to the eastern part of the country. A number of declines have been recognized in France. However, even in the areas believed to be most affected, mortality has remained at low levels. On a local scale, considerable mortality could occur, but it was insufficient to affect the regional estimates. Differential patterns were observed. For example, conifers improved in the north and east of the country when they were deteriorating in the south.

In addition to the regional variations, different species have varied in their crown transparency dynamics. Silver fir in the Vosges Mountains and the northern and southern Alps improved between 1985 and 1990 whereas in the Pyrenees and the Massif Central it continued to deteriorate until 1990. In the Pyrenees, a marked recovery occurred in 1991, compared with a slight deterioration in the Vosges the same year. Norway spruce, showed a similar pattern to silver fir, although the changes were less marked. Scots pine transparency had a very irregular pattern, increasing in some areas and decreasing in others. Beech showed quite marked variations, with transparency mostly increasing between 1985 and 1987 and then decreasing in 1988. Between 1989 and 1990, a marked deterioration occurred (as in Great Britain) and this was followed by a slight recovery in 1991. In many regions, the transparency of oak crowns increased steadily between 1985 and 1991.

The French results are entirely consistent with the patterns that would be expected from the occurrence of localized declines related to site-specific (e.g. soil conditions) or region-specific (e.g. climate) factors (Landmann, 1992). There is no indication of a widespread and consistent decline of forest health throughout the country.

Germany

The national figures for Germany are complicated by several factors. These include differences in survey procedures between different states, the reunification of the east and west parts of the country and the absence of a survey in some states in 1990. However, it is possible to examine trends on a state-to-state basis. Because of uncertainties over past surveys, the following paragraphs only deal with states in the former Federal Republic of Germany, and accounts have only been given for selected states.

In Lower Saxony, the proportion of trees with 0–10% transparency has fluctuated between 46% and 63% during the period 1986–91 (Bruns, 1991). The lowest figure was obtained in 1990, but the overall condition of trees subsequently recovered in 1991. Norway spruce reflected this pattern, but beech showed a steady decline in condition between 1986 and 1990, followed by a recovery in 1991.

In Schleswig-Holstein, the proportion of trees with 0–10% transparency dropped between 1984 and 1987, rose in 1988 then dropped again in 1990 and 1991. Beech did not reflect this trend. The condition of trees dropped markedly between 1984 and 1987 (91% with 0–10% transparency in 1984 compared to 22% in 1987) but has since steadily recovered (Weiss, 1991).

In Hesse, the overall figures show a progressive reduction in the proportion of trees in the 0–10% transparency category between 1984 (58%) and 1991 (29%) (Nassauer, 1991). As elsewhere, individual species have shown variations from this trend. The proportion of Norway spruce in the healthiest category has declined progressively, although there was a particularly sharp jump in the proportion of trees with more than 25% transparency between 1990 and 1991. Scots pine trees were more variable, with no clear trend apparent until 1987, since when there has been a progressive deterioration. Beech has behaved similarly to Norway spruce, with a continuous decline in the proportion of trees in the 0–10% category since 1984. Oak trees were more variable. The proportion of trees in the 0–10% category dropped markedly between 1984 and 1985 and has declined more slowly since then. The proportion of trees with more than 25% transparency increased to 1987, dropped between 1988 and 1990, then rose sharply in 1991. In all four species, the proportion of trees with more than 25% transparency in 1991 was the highest since the start of the monitoring programme.

The results for Baden-Württemberg are of interest as the Black Forest, where many of the original reports of the most recent phase of declines originated, forms an important part of the forest area of this state. The results, which are probably more reliable than for any other state in Germany, suggest that the proportion of trees with 0–10% transparency has remained fairly constant since 1984. An improvement was noted in the period 1986–88 but, since then, there has been a marginal deterioration (Braun and Schröter, 1991).

Although part of a different network to the main national monitoring programme, Baden-Württemberg is interesting as it contains some silver fir plots that have been monitored for changes in health since 1979. The results from some of these are given in Fig. 6.7. The level of deterioration was variable, but in all three, a marked and irrefutable decline in health occurred. Mortality was recorded in all cases. Stabilization of the decline

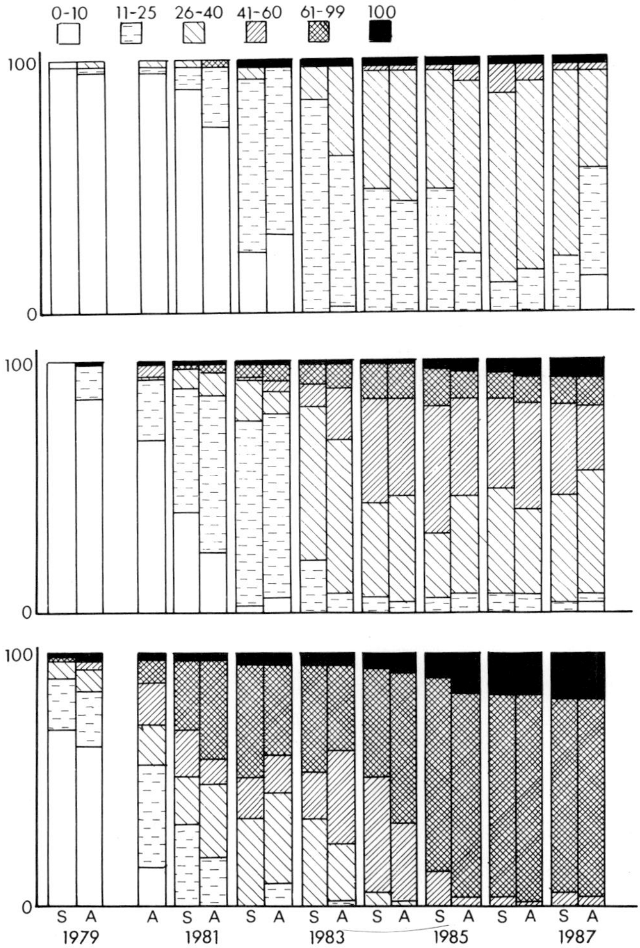

Figure 6.7 Development of crown transparency in three silver fir permanent monitoring plots in Baden-Württemberg. S: spring; A: autumn. From: Mettendorf *et al.* (1988).

occurred in 1984 or 1985, depending on the plot. Since then, there has been a recovery, although this pattern is not so readily apparent in other species (Fig. 6.8), with beech in particular continuing to decline. The results are important as similar data are unavailable for any other European country. They clearly imply that in Baden-Württemberg at least, the national monitoring started at a time when tree health was particularly poor. This has considerable implications for the use of tree condition in the mid- or late–1980s as a basepoint for the future monitoring of crown transparency and tree health in general.

The results for Bavaria are also of interest in view of the many

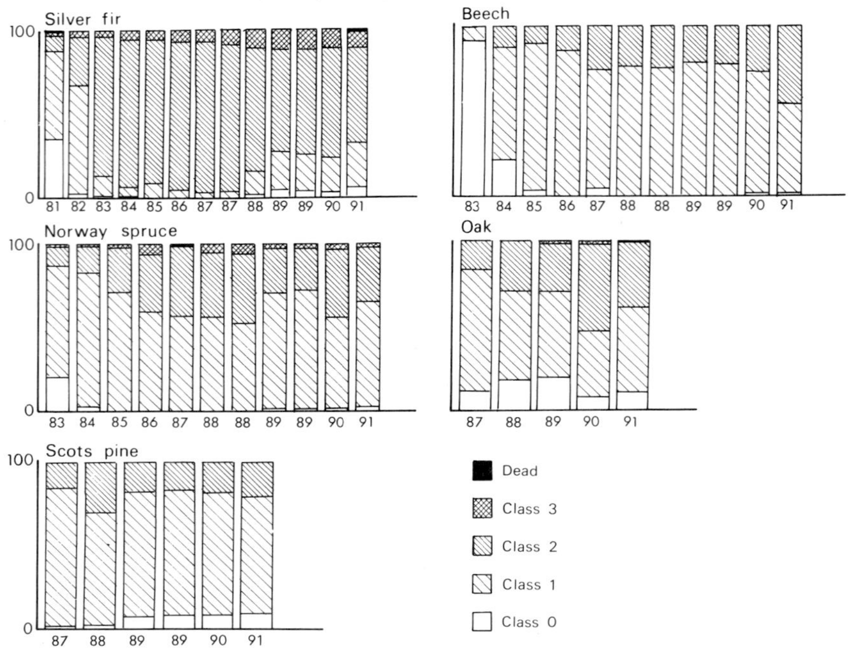

Figure 6.8 Development of crown transparency in permanent monitoring plots in Baden-Württemberg, 1981–91. Class 0: 0–10% transparency; 1: 11–25% transparency; 2: 26–60% transparency; 3: >60% transparency. From: Anon. (1991d).

reports of decline in the Bavarian Forest and the Bavarian Alps. The proportion of trees with no transparency rose between 1983 and 1987, then dropped slightly (Mößmer and Mayer, 1991). An increase was recorded between 1989 and 1991, although no survey was conducted in 1990. The proportions of trees with 11–25% and 26–60% transparency in 1991 were higher than in any other year. The pattern of change among individual species (Norway spruce, beech and oak) has been remarkably similar.

Bavaria also had a number of permanent monitoring plots during the critical period in the early 1980s. Within the state, there were quite marked regional variations in the development of the defoliation (Kennel, 1986). In south Bavaria, the proportion of Norway spruce with more than 25% transparency dropped between 1981 (32%) and 1982 (25%), then rose until 1984 (57%) before dropping again in 1985 (45%). In northwest Bavaria, there were no plots in 1981, but the proportion of Norway spruce with more than 25% transparency increased steadily from 4% in 1982 to 34% in 1985. The Oberpfälzer Forest along the Czechoslovakian border showed a completely different pattern. The proportion of Norway spruce with more than 25% defoliation decreased between 1981 (15%) and 1982 (11%), then increased dramatically to 63% in 1984 and 71% in 1985. To the south-east, in the Bavarian Forest, the proportion of Norway spruce with more than 25% transparency increased from 32% in 1981 to 77% in 1984 before dropping to 58% in 1985, and a similar pattern was evident in the Frankenwald, northwest of the Oberpfälzer Forest.

The results from Germany display a similar pattern to France, in that individual species in individual areas have behaved very differently. Even within the states, there have been substantial variations with, for example, beech in Baden-Württemberg showing different dynamics in different regions (Braun and Schröter, 1991). The results are not consistent with a widespread and pervasive decline of forests throughout the country. Instead, they again suggest the presence of localized declines controlled by site-specific and region-specific factors.

The Netherlands

Crown health is assessed in The Netherlands rather differently to many other countries. In conifers, defoliation is determined by assessing the needle retention and comparing this to what is considered normal. For example, fully foliaged Scots pine over 40 years old are considered to have on average two needle years, whereas younger ones have on average 1.5 years. Assessments are reported according to the 4-category system (0–10%, 11–25%, 26–60%, >60%). Assessments were started in 1983 and have continued since. As with many other countries, the techniques used in the first year of survey were inconsistent with later surveys and the effective starting date for the records was 1984. The survey network is one of the densest in Europe, involving 3400 sample plots, although not all of these are assessed every year.

Between 1984 and 1990, the proportion of trees classed as healthy (0–10% needle/leaf loss) ranged from 42.6–52.5% (Anon., 1990d). The condition of trees deteriorated between 1984 and 1987, but then improved. As with other countries, the general figures conceal several very marked trends. The proportion of Scots pine classed as healthy rose from 34% in 1984 to 72.2% in 1988 then dropped back to about 62% in 1989 and 1990. Austrian/Corsican pine has shown the reverse pattern, dropping from 57% healthy trees in 1984 to 13.2% in 1988 and then improving to 20.4% in 1990. Douglas fir also deteriorated markedly during the period, dropping from 50% healthy trees in 1984 to 6.7% in 1990. Despite these conspicuous changes, the proportion of Douglas fir with more than 60% needle loss only increased by 5% between 1984 and 1990.

Oak generally deteriorated between 1984 and 1987 (a drop in the percentage of healthy trees from 57% in 1984 to 19.9% in 1990) but the proportion of healthy trees has since increased. A similar pattern was evident for beech. These patterns are in marked contrast to the trends seen for these two species in Great Britain and France.

Finland

In Finland, an annual assessment of forest condition has been undertaken since 1986. Assessments are reported using a slightly different system to that used by many other countries. Instead of the uneven transparency classes (1–10%, 11–25% etc.), 10% classes are used. The survey is conducted using a systematic grid with two densities. In the north, a 32 km × 24 km grid has been adopted whereas in the south the grid is 16 km × 16 km. As well as crown transparency, several other indices are assessed, including transparency pattern, crown degeneration, fruiting, abiotic and biotic damage and the incidence of the pathogen *Ascocalyx abietina*.

Young stands of Scots pine deteriorated between 1986 and 1988, but have since been stable (Jukola-Sulonen *et al.*, 1990; Salemaa *et al.*, 1991). Older Scots pine showed higher levels of transparency in all years, with the proportion of trees with more than 10% transparency increasing between 1986 and 1988 and then showing a slight recovery in 1990 and 1991. Norway spruce has shown a similar pattern, deteriorating in condition between 1986 and 1988 and then remaining constant. The modal class of transparency was always 0–10%, contrasting markedly with the estimates for Great Britain shown in Fig. 6.4.

The highest levels of transparency seen were in the regions of Lapland, Kainuu and Ostrobothnia and it was in these regions, together with the southern coast, that the greatest increases in transparency were seen between 1986 and 1988. The causes of the defoliation differed regionally and between years. Stand age, fungal pathogens and climatic factors were all important, although air pollution was also considered a possible factor affecting tree condition along the southern coast.

An historical perspective to the pattern of defoliation has been provided by two independent studies. The first was conducted on Scots pine in northern Finland, using the vascular bundle method described in Chapter 4. In the four trees studied, there was no persistent change in needle retention between 1868 and 1988 (Jalkanen and Kurkela, 1990). The lowest levels were reached at the start of the twentieth century, and coincided with a period of reduced growth in northern Fennoscandia. Outside this period, there was no good relationship between needle retention and radial increment.

Another study involved the long-term litter fall in a Scots pine stand near Punkaharju Forest Research Station in the south of the country (Kouki and Hokkanen, 1992). The series ran from 1962 to 1986. Major variations were found during the period and no clear trend was apparent in the data. Litter fall decreased between 1983 and 1984, then increased in 1985 before dropping again in 1986. The annual variations are characteristic of Scots pine needle litter fall (Viro, 1955; Mälkönen, 1974; Albrektson *et al.*, 1977; Flower-Ellis, 1985) and suggest that crown transparency should be expected to vary substantially from year to year.

The surveys in Finland did not reveal any marked trends in the data. There was no evidence of a large-scale deterioration in forest health. However, the surveys were started relatively late (1986) and the results from more southerly countries suggest that this was when the worst of the recent phase of tree decline was nearly over. However, other studies in Finland with genuine historical perspectives do not support the hypothesis of a recent large-scale decline of forest condition in the country.

Switzerland

While some of the longest time series are available from the permanent monitoring plots in the south of Germany, the most reliable data are probably those collected in Switzerland. Unlike many European countries, the Swiss Sanasilva programme placed great emphasis on quality control. A set of standard photographs were produced (Sanasilva, 1986) and considerable emphasis was placed on training. Currently, all surveyors receive training at the beginning of the survey period and this is augmented regularly as the surveyors move through the main physiographic regions of the country. Quality control is ensured by two control teams who assess the reliability of the observations made by the six survey teams.

The survey design is different from most national monitoring programmes in that the plots, which are distributed on a 4 km × 4 km subset of the 1 km × 1 km national forest inventory grid are based on fixed areas (in the horizontal plane) rather than fixed numbers of trees. In addition, over-topped and suppressed trees are included in the assessments. Statistically, this results in a more representative assessment of the health of Swiss forests than other sampling designs, but its application to the monitoring of forest condition, which is more concerned with the changes through time of individual trees, is less clear.

Strictly comparable results are available for the period 1985 to 1991 (Sanasilva, 1991). Nationally, the proportion of trees with less than 10% transparency has declined from 64% in 1985 to 32% in 1991. The decline has been irregular, and trees showed some recovery in 1988 and 1989. The proportion of trees with more than 60% transparency has remained constant throughout the survey period. The national trend is closely reflected by the aggregated trends shown by conifers and broadleaves. Both groups have shown an overall increase in transparency between 1985 and 1991, with a period of lower transparency in 1988 and 1989.

As in other national programmes, individual species have not always followed the national trend. The patterns shown by Norway spruce, beech, maple and ash closely reflect the national trends. Scots pine also reflects the national pattern, but the changes have been less marked, probably because of the higher initial levels of transparency. The transparency of silver fir declined between 1985 and 1987, recovered a little in 1988, then continued to decline. A similar pattern was evident in oak. European larch shows the greatest departure from the national trend. A sharp increase in transparency between 1985 and 1986 was followed by a recovery to 1988 and then an increase in transparency.

Regional results are difficult to interpret as they are not presented by species and differences in the distribution of species mean that differences between regions may be more attributable to the species composition than any genuine regional trends in individual species. Most regions revealed the improvement in transparency levels in 1988 and/or 1989 and the subsequent deterioration. However, in the Alps, an improvement in tree condition occurred in 1991, in marked contrast to the deterioration seen in the Mittelland, Voralpen and Alpensüdseite regions.

The results from Switzerland appear to show more coherency between species than many other countries. However, as the country is small compared to Germany and France, the consistency might be expected. The differences between the trends seen in Switzerland and the neighbouring countries are more difficult to interpret. For example, it is not clear why Swiss silver fir in the Jura Mountains should be showing a different trend to French silver fir in the same mountain range. Similarly, the trends seen in the Mittelland region appear to differ from the trends seen in the adjacent Black Forest of Germany. These apparent discrepancies need to be resolved.

Conclusions

The results from Great Britain and the rest of Europe do not support the hypothesis of a widespread decline phenomenon spreading out across Europe during the 1980s. This scenario, which was advocated by several scientists who should have known better, is wrong. Instead, the evidence supports the idea of many local factors affecting the condition of individual tree species in different regions and countries. Consequently, the value of assessing changes in the condition of European forests as a whole seems limited.

The results reveal the importance of breaking down the data by species and region. This implies that considerable sample sizes are required to determine change reliably. The 16 km × 16 km grid adopted by the United Nations Economic Commission for Europe and the Commission of the European Communities would have a sufficient density to provide adequate sample sizes in areas with relatively homogeneous environments and forest compositions. However, over much of Europe, neither the environment nor the forest composition and structure are homogeneous at a scale of 16 km × 16 km. Consequently, greater sampling densities are required. These have been installed in most countries and, for example, are reported under the individual country sections of the UN-ECE reports.

The data from these more detailed grids reveal a complex pattern of improvements and deteriorations in individual species. The area and species affected by a particular deterioration in condition is largely controlled by the nature of the stress. For example, summer drought may affect large areas whereas damage by hail or wind tends to be much more localized.

It is now very unlikely that further standardization in the 'defoliation' assessments between countries will be achieved. Comparisons between absolute levels of defoliation/transparency in different countries have been avoided here as the results from the previous chapter show that such comparisons would be nonsensical. Instead, emphasis has been placed on comparing the changes occurring within countries. Even this approach is of questionable validity, as has been shown. There seems to be very little justification for aggregating data from different species and different regions if the causes of changes are being sought. A much more valuable approach would be to identify reasonably homogeneous regions and investigate the changes of individual species within these. Such comparisons would be based on the assumption that changes in transparency (as opposed to their absolute levels) are unaffected by national or regional differences in assessment, which is questionable. However, such an approach seems to be the most promising avenue for further research.

A final point that should be made concerns the reliance of several international bodies on the transparency figures generated from the inventories of forest condition. The studies that have been made clearly reveal that comparisons among the absolute levels of defoliation in different countries are invalid. However they continue to be published. While more difficult to collate, national data are available and are regularly published or otherwise made available in all countries. Future studies using transparency data should go back to the source material rather than relying on the politically-motivated syntheses published by the UN-ECE. Had this been done from the start, then much of the uncertainty surrounding the nature, timing and distributions of forest declines in Europe would have been avoided.

7 Assessment of Other Indices of Tree Condition

> A few years ago, when it became a matter of quantifying 'forest decline', scientists and foresters faced a major difficulty: there was no universally acknowledged health indicator.
>
> *Landmann, 1991*

As shown in Chapter 4, there are many indices of tree condition that can be used to assess the condition of a forest. In the British monitoring programme, a variety of indices have been adopted; these are listed in Table 7.1. Some of these indices are related to the nature or position of the tree in the stand and are not strictly indices of tree health. However, they have been summarized here as they have an influence on tree health parameters. The addition of extra indices of tree condition provides invaluable information for the interpretation of crown transparency data. However, very few countries include such data in their reports and this chapter deals almost exclusively with data from Great Britain.

Dominance

Four categories of tree dominance were recognized in the British programme, although sampling was restricted to trees in the first three classes. The results are presented in Table 7.2. As expected, most trees were in either the dominant or co-dominant classes. Trees in the intermediate category might be expected to suffer from a certain amount of shading from adjacent dominants and co-dominants, and this factor needs to be considered when examining the data. The absence of trees in the suppressed category reflects the deliberate exclusion of such trees. Interestingly, despite some trees being selected as early as 1984, no trees changed from being dominant/co-dominant/intermediate to suppressed during the monitoring period.

Table 7.1 Indices determined in the 1991 British assessment programme

Variable	Sitka spruce	Norway spruce	Scots pine	Oak	Beech
Branch density	x	x	x		
Branch pattern	x	x			
Browning of current needles	x	x	x		
Browning of leaves				x	x
Browning of older needles	x	x	x		
Butt damage	x	x	x	x	x
Canopy closure around tree	x	x	x	x	x
Crown transparency	x	x	x	x	x
Crown dieback extent				x	x
Crown dieback location				x	x
Crown dieback type				x	x
Crown form			x	x	x
Diameter at breast height	x	x	x	x	x
Defoliation type	x	x	x	x	x
Degree of leaf-rolling					x
Epicormics on branches				x	
Epicormics on the stem				x	
Flowering in lower crown			x		
Flowering in upper crown			x		
Frequency of leaf-rolling					x
Fruiting	x	x	x	x	x
Fungal presence	x	x	x	x	x
Insect damage	x	x	x	x	x
Leader condition	x	x	x		
Leaf size					x
Mechanical damage to crown	x	x	x	x	x
Needle retention	x	x	x		
Overall discoloration	x	x	x	x	x
Premature leaf loss					x
Secondary shoot frequency	x	x			
Secondary shoot location	x	x			
Shoot death extent	x	x	x		
Shoot death location	x	x	x		
Stem damage	x	x	x	x	x
Tree dominance	x	x	x	x	x
Type of leaf browning				x	x
Type of leaf yellowing				x	x
Type of needle yellowing	x	x	x		
Yellowing of current needles	x	x	x		
Yellowing of leaves				x	x
Yellowing of older needles	x	x	x		

Table 7.2 Percentages of trees involved in the assessment programme, 1991, in each of the dominance categories

		Category			
Species	N	1	2	3	4
Sitka spruce	1746	23	69	8	0
Norway spruce	1840	22	72	6	0
Scots pine	1944	20	73	7	0
Oak	1847	17	77	6	0
Beech	1464	17	80	3	0

Classes: 1: dominant trees; 2: co-dominants; 3: intermediate; 4: suppressed.

Canopy closure

Canopy closure was assessed by recording the number of quadrants of the crown that were touching the crowns of adjacent trees. This variable would not be expected to be equally distributed as crowns in the highest category (i.e. with all four quadrants having adjacent crowns) would only be visible with difficulty. In addition, the sampling design is biased towards trees growing on stand edges as it is often difficult to see the crowns of trees within stands.

The results in Table 7.3 suggest that most conifers were scored in classes 2 and 3, with few trees in classes 0 and 4. In contrast, significant proportions of oak and beech were in class 4. This reflects the nature of forest stands in Great Britain, with conifers generally occurring in dense blocks that can only be assessed on stand edges and broadleaves occurring in more open stands. As with dominance, the degree of canopy

Table 7.3 Crowding categories for trees involved in the assessment programme, 1991

		Category				
Species	N	0	1	2	3	4
Sitka spruce	1746	1	9	34	50	6
Norway spruce	1840	4	15	37	40	4
Scots pine	1944	8	21	38	29	4
Oak	1847	3	14	26	38	19
Beech	1464	2	11	26	42	19

Categories: 0: crown open-grown with no evidence of shading effects;
1: crown in contact with others on one side;
2: crown in contact with others on two sides;
3: crown in contact with others on three sides;
4: crown in contact with others on four sides.

closure may affect some other scores and this should be accounted for when examining the data.

Crown form and defoliation type

Crown form was assessed in Scots pine, oak and beech. Defoliation type was assessed in all species. Most Scots pine (71%) were scored as having vigorous apical dominance (class 1), as might be expected from the young age distribution of the sample trees. Reduced apical dominance (class 2) accounted for 22% of the trees and a further 5% had no apical dominance (class 3). Only 2% of the trees showed signs of developing the platform typical of over-mature trees (classes 4–6).

In broadleaves, crown form was assessed using the methods developed by Roloff (1985a, 1989, 1991). The reliabilty of these estimates is uncertain as I found that surveyors had considerable difficulties with the index. However, the problems had been largely overcome by the 1991 survey. In oak, half the trees were scored as having both side shoots and apical shoots with reduced growth. Crown development had stopped in 5% of the trees and was classed as vigorous in only 13% of the trees. These figures are not altogether surprising as the stands had a minimum age of 80 years. This means that most trees would have been past their most active phase of shoot extension growth. A similar distribution was seen in beech, with reduced apical and side shoot growth in 40% of the trees, vigorous growth in 20% and halted crown development in 5%.

In 1991, the most common form of defoliation in the two spruce species involved the loss of needles from the base of the crown upwards (class 6) (Fig. 7.1). This contrasts with previous years when the uniform loss of needles from throughout the crown (class 4) was the most common type of defoliation. In Finland and Sweden, the window (classes 1 and 2) is the commonest type of defoliation (Westman and Lesínski, 1986; Salemaa *et al.*, 1991), suggesting the causes of defoliation in Great Britain and Sweden may differ.

In Scots pine, the most common form of defoliation was the uniform loss of needles from throughout the crown (class 4), with defoliation of the lower crown (class 1) and gap-like defoliation (class 3) also being important (Fig. 7.2). As with the two spruce species, there was a marked increase in the proportion of trees scored as defoliated in the lower part of the crown. The amount of trees described as having gap-like defoliation (class 3) has also increased since 1989. These results are similar to those from Finland, where the most common type of defoliation in Scots pine recorded by Salemaa *et al.* (1991) was the gap-like category, closely followed by uniform defoliation.

The reasons for the changes in defoliation type are not readily apparent. One possibility is that the differences were due to changes

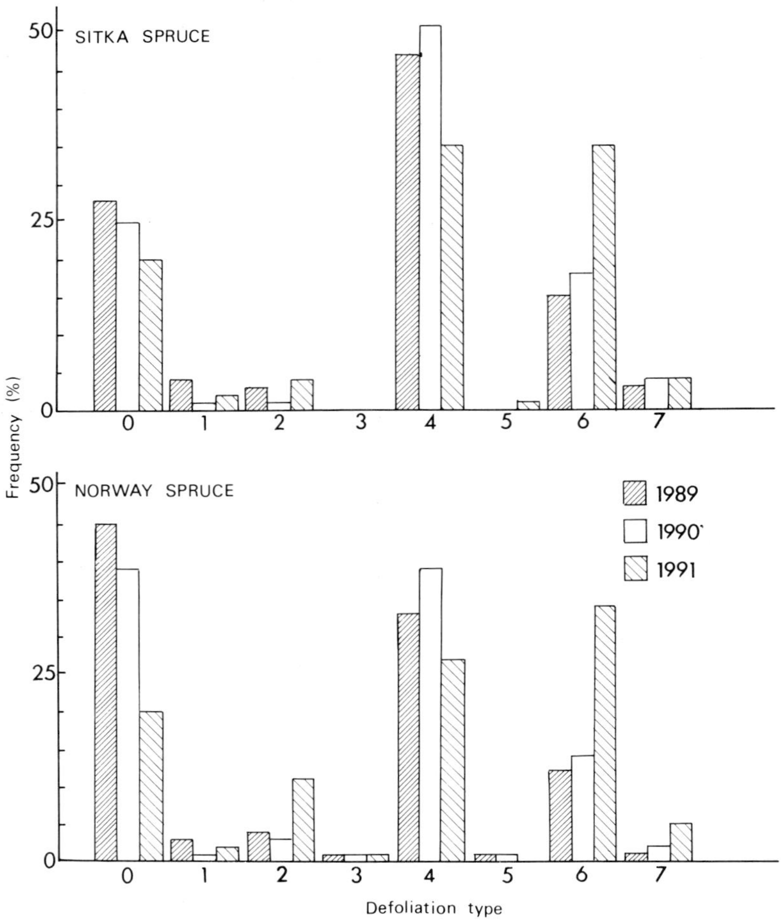

Figure 7.1 Defoliation types in spruce. 0: no obvious defoliation; 1: small window in upper crown; 2: large window; 3: top-dying; 4: uniform loss of needles throughout the crown; 5: peripheral defoliation; 6: loss of needles from base upwards; 7: other. (Figures are percentages of trees in each species in each class.)

in observer perception of defoliation type. However, it is unlikely that observer inconsistency would cause changes of the magnitude seen over the three years. Another possibility is that changes in the sample of trees caused the variations. This is also unlikely as only Sitka spruce has changed substantially over the three years (with an increase in the sample size in 1991).

With Sitka spruce (Table 7.4), the most frequent changes were between the no defoliation (class 0), overall thinning (class 4) and base of the crown (class 6) categories. A similar pattern was apparent with Norway spruce (Table 7.5), although many trees classed as having a large window (class 2) in either 1989 or 1990 were classified as having uniform loss of needles in 1991. The proportion of Norway spruce changing from basal thinning to uniform thinning and vice versa was smaller than with Sitka spruce. In Scots pine (Table 7.6), the greatest amount of interchange occurred between defoliation of the lower crown, gap-like defoliation and uniform loss of needles. There was a particularly marked change in the uniform defoliation category, with 209 and 140 of the 755 trees in this category in 1989 classed as having defoliation of the lower crown or gap-like defoliation, respectively, in 1991.

The most common forms of defoliation in oak and beech involved either small gaps in the crown (class 2) or gaps in the lateral branch systems (class 3) (Fig. 7.3). The largest changes over the period 1989–91 occurred in the lowest category (no defoliation), a trend which corresponds with the changes in crown transparency reported in the previous chapter. In line with these changes, the proportions of trees in classes 2 and 3 (small gaps, gaps in the lateral branch system) increased (Tables 7.6 and 7.7). In contrast to the defoliation types in the conifers, the defoliation types in broadleaves represented a progression from healthy to dying trees. However, there was no evidence of a gradual shift beyond class 3, suggesting that there was no trend of worsening condition among trees having some defoliation in 1989.

As with the conifers, there was evidence of a considerable interchange

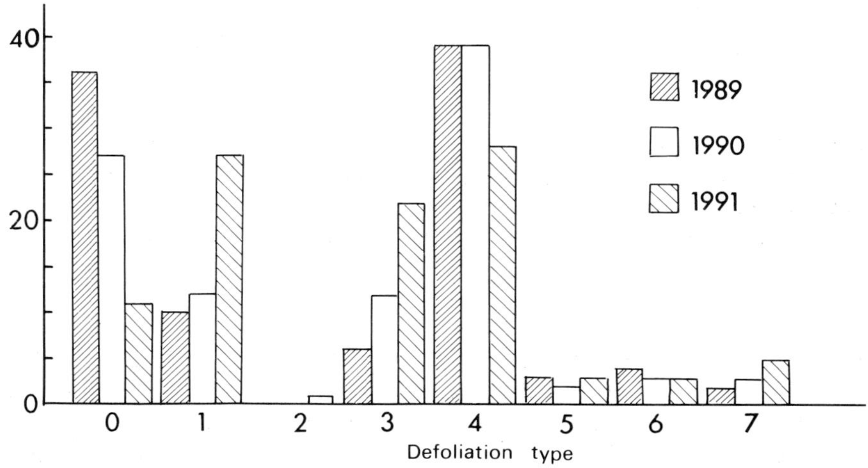

Figure 7.2 Percentages of Scots pine with different defoliation types. 0: no obvious defoliation; 1: lower crown only; 2: peripheral defoliation; 3: gap-like defoliation; 4: uniform loss of needles; 5: spot-like defoliation; 6: upper crown only; 7; other.

Table 7.4 Changes in defoliation type of Sitka spruce between 1989 and 1991, and 1990 and 1991

Defoliation type 1991	Defoliation type 1989								Defoliation type 1990							
	0	1	2	3	4	5	6	7	0	1	2	3	4	5	6	7
0	**163**	3	2	0	49	0	17	7	**164**	0	0	0	36	1	35	6
1	8	**4**	0	1	11	0	1	3	8	**1**	0	0	13	0	2	2
2	12	3	**4**	0	23	0	10	4	19	0	**2**	0	24	0	7	5
3	1	1	0	**0**	1	0	0	0	1	0	0	**0**	1	0	1	0
4	84	14	14	3	**298**	2	55	10	47	4	9	2	**340**	0	60	11
5	7	0	0	0	4	**0**	0	0	2	0	1	0	8	**0**	0	0
6	91	16	19	3	215	2	**97**	8	85	7	5	0	227	1	**125**	6
7	19	5	3	0	12	0	3	**8**	14	2	2	0	12	0	1	**15**

Categories are the same as in Fig. 7.1.

Table 7.5 Changes in defoliation type of Norway spruce between 1989 and 1991, and 1990 and 1991

Defoliation type 1991	Defoliation type 1989								Defoliation type 1990							
	0	1	2	3	4	5	6	7	0	1	2	3	4	5	6	7
0	**257**	2	3	2	34	2	19	0	**260**	5	2	2	32	2	15	3
1	17	**3**	1	2	7	0	1	1	12	**3**	0	3	9	0	3	2
2	65	7	**23**	1	60	0	23	2	46	0	**15**	1	92	0	26	2
3	8	0	4	**1**	4	0	5	2	9	0	5	**3**	2	0	5	0
4	100	15	27	4	**253**	2	33	5	62	8	12	6	**301**	1	42	11
5	2	0	0	0	2	**0**	1	0	1	1	0	0	3	**0**	0	0
6	226	23	13	3	200	8	**107**	5	197	3	5	5	214	6	**150**	5
7	59	2	1	0	8	0	5	**2**	54	2	2	0	14	0	0	**5**

Categories are the same as in Fig. 7.1.

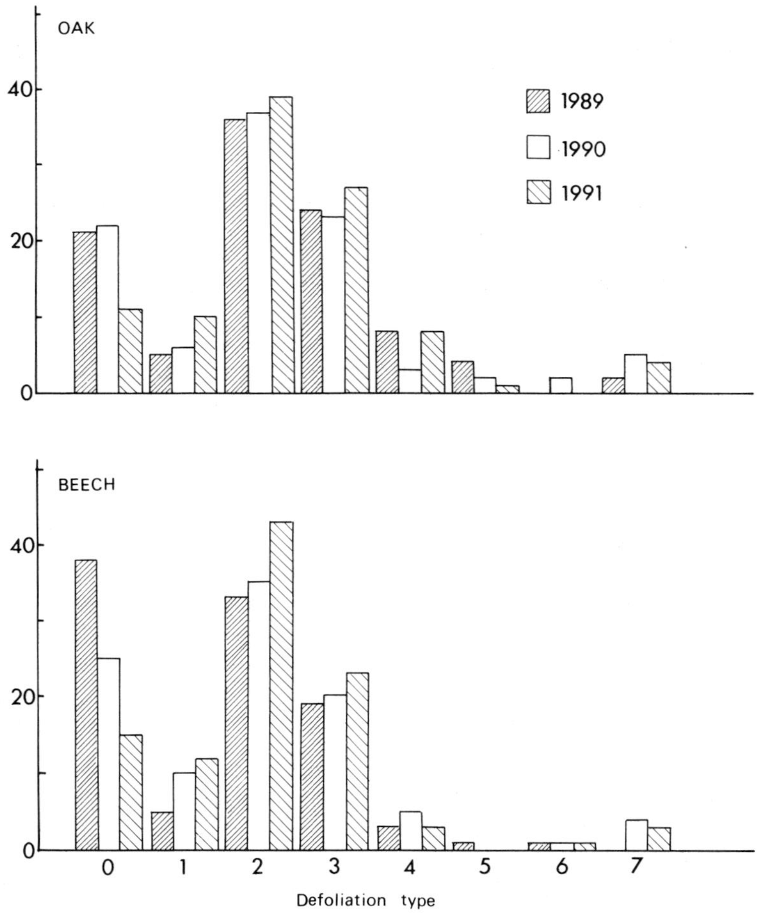

Figure 7.3 Percentages of broadleaves in specific crown pattern categories, 1991. 0: 0-15% loss of density; 1: no clear pattern; 2: small gaps; 3: gaps in the lateral branch system; 4: large gaps in lateral branch system; 5: predominantly large gaps; 6: whole or part of crown completely defoliated; 7: other. Figures are percentages of trees in each species in each class (i.e. rows sum to 100).

in the scores of individual trees in different years (Tables 7.7 and 7.8). Most of the oak trees with increasing levels of defoliation moved from class 0 to class 2, as might be expected. Many trees in classes 2 and 3 oscillated between these two classes over the three years. A similar pattern was evident for beech, although the numbers of trees moving between classes 2 and 3 were greater.

Table 7.6 Changes in defoliation type of Scots pine between 1989 and 1991, and 1990 and 1991

Defoliation type 1991	Defoliation type 1989								Defoliation type 1990							
	0	1	2	3	4	5	6	7	0	1	2	3	4	5	6	7
0	**146**	17	0	4	26	1	0	4	**163**	15	0	8	22	0	1	2
1	163	**61**	0	27	212	15	15	8	127	**100**	2	51	209	7	19	8
2	6	1	**0**	2	1	2	1	0	6	2	**0**	3	1	1	0	1
3	156	49	0	**46**	130	12	16	11	92	40	1	**115**	140	9	7	23
4	109	43	1	26	**263**	20	21	6	79	53	1	31	**325**	8	17	10
5	9	7	0	2	32	**7**	2	1	4	5	0	15	25	**9**	2	0
6	13	5	0	1	18	2	**10**	1	9	5	0	8	15	0	**14**	3
7	40	10	0	4	19	0	1	**0**	41	3	0	5	18	0	5	**19**

Categories are the same as in Fig. 7.2.

Table 7.7 Changes in the defoliation type scores for oak between 1989 and 1991, and 1990 and 1991

Defoliation type 1991	Defoliation type 1989								Defoliation type 1990							
	0	1	2	3	4	5	6	7	0	1	2	3	4	5	6	7
0	**110**	5	27	6	0	0	0	0	**111**	11	21	7	1	0	0	0
1	24	**5**	25	9	1	0	0	1	8	**18**	30	5	6	0	0	6
2	139	17	**137**	65	1	0	2	1	79	43	**181**	54	3	1	1	8
3	36	7	58	**57**	20	4	2	1	12	13	47	**83**	20	1	5	8
4	1	1	8	8	**3**	2	0	0	0	3	2	8	**8**	1	0	1
5	0	0	1	3	0	**0**	0	0	0	0	1	2	0	**0**	1	0
6	1	0	1	2	1	0	**1**	0	0	0	2	2	0	0	**0**	2
7	5	3	14	9	1	0	0	**0**	0	0	8	10	2	0	4	**11**

Categories as in Fig. 7.3.

Table 7.8 Changes in the defoliation type scores of beech between 1989 and 1991, and 1990 and 1991

Defoliation type 1991	Defoliation type 1989								Defoliation type 1990							
	0	1	2	3	4	5	6	7	0	1	2	3	4	5	6	7
0	**89**	12	38	7	1	1	0	1	**129**	12	35	10	0	1	1	4
1	29	**7**	47	24	7	0	0	1	44	**28**	60	20	4	0	0	11
2	96	30	**216**	104	24	5	0	3	139	49	**361**	104	6	7	7	13
3	33	12	124	**131**	43	18	1	9	29	12	136	**211**	18	9	9	16
4	11	1	20	29	**19**	5	0	1	16	3	30	38	**19**	13	5	3
5	0	2	2	3	2	**12**	0	1	0	2	1	3	3	**10**	4	1
6	0	0	0	1	0	0	**1**	0	0	0	0	3	0	0	**1**	0
7	9	4	18	5	1	4	0	**5**	9	2	14	3	2	3	0	**35**

Categories as in Fig. 7.3.

Dead shoots in conifers

Shoot death in conifers was assessed from the location within the crown, location on branches and frequency. In all three species, the extent of shoot death increased from 1989 to 1991 (Fig. 7.4, Tables 7.9 and 7.10).

Table 7.9 Location of shoot death on individual branches within the crowns of trees in the assessment programme

	Location					
	0	1	2	3	4	5
Sitka spruce						
1989	56	4	5	5	22	8
1990	46	5	2	7	25	15
1991	35	11	3	10	23	18
Norway spruce						
1989	53	5	3	6	19	14
1990	47	6	1	6	23	17
1991	33	17	3	5	22	20
Scots pine						
1989	52	16	3	1	6	22
1990	43	18	1	6	7	25
1991	31	37	1	2	8	21

Categories: 0: no shoot death; 1: outermost shoots only; 2: shoots on the middle of the branch; 3: shoots on innermost part of branch; 4: middle and inner parts of branch; 5: all over branch.

Table 7.10 Extent of shoot death in the live crowns of conifers, 1991

	Percentage frequency				
	Absent	Rare	Scattered	Common	Abundant
Sitka spruce					
1989	56	7	14	21	2
1990	46	8	16	28	2
1991	35	19	23	20	2
Norway spruce					
1989	53	8	18	19	1
1990	47	10	19	23	1
1991	33	23	22	20	2
Scots pine					
1989	51	10	19	10	1
1990	43	22	23	10	2
1991	31	38	21	9	1

The percentage of trees in each species in a given category is shown.

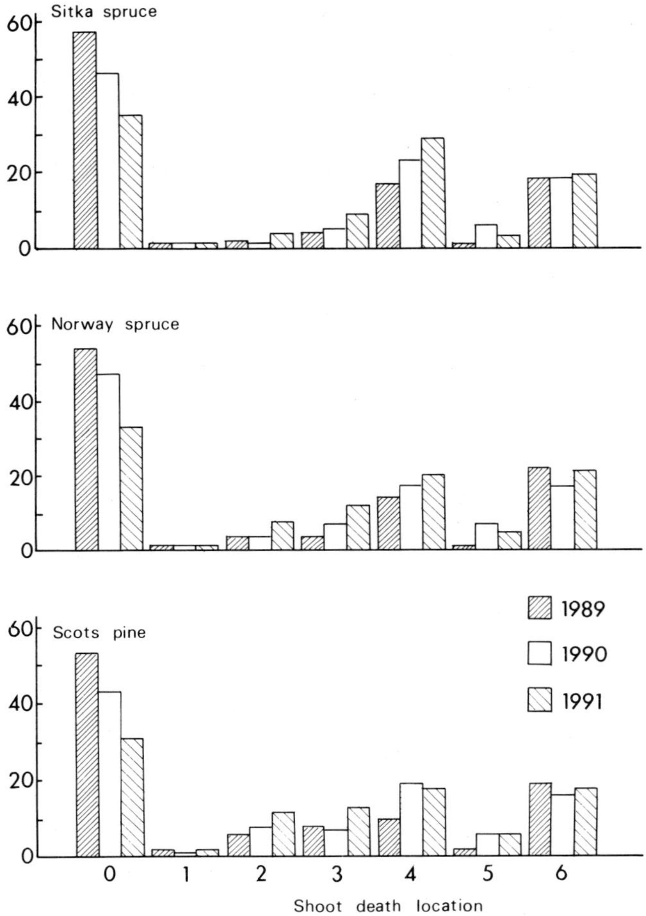

Figure 7.4 Location of shoot death within the crowns of trees in the assessment programme. Categories: 0: no shoot death; 1: top three branch whorls only; 2: top quarter of crown; 3: top half of crown; 4: bottom half of crown; 5: middle part of crown only; 6: all over crown.

Shoot death was recorded in over 65% of the trees assessed in 1991. Given the marked increase in the proportion of trees with shoot death since 1989, the figures for 1991 may be important. However, it is unknown whether the figures for 1991 lie within the range that might be expected given the year-to-year fluctuations in the various agents causing shoot death (e.g. insects, climatic extremes, fungal infections etc.). In Finland, shoot and/or branch death was recorded on 43% of the Scots pine and 76% of the Norway spruce assessed in the survey

(Salemaa *et al.*, 1991). These results suggest that the data from Great Britain may be within the range that occurs naturally.

The location of shoot death within the crowns was broadly similar from year to year. Mostly, shoot death was located either in the bottom half of the crown or all over the crown. However, in 1991, there was a marked increase in the proportion of Scots pine with shoot death in the upper half of the crown. As the proportion of trees in the other categories (except class 0) remained similar, the increase must have occurred primarily in trees with no previous shoot death. The increase is likely to be related to attacks by the pine shoot beetle. As with the location within the crown, the location of dead shoots on individual branches was reasonably consistent over the period. However, all three species had a noticeable increase in the proportion of outer shoots recorded as dead in 1991. The increase was marked for Scots pine and was also considerable in Norway spruce.

Although an increase in the proportion of trees with shoot death was recorded over the three years, the incidence of shoot death was not particularly high. The incidence did not show any major increase over the three years. The proportion of trees scored as having either common or abundant dead shoots remained approximately the same over the period, with the greatest increases occurring in the rare and scattered categories. Generally, the extent of shoot death was greater in the two spruce species than in Scots pine. The figures in Table 7.10 suggest that there is an endemic level of shoot death which is due to factors operating in most years (e.g. insect attack). Exceptional events will also cause shoot death in conifers, but no such event was recorded between 1989 and 1991.

Crown dieback in broadleaves

As with the occurrence of shoot death in conifers, the extent of crown dieback in broadleaves provides a good measure of the health of individual trees. As a result, there has been a considerable effort devoted to devising suitable ways of assessing dieback. The system used here is based on Westman (1989) and involves assessing the degree (Fig. 7.5), location (Fig. 7.6) and extent of dieback (Table 7.11).

Over the three-year period, the number of oak trees with dieback showed a consistent increase. The changes always involved moves from scores of no dieback to scores of minor amounts of dieback. There was no increase in the proportions of oak trees scored with medium to severe levels of dieback over the three years. The proportion of beech showing dieback increased dramatically between 1989 (53%) and 1990 (76%), but did not increase further in 1991. As with oak, dieback was minor, and medium to severe dieback was very rare.

Most dieback in both species was located in the top of the crown (class 1), the top and middle (class 3) or all over the crown (class 5). Suppressed branches were omitted from the assessments and this is reflected in the low scores for dieback in the lower part of the crown. There was no obvious change in the pattern of dieback over the three years.

The extent of dieback remained similar over the three years, with most trees developing the symptom being scored as having less than 20% of the crown affected. As with shoot death in conifers, there appears to be an endemic level of dieback that occurs in response to the normal range of stresses affecting trees. Severe dieback, although recorded, did not occur in enough trees for a percentage score to be recorded.

Considerable variation was apparent when the changes in individual trees were examined (Table 7.12). Few trees remained in the same

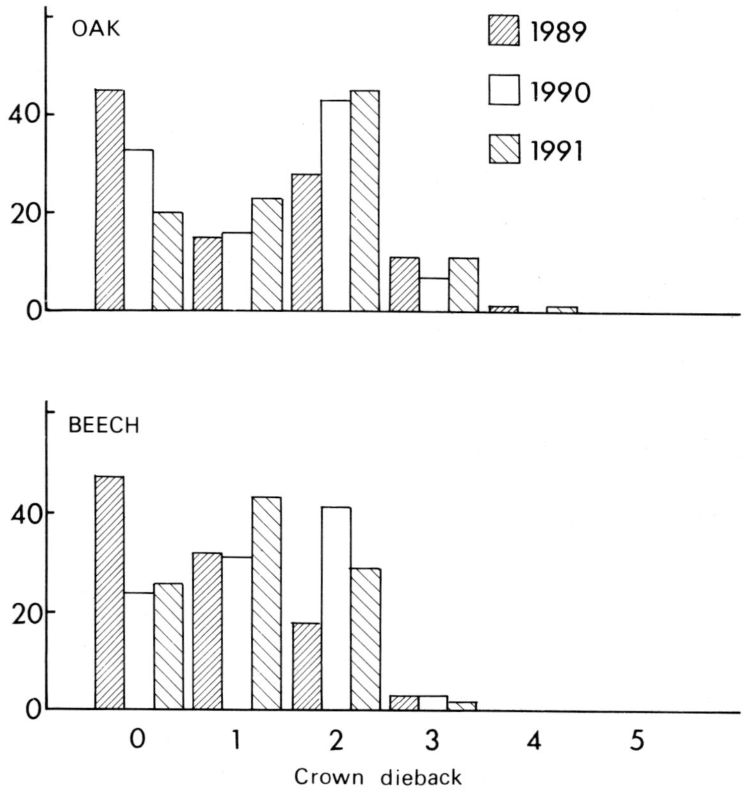

Figure 7.5 Percentages of oak and beech in each crown dieback class. 0: no dieback; 1: leaf loss only; 2: breaks to thin branches; 3: breaks to thick branches; 4: stem broken; 5: other. The percentage of trees in each species in a given category is shown.

dieback class from one year to the next, with substantial changes occurring in category 1 (leaf loss only). This was expected because of the transient nature of class 1. It is not clear whether this class represents a definite dieback category; leaves may be lost from shoots without the shoots necessarily dying. Although a general trend is apparent in the overall figures for dieback, the results in Table 7.12 reveal that individual trees are much more dynamic than might be expected and that both improvements and deteriorations occur in any given year.

The location of dieback in the crowns also varied (Table 7.13). As expected from the figures presented above, many trees without dieback in 1989 were subsequently affected. In beech, the greatest increases were related to dieback in the tops of the crowns. However, in oak, there was a much more uniform spread of dieback throughout the crown, with many trees being placed in this category. In beech, most changes occurred during 1989–90 (not shown in Table 7.13), reflecting the

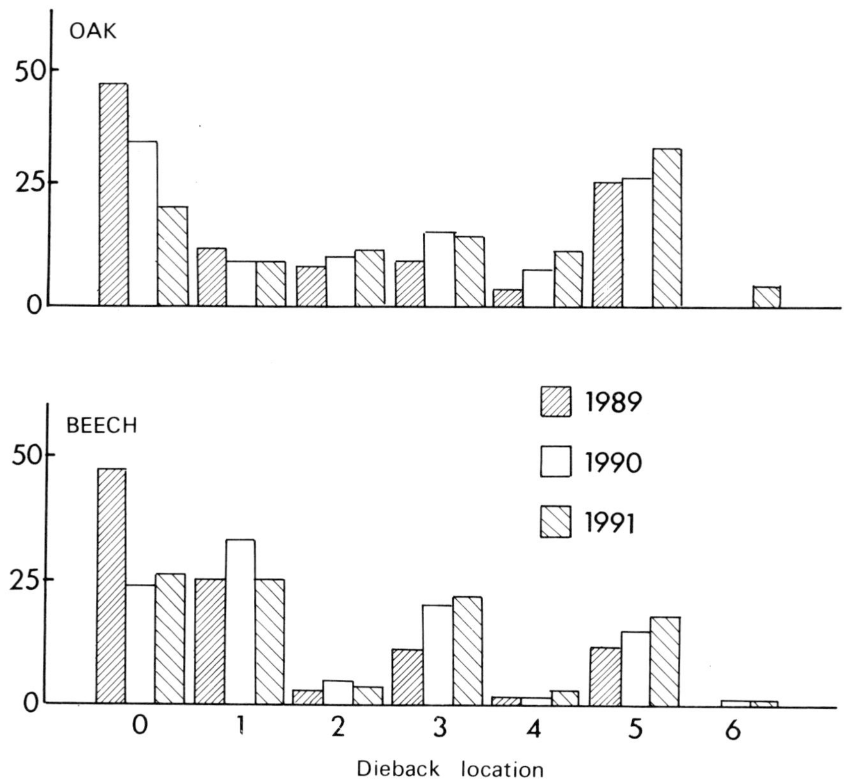

Figure 7.6 Location of dieback within individual crowns. Categories: 0: no dieback; 1: top of the tree only; 2: middle parts of the crown; 3: top and middle of the crown; 4: base of the crown; 5: throughout the crown; 6: other.

Table 7.11 Extent of dieback (in 10% classes)

	0	1–10	11–20	21–30	31–40	41–50	51–60	61–70	71–80	81–90	91–100
Oak											
1989	45	33	14	5	2	1	0	0	0	0	0
1990	33	53	10	2	1	1	0	0	0	0	0
1991	20	56	18	4	2	0	0	0	0	0	0
Beech											
1989	47	36	14	2	1	0	0	0	0	0	0
1990	24	60	12	3	1	0	0	0	0	0	0
1991	26	59	11	3	1	1	0	0	0	0	0

The percentage of trees of each species in a given category is shown.

importance of the 1989 drought (Innes, 1992a). In contrast, the changes in beech in 1990–91 were relatively minor, although there was a clear trend for the progression of dieback in some trees affected in 1990. Oak showed the reverse, with greater changes in 1990–91 than in 1989–90.

Although there has been little interest in the dieback of beech (as opposed to changes in branch structure), dieback of oak has received considerable attention (Eichholz, 1985; Balder and Lakenberg, 1987; Donaubauer, 1987; Jakucs, 1988; Hartmann *et al.* 1989). These studies suggest that the incidence of dieback is increasing in some areas. The levels recorded in Great Britain between 1989 and 1991 were not particularly unusual, although there is still a major lack of base-line information.

Discoloration

When forest health surveys were initiated, great emphasis was placed on the incidence of discoloration (see Chapter 3). This was because discoloration was one of the main symptoms associated with the observed declines. Foliage discoloration in trees in Great Britain was generally extremely rare between 1987 and 1991. Both browning (Table 7.14) and yellowing (Table 7.15) involved less than 10% of the conifers assessed, and the discoloration was primarily related to older needles. Sitka spruce

Table 7.12 Changes in the degree of dieback between 1989 and 1991

Score in 1991	Score in 1989					
	0	1	2	3	4	5
Oak						
0	**197**	32	24	8	0	0
1	135	**48**	107	16	0	0
2	198	100	**192**	75	2	1
3	34	14	35	**41**	4	0
4	1	0	1	2	**1**	0
5	0	0	1	0	0	**0**
Beech						
0	**166**	49	4	0	0	0
1	122	**124**	57	4	0	0
2	95	81	**82**	10	2	0
3	4	1	13	**6**	0	1
4	0	2	0	2	**0**	0
5	0	0	0	0	0	**0**

0: none; 1: leaf loss only; 2: dieback restricted to relatively thin branches; 3: several large branches involved; 4: main stem involved.
Numbers of trees in each category are given.

showed no marked annual fluctuations in browning during the period, but the yellowing scores showed some variation. In current-year needles, there was more yellowing in 1990 than in any other year. In contrast, the older needles showed the least yellowing in 1990 and the greatest amount in 1987.

In Norway spruce, there was an increase in the incidence of browning of current-year foliage in 1991. This was associated with the development of the needle reddening syndrome in some places (Innes and Boswell, 1991). An increase in browning was also apparent in older foliage. Needle yellowing remained fairly constant during the period. However, an increase in the incidence of yellowing of older needles was recorded in 1991. This may well reflect better recording of yellow-tipping (see below) than any real change.

The amount of discoloration of Scots pine was also reasonably constant except in 1991, when marked increases in the extent of browning were reported. This primarily occurred in current-year needles and is believed to reflect a major increase in the incidence of shoot death caused by the pine shoot beetle. The amount of yellowing of current-year foliage was marginally greater than in previous years.

In oak (Table 7.16), the amount of leaf-browning remained similar between 1987 and 1990, but a marked increase was noted in 1991. This

Table 7.13 Changes in the location of dieback between 1989 and 1991

Location in 1991	Location in 1989						
	0	1	2	3	4	5	6
Oak							
0	**197**	20	15	4	6	18	0
1	67	**23**	3	16	2	11	0
2	69	16	**20**	8	8	14	0
3	76	26	15	**29**	3	29	0
4	86	17	9	7	**8**	30	0
5	60	28	31	47	11	**196**	0
6	10	4	4	9	3	14	**0**
Beech							
0	**166**	23	11	10	6	2	0
1	91	**87**	5	15	5	13	0
2	24	6	**4**	2	1	4	0
3	58	60	7	**38**	2	38	0
4	10	0	1	4	**3**	11	0
5	33	19	2	24	4	**29**	0
6	5	0	0	0	0	2	**0**

0: none; 1: top of tree only; 2: middle parts of crown; 3: top and middle; 4: base of crown; 5: throughout crown; 6: other.
Numbers of trees in each category are given.

Table 7.14 Percentage of conifers in each needle-browning class, 1987–91

	1st year					Older				
	0-10%	11-25%	26-60%	>60%	dead	0-10%	11-25%	26-60%	>60%	dead
Sitka spruce										
1987	94	6	0	0	0	91	8	1	0	0
1988	99	1	0	0	0	89	8	3	0	0
1989	98	2	0	0	0	97	2	1	0	0
1990	99	1	0	0	0	99	1	0	0	0
1991	99	1	0	0	0	98	2	0	0	0
Norway spruce										
1987	95	4	1	0	0	94	5	1	0	0
1988	99	1	0	0	0	88	8	4	0	0
1989	99	1	0	0	0	95	4	1	0	0
1990	98	1	0	0	0	98	1	1	0	0
1991	91	4	1	4	0	90	4	2	4	0
Scots pine										
1987	94	5	1	0	0	92	7	1	0	0
1988	96	4	0	0	0	93	6	1	0	0
1989	93	7	0	0	0	92	7	1	0	0
1990	91	8	0	0	1	91	7	1	0	1
1991	77	12	2	8	1	89	3	0	7	1

Table 7.15 Percentage of conifers in each needle-yellowing class, 1987–91

	Current					Old				
	0-10%	11-25%	26-60%	>60%	dead	0-10%	11-25%	26-60%	>60%	dead
Sitka spruce										
1987	97	2	1	0	0	89	9	2	0	0
1988	96	4	0	0	0	93	5	2	0	0
1989	98	1	1	0	0	92	6	2	0	0
1990	94	3	2	1	0	98	1	0	0	0
1991	97	2	1	0	0	94	3	2	1	0
Norway spruce										
1987	98	2	0	0	0	98	2	0	0	0
1988	97	3	0	0	0	95	4	1	0	0
1989	98	2	0	0	0	97	2	1	0	0
1990	98	1	0	0	0	99	0	0	0	0
1991	97	2	1	0	0	90	5	4	1	0
Scots pine										
1987	98	2	0	0	0	96	4	0	0	0
1988	97	3	0	0	0	94	5	1	0	0
1989	97	3	0	0	0	89	8	3	0	0
1990	97	3	0	0	1	95	4	0	0	1
1991	94	4	1	1	0	96	3	0	1	0

may be related to a modification in the observations procedures. Until 1991, observers estimated the proportion of the crown that consisted of brown leaves, whereas in 1991, observers noted the proportion of leaves with browning present. Given the incidence of insect attack in oak, the frequency of browning that was recorded was expected. In contrast, yellowing is relatively rare in oak, and the scores remained similar throughout the period. Yellowing tends not to occur as patches on leaves and it was therefore not influenced by the modification to the observation procedure.

In beech, browning was quite widespread (Table 7.16), although it was very rare in 1990. A similar modification to the observation procedure to that made for oak was adopted in 1991. This is believed to have caused the increase in the proportion of trees with moderate to severe levels of browning. The browning is largely caused by the beech leaf miner and varying proportions of the leaf are affected. In this context, the 1990 figures are of interest since they suggest that damage caused by the beech leaf miner was very low. This might be explained by low nutritional contents of the foliage (as indicated by the high frequency of abnormally small leaves) associated with the excessive flowering. However, this hypothesis remains to be tested.

The amount of yellowing varied markedly from year to year. No clear pattern has yet emerged that might point to the cause of the fluctuations in yellowing. Yellowing appears to be related to soil chemistry, and an interaction between climate and soil chemistry is probably the factor determining the amount of yellowing at any given site in a specific year.

Table 7.16 Percentage of oak and beech showing leaf discoloration, 1987–91

	Browning					Yellowing				
	0-10%	11-25%	26-60%	>60%	dead	0-10%	11-25%	26-60%	>60%	dead
Oak										
1987	98	2	0	0	0	97	3	0	0	0
1988	95	5	0	0	0	92	7	1	0	0
1989	96	4	0	0	0	92	5	2	1	0
1990	95	4	0	0	0	92	6	1	1	0
1991	81	16	3	0	0	92	4	3	1	0
Beech										
1987	74	22	4	0	0	86	13	1	0	0
1988	72	24	4	0	0	47	35	16	2	0
1989	75	20	4	1	0	70	22	7	1	0
1990	91	5	4	0	0	88	8	3	1	0
1991	71	17	10	2	0	59	21	13	6	0

Nature of the discoloration

In contrast to most forest health inventories undertaken in Europe, the nature of any observed discoloration is recorded in the British monitoring programme. The exact distinction of different types of discoloration and its location both on needles and within the crown is an important but little used diagnostic tool (c.f. Friedland *et al.*, 1984; Redfern *et al.*, 1987; Niehaus, 1989). In conifers, only the different forms of yellowing were recorded whereas the nature of both browning and yellowing was recorded for broadleaves. Because of the changes in assessment procedures, only the data for 1991 have been summarized here. In contrast to the scores for the amount of discoloration, the scores for discoloration type are based on presence-absence. This means that the proportions of trees scored as having no discoloration type will be lower than the proportions of trees scored as having 0–10% discoloration.

As indicated above, yellowing of current-year needles of conifers was relatively rare. Where it did occur, in spruce it was restricted to trees having an overall yellowish appearance. In about half the cases for Sitka spruce and two-thirds of the cases for Norway spruce, the chlorosis was quite marked. In Scots pine, the yellowing either took the form of yellow tipping to the needles or marked chlorosis of the whole tree.

Discoloration of older needles was more obvious, but was almost certainly under-recorded as the needles become progressively more difficult to see as they become older. In both Sitka spruce and Norway spruce, the most frequent type of discoloration was yellow-tipping to the needles. The significance of yellow-tipping is uncertain, although it has been widely reported from continental Europe. For example, in Finland, it was reported on 40% of the Norway spruce assessed (Salemaa *et al.*, 1991). Its visibility is very dependent on the weather conditions and direction of sunlight at the time of observation and its actual frequency is likely to be higher than was recorded. Yellowing of the upper surfaces of needles was recorded on several occasions (4% of Norway spruce and 2% of Sitka spruce). This type of discoloration has also been reported from the continent, where it has been particularly associated with Type 1 Norway spruce decline. Its causes are unclear (although the mechanisms are understood). Widespread yellowing of needles throughout the tree was relatively common in Norway spruce. In Scots pine, yellowing of older needles was consistently associated with overall yellowing of the tree.

Various discoloration types were noted in broadleaves (Table 7.17). In oak, scattered brown leaves were recorded on over 20% of the trees, with marginal necrosis being recorded in 10% of the cases. In beech, marginal and tip necrosis were by far the most common types of discoloration, strongly suggesting that damage by the beech leaf miner was responsible. However, damage by the fungus *Apiognomonia errabunda* (Roberge) Hohnel may also have been important, as the two types of damage are visually similar and are consequently difficult to separate without close inspection of the leaves. Scattered brown leaves were also recorded, but were less frequent than in oak. Most yellowing in oak involved marginal discoloration. In beech, yellowing was predominantly recorded on the tips of leaves and as spots on leaves. The latter would be the yellowing type given to interveinal chlorosis.

Table 7.17 Nature of the discoloration in oak and beech, 1991

	n	0	1	2	3	4	5	6	7
Browning									
Oak	1847	57	10	2	7	21	2	0	0
Beech	1464	47	27	12	3	8	3	0	0
Yellowing									
Oak	1847	87	8	3	1	1	0	0	
Beech	1464	52	8	22	16	3	0	0	

Browning: 0: no discoloration; 1: leaf margins only; 2: leaf tips only; 3: spots on leaves; 4: scattered leaves entirely brown; 5: occasional shoots with leaves entirely brown; 6: occasional branches with all leaves entirely brown; 7: other.
Yellowing: 0: no yellowing; 1: leaf margins yellow; 2: some shoots yellow, the rest of the crown being of normal colour; 3: entire tree yellowish-green; 4: entire tree greenish-yellow; 5: entire tree chlorotic; 6: other.
The percentages of trees in each category are given.

Tendency for individual trees to show discoloration in successive years

As indicated in Chapter 5, foliage discoloration was one of the indices that showed the greatest amount of change during each annual survey period. The amount of discoloration recorded at some sites was therefore partially dependent on the date of assessment.

Sitka spruce

In Sitka spruce, the browning recorded on trees in 1989 was absent in 1990, although a few new trees developed a small amount of browning. Simultaneously, there was a substantial increase in the yellowing of current-year needles. This mainly involved trees scored as without yellowing in 1989; those that had yellowing in 1989 had largely recovered by 1990. The increase was restricted to current-year needles; very few trees developed yellowing of older needles and trees with yellow older needles in 1989 mostly recovered. The overall discoloration values reflect both the development of yellowing in some trees and the general recovery of those trees affected in 1989.

There was no indication that trees with browning in 1989 were more prone to browning in 1991. The amount of yellowing recorded in the two years was similarly independent. Consequently, it appears that the 'population' of discolored trees varies from one year to the next rather than being related to specific individuals.

Norway spruce

Very little discoloration of current needles was recorded on Norway spruce in either 1989 or 1990. Trees with browning of older needles in 1989 generally recovered, although some new trees developed browning. The yellowing of older needles seen in 1989 was absent in 1990. The changes were reflected in the overall discoloration scores, with a few trees developing discoloration and most of those with discoloration in 1989 recovering by 1991.

Between 1989 and 1991, there was very little indication that the same trees were involved in the discoloration. Only four trees with browning of current needles and seven trees with browning of older needles in 1989 were scored as having browning in 1991. The proportion of trees retaining yellowing over the period was equally small.

Scots pine

Discoloration was more frequent on Scots pine than on either of the spruce species. This is probably normal, reflecting the onset of autumnal senescence, the effects of which are much more marked in Scots pine than in the spruces. In 1990, there was an increase in browning. This occurred through an increase in the number of trees affected rather than an increase in the degree of browning on individual trees. As with the spruces, trees with discoloration in 1989 mostly recovered and the observed discoloration in 1990 was on trees previously classified as without any.

As with the two spruce species, there was very little indication that particular trees were more prone to discoloration than others. The proportion of trees having discoloration in both 1989 and 1991 was very small, despite the relative frequency of discoloration.

Oak

Oak showed relatively little change in 1989–90. A similar pattern to the conifers occurred, with discoloration developing on trees that had previously been unaffected and affected trees recovering. Some trees recorded as having limited (11–25% of the foliage) yellowing in 1989 deteriorated further. Very few trees remained in the same discoloration categories between 1989 and 1991, showing that discolored trees are capable of recovery.

Beech

The greatest changes between 1989 and 1990 occurred in beech, with marked reductions in both browning and yellowing. There was some evidence for the development of browning in trees, although the proportion involved was relatively small, particularly in comparison to the number of trees recovering.

Many trees that were discolored in 1989 were also discolored in 1991. This trend was apparent for both browning and yellowing, but the proportion of trees remaining with yellowing was much higher than with browning. Yellowing appears to be strongly related to site characteristics and it seems likely that if it occurs at a site in one year, it is likely to occur there again in future years. Weather conditions, however, may also play a role. In Table 7.18, the changes between 1990 and 1991 are shown, as the proportion of trees discolored in 1990 was much lower than in either 1989 or 1991. There was a tendency for trees classed as having slight yellowing in 1990 to have increased levels in 1991, although even in a year when there was a substantial overall increase in yellowing, some individual trees showed an improvement on their 1990 yellowing scores.

Table 7.18 Changes in discoloration of beech between 1990 and 1991

	Browning in 1990				Yellowing in 1990			
1991 score	0-10%	11-25%	26-60%	>60%	0-10%	11-25%	26-60%	>60%
0–10%	**567**	26	7	4	**461**	22	5	6
11–25%	128	**15**	3	0	156	**22**	7	2
26–60%	63	4	**15**	0	92	21	**6**	1
>60%	12	0	6	**0**	37	7	5	**0**

Numbers of trees in each category are given.

Needle retention

Needle retention provides an index of the number of years that needles are retained. It is only useful as a measure of needle loss if the loss occurs progressively from the oldest to the youngest needles. In Great Britain, uniform loss of needles from shoots of all ages is the most common cause of crown transparency and therefore needle retention estimates cannot be used as a surrogate for transparency estimates.

Using binoculars, it is usually difficult to count back more than about seven years in the upper crowns. Consequently, an upper limit of greater than seven years was adopted. When there are more than seven years of needles, there may be substantial variation between trees, but the assessment errors are proportionally greater. The index is subject to error as needle retention varies throughout the crown and observers have to make an estimate of the average length of retention for the whole crown. Consequently, increases of more than one year retention from one year to the next (which are obviously impossible) may be recorded. Although quite detailed indices of needle retention have been proposed, the errors associated with the estimates of retention mean that these are of relatively little significance. Consequently, average values have been used (Table 7.19).

Needle retention in Sitka spruce remained fairly constant except in 1990, when retention was noticeably less than in other years. In 1989 and 1990, some trees were recorded as having very low retention values, although these subsequently recovered. A similar pattern was evident in the transparency estimates, reflecting the relationship between needle retention and transparency estimates when the retention is very low. The reduction in retention can be directly attributed to damage caused by the green spruce aphid.

Retention figures for Norway spruce were more variable than for Sitka spruce. Both 1987 and 1990 were years in which the proportion of trees with more than seven years of needle retention was very small. In 1991, there was an increase in the proportions of trees with very low needle

Table 7.19 Percentage of trees with needle life (needle retention) for a given number of years, 1987–91

	Number of years								
	0	1	2	3	4	5	6	7	>7
Sitka spruce									
1987	0	0	0	1	4	15	18	18	44
1988	0	0	0	1	5	15	16	18	45
1989	0	2	2	4	7	13	15	18	39
1990	0	1	2	4	7	11	19	24	31
1991	0	0	2	3	8	10	10	17	49
Norway spruce									
1987	0	0	0	1	6	17	23	20	33
1988	0	0	0	1	3	11	20	24	41
1989	0	0	0	2	4	11	15	21	47
1990	0	0	0	2	6	13	21	24	32
1991	0	1	1	5	9	10	13	15	46
Scots pine									
1987	0	2	49	44	5	0	0	0	0
1988	0	2	59	37	2	0	0	0	0
1989	0	2	52	44	2	0	0	0	0
1990	1	3	50	45	1	0	0	0	0
1991	1	1	40	53	4	0	0	0	0

retention. This seemed to be related to the development of the needle reddening phenomenon, which was often associated with the loss of older needles from the trees.

Needle retention figures for Scots pine are usually easier to interpret as trees in Great Britain generally retain their needles for less than four years. However, substantial needle fall occurs in late summer and plots assessed at the end of the survey period may have lower retention estimates than they would have had had they been assessed at the start of the period. The majority (>90%) of trees had either two or three years needle retention in all five years. In the period 1987 to 1990, more trees had two years of needles than three. However, in 1991, the position was reversed. These figures are very different from northern provenances of Scots pine, which hold their needles for six more years or more (e.g. Yarmishko, 1989; Jalkanen and Kurkela, 1990; Salemaa *et al.*, 1991), and comparisons are therefore difficult. The retention in Great Britain is closer to that seen in eastern Europe. In eastern Europe, a reduction from a mean of 3.5 years to 2.6 has been recorded between 1950 and the late 1980s (Skuhravy, 1990). No data are available from visual estimates of needle retention in Great Britain that might suggest whether a similar trend has occurred.

Data concerning the changes in the needle retention of individual trees are only available for the period 1989–91. However, as with many other indices, considerable changes occurred on individual trees during this period (Table 7.20). The pattern of change was extremely complex and appeared to depend on the initial retention score. For example, between 1989 and 1990 (not shown in Table 7.20), Sitka spruce with six or fewer years of needles tended to increase their retention. This trend was even more marked in the period 1990–91. The changes were most apparent in the trees with higher needle retention. For example, 212 (37%) of the trees recorded as having more than seven years needle retention in 1989 had seven years or less in 1990. Similarly, whereas 48 (20%) of the trees recorded with seven years needle retention in 1989 had more in 1990, 79 (33%) had less.

Norway spruce had a different pattern, with similar numbers of trees showing increased and reduced retention during the period. However, in 1991, there was a marked increase in the number of trees with more than seven years of needle retention. The changes between 1989 and 1990 were more marked in Norway spruce than Sitka spruce, the former not being affected to the same extent by the 1989 aphid defoliation episode and therefore having fewer severely defoliated trees available for recovery. Trees with only two to four years needle retention in 1989 generally recovered whereas many trees with relatively high levels of needle retention deteriorated. Several trees with six or more years of foliage in 1989 had three or fewer years in 1990.

In Scots pine, large numbers of trees varied between two and three

Table 7.20 Changes in needle retention of Sitka and Norway spruce between 1989 and 1991

1991 retention	Needle retention in 1989 (years) 1	2	3	4	5	6	7	>7
Sitka spruce								
1	**0**	0	0	0	0	0	0	0
2	19	**4**	1	1	1	1	2	0
3	9	7	**11**	8	7	3	1	2
4	0	8	26	**23**	27	22	11	8
5	0	2	8	23	**49**	32	23	10
6	0	1	4	8	21	**30**	22	38
7	0	2	3	11	19	22	**51**	81
>7	0	0	0	1	24	64	127	**435**
Norway spruce								
1	**2**	0	2	1	2	3	1	2
2	1	**0**	1	3	8	2	3	1
3	0	0	**10**	7	12	25	16	14
4	0	1	5	**28**	44	29	28	29
5	0	0	4	15	**47**	42	53	23
6	0	0	0	12	46	**46**	45	65
7	0	0	0	1	13	48	**41**	119
>7	0	0	1	2	9	57	150	**535**

years needle retention. Approximately equal numbers of trees moved from two years to three years and vice versa in both the periods (Table 7.21). Even with these trees, there was evidence of observer error, with one tree scored as having only one year of needles in 1990 scored as having four years of needles in 1991. There was no indication of a progressive reduction in the needle retention of individual trees over the three years.

Needle retention in Scots pine using the vascular bundle method

Needle retention in Scots pine at two English sites was assessed using the vascular bundle method described in Chapter 4. Samples were obtained from Thetford (nine trees) and Alice Holt (five trees) Forests in south-east England. At Thetford, the mean survival of needles after one, two and three years was 85%, 59% and 29%, respectively. At Alice Holt, the mean survival rates were 89%, 66% and 22%, respectively. Substantial year-to-year variations in needle retention were found, as well as variations between the two sites and between trees within sites. Generally a decrease in needle survival rates has occurred since about 1980 (Fig. 7.7). Although there are still several unresolved problems related to this technique when it is applied to densely growing plantation trees with relatively short needle retention periods, the results suggest that a progressive change in needle retention patterns has occurred at the two sites during the last ten years. This clearly warrants further investigation.

Leader condition

The results of the assessments of leader condition in conifers are presented in Table 7.22. Bent or twisted leaders were fairly frequent in Sitka spruce but rare in the other two species. Both Sitka and Norway spruce had high proportions of trees with missing leaders, a form of damage that can be caused by a variety of agents. In Norway spruce, severe mechanical damage caused during a storm in 1990 resulted in many trees having the top third or so of the crown being snapped off. In such cases, the leader would be scored as missing. Usually, missing leaders will be replaced by a side shoot, and all three species had individual trees where side shoots had taken over.

Flowering in Scots pine

In Scots pine, male flowering replaces needle development and can result

Table 7.21 Changes in needle retention of Scots pine between 1989 and 1990

	Previous needle retention (years)				
	1	2	3	4	5
1990 vs. 1989					
1	**19**	22	5	0	0
2	20	**684**	213	4	0
3	0	236	**589**	21	0
4	0	1	18	**6**	0
5	0	0	0	0	**0**
1991 vs. 1990					
1	**11**	17	0	0	0
2	26	**562**	180	0	0
3	10	364	**604**	21	0
4	1	8	68	**3**	0
5	0	0	1	0	0
1991 vs. 1989					
1	**4**	12	0	0	0
2	21	**481**	218	2	0
3	9	409	**530**	22	0
4	0	24	48	**7**	0
5	0	0	1	0	**0**

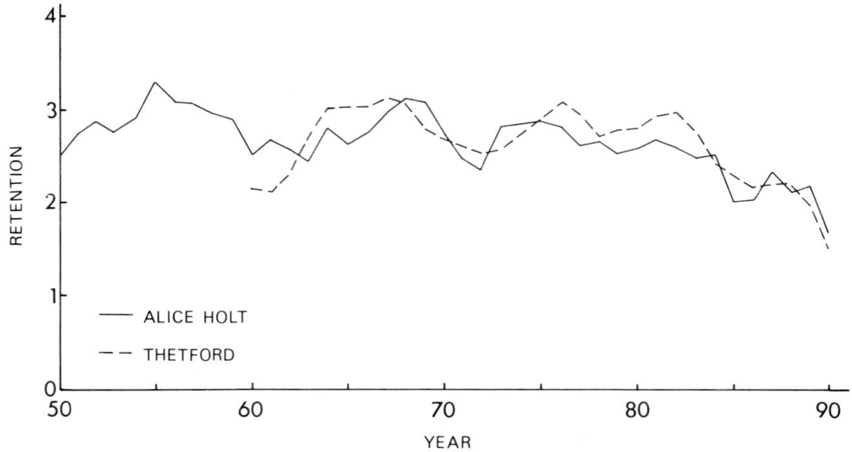

Figure 7.7 Needle retention (in years) of the leading shoots of Scots pine at Thetford and Alice Holt Forests, 1950–91.

in a thinner-looking crown. The effect may be particularly noticeable when looking up into a crown as flowering tends to be much heavier on shaded branches in the lower crown than in the upper half of the crown. In the assessment programme, the gaps on the shoots left after the flowers have dropped off were assessed on each tree, with assessments being divided between the upper and lower crowns (Fig. 7.8). In all years, flowering was much more abundant in the lower part of the crown than in the upper part.

The overall figures for 1989–90 were similar, although the amount of flowering in the upper crown was less in 1990. In 1991, there was a marked increase in flowering in both parts of the crown. Over half the trees had flowering present in the upper part of the crown (compared with only 37% in 1990). There was a good relationship between the amount of flowering in the bottom and top halves of trees (Table 7.23). Very few trees had more flowering in the top half than in the bottom half, and trees with high levels of flowering in the top half normally had high levels of flowering in the bottom half.

There was considerable annual variation in the flowering levels of individual trees (Table 7.24). Between 1989 and 1991, approximately one third of the trees had no flowering in the upper crown. There was little indication that trees with high levels of flowering in the top half of the crown in 1989 were likely to have high levels in subsequent years. Very few trees (<5%) had no flowering in the bottom half of the crown in all of the three years. However, 207 trees were recorded as having abundant flowering in the bottom half in both 1989 and 1991. There

Table 7.22 Percentages of trees with the leader in a given condition, 1990

	Type								
	0	1	2	3	4	5	6	7	8
Sitka spruce									
1989	64	2	12	10	4	2	0	6	0
1990	65	4	9	10	4	0	0	6	0
1991	75	2	7	9	2	1	1	3	0
Norway spruce									
1989	68	6	11	3	6	1	1	4	0
1990	65	6	16	4	4	1	1	5	0
1991	67	8	15	2	2	1	1	4	0
Scots pine									
1989	60	3	5	2	12	0	1	8	9
1990	62	4	5	1	5	0	1	11	10
1991	67	4	8	1	3	0	2	9	6

0: normal; 1: shorter than current-year side shoots; 2: missing; 3: bent or twisted; 4: double; 5: broken; 6: bare; 7: side shoots taken over; 8: complete loss of apical dominance.

Table 7.23 Numbers of Scots pine with given scores for flowering in the top and base of the crown

1991 flowering score	Top half				
Bottom half	0	1	2	3	4
0	**196**	13	7	0	0
1	259	**49**	11	2	0
2	278	125	**29**	2	2
3	152	161	140	**10**	0
4	46	73	164	194	**31**

Data refer to the 1991 assessment programme.
Classes: 0: none; 1: rare (1–10% of branches); 2: infrequent (11–25% of branches); 3: common (26–60% of branches); 4: abundant (>60% of branches).

were no readily discernable differences in the patterns of changes in 1989–90 and 1990–91.

Fruiting

The amount of fruiting (coning, acorns and mast) recorded in 1989–91 is presented in Table 7.25. In the conifers, only fresh cones were assessed. In all cases, the figures refer to the amount of visible fruit, and no account was taken of the quality of the seed. No data have been presented for oak as it was apparent from the repeat assessments undertaken in 1990 and 1991 (see Chapter 5) that evaluations of the numbers of acorns early on in the season may seriously underestimate the actual numbers.

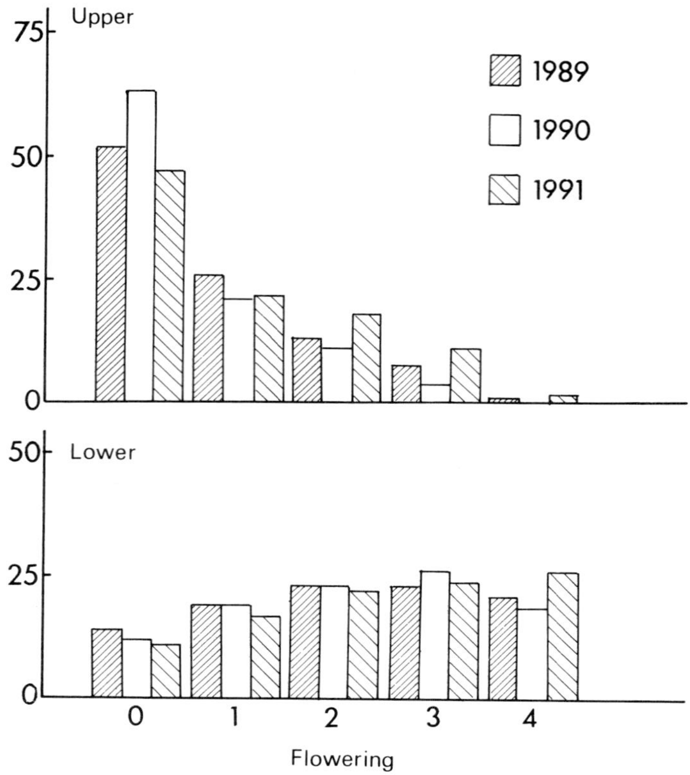

Figure 7.8 Percentage of pine in each flowering class, 1989–91. 0: none; 1: rare; 2: infrequent; 3: common; 4: abundant.

In Sitka spruce, Norway spruce and beech, 1990 was a year with very high levels of fruiting. The increases involved both trees without cones in 1989 and trees with rare or scattered cones. In 1991, in marked contrast, there was very little evidence of fruiting in these three species. The fruiting in 1990 was related to the very dry and warm summer of 1989, although it is interesting that the dry summer of 1990 did not result in high levels of fruiting in 1991. Scots pine showed a different pattern, with similar levels of coning in all three years. A relationship between the fruiting of beech and spruce and the temperature during the previous summer has been known for some time (Matthews, 1955; Eklund, 1957), and the results from Great Britain were therefore anticipated.

The large amount of beech mast was particularly noticeable in 1990. It was much more frequent than in 1989, with many trees having mast in 1990 that were without it in 1989. In addition, trees with light levels of mast in 1989 had higher levels in 1990. In 1991, masting was virtually absent in all the trees, despite the hot summer of 1990. This strongly

Table 7.24 Changes in the extent of flowering on individual Scots pine between 1989 and 1991

Flowering class in 1991	Flowering class 1989				
Upper crown	0	1	2	3	4
0	**537**	213	79	24	6
1	207	**108**	53	24	5
2	134	94	**62**	39	4
3	50	35	47	**53**	9
4	10	8	7	6	**0**
Lower crown					
1	**51**	54	41	18	15
2	72	**81**	74	50	23
3	66	80	**112**	106	44
4	37	86	110	**120**	91
	22	46	85	124	**207**

The classes are the same as in Table 7.23.

suggests that the weather conditions of the previous summer cannot be considered on their own; it is necessary to look at several years before the

Table 7.25 Percentages of trees with a given level of fruiting

	Frequency			
	None	Scarce	Common	Abundant
Sitka spruce				
1989	69	20	8	3
1990	36	34	25	6
1991	95	5	1	0
Norway spruce				
1989	90	8	2	0
1990	51	25	20	5
1991	96	3	1	0
Scots pine				
1989	17	38	30	15
1990	24	36	32	8
1991	25	34	29	12
Beech				
1989	59	29	9	3
1990	6	11	33	50
1991	94	5	1	0

For spruce, only fresh cones were included, 1989–91.
For Scots pine, only unopened second year cones were included.

masting response in any particular year can be predicted. This is borne out by detailed research undertaken in Scandinavia, which has not determined any clear relationships between specific environmental variables and the incidence of fruiting.

Secondary shoots in spruce

The pattern of the occurrence of secondary shoots provides a valuable insight into the health of a tree. The proportion of secondary shoots increases with tree age (Westman and Lesínski, 1985) as gaps left by the fall of needles from older shoots are filled in (see Fig. 4.12). This means that under natural conditions, secondary shoots should start to develop from the interior of the branches and spread outwards (Westman and Lesínski, 1986). However, shading of the interior part of the crown may prevent the shoots from fully developing in this area.

Secondary shoots were assessed in all three years. During the quality assurance work in 1989, there were problems with variation in the ways that secondary shoots were being assessed. Consequently, no figures are available for 1989. Secondary shoots were assessed by abundance and location, following the recommendations of Lesinski (1989). The relative frequency of location is given in Fig. 7.9 and their abundance is given in Fig. 7.10. Differences were apparent between 1990 and 1991, with the proportion of Norway spruce recorded as having secondary shoots increasing in 1991. In both Sitka spruce and Norway spruce, the most common category recorded revealed that a few secondary shoots were present on many branches. Differences in the location of the secondary shoots also occurred, but I strongly suspect that this was attributable to observer inconsistency. As the problem was addressed during the 1991 training programme, the figures for 1991 are more reliable than for 1990.

The patterns for the two species were remarkably similar, despite the differences in transparency levels. The presence of secondary shoots is believed to be a good index of vigour, indicating that the tree is capable of recovering following defoliation (Gruber, 1987; Lesinski and Westman, 1987; Lesinski and Landmann, 1988; Rehfuess and Rehfuess, 1988; Lesinski, 1989; Rehfuess, 1989). Consequently, the relationship between crown transparency and the occurrence of secondary shoots is of interest, and has been presented in Fig. 7.11.

Again, the two sets of results are fairly similar, although the proportion of Norway spruce in the lower transparency categories without secondary shoots was lower than for Sitka spruce. In both cases, there appeared to be no relationship between lower levels of transparency and the proportion of trees with secondary shoots. At the highest levels of transparency, an increase in the proportion of trees with secondary shoots was apparent, but the sample sizes were very small. As much of the defoliation is known to have occurred in 1988–9, it is possible that secondary shoots had insufficient time to develop on many of the trees.

Epicormic branches on oak

To a certain extent, epicormic shoots on oak serve a similar function to secondary shoots on spruce. The proportions of trees with branch and stem epicormics are given in Fig. 7.12. Epicormic shoots tended to be more frequent on branches than on stems. In a few cases, this might reflect stem pruning, but this seems to be quite a rare practice in Great Britain. There was a strong suggestion of a decline in the proportion

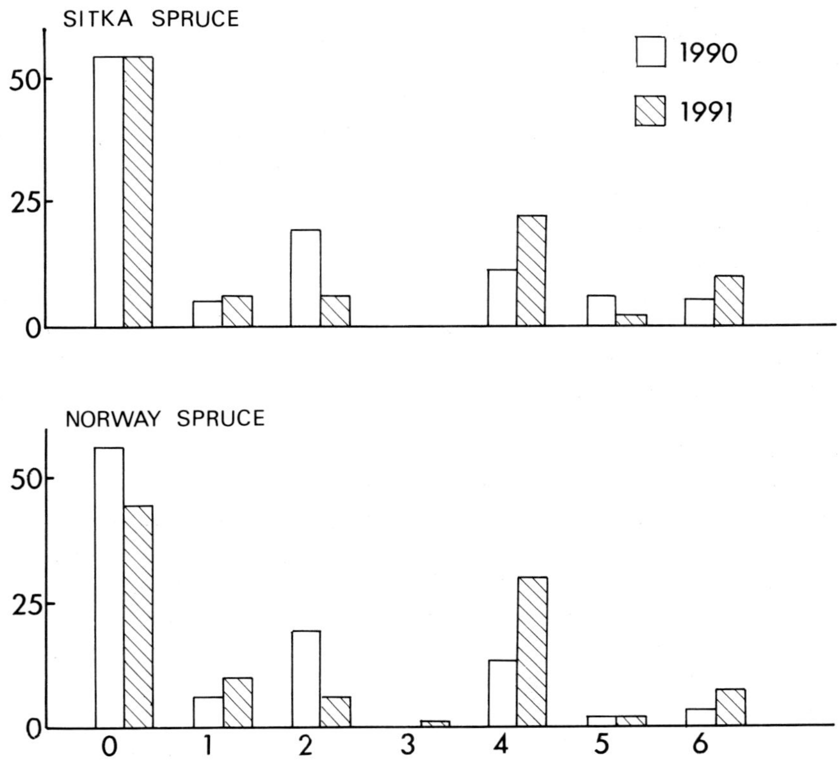

Figure 7.9 Location of secondary shoots on Sitka and Norway spruce. 0: none; 1: outer parts of branches only; 2: middle of branches; 3: inner parts of branches; 4: outer and middle parts; 5: middle and inner parts; 6: all along branch.

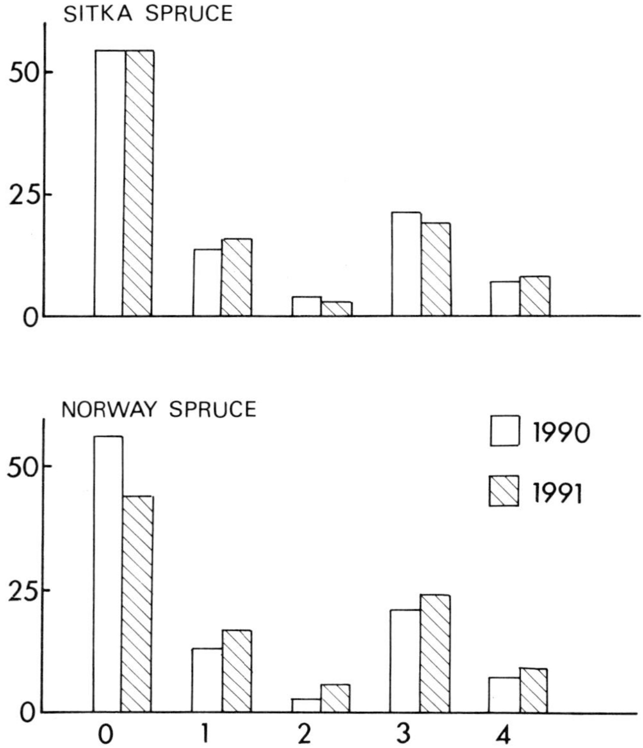

Figure 7.10 Abundance of secondary shoots on Sitka and Norway spruce in 1990 and 1991. 0: none; 1: a few shoots on less than a quarter of the branches; 2: many shoots on less than a quarter of branches; 3: a few shoots on many branches; 4: many shoots on many branches.

of trees with abundant stem epicormics over the three years. The reason for this is unknown, but a possible explanation is drought stress. There was no readily discernible pattern in the occurrence of stem and branch epicormic shoots on individual trees (Table 7.26), although trees with high numbers of epicormics on the stem also tended to have high numbers on the branches.

Leaf-rolling in beech

Leaf-rolling is believed to be associated with water stress. This appears to be confirmed by both the repeat assessments described in Chapter 5 and the general figures for 1990 (Table 7.27), a year of marked drought stress. However, high levels of rolling were also seen in 1991, despite the

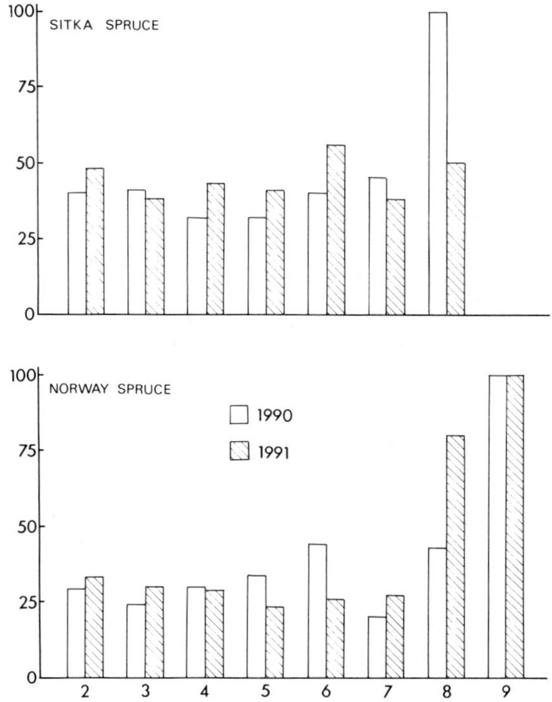

Figure 7.11 Relationship between crown transparency and the presence of secondary shoots, 1990. The proportion of trees without secondary shoots in each 10% transparency category is given. The occurrence of secondary shoots was not assessed on trees with 20% or less loss of transparency. Classes: 2: 21–30% transparency; 3: 31–40%; 4: 41–50% ... 9: 91–100%.

relatively favourable climatic conditions during the growing season. One possibility is that the leaf-rolling occurs in response to water stress caused by damage to the fine root system. This could explain the presence of rolling in 1991.

The 1990 assessments revealed an increase in the incidence of leaf-rolling since 1989 but, as with other indices, individual trees showed considerable variation. Almost half the trees without rolling in 1989 showed rolling in 1990, although rolling was also absent on many (34%) trees that were affected in 1989. The degree of leaf-rolling did not change substantially between the two years, although there were considerable changes among individual trees. Over the three-year period, there was no indication that certain trees are more likely to show leaf-rolling than others (Table 7.28). However, the spatial distribution of stresses such as drought also varied among the three years and this may have contributed to the year-to-year variations.

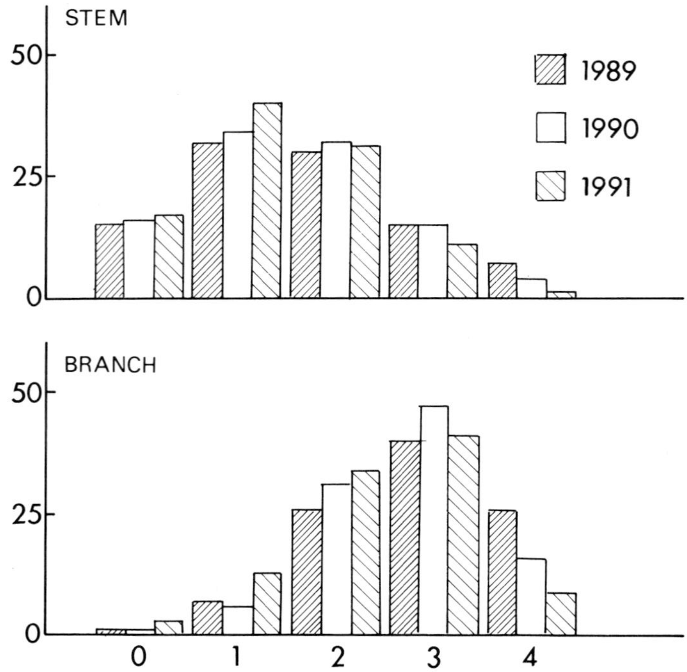

Figure 7.12 Percentages of oak with branch and stem epicormics, 1990. 0: none; 1: rare; 2: scattered; 3: common; 4: abundant.

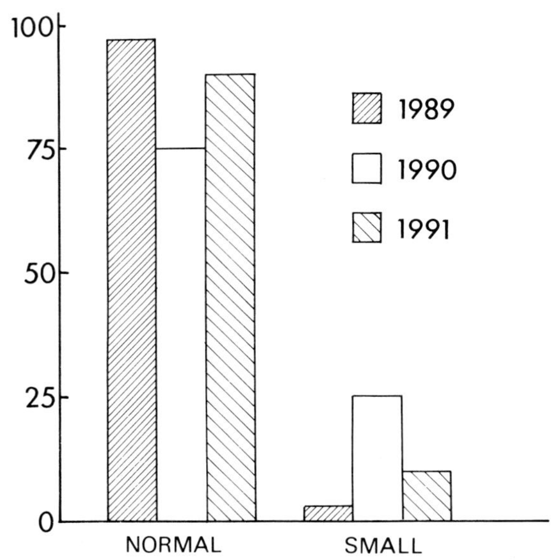

Figure 7.13 Leaf size in beech. The proportions of trees with normal- and small-sized leaves in each of the three years are given.

Table 7.26 Relationship between stem and branch epicormics in oak, 1991

Branch frequency category	Stem frequency category 0	1	2	3	4
0	**46**	7	6	0	1
1	84	**124**	30	1	0
2	105	306	**180**	34	0
3	69	253	301	**124**	4
4	10	44	55	53	**11**

Leaf size in beech

Leaf size in beech was recorded (small – normal – large) in all three years (Fig. 7.13). A particular feature of 1990 was the large number of very small leaves on some trees, particularly in the upper crown. In 1991, the percentage of trees with small leaves in the upper crown dropped, but it

Table 7.27 Percentages of beech with leaf rolling in the crown, 1989–91

Degree	Extent 0	1	2	3	4	5	6	7
1989								
None	49	0	0	0	0	0	0	0
Slight	0	13	1	12	3	6	5	0
Medium	0	2	0	3	0	3	1	0
Severe	0	0	0	0	0	0	0	0
1990								
None	45	0	0	0	0	0	0	0
Slight	0	7	2	12	4	7	2	1
Medium	0	4	0	4	1	5	3	1
Severe	0	0	0	0	0	1	0	0
1991								
None	44	0	0	0	0	0	0	0
Slight	0	9	1	12	5	6	6	1
Medium	0	1	0	3	1	5	4	1
Severe	0	0	0	0	0	0	0	0

Extent classes: 0: none; 1: leaves on a few shoots in the upper crown; 2: leaves on a few shoots elsewhere; 3: half of upper crown affected; 4: half of shoots throughout crown affected; 5: most shoots in the upper crown affected; 6: most shoots throughout crown affected; 7: virtually all leaves affected.

Table 7.28 Changes in the frequency of leaf rolling in beech between 1989 and 1991

Leaf rolling frequency 1991	1989							
	0	1	2	3	4	5	6	7
0	**253**	47	2	37	6	24	9	0
1	30	**23**	3	15	5	9	6	0
2	5	6	**2**	2	1	0	1	0
3	52	12	0	**24**	9	15	11	1
4	15	10	0	4	**3**	7	9	2
5	21	6	6	19	5	**17**	7	0
6	25	17	1	17	0	9	**8**	0
7	0	0	0	2	0	0	4	**1**

Extent classes: 0: none; 1: leaves on a few shoots in the upper crown; 2: leaves on a few shoots elsewhere; 3: half of upper crown affected; 4: half of shoots throughout crown affected; 5: most shoots in the upper crown affected; 6: most shoots throughout crown affected; 7: virtually all leaves affected.

remained higher than in 1989. This trend was rather surprising as the 1990 drought might have been expected to have had an effect on the size of leaves the following season. However, the marked drop in fruiting levels would have resulted in more assimilate being available for leaf production. Of the 852 trees examined in the two years, 164 of the 210 trees recorded as having small leaves in 1990 had normal-sized leaves in 1991. Only 30 of the 639 trees with normal-sized leaves in 1990 had small leaves in 1991.

In the field, there appeared to be an association between the amount of mast and the presence of small leaves (see Fig. 4.3). This was confirmed by the data, with 76% of the trees with small leaves having abundant mast. Only 41% of the trees with normal-sized leaves had abundant mast. There was also evidence that the small leaves were associated thin-crowned trees in both 1990 and 1991 (Fig. 7.14). The nature of this relationship changed between the two years, with small leaves occurring more on defoliated trees in 1991 than in 1990.

Premature leaf loss in beech

Premature leaf loss, revealed by green leaves on the ground under trees, was very similar in all three years (Fig. 7.15). Infrequent leaf loss was noted under less than 20% of the trees in each of the three years, and the drought-related changes seen in some other indices were not apparent for premature leaf loss.

Presence of insects and fungal pathogens

The method of reporting insect and fungal damage on trees was changed in 1990 following the problems experienced in 1989. In 1990, assessments were standardised between species, with five categories being recognized: none, rare, infrequent, common and abundant. In each case, the category referred to the proportion of foliage affected. Consequently, reliable comparisons can only be made between 1990 and 1991. The scores were dependent on the ability of the observer to clearly see the canopy and were averaged across the canopy. In broadleaves, poor views of the upper canopy may be less important than for other indices, as insect damage is often greater in the lower canopy than in the upper part (Phillipson and Thompson, 1983; Moore *et al.*, 1991).

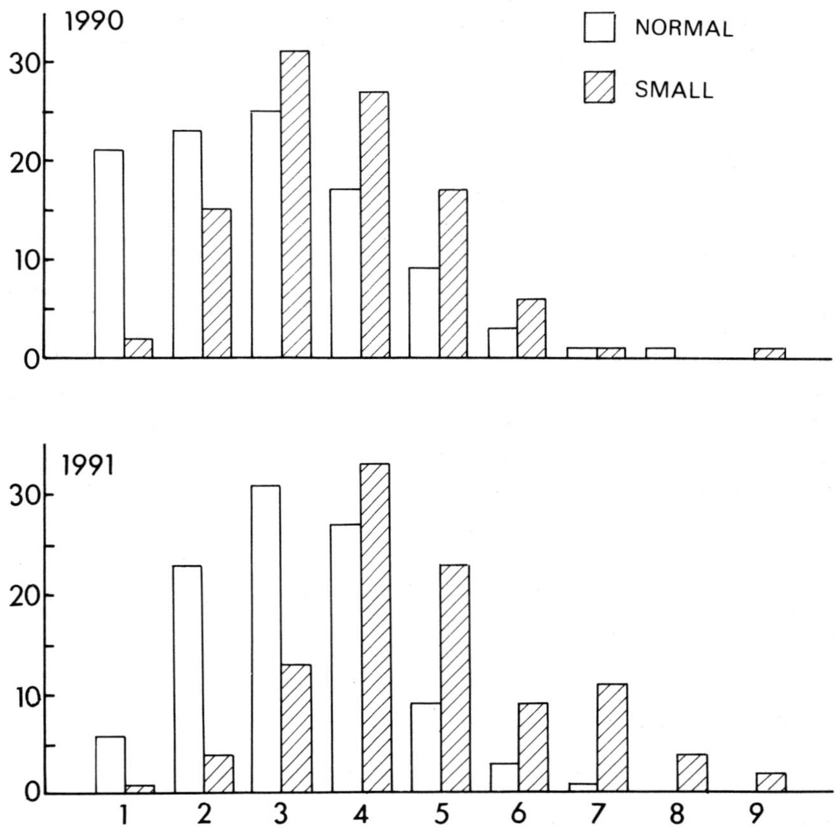

Figure 7.14 Proportions (%) of beech trees in each crown transparency class. The distributions for trees with normal- and small-sized leaves in 1990 and 1991 have been distinguished. Classes: 1: 0–10% transparency; 2: 11-20%; 3: 21–30%; 4: 41–50% ... 9: 81–90% transparency.

Sitka spruce and Norway spruce both had similar levels of insect damage, whereas in Scots pine it was much less frequent (Table 7.29). Oak had considerably more damage than any other species, reflecting the wide range of insects that feed on this species. The incidence of insect activity was similar in the two spruce species in 1990–1. However, activity in 1991 was markedly higher in both Scots pine and beech. In Scots pine, this was probably due to increased numbers of the pine shoot beetle; in beech, the damage was primarily by the beech leaf miner. Landsberg and Ohmart (1989) found that defoliation of beech by insects was a regular occurrence, occurring four years out of every eight, suggesting that the damage seen in the British beech trees was not unusual.

The pine shoot beetle was largely responsible for the brown first-year needles and the dead outer shoots recorded on Scots pine during the monitoring programme. Most feeding occurs in late summer during maturation feeding of the new beetle generation, although significant feeding may also occur in early summer by the parent generation. Growth responses may occur, but they are usually minor (Långström *et al.*, 1990; Långström and Hellqvist, 1991, 1992) at the levels of damage normally seen in pine forests. When damage increases beyond this background level, substantial growth losses may occur (Långström and Hellqvist, 1991). Because it is the first-year shoots in the upper crown that are lost, the effects on growth may be greater than those caused by insects or pathogens damaging older shoots.

Fungal damage was also recorded. Only visible damage to foliage was assessed, fruiting bodies on the stems or butts of trees being reported under damage to stems and butts. As with insects, reports only assess visible evidence of recent damage at the time of the assessment and it is likely that the observations severely underestimate the true frequency of occurrence. In all species, and in both 1990 and 1991, fungal problems were reported on less than 10% of the trees. Although collection of foliar and root samples has been possible in some monitoring programmes, it has not yet been feasible in the British programme.

Mechanical damage to the crown

Mechanical damage to the crown by wind represents an important cause of crown thinning in Great Britain. It is present in most years, but was particularly noticeable in 1990, when the effects of storms in October 1987 and January 1990 were still very evident. The proportions of trees in different damage categories are given in Table 7.30.

In 1990, damage was most apparent in oak, being recorded on 39% of the trees and least apparent in Norway spruce, with only 17% affected. Abrasion of peripheral shoots (caused by wind) and broken branches caused by wind and/or snow were by far the most common types of

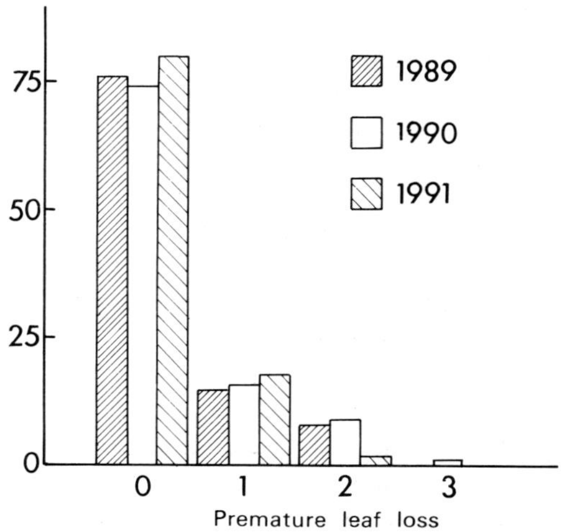

Figure 7.15 Occurrence of premature leaf loss in beech 1989–91. Categories: 0: none; 1: infrequent; 2: common; 3: abundant.

Table 7.29 Percentage of trees with insect damage to the foliage, 1990

	Frequency				
	None	Rare	Infrequent	Common	Abundant
Sitka spruce					
1990	56	14	14	12	5
1991	57	17	14	10	2
Norway spruce					
1990	62	12	13	11	2
1991	66	15	12	7	0
Scots pine					
1990	70	16	11	3	0
1991	47	35	15	3	0
Oak					
1990	20	31	31	16	2
1991	16	29	34	18	3
Beech					
1990	42	23	16	18	2
1991	26	30	21	20	3

The frequency scores refer to the percentage of the foliage affected.

damage. In 1991, there was a general increase in mechanical damage. Some of this was attributable to a period of heavy snow followed by strong winds that affected parts of the country in December 1990 (Innes and Boswell, 1991).

Butt and stem damage

The scoring for butt and stem damage was revised in 1990 to take into account experience gained in the 1989 programme, when the assessments were first introduced. Assessments were changed so that they were the same for all species, but this involved some reductions in the amount of information specific to particular species (e.g. past *Cryptococcus* infestation on beech). The results are presented in Table 7.31.

The most frequent type of stem damage on Scots pine was caused by squirrels (5% of trees) and involved bark stripping. The cause for the stripping is unknown, but the damage in beech can be correlated with phloem width, only occurring where phloem width exceeds 0.3 mm (Kenward, 1989). Tarry patches on the stems were also recorded in 1990

Table 7.30 Percentages of trees with specific types of mechanical damage within the crown in 1990 and 1991

	Type								
	0	1	2	3	4	5	6	7	8
Sitka spruce									
1990	79	17	3	0	1	0	0	0	0
1991	57	39	3	1	0	0	0	0	0
Norway spruce									
1990	83	14	2	0	0	0	0	0	0
1991	59	37	7	1	0	0	0	0	0
Scots pine									
1990	69	20	9	1	0	0	0	0	0
1991	52	29	19	3	1	0	0	0	0
Oak									
1990	61	24	13	1	1	0	0	0	0
1991	56	28	19	0	0	0	0	0	0
Beech									
1990	74	20	5	0	1	0	0	0	0
1991	63	27	6	2	0	1	0	0	0

0: none; 1: abrasion of peripheral shoots; 2: wind and snow damage; 3: prevailing wind effect; 4: damage caused by adjacent trees leaning into target tree; 5: hail; 6: lightning; 7: damage caused by harvesting operations; 8: other.
Values do not necessarily add up to 100 as a tree may have more than one type of damage.

Table 7.31 Percentage of trees with specific types of butt (below 1.3 m) and stem (above 1.3 m) damage, 1990

	Type														
	0	1	2	3	4	5	6	7	8	9	10	11	12	13	14
Sitka spruce															
Stem – 1990	97	0	0	2	0	0	0	0	0	0	0	0	0	0	0
1991	95	0	0	1	0	0	0	1	0	0	0	3	0	0	0
Butt – 1990	89	0	0	2	0	0	0	5	0	0	1	2	0	0	0
1991	88	0	1	4	0	0	0	2	0	0	1	4	1	0	0
Norway spruce															
Stem – 1990	94	0	0	2	0	1	0	2	0	0	0	1	0	0	0
1991	93	0	0	3	1	1	0	1	0	0	0	1	0	0	0
Butt – 1990	81	0	1	3	0	1	0	9	0	0	2	1	2	0	0
1991	81	0	3	6	2	2	0	4	0	0	0	3	2	0	1
Scots pine															
Stem – 1990	89	0	1	0	1	0	0	1	0	3	0	0	5	0	0
1991	91	0	1	0	1	0	0	1	0	0	0	0	5	0	0
Butt – 1990	94	0	1	0	0	0	0	4	0	0	0	0	0	0	0
1991	94	0	2	0	0	0	0	3	0	0	1	0	0	0	0
Oak															
Stem – 1990	96	0	1	0	0	2	0	0	0	0	0	1	0	0	0
1991	95	0	2	0	0	1	0	0	0	0	0	0	0	0	1
Butt – 1990	90	1	2	0	1	2	0	3	0	0	1	0	0	0	1
1991	91	1	4	0	1	1	0	1	0	0	0	0	0	0	1
Beech															
Stem – 1990	72	1	3	11	2	2	0	0	5	0	1	1	1	2	0
1991	74	1	8	0	2	4	0	1	6	1	3	2	1	0	1
Butt – 1990	56	2	8	1	2	5	1	3	5	1	3	1	12	0	0
1991	56	3	13	0	4	7	0	3	8	2	4	2	9	0	3

0: none; 1: dead area with bark still present; 2: area of exposed wood; 3: sap/resin flow from cracks or holes; 4: swelling; 5: crack at least 5× long as wide; 6: fungus; 7: extraction damage; 8: insect activity; 9: tarry spots on bark; 10: vandalism; 11: pruning or brashing wounds; 12: animal damage; 13: fire; 14: other.
Values do not necessarily add up to 100 as a tree may have more than one type of damage.

(3% of trees), but not in 1991. Sap flow from cracks or holes was recorded on 11% of beech stems in 1990, but was absent in 1991 and insect activity was seen on 5–6% of the stems. Sitka and Norway spruce and oak had no particularly frequent forms of stem damage.

Butt damage was most frequent on beech, with 56% of trees affected. Damage by grey squirrels and rabbits (*Oryctolagus cuniculus*) was particularly frequent, affecting 9–12% of trees. The incidence of beech bark disease was low, probably reflecting the absence of suitable precursors during the previous years. Areas of exposed wood, caused by unknown factors, were present on 8% of trees in 1990, rising to 13% in 1991. All species showed relatively high levels of extraction damage to the butts, with Norway spruce being the most affected. Norway spruce is particularly sensitive to damage during thinning operations (Kovbassa, 1988) and this is reflected in the results.

The importance of butt damage is uncertain. Exposure of the xylem can provide an opportunity for infection by various forms of rot. Often however, trees can partition the damaged area, limiting the spread of any rot. With severe damage, physiological processes may be disturbed, but trees seem to be able to withstand large amounts of damage with only limited effects on the crown (Fig. 7.16). The relationship between butt and stem damage and crown condition requires further investigation.

Relationships between transparency, discoloration and the other indices

It is extremely difficult to assess the relationships amongst the different variables. This partly arises because of the sampling design and the different scales of the variables, but a greater problem is the high level of multicollinearity in the data. The large data sets (ranging from about 1300 to 2000 trees of each species) mean that very low correlation coefficients may nevertheless be significant, leading to difficulties of interpretation. In addition, there is the strong possibility that some of the relationships between variables may be site-specific. The statistical problems associated with the analysis of forest monitoring data are currently being examined, and further developments in this area are likely.

At the simplest level, non-parametric correlation techniques (e.g. Spearman's Rank test) can be used to examine bivariate relationships. Spearman rank correlations were calculated for the British data. The results reveal some of the the more important trends in the data. In Sitka and Norway spruce and Scots pine, there was a strong correlation between the extent of shoot death and crown transparency. Branch density (a subjective assessment of the height increment) was also positively correlated with transparency. A strong negative correlation occurred between transparency and needle retention. The other strong correlation present in both spruce species was between transparency and secondary shoot abundance. The relationship indicates that as transparency increases, the incidence of secondary shoots increases. This may indicate that trees will usually recover following defoliation.

In oak and beech, transparency was primarily correlated with dieback, crown form and the extent of insect and mechanical damage in the crown. There was no relationship between the occurrence of discoloration and crown transparency.

These results point to a possible reason for the much higher figures for

Figure 7.16 Severe butt damage to a giant sequoia (caused by fire). Despite the loss of heartwood and xylem tissue, there was no obvious effect on the crown condition. Giant Forest, California.

transparency found for Great Britain in comparison with other European countries (Table 6.13). In many countries, transparency caused by factors such as dieback, insects and mechanical damage is excluded from the final transparency assessment. If this practice were adopted in Great Britain, then it is clear that the reported levels of transparency would be substantially lower. Revised estimates of transparency, taking into account various mitigating factors, have been presented by Innes (1992d).

Conclusions

The results collected from Great Britain for 1987–91 reveal the importance of obtaining data on a variety of indices in addition to crown transparency and discoloration. The extra data provide an invaluable supplement to the transparency data, and enable some of the trends to be interpreted.

The data for Great Britain do not point to any consistent trends in forest condition over the period. This supports the results of the crown transparency assessments. However, the droughts of 1989 and 1990 clearly had an effect on many indices. It is not clear how long these effects will last. Some may be transient, others may be more permanent. Only continued monitoring will resolve this.

8 Soils and Forest Decline

We must remember that the soil not only supports all plants growing in it but also supports myriad hidden processes that are necessary for its fertility and for healthy forests.
Maser, 1988

Introduction

Since the 1970s, there has been increasing concern that forest soils have been damaged by acidic deposition, due in part to sulphate, nitrate and ammonia. Other aspects of forest ecosystems, including forest health (e.g. Schröder and Vetter, 1989) are affected, particularly the ground flora of forests, the availability of certain essential nutrients such as calcium, magnesium and manganese and the concentrations of potentially toxic chemical species. There is a considerable literature to support these assertions, some of which has been reviewed below. Several detailed reviews of forest soil chemistry exist (e.g. Reuss and Johnson, 1986; Binkley *et al.*, 1989a; Lucier and Haines, 1990; Ulrich and Sumner, 1991) and these should be examined for a full discussion of the subject.

Soil quality has long been known to influence the productivity of forests. It is therefore not surprising that interest should turn to soils, following the 'discovery' that forest soils have acidified over the past 30–60 years. However, links between soil acidification and tree health have been difficult to establish in the field.

Acidification

Ulrich (1991) defined soil acidification as a decrease in the acid neutralizing capacity and/or an increase in the base neutralization capacity and/or an increase in the acid strength as indicated by a decrease in pH. Van Breemen *et al.* (1983) defined soil acidification as a decrease in the acid neutralizing capacity of the soil. Van Breemen *et al.* (1983) emphasized the important distinction between intensity, as measured by pH, and capacity, as measured by the moles of strong acid present. They defined the acid neutralizing capacity as the amount of strong acid required to reduce the pH to a reference pH value and suggested that the best way to estimate the acid neutralization capacity of forest soils (ANC_s) was by component composition:

$$(ANC_s) = 6(Al_2O_3) + 2(CaO) + 2(MgO) + 2(K_2O) + 2(Na_2O) + 2(MnO) + 2(FeO) - 2(SO_3) - 2(P_2O_5) - (H_2Cl)_5$$

where brackets denote molar concentration. They considered a pH of 3.0 to be a suitable pH reference value for forest soils.

Acidification of soils occurs when the rate of proton input to the soil exceeds the rate of proton consumption by the buffering system (Matzner and Ulrich, 1985). In lime-free soils, the only process that buffers protons without simultaneously producing cation acids is the weathering of silicates and the associated release of basic cations such as calcium, magnesium, sodium and potassium. If the buffering capacity represented by the weathering of silicates is insufficient to buffer all the available protons, buffering takes place by reactions with the clay mineral lattice and monomeric or polymeric aluminium ions are released. Amorphous aluminium hydrous oxides play a critical role in this process. The release of aluminium ions, in turn, leads to the displacement of exchangeable cations, mainly calcium and magnesium (Matzner and Ulrich, 1985). The amount of acidity that most forest soils can buffer through silicate weathering varies from about 0.2–1.0 keq ha^{-1} yr^{-1}. As the deposition of acidity is usually considerably greater than this, acidification follows.

The role of sulphur

A difficult issue to resolve is the fate of sulphur deposited from the atmosphere. Sulphur is an essential nutrient for plant growth and, in some areas such as the Pacific Northwest, forestry problems have been encountered due to a lack of sulphur (Blake *et al.*, 1988). In arable crops, sulphur requirements vary from 15 to 90 kg S ha^{-1} yr^{-1} (Johnson and Reuss, 1984), although forest requirements are believed to be much lower (<5 kg ha^{-1} yr^{-1}) due to recycling (Reuss and Johnson, 1986). Generally, sulphur is not considered a nutrient limiting the growth of forests. Sometimes, sulphate may not accumulate under present deposition loads (e.g. von Freiesleben, 1988), and it has even been suggested that under certain acidification conditions, the amount of sulphate in streams may actually decrease because of increased adsorption of sulphate in the soil (Fuller *et al.*, 1987). Consequently, inputs of sulphur do not necessarily lead to further acidification.

A variety of mechanisms have been proposed for sulphate retention in soils. The three most frequently cited are the specific adsorption of sulphate onto iron and aluminium oxide surface materials, involving ligand displacement (Rajan, 1979), the immobilization of sulphate by soil organic matter through the conversion of inorganic sulphate to organic sulphate compounds (Fitzgerald *et al.*, 1982; Swank *et al.*, 1984) and the precipitation of basic aluminium sulphate minerals (Nordstrom, 1982).

Autry *et al.* (1990) examined the sulphur retention mechanisms in forest soils. They found that organic sulphur accounted for more than 78% of the total sulphur content in the uppermost (0–20 cm) mineral soil horizons. Adsorbed sulphate was generally a minor component, accounting for less than 13% of the total sulphur (Fig. 8.1). Carbon-bonded sulphur was the most prevalent form of organic sulphur. From these observations, they argued that the most important retention mechanism was organosulphur formation. Nodvin *et al.* (1986) found that 93% of the soil sulphur was held in organic forms, 74% being carbon-bonded and 18% being ester sulphate and similar findings have been reported by Bettany *et al.* (1979). Strickland *et al.* (1986) found that inorganic sulphur only accounted for 2–18% of the total sulphur in the upper 20 cm of the forest litter and soil, with additional inorganic sulphur added to the soil rapidly being converted to organic forms. While other studies have not always supported these conclusions (e.g. Meiwes *et al.*, 1980), microbial immobilization of sulphur inputs can evidently be both rapid and important in some forest soils (Hern *et al.*, 1985).

In other soils, adsorption of sulphates is an important mechanism influencing the dynamics of atmospherically-deposited sulphur (Singh *et al.*, 1980; Johnson and Reuss, 1984; Hern *et al.*, 1985; Reuss and Johnson, 1986; Singh and Johnson, 1986; Harrison *et al.*, 1989; Fernandez and Rustad, 1990). The predominant adsorption mechanism appears to be OH^- ligand exchange, which occurs primarily on iron and aluminium hydrous oxide coatings on soil and clay particles. Harrison *et al.* (1989) found that the presence of hydrous oxides of aluminium was one of the best predictors of the sulphate adsorption potential of a soil, suggesting that these are particularly important adsorption substrates. A detailed account of aluminium and iron oxides in acidic soils has been given by Paterson *et al.* (1991).

Formation of basic aluminium sulphate minerals may also be involved in the retention of sulphate in forest soils. For example, Khanna *et al.* (1987) argued that the retention of sulphate in an acid forest soil was related to the precipitation and dissolution of jurbanite and gibbsite, a hypothesis supported by studies of sulphate retention in Danish forest soils (Von Freiesleben, 1988; Gundersen and Beier, 1988). Courchesne and Hendershot (1990), working on two Humic Cryorthods in Quebec, have suggested that sulphate retention may be related to the formation of alunite, although they failed to find this mineral in their soils and Süsser and Schwertmann (1991) have suggested the involvement of an aluminium hydroxy-compound.

The capacity for adsorption is dependent on pH because of the amphoteric nature of iron and aluminium oxides. It increases with decreasing pH (Fig. 8.2) (Chao *et al.*, 1962; Fuller *et al.*, 1985; Cosby *et al.*, 1986). The adsorption capacity is also concentration dependent. The forest floor leachate contains organic ligands and these may compete with sulphate on the exchangeable positive charge sites, resulting in either reduced or non-existent sulphate retention (Gobran and Nilsson, 1988). At low pH, the surfaces of colloids become positively charged because of

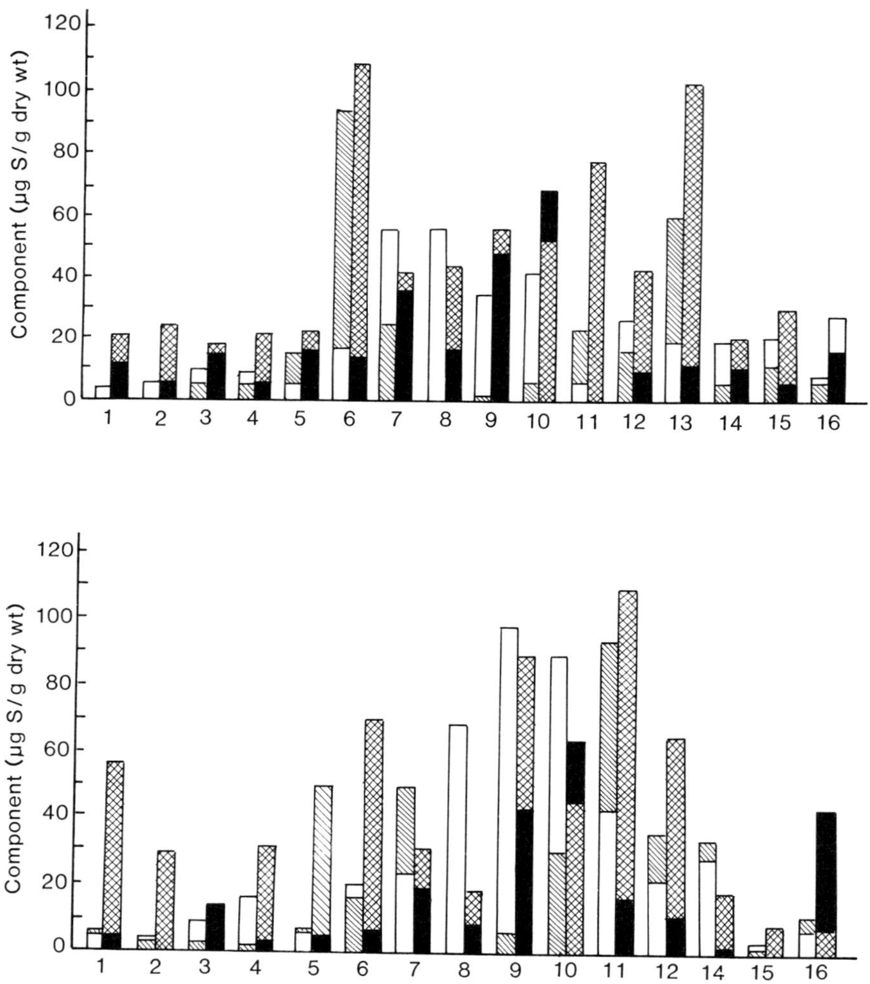

Figure 8.1 Soluble and adsorbed sulphur components of O1 (upper diagram) and O2 (lower diagram) horizon samples collected in 1985 from various forested sites in the United States. Solid: adsorbed sulphate; cross-hatching: ester sulphate; open: soluble sulphate; hatching: ester sulphate. From: Autry *et al.* (1990).

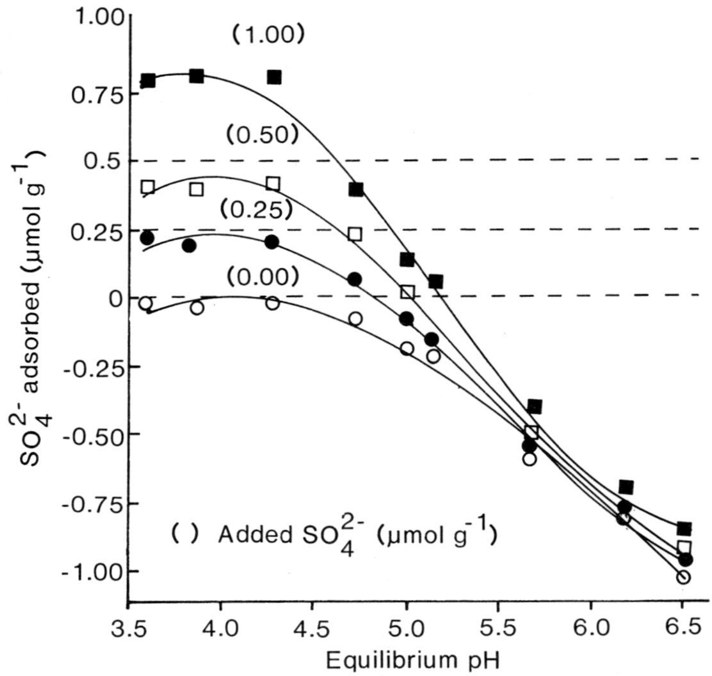

Figure 8.2 Sulphate adsorbed by the Bs2 horizon soil from Hubbard Brook Experimental Forest, U.S.A., at different concentrations of added sulphate across a range of equilibrium solution pHs. Negative adsorption numbers were due to desorption of ambient sulphate. From: Fuller *et al.* (1987). Reprinted by permission for Nature vol. 325, pp. 707–710; Copyright (c) 1987 Macmillan Magazines Ltd.

H^+ adsorption, enhancing anion adsorption. At high pH, the surfaces become negatively charged due to H^+ dissociation or OH^- adsorption (Johnson *et al.*, 1986). Other factors include the dissolution of aluminium below pH 4 (Chao *et al.*, 1962) and the successful competition for sulphate anions by aqueous alumino-sulphate complexes against sesquioxides at low pH (Nodvin *et al.*, 1986). Johnson and Reuss (1984) defined a lime potential (K_L) of the soil:

$$K_L = pH - 0.5p(Ca^{2+} + Mg^{2+})$$

If the lime potential of the incoming solution exceeds that of the soil, then the base saturation of the soil will be increased when equilibrium is reached. This has been confirmed in several field experiments (e.g. Khanna *et al.*, 1986). Meiwes *et al.* (1980) have argued that the maximum amount of sulphate adsorption occurs at about pH 4.0. This value has also been obtained in studies of the soils at Hubbard Brook: Nodvin in the soil, whereas 52% was retained at pH 4.8 and 13% at pH 6.3 (Fig. 8.2).

The role of ammonium

In several areas, particularly The Netherlands, atmospheric ammonium plays an important part in the acidification of forest soils. The reactions involved can be described as:

$$NH_4^+ + 2O_2 \rightarrow NO_3^- + H_2O + 2H^+$$

Denitrification results in a reversal of the acidification:

$$2NO_3^- + 5H_2 + 2H^+ \rightarrow N_2 + 6H_2O$$

The repeated application of ammonium fertilizers to agricultural soils has been known to result in acidification problems for some time and, on occasion, they have been used to deliberately acidify the soil to promote the growth of acidophilic crops such as blueberries and cranberries (Haynes and Swift, 1986). Atmospheric deposition frequently occurs as ammonium sulphate, following a reaction between ammonium and sulphur dioxide (Van Breemen *et al.*, 1982). On reaching the soil, the ammonium sulphate rapidly oxidizes to form nitric and sulphuric acids. In calcareous soils, the acids dissolve calcium carbonate, resulting in a predominance of dissolved calcium nitrate. In acidic soils, the acids result in the formation of aluminium nitrate (Van Breemen *et al.*, 1988). Van Breemen *et al.* (1982) estimated that the deposition to some forests in The Netherlands amounted to about 63 kg N ha^{-1} yr^{-1}, and subsequent studies in The Netherlands have suggested that the deposition is considerably higher in the worst affected areas.

The acidification problems caused by ammonium have been greatest on sandy soils and this is where the greatest effects on trees have been reported (see Chapter 3). Changes to the soil flora have occurred, with declines in terrestrial saprophytic fungi and increases in humus-living saprophytic fungi (Van Breemen and Van Dijk, 1988). The terrestrial flora has also changed, with nitrophilous species, such as broad buckler fern, climbing corydalis, common hemp-nettle and various roses (*Rubus* spp.) increasing in abundance (Ellenberg, 1979). Similar changes have been reported in the Belgian Ardennes (Muys, 1990), with broad buckler fern, bramble (*Rubus fruticosus* L.), bracken (*Pteridium aquilinum* (L.) Kuhn), climbing corydalis and common cleavers (*Galium aparine* L.) all increasing in abundance.

Leaching and other effects

The relative importance of the different mechanisms and effects of the documented acidification is still being debated. Bouman and Lohmann

(1987) attributed acid toxicity in the Harz Mountains of Germany to base deficiency, which may be caused by uptake and immobilization of exchangeable cations from the soil pool. However, production of acidic decomposition products in spruce and pine litter may also be an important mechanism (Tamm, 1950).

Natural soil acidification mainly occurs because of the dissociation of weak acids (De Vries and Breeuwsma, 1984). In acidic soils, natural acidification is about 0.1–2.0 kmol H^+ ha^{-1} yr^{-1} (Ulrich and Matzner, 1983; Van Breemen *et al.*, 1984; Fölster, 1985; Van Miegroet and Cole, 1985). This rate is similar to the level of acidification caused by the removal of vegetation, which is about 0.5–2 kmol H^+ ha^{-1} yr^{-1}. Although acidification of acidic soils is primarily the result of atmospheric deposition, natural acidification by organic acids can also occur, leading to the formation of podzols.

Leaching of base cations is both an effect of acidification and another important mechanism leading to soil acidification (e.g. Johnson *et al.*, 1985). The amounts of cations leached from the forest floor can be directly related to the acidity of the throughfall (Cronan, 1980). Cronan, using microcosms in a balsam fir ecosystem in the northeast United States, found that leaching of calcium, magnesium, potassium and ammonium increased with increasing throughfall acidity. Increased losses of calcium, potassium, ammonium, sulphate, nitrate and organic ions occurred as temperature increased. However, in a study of a forest in eastern Tennessee dominated by chestnut oak (*Quercus prinus* L.) and yellow poplar, uptake and cycling of potassium and calcium by the vegetation resulted in considerable conservation of these elements (Johnson *et al.*, 1985). Here, losses of calcium from the soil pool were due to uptake by biomass, whereas losses of magnesium were attributable to leaching (Johnson and Todd, 1990). Similar results have been documented in Switzerland (Blaser, 1992). Consequently, Johnson *et al.* (1985) considered it unlikely that net exports of cations would significantly reduce soil cation reserves, with the exception of calcium. In contrast, Mulder *et al.* (1989) argued that organic aluminium is currently being rapidly depleted by leaching in some soils in The Netherlands and that this loss may cause a drop in soil pH due to the lack of buffering.

Several European studies have suggested that acidification has led to the loss of nutrients from the soil through leaching. For example, Aune *et al.* (1989) compared forest soils in Høylandet (Trondelag, Norway), Aust Agder (Sørlandet, Norway) and the Black Forest (Baden-Württemberg, Germany), where pollution was believed to be low, intermediate and high, respectively. The sites had similar soils (podzols) and soil parent materials, but differed in climate, with the annual mean temperature at the Høylandet site being 1.5°C lower than at the Black Forest site. Potassium, sodium, magnesium and calcium concentrations in the Black Forest were about 20% of those at the Høylandet site, and the Black Forest soils were more acidic and had higher nitrogen contents. The lower sodium, calcium and magnesium values were attributed to acidic deposition, although it was acknowledged by the authors that differences in biomass storage, management and site conditions might play a part.

Bergkvist (1987a, b) and Bergkvist *et al.* (1989), working with podzols and brown forest soils in mixed coniferous and beech forests in southwest Sweden, found net losses of sodium, potassium, magnesium, calcium, aluminium, zinc, cadmium and nickel. In addition, there were significant movements within the soils, with almost all the iron and lead leached from the A horizon being precipitated in the B horizon. In addition, organic forms of aluminium were being leached from the A horizon and precipitated in the B horizon. Leaching of base cations was greatest from the brown forest soil, whereas leaching of aluminium was greatest from the podzol. In some areas, the rate of leaching exceeds the supply of important elements through weathering (Hofmann-Schielle, 1991).

The effects of acidification and liming on soil organisms and decomposition in a Scots pine forest in north Sweden were compared by Bååth *et al.* (1980a, b). Acidification treatments involved doses of 50 and 150 kg ha^{-1} H_2SO_4. The acidic treatments resulted in reduced decomposition rates and soil micro-organism populations and changes to the soil fauna composition. Conversely, liming did not affect decomposition, although the abundance of Enchytraeidae decreased markedly under both acidic and liming treatments. This contrasts with work done in Berlin, Germany, where liming of acidic soils under Scots pine led to increases in macrofaunal and microbiological activity (Marschner and Wilczynski, 1991). The effects of acidification on the rates of decomposition of Scots pine litter appear to depend on the initial rate of decomposition. Working at Kevo, in northern Finland, where decomposition rates are naturally very slow, Neuvonen and Suomela (1990) found no effect of treatments with pH 3 and pH 4 on the decomposition rates of litter.

Reductions in decomposition rates associated with acidic treatments have also been recorded for Norway spruce litter (Hovland *et al.*, 1980; Berg, 1986b), red spruce litter (Moloney *et al.*, 1983) and birch leaves (Fig. 8.3) (Hågvar and Kjøndal, 1981a, b), although the extent of the reduction may be dependent on soil types and other factors (e.g. Berg, 1986c). Liming had no effect on the rate of Norway spruce litter loss (Berg, 1986b).

Changes in forest soils, particularly those that have changed from a non-forested form of landuse to forest would be expected because of the uptake of base cations from the soil associated with the increase of biomass in an afforested stand (Williams *et al.*, 1979; Nilsson *et al.*, 1982; France *et al.*, 1989). As electroneutrality must be maintained, uptake of nutrient cations is accompanied by the uptake of an equivalent amount of anions or by the release of cations. Anion uptake is balanced by a flux in OH^- which, due to the high partial pressures of carbon dioxide in the soils,

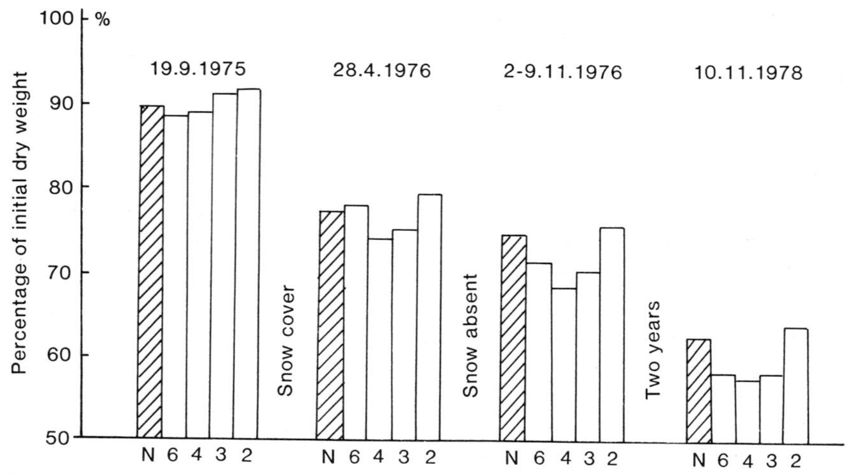

Figure 8.3 Dry weight reduction of birch leaves in litter bags (laid out 28.7.1975). Hatched columns: non-watered plots. Figures at the bottom indicate the pH of the artificial rain. From: Hågvar and Kjøndal (1981a).

accumulates as HCO_3^-. Changes in the chemical status of soils under conifers are well-known (e.g. Page, 1968; Grieve, 1978; Williams *et al.*, 1979; Messenger, 1980; Nys, 1981; Rehfuess, 1981; Silveira Lopes *et al.*, 1985; Sohet *et al.*, 1988). They are more important on base-poor, podzolic soils than on richer soils. Soil acidification has also been recorded when broadleaved forest is converted to coniferous forest (Alban, 1982) and broadleaved forest itself is quite capable of acidifying the soil, even to the extent of causing podzolization (Hornung, 1985). However, acidification of soils under broadleaves is frequently ascribed to acidic deposition (e.g. Beyer *et al.*, 1991). Bergkvist (1987b) found that Norway spruce increased soil acidification compared to beech.

Many forest soils have evolved from non-forest soils in relatively recent times. Much of the forested area of eastern North America has developed on abandoned agricultural land and the extent to which this has occurred in Europe has probably been underestimated. This has important implications for some studies of changes in forest soils (e.g. Binkley *et al.*, 1989b), as the chemical status of the soils may not have approached equilibrium with the new landuse type (Ulrich, 1983a). Consequently, any inferences about the role of acidic deposition should be drawn with care.

In The Netherlands, most forests have been planted on former heathlands (Van den Burg, 1991). The original soils in these areas were very acidic, and Scots pine was therefore chosen for the afforestation, as this species is generally tolerant of poor nutrient status and high soil acidity (Carter, 1987). However, the Scots pine failed in the north-east of

The Netherlands due to repeated attacks by needle-cast fungi (*Lophodermium* spp.). As a result, other species were planted, predominantly Japanese larch, Norway spruce, Douglas fir and pedunculate oak. These have higher nutritional requirements than Scots pine and liming and/or fertilization was often necessary.

In Great Britain, a vigorous afforestation programme has occurred during the 20th century. Much of this has been in upland areas, where the soils were already degraded from past agricultural activities (e.g. Innes, 1983a, b). Many soils had been subjected to severe grazing pressure, and erosion was quite widespread (Innes, 1983c, 1986, 1989). Consequently, these soils and also the peats that have been planted are not ideal for forest growth. In lowland areas, replacement of semi-natural broadleaved woodland with coniferous plantations has been widespread and this is also likely to have had major effects on the soils.

The nature of humus formation is particularly important in aggrading forest ecosystems. Where the soil is nutrient-rich, basic and well-aerated, organic matter derived from foliage litter is mixed into the mineral soil and humified. Highly polymerized, base-saturated humates are formed and these are bound to the mineral fraction of the soil, leading to the formation of humic mineral horizons, with mull as the humus form. If the biological activity is reduced for any reason, there is a corresponding decrease in the rate of formation of humus. The result is the formation of a humus layer of mor. The intermediate stage between mull and mor is moder, and this may develop in situations where acidification is occurring (Beyer *et al.*, 1991).

Various nutritional problems associated with soil acidification have been seen (see Chapter 3). Phosphorus deficiency is particularly important in some soils. Mohren *et al.* (1986) postulated that an abundance of nitrogen may result in a reduction in the growth of roots compared with shoot growth, resulting in a reduction in phosphorus uptake relative to nitrogen uptake. Adverse phosphorus-nitrogen interactions have also been suggested in Switzerland (Flückiger and Braun, 1992). An important process leading to a reduction in the availability of phosphorus in the soil is the precipitation of Al- and Fe-phosphates caused by low soil pH. The negative association between nitrogen and phosphorus availability has been confirmed by fertilizer trials (Mohren *et al.*, 1986), with nitrogen fertilization leading to phosphorus deficiency in a dry humus soil on former heathland at Kootwijk and a positive response to phosphorus fertilization being seen on a humus podzol at Ulvenhout in The Netherlands.

Measured changes in forest soils

Several studies have suggested that long-term (20–70 years), significant changes in forest soils have occurred. However, as with changes in the nutrition of trees discussed in the following chapter, it is essential that various potential problems are considered when making comparisons between recent and old samples (Ilvesniemi, 1991). These include the marked variations that occur in soil pH over short horizontal and vertical distances and through time and the importance of successional processes within developing forest stands.

Working on a variety of soil types in the Alltcailleach Forest in north-east Scotland, Billett *et al.* (1988) found substantial changes in the soils between initial sampling in 1949–50 and re-sampling in 1987 (Figs 8.4 and 8.5). Acidification of the surface organic layers by 0.07 to 1.28 pH units had occurred, and 73% of the soils had acidified at a depth of 40 cm. The acidification was greatest under older, open-canopy trees and the authors attributed most of the acidification to cation depletion and litter accumulation. Sites previously dominated by heather and converted to forest showed increases in soil pH, reflecting the highly acidic nature of heather soils.

Looking at other changes in the soils, Billett *et al.* (1990) found that nitrogen and organic carbon concentrations in the soil organic horizons had decreased over time, although an increase in the O horizon thickness had resulted in a large increase in the storage capacity of the soils (Fig. 8.6). The reductions in the nitrogen and organic carbon concentrations were attributed to the mineralization of what were the surface horizons in 1949–50. Patterns of heavy metal accumulation were variable, with copper and lead increasing and zinc decreasing (Billett *et al.*, 1991).

Elsewhere in Great Britain, Johnston *et al.* (1986) found evidence of soil acidification under both grassland and woodland in central England and subsequent analyses have shown that the acidification was much greater under the woodland than in the ungrazed grassland soils (Goulding *et al.*, 1988). These analyses from Rothamsted Experimental Station are based on some of the longest available records of soil chemistry and worth examining in detail. The grassland experiments started in 1856 and the naturally regenerating deciduous woodland began to develop in the 1880s. The woodland has developed on abandoned agricultural land, with one area having been heavily limed approximately two hundred years ago (Goulding *et al.*, 1988). The pH of the soil in the limed area has remained at about pH 7.9–8.0 throughout the study period. In contrast, the unlimed area has shown a decrease in pH and the surface layers are now considered by Goulding *et al.* (1988) to be reaching an equilibrium value of about pH 3.6 (Fig. 8.7).

Other British studies include that of Kuylenstierna and Chadwick (1991), who documented the acidification of upland grassland soils by 0.1–1.7 pH units over the period 1957 to 1990 in north-west Wales. Peats in Scotland also appear to have been acidified over the last 50 years (Skiba *et al.*, 1988).

Similar results have been obtained from other parts of the world, with Brand *et al.* (1986) documenting long-term increases in soil acidity over a

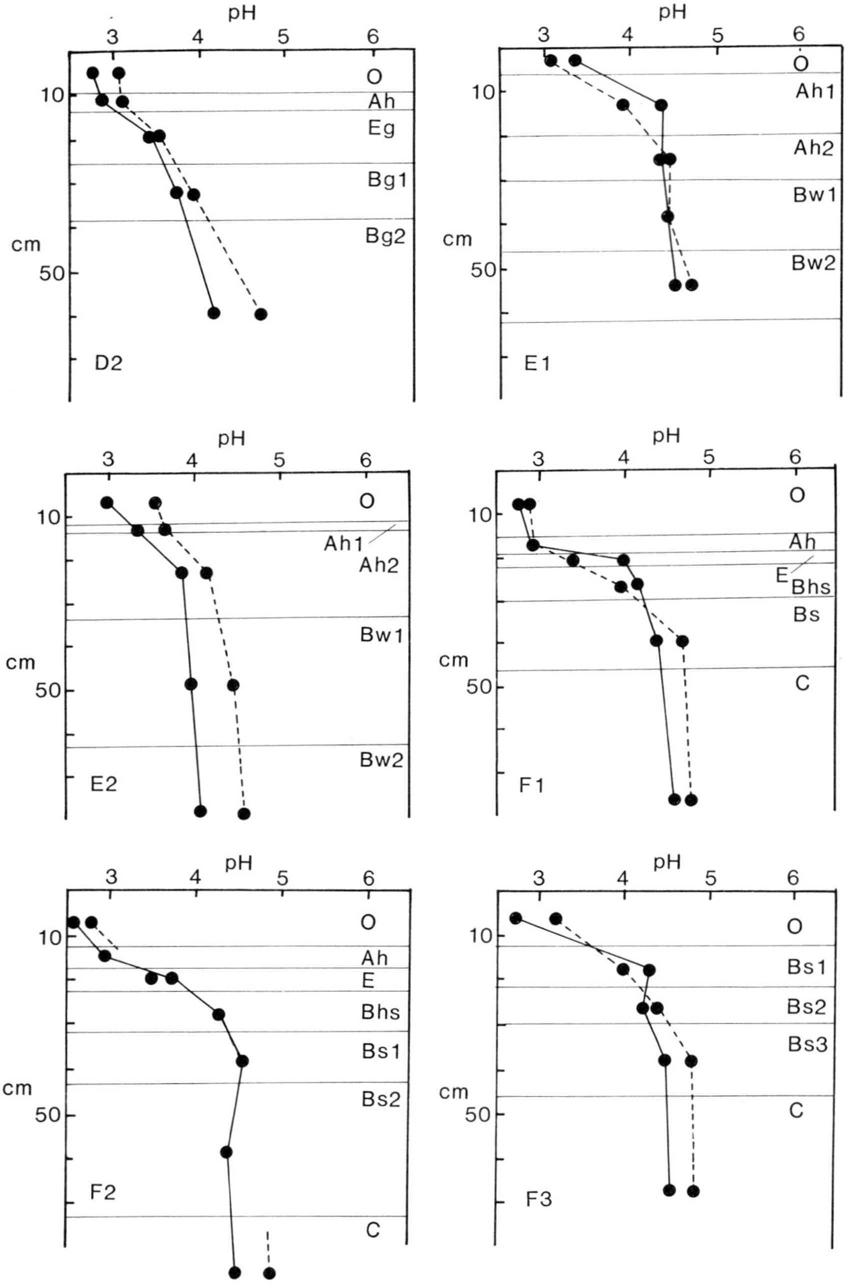

Figure 8.4 Changes in soil pH 1949–50 (dashed lines) and 1987 (continuous lines) at sites with freely draining soils and continuous forest cover since the 19th century. The depth of the 1949–50 horizons has been adjusted to enable comparison with the 1987 soil profiles. From: Billett *et al.* (1988).

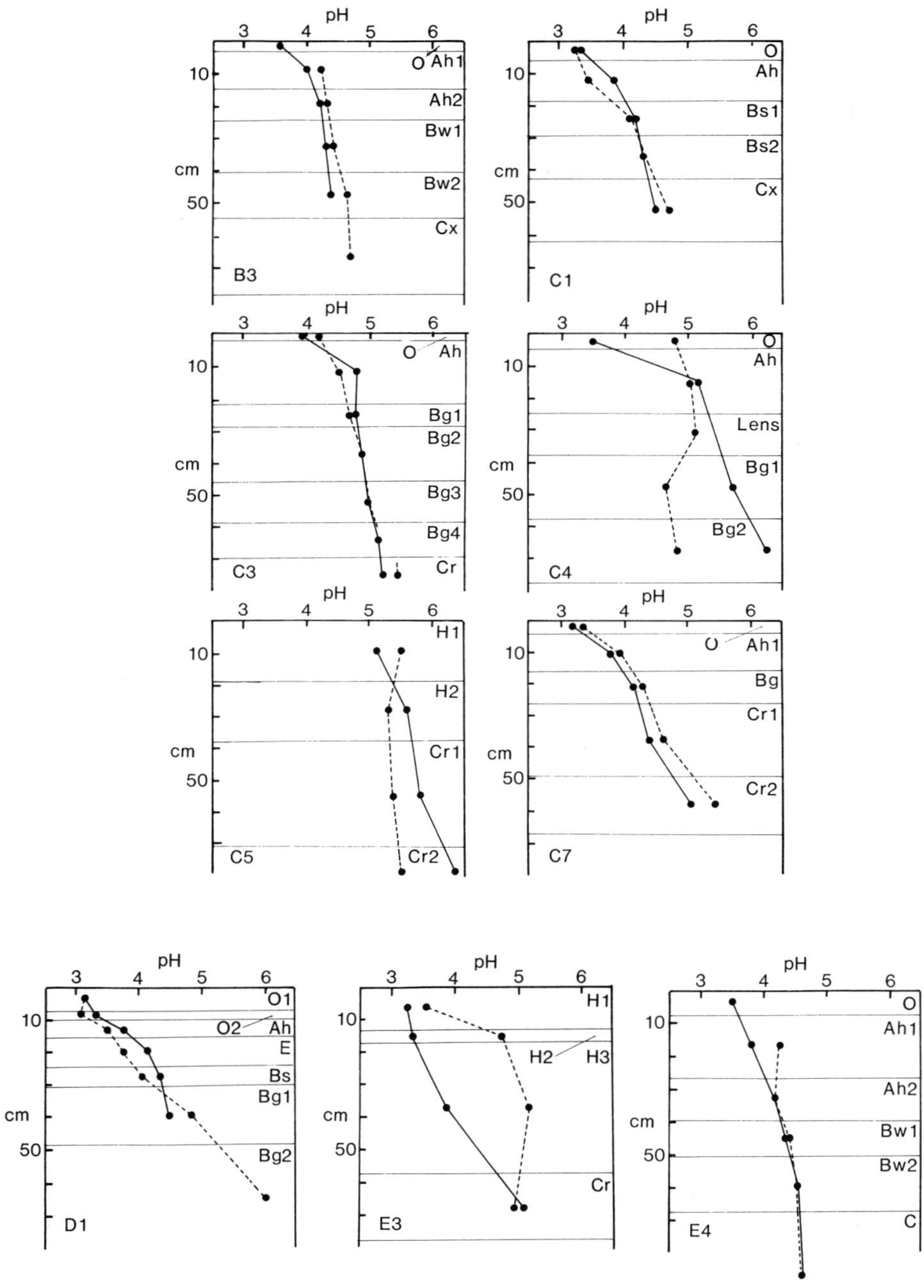

Figure 8.5 Changes in soil pH 1949–50 (dashed lines) and 1987 (continuous lines) at sites with differing drainage status and planted within the last 50 years, either after clear-felling or directly from heather moorland. The depth of the 1949–50 horizons has been adjusted to enable comparison with the 1987 soil profiles. From: Billett *et al.* (1988).

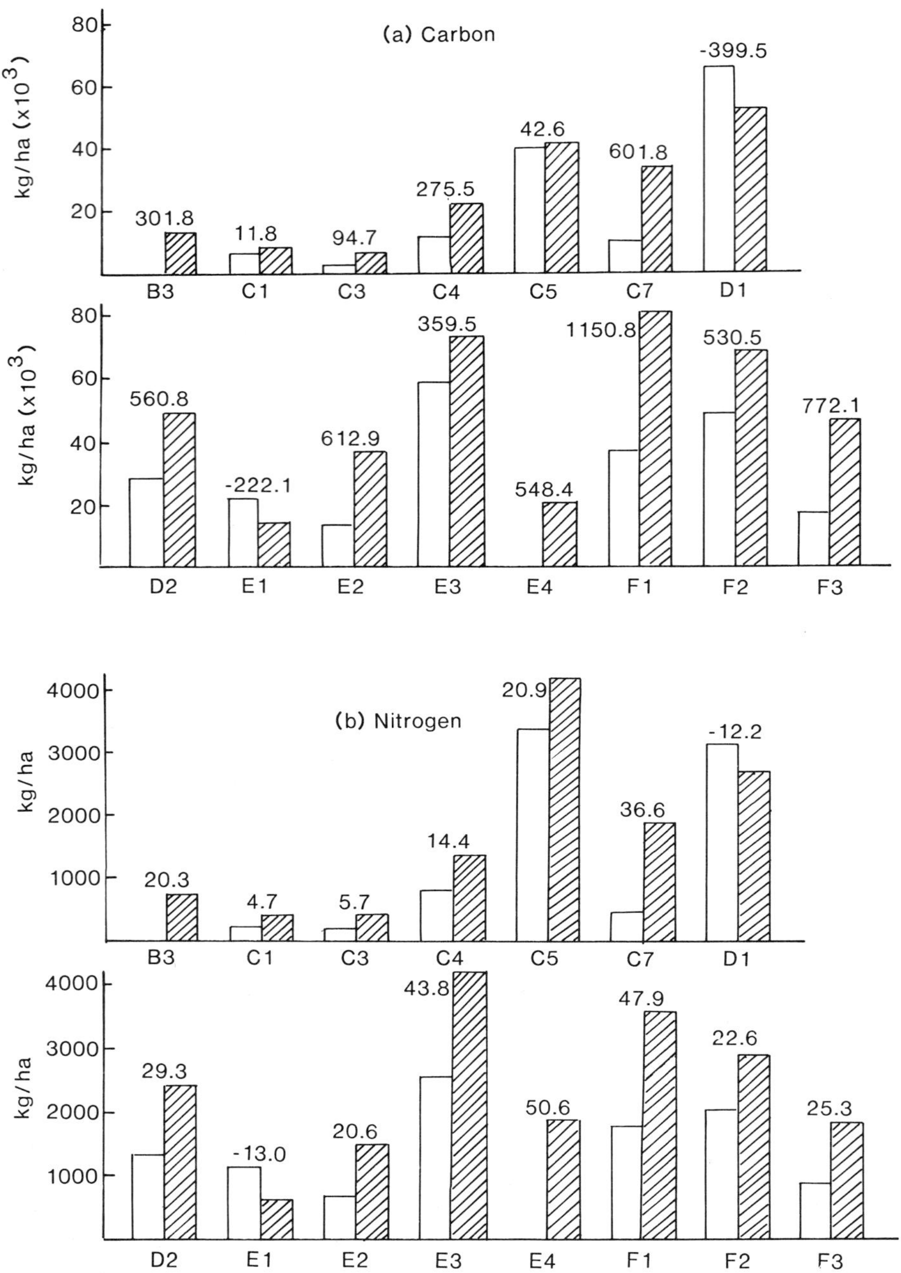

Figure 8.6 Carbon and nitrogen pools for each of 15 soil organic horizons (horizontal axes) from Scotland. Open bars represent the soils in 1949–50 and hatched bars represent the soils in 1987. Accumulation rates for C and N (in kg ha^{-1} yr^{-1}) are given for each site. From: Billett *et al.* (1990).

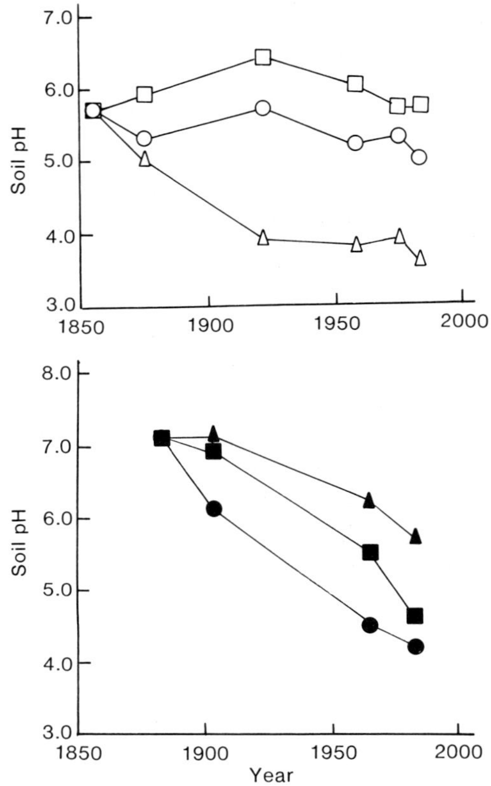

Figure 8.7 Upper diagram. pH values of soil samples taken from the unmanured (circles), ammonium-N treated (triangles) and nitrate-N (squares) plots of the Park Grass Experiment at Rothamsted, England. Lower diagram: pH values of soil samples taken from the 0–23 cm (circles), 23–46 cm (squares) and 46–69 cm (triangles) horizons of the Geescroft Wilderness at Rothamsted, England. From: Goulding *et al.* (1988).

46-year period under red pine and white spruce forest in the Ottawa Valley, Ontario, and France *et al.* (1989) finding changes over a 27-year period in a variety of woodland types in Ontario. In contrast, Roberge (1987) found either no change in pH or a slight increase over 10–20 years in the organic horizons of soils under sugar maple, balsam fir, black spruce and jack pine. In the United States, Anderson (1988) (reported in Johnson *et al.*, 1991b) documented changes in Adirondack Mountain forest soils over the last 50 years. Soils with acidic (< pH 4.0) organic horizons showed little change over the period. Less acidic organic horizons decreased in pH by between 0.3 and 0.5 pH units. Similar results have been obtained elsewhere (e.g. Wittmann and Fetzer, 1982; von Zezschwitz, 1982).

Several studies based on repeated measurements over 30 years or more have been reported from continental Europe. In Germany, acidification of forest soils has been reported from the Odenwald and Black Forest (Hildebrand, 1986), Oberpfalz (Grimm and Rehfuess, 1986), the Erzegebirge (Dässler and Ranft, 1989), the Harz Mountains (Hauhs, 1989), Lower Saxony (Ulrich *et al.*, 1989), Saarland (Wagner, 1986), the Rothaar Mountains of Sauerland (von Zezschwitz, 1982, 1985), Northrhine-Westfalia (Butzke, 1981, 1984; Gehrmann *et al.*, 1987), Bavaria (Wittmann and Fetzer, 1982) and north-west Germany (Rost-Siebert, 1988). Von Grenzius (1984) and Buck-Feucht (1986) also documented long-term acidification of German forest soils. Härtel and Cerny (1981) recorded changes in Norway spruce forest soils in Austria (near a pollution source) and soil acidification has occurred in uncultivated soils in Belgium (Ronse *et al.*, 1988), accompanied by an increase in soil organic matter.

Some of the most detailed work on soil acidification has been done in Scandinavia (Fig. 8.8) (Falkengren-Grerup, 1986, 1987, 1989a; Hallbäcken and Tamm, 1986; Tamm and Hallbäcken, 1988; Jacks *et al.*, 1989; Falkengren-Grerup and Eriksson, 1990; Ljungström *et al.*, 1990; Hallbäcken, 1992). Although it was considered in the early 1980s that the chemical changes seen in forest soils could not be unequivocally related to the effects of acidic deposition (Nilsson, 1983; Petersen, 1986), the consensus today is that deposition-driven acidification has occurred. However, Rosén (1989) emphasized the potentially adverse nutritional effects of trees removed during harvesting, and the role of leaching following nitrogen fertilization or clear-felling. These effects have not received as much attention as they deserve.

Neutral soils under beech in southern Sweden studied at different intervals between 1935 and 1983 decreased in pH by about 0.8 units in the upper horizons. Under a variety of forest types in Skåne, the pH of the humus layer was significantly more acidic in the 1980s than in the 1950s (Fig. 8.9). The decreases in pH were smallest in the soils that were originally the most acidic, but these soils showed a big increase in the hydrogen ion concentrations (the apparent discrepancy being due to the logarithmic scale used for pH measurements). In podzols characterized by the lowest pH values in the humus layer, pH decreased less over time in the upper horizons than at greater soil depths. In grey-brown forest soils, characterized by an intermediate humus layer pH, substantial decreases in pH were recorded throughout the profiles. Conversely, the grey-brown forest soils with the highest pH in the humus layer and with small increases in pH with depth showed substantial decreases in the humus layer pH but smaller changes in pH with increasing depth. Decreases in pH were recorded in both forest soils and in the soils under

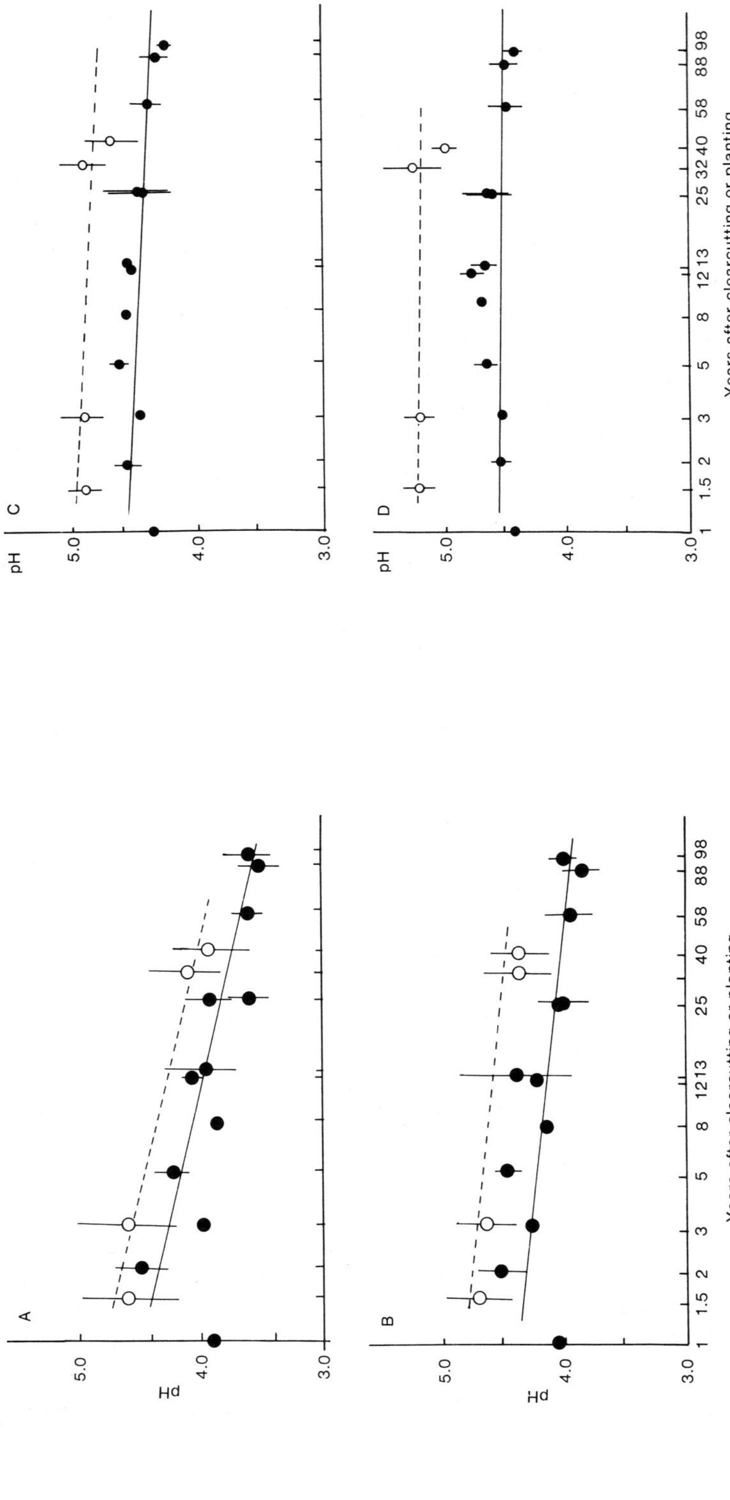

Figure 8.8 The pH of field moist soil samples plotted against the logarithmic age for different spruce stands at Tönnersjöheden Experimental Forest. Vertical bars represent the standard deviation of the measurements where were three or more profiles in stands of the same age. Dashed lines: samples collected in 1927. Solid lines: samples collected in 1984. A: Humus layer (A_0); B: A_2 horizon; C: B horizon; D: C horizon. From: Tamm and Hallbäcken (1986).

a heather-dominated heath. The wet deposition of sulphate and nitrate in the area of Skåne studies amounted to 10 kg ha^{-1} yr^{-1}, with total deposition probably being twice this figure (Falkengren-Grerup, 1989a). The changes in pH were accompanied by losses of exchangeable sodium, potassium, magnesium and calcium, as well as zinc and manganese and an increase in the concentrations of exchangeable aluminium (Falkengren-Grerup and Eriksson, 1990; Falkengren-Grerup and Tyler, 1991a). Similar changes have not been documented over the same period in forest soils at Kulbäcksliden in northern Sweden where levels of acidic deposition are low (Hallbäcken and Tamm, 1986; Tamm and Hallbäcken, 1988), suggesting that acidic deposition may have influenced the changes in Southern Sweden.

Hallbäcken (1992) calculated that the exchangeable storage of sodium, potassium, calcium and magnesium in the soil down to 70 cm depth had decreased between 1927 and 1984 by 57–60% under beech forest and by 56–74% under Norway spruce forest at Tönnersjöheden Forest in south-west Sweden. Biomass uptake was estimated to account for 41–43%

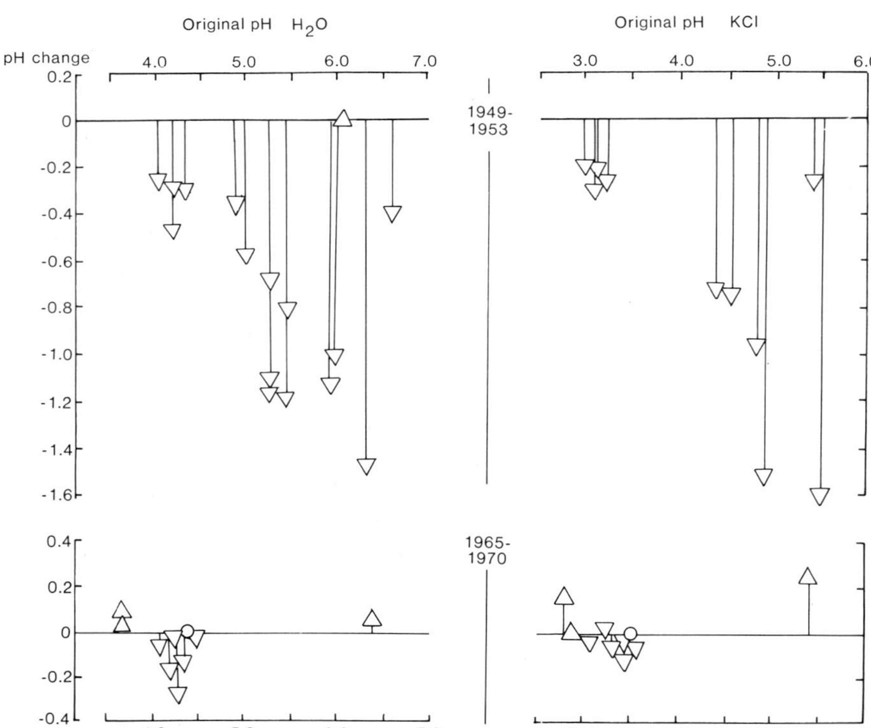

Figure 8.9 pH changes in the humus layer in relation to original pH of sites investigated in 1949–53 or 1965–70. pH_{KCl} was not measured for some sites. From: Falkengren-Grerup (1987).

of the base cation depletion under beech and 34–45% of the depletion under Norway spruce.

Sometimes, the changes in soil chemistry have been accompanied by changes in the ground flora of the forest, in others, no changes in species composition (e.g. Wittig *et al.*, 1985; Buck-Feucht, 1986) or tree health (Ljungström *et al.*, 1990) have been observed. Falkengren-Grerup (1989a) recorded increases in the numbers of species over time and attributed this to nitrogen enrichment, and increases in the numbers of nitrophilous species on well-buffered beech soils in Holstein, Germany, have been documented by Rost-Siebert and Jahn (1988). Tyler (1987a) recorded nitrophilous species as being more abundant in Skåne than in north-east Småland, nitrogen deposition being considerably higher in Skåne. However, several other environmental factors may have been involved in the difference. Kuhn *et al.* (1987), working in coppiced communities in northern Switzerland, recorded species losses in the herb and shrub layers over time and an increase in the incidence of nitrophilous species. However, some changes may have been attributable to changes in the coppicing regime and changes in the canopy of mature beech forest may have had an influence on observed changes in the ground flora (Wilmanns, 1989). Elsewhere in the Alps, stands with low vitality have shown no evidence of changes in their phytosociological composition (Grabherr and Mucina, 1989).

Falkengren-Grerup and Tyler (1991b) identified many changes in the flora of Swedish beech forests between 1979–81 and 1989–90. Changes of individual species could be divided into several groups, depending on their ecological requirements. Species that require low soil acidity, such as woodruff (*Galium odoratum* (L.) Scop.), dog violet (*Viola riviniana* Reichenb. / *Viola reichenbachiana* Boreau), wood melick (*Melica uniflora* Retz.), common Solomon's seal (*Polygonatum multiflorum* (L.) All.), dog's mercury (*Mercurialis perennis* L.) and Norway maple (*Acer platanoides* L.), had decreased in abundance. However, management practices were greatly affecting the species composition, with heavy thinning having particularly marked effects.

Most studies of chemical change have emphasized either the changes in soil chemistry occurring because of nutrient uptake by trees or the changes caused by acidic deposition. However, it is important to remember that many forest soils have been chemically treated, with nitrogen, potassium and phosphorus fertilization commonly being used. The effects of these are examined in a subsequent section.

The importance of sampling design

Studies of changes of soils depend on adequate sampling. This may be difficult as the pH of soils in forests is highly variable in both space and time (e.g. Beckett and Webster, 1971; Crampton, 1980; Yost *et al.*, 1982;

Federer and Hornbeck, 1985; Thompson and Loveland, 1985; Skyllberg, 1990; Grigal *et al.*, 1991; Rampazzo, 1991). In particular, many studies have shown that stemflow has an important impact on the acidity of the soil close to the trunk (e.g. Zinke, 1962; Gersper and Hollowaychuk, 1971; Skeffington, 1983; Cloutier, 1985; Glavac *et al.*, 1985; Hornung, 1985; Wittig and Neite, 1985; Riha *et al.*, 1986a, 1986b; Wittig, 1986; Wolfe *et al.*, 1987; Bringmark, 1989; Falkengren-Grerup, 1989b). The effect is strongly species-dependent (Figs 8.10 and 8.11), with the soil around oak trees showing little acidification towards the trunk (Beniamino *et al.*, 1991) and this genus causing less soil acidification than many other species (Moffat and Boswell, 1990). Pallant and Riha (1990), working with red pine and Norway spruce, have argued that the increased acidification of soils towards the bases of trees may be related to the occurrence of specific oxalate-producing soil fungi. However, the relative importance of fungi as a source of the acidification remains unclear, although they appear to be minor in comparison to stemflow.

The spatial differences in the pH of forest soils (Fig. 8.12) mean

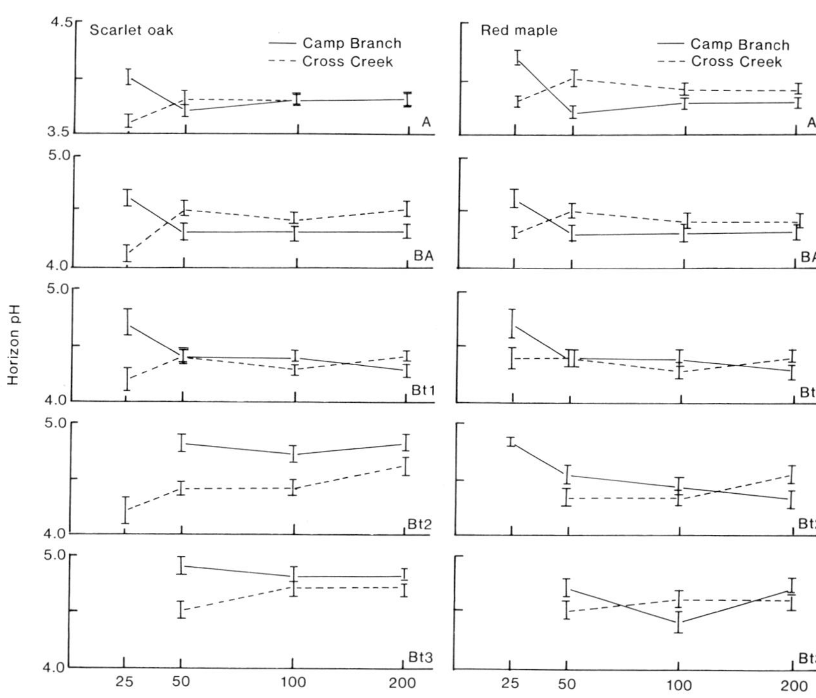

Figure 8.10 Mean pH (± one standard error) at four distances from the stem of red maple and scarlet oak at Cross Creek and Camp Branch, Tennessee. From: Wolfe *et al.* (1987).

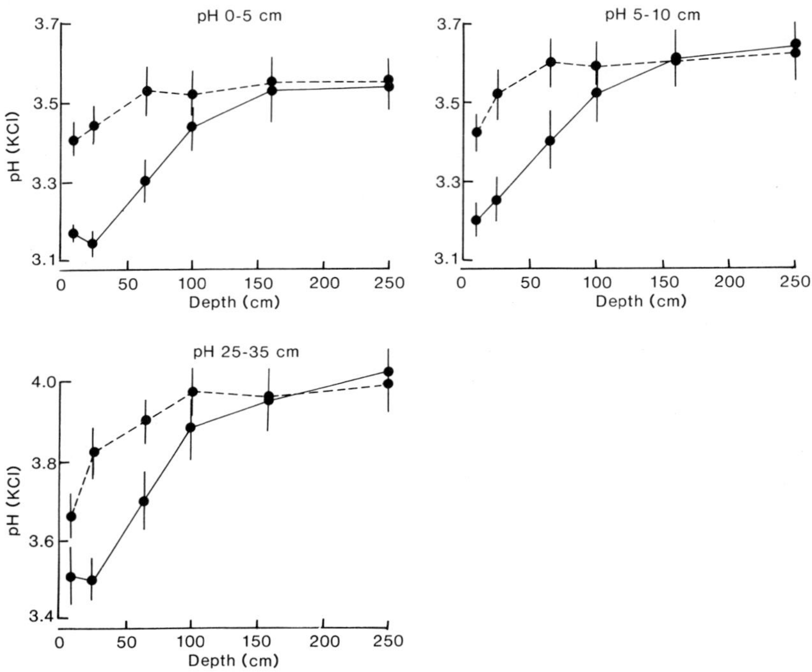

Figure 8.11 Mean soil pH (in KCl) and standard error at three soil depths with increasing distance from European beech tree stems. The transects were chosen along the most obvious stemflow track (solid line) from twenty trees, and a control transect (dashed line) was taken at right angles. From: Falkengren-Grerup (1989).

that sampling is much more difficult than in other soils. The 20 bulked samples per hectare that have been recommended by Davey and Conyers (1988) are clearly insufficient to adequately characterize the soil over such an area, yet many studies use a smaller sample. In addition, the very steep gradients in pH that occur in the rhizosphere (Marschner *et al.*, 1991) and the vertical gradients within the mor layer (Skyllberg, 1990) throw into question the use of soil pH measurements as an index of forest condition.

In one of the most detailed investigations undertaken to date on chemical changes in forest soils, Ilvesniemi (1991) reported the evolution of pine forest soils between 1982 and 1988. As might be expected over such a short period, a substantial amount of noise was apparent in the data. Big differences were found in the samples taken from the same plots in the same years, suggesting that substantial changes in the soil chemical status would be required before any reliable signal was detected. The number of samples required to identify statistically significant differences for a 10%

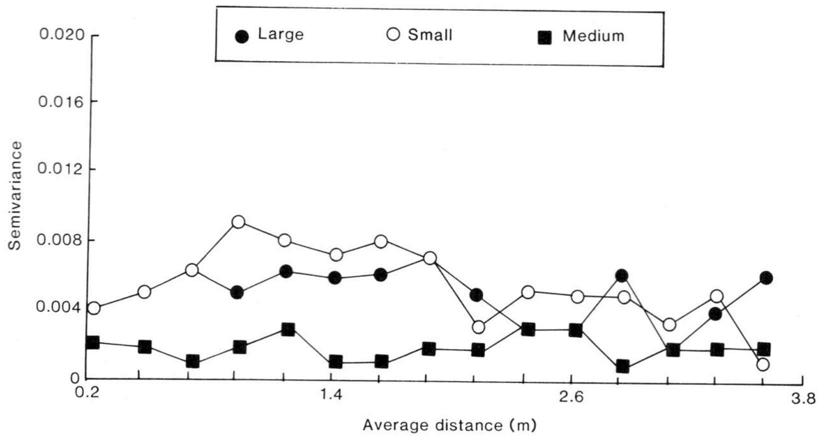

Figure 8.12 Semi-variogram of surface mineral soil pH with lag distances of 20 to 360 cm, used to study small-scale spatial variability in the pH of forest soils. Results from different sizes of cores (small, medium, large) are indicated. From: Riha *et al.* (1986b).

change in the measured value at the 5% risk level and at 90% certainty generally varied between five and 25 for humus samples. For mineral soil samples, the numbers were much higher, ranging from 20 to over 300. Given the high cost of sampling and analysing soils, these numbers suggest that regular monitoring of mineral soils may be impractical.

Other studies have found similar levels of variation, throwing serious doubts on some published results. For example, Skyllberg (1990) recorded a variation of 0.6 pH units in the top four centimetres of soil in a Norway spruce stand and Aaltonen (1925) recorded differences of more than 1 pH unit within individual stands. When such variations have been found, it is difficult to see how some authors have managed to identify a decrease in pH of 0.5 to 1.0 pH units on the basis of a small number of repeated samples.

Soil variation within two oak plots in southern England

The extent of variation in forest soil can be illustrated by a case study from southern England. Two plots were examined, one at Highclere Castle and the other at Goose Green Enclosure in Alice Holt Forest, both in Hampshire. The Highclere Castle stand was a pure, 90-year-old, pedunculate oak woodland at 168 m ASL on a shallow south-facing slope. The Alice Holt stand consisted of pure, 170-year-old pedunculate oak woodland on flat ground at an altitude of 76 m.

The Highclere Castle site was located on soils of the Givendale and

Winchester Series of the Carstens Association. These are chalky soils, usually covered by a layer of variable thickness of loess and clay with flints. At the site, the loess layer was absent and the soils had developed on the clay with flints. They were deep, well-structured and freely-draining to the underlying chalk. Despite the presence of the underlying chalk, the Carstens Association soils are naturally acidic, containing little magnesium but adequate amounts of phosphorus and potassium.

At Alice Holt, the soils belonged to the Denchworth Association. These are permeable, seasonally (during winter) waterlogged clayey soils which have developed on Jurassic and Cretaceous clays. The soils are stoneless and strongly mottled and are naturally acidic, with good levels of potassium but low phosphorus contents.

Samples from the Highclere stand were obtained in late November 1989 (120 samples) and late February 1991 (40 samples). Samples (160) from Alice Holt Forest were all obtained in late March 1991. At each site, soils were sampled in the mid-point of the six trees used in each sub-plot of the forest monitoring programme. Five pits were excavated at each of the four sub-plots. Samples were obtained from 10 cm and 40 cm depths within the pits, with samples being collected from each of the four 50-cm sides of the pits. This resulted in eight samples per pit, 40 samples on each side of the plot and 160 samples for the plot as a whole.

The summary statistics for samples for the two depths at the two plots

Table 8.1 Summary statistics for the two depths at the two plots

	Highclere Castle					Alice Holt Forest				
Variable	Mean	S.d.	S.e.	Min	Max	Mean	S.d.	S.e.	Min	Max
10 cm depth										
pH	4.55	0.6	0.07	3.85	6.00	4.21	0.24	0.03	3.6	4.65
Conductivity	46.9	10.4	1.16	29.9	77.6	43.6	8.87	0.99	30.4	72.4
Buffer capacity	3.07	0.76	0.09	1.55	6.25	4.03	1.4	0.02	1.90	7.35
Water content	18.4	4.93	0.55	11.5	28.9	31.8	3.14	0.35	25.1	39.0
Organic content	4.82	2.09	0.23	0.55	10.26	10.1	2.25	0.25	3.06	15.24
Stone content	36.52	8.91	0.99	19.47	59.72					
Nitrogen content	31.6	46.5	5.2	0.0	250.0					
40 cm depth										
pH	5.02	0.69	0.08	4.0	6.35	4.46	0.21	4.00	4.80	
Conductivity	39.8	12.0	1.35	18.4	71.7	34.8	6.95	0.78	26.1	63.1
Buffer capacity	1.94	0.57	0.06	0.83	3.5	2.87	0.75	0.08	1.2	4.95
Water content	14.8	3.08	0.34	8.73	20.17	26.6	3.17	0.35	18.9	41.2
Organic content	2.91	1.44	0.16	1.52	6.77	6.51	1.18	0.13	4.67	9.61
Stone content	49.6	15.9	1.8	16.6	83.6					
Nitrogen content	12.3	23.0	2.57	0.0	100.0					

S.d. = standard deviation; S.e. = standard error.

are presented in Table 8.1. As indicated above, the samples from the east side of the site at Highclere Castle were collected at a different time to the rest. The variable that is most likely to differ seasonally is water content, and this, in turn, may affect some other variables. Consequently, the water contents for the four sides have been compared in Table 8.2. Water contents were generally high at the time of the second sampling. At 10 cm depth, the water contents were significantly higher than at either the south or west sides, but only marginally higher than the north side of the plot. The water contents at 40 cm depth on the east side were less than those recorded on the north side, but greater than those on the south or west sides. Comparisons in the properties of soils from the different sides of the plot at Highclere Castle should therefore be made with care.

A nested analysis of variance showed that in general, the extent of variation was greater at the smaller scale than between sides of plots. This was unexpected as variation in soils generally increases with distances between sample points. However, most studies that have looked at variation in soils have dealt with non-woodland soils. Soils located under trees are subjected to many influences (such as rooting and throughfall) which vary at the microscale and they are therefore much more variable than other soils. There were very few variables which showed non-significant variation within pits. The organic content of the soils at Highclere Castle was the only variable that showed non-significant variation at both depths. However, despite the same broad type of vegetation, the organic content of the soils at both depths at Alice Holt showed significant high levels of variation within individual pits, as expected. Several studies have shown that pH is highly variable in forest soils, For example, Bringmark (1989) working with brown forest soils under beech in southern Sweden, found several levels of spatial variation.

Table 8.2 Water contents from each of the sides of the plots at Highclere Castle

	Mean	S.d.	S.e.	Min	Max
10 cm depth					
North side	21.7	2.7	0.6	17.1	26.6
South side	13.3	1.5	0.4	11.5	16.7
West side	15.0	1.9	0.4	12.3	19.3
East side	23.8	2.6	0.6	19.4	28.9
40 cm depth					
North side	17.6	1.3	0.3	15.5	20.2
South side	11.9	1.8	0.4	8.7	15.0
West side	12.3	1.4	0.3	9.8	15.0
East side	17.3	1.6	0.4	13.0	19.3

The east side was sampled in February 1991 and the other sides in November 1990

Marked thresholds occurred at distances of 0.05 m, 0.1–0.7 m, 1.0 m, 1.5–2.5 m and 4.0 m. Similarly, Nykvist and Skyllberg (1989) found that the pH of humus in coniferous forest soils in northern Sweden could vary by as much as 0.6 pH units within a 2 m^2 plot. The variation not only occurs horizontally but also vertically. For example, Skyllberg (1990) found that there could be more than 0.7 pH units difference between the pH of the top and bottom of the mor layer.

Although the stemflow of oak is considered acidic (Carlisle *et al.*, 1967), the rough and porous nature of oak bark means that little stemflow reaches the ground (Beniamino *et al.*, 1991), and a marked trend of decreasing soil pH towards individual trunks would therefore not be expected.

At Highclere, the variations in pH within the stand were very large, with a range of 2.15 pH units at 10 cm depth and 2.35 pH units at 40 cm depth. The variation at Alice Holt was considerably less, being 1 pH unit at 10 cm depth and 0.8 pH units at 40 cm depth. The high level of variation at the Highclere plot is almost certainly a function of the local geology and soil type. The high pH recorded in some pits probably reflected local variations in the depth of the clay-with-flints layer, the high pH occurring where the chalk was close to the surface. Although this soil type is quite widespread in some parts of England, the variation in pH should not be extrapolated to other soil types.

The variation in pH at the Alice Holt stand (clay soils) may be more representative of the variation that occurs within forest stands. However, clay soils are largely restricted to lowland sites in Great Britain whereas many plantation forests are situated on podzols and peats in the uplands. Consequently, a similar study of the variation in the pH of upland soils is required before any wider conclusions can be reached.

Conductivity measurements showed a similar range at both the Highclere and Alice Holt sites, although the standard deviations and standard errors at both depths were higher at Highclere than at Alice Holt. At both sites, conductivity was greater at 10 cm depth than at 40 cm depth. The F-ratios differed markedly at the differing sampling intervals. Significant F-ratios were found for three of the four depth-site combinations at the smallest sampling scale (between the different sides of the pit). They suggested that while variation may be considerable over distances of about 1 m, it is not so apparent over larger distances.

The buffering capacity of the soil at Highclere was generally lower than at Alice Holt, despite the higher mean pH of the soils at Highclere. The buffering capacities at both sites and at both depths were very variable, with coefficients of variation ranging from 25% to 35%. The results of the nested analysis of variance revealed considerable variation between site-depth combinations. The F-ratio for the soil at 10 cm depth at Alice Holt was the only one that was significant at the between-pit level. In contrast, the buffering capacities of the soil at this depth at Alice Holt did not have a significant *F*-value for the within-pit level of variation. The other three site-depth combinations had significant *F*-values.

Similar to the pattern seen in most forest soils, the organic content of the soils decreased with depth at both sites. The values obtained were very variable, with the levels of variation being similar at both sites. The analysis of variance provided rather inconsistent results. Mostly, the *F*-ratios were significant and the variation was generally greater between pits than within pits.

The variation in stone content is always difficult to assess as, during sample collection, the larger rocks are usually removed. A subjective element introduced is therefore introduced. The stone content was only assessed at Highclere. Predictably for a clay-with-flints soil, the stone content was very variable, ranging from 19 to 60% at 10 cm depth and from 17 to 84% at 40 cm depth. The variation is reflected in the results of the analysis of variance, which indicated a significant amount of variation between samples at all levels.

The nitrogen content of the soil was only assessed for the Highclere soil. The degree of variation is difficult to interpret because of the crude method used to derive the nitrogen contents. The estimates of the nitrogen content were only semi-quantitative and should therefore be treated with care. The results of the analysis of variance showed that the variation between pits was much greater than within pits. The *F*-ratio for the within-pits level at 10 cm depth was not significant while at 40 cm depth, it was.

Cation exchange capacities and calcium-magnesium contents were only determined for a few samples at Highclere. The analysis of the variance of the cation exchange capacities indicated that the level of variation within pits did not form a significant component of the between-pit variation. In contrast, the *F*-ratios for the calcium-magnesium contents showed a significant amount of variation at the within-pit level. This corresponds with the results of Grigal *et al.* (1991), who found that calcium contents varied markedly over relatively short distances and recommended that about 50 samples be taken to adequately characterize the soil calcium levels in a stand.

Several other studies have found marked variations in the soil properties within forest stands. David *et al.* (1990) examined the variation of spodosols in Maine. The soils were characterized by acidic forest floors (pH 3.6) and upper mineral soil horizons (pH 3.4–4.0), low concentrations of base cations in the mineral horizons and enrichment of the spodic horizon with carbon, nitrogen and sulphur. All chemical properties investigated, except pH, showed high levels of variation, with the relative standard deviation of the Ca^{2+} concentration being 42–95% over an area of 2.2 ha. Grigal *et al.* (1991) recorded coefficients of variation for forest floor pH of 5%, for organic fractions of 25% and exchangeable cations of 28%. In the mineral soil, the figures were 5%, 20% and 35%, respectively. Within a stand, Grigal *et al.* (1991) found that the coefficients of variation for forest floor properties were the same over about 2.5 m as over 30 m.

These studies indicate that the chemical and physical composition of forest soils is extremely variable. This poses considerable problems in studies of change as soils cannot be resampled at the same three-dimensional point (as the sample has already been taken). Consequently, many samples are required from each site before changes can be reliably identified, even with rigorous temporal controls. The majority of published studies of changes in soil parameters are not based on a sufficient sample size and must therefore be viewed with caution.

Buffering and toxicity

The availability of toxic elements within the soil depends on, among other things, the reaction status of the soil. Generally, toxicity problems are rarely encountered with soils of pH > 5.5 (Sumner *et al.*, 1991). Ulrich (1983a, b, 1991) has identified five buffer ranges in the soil, dependent on pH. These can be listed as:

Buffer range	*pH range*	*Base saturation*	*Buffer reaction*
Carbonate	8.0–6.2	1.00	$CaCO_3 + H_2CO_3 \rightarrow Ca^{2+} + 2HCO_3^-$
Silicate	6.2–5.0	1.00–0.70	$CaAl_2Si_2O_8 + 2H_2CO_3 + H_2O \rightarrow$ $Ca^{2+} + 2HCO_3^- + Al_2Si_2O_5(OH)_4$
Cation exchange	5.0–4.2	0.7–0.05	clay mineral = $Ca + 2H^+ \rightarrow$ H-clay mineral-H + Ca^{2+}
Aluminium	4.2–3.0	0.05–0.00	$ALOOH + 3H^+ \rightarrow Al^{3+} + 2H_2O$
Iron	<3.8	0.00	$FeOOH + 3H^+ \rightarrow Fe^{3+} + 2H_2O$

The various ranges have been described in detail by Ulrich (1983a) and the following summary is drawn primarily from this account. For soils to remain in the calcium carbonate buffer range, they must contain a significant proportion of $CaCO_3$ in their fine fraction. Consequently, such soils are mainly restricted to those developed on chalk and limestones. The dominant acid is generally carbonic acid, which is mostly produced by root and decomposer respiration. The dominant cation in the soil solution is calcium. Forest soils in this buffering range can buffer atmospherically derived acidity very efficiently.

In the silicate buffering range, calcium carbonate is not important in the buffering process. Instead, buffering occurs through the weathering of silicate minerals. The weathering process is complex and may involve several stages. The precise nature of the reactions depends on the particular silicate minerals present in the soil. The buffering rate is generally about 0.2 to 2 keq H^+ ha^{-1} yr^{-1}, although deposition rates of acidity are frequently much greater than this. When deposition rates exceed this buffering capacity, the soil is likely to move into the cation exchange buffer range.

Soils in the cation exchange buffer range rely on the presence of exchangeable base cations for buffering. If a net production of a strong acid occurs in the soil, the pH is buffered by proton exchange with Ca^+ and other bases at the exchange sites in a reversible reaction. The process works as long as the percentage of exchangeable calcium and other bases in the soil remains above 5–10% of the cation exchange capacity. Unlike the two preceding buffer ranges, release of aluminium ions may take place. Within the pH range 4.2–5.0, the aluminium ions tend to form nonexchangeable polymeric Al-hydroxy cations or Al^{3+} ions. The former result in a reduction in the cation exchange capacity, the latter in a loss of base saturation. At this level, significant manganese mobilization may occur, allowing either high manganese uptake by plants or losses of manganese via leaching (or both).

Many forest soils are more acidic than the above soils and lie within the aluminium buffering range, where aluminium is the dominant exchangeable cation. The exchangeable calcium and magnesium content in these soils is less than 5–10%. The aluminium is primarily derived from the dissolution of aluminium oxides and hydroxides and the weathering of clay minerals.

In the most acidic soils, the iron buffer range predominates. The pH of these soils is less than 3.2. At this level of acidity, iron oxides become highly soluble and iron concentration levels in the soil solution may reach very high levels until they are removed (e.g. in the E horizon of podzols).

Soils with a pH of greater than 4.2 mainly buffer protons by the release of base cations (Ca, Mg, Na and K) whereas at less than pH 4.2, aluminium compounds become active. By far the greatest emphasis has been placed on the aluminium buffer range, primarily because certain forms of aluminium are highly toxic to plants. This is discussed in the following section.

Aluminium toxicity

Aluminium toxicity has frequently been associated with the acid rain debate. It was initially identified as an important buffer system in catchments exposed to acidic deposition (Norton, 1976; Cronan and Schofield, 1979; Cronan and Goldstein, 1989). Ulrich *et al.* (1980) later associated the increased aluminium concentrations in the soil with toxic effects on trees.

The dynamics of aluminium mobilization and leaching have been examined by several authors. A particularly interesting study compared the ability of nitric and sulphuric acids to leach aluminium from organic soil horizons (James and Riha, 1989). In this laboratory-based study, the leachates from soil columns containing a forest-floor spodic horizon sequence were examined. Increasing the nitric acid from pH 5 to pH 3 resulted in an increase in the total aluminium from 0.7 to 0.85 mM. Increasing the sulphuric acid by the same amount had no effect. Neither acid had any effect in the E and A horizons. The study therefore strongly suggested that the forest floor represented an important source of aluminium and that nitrogen deposition may be more important than previously thought in relation to aluminium mobilization.

Adams *et al.* (1990) examined the relationship between aluminium concentrations in drainage waters and landuse. Concentrations of Al^{3+} were consistently greater in streams draining Sitka spruce plantations in Great Britain than in streams draining adjacent catchments dominated by rough grazing (Fig. 8.13). The concentration of Al^{3+} was not

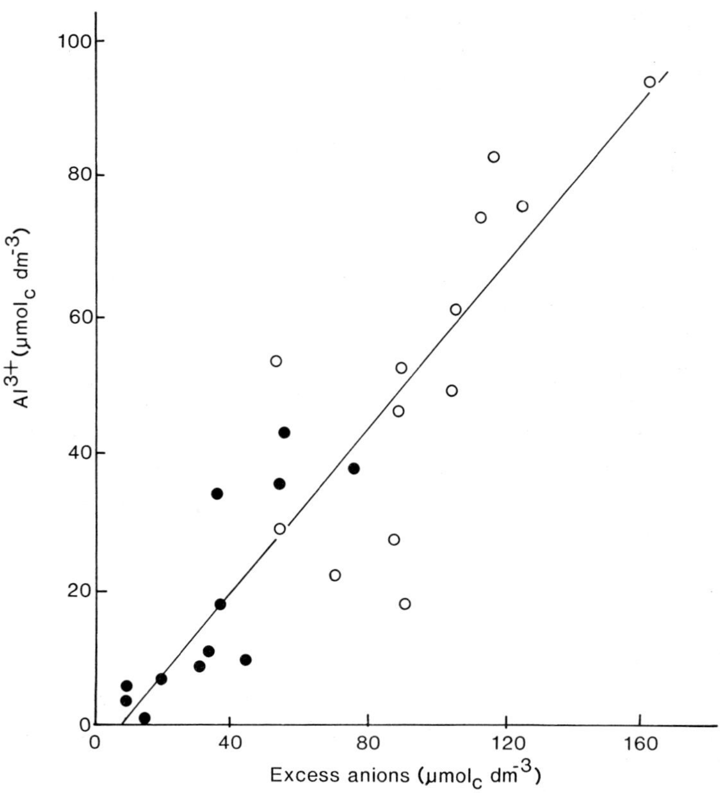

Figure 8.13 The relationship between Al^{3+} and excess anions in stream water samples (streams draining Sitka spruce plantations – open circles; streams draining rough grazing – filled circles). From: Adams *et al.* (1990).

related to the solubility product of gibbsite, and instead was controlled by the concentration of excess anions. The source of the Al^{3+} was believed to be exchangeable aluminium, with displacement of exchangeable aluminium being the main process determining the occurrence of Al^{3+} in streams.

The mobilization of some forms of aluminium occurs over a relatively narrow pH range at about 4.4 to 4.2 (Bergkvist, 1987b), with the dominant species at this acidity being Al^{3+} (Eldhuset, 1988). Above pH 4.5, aluminium toxicity is not expected because of aluminium hydroxide precipitation (Kelly *et al.*, 1990). Various factors influence the mobilization, although pH appears to be by far the most important. Evans and Zelazny (1987) have suggested that sulphate additions to the soil solution may increase aluminium mobilization by influencing the concentrations of dissolved organic carbon. However, this hypothesis requires further testing. The interpretation of these results is complicated by the suggestion that natural forest soils behave very differently to soils reconstituted in the laboratory (Matzner, 1992). The importance and reasons for the differences remains to be fully evaluated.

Aluminium mobilization occurs naturally in acidic soils and many plants have developed tolerance mechanisms (Eldhuset, 1988). For example, organic acids, such as malate and citrate may be excreted to chelate the aluminium. Alternatively, the pH of the rhizosphere may be raised so as to precipitate or polymerize aluminium. In yet another mechanism, aluminium can be actively excluded from the plasmalemma of the root cells. In looking at toxicity and mobilization, it is important to remember that most tree roots are concentrated in the O and A horizons of the soil, where a substantial proportion of the aluminium may be complexed to organic matter.

One of the most detailed studies done to date on the dynamics of aluminium in forested ecosystems is the ALBIOS study. Among other aspects, this project involved the comparison of the aluminium chemistry of 14 different catchments, 10 in North America and 4 in Europe. Sulphur deposition varied markedly between the sites, ranging from 4 kg ha^{-1} yr^{-1} to more than 80 kg ha^{-1} yr^{-1} (Cronan and Goldstein, 1989). Interestingly, the comparison showed that sulphur deposition at the Solling site in Germany was almost double that at any other site, suggesting that the extensive research that has been done there may have little application elsewhere. Total monomeric aluminium (MAL) concentrations in soil waters varied widely, ranging from less than 1 μmol l^{-1} in parts of the US to >240 μmol l^{-1} in German Norway spruce stands. However, activities of Al^{3+} were much lower than either the MAL or total aluminium (TAL) concentrations (Cronan and Goldstein, 1989).

The ALBIOS project showed that increasing MAL concentrations in soil water samples were generally associated with a gradient of increasing soil solution sulphate and ionic strength. However, streamwater samples showed no such trend. As other studies have suggested, the speciation of the aluminium between sites varied markedly.

The ALBIOS project was initially set up to test the hypothesis that aluminium mobilization and transport in forest soils were strongly influenced by acidic deposition and that aluminium leaching increased with increasing soil inputs of strong acids (Cronan and Goldstein, 1989). However, the study revealed that complex relationships existed between sulphate and nitrate inputs and aluminium outputs from the catchments. Aluminium leaching was a function of the interactions between several factors, two of the most important being the mechanisms and rates of the formation of alkalinity and the hydrological characteristics of the catchment. Soils dominated by the aluminium buffering system and having low retention rates of strong acid anions had a strong correlation between increased fluxes of sulphate and/or nitrate and increased aluminium leaching. These soils were characterized by pH values of less than 4.9 and base saturation levels of less than 15%. On the other hand, soils with high base saturation showed little or no leaching of aluminium.

The establishment of aluminium toxicity thresholds in the soil is difficult as they are extremely variable. Aluminium tends not to be toxic in the litter and humus layers as it complexed with organic matter (Murach, 1984; Bengtsson *et al.*, 1988), the capacity of the organic matter to bind aluminium increasing with increasing pH. The organically-complexed aluminium is non-toxic and it is this form that may provide trees with a source of aluminium (Arp *et al.*, 1989). The ALBIOS programme, concluded that aluminium toxicity to trees was influenced by a wide range of factors, including solution pH, aluminium speciation, concentration of calcium and magnesium in the soil solution, overall ionic strength of the solution, the nature of any inorganic nitrogen in the solution, mycorrhizal interactions, temperature, soil moisture levels, initial tree vigour, the nutrient status of the tree and the species and provenance (Cronan and Goldstein, 1989; Cronan *et al.*, 1989). It is therefore not surprising that different studies have come up with different results (Table 8.3), with some of the more marked differences occurring between soil and solution cultures (Cronan *et al.*, 1989). The most toxic forms of aluminium are some monomeric species: Al^{3+}, $AlOH^{2+}$ and $Al(OH)_2^+$ (Pavan and Bingham, 1982; Pavan *et al.*, 1982; Alva *et al.*, 1986; Cronan and Goldstein, 1989; Sucoff *et al.*, 1990), with the activity of Al^{3+} (as opposed to its concentration) being particularly important. Although polymeric aluminium may be equally or even more phytotoxic than the monomeric species, it is not known how common this form is in the rhizosphere (Cronan and Goldstein, 1989).

The aluminium toxicity threshold for individual trees varies from less than 50 μmol l^{-1} to more than 3000 μmol l^{-1} (McCormick and Steiner, 1978; Göransson and Eldhuset, 1987; Eldhuset, 1988; Cronan *et al.*, 1989; Kelly *et al.*, 1990), with honeylocust (*Gleditsia triacanthos* L.)

Table 8.3 Thresholds for statistically significant reductions in root growth for pine and spruce species as determined in solution, sand and soil culture experiments with seedlings

Study	Species	Threshold mmol l^{-1}	Medium
Hutchinson *et al.* (1986)	*Pinus banksiana*	1.48	Sand
	Pinus strobus	2.96	Sand
Eldhuset *et al.* (1987)	*Pinus sylvestris*	3.0–5.0	Solution
Humphreys and Truman (1964)	*Pinus radiata*	>0.74	Solution
McCormick and Steiner (1978)	*Pinus rigida* Mill.	4.44	Solution
	Pinus sylvestris	4.44	Solution
	Pinus virginiana	4.44	Solution
Williams (1982)	*Pinus clausa* Sarg.	1.22	Solution
	Pinus taeda	1.22	Solution
Thornton *et al.* 1987	*Pinus taeda*	>3.0	Solution
	Pinus taeda	1.5	Solution
Paganelli *et al.* 1987	*Pinus taeda*	0.19	Sand
Joslin and Wolfe 1989	*Pinus taeda*	1.3	Soil
Evers (1983)	*Picea abies*	>1.5	Solution
Rost-Siebert (1983)	*Picea abies*	0.3	Solution
Abrahamsen (1984)	*Picea abies*	0.74	Solution
Van Praag *et al.* (1985)	*Picea abies*	3.33	Sand
Makkonen-Spiecker (1985)	*Picea abies*	2.96	Sand
Ingestad *et al.* (1985)	*Picea abies*	1.0	Solution
Hutchinson *et al.* (1986)	*Picea mariana*	0.37	Sand
	Picea glauca	0.37	Sand
Nosko *et al.* (1988)	*Picea glauca*	0.05	Sand
	Picea rubens	0.37	Sand
Schier (1985)	*Picea rubens*	3.7	Solution
Thornton *et al.* (1987)	*Picea rubens*	0.25	Solution
Ohno *et al.* (1988)	*Picea rubens*	>0.54	Soil
Joslin and Wolfe (1988)	*Picea rubens*	0.25	Soil

A greater than symbol (>) indicates no significant reduction was observed at the highest aluminium level tested. (From Raynal *et al.*, 1990)

being one of the most sensitive tree species studied to date (Sucoff *et al.*, 1990). Generally, these concentrations are well above those found in soil solutions (Cronan *et al.*, 1987; Göransson and Eldhuset, 1987; Eldhuset, 1988; Kelly *et al.*, 1990), although Godbold *et al.* (1988a, b) and Cronan *et al.* (1989) considered that conditions for aluminium toxicity might exist at some German sites, particularly during climate-driven acidification 'pushes'. In addition, high aluminium concentrations have been recorded in some North American spruce-fir forests subject to frequent cloud immersion (Joslin and Wolfe, 1992). Red spruce, beech and sugar maple are moderately sensitive whereas American beech and loblolly pine appear to be relatively insensitive. Asp and Berggren (1990) found that aluminium complexed with fulvic acids did not affect the uptake of calcium. The calcium : aluminium ratio is thought to be an important index of the likelihood of damage to trees, yet Thornton *et al.* (1989) found no variation in root growth of European beech over a calcium : aluminium ratio range of 60–0.016, despite this species being the most sensitive of three species investigated (red oak, American beech and European beech), with effects being apparent at 500 μmol l^{-1} Al. The calcium : aluminium ratios in Solling have been reported to be in the range 0.05–0.06 and in Skåne, south Sweden, to be less than 0.02 (Tyler, 1987b).

A frequently cited factor that is believed to affect the toxicity of aluminium is the amount of calcium in the soil. The calcium : aluminium ratio is one of the main criteria used in deciding the critical load of acidity for forest soils (e.g. de Vries, 1988; Sverdrup *et al.*, 1992). Some studies have suggested that damage to the fine roots occurs when calcium : aluminium ratios are low (e.g. Junga, 1984; Jorns and Hecht-Buchholz, 1985; Rost-Siebert, 1983, 1985). However, other studies have failed to find any good relationships between tree growth and the calcium : aluminium ratio (e.g. Göransson and Eldhuset, 1987; Kazda and Zvacek, 1989; Keltjens and Van Loenen, 1989), and the calcium : aluminium may on occasion be a poor index of the risk of aluminium toxicity. Other studies have failed to find any correlation between soil aluminium concentrations and growth of trees, such as those with black spruce (Arp *et al.*, 1989).

Many studies have failed to find a good relationship between foliar aluminium concentrations, root aluminium concentrations and readily available aluminium in the soil (e.g. Arp *et al.*, 1989). Sapwood concentrations of aluminium also appear to be unrelated to soil levels (DeWalle *et al.*, 1991). However, some of these results may be due to the aluminium species examined and these are not fully detailed in many publications.

Partitioning and isolation of aluminium in trees is particularly effective. Much of the aluminium that is taken up by trees through the root cap and the mucilaginous secretion covering the epidermal cells remains in the fine roots (Arp *et al.*, 1989) or even in the mucilage itself (Eldhuset, 1988). The aluminium binds very specifically to the mucilage by exchange adsorption on the polyuronic acid, by complexing with pectic substances and by the formation of polyhydroxy forms (Roy *et al.*, 1988). However, Cronan and Goldstein (1989) listed several effects that might occur given that *in vitro* studies have shown that the main mechanisms of toxicity are related to strong interactions of aluminium with orthophosphate groups of organophosphates and the interference of aluminium with calcium metabolism. They argued that these could have several effects:

- aluminium may bind to nucleotides and nucleic acids, inhibiting cell division;
- aluminium may bind to ATP, ADP or membrane-bound ATPases, interfering with energy transfer;
- aluminium may interfere with enzyme systems such as the acid phosphatases;
- aluminium may bind to calmodulin and other calcium peptides, seriously disrupting cellular control and contractile processes;
- aluminium may bind to phospholipid groups and alter membrane permeability; and
- aluminium may interfere with membrane selectivity for ions, including zinc, copper, calcium, magnesium, manganese, potassium, phosphorus and iron.

Experimental studies have suggested that the effects of aluminium are first seen in the roots, with reduced root length, root thickening and dead root tips occurring (Eldhuset, 1988). Prolonged exposure leads to symptom expression in the shoots, with reduced growth, alterations to the chlorophyll content of the foliage and discoloration (yellowing and purpling), wilting and death. Eldhuset (1988) described five main physiological effects of aluminium toxicity. Cation uptake was reduced because of competition for uptake sites and because of effects on membrane proteins such as calmodulin and ATPases. The amount of available phosphorus was reduced due to the precipitation of aluminium phosphate. Cell length growth was inhibited by aluminium binding to pectins. DNA synthesis, and therefore cell division, was inhibited due to aluminium binding to phosphate groups in the DNA. In addition, mycorrhizal fungal associations could be affected.

Ulrich *et al.* (1979) hypothesized that aluminium could be responsible for the inhibition of the growth of Norway spruce roots, although subsequent studies have shown that the inhibition occurred before aluminium concentrations rose (Rehfuess, 1981). The most likely explanation for the decline in root vitality was drought. This is disputed by Ulrich *et al.* (1989) who have since argued that the root damage has been present since the late 1950s. The argument has not been resolved and further work is necessary before any reliable conclusions can be drawn.

Summary

The role of aluminium remains equivocal. Davis (1990) found higher levels of total aluminium and inorganic monomeric aluminium in asymptomatic stands than in symptomatic stands, suggesting that aluminium is not always associated with toxicity. The experimental studies clearly suggest that aluminium can have toxic effects on tree seedlings, but the evidence for this in the field is very limited. Work done at Solling in Germany suggests such effects, but this site had abnormally high concentrations of soil aluminium. Speciation is a critical factor, as is the interaction between aluminium and other elements, particularly calcium. However, much more work is required in this area.

Other elements

While most studies have focused on the direct effects of soil acidification on tree roots, other aspects of the forest ecosystem may be equally important. Sulphur dioxide is highly toxic to soil micro-organisms, due to the combination of the solubility products of sulphuric acid (Bewley and Stotzky, 1983a; Bewley and Parkinson, 1984), and it may inhibit microbial activity (Grant *et al.*, 1979a, b). However, acidic deposition and the resulting soil acidification appears to have less of an effect (Gray and Ineson, 1981; Bewley and Stotzky, 1983a), with reductions in the soil pH to 2.9 having little effect on either the duration of the lag in respiration or the total amount of carbon evolved as carbon dioxide. One factor mitigating the toxicity may be the presence of the clay mineral montmorillonite (Bewley and Stotzky, 1983b).

Most metals, including zinc, cadmium, manganese, copper and lead, become increasingly soluble as pH drops. However, as with aluminium, the presence of organic matter may result in the formation of soluble metal organic complexes, reducing the potential toxicity of many of these metals. For example, copper can be organically bound in both soluble and insoluble complexes and its availability may be so reduced as to induce deficiency in plants (Sumner *et al.*, 1991). However, Neite *et al.* (1992) have argued that soil acidification may have resulted in the increased potential for mobilization of heavy metals stored in the upper soil horizons.

Manganese reaches potentially toxic levels in many acidified soils (Cassens-Sasse, 1987) but there is little evidence for it having an adverse effect on trees at the concentrations normally found in acidic soils. Instead, its mobilization at low pH may result in it being leached from the soil, resulting eventually in manganese deficiency. High levels of manganese may induce calcium deficiency (e.g. Keil *et al.*, 1986), although Norway spruce is generally highly tolerant to manganese (Kreutzer and Bittersohl, 1986; Nebe, 1987; Kazda and Zvacek, 1989).

The role of heavy metals remains very uncertain in soil studies (c.f. Glatzel, 1985). As with trees, insufficient attention has been given to the soil as an ecosystem component and the various interactions between the mineral and biotic phases of the soils are only now beginning to be investigated. Much work relies on the toxicity of heavy metals in agricultural situations, which may not be relevant to forest ecosystems. Consequently, this is an area where further work is needed.

Nitrogen saturation of forest soils

Nitrogen has long been known to be a critical element in some forests (Mälkönen, 1990). Temperate and boreal forests appear to be more prone to nitrogen limitation than tropical forests, as the latter contain many species that can fix atmospheric nitrogen. The loss of these species is one reason for the low fertility of soils in areas cleared of tropical rain forest.

A phenomenon that has been causing increasing concern is the possibility that forest ecosystems are becoming saturated with nitrogen from atmospheric sources. Nitrogen saturation has been defined as occurring when losses of nitrogen from the ecosystem equal or exceed nitrogen inputs (Ågren and Bosatta, 1988). Alternative definitions exist, such as ecosystems unable to absorb any more nitrogen (Ågren, 1983) and ecosystems where further nitrogen supply will not increase primary production (Nilsson, 1986). As nitrogen inputs and exports vary through time, any balance must be calculated over a considerable period, Ågren and Bosatta (1988) recommending at least 20–50 years.

In the past, forests were often thought to be nitrogen-limited (Gundersen, 1991) and many fertilization trials have shown increased growth following the application of nitrogen (e.g. Miller and Cooper, 1973; Miller and Miller, 1988; Miller *et al.*, 1977; McIntosh, 1984). However, losses of nitrogen from forest ecosystems have been reported (e.g. Nilsson and Grennfelt, 1988; Hauhs *et al.*, 1989; Förster, 1990; Gundersen and Rasmussen, 1990) and atmospheric deposition, with other processes, may be leading to an excess of nitrogen availability over demand (Aber *et al.*, 1989). When this occurs, nitrate losses from the soil may result in acidification of drainage waters and potentially toxic aluminium concentrations (Cronan, 1985). In addition, increased nitrogen availability may result in increased uptake, which can have adverse effects on trees.

The situation appears to have been exacerbated by the reduced removal of nitrogen from forests; in the past, removal of litter and foliage would have represented a significant pathway for nitrogen losses from forest ecosystems (Emanuelsson, 1988). Whole-tree harvesting still represents an important anthropogenic drain on forest nutrients (Nykvist and Rosén, 1985). The importance of litter removal has probably been underestimated in recent studies: the soil and litter represents a major sink for nitrogen in forest ecosystems (Heilman and Gessel, 1963; Miller *et al.*, 1976; Raison *et al.*, 1990). Natural losses of nitrate are estimated to be less than 1 kg ha^{-1} yr^{-1} from coniferous forests and less than 2–3 kg ha^{-1} yr^{-1} from broadleaved forests (Gundersen and Rasmussen, 1990). These values are considerably less than the deposition levels encountered in many forests.

The ability of an ecosystem to absorb incoming nitrogen depends on a variety of factors, including the soil type, the nature of the litter layer, the temperature and the contact time (Ågren and Bosatta, 1988; Duckworth and Cresser, 1991). Trees require nitrogen for their metabolism and the uptake of nitrogen in undisturbed coniferous and broad-leaved forest ecosystems is estimated as 30–75 and 50–100 kg ha^{-1} yr^{-1}, respectively (Cole, 1981; Gosz, 1981; Melillo, 1981; Nadelhoffer *et al.*, 1984; Becquer *et al.*, 1990). Generally, nitrate made available during the summer is taken up, whereas nitrate made available during winter may be leached from the system (Becquer *et al.*, 1990). This is particularly true during mild autumns and winters, when increased nitrification may result in the production of substantial amounts of nitrate. Nitrification may also be enhanced during warm conditions (Ulrich, 1981, 1983a, b; Matzner and Ulrich, 1985), resulting in the climate-driven acidification pulses described above.

Emmer and Tietema (1990) found a linear relationship between temperature and net nitrogen mineralization of oak and beech litter. This took the form of:

$$\text{NMIN} = 68.5T\ 520.8$$

where NMIN: net nitrogen mineralization in mg N kg^{-1} over four weeks and *T*: temperature (°C). As warm conditions may reduce uptake by trees because of drought stress, acidification pulses can occur (e.g. Matzner and Thoma, 1983). The process has been termed 'temporal discoupling' by Ulrich (1983a). The warm summers of the late 1980s may have resulted in improved nitrogen supply at many locations because of this temperature relationship (Hippeli and Branse, 1992).

Nitrification involves the biological oxidation of $NH4^+$ to NO_3^-, with NO_2^- as an intermediate stage. It also involves the conversion of reduced forms of nitrogen to NO_2^- or NO_3^-. The most important microbial species involved in the transformation belong to the *Nitrosomonas* and *Nitrobacter* groups. In very acidic soils, heterotrophic nitrification forms the main pathway, with the process carried out by fungi such as *Absidis cylindrospora* and *Verticillium lecanii* rather than bacteria (Lang and Jagnow, 1986; Gundersen and Rasmussen, 1990), although Vonk *et al.* (1988) have isolated several heterotrophic nitrifying bacteria from acidic Douglas fir forest soils in The Netherlands. Nitrification tends to be higher under broadleaves than conifers (Gosz, 1981). It increases with increasing temperature, up to about 30°C and decreases with increasing moisture (Haynes, 1986), although this relationship does not apply to heterotrophic nitrification conditions. Kriebitzsch (1978) argued that soil pH was a poor indicator of its nitrifying ability, and experimental work by Like and Klein (1985) and Salonius (1990) found that treatment of soil columns with simulated acidic precipitation ranging in pH from 3.6 to 4.6 did not affect nitrification rates. The effects of calcium, potassium and phosphorus contents were greater, with the nitrifying ability increasing with increasing nutrient supply (Kriebitzsch, 1978). Similar results have been described by Farrell (1990). Soils with calcium : nitrogen ratios greater than 25–30 in the organic horizon are unable to nitrify (Gundersen and Rasmussen, 1990). Nitrification rates are also related to soil acidity, with the activity of ammonium relative to nitrate increasing with increasing acidity (Hern *et al.*, 1985; Haynes and Swift, 1986). Nitrate tends to be extremely mobile in soils whereas ammonium is fixed on the cation exchange complex or in clay minerals.

When nitrogen is scarce, it may be cycled tightly within trees. Aronsson and Elowson (1980) documented about 30% of the nitrogen content of older Scots pine needles being translocated to new shoots and needles. As the nitrogen content of the litter dropped, the carbon : nitrogen ratios increased, leading to falling nitrification rates. Increased nitrogen availability, conversely, results in higher circulation of nitrogen through the litter pathway. At Alltcailleach Forest in northeast Scotland, long-term records of nitrogen concentrations showed that there has been no change in the calcium : nitrogen ratio, despite the site exceeding the critical load of 5–20 kg N ha^{-1} yr^{-1} (Billett *et al.*, 1990).

Nitrogen additions are known to result in changes to the soil chemical status. Below pH 5.5, application of ammonia-fertilizers resulted in an increase in exchangeable aluminium concentrations in the soil and simultaneous increases in the levels of calcium, magnesium, potassium and sodium in the soil solution (Haynes and Swift, 1986).

The balance between nitrate and ammonium as sources of nitrogen has been carefully investigated, particularly in relation to soil problems in The Netherlands. Trees generally take up ammonium in preference to nitrate (Boxman and Roelofs, 1988; Scheromm and Plassard, 1988; Marschner *et al.*, 1991; Arnold, 1992), a process that leads to acidification. This process has been termed by Ulrich (1986) the 'spatial discoupling of ammonification and nitrogen uptake'. However, as the soil acidity increases, the preference for ammonium declines (Arnold and Van Diest, 1991).

The amount of nitrogen in the litter and mineral soil is important as it influences several processes. For example, the turnover of lignin, which exerts an important control on nutrient release, is, among other factors, dependent on the nitrogen concentration, with higher levels of nitrogen retarding lignin decomposition due to adverse effects on the microbially formed lignolytic enzyme system (Keyser *et al.*, 1978; Berg, 1986a).

The role of nitrogen deposition in increasing forest yields in the 20th century is difficult to ascertain. Yield increases have been documented for many areas with, for example, Norway spruce (Björkdahl and Eriksson, 1989) and beech (Falkengren-Grerup and Eriksson, 1990) increasing in southwest Sweden, and the question of growth increases caused by atmospheric nitrogen fertilization has been examined in detail in Chapter 10. The attribution of the growth increases entirely to nitrogen deposition is questionable as fertilizer trials in many areas have revealed a fairly small and temporary response of tree growth to nitrogen (e.g. Miller and Miller, 1988; Farrell, 1990). However, there may be a strong interaction between

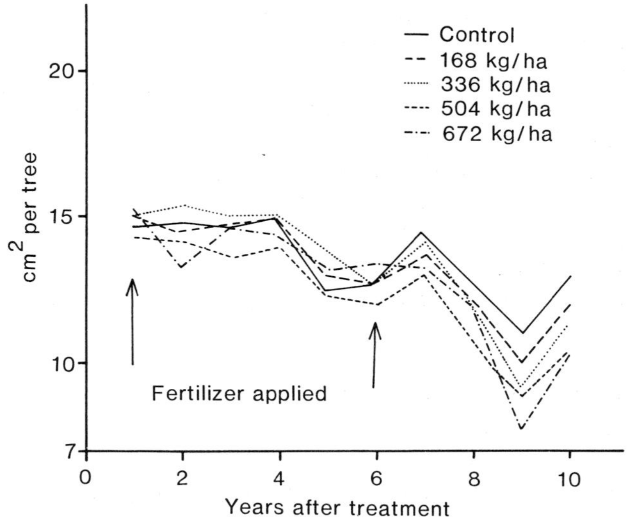

Figure 8.14 Annual growth response of sugar maple basal-area increment to added nitrogen; pooled data from seven locations in New York State. From: Stanturf *et al.* (1989).

climate and nitrogen (Miller *et al.*, 1977), a response that is markedly site-dependent, and many past trials have been on soils where a response would not be expected. This is confirmed by those studies which have shown positive responses to nitrogen fertilization (Fig. 8.14) (e.g. Mitchell and Chandler, 1939; Stanturf *et al.*, 1989). Trials are also complicated by many other mitigating factors, including the age of the trees being fertilized. For example, Miller and Miller (1988) pointed out that a tree crop may respond to nitrogen fertilization when young, but may then show no such response once canopy closure has been achieved. Late in the development of the stand, the trees may again become responsive to nitrogen fertilization.

In Scandinavia, nitrogen fertilizers have been widely used to improve growth rates of trees. For example, in Sweden, between 140 000 and 150 000 ha of forest were fertilized each year in the recent past (Hansson, 1984), although forest fertilization has now been severely curtailed in southern Sweden. Some of the experiments provided useful information on the effects of nitrogen on forests. The majority (>70%) of nitrogen applied as fertilizer in areas with low levels of available nitrogen is generally retained within the forest ecosystem (Melin, 1986; Melin and Nömmik, 1988). Nitrogen fertilization occasionally resulted in damage to trees in northern Finland and northern Sweden, where it induced boron deficiency (Aronsson, 1983; Lipas *et al.*, 1983; Möller, 1983). In contrast, fertilization in mid- and southern Sweden increased stemwood production by 30–50% in the 10 years following fertilization (Staland, 1985). The amount of damage caused by nitrogen fertilization depended on the site quality, with fertilization of nitrogen-deficient stands showing no long-term harmful consequences and generating increased growth (Mälkonen and Kukkola, 1991). Nitrogen fertilization of better sites needed to be accompanied by additions of phosphorus to get a positive response (Mälkönen, 1990). Fertilization with ammonium nitrate generally resulted in an increase in the H^+ activity in the soil, lasting up to three years (Nohrstedt, 1989). Over longer periods, fertilization with up to 1800 kg N ha^{-1} did not affect the pH of the humus and eluvial layers of the soil. However, at intermediate application rates, a pH decrease of 0.2 to 0.3 units was identified in the spodic horizon (Nohrstedt, 1990). The amounts of exchangeable cations were not affected by the nitrogen applications.

In Finland, nitrogen fertilization resulted in significant increases of organic matter and nitrogen in the humus layer (Mälkönen, 1990; Mälkönen *et al.*, 1990), although no effects on soil pH or the concentrations of base cations were noted. As in Sweden, increased growth occurred on the least fertile sites, with little response being obtained on the more productive sites. This may be an explanation for the different effects between species shown in some studies. In contrast, Fenn (1991) argued that nitrogen deposition in the San Bernadino Mountains of California may actually enhance litter decomposition rates.

Outside Scandinavia, the application of nitrogen fertilizers to mature forest is rare, although Rehfuess *et al.* (1991) documented improvements in the growth of Scots pine in an impoverished soil in north-eastern Bavaria. Both the treatments and the controls showed improved growth and Rehfuess *et al.* (1991) attributed the increased growth in the control plots to atmospheric deposition of nitrogen. In many continental forests, litter raking has led to the loss of calcium and magnesium from the forest ecosystems. Because of this, additions of atmospheric nitrogen are causing greater nutritional disturbances than they would have otherwise done (Kazda, 1990).

Nitrogen fertilization in Great Britain has generally produced very limited responses. This appears to be because other nutrients tend to become rapidly limiting (Miller and Miller, 1988). There is no evidence from British nitrogen fertilization trials that growth abnormalities or site deterioration will occur (Miller and Miller, 1988). In Belgium and France, nitrogen fertilization on base-poor soils in the Ardennes resulted in the development of needle chlorosis due to magnesium and/or calcium deficiency and needle and increment losses (Nys, 1989; Muys, 1990). On richer soils, growth increases occurred.

Conclusions

Skeffington and Wilson (1988) emphasized that nitrogen fertilization experiments should not be used as a surrogate for studies of the effects of atmospheric deposition of nitrogen, because of major differences in the rates, timing and nature of the exposures. Johnson (1992) has also drawn attention to the major differences in the forest ecosystem response between pulse fertilization and the long-term atmospheric deposition of nitrogen. However, many supposed effects of atmospheric nitrogen deposition are based on fertilizer experiments. Atmospheric nitrogen is clearly affecting forest soils, although it is uncertain whether it is adversely affecting the forest ecosystems, except in the most extreme cases. Fertilization with nitrogen has been a widely used technique in forestry, although its cost-effectiveness is limited. Atmospheric deposition of nitrogen will stimulate forest growth, if other elements are in sufficient supply. However, when the deposition rates are very high, as for example in The Netherlands, then adverse effects occur and nitrogen saturation of the ecosystem may ensue.

Soil amelioration

Many European forest soils have been severely degraded by management practices. In particular, the widespread use of litter raking led until quite recently to the large-scale transfer of nutrients from forests to fields, resulting in severe disruption of the nutrient cycle (Ebermayer, 1876; Ramann, 1883; Wittich, 1951; Kazda, 1990; Evers, 1991; Facelli and Pickett, 1991; Feger *et al.*, 1991). Losses attributed to raking have been estimated at 20–50 kg nitrogen ha^{-1} yr^{-1} and values for other nutrients (in ha^{-1} yr^{-1}) were 2–4 kg of phosphorus, 12–25 kg of potassium, 15 to more than 40 kg of calcium and 3 to more than 10 kg of magnesium (Ebermayer, 1876; Kreutzer, 1972; Krapfenbauer, 1983; Glatzel, 1991). These losses would have reduced the acid neutralizing capacity of the soil by from 2.4 to more than 5 kmol ha^{-1} yr^{-1}. Removal of biomass for fodder and timber for fuel, together with the use of woodlands for grazing, led to further depletion of the forest nutrient pools (Fig. 8.15). These reductions may have been greater than those caused by recent acidification related to acidic deposition (Feger *et al.*, 1991; Glatzel, 1991; Feger and Raspe, 1992).

Much of the current area of European forest has been planted within the last 200 years. In addition, large areas of deciduous woodland have been converted to coniferous woodland, often using fast-growing species unsuitable for the site (van Goor, 1985; Nys, 1989). Elsewhere, forests have been planted on low-quality soils formerly used as grazed heathlands (Nihlgård, 1990). All these factors led to a situation at the beginning of this

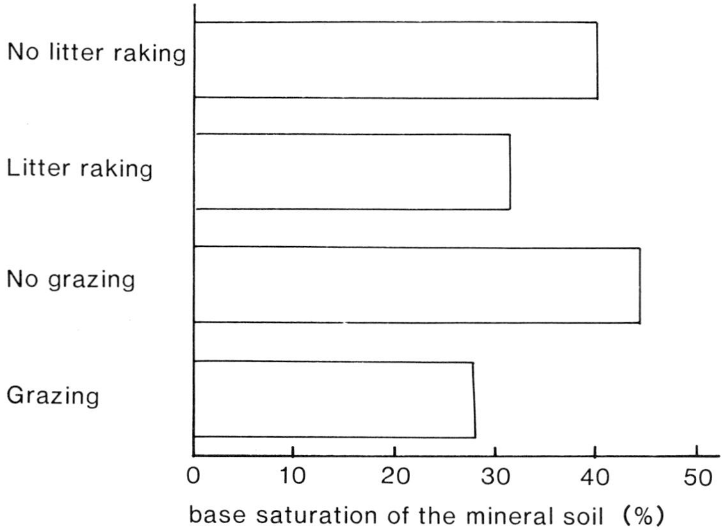

Figure 8.15 Base saturation of forest soils on siliceous parent material in Tyrol, Austria, as influenced by litter raking and grazing. From: Glatzel (1991).

century in which many forest soils were severely degraded. The recent stress created by acidic deposition may have hindered the recovery of these ecosystems (Glatzel, 1991).

The degradation was recognized at an early stage, as were the problems of afforesting areas with poor soils. For example, the Count of Waterloo-Merode used a variety of amelioration techniques, including the application of slag and lime and the use of nitrogen-fixing lupins (*Lupinus* spp.), to improve heathland soil before planting it with Scots pine and Norway spruce (Heyer and Hess, 1906). Similarly, positive responses of Norway spruce to fertilizers (slag) were found in 1888 in the Vosges Mountains (Giersberg, 1901). Various treatments have been suggested, the two most common being liming and fertilizer application.

Liming

There has been considerable interest in the use of liming and various fertilizers for the improvement of soil conditions in forests. Liming in particular has received a great deal of attention, both in recent years and in the past. It has been widely assumed that liming will offset the adverse effects of soil acidification caused by acidic deposition. While this may be true of acidification, liming with calcium carbonate has failed to offset the magnesium-deficiency symptoms seen in areas with Type 1 Norway spruce decline (Wachter, 1986; Aldinger, 1987).

Considerable care should be taken when using liming with some species. For example, Scots pine is intolerant of alkaline conditions and growth reductions have been recorded associated with an increase in soil pH from pH 5.5 to 6.2 (Carter, 1987). However, this may have been a site-specific response as other studies (e.g. Olsson, 1986; Spiecker, 1991) have failed to reveal an adverse effect of liming on increment. One problem associated with liming is that an increase in calcium uptake is associated with a decrease in phosphorus uptake. This results in the development of nutritional imbalances. In addition, the application of lime may trigger losses of nitrate from the soils, with consequent effects of the quality of drainage water. However, there appears to be no danger of nitrate losses if low applications of lime (2–3 tonnes ha^{-1}) are used (Matzner, 1985; Krieter, 1986; Evers, 1989) and the pH of the organic matter remains below about 4.2 (Evers, 1991). On sites where the humus is susceptible to nitrification, particularly those that have experienced past changes in tree species, liming with 2–4 tonnes ha^{-1} may be sufficient to stimulate nitrification and subsequent losses of nitrogen (Kreutzer, 1989).

Liming has been widely used in Finland. The results of several trials have been described by Derome *et al.* (1986) and Derome and Pätilä (1989, 1990). Based on 96 experiments spread throughout Finland, Derome *et al.* (1986) concluded that calcium was retained for a long time in the surface layer of forest soils, with 65–75% of applied limestone still present in the humus and uppermost 20 cm of the mineral soil horizons almost 25 years after the initial application. Liming reduced the extractable aluminium in the humus layer by half and in the mineral soil horizon by 20%. The depth of the humus layer strongly affected the extent to which liming increased the pH of the soil. Liming also resulted in an increase in the organic matter content of the humus layer and a corresponding decrease in the thickness of the uppermost layer of the mineral soil. The effect of liming on tree growth was to reduce long-term growth of Scots pine by 3% and Norway spruce by 10%.

In a later study, Derome and Pätilä (1989) showed that the effects on the soil reaction were still apparent 60 years after the initial treatment. Long retention periods have also been noted in studies elsewhere (e.g. in Germany by Grimm and Rehfuess, 1986). Growth reductions occurred with as little as 500 kg ha^{-1} application, whereas 10 tonnes ha^{-1} were required to change the base saturation of the soil. Derome and Pätilä (1989) found that the effects differed between soil types and species. On mineral soils, liming had no effect on Scots pine and reduced the growth of Norway spruce. Overall, there was a long-term increase in soil pH. The amount of organic matter in the humus layer increased after liming. On peat soils, liming had either insignificant or negative effects on growth, possibly because of nutrient uptake by the microflora after liming.

Derome and Pätilä (1990) examined the results of about 40 liming trials in the southern half of Finland. On mineral soils, liming led to a significant increase in the organic matter in the litter layer under both Norway spruce and Scots pine, with the increase being greatest on the least productive sites. The organic matter content of the upper mineral soil was unaffected. They attributed the increase in organic matter to increased litter fall and root biomass of the ground vegetation, rather than increased litter fall from the trees, as liming either had no effect or a slightly negative effect on stand growth (Derome *et al.*, 1986). In Germany, Marschner and Wilczynski (1991) found very different results. Liming led to an increase in soil pH from 3.3 to pH 5.5. This was associated with severe losses of organic matter due to increased biological activity, involving a change from slow growing fungi towards bacteria.

Liming (with kieserite – $MgSO_4.H_2O$) is one treatment that has been applied in the ARINUS research project being conducted in the Black Forest, Germany (Feger *et al.*, 1991). The ARINUS site is of interest as it is typical of those showing Type 1 Norway spruce decline. Growth of trees at the site was slow, amounting to only 7 m^3 ha^{-1} yr^{-1}, and was controlled primarily by the temperature during the growing season. Magnesium deficiencies were evident, although magnesium was deposited in precipitation at a rate of about 0.9 kg ha^{-1} yr^{-1}. In throughfall, this increased to about 1.2 kg ha^{-1} yr^{-1}. The average uptake of magnesium by the trees amounted to about 1.1 kg ha^{-1} yr^{-1}. Leaching of magnesium from the upper soil layers was calculated to be 1.7 kg ha^{-1} yr^{-1}. A total of 750 kg ha^{-1} kieserite was applied, resulting in the introduction of 130 kg ha^{-1} of magnesium and 170 kg ha^{-1} of sulphur. More than 75% of the applied magnesium has remained in the forest ecosystem. The initial response to the treatment was the depression of soil pH and the mobilization of aluminium. Although Al^{3+} was released, the fine roots and mycorrhizae were unaffected (Haug and Feger, 1990). The treatment led to an increase in foliar magnesium, particularly in older needles, and the magnesium : aluminium ratio in the soil and soil solution improved following the fertilizer applications. The application of magnesium in this situation clearly relieved a very tight cycling system which was experiencing net losses.

The effects of liming on Scots pine in Germany have been studied by Marschner *et al.* (1989). Granulated lime was applied at a rate of 6.1 tonnes ha^{-1} in April 1986. The lime was supplemented with potassium (145 kg ha^{-1} K_2SO_4) to reduce the risk of calcium : potassium antagonism in root uptake. A short-term increase in aluminium, manganese and cadmium concentrations was recorded because of the sulphate in the potassium treatment. However, two years after the liming treatment, 80% of the calcium and magnesium was still present in the forest floor, although there was no trace of the added potassium. The amount of organic carbon stored in the forest floor had decreased. In the mineral horizon, pH had significantly increased and exchangeable aluminium was significantly reduced.

Liming trials in the Black Forest and elsewhere in southern Germany have generated increased growth of Norway spruce (Spiecker, 1991). However, the responses only became evident more than six years after the treatments. Treatments including nitrogen resulted in much more rapid responses and the longest-lasting and most marked growth increases were in response to combinations of liming and nitrogen and phosphorus fertilizers. However, Kenk *et al.* (1984a) found that nitrogen-phosphorus-potassium fertilization on sandstone sites in the Black Forest resulted in substantial decreases in the available magnesium and calcium.

In 1989, more than 176 000 ha of forests in Germany were either fertilized or limed primarily to reduce 'damage caused by air pollution' (Schierl and Kreutzer, 1991). Applications of kieserite, dolomite and other sources of magnesium generally resulted in the disappearance of magnesium deficiency symptoms (Bauch *et al.*, 1985a; Hüttl, 1985a; Kenk *et al.*, 1985; Zöttl, 1985a, b; Aldinger, 1987; Isermann, 1985, 1987; Kaupenjohann *et al.*, 1987). Cytological damage, in the form of hypertrophied sieve and Strasburg cells, disappeared following fertilization and cambium cells were again able to differentiate efficient sieve cells (Fink, 1986a; Hüttl and Fink, 1988). The changes were accompanied by a return to normal photosynthesis (Lange *et al.*, 1989).

Other treatments

In early studies, fertilizer application affected the responses of trees to air pollution, particularly 'smoke damage', in several areas (Ranft, 1984; Materna, 1987; Paternak *et al.*, 1987). The responses were probably related to the increased sensitivity of nutrient-deficient trees to sulphur dioxide (Jung and Winter, 1992). The connection appears to lie in the ability of trees to detoxify the sulphur dioxide when they are supplied with adequate amounts of nutrients (Klein and Jäger, 1976). The recent deterioration of tree condition was first apparent in the Harz mountains on poor soils (Stock, 1988) and the history of management and exploitation have been cited as important additional factors by Zöttl (1985a, b). In long-term fertilizer trial plots in Germany, there have been marked differences in the responses of trees in different areas. For example, Evers (1991) reported little difference in the needle losses of trees in fertilized and unfertilized plots, although mortality was greater in fertilized plots, whereas Kenk *et al.* (1984) found big differences in the damage between fertilized and unfertilized plots in the Odenwald.

At Elzach, Germany, magnesium oxide (MgO) and ammonium sulphate were applied to reduce the visible effects of magnesium deficiency (Hüttl, 1990). Soil analyses in a stand of 12-year-old chlorotic Norway spruce had revealed low levels of magnesium and calcium, low base saturation and low magnesium : aluminium and calcium : aluminium ratios. One treatment consisted of applying 150 g of magnesium oxide to each tree. The deficiency symptoms disappeared within four to five months of the treatment, and had not re-appeared four years later. However, the response of the trees may have been modified by extraneous factors. Nitrogen concentrations in both fertilized and unfertilized trees peaked in the autumn of 1985 (the initial treatment was in spring 1983), and then decreased. Calcium contents showed marked variations from year to year, and the magnesium contents in the unfertilized trees also varied markedly. In the ammonium sulphate trial, the equivalent of 126 kg N ha^{-1} was applied in spring 1985. In the autumn of that year, the trees showed marked chlorosis. Nitrogen : magnesium ratios were very high in 1985 and rose even further in 1986 before recovering in 1987.

In another trial, at Forbach, Germany, magnesium and nitrogen-phosphorus-potassium were applied in separate treatments to 10–15-year-old naturally-regenerated chlorotic Norway spruce (Hüttl, 1990). The magnesium treatment resulted in the recovery of the trees, whereas the NPK treatment led to a decrease in magnesium contents of needles.

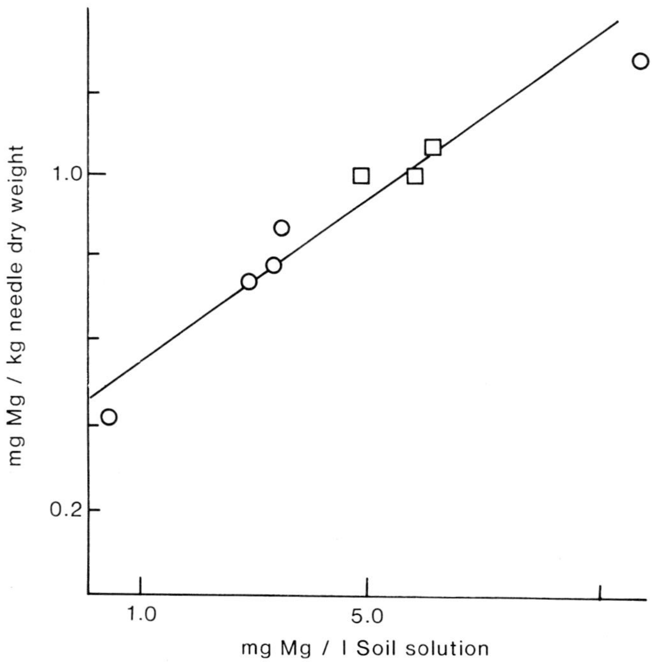

Figure 8.16 Magnesium content of spruce needles at Oberwarmensteinach, Germany (circles), and Wülfersreuth, Germany (squares), as a function of the magnesium content of the soil solution (P80 suction cups, mean value for the season). From Kaupenjohann *et al.* (1987).

The form of the applied magnesium is very important. Kaupenjohann *et al.* (1987) found that magnesium sulphate rapidly increased the magnesium contents of deficient trees (Fig. 8.16), with fertilized trees having magnesium contents double those in the controls. Fertilization with $MgO.CaCo_3$ worked less quickly, but induced higher magnesium contents in the needles. However, Monchaux and Thivolle-Gazat (1990/1991) found no differences in the effects of five different types of magnesium fertilizers applied to 12–15-year-old Norway spruce stands at the Col du Donon in the Vosges Mountains of France, again suggesting the importance of site-specific factors controlling responses.

Potassium deficiency has been associated with one form of decline identified in Norway spruce in Germany. It has been reported from a variety of areas, including low-altitude stands in northern Bavaria (Zech *et al.*, 1983), morainic loams and moorland soils in the alpine foothills (Zöttl and Hüttl, 1985), the limestone Alps of Bavaria (Rehfuess, 1987), the slate and quartzite rocks of upper Franconia (Hüser, 1983) and in the Erzgebirge (Fiedler, 1986). In recent years, it has also appeared on loamy soils in south-west Germany (Evers and Schöpfer, 1988). The deficiency is primarily due to anomalous soil structure and a reduction in the exchange capacity of the surface soil horizons (Hildebrand, 1988). As with other deficiency problems, remedial fertilization has been successful at alleviating the symptoms.

Conclusions

Unlike forest condition, there is clear evidence that the chemical status of forest soils has changed over the last 100 years. The causes of these changes are complex; acidification caused by forest growth has certainly occurred, but acidification because of acidic deposition also appears to have been important. The acidification has been accompanied by losses of base cations, to the extent that some forest soils are now deficient in these. Often, the effects of these losses have been aggravated by the poor initial chemical status of the soils. Many forest soils have been mismanaged in the past, and poor matching of species to site has exacerbated the problem.

There is good evidence of a strong association between the condition of the soil and the status of forests at particular sites. This has been clearly illustrated by studies in the Harz Mountains of Germany and elsewhere. In particular, magnesium deficiency is closely correlated with low soil magnesium levels. Elsewhere, exhaustion of nutrients has been associated with the decline in the growth rates of trees and even in a decline in the health of the trees themselves. This phenomenon is not restricted to the most recent reports of forest declines. It is widespread and has been reported from a number of diverse environments.

The problem of soil toxicity, particularly as related to aluminium, remains largely unresolved. The original work at Solling, near Göttingen in Germany, showed that potentially phytotoxic concentrations of aluminium can occur in the soil. However, the ALBIOS programme revealed that the conditions at Solling were very unusual, and it is unclear how widespread the effects seen at Solling are. The detailed work that has been conducted on aluminium toxicity has shown that there are many mitigating factors which influence the relationship. The early threshold values and the calcium : aluminium ratio were clearly inappropriate as indices of toxicity. However, the latter has been widely adopted in critical loads modelling.

Nitrogen additions to soils have been occurring for some time. In areas such as The Netherlands, the deposition rates are sufficiently high to seriously disturb soil processes, with important implications for trees and other elements of the flora. The nitrogen nutrition of trees elsewhere appears to be disturbed by low levels of deposition, and it is possible that the nitrogen dynamics of forest ecosystems are more sensitive to perturbations than previously recognized. Nitrogen enrichment is involved in some cases of nutrient deficiency, as indicated in the following chapter.

Soils are clearly a critical part of forest ecosystems, and their condition needs to be monitored as carefully as the trees themselves. However, there are many pitfalls associated with soil analyses, not least the very high levels of variation that occur at small scales. Cause–effect inferences consequently need to be made with great care.

9 Foliar Analyses

> There is then, for a given element and a given species, an optimal level (maximal growth), a critical level (lower growth) and a deficiency level (strong growth reduction and some decay symptoms).
> *Fromard et al.* (1991)

Introduction

The nutritional status of trees provides an important diagnostic tool for establishing their condition. While most studies have concentrated on identifying deficiencies, excesses may also occur, particularly with nitrogen, sulphur and some heavy metals. Recently, it has been increasingly recognized that another form of nutritional disturbance may occur. This involves the increased supply of one nutrient (usually nitrogen) relative to others. All three types of disturbance have been identified in relation to forest declines.

Major problems occur with the comparison of foliar nutrient contents across space and through time because of the spatial and temporal variation of nutrient contents. Nutrient concentrations of coniferous foliage vary according to the nutrient content of the soil (Bell and Ward, 1984), with the needle and tree age (Florence and Chuong, 1974; Mälkönen, 1974; Miller *et al.*, 1981; Madgwick *et al.*, 1983; Hüttl, 1985b; Huttunen *et al.*, 1985; Knabe *et al.*, 1989; Santerre *et al.*, 1990), with the needle position in the crown (Will, 1957; Wells and Metz, 1963; Mead and Will, 1976) and with the annual physiological cycle (Tamm, 1955; Van den Driessche, 1974; Fife and Nambiar, 1984, 1987; Kazda and Weilgony, 1988; Helmisaari, 1990). Consequently, it is very difficult to control adequately the inherent variation in the sampling. In addition, the exact nature of the variation has only been established for some species and the nutrient dynamics of many tree species remain virtually unknown.

Several nutrients, including potassium, nitrogen, phosphorus, boron and magnesium, are very mobile and may therefore be prone to variation through time (Figs 9.1 to 9.4) (Helmisaari, 1990). In contrast, manganese increases in needles through time and is therefore relatively immobile. The

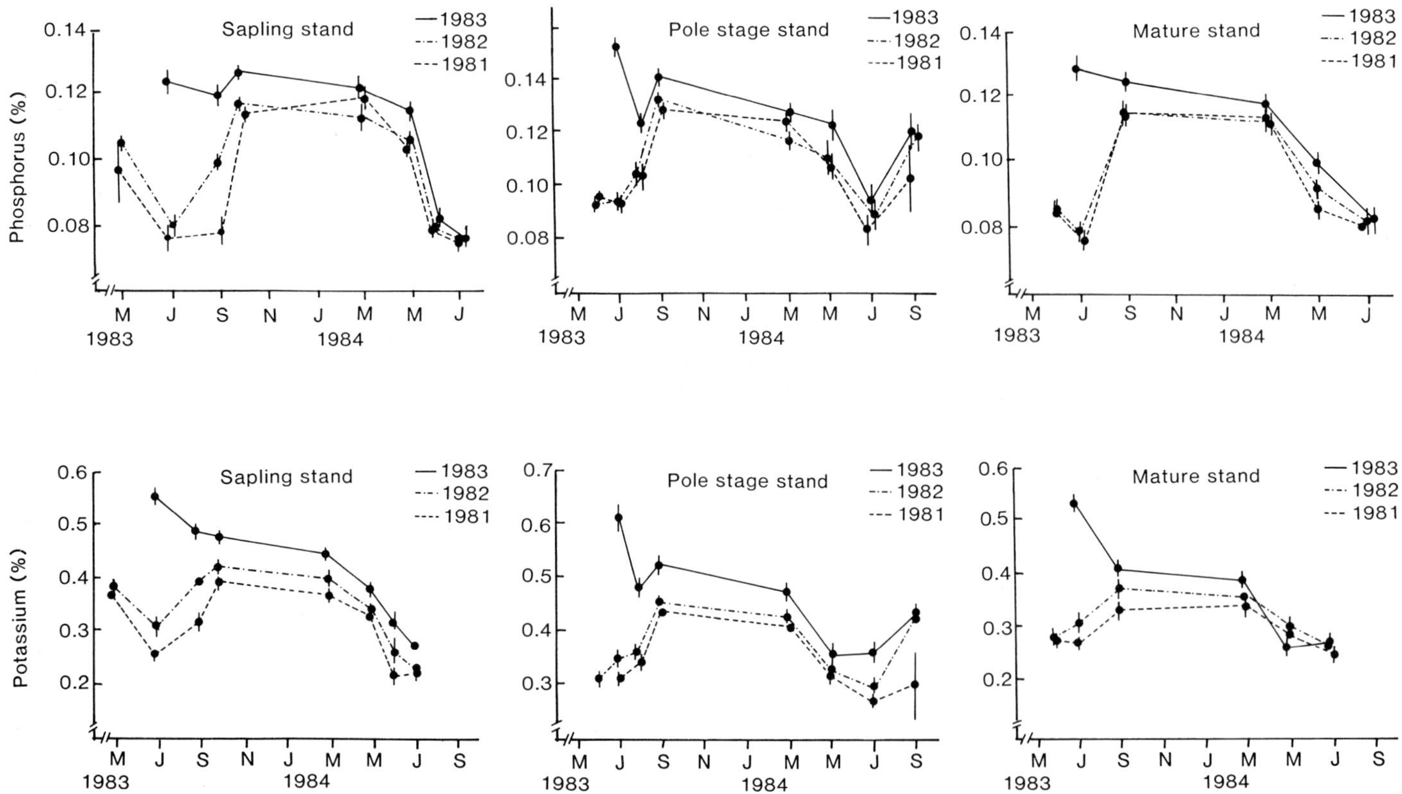

Figure 9.1 Annual variation in the potassium and phosphorus contents of Scots pine needles in stands of various ages. The contents of the three different needle years present are given. Vertical bars indicate the standard errors. From: Helmisaari (1990).

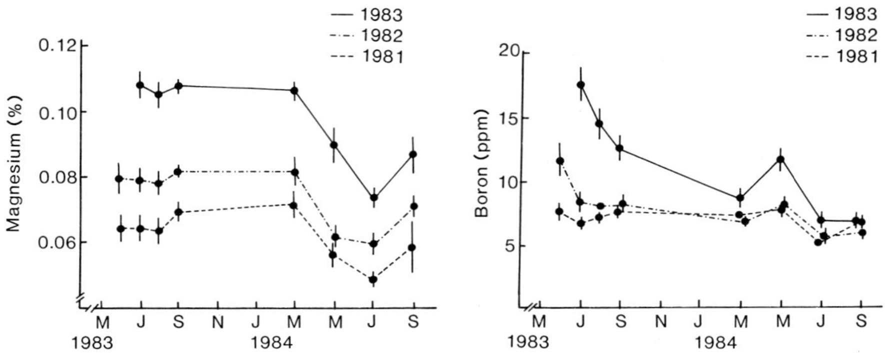

Figure 9.2 Annual variation in the magnesium and boron contents of Scots pine needles in a pole stage stand. The contents of the three different needle years present are given. Vertical bars indicate the standard errors. From: Helmisaari (1990).

seasonal variation in the nutrient concentrations means that the time of sampling is critical.

An added problem is that different analytical techniques may produce different results. For example, Jarva and Tervahauta (1989) have reported major differences in the sulphur contents of foliage depending on the method of analysis used. Between 24 laboratories, estimated total sulphur contents for the same Norway spruce foliage ranged from 0.07% to 0.16% and for the same Scots pine foliage from 0.09% to 0.19%. Many similar problems appear to exist with the analysis of broad-leaved foliage (Zimmermann, 1989).

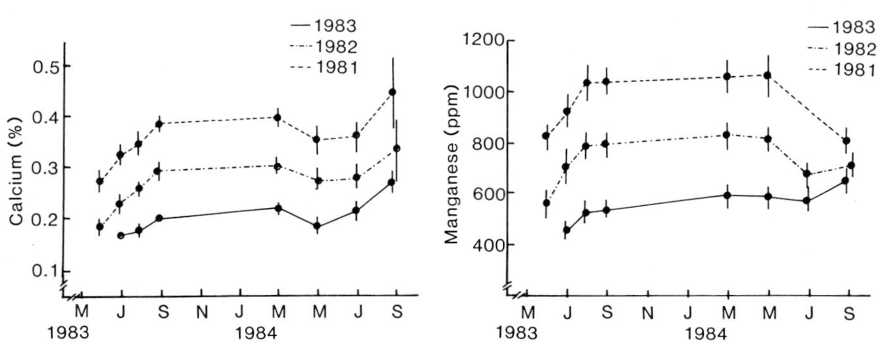

Figure 9.3 Annual variation in the calcium and manganese contents of Scots pine needles in a pole stage stand. The contents of the three different needle years present are given. Vertical bars indicate the standard errors. From: Helmisaari (1990).

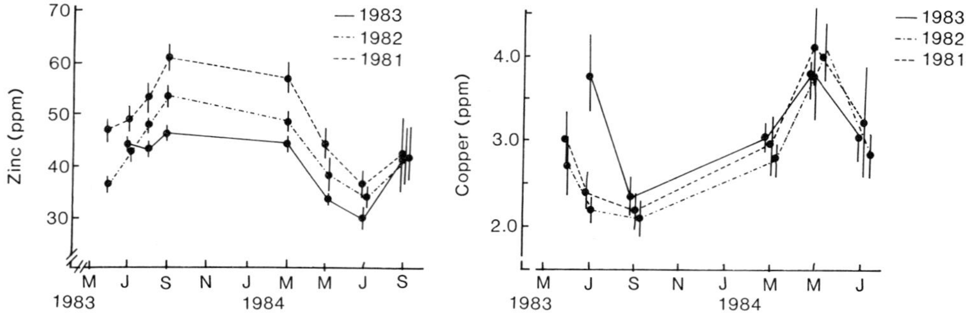

Figure 9.4 Annual variation in the zinc and copper contents of Scots pine needles in a pole stage stand. The contents of the three different needle years present are given. Vertical bars indicate the standard errors. From: Helmisaari (1990).

The various problems can be specified as follows:

- differences in the standards between analytical techniques and between different laboratories;
- effects of sampling time and sampling position;
- differences between trees;
- differences between sites; and
- differences between the nutrition of young and older trees.

Changes in nutrition through time

Many of the declines described in Chapters 1–3 can be related to changes in the nutritional status of the trees. Over the longer term, there may be a net transfer of certain elements from soil reserves to timber, resulting in a permanent loss of those nutrients when the timber is harvested. There are many studies which have looked at element budgets in this way and they need not be reviewed here (e.g. Matzner and Ulrich, 1981). Nutritional problems can occur when the relationship between the availability of and the demand for specific nutrients changes through time. This may arise because of depletion of the soil reserves or from the increased availability of certain elements relative to others. For example, the nutritional problems associated with Type 1 Norway spruce decline may be associated with a simultaneous reduction in the availability of magnesium and an increase in the availability of nitrogen.

Many problems listed in the introductory section of this chapter are applicable to studies of nutritional changes over time. In particular, as with studies of soil changes, it is important to ensure that the methods used are exactly duplicated. The degree to which this has been achieved is not always reported in studies purporting to describe changes in forest nutrition.

Magnesium deficiency has been widely associated with Type 1 Norway spruce decline and there is evidence for changes in the magnesium status of trees through time (Prinz *et al.*, 1985; Hüttl and Wisniewski, 1987). As suggested in Chapter 3, the causes of the changes remain uncertain. They are clearly related to changes in the rate of uptake, as shown by the correlation between the xylem sap and needle magnesium contents (Oren *et al.*, 1988). Needle magnesium contents can also be related to soil magnesium contents (Fig. 8.16) (Kaupenjohann *et al.*, 1987; Hantschel *et al.*, 1988). The behaviour of magnesium in needles varies according to its abundance. In the Fichtelgebirge, translocation of magnesium from older to new needles accounted for 75% of the total content in asymptomatic trees. In trees with magnesium deficiency, translocation from older needles only accounted for 50% of the magnesium, the rest being derived by uptake from the soil (Oren *et al.*, 1988b).

Changes in the magnesium nutrition of trees appear to be related to changes in magnesium availability. A similar conclusion has been reached for a variety of other important macro- and micro-nutrients. The changes in availability appear to be a direct consequence of silvicultural activities. If the tree requirements exceed the supply in the soil and the rate at which the the soil supply is recharged through weathering and atmospheric deposition, then deficiency ensues. Fertilization will normally relieve this problem, but the environmental impact of fertilization may be unacceptable under certain circumstances. There appear to have been few mass balance studies that have looked at the long-term sustainability of specific soil types under different tree species. If long-term forest sustainability is to be a goal of modern forestry, then this is an area where more work is required.

Excess nitrogen

As discussed in the previous chapter, forest fertilization experiments have shown that typically the growth of trees increases in response to nitrogen fertilization. Nitrogen depletion because of management practices such as litter raking appears to have been a problem in the past and has now mostly stopped. Most areas in central Europe are currently experiencing increasing levels of nitrogen deposition and this, together with the reduction in nitrogen removal, is resulting in increased nitrogen availability.

Studies of nutrient concentrations of foliage in the severely affected area of the Peel in The Netherlands have provided some useful insights into the problem of excess nitrogen. Needles of Scots pine infected by *Gremmeniella abietina* and *Sphaeropsis sapinea* had higher nitrogen concentrations and lower magnesium concentrations than unaffected needles (Roelofs *et al.*, 1985), although subsequent studies did not consistently find this pattern (Van den Burg, 1988). Nitrogen : magnesium ratios in the needles ranged from 2 to 4. Elsewhere, Weissen *et al.* (1990) attributed nitrogen : magnesium ratios greater than between 30 and 40 as being indicative of a nutritional problem.

Increases in the nitrogen concentrations of older needles of Douglas fir have been reported from The Netherlands, with nitrogen contents of unhealthy trees being much higher than those of healthy trees. The increases in nitrogen content with age are the reverse of those expected for healthy trees (Van den Burg, 1988). In Douglas fir, nitrogen : phosphorus ratios have been found to be a useful tool for recognising nutritional disturbance (Mohren *et al.*, 1986), with ratios being higher in defoliated trees. As with Scots pine, the increased nitrogen contents are associated with a number of secondary effects, including deficiencies of other nutrients and increased insect activity.

Studies of Scots pine (Van Dijk and Roelofs, 1988) in areas with excessive ammonium deposition have shown that the nitrogen contents of chlorotic trees were higher than normal trees. Calcium and zinc were also higher in chlorotic trees. The chlorotic needles had very high levels of arginine, accounting for 35% and 29% of the total nitrogen in young and old needles, respectively. In green trees, arginine accounted for 2% of the nitrogen in young needles and 8% of the nitrogen in old needles. Histidine, lysine, leucine and isoleucine were all higher in chlorotic trees. However, the implications of these changes for tree health are not fully understood.

Association between soil and foliar chemistry

Balsberg Påhlsson (1989) made a detailed study of the mineral composition of beech foliage. It involved 89 sites in Scania, southern Sweden. Element concentrations in the foliage were related to element concentrations in the soil. Only a weak association was found between foliar and soil chemistry. Foliar nitrogen levels were optimal or above-optimal whereas potassium, magnesium and phosphorus levels were near or below minimum levels. Insect damage was not related to foliar chemistry. The lower crowns generally had higher concentrations of elements, although calcium and zinc were exceptions to this. Leaf chlorosis was negatively correlated with nitrogen, magnesium and potassium concentrations, as also found by Zech *et al.* (1985) in Germany.

Hüttl (1985b) examined the nutritional status of Norway spruce from 150 sites in south-west Germany. The sites mainly consisted of stands more than 60 years old and the trees showed weak to moderate levels of damage. The soil types were generally poor, having developed on gneiss/granite, sandstone, poor limestone and moraine. Most stands were well-supplied with nitrogen and phosphorus levels were adequate. Potassium, calcium, magnesium, manganese and zinc ranged from optimal to deficient. The mineral nutrition of the trees closely reflected the availability of specific nutrients in the soil which, in turn, was determined by the composition of the parent rock.

In Norway, Aamlid *et al.* (1992) found that the concentrations of some nutrients, such as calcium, nitrogen and manganese, were significantly related to the soil concentrations of these elements. However, no such relationship was found for potassium, although potassium deficiency is one of the most widespread causes of yellowing seen in Norway.

Those studies that have been done clearly reveal an association between the soil nutritional status and the nutritional status of the trees. This link has been used to explain, for example, the distribution of Type 1 and Type 2 Norway spruce decline in the Harz Mountains of Germany (Hauhs, 1985, 1989). One problem that has not been sufficiently addressed is the relationships between total and available nutrients within forest soils. Many studies have looked at the total element content, but much of this may be in forms that are unavailable to plants. When studies have investigated available nutrients, most deficiencies can be directly attributed to soil deficiencies. There is no evidence that foliar leaching induced by acidic deposition has directly caused nutritional deficiencies.

Heavy metals

Most effects of heavy metals appear to occur through interactions in the rhizosphere. Consequently, they have been mainly discussed in the previous chapter. There are relatively few studies that have looked at foliar heavy metals, probably because trees can normally exclude or isolate excess quantities (many heavy metals are required in small quantities for normal physiological processes).

One study in Great Britain examined the effects of heavy metals on the growth of Sitka spruce in the south Wales coalfield (Burton *et al.*, 1983). This area was highly polluted by a number of industrial complexes and major problems have been encountered with the growth of some Sitka spruce stands. Burton *et al.* (1983) defined the upper critical concentration (the point at which yield starts to decline) for lead as 19 mg kg^{-1}, cadmium as 3.49–6.22 mg kg^{-1}, depending on plant age, copper as 61–104 mg kg^{-1}, nickel as 5.83 mg kg^{-1} and zinc as 226 mg kg^{-1}. Heavy metals were present in sufficient quantities in the foliage to pose a possible risk to growth. The figures for lead and cadmium were remarkably similar to the foliar concentrations of these elements that Breckle and Kahle (1992) identified as having an impact on the growth of beech.

Air pollution and nutrition

Gradient studies

There have been very few attempts to examine nutrient concentrations across large-scale gradients, possibly because the role of site factors throws into question the usefulness of such studies. However, a study by Cape *et al.* (1990) examined the element contents of three year classes of foliage of Norway spruce from 12 European sites. Sampling was restricted to green needles from the middle third of the canopy. Epicuticular waxes were removed before the analysis. At sites with poor overall crown condition, poor trees had significantly less magnesium and calcium and more potassium in the second-year foliage than 'healthy' trees (defined as having less than 25% needle loss). Higher levels of potassium in damaged (chlorotic) needles were also recorded by Lange *et al.* (1989). In contrast, Krivan *et al.* (1986) found that potassium contents of damaged trees were frequently, but not consistently, lower than healthy trees.

In the study by Cape *et al.* (1990), poor crowns were associated with larger ratios of nitrogen : magnesium, potassium : magnesium, sulphur : magnesium and potassium : calcium and smaller ratios of sulphur : potassium and nitrogen : potassium. In poor trees, calcium showed no increase with needle age and nitrogen and magnesium contents decreased through time, the latter markedly. Sulphur and calcium contents of the foliage increased with increasing needle age at all sites whereas magnesium and potassium decreased. Nitrogen was relatively constant, except at The Netherlands sites where it increased with needle age. These results are, however, difficult to extrapolate beyond the study as so few sites were involved. Given the inherent variation in soil and tree nutrition, a much larger sample would be required before any realistic conclusions could be drawn.

It is not clear how valuable such gradient studies are. The overwhelming influence of the soil nutrition strongly suggests that gradient studies should concentrate on forest soils rather than the composition of the foliage. Within a site, it may be possible to identify gradients associated with the sharp gradients in the deposition of atmospheric pollutants that occur at forest edges, but such studies are rare. Given the number of factors affecting foliar nutrition, it seems that gradient studies are unlikely to provide particularly useful information, as confirmed by the study by Cape *et al.* (1990).

Nutrition as a diagnostic tool

Sample size requirements

There is little information of the sample sizes required to adequately characterize the nutrient contents of the trees within a stand. Landolt *et al.* (1989) found that in Norway spruce, the most variable elements within a stand were boron, manganese, sodium and nickel. Copper, iron, potassium and sulphur were the least variable. Manganese was also found to be very variable by Påhlsson (1989), working with beech in southern Sweden. Nihlgård (1989) found very variable levels of potassium and magnesium in Scots pine and Norway spruce stands.

Nihlgård (1989) took foliar samples from 42 Norway spruce and Scots pine stands in southern Sweden in 1984–5 and again in 1986–7. During this relatively short period, nitrogen, phosphorus, sulphur and zinc concentrations in current-year needles increased while aluminium decreased. Simultaneously, the amount of yellowing decreased, as did the incidence of frost damage, fungi and green algae (*Pleurococcus vulgaris*) on the surfaces of the needles. These results strongly suggest that quite marked fluctuations in nutrient concentrations can occur from one year to the next. Other work (Evers, 1972) has shown that nitrogen, phosphorus and potassium levels fluctuate according to the climatic conditions, decreasing after dry years and increasing after wet years. Nihlgård found that the pH of the humus layer was a better indicator of foliar status than the pH of the mineral soil. However, aluminium in the mineral soil was a better predictor of the foliar phosphorus content and phosphorus : nitrogen ratio in the foliage than the aluminium content of the humus layer, probably because of binding to water soluble organic matter in the humus. There was a positive correlation between the humus pH and the foliar calcium content.

One of the best studies of the variation in the foliar composition of Norway spruce needles has been reported in a series of papers by Wyttenbach and colleagues (Wyttenbach and Tobler, 1988; Wyttenbach *et al.*, 1985, 1989a, b). Working in the Winterthur area, near Zürich, Switzerland, Wyttenbach *et al.* (1985) found that the only elements that clearly reflected environmental influences were sodium, chlorine and bromine. Given the proximity of some sample trees to a motorway, the influence of salt was predictable (Dobson, 1991), and this was reflected in the results (Wyttenbach *et al.*, 1989b). Analysis of the implanted material (Wyttenbach *et al.*, 1989a) showed that the dust load was a significant component for some sites illustrating the dangers of using the common practice of analysing unwashed needles during assessments of tree nutrition.

Wyttenbach and Tobler (1988) looked at the seasonal variations in a variety of elements in first and second year Norway spruce needles.

They identified three main groups of elements: calcium, strontium, barium and manganese in the first group, aluminium, bromine, cobalt, iron, mercury, lanthanum, scandium, antimony and zinc in the second and potassium, rubidium, caesium, phosphorus and chlorine in the third. The elements in the first and third groups showed the greatest amount of variation during the growing season, with the variation being greater in first-year needles than in second-year needles. In marked contrast, elements in the second group showed a near-linear increase in concentrations during the year in both younger and older needles, suggesting that these elements accumulate in needles.

These studies all suggest that very large sample sizes are required before the nutritional status of a forest stand can be assessed with any degree of confidence. This is unfortunate, as most studies have used relatively small sample sizes (see, for example, the study undertaken in Britain described below). The large variation that has been found even in very carefully controlled sample collections means that most surveys of nutritional status must be treated with caution.

Sulphur

As forest declines in Europe have been widely associated with air pollution, it is predictable that the sulphur contents of tree foliage should have been studied in some detail. The total sulphur content is often analysed, but Gasch *et al.* (1988) and Legge *et al.* (1988) argued that the ratio of organic to inorganic sulphur provides a better index of the impact of sulphur dioxide. Under heavy pollution loads, sulphur contents of conifer foliage increase through time (Materna, 1982; Molski and Dmuchowski, 1984; Hager and Kazda, 1985; Penka and Cervená, 1985) and the age of the foliage is therefore critical when making comparisons.

At a single site in Austria, foliar sulphur contents of Norway spruce ranged from 0.106% to 0.222%, depending on the position of sampling, the age of the needles and the dominance class of the tree (Hager and Kazda, 1985). Higher sulphur contents were found in older needles in exposed parts of the upper canopy. The sulphur contents of needles from a thinned stand were about 14% higher than in an unthinned stand, suggesting that crown exposure is an important factor influencing the sulphur contents of the foliage. However, thinning also results in increased growth which may be related to increased sulphur uptake. Outside areas with exceptionally high pollution loads, sulphur contents vary markedly from one year to the next (Stefan, 1989). Van der Stegen and Myttenaere (1991) found that the sulphur content varied according to the way contamination occurred. Atmospheric deposition resulted in high levels of surface contamination whereas sulphur absorbed through the roots tended to be localized in the inner parts of the leaves. However, a considerable amount of translocation also occurred, with sulphur being transferred from exposed leaves to non-exposed leaves.

Sulphur concentrations have been reported from a study of elements in the needles of Norway spruce in Baden-Württemberg (Evers and Schöpfer, 1988). Levels were generally high, in excess of 0.168% and local sources of sulphur were believed to have significantly affected the results. The highest value recorded was 0.331%. The pattern of concentrations in needles coincided approximately with sulphur dioxide concentrations, and a similar pattern has been found in other species elsewhere (Li and Zhang, 1989). Problems are expected to occur at sulphur concentrations in excess of 0.17% (Evers, 1984) or 0.2% (Zöttl, 1985).

In a study of the sulphur contents of the foliage of Norway spruce, silver fir and beech in southern Poland, Greszta *et al.* (1989) were able to map the sulphur contents across the area. All three species were considered good bioindicators of pollution, as sulphur contents increased towards point sources. In Norway spruce, the sulphur contents of second year needles ranged on average from 0.06 to 0.29%, rising to 0.36% near point sources. Contents of sulphur in beech leaves ranged from 0.15% to 0.27%. The plots at the highest elevations had the highest concentrations of sulphur, despite being the furthest away from sources of sulphur dioxide. This strongly suggests that enhanced deposition of sulphur was occurring at higher altitudes or that soil-derived sulphur was making an important contribution.

Malcolm and Garforth (1977) have recommended the use of the sulphur : nitrogen ratio in foliage to assess the possible impact of atmospheric pollution. They found that the sulphur : nitrogen ratios in foliage from relatively unpolluted sites in Scotland were about 0.034 in Sitka spruce and 0.028 in Scots pine, although these ratios varied slightly during the year. Sulphur contents of foliage of Sitka spruce in its natural (pristine) range were 0.12–0.14% dry weight (Malcolm and Garforth, 1977). In Scotland, in the mid–1970s, the sulphur contents ranged from 0.08% to 0.14% dry weight (Malcolm and Garforth, 1977). Considerably higher values were found in a Forestry Commission survey in the mid–1980s (see below).

Although sulphur contents provide a good index of the level of sulphur dioxide pollution (Materna, 1982; Tesar and Temmlová, 1982; Molski and Dmuchowski, 1984; Mankovská, 1988; Innes and Boswell, 1989b; Xiaozhu and Youbiao, 1989), the sulphur contents do not provide a good indication of the extent of defoliation (Materna, 1988; Innes and Boswell, 1989b). Generally, the degree of transparency bears little relation to the sulphur contents, suggesting that sulphur content *per se* is not a good index of tree condition. However, Landolt *et al.* (1984) found significant correlations between the extent of dieback in Norway spruce and the needle contents of sulphur, chlorine and fluorine. The available results suggest that the absolute concentrations of total sulphur do not provide a reliable index of stress, except at very high concentrations. Consequently, they are unlikely to be particularly useful in monitoring studies, except near sulphur sources.

Aluminium, manganese and calcium

The concentrations of aluminium, manganese and calcium in tree foliage appear to be closely linked and have been treated together. Kazda and Zvacek (1989) examined the foliar aluminium, manganese and calcium contents in Norway spruce in the Bohemian Forest of Upper Austria. Four- to six-year-old branches from the seventh whorl of four trees from each plot were sampled monthly during the growing season. There was no relationship between either the calcium or the aluminium contents of the needles and soil calcium : aluminium ratios. However, mean soil calcium levels were correlated with mean needle calcium contents. Similarly, manganese levels in the soil were correlated with foliar manganese contents, although there were major between-tree variations in the foliar manganese contents. Zöttl (1983, 1985a) found no relationship between the aluminium contents of the foliage and fine roots and the amount of damage in the stands and Krivan *et al.* (1986) also found no relationship between nutrient composition and the occurrence of damage symptoms (in the Black Forest). Gärtner (1985) recorded a strong correlation between soil and foliar manganese, as did Zvacek (1988).

Gärtner (1985) looked at manganese levels in foliage of Norway spruce in Hesse, Germany. Manganese contents were consistently negatively correlated with the degree of defoliation. Consequently, manganese contents were recommended as an indication of damage. However, this study has not been supported by others, and manganese contents are generally seen as a critical factor in the recent tree declines seen in Europe.

Other elements

The potassium : zinc ratio has been recommended by Liu and Hüttl (1991) as a useful indication of the likelihood of yellowing occurring. As the ratio rises, so does the risk of yellowing.

Decreases in phosphorus contents with increasing defoliation and increases in potassium have been reported by Heinsdorf *et al.* (1988). However, there are usually very poor correlations between the nutritional status of needles and the degree of defoliation (Liu and Hüttl, 1991), although needle loss in Norway spruce can be expected when calcium levels fall below 5.0 mg g^{-1} and iron levels exceed 80 μg g^{-1}.

Table 9.1 Deficiency levels for specific elements in Norway spruce

Element	Visual symptoms	Deficiency range for growth	Normal	Optimum
N	0.8–1.3	1.0–1.6	1.08–2.4	1.7–2.4
K	0.14–0.4	0.2–0.7	0.4–1.21	0.5–1.3
Ca	0.02–0.3	0.13–0.3	0.1–1.26	0.03–0.68
P	0.05–0.13			0.11–0.3
Mg	0.02–0.08	0.03–0.07	0.05–0.31	0.07–0.14
S	—	—	0.03	0.12–0.21
Zn	0.001–0.003			
Mn	0.0002–0.002			
Fe	0.0002–0.002			

From Ingestad, 1959; Zöttl and Mies, 1983; Fiedler, 1988; Landolt *et al.*, 1989; Materna, 1989; Liu and Hüttl, 1991.
Expressed as % dry weight unless otherwise stated.

Iron, however, is difficult to assess as a high proportion is normally present as dust on the needle surface and studies that fail to wash off the dust will include this source of contamination. Iron, as well as cadmium and lead, is important around point sources, but does not appear to be transported long distances (Mankovská, 1988).

Recognition of deficiency levels

Several authors have proposed deficiency levels. In Table 9.1, data are presented from a variety of sources. Cape *et al.* (1990) defined a number of risk thresholds, based on the presence of a mix of healthy and symptomatic trees at a particular site. They argued that these values could be used to calculate a risk index for the site, based on the following relationship:

$$\text{Risk index} = \frac{1}{8N} \sum_{i=1}^{n} (\text{number of thresholds exceeded})$$

where N is the number of trees assessed. The thresholds are given in Table 9.2.

The value of this index has still to be determined. Many studies have revealed the importance of foliar ratios and their inclusion in the risk index is useful. However, it is likely that other ratios may be just as important and much more work is required before the the risk index of Cape *et al.* (1990) can be universally accepted.

Table 9.2 Risk thresholds, as specified by Cape *et al.* (1990) for element concentrations and ratios in Norway spruce and Scots pine

Element or element ratio	Threshold
Calcium	<0.3
Magnesium	<0.05
Sulphur:Calcium	>0.4
Sulphur:Magnesium	>3
Nitrogen:Calcium	>5
Nitrogen:Magnesium	>30
Potassium:Calcium	>2
Potassium:Magnesium	>15

All values are for 1–2 year-old needles.

Analyses of foliar composition in Forestry Commission survey plots

In 1985, one tree from either in or close to each subplot then in existence was randomly selected for foliar analysis (Binns *et al.*, 1986). The sample comprised 34 Sitka spruce plots, 35 Norway spruce plots and 39 Scots pine plots. All stands were even-aged, with the ages of the plots ranging from 30 to 50 years. Foliage from trees less than 20 m high was sampled by squirrel pruners. Taller trees were either felled (in which the case the tree felled was not a tree in the plot) or climbed. One sample branch was cut from the seventh whorl from the top of each sample tree. The one-year-old and two-year-old internodes were selected from the main axis of the branch and the four first-year and four second-year shoots obtained from each plot were then bulked to form two separate samples.

Epicuticular wax was not removed from the needles before the analysis, although needles were washed in distilled water to remove surficial material. Needle tissue was digested using perchloric acid. Nitrogen and phosphorus were determined using continuous flow analysis. DC-plasma emission spectrophotometry was used for calcium, magnesium and potassium determinations. Inductively-coupled plasma emission spectrophotometry was used to determine the concentrations of sodium, iron, aluminium, titanium, zinc, copper, molybdenum, manganese, boron, sulphur, nickel, cadmium, chromium, lead and cobalt. Values for titanium, molybdenum, nickel, cadmium, chromium, lead and cobalt were all below the reliable detection limit of the equipment used.

As discussed above, there are many ways of presenting foliar analysis data. The simplest method is to present the absolute concentrations. These are shown in Tables 9.3 to 9.5. There are no particularly

Table 9.3 Elemental concentrations of Sitka spruce needles sampled in 1985

		Current year			Second summer		
Element	Unit	Min	Mean	Max	Min	Mean	Max
Nitrogen	%	0.94	1.34	1.76	1.08	1.37	1.92
Phosphorus	%	0.12	0.17	0.26	0.10	0.16	0.22
Potassium	%	0.73	1.00	1.27	0.66	0.93	1.25
Magnesium	%	0.06	0.09	0.15	0.04	0.08	0.14
Calcium	%	0.20	0.32	0.50	0.23	0.38	0.65
Chlorine	%	0.50	1.30	2.30	0.40	1.35	2.20
Aluminium	ppm	37	112	225	48	134	290
Sodium	ppm	23	65	151	27	101	390
Iron	ppm	44	93	205	47	91	172
Zinc	ppm	20	33	51	21	32	60
Copper	ppm	3.00	4.73	12.8	3.00	4.58	16.8
Boron	ppm	13	23	45	11	24	63
Sulphur	ppm	930	1466	2160	980	1554	2230
Manganese	ppm	260	807	2270	180	883	2050

obvious features about the ranges shown in these tables. As might expected from the lack of discoloration seen in the trees, magnesium deficiency was not apparent, although some trees were in the range where deficiency symptoms might be expected. Trees in Britain are generally supplied with adequate amounts of magnesium because of the atmospheric deposition of marine magnesium (Roberts *et al.*, 1989).

Table 9.4 Element contents of Norway spruce needles sampled in 1985

		Current year			Second summer		
Element	Unit	Min	Mean	Max	Min	Mean	Max
Nitrogen	%	1.07	1.27	1.74	1.01	1.33	1.64
Phosphorus	%	0.11	0.17	0.21	0.08	0.15	0.22
Potassium	%	0.29	0.74	1.02	0.49	0.74	1.11
Magnesium	%	0.06	0.08	0.14	0.04	0.08	0.11
Calcium	%	0.15	0.34	0.63	0.22	0.48	0.86
Chlorine	%	0.40	1.53	2.70	0.40	1.73	3.40
Aluminium	ppm	45	112	311	56	139	542
Sodium	ppm	24	65	153	26	90	189
Iron	ppm	45	75	184	40	84	169
Zinc	ppm	25	37	63	25	36	55
Copper	ppm	3.20	5.73	63.2	2.80	4.21	27.5
Boron	ppm	12	24	65	11	22	78
Sulphur	ppm	730	1108	1960	750	1208	1620
Manganese	ppm	140	827	3970	100	1080	5540

Table 9.5 Element contents of Scots pine needles sampled in 1985

		Current year			Second summer		
Element	Unit	Min	Mean	Max	Min	Mean	Max
Nitrogen	%	1.10	1.57	2.40	1.25	1.74	2.32
Phosphorus	%	0.11	0.16	0.20	0.09	0.14	0.26
Potassium	%	0.60	0.71	0.96	0.52	0.68	0.81
Magnesium	%	0.05	0.07	0.11	0.05	0.07	0.09
Calcium	%	0.08	0.21	0.58	0.17	0.30	0.50
Chlorine	%	0.30	0.84	1.70	0.60	1.30	2.10
Aluminium	ppm	74	213	464	113	309	532
Sodium	ppm	34	139	593	76	379	1152
Iron	ppm	44	102	308	77	162	373
Zinc	ppm	29	40	53	27	41	68
Copper	ppm	3.20	5.16	9.90	3.00	4.39	11.5
Boron	ppm	14	23	44	14	24	37
Sulphur	ppm	850	1236	1950	970	1390	1950
Manganese	ppm	120	362	1260	150	590	1740

Calcium contents were well above levels at which deficiency symptoms would be expected.

Nitrogen levels were generally in the normal range for Sitka spruce and Norway fir. However, very high levels of nitrogen were seen in some Scots pine plots and these may reflect deposition of atmospheric nitrogen. Nitrogen levels were correlated with sulphur contents, and there is a strong possibility that the high concentrations were the result of the co-deposition of sulphur and nitrogen as ammonium sulphate. Analysis of the spatial variation of the sulphur contents revealed that the needle concentrations were related to atmospheric concentrations of sulphur dioxide (Innes and Boswell, 1989b) rather than with the wet or dry deposition of sulphur and this area requires further investigation.

The relationships between the different nutrient levels have been examined using principal components analysis (Table 9.6). Analyses were undertaken for the first-year and second-year needles separately and then for both years combined. In general, the amount of variance accounted for by the first two components was low, with the highest level being 47% (for first-year Scots pine needles). In all three species, sulphur contents had an important influence on the first component. Other important elements included potassium and zinc in Sitka spruce, aluminium, sodium and iron in Norway spruce and iron and manganese in Scots pine.

Table 9.6 Results of a principal components analysis of the needle element contents

	Sitka spruce			Norway spruce			Scots pine		
	1st	2nd	both	1st	2nd	both	1st	2nd	both
First component									
N	–0.21	0.28	–0.27	–0.27	–0.21	–0.26	–0.28	–0.37	–0.29
P	0.00	0.15	0.04	0.04	0.08	0.17	–0.06	–0.21	0.05
K	–0.31	0.31	0.27	–0.13	–0.11	–0.09	–0.01	–0.26	0.03
Mg	–0.20	0.11	0.08	–0.22	0.00	–0.04	0.01	0.11	0.16
Ca	–0.33	0.17	0.23	–0.29	–0.22	–0.32	–0.27	–0.27	–0.33
Cl	–0.36	0.25	0.30	0.05	–0.32	–0.22	–0.30	–0.13	–0.34
Al	–0.22	0.25	0.27	–0.41	–0.24	–0.28	–0.35	–0.18	–0.34
Na	–0.22	0.34	0.33	–0.33	–0.39	–0.41	–0.33	0.01	–0.28
Fe	–0.11	0.37	0.29	–0.36	–0.45	–0.39	–0.33	–0.37	–0.36
Zn	–0.34	0.30	0.29	–0.18	–0.17	–0.18	–0.16	0.03	–0.08
Cu	–0.24	0.23	0.25	0.13	–0.17	0.03	–0.24	–0.30	–0.09
B	–0.24	0.27	0.29	–0.33	–0.31	–0.26	–0.25	–0.25	–0.25
S	–0.41	0.42	0.44	–0.43	–0.45	–0.45	–0.36	–0.46	–0.35
Mn	–0.28	0.04	0.14	–0.17	–0.15	–0.22	–0.37	–0.34	–0.37
VAF	25	26	26	24	24	23	35	23	31
Second component									
N	–0.35	–0.14	–0.25	–0.31	–0.35	–0.33	–0.34	0.24	–0.29
P	0.18	–0.47	0.21	–0.32	–0.52	–0.51	–0.45	0.26	–0.48
K	0.13	–0.07	0.25	–0.17	–0.26	–0.30	–0.01	0.23	–0.40
Mg	0.34	–0.55	0.45	0.23	–0.27	–0.23	–0.19	0.23	–0.32
Ca	0.31	0.22	0.24	0.37	0.28	0.28	–0.02	–0.08	–0.04
Cl	0.11	0.13	0.28	0.24	0.27	0.22	–0.22	–0.49	0.23
Al	–0.44	0.18	–0.44	–0.24	–0.17	–0.14	0.38	–0.06	0.07
Na	–0.18	0.09	–0.11	0.17	0.11	0.12	–0.17	–0.51	0.29
Fe	–0.37	0.14	–0.34	–0.26	–0.02	–0.16	0.27	–0.11	–0.02
Zn	0.29	–0.12	0.39	0.31	–0.05	–0.01	0.28	–0.02	–0.09
Cu	–0.13	0.12	–0.03	0.12	–0.22	–0.01	0.42	0.09	–0.42
B	–0.25	–0.18	–0.05	–0.05	–0.10	–0.22	–0.09	–0.40	–0.07
S	–0.17	0.07	–0.09	–0.07	–0.15	–0.20	–0.29	0.21	–0.31
Mn	0.21	0.51	0.13	0.51	0.44	0.47	0.01	–0.19	0.00
VAF	19	15	15	16	17	15	12	16	14

Conclusions

Considering the importance of nutrition in several types of forest decline, there is surprisingly little information available about the nutrition of large trees. Most studies have concentrated on seedlings or young trees but, as with other experiments, it is uncertain whether such results can be applied to mature forests. There have been several large-scale surveys of nutritional status undertaken (e.g. in Switzerland, Austria and Baden-Württemberg, Germany). However, the results of these are as difficult to interpret as the British results.

It seems fairly clear that the absolute concentration of a particular element is sometimes of little value as an index of nutrition. Instead, the relationships between different elements should be examined. This will be particularly important as elevated nitrogen concentrations increase in importance. Numerous different ratios are available and there is an urgent need for a complete review of all the available material. Until this is done, considerable uncertainty will surround the interpretation of the nutritional status of mature forests.

10 Forest Decline and Increment in Europe

Descriptions of decline often neglect the most critical aspect, which is the decreased radial growth of affected trees.
Treshow and Anderson, 1989

Introduction

One of the most important measures of forest condition is annual volume increment. Annual volume increment determines the economic value of the stand and also provides a measure of its vigour. Changes in annual volume increment provide an indication of changes in the condition of the stand. The greatest problem with the assessment of changes in increment associated with forest decline is how to find out what growth would have been in the absence of pollution. Considerable confusion has arisen over this issue because of the tendency to attribute any transparency in the crown to damage by atmospheric pollution. While it is now usually admitted that up to 25% transparency does not necessarily indicate any problem, even this figure is probably too low. In practice, it is very difficult to attribute a cause to transparency. Even when it can be done, Athari and Kramer (1983) pointed out early on in the debate that simple comparisons between transparent and asymptomatic trees are invalid because of the number of potentially confounding variables.

Studies around point sources

Various approaches have been used to try and identify the impact of atmospheric pollution on forest growth. A traditional method is to compare the growth of trees near a pollution source with the growth of trees further away. For example, Athari (1981) compared the growth of Norway spruce from 17 stands in an area affected by air pollution with the growth of trees in three unpolluted stands. Similarly, Serre-Bachet (1987) looked at the growth of trees along a transect extending away from a known source of fluoride pollution in Savoie, France. Bunce (1984) investigated the effect of a fluoride-producing aluminium smelter on the growth of western hemlock (*Tsuga heterophylla* Sarg.) and Pacific silver fir in British Columbia, Canada. Growth was compared close to the smelter and in an area virtually free of fluorides. He established the effect of the smelter on tree growth and also identified an increase in growth after 1973 which could be linked to a decrease in fluoride emissions from the smelter. Such studies can be enhanced by examining growth along a distance gradient from the smelter (e.g. Fox *et al.*, 1986), although this assumes that the environmental conditions remain relatively constant over the length of the gradient (Schweingruber, 1987).

Regional patterns of forest increment

An alternative to gradient studies is to map regional patterns of increment. A regional tree ring inventory was conducted in southern Norway in the early 1970s in relation to acidic precipitation (Strand, 1980a, b). Annual rings from Norway spruce and Scots pine from a total of 6150 systematically-distributed plots were examined for the period 1911–72. All trees

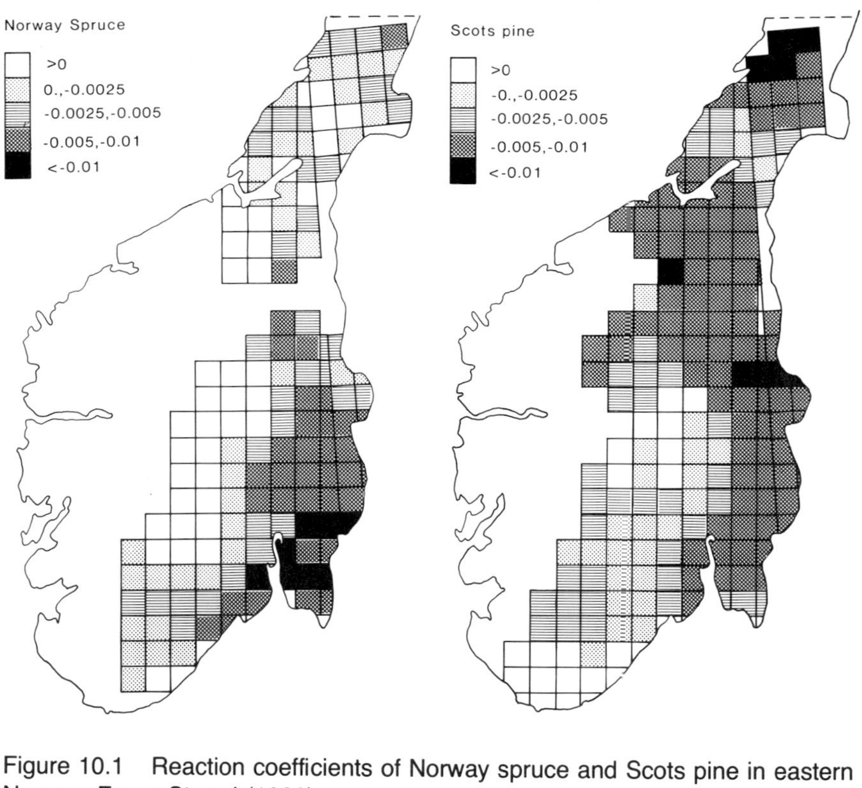

Figure 10.1 Reaction coefficients of Norway spruce and Scots pine in eastern Norway. From: Strand (1980).

used in the analysis were more than seventy years old. The series were averaged into clusters of 1 km × 1 km, resulting in a reduction to 2814 series. Using linear regression techniques, a reaction coefficient for each site was obtained. As these showed a great deal of site-to-site variation, they were averaged over 30 km squares (Fig. 10.1). A long-term decline in increment over the period was found, but this was attributed to ageing effects. Several indicator years were present, with 1928 being a poor year for both species at high altitudes, 1960–1 being years of poor growth in the south and 1965–70 being a period of definite decline in some areas, particularly those at low altitudes. The pattern of growth showed some marked similarities with the pattern of rainfall but, significantly, Strand was unable to attribute any of the growth trends to acidic deposition.

Nichols (1981) examined the growth patterns of several forest species in England and Wales. The analysis was based on records of estimated yields of individual stands throughout the area. As might be expected from the environmental variation in England and Wales, marked spatial patterns in the yield classes were identified. The areas with the highest growth rates for each species varied among species, reflecting the environmental requirements of those species. Although the author was not concerned with the possible effects of pollution, his maps reveal no indication of a correlation between patterns of growth and patterns of air pollution in Great Britain. Instead, climate and soil type appear to be the main factors affecting yield.

Traditional dendroecological approaches

Relatively few studies have used 'traditional' dendroecological methods to assess growth declines in Europe. Gradient studies seem to work well in cases where pollution is derived from an identifiable point source, but problems arise when regional-scale effects are being examined, particularly if there are no clear gradients. This is the normal situation in relation to forest decline studies. A variety of solutions have been proposed. Abetz (1985) suggested the use of reference trees, defined as theoretical trees whose growth has been determined by yield class estimates. For a given yield class, the diameter of the stem can be predicted at specific age intervals. The effect of age can then be eliminated by expressing the mean radial increment of the trees in question as a percentage of the reference tree. This approach was adopted by Eichkorn (1986) for analysing Norway spruce and by Gerecke (1986, 1990) for silver fir tree ring series from Baden-Württemberg (Fig. 10.2).

The most widely used technique is to compare actual growth versus predicted growth. Dong and Kramer (1986) attempted to assess the effects of pollution on Norway spruce in northern Germany (Type 5 decline) by comparing growth of individual trees over ten year periods. They claimed to have removed the ageing effect and then found that growth over the

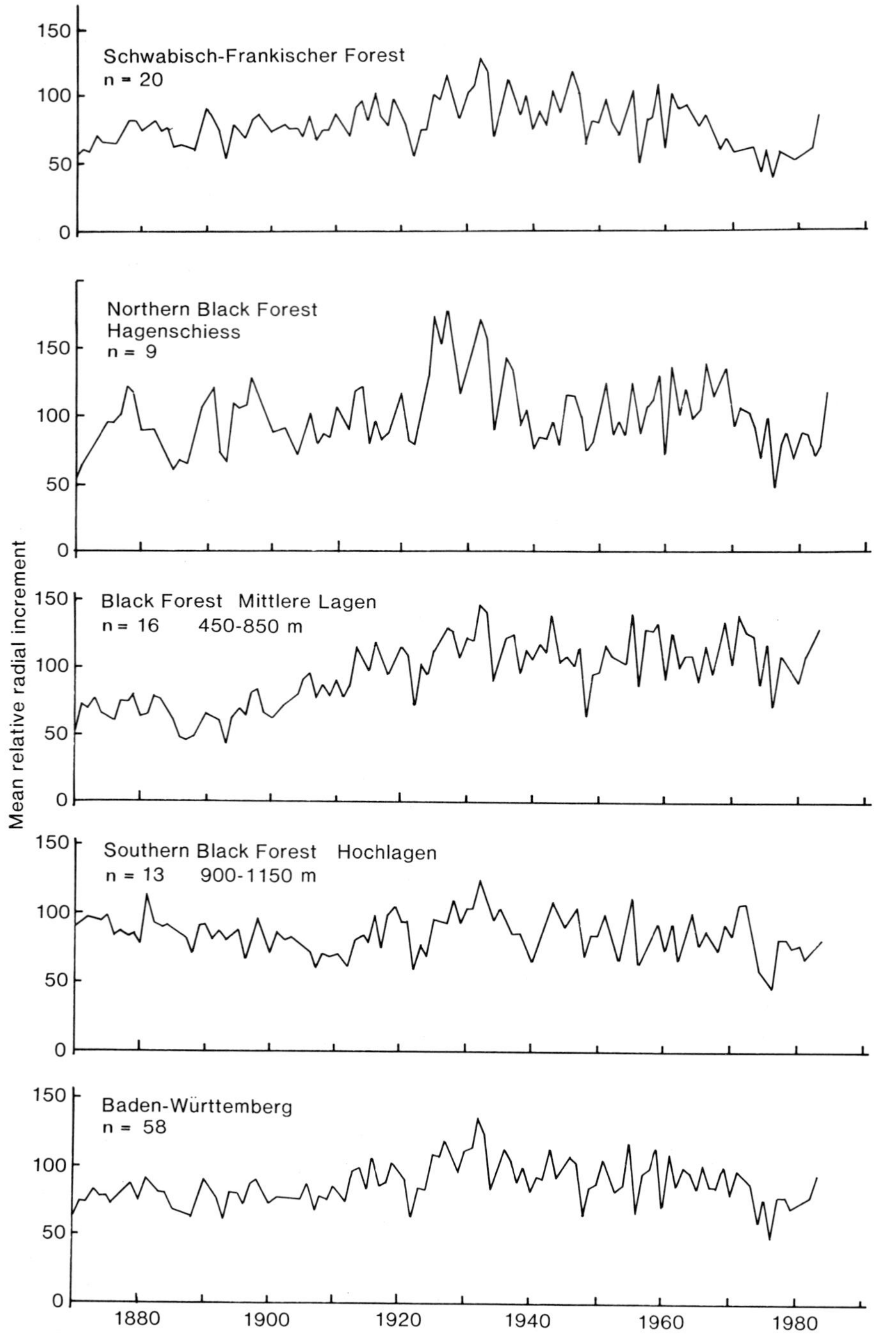

Figure 10.2 Mean relative radial increment of predominant silver firs for several locations in Baden-Württemberg, Germany. From: Gerecke (1986).

most recent 10-year period was markedly less than the growth during the previous 10-year period.

A similar approach was adopted by Eckstein (1985) and Eckstein *et al.* (1984), who examined the growth of beech in the Vogelsberg of Germany 'before' and 'after' the onset of pollution. In a sample of 386 trees (sampled in 1981), the effect of the 1976 drought was very marked, with many missing rings present after 1976. Eckstein argued that there had been several dry summers as severe as that in 1976 during the 20th century, yet the effects of the 1976 drought were more marked than on previous occasions. However, some trees showed a marked recovery following the drought and subsequently grew better than at any stage in the past. This was probably due to the removal of competition by drought-induced mortality, a phenomenon that may be more widespread than realised. Eckstein and Krause (1989) used the same argument in a study of Norway spruce around Hamburg. The apparently unprecedented effects of the 1976 drought suggested that another factor had made the trees more susceptible to drought. A basic problem with this approach is that the true date of the onset of pollution is unknown, unlike cases where point source pollution has been investigated (e.g. Liese *et al.*, 1975; Kienast, 1982). In addition, it is assumed that the growth response of an individual tree to climate is constant through time. This is unlikely given the physiological changes in a tree as it ages, and changes in the growth responses of individual trees to climate through time have been documented by several authors.

The approach used by Eckstein and co-workers, described above, could be interpreted as a fairly standard approach, depending on the methods used to remove the effects of age from the series. Heliotis *et al.* (1988) used a flexible cubic spline digital filter to remove the age trend in Greek fir from the Parnis National Park in Greece. They compared the resulting series with series obtained from a control site 3 km to the north. At the Mont Parnés site, growth declines first occurred in 1962, although the interpretation of this is very questionable (Fig. 10.3). No similar decline was recorded at the control site, but growth there has declined since 1970. The study illustrated several problems, including the removal of potentially useful information during de-trending and the problems of making comparisons between sites when regional pollution may be involved.

Traditional dendroecological approaches appear to offer considerable potential in the study of growth changes related to forest declines (Innes, 1991c; Innes and Cook, 1989). However, as discussed below, some of the techniques used in such approaches need to be carefully evaluated. For example, the use of mean chronologies is inappropriate when only some of the trees are declining. More emphasis needs to be placed on individual tree series in such situations. Also, the methods used to remove ageing effects may well remove some or all of any decline attributable to pollution and better techniques are required for detrending.

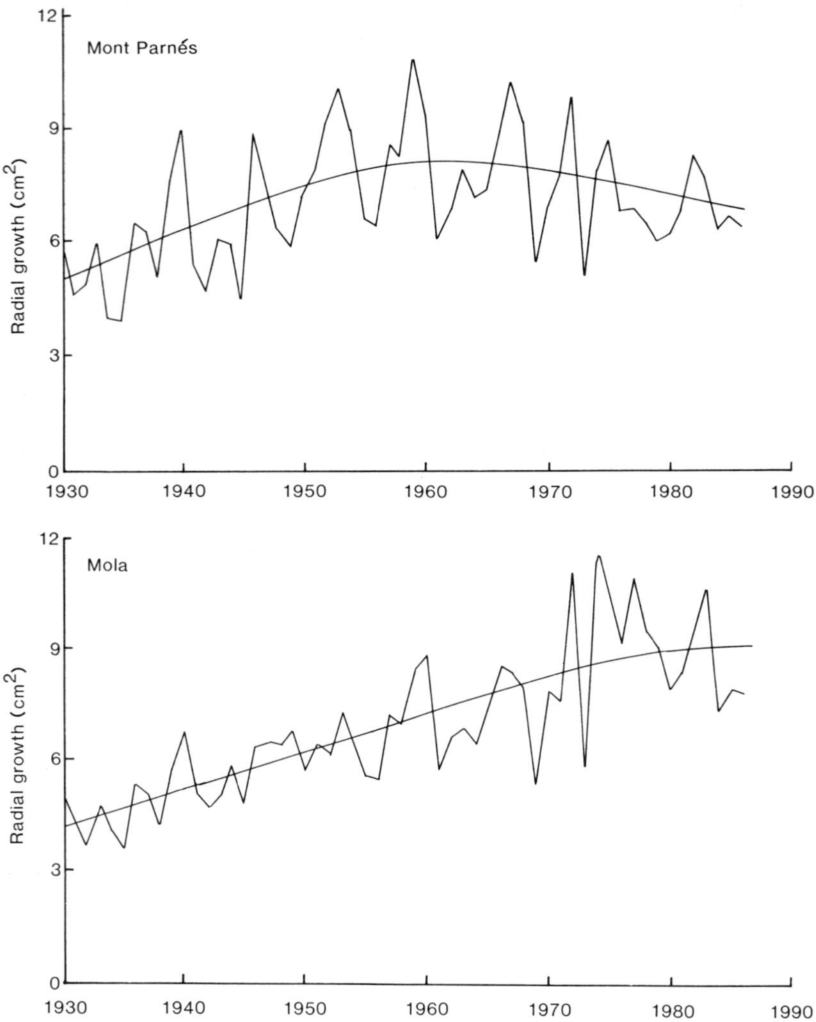

Figure 10.3 Annual growth of Greek fir at Mont Parnés and Mola, Greece. From: Heliotis *et al.* (1988).

Growth decreases

A common assumption is that the reports of forest declines throughout Europe have been accompanied by widespread reductions in forest growth. Such an assumption rests on the validity of the relationship between transparency and increment, a subject discussed below. However, since past increment can be determined by tree ring analysis whereas few techniques exist for establishing past transparency levels, growth analyses might provide evidence of a deterioration in the condition of European forests.

Norway spruce

Several studies have looked for growth decreases associated with reports of forest decline. Greve *et al.* (1986) documented decreases in the growth of Norway spruce in north-east Bavaria extending back to the 1940s. The decreases were identified by comparing actual against predicted standardised growth rates (Fig. 10.4).

Worbes (1989) suggested that increment reductions in Norway spruce were widespread in the Bruchberg and Sonnenberger areas of the Harz Mountains, Germany. He found very little relationship between precipitation and growth, but temperature was an important variable affecting growth. Of the trees sampled, between 50% and 90% showed a reduction in growth after 1960. While some of these declines may be related to release-suppression sequences, the number of growth reductions appears to have increased in recent years, suggesting a link with the reports of declines.

Increment analysis has revealed that the recent phase of Norway spruce declines began some time ago. The onset of the decline seems to have varied between areas (Aszmutat and Knigge, 1987; Athari, 1983; Greve *et al.*, 1986; Murri and Schlaepfer, 1986; Schneider *et al.*, 1988; Schweingruber *et al.*, 1983). Most of these authors agree that the initial divergence in growth rates occurred following an external forcing event, such as a drought or prolonged dry period. The 1976 drought is documented as severely reducing increment (Eckstein and Krause, 1989; Eichkorn, 1986; Spiecker, 1987) and differences in annual increment between symptomatic and vigorous trees appears to have become particularly marked after the drought (Evers *et al.*, 1986). Despite the apparent decline, increment analyses suggest that growth of Norway spruce in the Black Forest is above normal (Eichkorn, 1986).

These results suggest that in the trees surveyed, growth reductions have occurred. However, all the studies were of purposefully selected trees. The studies did not attempt to determine growth changes systematically. Instead, as in many American studies, assessments were concentrated on trees in areas where growth decreases would have been expected.

Silver fir

In the Beskedy Mountains of southern Poland, annual ring widths of silver fir measured from 1925 to 1980 revealed that the average width of rings was 50% less in the period 1971–80 than in the period 1941–50 in relatively young stands, and 40% less in older stands. A steady decline in radial increment has occurred since about 1960 (Jaworski and Skrzyszewski, 1987), although this study was probably completed too early to pick up the widespread recovery of silver fir increment that was seen in many parts of Europe in the mid- to late-1980s. A further problem with this study is the role of ageing. Trees would be expected to have reduced increment through time, and the effects of this natural reduction in growth were not taken into account.

Using reference trees, Eichkorn (1986) found no evidence of long-term reductions in the growth of silver fir in Germany, although droughts had a marked influence on annual increment. The strongest decrease in the

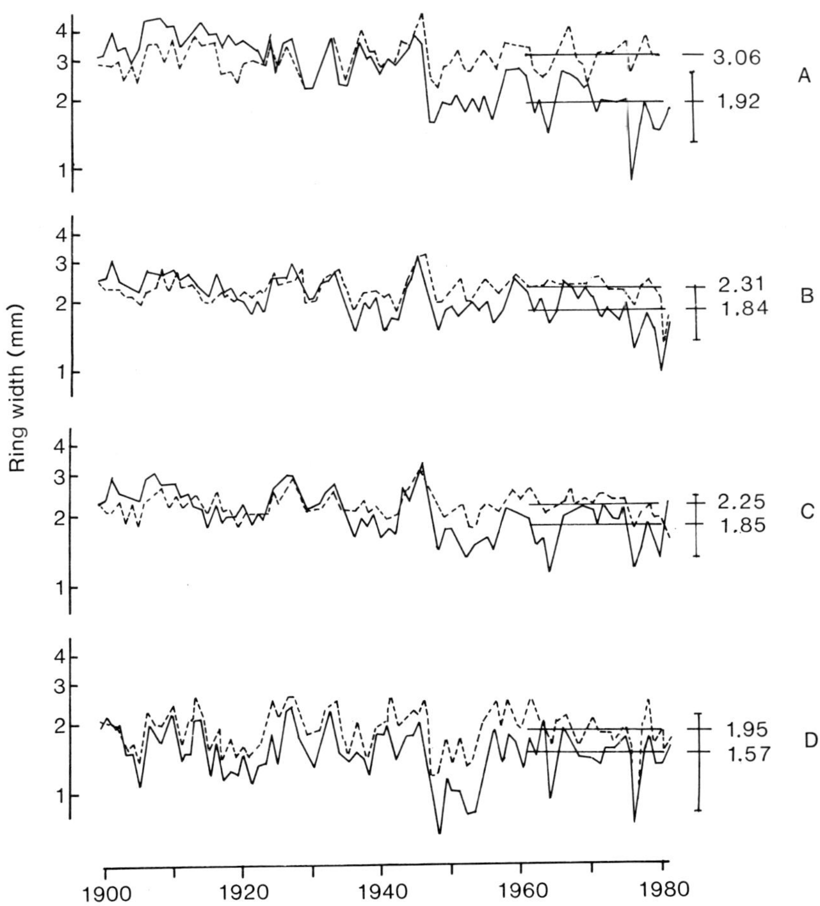

Figure 10.4 Growth of Norway spruce at several sites in north-east Bavaria, Germany. The solid lines show actual growth and the dashed lines show predicted growth. From: Greve *et al.* (1986).

growth of symptomatic trees occurred in 1976. Trees currently with more than 25% needle loss started to have reduced increment in the mid–1960s. Similar results were obtained by Bräker (1990, 1991) for Norway spruce and silver fir in Switzerland. Gerecke (1986) concluded that drought, management and site influences were all extremely important in determining the growth of individual trees relative to reference trees. A later study by Gerecke (1990), again using reference trees as a standard, showed that the anomalous increment could largely be related to differences in the growing space of individual trees. Trees with reduced increment since the 1960s were mostly those with short crowns. Short-term extremes were caused by both droughts and frosts, the latter displacing radial increment upwards on the stem. Towards the end of the 1980s, a recovery in the annual increment was identified. These results support those obtained in the Black Forest, Germany, by Spiecker (1986). He found that both Norway spruce and silver fir with longer crowns had faster diameter and height growth and a better performance during dry periods.

In Bavaria, silver fir from four sites showed a striking growth depression during the period 1965–82 (Eckstein and Saß, 1989). Since then, a rapid recovery has occurred, matching the pattern seen in Baden-Württemberg (Konnert and Mettendorf, 1990), the Vosges (Becker, 1989) and Switzerland (Sell *et al.*, 1988; Schmid-Haas, 1989a). The annual coefficient of variation increased from about 1925, peaked in 1976 and 1981 and subsequently decreased (Fig. 10.5). A similar trend of increasing climatic sensitivity has been noted in Norway spruce and silver fir in Germany (Widmann and Avemark, 1988). However, according to Eckstein and Saß (1989), the growth decline was independent of climate.

Keller and Imhof (1987) described growth declines in symptomatic

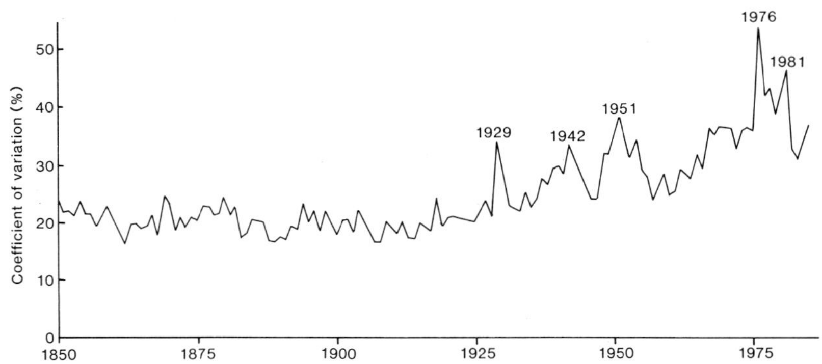

Figure 10.5 Annual coefficient of variation of the tree ring width of silver fir trees from Bavaria. From: Eckstein and Saß (1988).

silver fir in Switzerland starting in the 1960s. In Baden-Württemberg, where high-altitude silver fir is known to have become increasingly responsive to dry and warm summers (Mitscherlich, 1983), Kenk (1983) and Kenk *et al.* (1984b) found that the increment of symptomatic trees was about 33% of that of asymptomatic trees in the late 1970s and early 1980s. Mostly, increment reductions began in the mid 1950s and differentiation occurred after a year of particularly low increment in 1955–6. This probably reflects the sensitivity of silver fir to climatic extremes (Eckstein *et al.*, 1983). The extent of the volume loss was dependent on the site quality, the proportion of symptomatic trees and the degree of crown transparency. Fertilization had an initial effect, but growth quickly reverted to its former pattern. However, Kenk *et al.* (1983) found that fertilization in 1956 and 1957, with a partial repetition in 1966, resulted in increased growth of silver fir. Symptomatic and asymptomatic Swiss silver fir showed a similar pattern of growth until the early 1950s, since when symptomatic trees have had reduced growth (Lenz *et al.*, 1988). The dry period of 1941–53 and the severe frost of February 1956 are believed to have triggered the differentiation. Silver fir is particularly sensitive to warm summers and cold winters in some areas. In Baden-Württemberg, cold temperatures and the frost in 1956 greatly affected the growth of trees (Kenk, 1990).

In the former Yugoslavia, symptomatic silver fir showed a reduction in growth from 1950 onwards, with missing rings being present in many of the most transparent trees (Torelli *et al.* 1986). The symptomatic trees tended to show an upward movement of the zone of maximum increment.

Beech

Beech trees in the Swabian Alb, Germany, showed heavy depressions in annual increment associated with the drought years of 1976 and 1983 (Fig. 10.6) (Abetz, 1988) and recovery was slower on drier soils. In the study by Abetz, young trees at the time of assessment were growing better than older ones when they were young. Given the climatic sensitivity of beech, systematic studies of its growth would undoubtedly produce extremely useful information.

Abrupt growth changes

An alternative approach to searching for a decline is to look for evidence of abrupt growth changes (either decreases or increases). Avemark and Schöpfer (1988) described the use of abrupt growth changes to assess increment in Norway spruce and silver fir in Germany. Their definition of an abrupt change was based on that of Schweingruber (1985, 1986) and Schweingruber *et al.* (1986) and was defined as occurring when radial

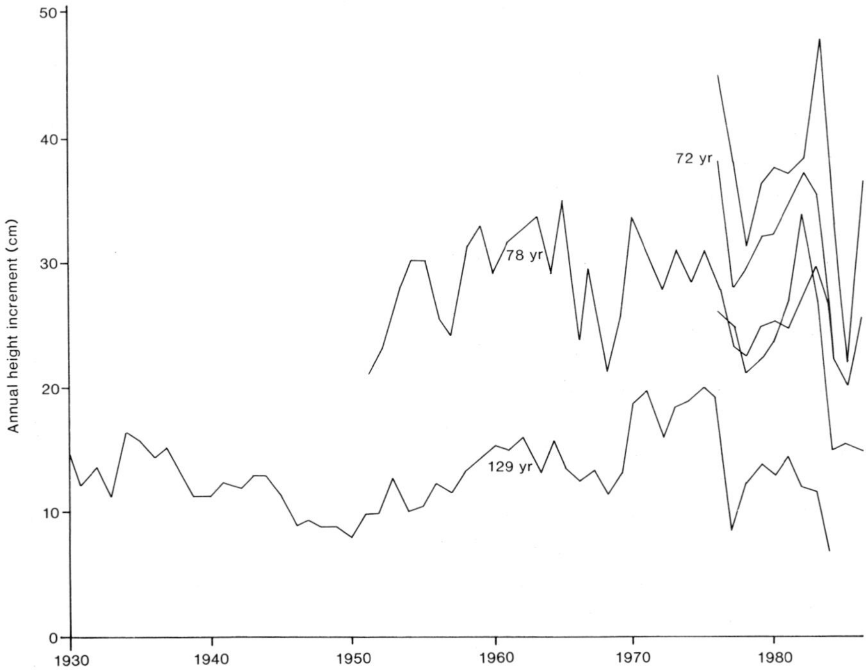

Figure 10.6 Mean annual increment of 42 old beech trees and 37 young beech trees. From: Abetz (1988).

increment was reduced over a period of at least four years by at least 40% when compared with the previous four years. A similar definition was adopted by Schweingruber *et al.* (1990). The percentage of trees reacting to an event causing a growth change was considered a measure of the effectiveness of that event. The sample analysed by Avemark and Schöpfer (1988) involved 1591 Norway spruce and 369 silver fir. In Norway spruce, a pattern of reductions followed by recovery phases was found. In the 1940s, growth reductions occurred, followed by recovery in the 1950s, reductions up to 1965 and recovery to 1970. Silver fir showed a similar pattern, but the swings from reduction to recovery became more marked after 1960. There was no evidence of abrupt changes in the first 40–50 years of the trees' life. Extreme years for Norway spruce were noted in 1921, 1929, 1934, 1948 and 1976. In silver fir, the most extreme years were 1922, 1934, 1948, 1956, 1960 and 1976. The years 1921–2, 1934, 1948 and 1976 appear to have been particularly important, as both silver fir and Norway spruce were affected. However, several years appear to have been important for one species and not the other, a pattern that has been found for other species elsewhere (e.g. Bauch, 1983; Barzut, 1984; Bridge and

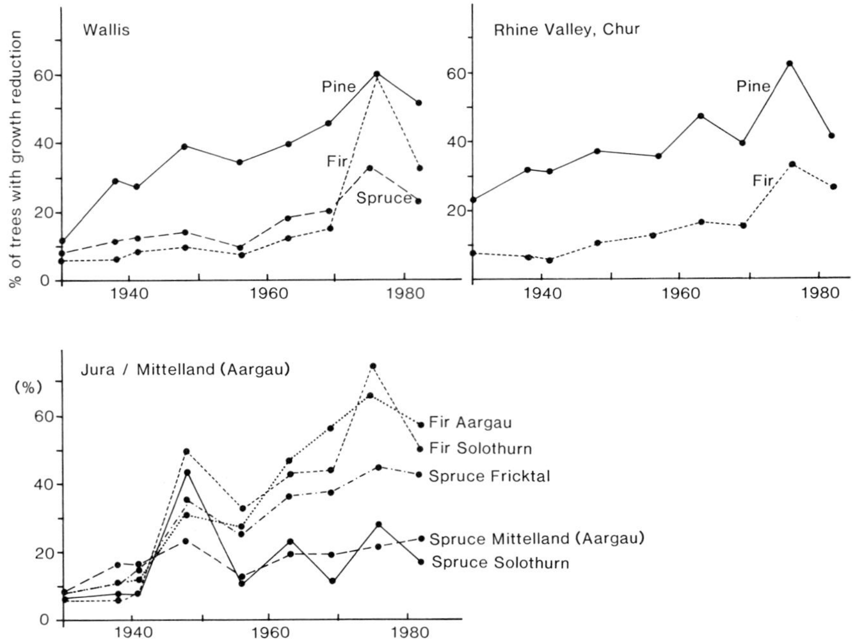

Figure 10.7 Percentage of trees with growth reductions in different regions of Switzerland. Pine, fir and spruce from Wallis, the Rhine Valley near Chur, and the cantons of Solothurn and Aargau. From: Schweingruber (1986).

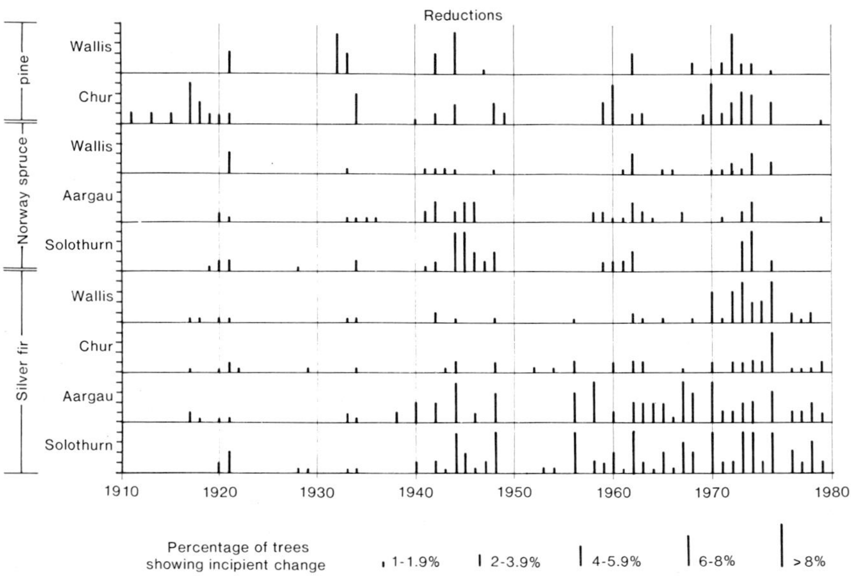

Figure 10.8 Onset of persistent growth changes (reductions) in particular species and regions; only growth reductions evident in 1–8% of the trees are shown. From: Schweingruber (1986).

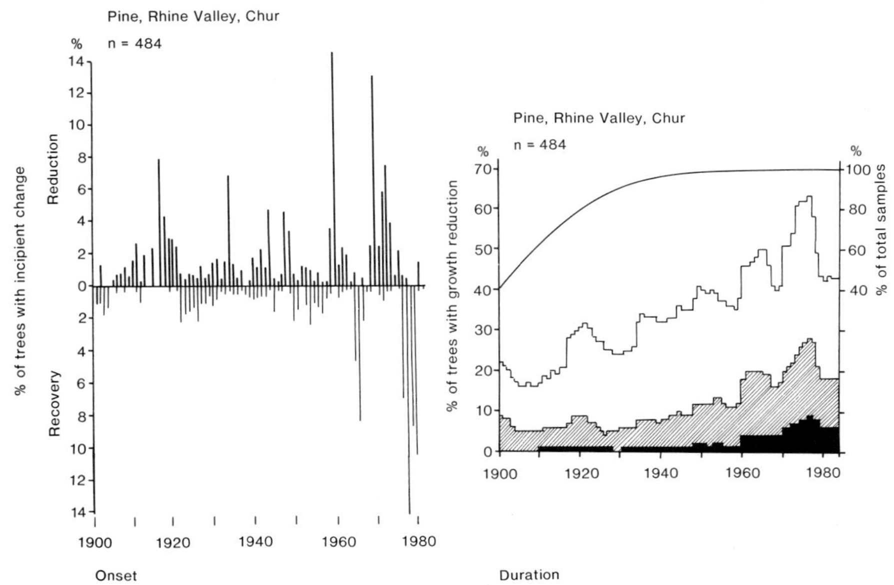

Figure 10.9 Dates of onset of abrupt changes in growth (left) and percentage of trees with growth reduction (right) in pines from the upper Rhine Valley near Chur, Switzerland. Left: upward lines indicate the percentage of trees with incipient growth reduction and downward lines indicate the percentage of trees with incipient growth recovery. Right: blank: 40–55% reduction; hatched: 45–70% reduction; black: >70% reduction. The curve at the top shows the percentage of the total number of samples available for each interval. From: Schweingruber (1986).

Ni Bhreartuin, 1988). Consequently, it seems that data from different species should only be compared with care.

In Switzerland, a detailed study of abrupt grow'h changes in Scots pine, silver fir and Norway spruce has been made in the Rhône Valley (Kontic *et al.*, 1986; Schweingruber et al., 1985, 1986). In 1976, 61% of the pine samples, 58% of the fir and 33% of the spruce showed an abrupt reduction in growth, suggesting the importance of the drought in that year. In an earlier study, Schweingruber *et al.* (1983) examined the growth of trees in Valais (Wallis), Emmental, the northern Jura and the Swiss Central Plateau area (Figs 10.7, 10.8 and 10.9). They found that symptomatic trees generally showed abrupt reductions in increment, amounting in most cases to more than 50% of the previous increment. There was poor agreement between the transparency category and increment reduction in individual trees, although this was not examined statistically. The date of the initial growth reduction varied markedly between trees. The reductions appeared to be unprecedented in the tree

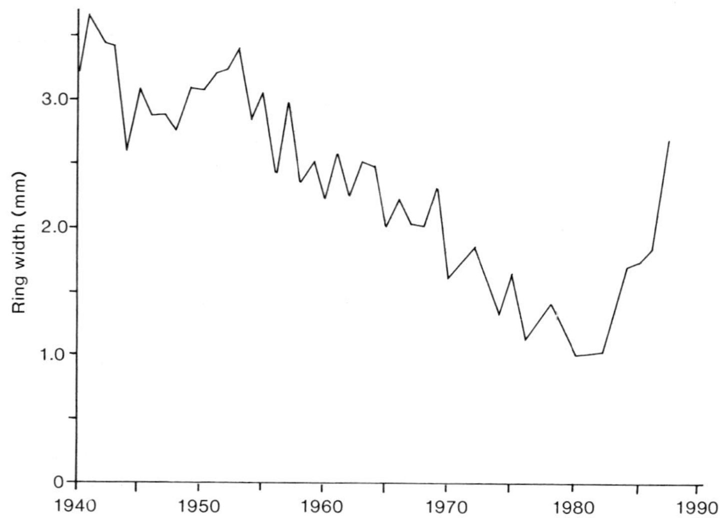

Figure 10.10 Pattern of mean diameter increment at breast height of 49 silver fir trees at Bremgarten, Switzerland. From: Schmid-Haas (1989).

ring record. However, recovery was present in all of the firs and most of the pines in the Valais region. Using the same approach, Schmid-Haas (1989b) looked at growth changes of silver fir in the Reuss Valley, south-west of Zürich, Switzerland. A local source of pollution in Bremgarten was responsible for the depression of growth in surrounding silver fir stands. During the 1980s, substantial growth increases were recorded, reversing the apparently steady decline in the growth rate of the trees (Fig. 10.10).

The analysis of abrupt growth changes offers considerable potential, much of which is still to be exploited. There is a considerable volume of literature on the statistical analysis of time series and much of this has yet to filter through to tree ring studies. Analytical techniques are becoming increasingly sophisticated and these, together with the emphasis on changes in individual tree ring series, mean that the analysis of growth changes is likely to be one of the most important areas of future research involving forest declines.

Growth increases

Changes in growth are not restricted to growth reductions. Growth increases have been reported from a variety of areas and from several different species. These are not confined to high altitude or high latitude areas, although the increases appear to have been particularly marked in these areas (Innes, 1991a). Growth increases can be defined as both absolute increases in growth (which are the easiest to identify) and increased growth relative to what would be expected (i.e. slower rates of decline). As the radial growth of most trees declines through time (until a minimum point is reached), the latter may be particularly important.

In Finland, Arovaara *et al.* (1984) examined inventory data and found an increase in annual growth. They were unable to attribute the increase to stand dynamics. Between the 3rd and 7th Finnish forest inventories, a decline in growth was predicted from changes in stand characteristics. However, over this period, an increase in growth of about 16% occurred. The increase in the forest resource is now well-established in Finland (Mielikäinen and Kauppi, 1991) and is matched by similar patterns in eastern North America (see Chapter 2).

Briffa (1991b) examined the growth of conifers from sites throughout Europe. A variety of species were investigated, depending on the region. Analysis of the series revealed that since the 1950s, ring widths have generally increased, particularly in central and southern Europe. When the data were corrected for age and expressed as basal area increment, unprecedented increases in growth were found during the last 100 years. In spruce from central Europe, the increase mainly occurred in the second half of the 19th century. In northern Europe, the increased growth of spruce and pine occurred about 1900. These results are important as they clearly show that on a European scale, growth reductions have not occurred. Instead, the evidence from long-lived trees growing in essentially natural conditions points to a growth increase during the 20th century.

Norway spruce

Studies on Norway spruce in Baden-Württemberg (Eichkorn, 1986; Kenk, 1990) revealed that growth in the mid and late 1980s was greater than would be predicted from classical yield models. Similarly, in North Rhine-Westphalia, growth of Norway spruce was higher than predicted from reference curves (Spelsberg, 1988). Although defoliation of individual trees resulted in reduced growth, Hocevar (1990) reported that the stand-level growth of montane Norway spruce on the Pokljuka-Jelovica plateau of Slovenia was better in the late 1980s than ever before. These results agree with those of Briffa (1991b), who found increased growth of Norway spruce in central and northern Europe.

Silver fir

In central Europe, increases in the growth of silver fir occurred on poor and intermediate sites during the 1980s (Becker *et al.*, 1990a, b). The

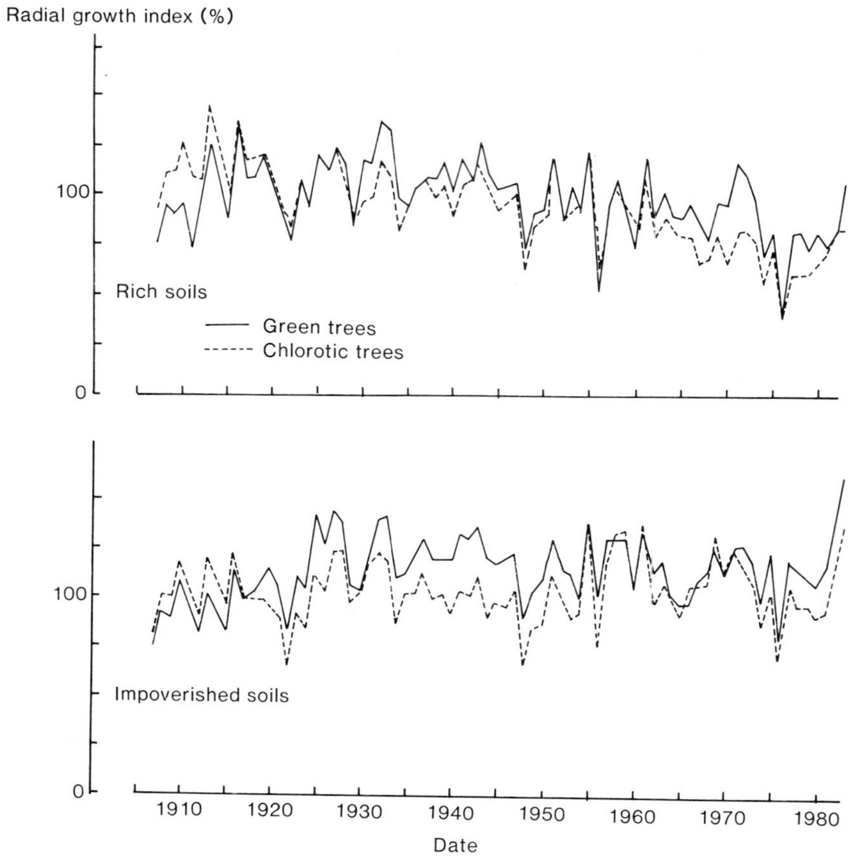

Figure 10.11 Growth indices of silver fir in the Vosges growing on rich (above) and poor (below) soils. In each case, growth of normal and chlorotic trees has been separated. From: Becker *et al.* (1989).

increases were much more marked than on more favourable sites (Fig. 10.11), suggesting that growth is more sensitive on the more acidic sites (as might be expected). The increases in the growth of silver fir were anticipated in view of the recovery of this species in the mid- and late-1980s (see Chapter 6).

Scots pine

Pretzsch (1989) has reported a major increase in the increment of Scots pine in north-east Bavaria that has occurred since the 1950s, whatever the degree of defoliation. The reasons for the increase in growth are unknown, there are appear to be no similar records from elsewhere.

However, Briffa (1991b) recorded increased growth of pines in the series that he investigated, suggesting that growth increases may have been quite widespread.

Crown transparency and growth

A major area of concern has been the relationship between crown transparency and growth. It has been widely assumed that increases in transparency result in reduced growth. However, this relationship is not always clear. For example, Hartmann *et al.* (1987) argued that the relationship between transparency and increment was more variable for Norway spruce than it was for Scots pine, and Frühwald *et al.* (1984b), Bauch *et al.* (1985b), Göttsche-Kuhn (1988), Sterba and Eckmüller (1988) and Utschig (1989) all found that depression of radial growth was a poor indicator of high transparency levels. Conversely, Schmid-Haas (1991a) argued that the recent radial increment of beech, Norway spruce and silver fir was related to the transparency of the trees.

The differences may be related to the timing of the defoliation event. When assessing crown transparency, it is usually impossible to determine when, if ever, the transparency changed to its current level. In trees that have high levels of transparency because of, for example, their growth form, a poor relationship with increment would be expected. Trees that have actually lost foliage might, however, be expected to show a growth reduction soon afterwards. The date of the increment reduction would then depend on the date of defoliation.

Norway spruce

Data from Norway spruce stands in northern Germany are given in Fig. 10.12 (Dong and Kramer, 1987). Although a relationship between

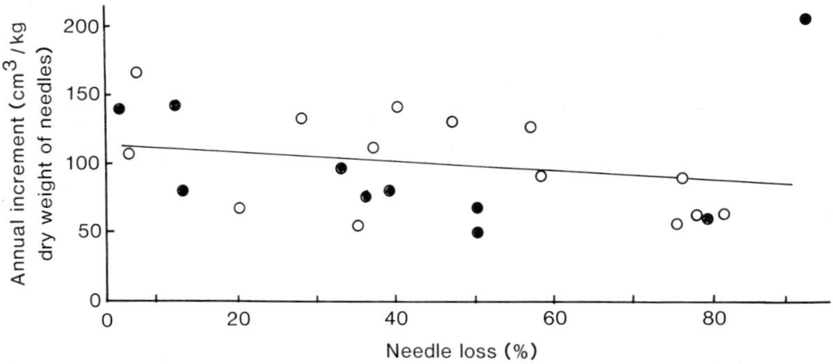

Figure 10.12 Annual increment in cm^3 per m^2 crown surface area in relation to needle loss in Norway spruce from northern Germany. Solid circles: Wingst, open circles: Clausthal-Schulenberg. From: Dong and Kramer (1987).

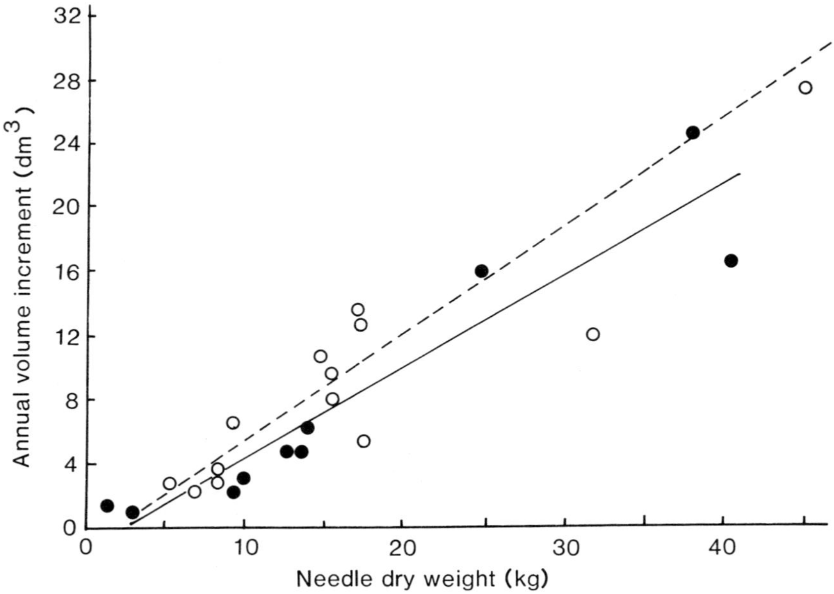

Figure 10.13 Annual volume increment in dm^3 in relation to dry weight of needles (kg) in Norway spruce from northern Germany. Solid circles: Wingst, open circles: Clausthal-Schulenberg. From: Dong and Kramer (1987).

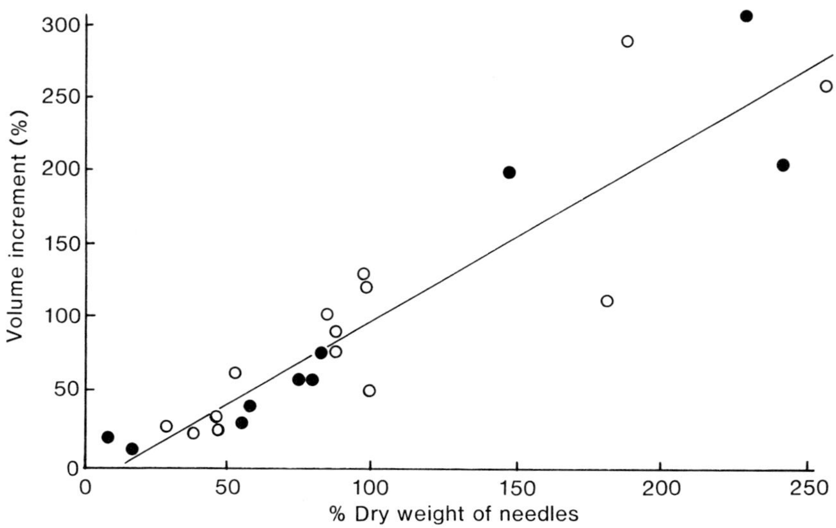

Figure 10.14 Volume increment (%) in relation to percentage dry weight of needles in Norway spruce from northern Germany. Solid circles: Wingst, open circles: Clausthal-Schulenberg. From: Dong and Kramer (1987).

transparency and increment existed, the trend was very weak. Much better relationships were obtained when annual volume increment was plotted against the dry weight of needles (Fig. 10.13) and when volume increment as a percentage was plotted against percentage dry weight of needles (Fig. 10.14).

In north Germany, Athari (1983) found a reasonable relationship between transparency and increment reduction, with trees with 25–60% transparency, 60–100% transparency and dead trees showing increment reductions of 10%, 59% and 70%, respectively, during the 1970s. In this area, it was noticeable that it was the dominant trees in stands that were first affected by yellowing (Hauhs and Wright, 1986). However, working in the Black Forest, Kenk (1990) argued that consistent increment reductions did not occur in Norway spruce below 30% transparency. Differences in the growth rates of Norway spruce according to the amount of transparency have been reported by Spelsberg (1988) and are shown in Fig. 10.15. As in many other cases, trees classed as healthy at the time of assessment were growing more slowly than other trees until a period of stress reversed the situation.

Increment studies in the Göttingen area of Germany revealed that increment reductions in very transparent Norway spruce started in 1964 (Aszmutat and Knigge, 1987). In the Harz Mountains and other areas in northern Germany, trees in all transparency classes have shown an almost uniform recovery since the 1976 drought, despite varying levels of crown transparency (Frühwald *et al.*, 1984; Lorenz and Eckstein, 1988). However, on at least one site, growth has failed to recover from the 1976 drought (Lorenz and Eckstein, 1988) and the drought appears to have triggered prolonged increment declines in some areas (Evers *et al.*, 1986). In Baden-Württemberg, increment of transparent trees has been declining since the end of the 1960s and significantly so since

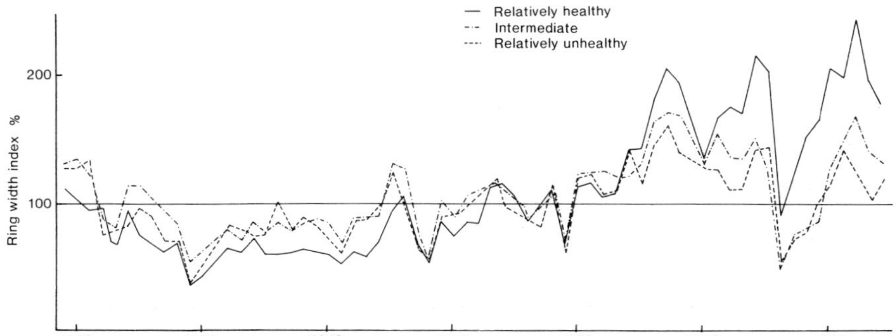

Figure 10.15 Relative annual ring width development of Norway spruce trees in different health classes in Nordrhein-Westfalen, Germany. From: Spelsberg (1988).

1972 (Kenk *et al.*, 1984b; Evers *et al.*, 1986). The divergence in growth patterns was aggravated by the 1976 drought. 1976 was also a year of markedly reduced increment in Norway spruce at Tharandt, Germany (Vogel, 1989).

Working in northern Germany, Dong and Kramer (1986, 1987) found that trees in the same stand having largely similar stem dimensions and similar levels of transparency often had very different rates of growth. However, once a variety of confounding factors were considered, including the stem dimensions, degree of transparency, various crown parameters, the available growing space and the adjacent competition, a good relationship was obtained between relative volume increment (cm^3 m^{-2} crown canopy area) and transparency in Norway spruce, Scots pine and silver fir. Their results are presented in Table 10.1. In interpreting this table, it is important to remember that these values take other factors into account and that the growth period refers to the last 10 years only.

Working in the same area, Kramer (1986a, b) found that increment of Norway spruce was lower in the period 1975–85 than during the previous ten-year period, even when ageing and climatic factors had been taken into account. Although there was a general agreement between transparency

Table 10.1 Relationship between needle loss (%) and relative volume increment (cm^3 m^{-2} crown canopy area) for various species in northern Germany

Species	Site	Needle loss	Relative volume increment
Norway spruce	Neuenburg	10	213
		20	175
		30	138
		40	100
		50	62
Scots pine	Neuenburg	10	308
		20	260
		30	212
		40	164
		50	116
Silver fir	Neuenburg	10	344
		20	290
		30	236
		40	182
		50	128
Norway spruce	Wense'sche	10	199
		20	159
		30	120
		40	80
		50	41

From Dong and Kramer (1986)

and increment loss, individual trees showed substantial discrepancies. Drought greatly affected increment and the reaction to drought was stronger in high-increment trees than in low-increment ones. Increment per square meter of crown surface area decreased with increasing transparency, but the use of normal forest health survey data to infer increment changes was invalid. Transparency was an important variable but was outweighed by factors such as stem and crown dimensions and the growing space of the sample tree.

A different approach was used by Gärtner and Hradetzky (1988). They divided approximately 2000 diameter growth curves into three types (Fig. 10.16). The first consisted of steeply rising curves without any noticeable break in diameter. The second type was similar, but involved a delayed rise in the initial growth. The last type involved a steep rise at first, followed by a rapid flattening off. The curves were not universally applicable, with 21% of trees being impossible to attribute to any of the three categories. By far the most common form was the first type. They found that transparency characteristics were unrelated to any particular type of growth curve.

Göttsche-Kuhn (1988) examined the timing of growth in asymptomatic and symptomatic Norway spruce in Germany. There was no evidence of differences in the fine structure of the cambium of the trees. However, in very transparent trees, cell division started later and radial increment finished earlier than in asymptomatic trees. The duration of cell division was reduced by about 60% and the number of cell divisions by about 85%. Primary wall extension of the earlywood tracheids was reduced in the

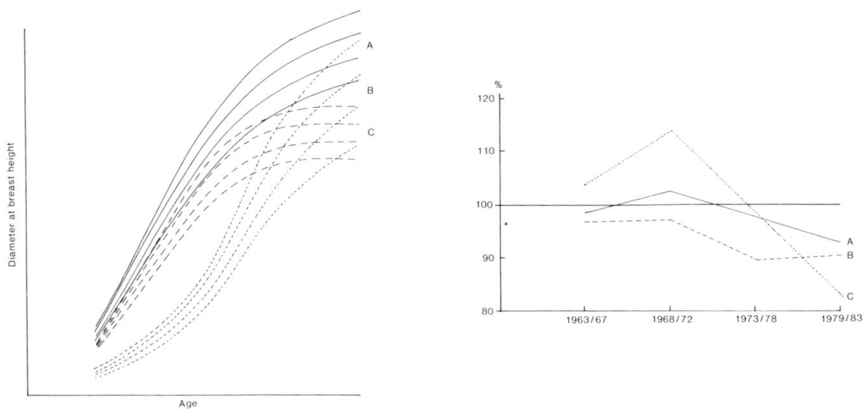

Figure 10.16 Diameter development through time for three types of Norway spruce. The left-hand curves show the three growth types and the right-hand diagram shows the increment of transparent trees relative to asymptomatic trees (horizontal line). From: Gärtner and Hradetzky (1988).

radial direction and, to a lesser extent, in the tangential direction in defoliated trees.

With Norway spruce in Switzerland, Hartmann *et al.* (1987) found that there was a reasonable amount of agreement between transparency and increment reduction. Trees with transparent crowns at the time of assessment grew better in the period 1935–55 than trees with no defoliation at the time of assessment. Obvious stress years included 1962 and 1976, with differences in increment between symptomatic and asymptomatic trees being more marked after these dates. Murri and Schlaepfer (1987) looked at the relationships between transparency and increment in 670 Norway spruce from two sites in the Bernese Oberland. A significant relationship was found between basal area increment and transparency (Fig. 10.17). Increment was generally low in the period 1970–84. On average, a reduction in radial increment of 11–13% was found for every 10% increase in transparency.

Norway spruce in the Jura Mountains of France have not shown an overall change in vigour during this century (Badot *et al.*, 1990). Symptomatic trees today have had lower increments than asymptomatic trees since 1949, with lower increment being particularly noticeable in the periods 1949–59 and 1973–87. Both periods corresponded with periods of below-average rainfall, and the authors suggested that the drought was an important factor affecting those trees with crown transparency. Schneider *et al.* (1988) also examined the increment of Norway spruce in this area. They found that the radial increment of trees with 26–70% transparency decreased in relation to asymptomatic (0–10% transparency) trees from the end of the 1950s. Significant differences in increment were apparent from the mid–1960s, with increment differing by about 20%.

In southern Sweden, a reasonable relationship has been established between increment and transparency, although the nature of the relationship varied between localities. The divergence in growth extended back to the 1950s. An increment reduction of approximately 10% was associated with every 20% reduction in transparency (Björkdahl and Eriksson, 1989).

The effects of drought on Norway spruce were also apparent in Denmark (Münster-Swendsen, 1987a). There, a relationship between drought, tree growth and the population dynamics of the spruce bell moth (*Epinotia tedella* Clerck) was found. Drought, which explained up to 74% of the annual variation in growth, inhibited the growth and the resistance of the trees, making them more susceptible to infestation by the bell moth larvae and other pests.

Few studies have attempted to investigate the problem of increment change on a stand-level basis. However, Röhle (1986) compared average stand transparency with the decrease in basal area increment. The relationship varied with stand age, and increment losses became particularly apparent at ≥50% transparency.

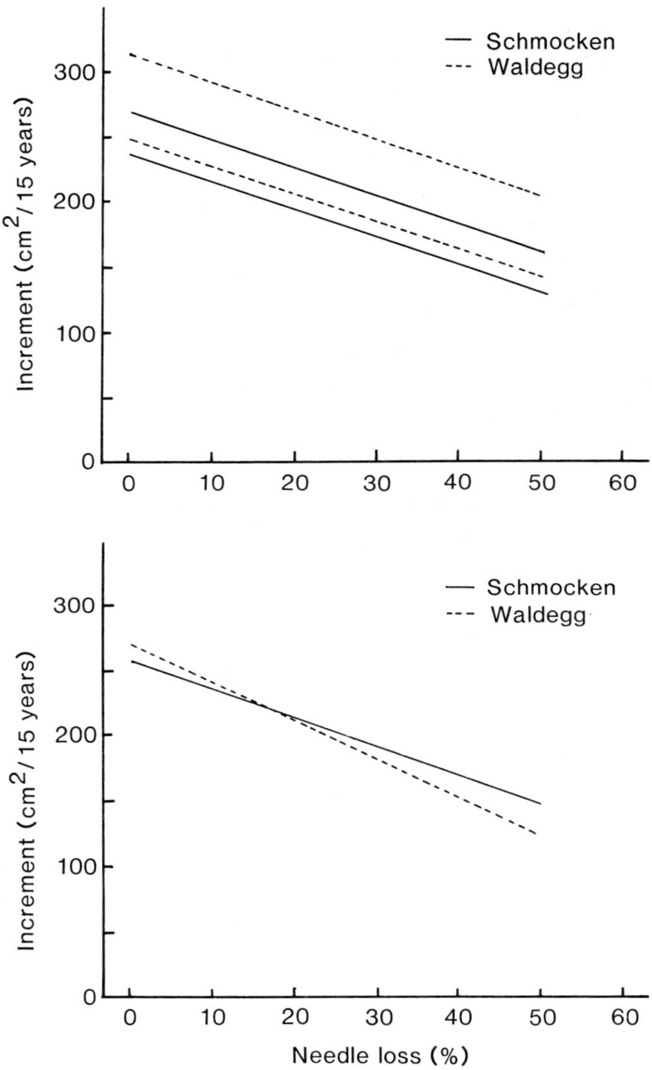

Figure 10.17 Relationship between basal area increment and needle loss for trees of differing crown diameter. In the upper diagram, the upper two lines are for crown diameters of 6 m and the lower two lines for diameters of 3 m. The lower diagram shows the average relationship between needle loss and basal area increment. From: Murri and Schlaepfer (1987).

The studies that have been conducted with Norway spruce reveal a complex situation. As with other species, many factors other than transparency affect increment, and this is reflected in the diversity of the results. It seems unlikely that any general relationship can be derived, but it may be possible to specify site-specific relationships for carefully controlled trees. No threshold of crown transparency at which increment reductions will occur can be specified, throwing into doubt the widely used 25% transparency threshold. Given the variation in the results, it is unlikely that further work in this area will significantly clarify the situation.

Sitka spruce

Sitka spruce in Great Britain appears to follow much the same pattern as Norway spruce, with recent growth of more transparent trees being lower than that of asymptomatic trees (Innes and Neumann, 1991). Differentiation of growth in the different classes of tree probably occurred because of environmental stress (in this case summer drought) (Fig. 10.18). As in many other studies, it was the faster growing trees that were most affected by the stress and which subsequently slowed in their growth (Fig. 10.19).

Silver fir

Schmid-Haas (1989b) reported relatively good relationships between transparency and increment reductions in silver fir, even at transparency values as low as 10–15%. This agrees with other work in Switzerland (Bräker and Z'Graggen, 1987) which suggested that significant growth reductions occurred at >15% transparency in silver fir. However, Schmid-

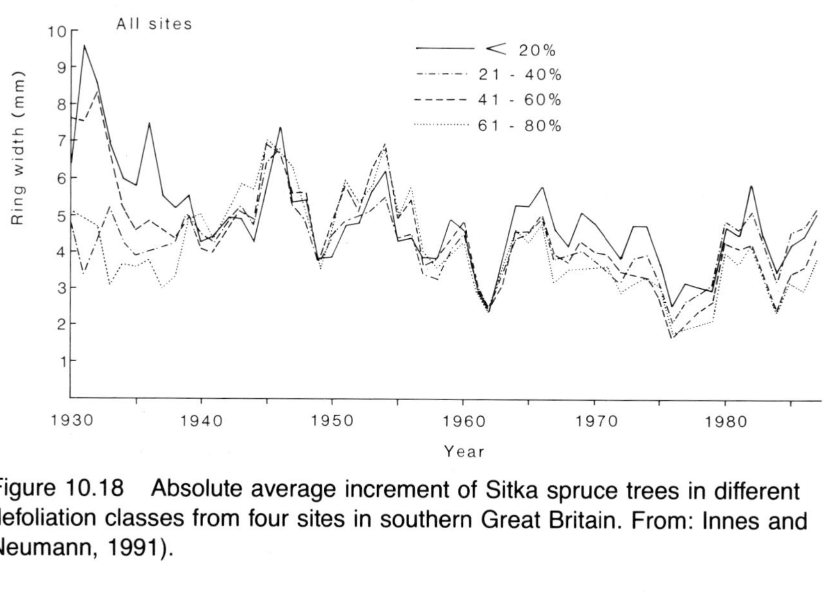

Figure 10.18 Absolute average increment of Sitka spruce trees in different defoliation classes from four sites in southern Great Britain. From: Innes and Neumann, 1991).

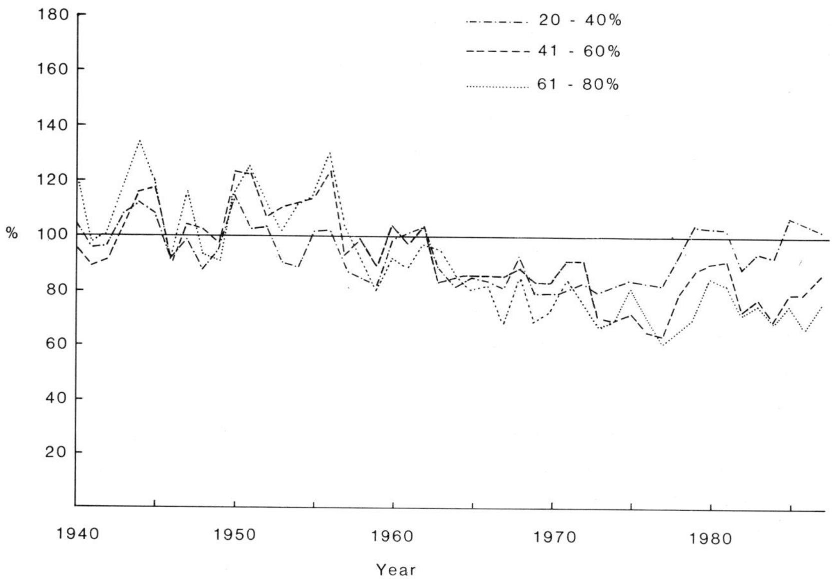

Figure 10.19 Average increment of Sitka spruce trees in differing crown density classes relative to trees with less than 20% reduction in crown density. Data drawn from four sites in southern Great Britain. From: Innes and Neumann (1991).

Haas acknowledged that the relationship between needle loss and increment varied between sites. Eichkorn (1986) also found a relationship between increment loss and transparency in silver fir. However, Konnert and Mettendorf (1990) found no indication of reduced increment in silver fir in Baden-Württemberg at defoliation levels of less than 20–30%. Kontic (1987) argued that it was impossible to distinguish trees with or without a reduction growth by their transparency characteristics, even when growing in the same habitat. The amount of transparency necessary to cause a growth reduction in silver fir is uncertain, with Kenk (1990) reporting that trees with 40–50% needle loss showed no reduction in increment.

The relationships between transparency and radial increment in silver fir appear to differ from other species. By far the most detailed studies of silver fir have been undertaken in the Vosges Mountains of France. Becker (1987) found no significant trend in radial growth variation since the start of the 20th century, nor was there any evidence that particular populations were behaving significantly differently from the mean. Most stands suffered a period of severe stress beginning in 1973, peaking in 1976 and almost ending by 1983. Comparable stress periods occurred in 1917–25 and 1943–51, both of which periods were correlated with severe droughts. Generally, there was a poor relationship between crown condition and increment, with growth curves deviating significantly from average only at transparency values over 40%. However, subsequent work in the Vosges and elsewhere (Becker *et al.*, 1990a, b) showed that this relationship was very variable, with some trees showing a reasonable relationship and others not showing any. An important factor influencing the relationship was social stratification within the stands. Where differences in growth occurred, they were apparent as far back as 1925, when they followed a period of markedly reduced growth. Differentiation also occurred in 1956 and 1976.

Growth of silver fir in the Jura Mountains of France shows a rather different pattern (Fig. 10.20), with mean radial growth being almost twice that recorded in the Vosges (Bert and Becker, 1990). Several periods of sustained growth decline were present in the tree ring record, all associated with severe climatic stress, the key years being 1922, 1948, 1976. Superimposed on this pattern was a long-term trend of increasing growth over the period 1870–1930. A good relationship was found between transparency and growth, with the growth curves diverging in the period 1915–33.

Douglas fir

Studies with Douglas fir have also shown a relationship between transparency and annual increment (de Kort, 1986; de Kort and Baas, 1986). In a study of this species in the Peel area of The Netherlands, where pollution from ammonia is a major problem, asymptomatic trees had relatively narrow (2.5–5 mm) rings in their juvenile stage and currently show no growth decrease. Trees with severe symptoms of decline

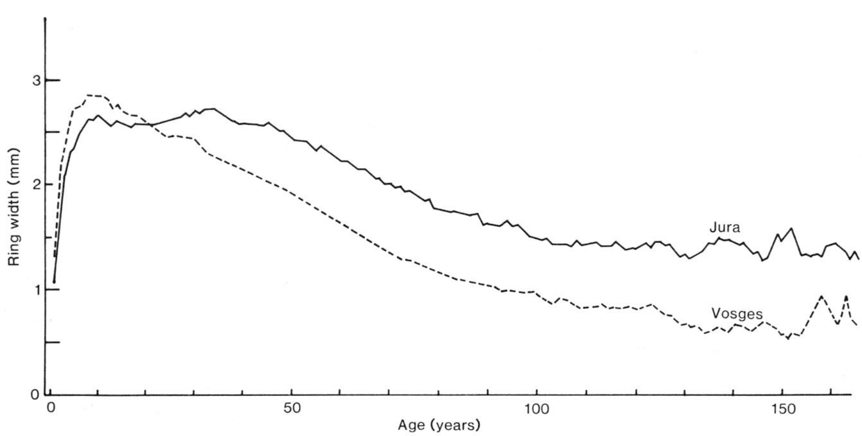

Figure 10.20 Radial growth of silver fir in the Jura and Vosges. From: Bert and Becker (1990).

(transparency and chlorosis) had wide (5–8 mm) rings in the juvenile stage but then had a decline in ring width to less than 0.5 mm. The reduction started between 1958 and 1968. In particularly affected trees, no wood formed at breast height in the 10 years following 1976. Growth was particularly poor in 1976, suggesting that drought was a major factor influencing growth.

Scots pine

In Scots pine, several studies have reported relationships between diameter increment and crown condition (e.g. Kenk, 1983, 1986; Schöpfer, 1986; Kontic *et al.*, 1986; Schmid-Haas *et al.*, 1986; Hartmann *et al.*, 1987). Studies of the impact of the pine shoot beetle have revealed that defoliating events can result in reduced radial increment for six years or more (Långström and Hellqvist, 1991). As this insect is quite common in the northern temperate forests of Europe, increment depressions associated with attacks seem likely.

Broadleaved species

Very little work has been done on the relationship between transparency and growth in broadleaved species. Gao *et al.* (1985) found that poplars could withstand up to 50% defoliation in any given year with very little effect on growth. Similarly, Hilton *et al.* (1987) found that experimental removal of 30% of the foliage of pedunculate oak did not affect growth. However, defoliation resulted in the earlier production of growth spurts and the formation of more lateral branches and the production of more abundant, smaller leaves. Both of these could affect crown defoliation estimates made later in the season. These studies dealt with true defoliation (leaves were removed experimentally) and their relevance to studies of transparency is therefore questionable.

The growth of oak trees responds to climate in western Europe (Kelly *et al.*, 1989) and the lack of any detailed study is therefore surprising. Oak is difficult to work with in comparison to many other species, but there is great potential for the analysis of long ring series from throughout Europe. Given the recent deterioration in the crown condition of oaks in many areas, studies of increment are urgently required.

Athari and Kramer (1989b) examined the relationships between crown classification of beech and increment. Basal area increment was related to crown projection area and crown surface area. However, when these two variables were controlled, there was no apparent relationship between increment and damage class, as determined by past twig extension. Generally, the relationship between transparency and basal area increment was much weaker in beech than in conifers. Koltzenburg and Knigge (1987a, b) found that the recent increment of beech with transparent (25–60%) crowns was greater than that of trees with more foliage. Similar results have been obtained by Rademacher (1986) and Wahlmann *et al.* (1986). Frühwald (1986) found that there was no significant relationship between damage in beech, as assessed by the Roloff method, and annual increment.

Wood quality

There has been concern that the deterioration in the crown condition of trees might be accompanied by a deterioration in the quality of the wood. This might have been important given the widespread practice of sanitation felling, whereby trees with very transparent crowns are preferentially felled. A considerable number of investigations have been made, and some of these are reviewed in the following species-by-species account.

Norway spruce

Aszmutat and Knigge (1987) found no significant reduction in the wood quality of very symptomatic Norway spruce from the Göttingen area in Germany, although the moisture content and percentage of sapwood decreased with increasing damage. Similarly, in Bavaria, Frühwald *et al.* (1986) found slightly reduced sapwood widths and moisture contents in symptomatic trees, but no effect on either wood density or other strength properties. Reduced moisture contents and sapwood areas have been reported in symptomatic Norway spruce from several other German sites (Rademacher *et al.*, 1986; Göttsche-Kuhn, 1988).

The bending strength of beams and sawn timber samples were the same for wood from symptomatic and asymptomatic trees from a variety of German sites affected by crown thinning (Glos and Schulz, 1986; Göttsche-Kuhn and Frühwald, 1986; Kufner and Schulz, 1986). The samples showed the normal variation in wood density, knot size, ring width and strength properties. There was no evidence of a variation in the fracture type, bending or tensile strength of the wood from the two sources. In Baden-Württemberg, there was no relationship between the amount of defoliation and wood quality, although Norway spruce with more than 70% transparency showed reduced sapwood moisture content (Grammel *et al.*, 1986).

Density might be expected to be affected as the proportion of latewood increases with defoliation. However, there is a simultaneous decrease in cell wall thickness, which largely compensates for the changing proportions of earlywood and latewood (Frühwald *et al.*, 1984b; Schulz, 1984).

Chemical contents of the wood also appear to be unaffected, with the percentage of extractives and soluble polysaccharide and dioxane-soluble lignin contents being the same in asymptomatic and symptomatic trees

(Fengel, 1988). However, Puls and Rademacher (1986) found that the wood of symptomatic trees had higher concentrations of sugar and a higher dry matter content of cold water methanol extracts.

Particle boards (V100-grade) have been made from asymptomatic and symptomatic Norway spruce. There was no significant difference in the bending or tensile strength or in the thickness swelling of boards made from symptomatic and asymptomatic spruce (Buchholzer and Harbs, 1986).

After four weeks of storage, Frühwald (1986) found that the moisture content of asymptomatic trees was about 110% of the dry weight and still too high for any fungal attack to occur. Wood from symptomatic trees stored for the same amount of time had moisture contents 95% of the dry weight. After seven months of storage the moisture content of wood from asymptomatic trees had dropped to less than 60% of the dry weight, making the wood susceptible to fungal attack. The moisture content of the wood from symptomatic trees had dropped to 42%, below the level suitable for fungal attack. Göttsche-Kuhn and Frühwald (1986) reported that the incidence of fungal attack was greater in stored wood after four months in the wood from symptomatic trees. However, after seven months, the attack was greater in the wood from asymptomatic trees. Weiss and Knigge (1989) concluded that there are no special storage requirements for wood from symptomatic trees.

Although the quality of the wood appears to be unaffected, there have been reports of reduced penetration of Cu-Cr-F preservative, possibly because of the reduced amount of sapwood (Sell *et al.*, 1989).

Scots pine

Several studies conducted in north-east Bavaria, Germany, the wood quality, including density, bending strength, percentage of sapwood and moisture content, of asymptomatic and symptomatic Scots pine trees showed no significant differences (Kufner and Schulz, 1986; Bues and Schulz, 1988; Bues *et al.*, 1989). Similar results have been obtained from elsewhere in Germany (Glos and Schulz, 1986). However, Hapla and Kottwitz (1987) found that the compressive and bending strength of wood from symptomatic trees in Lower Saxony was higher than that of asymptomatic trees. They attributed this trend to the more uniform growth ring pattern in symptomatic trees.

As with Norway spruce, the percentage of extractives and the soluble polysaccharide and soluble and dioxane-soluble lignin contents of symptomatic and asymptomatic trees were the same (Fengel, 1988).

Detailed studies of the wood structure have failed to reveal any differences in the tracheid length, diameter of the cell walls or the width of the lumina in timber from symptomatic and asymptomatic trees (Nimman and Knigge, 1989).

Silver fir

Torelli *et al.* (1986) looked at the properties of symptomatic silver fir in Slovenia. They found that tracheid length and cell wall thickness were unaffected. However, symptomatic trees showed many characteristics typical of suppressed trees suffering from starvation of the cambium or from internal water stress. Trees showed pronounced wetwood formation, with wetwood developing in previously-formed light heartwood and irregular or ligniform wetwood extending into the sapwood. The formation of the wetwood was usually preceded by visible dehydration, suggesting that the wetwood may be related to the sapwood being intensively dehydrated rather than it turning into wetwood. Asymptomatic and symptomatic silver fir from Baden-Württemberg showed no differences in their dioxane-soluble lignin contents (Fengel, 1988).

Frühwald (1986) found that the moisture content of the sapwood of symptomatic trees was less than in asymptomatic trees. However, the mechanical properties of the wood were unaffected, and a similar conclusion was reached in relation to silver fir from the Black Forest (Grammel *et al.*, 1986).

Douglas fir

There has been very little work done with the wood quality of Douglas fir in relation to declines. However, reduced sapwood widths have been found in symptomatic Douglas fir in The Netherlands (De Kort, 1986; De Kort and Baas, 1986). De Kort *et al.* (1991), in a more recent study, examined the wood quality of symptomatic Douglas fir from near Delden in the eastern Netherlands. A consistent, positive correlation occurred between density, cell wall percentage and latewood percentage. Symptomatic trees had narrower rings (as also found by Van den Ancker *et al.*, 1987) and in these, the proportion of latewood to earlywood was reduced. However, the sample size was too small to infer any conclusions about relationships between transparency and wood quality.

Oak

Asymptomatic and symptomatic trees sampled in Bavaria had wood density and strength within the normal range for these properties, irrespective of the degree of damage. The percentage of sapwood and the moisture content in the trees were not significantly different and there was no apparent effect on any other wood properties (Bues and Schulz, 1990).

Beech

In contrast to the above studies, Mehringer *et al.* (1988) found that the degree of transparency in beech in Germany did not affect the proportion of sapwood versus heartwood, the wood moisture content or the extent of heartwood discoloration. Similarly, Frühwald (1986) found that transparency was unrelated to the sapwood moisture content of beech.

Koltzenburg and Knigge (1987b) compared the fibre length, water conducting tissue and the distribution of cell types in the annual rings of beech from Lower Saxony for the period 1975–85 with a similar sequence for 1945–55. No changes were found associated with the transparency. Similarly, Mahler *et al.* (1988a, b) found that the degree of transparency did not affect the moisture content, density, bending strength or the compressive strength parallel to the grain of beech timber from Baden-Württemberg.

Wood from symptomatic and asymptomatic trees was used to manufacture 20 mm 3-layer PF laboratory particle boards, with a density of 710 kg m^{-3}. There was no difference in the storage, chipping or board manufacturing properties or for the strength or swelling properties of finished boards from the two types of timber (Buchholzer, 1988).

Forest decline, air pollution and the forest industry

The general impression resulting from the above information is that, except for the area encompassed by southern Poland, northern Czechoslovakia and eastern Germany, and small-scale enterprises in certain areas, the effects of the recent forest declines on timber supply and quality were small. However, certain reports have received extensive media coverage, particularly those published by the Timber Committee of the United Nations Economic Commission for Europe. For example, Nilsson (1991, 1992) estimated the volume of standing timber showing signs of defoliation and therefore at risk, or potentially at risk. The estimates were presented for all countries submitting data to the International Co-operative Programme for the assessment of forest health. The report, which implied but failed to find major financial losses caused by air pollution, had three major problems. First, it assumed that the transparency data from different countries could be compared. The results presented in Chapter 5 clearly show this assumption to be untenable, yet Nilsson (1991) and others (e.g. Giese, 1991) have done precisely this. Second, it assumed that there is a link between transparency and air pollution. While this is undoubtedly sometimes true, it is impossible to separate the effects of air pollution from other factors when overall national data are being considered. The assertion that in non-coniferous trees, 'there is no evidence in the literature reviewed that loss of 10–25%

of leaves should primarily be generated by natural stress factors' is refuted by numerous published studies. Lastly, the study assumed a link between transparency and increment reduction, a link which is also unsupportable. Consequently, the inferred link between air pollution and reduced increment on an international scale cannot be substantiated.

Nilsson (1991) acknowledged that factors other than air pollution may cause defoliation and stated (p. 3) that 'damage class 1 (a loss of 10–20% to 25% of foliage of coniferous species) is not regarded as explicitly due to damage by air pollutants'. He also indicated that defoliation may be caused by many other factors, but argued that transparency of more than 10% in broadleaves and more than 25% in conifers was caused primarily by air pollution. The omission of trees in the 10–25% defoliation class was important, as many trees fall into this category. Nilsson (1991) argued that trees in this class were more susceptible to air pollutants and therefore should be considered as at risk.

The problems with the analysis are reflected in the lack of any observed market effect of the purported decline in production. For example, Nilsson (1991) found the lack of any reduction in the production of sawlogs during a period when forest decline is supposed to have been increasing 'remarkable'. Similarly, those countries with the greatest amounts of decline have not shown any major increases in their exports of sawlogs. Nilsson (1991) attributed the lack of any observed effect to a lack of suitable data. However, it is clear that there were major problems with the figures used in the studies and that other factors, such as storm damage, were likely to have a much greater impact on the market than forest decline.

Much the same problems apply to other studies of purported economic losses attributable to air pollution. For example, Brabänder and Bitter (1989) suggested that losses in the mid 1980s in Germany amounted to DM 30–100 per hectare, purely through increased costs and decreased returns. Reductions in growth were not included in the estimates. However, Kroth *et al.* (1989) put the costs rather lower, at DM 22 per hectare. Most other studies of the economic losses caused by forest declines can also be criticized because of their basic assumptions and the methods by which the purported losses were calculated.

Some studies have shown that major economic losses in forests associated with air pollution are unlikely outside certain badly affected areas. For example, Mielikäinen and Kauppi (1991) argued that as forest production in Finland has been increasing in recent years, air pollution is unlikely to have had any effect on wood supply or markets in Finland. Similarly, Seppälä (1992) found little evidence that forest declines associated with air pollution were having any significant effect on the Finnish forestry sector.

The problems with most studies is that they have attempted to use questionable data in even more doubtful models. It is important to break down the costs of forest declines into different components (Brandl and Matthies, 1984; Altwegg, 1989), depending on the nature and timing of the decline. When this is done, then better figures may become available.

Conclusions

Considerable uncertainty has surrounded the relationships between forest growth and forest declines in Europe. In badly affected stands, radial growth can slow or even stop. Similarly, on badly affected trees, radial growth may stop. However, these results cannot be extrapolated to the European forest as a whole, despite attempts to do this.

Although growth decreases have been recorded in several years, the period of increased reports of forest declines in the late 1970s and early 1980s was followed by a period of recovery and increased growth. By the end of the 1980s, increment at many sites had reached levels higher than any previously recorded. The reasons for this are unclear. Fertilization by nitrogen and/or carbon dioxide, favourable climatic conditions and better stand management may all have contributed. In some stands that had previously been affected by mortality, competitive release may have been an additional factor leading to increased growth.

The relationship between crown transparency and growth depends on many interacting factors, and crown transparency on its own is a poor indicator of increment. When all other factors are controlled, a relationship exists, but such studies are relatively rare. Instead, several assumptions have been drawn based on the observed relationships in small samples of purposefully selected trees.

The absence of any apparent effect of transparency on wood quality is curious, given the changes in latewood density that have been documented. One possibility is that the changes in density were restricted to a few years at the periphery of the stem and that these do not significantly affect the overall properties of the stem. However, the conclusions of many different studies are almost unanimous. The increased reports of forest decline in the early part of the 1980s were not accompanied by any appreciable market effects at a European scale.

11 Forest Condition in Great Britain and Continental Europe

> It is worthy of note that decades of scientific research have concentrated on every possible cause of forest decline except that it might be the direct result of intensive plantation management based on ignorance of forest processes.
> *Maser, 1988*

Introduction

In the preceding chapters, several types of forest decline were described. Sometimes, the declines were clearly defined, involving reductions in crown vigour, growth reductions and even mortality. In other cases, it was much more difficult to establish whether the observed poor condition of a tree, stand or forest represented part of a decline or whether the poor condition was either the normal situation for that stand or simply represented the effects of a transient stress. This problem can only be resolved through long-term monitoring and, so far, no 'short-cuts' have been devised. However, the tree ring analyses described in the previous chapter provide a useful means of establishing the impact of poor crown condition on growth. They also provide an indication of the uniqueness of any increment reduction. Curiously, tree ring analysis appears to have been almost completely ignored as a means of investigating forest vigour in Great Britain.

Environmental factors affecting crown condition in Great Britain

Consistent surveys of forest condition have been conducted in Great Britain between 1987 and 1991. Consequently, this period has been examined in detail. However, it is important to consider earlier environmental factors that might have affected tree condition at the start of the monitoring programme. In particular, the drought of 1976 is known to have caused extensive mortality in some areas, particularly among beech.

Stands opened up by this drought were unlikely to have filled again by the time of the first surveys of beech in 1987. The crown condition of beech usually reacts adversely to being suddenly exposed and lag effects from the 1976 drought might therefore be expected.

Redfern *et al.* (1986) described widespread damage to Sitka spruce and other species in Great Britain during the winter of 1983–4. Damage typically involved shoot death, with foliar injury alone being rare. The damage frequently extended to three- to five-year old branches. Damaged shoots were concentrated in a height zone of varying extent, from the upper few whorls of the tree to the lower half or three-quarters of the tree. The top whorls were only damaged in the most extreme cases. Dead shoots tended to occur in the central region of individual branches and rarely involved apical shoots. Up to 1991, the damage had not been repeated at the same scale. The severity of the damage increased with altitude and with tree age, being most severe above 250 m ASL and on trees older than 45 years. The damage was frequently associated with stand openings, crop edges and similar situations, although some stands showed damage throughout the crop.

Redfern *et al.* (1986) believed that temperature fluctuations were important as a cause of the damage rather than low temperatures *per se* and this is consistent with more recent studies in North America. The winter of 1983–4 was characterized by a very mild December, a cold and windy January, a mild February and a cold March. In addition, the dry summer of 1983 might have predisposed the trees to damage. The temperature fluctuations were less than those that normally cause winter injury but, in its natural environment, Sitka spruce is not exposed to the type of temperature fluctuations experienced in 1983–4.

1986–7

In 1986, severe infestations by the needle cast fungus *Lophodermium seditiosum* were noted in Scots pine in many areas. These would have reduced the crown densities of Scots pine assessed in 1987. In early 1987, there were widespread reports of the green spruce aphid on Sitka and Norway spruce (Innes and Boswell, 1988), with the damage to Sitka spruce being the worst since the mid–1970s. Insects were also active in oak and beech, with the oak leaf roller moth being abundant in some parts of the southeast and severe attacks by the beech leaf miner being widespread. Discoloration of beech may have been further enhanced by an above average incidence of the fungus *Gloeosporium fagicolum* Pass.

1987–8

Following the 1987 survey, severe gales in October caused widespread destruction in southern Britain. Several sample plots were destroyed and severe damage to many of the remaining trees occurred. Damage was particularly apparent in oak, which tended to lose branches. In contrast, conifers tended to be blown over or snapped. This affected the results of the monitoring programme, as trees blown over were replaced whereas standing, but damaged, trees were assessed. Similar damage was associated with gales in Scotland during July 1988.

As in 1986–7, insect populations were high during 1987–8. The green spruce aphid was widespread among Sitka spruce and the oak leaf roller moth was locally abundant in some parts of England. The beech leaf miner was again common and browning of beech leaves by *Discula umbrinella* (Berk. and Br.) Sutton was noted on several occasions.

1988–9

September 1988 to September 1990 was marked by several notable meteorological events. The winters in 1988 and 1989 were unusually mild, although in 1989, a cold spell in November was probably responsible for reducing many populations of over-wintering insects in the south and west of Great Britain. Overall, the 1988–9 winter was 2.3°C above normal (Morison and Spence, 1989). Major outbreaks of the green spruce aphid were noted, causing severe defoliation in some spruce stands (Carter, 1989).

In both 1989 and 1990, marked summer droughts occurred. The drought in 1990 differed considerably from that of 1989, when the drought developed early in the summer and extended into August. In 1990 the onset was much later, with many affected areas only experiencing unusually dry conditions from the beginning of July onwards. In 1989, relatively few drought symptoms were seen in trees assessed in the survey plots, in marked contrast to trees growing in open situations (which are excluded from forest surveys) (Innes *et al.*, 1989).

1989–90

Ozone levels were relatively high during the summer of 1990, with the number of episode days (defined as occurring when two or more national network sites experience maximum hourly average ozone concentrations exceeding 60 ppb) being higher than in either 1987 or 1988. The number of 60 ppb hourly average exceedances increased by a factor of three and the corresponding exceedances of 80 ppb and 100 ppb increased tenfold (Bower *et al.*, 1991). The exposures to ozone were greatest in the south and south-east of the country. Despite the number of episodes, no visible symptoms of ozone damage were seen on trees assessed in the survey.

Severe winds occurred in January and February 1990, resulting in the loss of many trees in the southern part of the country. New growth in the spring of 1990 occurred earlier than usual and was badly affected by frosts in early April and late May. In relation to the survey, one effect of the frosts was the death of flowers on oak, which resulted in the low frequency of acorns in the autumn of 1990.

1990–1

The 1990 drought persisted in many areas of southern England, with severe drops in the groundwater tables being recorded. However, the summer of 1991 was generally much more moist than the previous two summers and there was little evidence of any direct moisture stress. Major damage occurred in eastern England in December 1989 because of a combination of snow and wind conditions, with severe mechanical damage being recorded on many sample trees. Further climatic damage occurred in some areas because of a late spring frost. Insect and fungal problems were generally few, probably as a result of the cooler winter temperatures.

General factors affecting trees in Great Britain

Anderson (1930) argued that growth rates of conifers in Scotland was dependent on exposure and soil conditions. Altitude was not considered a good measure of exposure as it depends on the shelter provided by neighbouring hills. Mayhead (1973) and Worrell (1987) considered exposure important, but also included altitude as an important variable. These results are consistent with other British studies (e.g. Hiley and Cuncliffe, 1922, 1923) and those obtained in many other countries and they can also be directly related to crown condition. For example, Neumann (1989a) found that the only factors significantly affecting defoliation scores in Austrian trees were stand age, canopy density and elevation. Crowns were thinner on less suitable sites. Similar results have been obtained elsewhere.

Individual species responses

The responses of individual species to specific environmental conditions are variable. Consequently, each of the five species assessed in the British monitoring programme has been summarized separately.

Sitka spruce

The natural range of Sitka spruce is the western seaboard of North America, where it occurs from northern California to Alaska. It is restricted to a narrow coastal belt, never extending more than about 210 km inland (Lines, 1987). In Great Britain, seed was initially derived from a variety of sources, although it seems likely that much came from Washington State. Most younger trees planted in Great Britain are of the Queen Charlotte Islands provenance and therefore strictly maritime in origin. The sample plots used in the programme were all located within 90 km of the sea. They were generally absent from the south-east of England, which is characterized by a more continental climate that silviculturalists consider unsuited to Sitka spruce. However, environmental gradients can be very steep in Great Britain, and distance from the sea is not a particularly good measure of the continentality of a given site.

Climate is by far the most important factor influencing the productivity of Sitka spruce at higher elevations in Scotland (Worrell and Malcolm, 1990a). In a study using 187 temporary plots spread over 37 sites, there was an average decrease in productivity with altitude of 4.3 m^3 ha^{-1} yr^{-1} per 100 m increase in altitude. Generally, inland sites were more productive than coastal ones and productivity declined northwards. These trends reflected trends in windiness and growing season temperatures. The tree-line varied from 350 m in the north to 650 m in southern inland areas. Regression analysis suggested that elevation, temperature and windiness accounted for 72–78% of the variability in productivity (Worrell and Malcolm, 1990b). Inclusion of soil types as a regression variable only increased the explanatory power of the model by 2–4%, with brown earths being the most productive soil types. A particularly interesting discovery was that the general yield class increased by about 1.3 m^3 ha^{-1} yr^{-1} for every 10 ten years decrease in crop age. This suggested that young crops were more productive in the 1980s than previously. The authors attributed the increased productivity to improvements in silvicultural practices.

During the monitoring period, the most important short-term factor affecting the crown condition of Sitka spruce was defoliation by the green spruce aphid. Severe outbreaks of this pest, associated with the mild winter of 1988–9 and, to a much lesser extent, the mild winter of 1989–90, caused significant defoliation in many areas. Severe cases involved the loss of all needles, particularly in the lower crown, although trees usually produced new growth in the spring following the defoliation. The defoliation in 1989 was mainly confined to the loss of older needles, although shoot death was recorded in some extreme cases. The defoliation primarily occurred from the base of the crown upwards, or involved the gradual loss of needles from throughout the crown. In 1990, many trees were showing signs of recovery, and this trend continued into 1991.

Another frequent cause of damage was late spring frosts. While these primarily affected younger trees, I saw effects on 30-year-old trees in Wales in the summer of 1991. The trees showed the tasselling that is characteristic of this form of injury (Gregory and Redfern, 1987; Innes and Boswell, 1991b).

The transparency of individual Sitka spruce fluctuated considerably between 1987 and 1991, although there was a general tendency for lower scores towards the end of the period. The cause of this general improvement is unknown. However, it seems likely that the absence of any major climatic stresses in the area in which Sitka spruce occurs may have played a major part. The reason for the high transparency scores at the start of the programme is also unknown. Dendroecological work has demonstrated the importance of severe drought stress in particular years, which tends to affect the larger trees within a stand more than it does the smaller ones (Innes and Neumann, 1991). However, major drought effects were not seen during the monitoring period, except in some isolated stands in the south of the country.

Dead shoots were recorded on over half the trees in the programme, being common or abundant on 30%. The cause of death was not recorded, but a variety of factors could be involved. As already indicated, frost tasselling and damage by the green spruce aphid could both kill shoots. In addition, abrasion was important on many trees.

In all years, discoloration was rare. Browning was present on about 1% of the trees, although yellowing was slightly more frequent, with about 5% of trees having yellowing of current-year needles. Yellowing of older needles generally involved about 5% of trees, although more were affected in 1989. The increase in the yellowing of current-year needles in 1990 is of interest. There was no apparent pattern to the increase, but sites in Cornwall (south-west England), Anglesey (north-west Wales) and Speyside (north-east Scotland) showed the greatest increases. A variety of nutrient disorders can cause yellowing and, without foliar analysis, it was impossible to assign a cause. There appeared to be some recovery in 1991, when fewer current-year needles showed signs of yellowing.

A marked increase in coning was seen in Sitka spruce in 1990. Only 36% of trees were without coning, compared to 69% in 1989. In 1991, the amount of coning dropped markedly, with only 5% of trees recorded as having fresh cones. There was no indication that the amount of coning was related to crown density, although severely defoliated trees were generally without cones.

The results for Sitka spruce do not suggest a long-term decline in tree condition, although the substantial fluctuations that occur from year-to-year clearly warrant careful monitoring. In the short span of data that is available, there is no indication of any downward trend in condition. However, such a trend would be difficult to identify in view of the effects of the green spruce aphid. Examination of anomalous plots did not reveal any unexplained problems, although the precise cause–effect mechanisms were not evaluated for the two plots in worst condition.

Norway spruce

Norway spruce is the species of greatest interest in central Europe in relation to the problem of forest decline. As suggested in Chapter 3, the most important form of decline is known as Type 1 decline and is associated with yellowing of older needles, caused by magnesium deficiency, and needle loss. These symptoms are characteristic and the yellowing in particular can easily be recognized. Innes and Boswell (1990a) reported that this type of decline, and most of the other types recorded in Germany, has not been detected in Great Britain. One possible reason for this is the differences in crown structure that are apparent between British Norway spruce and those elsewhere. In particular, comb-type trees are rare in Great Britain, comprising only 1.6% of the trees assessed in the programme.

In Great Britain, the crown transparency in Norway spruce took the form of a uniformly distributed absence of needles throughout the crown, with transparency from the base of the crown upwards being the second most frequent form. The unexplained syndrome known as 'spruce-reddening' was particularly noticeable in 1990 and 1991 and several sample plots were affected. As 'spruce-reddening' does not necessarily involve dying from the top downwards (reddening or browning of the foliage is often the most obvious symptom), the presence of the problem in sample plots is probably better judged from records of discoloration.

Between 1987 and 1991, the crown transparency of Norway spruce remained fairly constant, with individual trees being more stable than Sitka spruce. Norway spruce is not subject to defoliation episodes by the green spruce aphid to the same extent as Sitka spruce. It also tends to be planted in more sheltered situations, making it less susceptible to severe climatic stress. It is less sensitive to drought than Sitka spruce and a complex set of conditions are sometimes necessary for drought to trigger a decline in condition (Cramer and Cramer-Middendorf, 1984; Münster-Swendsen, 1987b). Despite this, Norway spruce is often very responsive to drought stress. The number of common sample trees for the period 1987–91 was the highest for any of the species and the figures are therefore likely to be the most reliable. However, the 1987 storms resulted in the loss of most Norway spruce sites in south-east England, and this area remained under-represented up to and including 1991 (Fig. 6.2).

Shoot death occurred at about the same frequency as in Sitka spruce. Approximately half the sample trees had shoot death present in 1989 and 1990, rising to two-thirds in 1991. Although the proportion of trees with shoot death increased in 1991, the proportion of trees on which shoot death was common or abundant remained the same. This suggests that the increase may have been transient.

Shoot death is likely to occur naturally through suppression but in that case should have been excluded from the assessment. There will also be some turnover of shoots within the crown. Little is known about the phenomenon and the monitoring programme is likely to reveal the nature of any turnover of shoots within the crown and how this affects other parameters of crown condition. The cause of the shoot death is unknown but, sometimes, it may have been related to spruce-reddening. This was confirmed by the presence of significant ($p < 0.001$) correlations between the extent of shoot death and both the browning of current needles and the browning of older needles.

Needle discoloration is an important symptom associated with most of the different types of Norway spruce decline that have been identified. In Great Britain, very little discoloration was noted in Norway spruce between 1987 and 1990, but a sharp increase in browning occurred in 1991. This was related to the occurrence of 'top-dying' or 'spruce-reddening' and appeared to be triggered by the weather conditions in 1989 and 1990. Whether the change was attributable to the mild winters or the summer droughts or a combination of the two remains unclear. Needle pathogens were not considered to have been involved, in contrast to the reports of reddening associated with Type 3 Norway spruce decline in Germany (Rehfuess, 1987, 1988b; Rehfuess and Rodenkirchen, 1984, 1985). The small number of records of yellowing may be related to the majority of sites being located away from the areas affected by the 1990 drought. It also reflects the good nutritional status of most Norway spruce plantations, with magnesium concentrations of first-year needles being 0.06–0.14% and second-year needles having 0.04–0.11%. As indicated in Chapter 9, the adequate magnesium concentrations in British trees indicate the importance of marine-derived magnesium and its frequency in the soil parent materials.

Needle retention was lower in 1990 than in 1991 or between 1987 and 1989. Many trees scored in 1989 as having more than seven years needle retention had seven years or less in 1990. This may be related to the drought conditions in 1989. However, the greater proportion of trees with more than seven years of needle retention in 1991 suggests that the relationship between drought and retention may be complex. The higher number of trees with very low retention values was associated with the needle discoloration described above. It is unknown how many years of needles should be retained in Norway spruce and the figure is likely to vary with provenance and environment. Consequently, changes in retention are likely to be more informative than the absolute levels.

The climatic conditions in 1989 and 1990 were undoubtedly responsible for the increase in coning recorded in 1990. Almost 46% of the trees recorded in 1989 without coning had cones in 1990 and, in 3% of the cases, coning was abundant. The amount of coning was less than in Sitka spruce, but the extent was still of note. As with Sitka spruce, coning was extremely rare in 1991.

The general picture for Norway spruce was one of a fairly stable situation, with only a few indices fluctuating to any degree. There was no indication of any major nutrient problems in the sample trees, which would be apparent from foliage discoloration. A major cause of concern is the widespread occurrence of needle-reddening. The causes of this remain unknown, yet the problem can be very severe in some stands, leading to extensive mortality. It appears to be related to disturbed water movement in the trees. This may be caused by drought, unusual winter weather, management activities (particularly late thinning), or a combination of the three. An interaction between management practices and drought has occurred in silver fir in the Vosges Mountains of France, but the problem with Norway spruce appears to be different. Some good quality research is needed on the problem.

Scots pine

Scots pine is the most important native conifer in Great Britain, although many stands planted in England and some stands planted in Scotland and Wales are derived from overseas sources (Lines, 1987). The origin of many stands is unknown as a principal source of seed; East Anglia has a mixture of stands originating in Great Britain and the continent. Experimental work has revealed major differences in the performance of different provenances (Lines and Mitchell, 1965; Lines, 1987; Worrell, 1992). Provenances from Scotland planted in Scotland out-performed other provenances whereas some provenances from Scandinavia, southern Spain and the Caucasus failed. Other provenances, including some from Scandinavia, Latvia, Poland, northern Germany and northeast France grew more slowly than the Scottish trees. Provenances from the northern European Plain and central Germany grew better than Scottish provenances when planted in East Anglia, but worse when planted in Scotland (Kinloch *et al.*, 1986). The rate of failure was directly associated with the site conditions, being highest on poor, exposed sites (Worrell, 1992).

The correct selection of provenance appears to be a major factor affecting the success of Scots pine plantations in Great Britain. Many pines in East Anglia and southern England are hybrids between Scottish and Belgium-France types (Wright and Bull, 1963). They appear to be relatively successful in this area. Problems arise when these trees and provenances derived from continental Europe are planted in upland sites, with complete crop failures being documented. Continental trees seem much less able to withstand the adverse climatic and edaphic conditions present in many upland areas in Great Britain. For example, trees from Wester Ross were much more resistant to winter blast than other provenances (Kinloch *et al.*, 1986).

Problems with Scots pine have been identified in several areas in continental Europe, notably The Netherlands and northern Germany. Excessive deposition of ammonia is believed to be causing the disruption of many physiological processes in these areas (den Boer and van den Tweel, 1985; van Breemen and van Dijk, 1988; van Dijk and Roelofs, 1988; Roelofs *et al.*, 1987; de Temmerman and Coosemans,

1989). Growth disturbances, foliage discoloration and increased insect activity have all been reported from affected trees. Ammonia deposition is also high in some parts of Great Britain, although precise figures remain elusive because of difficulties in the measurement of some forms of deposition.

Of the five species examined in the British monitoring programme, the data for Scots pine showed the least consistency between observers. This reflects the general difficulty experienced with the assessment of Scots pine. Care should therefore be taken in interpreting the following summary.

Between 1988 and 1989, there was a marked improvement in the crown transparency of many trees. The improvement continued into 1990, but was followed by a deterioration in 1991. Scots pine had the highest recorded mortality in the programme, with 0.8% of the sample being recorded as dead in 1991. Six of these were at a single site, which was previously identified as having a major problem (Innes and Boswell, 1989a). The most likely explanation for the decline and death of the trees at this and other similar sites is inappropriate provenance selection. The vigour of the trees is affected by adverse climatic conditions and low vigour, combined with damage by the pine shoot beetle and *Brunchorstia* and *Lophodermium* fungi, resulted in the deaths of the trees. Together with these site-specific problems, an increase in mortality from 1989 onwards occurred among the survey trees. Single trees died at several sites from unknown causes.

The crown transparency data for Scots pine illustrate the complex patterns of improvement and deterioration that make up the annual figures for forest condition. For example, although an increase in mortality was identified in 1990, some trees with moderate to severe transparency improved. Until 1991, there was a general improvement in transparency. Sometimes, this was very marked, with trees scored as having 71–80% transparency in 1987 being scored as having less than 10% transparency in 1990. In 1991, a marked increase in transparency occurred. These rapid changes would be expected in Scots pine as a single year of needles makes up 33–50% of the foliage in the crown and, over a four-year period, a complete turnover of the foliage occurs.

As with the spruce species, the most frequent forms of defoliation involved uniformly distributed transparency throughout the crown and transparency from the base upwards. Shoot death was recorded marginally more frequently in Scots pine than in the two spruce species, but the numbers of dead shoots on individual trees were generally less. The patterns of increase mirrored that of the two spruce species. A frequent cause of shoot death was the pine shoot beetle, which was present in considerable numbers in some years. One factor affecting the frequency of this species is the availability of breeding substrate. The large volume of wind-blown timber in some forests provided an ideal opportunity for the development of large populations of this insect during the monitoring period.

Needle discoloration was more frequent on Scots pine than on the spruce species. A certain level of discoloration of older needles would be expected as part of the normal senescence process. However, the marked increase in the frequency of records for browning of current needles that was recorded in 1991 cannot be attributed wholly to senescence. Instead, the most likely cause was damage by the pine shoot beetle, as most records of browning were for the majority of needles on a few individual shoots rather than for a few needles on each shoot. This is typical of damage caused by the pine shoot beetle which, in mid to late summer, bores out the centres of current-year shoots, resulting in the death of the shoots.

Needle retention has fluctuated markedly from year to year, as expected from the historical variations shown in Fig. 7.7. As with some other indices, the variation recorded during the field assessments masked considerable changes in individual trees. Needle retention generally increased in 1991, possibly reflecting the lack of accelerated needle senescence associated, in 1989 and 1990, with the summer droughts. The reduction in needle retention recorded in the historical studies at both Thetford and Alice Holt should be of some concern. Similar reductions have been documented elsewhere (e.g. in southern Bohemia: Skuhravy, 1990), and there is a real possibility of a long-term change. Similar studies from other sites are urgently required.

Considerable changes in the amount of flowering occurred on individual trees, making the identification of any pattern difficult. In all years, heavy flowering was associated with lower crown densities. The amount of coning did not fluctuate to the same extent as in the two spruce species, and no clear effects of the droughts were observed on this index.

Scots pine showed much more mechanical damage to the crowns than either of the two spruce species. Damage mainly took the form of abrasion to peripheral shoots and, as with the spruces, the amount of abrasion increased between 1989 and 1991. This was hardly surprising given the storms of early 1990 and subsequent periods of strong winds. Structural damage to the crowns caused by wind was also more apparent in Scots pine than in the spruces, with 9% of trees affected in 1990 and 19% in 1991. This was more likely to be the result of the branch form of Scots pine than due to trees being located on sites more prone to damage.

Overall, the amount of mortality in Scots pine remained higher than in the other species, but half of this was accounted for by one site. Up to 1990, the crown transparencies of trees gradually improved on 1987 levels, although wind and insect damage caused local problems and there was a deterioration in 1991. There was little evidence that the trees were adversely affected by the two droughts, confirming the general impression that the provenances of Scots pine planted in south-east England are relatively drought-resistant. However, the marked increase in transparency seen in 1991 may have been related to the droughts. The progressively reduced needle retention during the 1980s deserves further attention as it is the first evidence from Great Britain for a possible long-term decline in crown transparency.

Oak

Oak is a common component of broadleaved woodland in Great Britain and, as such, it is of considerable commercial, ecological and aesthetic value. There has been some concern about the state of hedgerow oaks in Great Britain, but these were excluded from the British monitoring programme, which dealt only with forest condition. There has also been concern about an apparent increase in dieback in European forest oaks (see Chapter 3). From 1989, reports of similar dieback have occurred in Great Britain. The cause remains uncertain, but the timing, together with the timing of past dieback episodes, strongly suggests drought as a primary cause.

A major drawback of the British monitoring programme is that the different species of oak are not distinguished. The trees were either sessile or pedunculate oak, or hybrids between the two. The lack of distinction is important, as the two species have different ecological requirements and, when planted together, one may decline while the other remains healthy (Durand *et al.*, 1983; Guillaumin *et al.*, 1983; Bonneau and Landmann, 1988; Landmann, 1992). In Great Britain, sessile oak tends to be more common on poor sites in the west of the country whereas pedunculate oak occurs throughout the country (Perring and Walters, 1962). Most of the oak trees planted before 1927 (i.e. those included in the monitoring programme) were derived from British seed stands, but the origin of these trees is unknown. The performance of different provenances of oak is largely unknown, mainly because of the poor experimental design of some Forestry Commission provenance trials (Lines, 1987).

Of all the species investigated in Great Britain, oaks had the poorest scores for crown transparency. This does not necessarily mean that oaks were more damaged; the amount of ground vegetation under oak is a common feature of stands and reflects the amount of light that is normally able to penetrate the canopy. The transparency of the trees was variable between 1987 and 1991. More trees improved than deteriorated between 1987 and 1989, although between 1987 and 1988, approximately 30% of the trees had increased transparency, some markedly so. This was probably the result of storm damage in October 1987. Little change was seen between 1989 and 1990, but a marked increase in transparency occurred in 1991, affecting over 40% of the trees. Over 15% of the trees deteriorated by more than 10%. Most transparency took the form of small gaps in the crown or small gaps in the lateral branch system, although many trees had large gaps present. These types of transparency seem to be characteristic of oak.

Dieback provides a good indication of the condition of a broadleaved tree. Only 20% of oaks were without dieback in 1991, and the proportion of trees with dieback steadily increased between 1987 and 1991. Often, the dieback was noted throughout the crown. Most involved a relatively small proportion of the crown. About 95% of the trees had less than 20% of the crown affected in 1991. The increases in dieback were probably associated with the summer droughts of 1989 and 1990, but it has not been possible to confirm this.

Discoloration of foliage was generally relatively rare, involving less than 10% of the trees. However, in 1991, there was a substantial increase in the proportion of trees with leaf-browning. The cause of this was unknown as a variety of problems can induce browning. In most years, yellowing was more frequent than browning, and there were several instances of moderate to severe discoloration. As with the conifers, the trees affected were different in each year, suggesting that the discoloration was not caused by mineral deficiencies.

Levels of mechanical damage to the crowns of oak were very similar in 1989 and 1990, but increased in 1991. The proportion of trees recorded as having wind damage increased between 1989 and 1991. Much of the increase can be attributed to storm damage in early 1990. Oak is particularly susceptible to branch damage during strong winds and, as already mentioned, mechanically damaged trees continue to be assessed in the British monitoring programme.

The condition of oak, more than other species, gives rise for concern. A deterioration in the general condition of the trees was seen between 1987 and 1991, although much of this was attributable to storm damage. However, independent work has documented an increase in the occurrence of dieback in oaks in Great Britain (Greig, 1992) during the monitoring period. The dieback is not unprecedented, having occurred previously in association with drought (Young, 1965). The results suggest a need for some species-specific research on the condition of oak, together with historical analyses of growth patterns of the two species in relation to climate.

Beech

In Great Britain, beech is native to the south and east of the country (Birks, 1989), although natural regeneration occurs in many stands where it has been introduced. High quality stands are present over much of the country and the largest trees in the country are well outside their supposed natural range (Mitchell *et al.*, 1990). It is likely that, given sufficient time, the natural range of beech would have extended northwards. As with oak, there has been a lack of good, long-term provenance trials (Lines, 1987). Those trials that have been completed suggest better height growth and form of trees from northern France, Belgium and The Netherlands than from Great Britain (Worrell, 1992).

There has been considerable concern over the condition of beech in Europe and this concern has extended to Great Britain. Much of the interest has arisen because of the work on shoot growth by Andreas Roloff (Roloff 1985a, 1985b). Studies in Great Britain (Lonsdale, 1986a, 1986b; Lonsdale *et al.*, 1989) suggested that the shoot growth depressions reported in Germany were also present in Great Britain, although the causes were uncertain. Those studies revealed that twig extension growth at some sites was severely depressed by the 1976 drought and that, sometimes, growth failed to recover to pre–1976 levels.

Despite the interest in beech and the purported evidence for its poor health, the Forestry Commission did not achieve anything like its sampling target for this species until 1991. This limited the analysis that could be undertaken and inferences concerning earlier changes in condition must be based on relatively small sample sizes. The small sample size, the marked changes in tree condition that occurred during each annual survey and the problems over the consistency of some of the data all combined to make any interpretation of the results tricky.

In 1987–89, the transparencies of beech crowns tended to improve, although local exceptions existed. In contrast to the other species, which either improved or remained constant, the crown transparency of beech generally deteriorated between 1989 and 1990. Sometimes, the deterioration was very marked. A further deterioration was evident in 1991. Often, the deterioration involved individual trees that had previously been scored as having relatively low levels of defoliation. Reductions on 1989 values of up to 70% were recorded, revealing that some trees were severely stressed (Innes, 1992a).

Crown dieback increased markedly in 1990, but there was a slight recovery in 1991. It primarily involved leaf loss or death of small branches and was mostly restricted to the top or the top and middle of the crowns. As with oak, crown dieback mainly involved a relatively small (< 20%) proportion of the crown. There was evidence of a progression of dieback, with almost half the trees scored in 1989 as having leaf loss only being scored in 1990 as having dieback involving small branches. However, many trees without dieback in 1989 moved to the small branch category in 1990. Many trees with some degree of dieback in 1989 showed a reduction in the proportion of the crown affected in 1990, and trees with severe dieback in 1989 had generally improved by 1991, suggesting that the changes were far from uniform. The reduction of dieback in 1991 probably reflects the recovery of trees at some sites following the 1989 and 1990 droughts.

Discoloration fluctuated markedly from year to year during the monitoring period. Browning was fairly consistent, although very little was recorded in 1990. In contrast, the proportion of trees affected by yellowing ranged from 12% to 53%. Although beech showed many drought symptoms in 1990, discoloration was largely absent in 1990. Yellowing was at its most marked in 1991, but the reasons for this are unknown. Although yellowing in beech has been widely recorded, occurring every year at some sites, the monitoring results reveal that the fluctuations are not driven by any readily discernible weather pattern. The cause of the discoloration at some sites was 'lime-induced chlorosis', but this did not explain the yellowing at all sites. A detailed investigation of the subject might reveal the causes.

There was a major increase in mast in 1990, followed by very low levels in 1991. In 1990, the mast was often associated with unusually small leaves, the frequencies of which were also much greater in 1990. Only 17 trees showed a reduction in mast on 1989 levels, whereas 707 (86%) showed an increase. The degree of masting was so high that it may have occasionally affected the crown density estimates. Masting was particularly high on moderately to highly transparent trees, as recorded in Germany (Gärtner, 1988), suggesting a link between the two. Heavy fruiting is known to occur in response to previous summer growing season drought (Matthews, 1955; Bernier *et al.*, 1989), and this seems to be the most likely stress involved in the triggering of the high mast load.

Leaf-rolling also increased in 1990, although not by the same extent as masting. Leaf-rolling is likely to reflect the degree of stress experienced by trees during the current growing season, whereas the potential amount of mast is determined during the previous growing season. The severity of leaf-rolling was very similar in 1989 and 1990. In 1991, 14% of the sample trees had rolling present in most or all of the crown. This was considerably more than in either of the previous years. The cause of the rolling was uncertain; drought conditions were not evident in 1991 and the rolling therefore cannot be directly attributed to this. However, if the two previous droughts had damaged the root systems of the trees, then the trees might have suffered from moisture stress in 1991, even if soil moisture conditions were adequate. Although insect damage in beech was much higher in 1991 than in 1990, no insects that might have induced the rolling were noted, and the cause remains unknown.

As with oak, the amount of mechanical damage in beech crowns increased between 1989 and 1991, with wind damage being the main factor involved. This again is likely to reflect damage caused by storms in early 1990.

Damage by the beech leaf miner was widespread in most years. It is uncertain whether this is a normal phenomenon. Hiltbrunner and Flückiger (1992) found that weevils preferred beech leaves exposed to ambient air to those exposed to filtered air. This suggests that ambient air at some locations may alter the leaves in such a way as to make them more favourable to weevils. Consequently, damage might be increased. Infestations by the felted beech coccus (*Cryptococcus fagisuga* Lind.) were limited in 1990 and 1991. In both years, less than 7% of the trees were affected. Active infection by the fungus *Nectria coccinea* (Pers.) Fr. was also uncommon, occurring on 0.8% of the trees in 1990 and 1.6% in 1991. The low levels of infestation are characteristic of this problem (Lonsdale and Wainhouse, 1987), and the restriction of the sampling to trees older than 60 years would have further reduced the likelihood of severe infestations being present.

The changes seen in beech during the monitoring period were relatively consistent and strongly suggested that climatic conditions, particularly drought, were important. The effects of the 1989 drought were reflected by increased levels of dieback, small leaves and masting, whereas the effects of the 1990 drought were evident from premature leaf loss. The reduction in leaf discoloration in 1990 appears to be anomalous. However, browning of beech leaves in Great Britain is primarily caused by the beech leaf miner and the fungus *Apiognomonia errabunda*, both of which appeared to be less frequent in 1990. Their low incidence may be associated with changes in the mineral composition of the leaves (particularly the nitrogen content) which are known to occur during years of heavy seed production (Bernier and Brazeau, 1988; Bernier *et al.*, 1989). Yellowing develops on particular sites in most years, and the reasons for the low levels in 1990 are unknown. The changes in the condition of individual trees were very variable, reflecting the known variation in the response of beech to environmental stress (Müller-Starck, 1985, 1989). The continued deterioration of trees into 1991 indicates that the effects of droughts extend beyond the year of the drought, as also shown by tree ring analysis (Eckstein *et al.*, 1984; Abetz, 1988). As with all other species in Great Britain, there is a need for some high quality work on the response of trees to climatic variations. Dendroecological methods provide a means for assessing the responses but, apart from a few studies, they have yet to be adopted seriously in Great Britain.

The role of air pollution in Great Britain

Air quality in Great Britain is extremely variable. It ranges from almost pristine in the north-west of Scotland to moderately to heavily polluted in the south-east of England. Superimposed on this trend are trends in precipitation quality, with the most polluted rainfall occurring in the east of England. However, the gradient in precipitation quality is complicated by marked gradients in precipitation amount, and the total deposition of some pollutants has a very different spatial pattern to the pollutant concentrations.

Monitoring of forest condition has not been conducted for a sufficiently long period to establish a temporal correlation between forest health and the levels of specific air pollutants. Instead, the only valid approach of establishing a link between the survey data and pollution levels at the present time is to seek a spatial correlation. Such an approach depends on several factors:

- the distribution of specific air pollutants or pollution episodes should be known;
- the distribution of specific air pollutants or pollution episodes should be distinguishable from the distributions of other factors likely to affect crown condition;
- the data for forest condition should be spatially reliable; and
- other factors affecting forest condition should be adequately controlled.

None of these requirements is satisfied. The distributions of specific air pollutants were correlated with the distributions of several important climatic variables and, so far, it has not been possible to factor these out. While the distribution of some air pollutants is well-known in Great Britain, major gaps exist in our knowledge, particularly in remote rural areas. As these are where forests tend to be located, the gaps are important. In addition, the relationships between the measured concentrations of pollutants and the exposures and doses actually experienced by the trees in forest situations remains very unclear.

Problems also exist with data on forest condition. As shown in Chapter 5, significant differences between observers were seen for most indices. Consequently, a spatial analysis of the data cannot be reliably undertaken. The last requirement is also unsatisfied. The sample sizes of all species were too small to adequately control variables such as soil conditions, provenance, tree age and stand density. Consequently, the data from the forest monitoring programme cannot be used to determine whether pollution is an important factor affecting the condition of trees in Great Britain.

Environmental factors affecting recent tree condition in Europe

As indicated in Chapter 5, there are major problems with the comparison of absolute levels of defoliation in different countries. In addition, problems exist over the definition of discoloration, the other main index used in the monitoring programmes. Very few other indices are regularly reported. These problems make the interpretation of reports from particular countries difficult.

A widely cited cause of damage in Europe is drought. The damage may take the form of drought stress or, in many countries, increases in the numbers and size of forest fires. Over the last 10 years, there have been several widespread summer droughts, including 1984, 1989 and 1990. In addition, widespread damage was caused by storms in early 1990. Other frequently reported causes of damage include late spring frosts and insect defoliation. Direct injury by air pollution is very rarely mentioned: it was recorded on only 0.1% of the 70 000 trees sampled in the Commission of the European Communities programme in 1990. Air pollution is frequently cited as an exacerbating factor, but the field evidence for this remains elusive.

Conclusions

The results from Great Britain support the suggestion that tree condition fluctuates around a mean, with annual variations depending on extraneous factors such as climate and insect activity (which themselves may be linked). Others have also reached this conclusion (e.g. Kandler, 1988, 1989). With the exception of the limited evidence for Scots pine, there is as yet no indication of any long-term deterioration in tree condition in Great Britain. However, it is highly unlikely that such a trend would become evident over the space of five years.

The successive droughts of 1989 and 1990 clearly had major impacts on the condition of some trees, although this was not reflected in the overall figures for four of the five species. The drought in 1990 occurred during the monitoring programme, and repeat assessments suggested that the condition of individual trees changed between early July and late August (Chapter 5). Consequently, the full effects of the 1990 drought are unlikely to have been recorded. While this presents a limitation for evaluating the effects of the 1990 drought, in the longer term, it means that the amount of noise associated with the 1990 drought will be lessened. As in 1989 (Innes *et al.*, 1989), the effects of the 1991 drought were more apparent in isolated trees than forest trees. The reasons for this have not been formally established although it seems likely that the microclimatic conditions in forests during drought periods are more favourable than around isolated trees.

The storms of January and February 1990 caused considerable damage to trees in southern Britain, but the damage to plots included in the monitoring programme was much less obvious than in 1987, when the storm occurred while broadleaved trees were still in leaf. A few sites were lost and structural damage and peripheral damage to shoots was recorded. The results from 1988 and 1989 suggest that the trees will gradually recover from this damage. The damage in central England in December 1990 was more severe, with many spruce trees being snapped two or three metres below the top of the crown. These trees will take some time to recover.

12 Conclusions

> The scientific ethic demands an open mind toward the evaluation of all possible causes of a problem as well as caution in arriving at interpretations. Scientists who objectively seek the truth independent of political considerations must freely admit to ambiguities in their results.
> *Terborgh, 1989*

Introduction

The existence of forest decline has increasingly been taken granted. This is quite correct, as there are numerous documented cases of forest decline in the literature. Many of these were examined in Chapter 1. What is much less certain is the nature of other declines. Those that have been described are restricted in area and often in the species that are affected. The presence of widespread, large-scale forest declines at the present time in Europe and North America has not been demonstrated. Before attempting to establish the role of air pollution in causing forest decline, it is important to decide whether there is any evidence of large-scale forest decline.

Evidence for large-scale forest declines today

Currently, there are few well-documented examples of large-scale forest declines. By this I mean the decline in condition of large (thousands of square kilometres) areas of forest. Instead, there is evidence of small-scale tree declines in several areas. In many such cases, a single species growing within a mixed forest is showing signs of decline. Examples include the decline of oak in Europe and the possible decline of red spruce at some sites in the northern Appalachians. These are not examples of forest decline, since other species are not directly affected. However, effects on one species may result in effects on others, as appears to be happening in spruce-fir ecosystems.

The case for a clear decline in forest condition away from point sources of air pollution is highly controversial. Whenever critical assessments are made, the evidence for a large-scale decline fails to materialize. For example, the surveys conducted across Europe by the Commission of the European Communities revealed that forest condition varies across the Community and from year to year. However, there is no indication of a substantial, long-term decline in the condition of forests. Similar results are apparent from, for example, the North American Maple Project. However, the time series that are available are quite short and may be insufficient to reveal a slowly-developing long-term decline in forest condition.

The absence of any examples of large-scale forest decline limits further debate. Inventories reveal that small-scale declines have occurred, although even the results of reliable inventories of forest condition should be interpreted with caution (Schmid-Haas, 1991b). Two additional points need to be considered. First, it is likely that most declines are limited to individual species and the condition of species therefore needs to be examined separately. Throughout this book, the importance of declines in individual species has been stressed. Second, it is possible that the issue of forest decline may be a much more subtle process than previously recognized. For example, a species may have reduced growth or increased transparency, but this may be more than offset by other factors. The global increase in carbon dioxide concentrations that has occurred over the last century may have offset growth reductions attributable to pollution. Establishing the presence of such an interacting complex is extremely difficult, particularly as forest ecosystems are characterized by a range of interacting factors.

The role of air pollution

A central issue in the debate about forest decline has been the putative effect of air pollution. The seeming absence of any long-term, large-scale decline in the overall condition of forests does not preclude any adverse effects of air pollution. Whereas acute injury is usually easy to diagnose, chronic effects are much more difficult to identify. This has not prevented some from making statements about air pollution. For example, Jönsson (1991, p. 153) claimed that 'in recent years forest damage has increased in Sweden because of air pollution, possibly in combination with climatic effects'. As with other such claims, the evidence supporting this assertion is very limited.

Although often described, the medical cause–effect prerequisites of Koch (1876) have seldom been applied in air pollution studies. They are:

- the infecting agent must be present in all patients showing symptoms of the disease;

- the infecting agent must be isolated from the patient; and
- the infecting agent must produce the disease under controlled laboratory conditions.

These criteria were expanded by the Committee on Biologic Markers of Air Pollution Damage in Trees (1989). Their five primary criteria for establishing cause–effect relationships were:

- strong correlation;
- plausibility of mechanism;
- responsiveness or experimental replication;
- temporality; and
- weight of evidence.

I have previously (Innes, 1991b) used these criteria to look at the evidence for an air pollution – forest decline interaction and the criteria have been re-examined in the following account.

A similar approach was adopted by Schlaepfer (1992). He identified six stages that were necessary when looking at the causes of forest decline. These were:

- detection and definition of the problem;
- description of magnitude, dynamics and variability of the phenomenon;
- detection of associations in space and in time between the symptoms and the hypothetical causes;
- experimental reproduction of the observed symptoms;
- explanation of the mechanism; and
- validation of the models.

Two extra stages that should be added to this list are the evaluation of historical accounts in the literature and the use of proper diagnostic techniques. Much of this book has been concerned with the first two stages of this approach. The identification and description should be one of the first stages of research, yet in forest decline investigations, the phenomenon has largely been taken for granted. As suggested in the previous chapters, this was a mistake and should be considered as such if continued in the future.

Strong correlation

This criterion was used by Koch (1876) and was also listed by Mosteller and Tukey (1977). It is based on the principle that there should be a strong association between the proposed cause and the observed effect(s). However, there is the assumption that both the cause and the effect can be measured and characterized.

The characterization of a biologically meaningful pollutant exposure is extremely difficult. Lee *et al.* (1988) reviewed a wide range of different exposure indices and concluded that no single index was best. Peak concentrations and accumulating (over time) concentrations were generally better than averaged concentrations. They also found that there was a need to account for phenological weights as the stage of development played an important part in the response of the plants to a specific level of a given pollutant.

Similarly, the characterization of forest condition is very difficult. As with pollution, no single characteristic is ideal and some form of multivariate index appears to be the optimum solution. However, as only two characteristics of crown condition (transparency and discoloration) are normally assessed in large-scale inventories of crown condition and the usual means of combining these is extremely crude, much more data will be required before such an index can be constructed for most areas.

There have been various attempts to draw conclusions from the distribution of pollutants and the incidence of decline symptoms or other indices of condition. For example, Benarie (1987) argued that there was a lack of evidence of forest decline in urban areas and that the hypothesis that air pollution was a cause of forest decline was therefore untenable. However, his argument was based on an incorrect interpretation of the regional forest health statistics in Germany (Innes, 1988c). Innes and Boswell (1988, 1989a) examined the distribution of defoliation and other indices of forest condition in Great Britain in relation to the distribution of several forms of air pollution. We found no significant relationships between poor forest condition and air pollution; if anything, the reverse was the case. However, there were many interactions among the data, making interpretation difficult. In addition, the reliability of some pollution data was questionable. Elsewhere, the importance of factors such as climate and stand management appear to override any effects that air pollution might be having (Bucher, 1988; Brooks, 1989; Hauhs, 1989; Neumann, 1989a).

The presence of a correlation alone is a poor indication of a cause–effect relationship (Schreuder and Thomas, 1991). A correlation does not necessarily imply causation and there are numerous instances which illustrate this general rule. Consequently, it is necessary to include other evidence before identifying a cause–effect relationship.

Plausibility of mechanism

Any observed association between cause and effect should have a reasonable biological explanation. There have been many experimental studies on the effects of air pollution on seedlings and young trees. The failure of many (but not all) such studies to reproduce the symptoms seen in forests suggests that a link in the cause-effect argument may be faulty or at least not as yet demonstrated. It is important that such experiments use the same pollutants and tree species as the cause–effect hypothesis; there is little point in showing an effect on one species and then using the experiment to 'explain' the effect of the same pollutant on a different species.

There is an increasing amount of empirical evidence that suggests that the results obtained from young trees cannot be readily extrapolated to mature trees (e.g. Borchert, 1976; Edwards *et al.*, 1992). This represents a major problem with the interpretation of experimental work on the effects of air pollutants on trees. The volume of literature that exists on the effects of air pollutants on seedlings and young trees clearly shows that under certain conditions, air pollutants can affect trees. The mechanisms of these effects are now being evaluated in some of the better experiments (e.g. Bender *et al.*, 1989; Matyssek *et al.*, 1991, 1992), but the results from these will only form a basis for understanding the mechanisms operating in the field. For example, many types of pollution are highly episodic, enabling periods of repair to occur. There have been very few studies of the factors affecting these repair processes.

Responsiveness or experimental replication

A major criticism of most experimental studies of 'forest decline' (as opposed to those studies dealing with the effects of air pollution on trees) is that there has been a tendency to generate symptoms under laboratory conditions and then to look for these in the field. In decline studies, the reverse process should apply, with experimental studies trying to replicate the field symptoms. An example of the former approach is provided by the work of Lucas (1989), who argued that exposure of Sitka spruce seedlings to 'ambient' ozone resulted in the development of chlorosis. However, chlorosis of the nature described by Lucas (1989) has not been reported from the large area of Sitka spruce forest growing under ambient ozone conditions in Great Britain.

A problem for experimental scientists is the poor quality of symptom description in inventories of forest condition. As already mentioned, defoliation and discoloration are generally the only phenomena described in the field, and these do not provide a satisfactory basis for experimentation. However, there are an increasing number of studies producing high-quality field data (see Chapter 4) which can be used in experimental replication studies. Examples include the work done in the Fichtelgebirge (Germany), the Col du Donon (France) and at Lägeren and Davos in Switzerland. In particular, cytological and biochemical studies of foliage and other tissues appears to offer great potential. This is clearly an area where better liaison between field and laboratory scientists would benefit both parties.

Temporality

A more controversial aspect of the air pollution–forest decline interaction is the timing of the declines. The increase in damage reported in the 1970s and early 1980s appears to have corresponded more with an increasing awareness of air pollution than with an increase in air pollution itself. The reputed correlation between the temporal incidence of forest declines and pollution has been cited as evidence of causation. The available information does not, however, support such a conclusion.

Much of the argument for a temporal association between forest decline and air pollution rests on the reports of forest decline published by international bodies. Throughout this book, it has been emphasised that these reports of forest decline consisted of a series of reports of individual tree declines, none of which appear to be unique. There are numerous reports of past declines in the historical literature, but scant attention appears to have been paid to these. Within such a framework, the argument for a temporal association becomes very weak. In addition, there has been widespread confusion over when transparency first occurred and when it was first described in inventories of forest condition. It is hardly surprising that the number of reports of transparency should have increased in the 1980s when surveys of transparency only started in the 1980s.

An example of the problems associated with a lack of temporal association is provided by Rehfuess (1981b), who re-examined data presented by Ulrich *et al.* (1980). The latter authors proposed that the fine root biomass of trees in the Solling area of Germany had decreased in parallel with an increase in the aluminium concentration of the soil solution. Rehfuess (1981b), however, showed that the decline of the fine root biomass had occurred before the rise in aluminium concentrations and that moisture stress was the primary cause of the decline.

It is now clear that the reports of forest decline in the late 1970s and early 1980s coincided with a period of considerable climatic stress. Climatic stresses have also occurred in the past, and an examination of the historical literature has revealed that most symptoms described in the recent period had previously been described. Consequently, it seems that what occurred in the latest phase of reports was merely a repetition of past events. Some of the localities involved may have been new and, in a few cases, the scale of individual declines may have been larger than previously recorded, but the declines were not without precedent. Although the declines were widely termed 'new forest damages', this was primarily to distinguish them from smoke damage rather than to imply that the phenomena had never previously occurred. Many have overlooked this. Some of these problems could be resolved by careful tree ring analyses (Cook and Innes, 1989), but the potential of this technique has yet to be fully realised.

Weight of evidence

The first four criteria of the Committee on Biologic Markers of Air Pollution Damage in Trees (1989) do not provide sufficient evidence of a cause–effect relationship when each is taken alone. Instead, all four criteria should be satisfied. There are very few reported cases where all four are met, the decline of pines in the San Bernadino Mountains being one of the few exceptions.

Several conclusions can be reached based on the 'weight of evidence'.

- Growth analyses, field surveys and intensive monitoring studies in Europe have not provided any indication of a widespread, large-scale decline.
- Individual components within forest ecosystems may change in response to small external forcing factors. However, the condition of tree crowns is probably one of the most complex indicators of ecosystem condition that could have been chosen. As tree crowns are affected by so many known pathogens, insects and abiotic stresses, they are of little value as indicators of pollutant stress, unless that stress is severe.
- Experimental work with young trees grown under artificial conditions shows that certain air pollutants, or combinations of air pollutants, can affect the metabolism of trees.
- These effects are likely to occur in forests but, mostly, they seem to be insignificant in relation to other forcing factors. It is not clear whether the pollutants represent an additional stress that will predispose trees to decline. Interactions between factors occur in more than one direction and pollutants may offset or reduce the effects of some stresses.
- The weight of evidence for the effects of air pollution on components of forest ecosystems other than trees is considerable. Lichen communities have been adversely affected throughout much of Europe. Forest soils have been acidified, although the relative contributions of acidic deposition and other processes are still debated. These effects alone justify action on pollution controls.

Specific pollutants

In examining the effects of air pollution on forests, it is necessary to distinguish between different pollutants. The term 'acid rain' has been avoided throughout this book as it has become corrupted by media overuse. Instead, where possible, specific pollutants have been described.

Acidic deposition

Throughout Europe and North America, the pH of rain in most areas appears to be insufficiently low to cause direct damage to the foliage of trees. A pH of less than 3.0 is generally required to cause foliar injury (see reviews by Morrison, 1984; Haines and Carlson, 1989). The pH of rainfall in Europe and North America is normally more than 10 times less acidic than this threshold, although the pH of water in individual cloud events may reach the critical threshold at which injury has been seen. However, it is important to remember that although individual cloudwater samples with a pH below 3.0 have been recorded, these form the exception rather than the rule, and most events at a site are less acidic. The experiments which have shown this threshold have generally involved persistent exposure to highly acidic water.

Acidic deposition affects the foliage of many tree species through its effects on the epicuticular wax. The wax changes in composition and structure through time, and acidic deposition (and some gaseous pollutants) may accelerate this process. However, damage to the epicuticular wax does not necessarily imply damage to any other part of the trees.

Sulphur dioxide

Sulphur dioxide has traditionally been one of the most important gaseous air pollutants affecting tree health. Its effects can be seen around point sources, although it is at its most damaging when combined with certain other stresses, such as frost. The mechanisms of sulphur dioxide damage are well understood in comparison with many other pollutants, but the role of very low concentrations (as now occur throughout much of the forested areas of western Europe and North America) is less certain. Ambient concentrations of sulphur dioxide in most forested areas are too low to cause direct effects on trees, but interactions with other pollutants and other forms of stress are possible.

Oxides of nitrogen

Unlike sulphur dioxide, concentrations of the oxides of nitrogen are increasing in many areas, although the gradual introduction of catalytic converters is limiting the increases in some countries. Generally, there is no evidence that the oxides of nitrogen are adversely affecting forest trees at the concentrations commonly encountered in forests. However, as a source of nitrogen, they may be combining with ammonia and ammonium to disturb the nitrogen dynamics of forest ecosystems.

Ammonia and ammonium

Ammonia emissions are likely to introduce patchy variation in acidic deposition, with the deposition of sulphur being particularly intensified (ApSimon *et al.*, 1987; McLeod *et al.*, 1990). There is clear evidence that ammonia is adversely affecting forest ecosystems in areas such as The Netherlands, where deposition levels are very high. However, the problem is often fairly local and it should not be confused with regional scale effects. Its importance may have been underestimated due to difficulties in the measurements of some forms of deposition.

Ozone

Ozone is probably the most controversial of the gaseous pollutants associated with the debate on forest declines. In recent years there has been a rapid improvement in our understanding of the ways in which the gas affects plant tissues. However, the occurrence of ozone is highly episodic in the field, and most early experiments failed to take this into account. In addition, as with early studies on other gases, many experiments have used unrealistically high exposures. For many, the importance of obtaining a significant result appears to have outweighed the value of undertaking realistic experiments. This situation is not helped by those studies which have claimed to use 'realistic' exposures which, under scrutiny, have turned out to be much higher than encountered in the field. Such claims verge on the fraudulent.

Good evidence for ozone effects on forests is currently limited to a few areas along the Pacific seaboard, although sensitive tree species elsewhere have been adversely affected. The symptoms seen on the trees have been reproduced in experiments and satisfactory cause-effect explanations have been devised. The evidence from elsewhere is much more controversial, since the evidence is not backed up by visible injury to foliage. This is hardly surprising as the exposures associated with foliar injury are much higher than those experienced in areas outside those mentioned above.

There is, however, the possibility that ozone is causing effects that are not readily visible. Sorting these out will be a major challenge for the next phase of research into the effects of air pollutants on trees. In addition, more work on the interactions between ozone and other stresses is required as is work on the effects of ozone on other components of forest ecosystems.

Other pollutants

Many other pollutants may reach forests from nearby emission sources. Some of these, such as fluoride, may cause extensive damage, even when present in very low concentrations. However, the role of others, such as the volatile organic carbons and heavy metals, is very uncertain. The weight of evidence suggests that in most areas, they are not particularly important as factors affecting tree condition. However, the impact of, for example, heavy metals on some components of forest ecosystems other than trees may be considerable.

The role of management

Loehle (1988) examined the idea that fast-growing trees are more susceptible to stress. He suggested that a trade-off occurred between slow-growing trees with high levels of defences and fast-growing trees with low levels of defences. Fast-growing trees may be able to shed and replace damaged leaves and may be more capable of healing wounds. However, once they reach maturity, water and nutrient access may slow them down, making them more susceptible to pathogens. As the tree ages, it loses vigour and is therefore less able to allocate energy to defences. A vigorous tree attacked by a pathogen may shift energy to defences. Older trees may have longer recovery periods (Loehle, 1988).

When a wound occurs, the tree may compartmentalize it to avoid further damage (Shigo and Marx, 1977; Tippett and Shigo, 1981; Shigo *et al.*, 1983). Trees that are good at doing this are all slower growing, defensive species.

A particularly important finding for silvicultural practices is the changes that occur in the dominance of individual trees during stand development. In some silvicultural systems, stands are selectively thinned by removing suppressed and sub-dominant trees. However, it appears that the dominant trees at any time may not always have been the dominant trees and may not continue to be so. The tree ring evidence suggests that although the dominant trees may be faster growing while they are dominant, they may also be more prone to stresses such as drought (e.g. Fischer and Rommel, 1989; Innes and Neumann, 1991). These important preliminary findings need to be explored in more detail.

Some forest management practices have clearly had an important impact on the condition of forest ecosystems. The following deserve particular attention:

- In the past, removal of litter led to the impoverishment of many forest soils. The effects of this are still being seen in some areas.
- Many forest areas today are first or second rotation crops on previously impoverished soils. Often, the forests are only there because the soils are of little use for any other form of agricultural activity. In such cases, nutritional problems would be the norm rather than the exception. Some areas that are unsuitable for trees have been planted, a problem that was particularly apparent

in Great Britain because of favourable tax concessions for afforestation.
- The silviculturalists' preference for the fastest growing species has often led to poor matching of species to site. Even where suitable species have been selected, the provenance may have been incorrect, leading to problems for the establishment and subsequent vigour of the trees. As forestry is a long-term operation, past mistakes may take a long time to rectify.

The role of climate

Throughout this book, a recurring theme has been the importance of climatic stress in triggering forest declines. Climate has been recognized as a cause of forest decline for as long as forestry has been studied. It is disquieting that so much early research linking forest declines to climatic stress should have been ignored in recent years. Whether this is merely typical of the lack of scholarship that characterizes much forestry research or whether it represents a deliberate attempt to focus research into fund-generating areas is unclear. A combination of the two seems likely as, with most environmental research money now being channelled into work on global change, the effects of climate on forest ecosystems has suddenly found its way into the research interests of many scientists previously involved purely with pollution research. Given the importance of climate, this change in emphasis is welcome.

The nature of the climatic stress seems to have varied between locations. In the case of northern red spruce in North America, winter injury was a trigger for, or even the cause of, the decline. In most of Europe, by far the most important factor appears to be drought, although winter injury was also involved with damage (but not necessarily decline) in several areas during the 1980s. The distinction between damage and decline in this context is important. Sometimes, the damage resulted in mortality that has not since been repeated. In such cases, a true decline did not occur. In other cases, the damage appears to have triggered or accelerated the decline of the trees. However, as discussed below, it is usually difficult to allocate different 'proportions of decline' to particular stresses.

The evidence for climate as a cause of many reported declines in forest condition is substantial and is consistent with the principles described above. There is a strong correlation between the occurrence of climatic stress, particularly drought, and the occurrence of declines, both in space and in time. The mechanisms by which drought affect trees are reasonably well understood, although the mechanisms of winter injury are still being sorted out. The mechanisms have been evaluated experimentally, and a high level of replication has been obtained. While there is insufficient information to develop models which reliably predict crown transparency under differing climatic conditions, successful response functions have been developed from tree ring work. From these, it is possible to predict growth for given climatic conditions. The weight of evidence for climate as having a major affect on trees and as being a cause of decline is overwhelming.

Much less certain is the importance of secondary interactions. Climatic stress influences a number of processes which are also affected by other factors. An example is the acidification pushes in soils that occur because of droughts. The extent to which factors such as air pollution predispose trees to damage from climatic stresses cannot be answered at the present time and should be the focus of future research on forest declines. This is particularly important given the uncertainty surrounding future climate-forest interactions (Innes, 1993).

Multiple stress

In recent years, the idea that trees are affected by combinations of stresses has received increasing attention. Contrary to what many have argued, the hypothesis is difficult to apply to the forest decline debate (Innes, 1992). Many stresses do not readily fall into the classifications proposed by Sinclair (1964) and Houston (1973). In addition, it is often very difficult to decide whether a particular stress has had a primary or a secondary effect.

The hypothesis has, to a large extent, been misused by some of those proposing air pollution as a cause of forest decline. As repeatedly indicated throughout this book, it is impossible to refer to forest decline as a single phenomenon. There are many different types of forest decline and some of these have no relationship to air pollution. The tendency towards the multiple stress hypothesis as an explanation for some types of forest decline appears to have arisen from the lack of evidence of direct effects of gaseous or other forms of air pollution on forest condition. In some situations, such as in parts of The Netherlands, the case for air pollution being involved in a multiple stress complex seems convincing. In others, the involvement of air pollution as a probable or even possible contributing factor is based on very limited evidence.

Landmann (1992) argued that many cases of forest decline are the result of a particular combination of silvicultural treatment, soil conditions and climate. His reasoning is supported by many field studies, both in France and elsewhere. He argued that the initial growth of a stand is primarily conditioned by the soil characteristics. The evolution of the crowns of the dominant trees occurs in a step-wise fashion, being determined by periods of marked stress. This is different from the traditional belief of many foresters, but is consistent with field data. The nature of the responses to these stress events is controlled by the structure of the stand, in particular the level of competition.

When applied correctly, the multiple stress hypothesis has much to commend it. However, it should not be used as a justification for including pollution as a stress affecting forests, as has been done in many cases described in the literature.

Remedial actions

In several areas in eastern Europe where forest decline has occurred, remedial action has been taken. This has proved necessary where mortality has occurred. However, a continuing problem is the artificially high game populations that exist in many parts of continental Europe. Numbers are so large that the adequate regeneration of declining stands, both by planting and by natural regeneration, is being severely compromised (e.g. Schäffner and Gürth, 1992). The situation today is a prime example of poor forest management leading to a lack of sustainability. An alternative to remedial actions such as liming and fertilization is to select provenances that are more resistant to pollution. Most species show marked variations in their sensitivity to pollution, and more careful matching of provenance to site might enable a more vigorous forest cover to develop. However, while provenance selection may help a forest cover develop, it does not deal with the source of the problem: the pollution.

Provenance selection is an area which has been mentioned on several occasions in the preceding chapters. Some reported declines can be directly related to poor matching of provenance to the physical site characteristics. Thus, problems with oak in the Forest of Tronçais can be related to the mixing of pedunculate and sessile oak when only sessile oak was really suited to the soil conditions encountered in some sections of the forest. The severe problems encountered with Scots pine in parts of upland Great Britain are probably related to provenance selection. Similarly, inappropriate provenance selection appears to be involved in the decline of large areas of maritime pine in France (Guyon, 1991). Many problems encountered today are the result of past mistakes; forest science has come a long way since the trees currently being felled were planted. However, the current overriding concern with increasing productivity may result in future problems, particularly in view of the increasing evidence of the greater sensitivity of faster-growing trees to environmental stress.

Summary

Forest and tree declines will no doubt continue to occur for as long as there are trees and forests. As in the past, a dynamic mosaic will occur in natural forests. However, the proportion of natural forests is rapidly shrinking, both because of deforestation and because of the increasing amount of afforestation with both native and exotic species that is currently occurring. Superimposed on this change in the nature of the world's forests are changes in the environmental conditions in which they are growing. These two trends present a real problem. The success of the increasing use of plantations throughout the world depends to a large extent on silvicultural experience. However, it seems that past experiences may not be particularly useful if environmental conditions are likely to change. For example, a site may have been suitable for one species 100 or even 50 years ago, but climatic change, changes in the soil conditions and other environmental changes may mean that that species is no longer suitable. Such changes are unlikely to be recognized by silviculturalists and the risks of subsequent declines may be increased. Despite allegations to the contrary, this situation already appears to have arisen in some fast-growing plantations in the Tropics.

Forest decline continues to be an emotive subject. As such, it attracts attention from the media and from politicians. Outside the scientific community, the general perception of the problem is still that major decline problems exist as a result of air pollution. Such a view is also held by some members of the scientific community. However, the increasing recognition that the problem has multiple causes represents a step in the right direction. As indicated above, further evaluation of the available information may convince more scientists that many stresses can be disregarded as of negligible importance. Convincing people outside the scientific community will take longer.

A move that provides some encouragement is the increasing recognition that forest trees are only one part of forest ecosystems. The forest decline debate has resulted in much basic research into various components of forest ecosystems, although recent attempts to apply the critical loads concept to forests reveals that this research is still woefully inadequate. Any understanding of the many different types of forest decline must involve an assessment of the processes operating within the forest ecosystem. The failure to recognize this represents a major shortcoming of past silviculture. As forest ecology is still hardly touched upon in most forestry textbooks and forestry courses, the failure to take proper account of the processes acting in forest ecosystems is likely to continue. The difficulties surrounding the assessment of forest health and the evaluation of the factors affecting it will only be resolved by adopting an ecosystem approach to forest research.

Bibliography

Aaltonen, V.T. (1925) Über den Aziditätsgrad (pH) des Waldbodens. *Communicationes ex Instituto quaestionum forestalium Finlandiae* **9,** 54 pp.

Aamlid, D., Venn, K. and Frogner, T. (1992) Sammenhang mellom næringsstoffer i nåler og jord på de faste intensive skogovervåkingsflatene i Norge. Rapport fra Skogforsk 11/92.

Aber, J.D., Nadelhoffer, K.J., Steudler, P. and Melillo, J.M. (1989) Nitrogen saturation in northern forest ecosystems. *BioScience* **39,** 378–86.

Abetz, P. (1985) Ein Vorschlag zur Durchführung von Wachstumsanalysen im Rahmen der Ursachenerforschung von Waldschäden in Südwestdeutschland. *Allgemeine Forst- und Jagdzeitung* **156,** 177–87.

Abetz, P. (1988) Untersuchungen zum Wachstum von Buchen auf der Schwäbischen Alb. *Allgemeine Forst- und Jagdzeitung* **159,** 215–23.

Abouguendia, Z.H. and Baschak, L.A. (1987) Response of two western Canadian conifers to simulated acidic precipitation. *Water, Air, and Soil Pollution* **33,** 15–22.

Ackermann, J. and Hartmann, G. (1992) Kronenschäden in Eichenbeständen Niedersachsens nach Farbinfrarot-Luftbildern aus den Jahren 1988/89. *Forst und Holz* **47,** 452–60.

Adams, D.F., Mayhew, D.J., Gnagy, R.M., Richey, E.P., Koppe, R.K. and Allan, I.W. (1952) Atmospheric pollution in the ponderosa pine blight area, Spokane County, Washington. *Industrial and Engineering Chemistry* **44,** 1356–65.

Adams, D.F., Shaw, C.G. and Yerkes, W.D. (1956) Relationship of injury indexes and fumigation fluoride levels. *Phytopathology* **46,** 587–91.

Adams, H.S., Stephenson, S.L., Blasing, T.J. and Duvick, D.N. (1985) Growth-trend declines of spruce and fir in mid-Appalachian subalpine forests. *Environmental and Experimental Botany* **25,** 315–25.

Adams, M.B., Campbell, R.G., Allen, H.L. and Davey, C.B. (1987) Root and foliar nutrient concentrations in loblolly pine: effects of season, site, and fertilization. *Forest Science* **33,** 984–96.

Adams, M.B. and O'Neill, E.G. (1991) Effects of ozone and acidic deposition on carbon allocation and mycorrhizal colonization of *Pinus taeda* L. seedlings. *Forest Science* **37,** 5–16.

Adams, W.A., Ali, A.Y. and Lewis, P.J. (1990) Release of cationic aluminium from acidic soils into drainage water and relationships with land use. *Journal of Soil Science* **41,** 255–68.

Addison, P.A. (1989) Monitoring the health of a forest: a Canadian approach. *Environmental Monitoring and Assessment* **12,** 39–48.

Ågren, G.I. (1983) Model analysis of some consequences of acid precipitation on forest growth. In: *Ecological effects of acid precipitation.* Report SNV PM 1636. Stockholm: SNV, pp. 233–44.

Ågren, G.I. and Bosatta, E. (1988) Nitrogen saturation of terrestrial ecosystems. *Environmental Pollution* **54,** 185–97.

Aikman, J.M. and Smelser, A.W. (1938) Structure and environment of forest communities in central Iowa. *Ecology* **19,** 141–50.

Alban, D.H. (1982) Effects of nutrient accumulation by aspen, spruce, and pine on soil properties. *Soil Science Society of America Journal* **46,** 853–61.

Albrektson, A., Aronsson, A. and Tamm, C.O. (1977) The effect of forest fertilization on primary production and nutrient cycling in the forest ecosystem. *Silva Fennica* **11,** 233–9.

Aldinger, E. (1987) Elementgehalt im Boden und in Nadeln verschieden stark geschädigter Fichten-Tannen-Bestände auf Praxiskalkungsflächen im Buntsandstein-Schwarzwald. *Freiburg Bodenkundliche Abhandlung* **19,** 266 pp.

Alexeyev, V. (1989) Forest health diagnosis and its application in air pollution impact studies. In: Noble, R.D., Martin, J.L. and Jensen, K.F. (eds), *Air pollution effects on vegetation including forest ecosystems. Proceedings of the second US-USSR symposium.* Broomall, Pennsylvania: Northeastern Forest Experiment Station, pp. 135–41.

Allen, R.B. and Rose, A.B. (1983) Regeneration of southern rata *(Metrosideros umbellata)* and kamahi *(Weinmannia racemosa)* in areas of dieback. *Pacific Science* **37,** 433–42.

Allison, J.R. (1982) *Evaluation of ozone injury on the Stanislaus National Forest.* Forest Pest Management Report 82–07. Berkeley: United States Department of Agriculture Forest Service Pacific and Southwest Region, 7 pp.

Allison, J.R. (1984a) *An evaluation of ozone injury to pines on the Eldorado National Forest.* Forest Pest Management Report 84–16. Berkeley: United States Department of Agriculture Forest Service Pacific and Southwest Region, 10 pp.

Allison, J.R. (1984b) *An evaluation of ozone injury to pines on the Tahoe National Forest.* Forest Pest Management Report 84–30. Berkeley: United States Department of Agriculture Forest Service Pacific and Southwest Region, 10 pp.

Allum, J.A.E. and Dreisinger, B.R. (1987) Remote sensing of vegetation change near Inco's Sudbury mining complexes. *International Journal of Remote Sensing* **8,** 399–416.

Alstad, D.W., Edmunds, G.F. and Weinstein, L.H. (1982) Effects of air pollution on insect populations. *Annual Reviews of Entomology* **27,** 369–84.

Altwegg, D. (1989) *Die Folgekosten von Waldschäden: Bewertungsansätze für die volkwirtschaftlichen Auswirkungen bei einer Beeinträchtigung der Schutzfunktion von Gebirgswäldern.* ETH, Zürich: Fachbereich für Forstökonomie und Forstpolitik.

Alva, A.K., Edwards, D.G., Asher, C.J. and Blamey, F.P.C. (1986) Relationships between root length of soybean and calculated activities of aluminium monomers in nutrient solution. *Soil Science Society of America Journal* **50,** 959–62.

Alvarado Rosales D., Hernández Tejeda, T., de Bauer, M. de L. de la I., and Galindo A., J. (1985) Declinación y muerte del bosque de oyamel en el sur del Valle de México: Síntomas, desarrollo y posibles causas. *Memoria XIX Congreso Nacional de Fitopatologia.*

Alvera, B. (1973) Estudios en bosques de coníferas del Pirineo Central. Serie A: Pinar con acebo de San Juan de la Peña: 1) « Producción de hojarasca ». *Pirineos* **109,** 17–29.

American Forestry Association (1987) An AFA white paper. The forest effects of air pollution. *American Forests* November/December 1987, 37–44.

Amman, G.D. and Speers, C.F. (1965) Balsam wooly aphid in the southern Appalachians. *Journal of Forestry* **63,** 18–20.

Ammer, U., Burgis, M., Koch, B. and Martin, K. (1988) Untersuchungen über den Zusammenhang zwischen Schädigungsgrad und Meereshöhe im Rahmen des Schwerpunktprogramms zur Erforschung der Wechselwirkungen von Klima und Waldschäden. *Forstwissenschaftliches Centralblatt* **107,** 145–51.

Ammer, U., Mössmer, R. and Bröker, U. (1983) Der Beitrag der Fernerkundung zur Erfassung der Waldschäden. *Forstwissenschaftliches Centralblatt* **102,** 149–57.

Van den Ancker, J.A.M., De Kort, I. and Van Oudenaarden, J. (1987) Vitaliteitsbeoordeling van Douglas met behulp van luchtfotografie, kroonaanzicht en jaaringanalyse. *Nederlandse Bosbouwtijdschrift* **59,** 309–17.

Andersen, C.P. and Hogsett, W.E. (1991) Ozone decreases spring root growth and root carbohydrate content in ponderosa pine the year following exposure. *Canadian Journal of Forest Research* **21,** 1288–91.

Andersen, C.P. and McLaughlin, S.B. (1991) Seasonal changes in shoot water relations of Picea rubens at two high elevation sites in the Smoky Mountains. *Tree Physiology* **8,** 11–21.

Andersen, C.P., McLaughlin, S.B. and Roy, W.K. (1991) Foliar injury symptoms and pigment concentrations in red spruce saplings in the southern Appalachians. *Canadian Journal of Forest Research* **21,** 1119–23.

Anderson, M.L. (1930). On the distribution of coniferous quality-classes in Scotland. *Scottish Forestry Journal* **44,** 45–54.

Anderson, R.L. (1959) Maple blight: a new problem in the Lake States. *The Northeastern Logger* May 1959, 28–9.

Anderson, R.L. and Belanger, R.P. (1987) A crown rating method for assessing tree vigor of loblolly and shortleaf pines. In: Phillips, D.R. (ed.), *Proceedings of the 4th biennial southern silvicultural research conference, 1986, November 4–6, Atlanta, Georgia.* General Technical Report SE-42. Asheville: United States Department of Agriculture Forest Service, Southeastern Forest Experimental Station, pp. 538–43.

Anderson, R.L., Brown, H.D., Chevone, B.I. and McCartney, T.C. (1988) Occurrence of air pollution symptoms (needle tip necrosis and chlorotic mottling) on eastern white pine in the Southern Appalachian Mountains. *Plant Disease* **72,** 130–2.

Anderson, S.B. (1988) *Long-term changes (1930–1932 to 1984) in the acid-base status of forest soils in the Adirondacks of New York.* PhD Thesis, University of Pennsylvania, Philadelphia.

Andersson, B. (1990) Defoliation of coniferous trees – assessments 1984–1987. *Environmental Monitoring and Assessment* **14,** 23–43.

Andersson, F.O. (1989) Air pollution impact on Swedish forests – present evidence and future development. *Environmental Monitoring and Assessment* **12,** 29–38.

Anonymous (1978) Symposium on remote sensing for vegetation damage assessment. Falls Church, Virginia: American Society for Photogrammetry.

Anonymous (1984) *Waldschadenserhebung 1984.* Bonn: Bundesministerium für Ernährung, Landwirtschaft und Forsten.

Anonymous (1985) *Waldschadenserhebung 1985.* Bonn: Bundesministerium für Ernährung, Landwirtschaft und Forsten.

Anonymous (1987) *Forest damage and air pollution. Report of the 1986 forest damage survey in Europe.* Geneva: Convention on Long-Range Transboundary Air Pollution, United Nations Economic Commission for Europe.

Anonymous (1988) *Forest damage and air pollution. Report of the 1987 forest damage survey in Europe.* Geneva: Convention on Long-Range Transboundary Air Pollution, United Nations Economic Commission for Europe.

Anonymous (1989a) *Forest damage and air pollution. Report of the 1988 forest damage survey in Europe.* Geneva: Convention on Long-Range Transboundary Air Pollution, United Nations Economic Commission for Europe.

Anonymous (1989b) *European Community. Forest health report 1987–1988.* Luxembourg: Office for Official Publications of the European Communities.

Anonymous (1989c) *Waldschadenserhebung 1989.* Bonn: Bundesministerium für Ernährung, Landwirtschaft und Forsten.

Anonymous (1990a) *European Community Forest Health Report 1989.* Luxembourg: Office for Official Publications of the European Community.

Anonymous (1990b) *Forest damage and air pollution. Report of the 1989 forest damage survey in Europe.* Geneva: Convention on Long-Range Transboundary Air Pollution, United Nations Economic Commission for Europe.

Anonymous (1990c) *La Santé des Forêts 1989.* Paris: Ministère de l'Agriculture et de la Forêts.

Anonymous (1990d) *1990 Forest Health Survey.* Utrecht: Ministry of Agriculture, Nature Management and Fisheries.

Anonymous (1991a) *Forest damage and air pollution. Report of the 1990 forest damage survey in Europe.* Geneva: Convention on Long-Range Transboundary Air Pollution, United Nations Economic Commission for Europe.

Anonymous (1991b) *European Community Forest Health Report 1990.* Luxembourg: Office for Official Publications of the European Community.

Anonymous (1991c) *La Santé des Forêts 1991.* Paris: Ministère de l'Agriculture et de la Forêts.

Anonymous (1991d) *Gesundheitszustand des Waldes in Baden-Württemberg 1991.* Mitteilungen der Forstlichen Versuchs- und Forschungsanstalt Baden-Württemberg 163.

Anonymous (1992a) *Forest condition in Europe. Report of the 1991 survey.* Geneva

United Nations Economic Commission for Europe, Brussels: Commission of the European Communities.

Anonymous (1992b) *La Santé des Forêts 1991.* Paris: Ministère de l'Agriculture et de la Forêts.

ApSimon, H.M., Kruse, M. and Bell, J.N.B. (1987) Ammonia emissions and their role in acid deposition. *Atmospheric Environment* **21,** 1939–46.

Armentano, T.V. and Menges, E.S. (1987) Air-pollution-induced foliar injury to natural populations of Jack and White pine in a chronically polluted environment. *Water, Air, and Soil Pollution* **33,** 395–409.

Armesto, J.J., Casassa, I. and Dollenz, O. (1982) Age structure and dynamics of Patagonian beech forests in Torres del Paine National Park, Chile. *Vegetatio* **98,** 13–22.

Armesto, J.J. and Figueroa, J. (1987) Stand structure and dynamics in the temperate rain forests of Chiloé Archipelago, Chile. *Journal of Biogeography* **14,** 367–76.

Armesto, J.J. and Fuentes, C.R. (1988) Tree species regeneration in a mid-elevation, temperate rain forest in Isla de Chiloé, Chile. *Vegetatio* **74:** 151–9.

Armstrong, J.K., Williams, K., Huenneke, L.F. and Mooney, H.A. (1988) Topographic position effects on growth depression of California Sierra Nevada pines during the 1982–1983 El Niño. *Arctic and Alpine Research* **20,** 352–7.

Arndt, U., Seufert, G. and Nobel, W. (1982) Die Beteiligung von Ozon an der Komplexkrankheit der Tanne (*Abies alba* Mill.) – eine prüfenswerte Hypothese. *Staub* **42,** 243–7.

ARNEWS (1992) *The acid rain national early warning system (ARNEWS): Canada's national forest health monitoring network.* Ottawa: Forestry Canada.

Arnold, G. (1992) Soil acidification as caused by the nitrogen uptake pattern of Scots pine *(Pinus sylvestris). Plant and Soil* **142,** 41–51.

Arnold, G. and Van Diest, A. (1991) Nitrogen supply, tree growth and soil acidification. *Fertilizer Research* **27,** 29–38.

Aronsson, A. (1983) Growth disturbances caused by boron deficiency in some fertilized pine and spruce stands on mineral soils. *Communicationes Institutales Forestales Fennica* **116,** 116–22.

Aronsson, A. and Elowson, S. (1980) Effects of irrigation and fertilization on mineral nutrients in Scots pine needles. In: Person, T. (ed.), *Structure and function of northern coniferous forests – An ecosystem study. Ecological Bulletin* **32,** 219–28.

Arovaara, H., Hari, P. and Kuusela, K. (1984) Possible effect of changes in atmospheric composition and acid rain on tree growth. *Communicationes Institutales Forestales Fennici* **122,** 1–16.

Arp, P.A., Akerley, C. and Mellerowicz, K. (1989) Distribution of aluminium in black spruce saplings growing on two New Brunswick forest soils with contrasting acid soil sulfate sorption. *Water, Air, and Soil Pollution* **48,** 277–97.

Arquillière, S., Filion, L., Gajewski, K. and Cloutier, C. (1990) A dendroecological analysis of eastern larch (*Larix laricina*) in subarctic Quebec. *Canadian Journal of Forest Research* **20,** 1312–9.

D'Arrigo, R.D. and Jacoby, G.C. (1991) A 1000-year record of winter precipitation from northwestern New Mexico, USA: a reconstruction from tree-rings and its relation to El Niño and the Southern Oscillation. *The Holocene* **1,** 95–101.

D'Arrigo, R., Jacoby, G.C. and Fung, I.Y. (1987) Boreal forests and atmosphere-biosphere exchange of carbon dioxide. *Nature* **329**, 321–3.

Asp, H. and Berggren, D. (1990) Phosphate and calcium uptake in beech *(Fagus sylvatica)* in the presence of aluminium and natural fulvic acids. *Physiologia Plantarum* **80,** 307–14.

Aszmutat, H. and Knigge, W. (1987) Auswirkungen von Immissionsbelastungen auf einige Eigenschaften des Holzes der Fichte (*Picea abies* (L.) Karst.). *Forstarchiv* **58,** 29–35.

Athari, S. (1981) Jahrringausfall, ein meist unbeachtetes Problem bei Zuwachsuntersuchungen in Rauchgeschädigten und gesunden Fichtenbeständen. *Mitteilungen der Forstlichen Bundesversuchsanstalt Wien* **139,** 7–27.

Athari, S. (1983) Zuwachsvergleich von Fichten mit unterschiedlich starken Schadsymptomen. *Allgemeine Forst Zeitschrift* **38,** 653–5.

Athari, S. and Kramer, H. (1983) The problem of determining growth losses in Norway spruce stands caused by environmental factors. In: Ulrich, B. and Pankrath, J. (eds), *Effects of accumulation of air pollutants in forest ecosystems.* Dordrecht: Reidel, pp. 319–25.

Athari, S. and Kramer, H. (1989a) Problematik der Zuwachsuntersuchungen in Buchenbeständen mit neuartigen Schadsymptomen. *Allgemeine Forst- und Jagdzeitung* **160,** 1–8.

Athari, S. and Kramer, H. (1989b) Beziehungen zwischen Grundflächenzuwachs und verschiedenen Baumparametern in geschädigten Buchenbeständen. *Allgemeine Forst- und Jagdzeitung* **160,** 77–83.

Atkinson, I.A.E. and Greenwood, R.M. (1972) Effects of the 1969–70 drought on two remnants of indigenous lowland forest in the Manawatu district. *Proceedings of the New Zealand Ecological Society* **19,** 34–42.

Auclair, A.N.D. (1989) An historical overview of forest decline in eastern Canada. In: *Proceedings of the 70th Annual Meeting of the Woodlands Section of the Canadian Pulp and Paper Association, March 21–22, 1989, Montreal, Canada.* 9pp.

Auclair, A.N.D., Martin, H.C. and Walker, S.L. (1990) A case study of forest decline in western Canada and the adjacent United States. *Water, Air, and Soil Pollution* **53,** 13–31.

Auer, V. (1958) The Pleistocene of Fuego-Patagonia. Part II. The history of the flora and vegetation. *Annales Academiae Scientiarum Fennicae* Series A, **50,** 1–239.

Auer, V. (1966) Climatic variations in Fuego-Patagonia. In: Blumenstock, D.I. (ed.), *Climatic variations in the Pacific area.* Honolulu: Bishop Museum, pp. 37–55.

Aune, E.I., Dahl, E. and Les, A.-K. (1989) Comparisons of forest soils in relation to acid precipitation in Central Norway, South Norway and Schwarzwald in West Germany. *Meddelelser fra Norsk Institut for Skogsforskning* **42,** 133–46.

Austarå, Ø. (1984) *Diametertilvekst og tredodelighet etter masseangrep av liten granbarvikler.* Norsk Institutt for Skogforskning Report 10.

Autry, A.R., Fitzgerald, J.W. and Caldwell, P.R. (1990) Sulfur fractions and retention mechanisms in forest soils. *Canadian Journal of Forest Research* **20,** 337–42.

Avemark, W. and Schöpfer, W. (1988) Abrupte Zuwachsänderungen von Fichten und Tannen. *Forst und Holz* **43,** 123–8.

Azevedo, H.D. and Morgan, D.L. (1974) Fog precipitation in coastal Californian forests. *Ecology* **55,** 1135–41.

Bååth, E., Berg, B., Lohm, U., Lundgren, B., Lundqvist, T., Rosswall, T., Söderström, B. and Wiren, A. (1980a) Soil organisms and litter decomposition in a Scots pine forest – effects of experimental acidification. In: Hutchinson, T.C. and Havas, M. (eds), *Effects of acidic precipitation on terrestrial ecosystems.* New York: Plenum Press, pp. 375–80.

Bååth, E., Berg, B., Lohm, U., Lundgren, B., Lundqvist, H., Rosswall, T., Söderström, B. and Wiren, A. (1980b) Effects of experimental acidification and liming on soil organisms and decomposition in a Scots pine forest. *Pedobiologia* **20,** 85–100.

Badot, P.-M. (1991) Sécheresse, pollution atmosphérique et dépérissement de l'épicéa dans le massif jurassien. *Revue forestière françaises* **43,** 27–34.

Badot, P.M., Garrec, J.P., Millet, B., Badot, M.J. and Mercier, J. (1988) Dépérissement et état hydrique des aiguilles chez le *Picea abies. Canadian Journal of Botany* **66,** 1693–701.

Badot, P.M., Perrier, P. and Badot, M.J. (1990) Dépérissement, déficit hydrique et variations de croissance radiale chez le *Picea abies. Canadian Journal of Botany* **68,** 1536–41.

Baillon, F., Dalschaert, X., Grassi, S. and Geiss, F. (1988a) Spruce photosynthesis: possibility of early damage diagnosis due to exposure to magnesium or potassium deficiency. *Trees: Structure and Function* **2,** 173–9.

Baillon, F., Dalschaert, X. and Grassi, S. (1988b) Possible use of chlorophyll fluorescence in early detection of forest damages: influences of different criteria. In: Mathy, P. (ed.), *Air pollution and ecosystems.* Dordrecht: Reidel, pp. 590–5.

Baker, E.A. and Hunt, G.M. (1986) Erosion of waxes from leaf surfaces by simulated acid rain. *New Phytologist* **102,** 161–73.

Balakrishnan, N. and Mueller-Dombois, D. (1983) Nutrient studies in relation to habitat types and canopy dieback in the montane rain forest ecosystem, Island of Hawaii. *Pacific Science* **37,** 339–60.

Balder, H. and Lakenberg, E. (1987) Neuartiges Eichensterben in Berlin. *Allgemeine Forst Zeitschrift* **42,** 684–5.

Ballantyne, D.J. (1984) Phytotoxic air pollutants and oxidative phosphorylation. In: Koziol, M.J. and Whatley, F.R. (eds), *Gaseous air pollutants and plant metabolism.* London: Butterworths, pp. 223–30.

Balsberg Påhlsson, A.-M. (1989) Mineral nutrients, carbohydrates and phenolic compounds in leaves of beech (*Fagus sylvatica* L.) in southern Sweden as related to environmental factors. *Tree Physiology* **5,** 485–95.

Balsberg Påhlsson, A.-M. (1992) Influence of nitrogen fertilization on minerals, carbohydrates, amino acids and phenolic compounds in beech (*Fagus sylvatica* L.) leaves. *Tree Physiology* **10,** 93–100.

Baltensweiler, W. (1985) Waldsterben: forest pests and air pollution. *Zeitschrift für angewandte Entomologie* **99,** 77–85.

Barbolani, E., Del Panta, S., Ferretti, M. and Pantani, F. (1986) Survey of acid deposition in coastal areas of Tuscany. *Annali di Chimica* **76,** 365–74.

Barbu, I. (1987) Cercetäri asupra proceselor de rarire si uscare a coroanelor la bradul din Bucovina. *Revista Padurilor* **102,** 195–9.

Barbu, I. (1991) *Moartea bradului – simptom al degradarii mediului.* Editura Ceres, Bucharest.

Barnard, J.E. and Scott, C.T. (1988) *Changes in tree growth rates in Vermont.* Research Note SE-350. Asheville: United States Department of Agriculture Forest Service, South-east Forest Experiment Station. (6 pp).

Barnard, J.E., Lucier, A.H., Johnson, A.H., Brooks, R.T., Karnosky, D.F. and Dunn, P.H. (1989) *Changes in forest health and productivity in the United States of America.* NAPAP State of Science/Technology Report No. 16. Washington DC: National Acidic Precipitation Assessment Program.

Barnes, J.D. and Davison, A.W. (1988) Interaction between ozone and frost sensitivity in clonal trees of Norway spruce. In: Bervaes, J., Mathy, P. and Evers, P. (eds), *Relationship between above and below ground influences of air pollutants on forest trees.* Air Pollution Report 16. Brussels: Commission of the European Communities pp. 77–82.

Barnes, J.D., Davison, A.W. and Booth, T.A. (1988) Premature degradation of epicuticular wax on *Picea abies* needles induced by ozone. In: Cape, J.N. and Mathy, P. (eds), *Scientific basis of forest decline symptomatology.* Air Pollution Report 15. Brussels: Commission of the European Communities, pp. 245–51.

Barnes, J.D., Eamus, D., Davison, A.W., Ro-Poulsen, H. and Mortensen, L. (1990) Persistent effects of ozone on needle water loss and wettability in Norway spruce. *Environmental Pollution* **63,** 345–63.

Barnes, R.L. (1972) Effects of chronic exposure of ozone on photosynthesis and respiration of pines. *Environmental Pollution* **3,** 133–8.

Barnhardt, S.J., McBride, J.R. and Warner, P. (1991) Oak seedling establishment in relation to environmental factors at Annadel State Park. In: *Proceedings of the symposium on oak woodlands and hardwood rangeland management, October 31 – November 2, 1990, Davis, California.* General Technical Report PSW-126. Berkeley: United States Department of Agriculture Forest Service, Pacific Southwest Research Station, pp. 25–30.

Bartholmess, H. (1989) Untersuchungen über den Zusammenhang zwischen Flechtenvegetation, Immissionsbelastung und Waldschäden in Baden-Württemberg. *Forstwissenschaftliches Centralblatt* **108,** 188–96.

Barzut, V.M. (1984) Features of the changes in the annual increment of pine and spruce growing together. *Lesnoi Zhurnal* **1984,** 113–5.

Basabe, F.A., Edmonds, R.L., Chang, W.L. and Larson, T.V. (1989) Fog and cloudwater chemistry in Western Washington. In: Olson, R.K. and Lefohn, A.S. (eds), *Effects of air pollution on western forests.* Pittsburgh: Air and Waste Management Association, pp. 33–49.

Batcheler, C.L. (1983) The possum and rata-kamahi dieback in New Zealand: a review. Pacific Science **37,** 415–26.

Battles, J.J., Johnson, A.H. and Siccama, T.G. (1988) Relationships between red spruce decline and forest characteristics at Whiteface Mountain, New York. In: *Proceedings of the US/FRG Research Symposium: effects of atmospheric pollutants on the spruce-fir forests of the eastern United States and the Federal*

Republic of Germany. General Technical Report NE-120. Broomall, Pennsylvania: United States Department of Agriculture Forest Service, Northeastern Forest Experiment Station, pp. 163–72.

Battles, J.J., Johnson, A.H., Siccama, T.G., Friedland, A.J. and Miller, E.K. (1989) Spatial patterns of red spruce decline on Whiteface Mt., New York. In: *Results of studies on red spruce decline at Whiteface Mt. New York. Report to the Forest Response Program*. Corvallis, Oregon: United States Environmental Protection Agency, pp. 18–39.

Bauce, E. and Allen, D.C. (1991) Etiology of a sugar maple decline. *Canadian Journal of Forest Research* **21,** 686–93.

Bauce, E. and Allen, D.C. (1992) Condition of the fine roots of sugar maple in different stages of decline. *Canadian Journal of Forest Research* **22,** 264–66.

Bauch, J. (1983) Biological alterations in the stem and root of fir and spruce due to pollution influence. In: Ulrich, B. and Pankrath, J. (eds), *Effects of accumulation of air pollutants in forest ecosystems*. Dordrecht: Reidel, pp. 377–86.

Bauch, J., Stienen, H., Ulrich, B. and Matzner, E. (1985a) Einfluss einer Kalkung bzw. Düngung auf den Elementgehalt in Feinwurzeln und das Dickenwachstum von Fichten aus Waldschadensgebieten. *Allgemeine Forst Zeitschrift* **40,** 1148–50.

Bauch, J., Rademacher, P., Bernecke, W., Knoth, J. and Michaelis, W. (1985b) Breite und Elementgehalt der Jahrringe in Fichten aus Waldschadensgebieten. In: Stratmann, H. (ed.), *Waldschäden: Einflussfaktoren und ihre Bewertung. Kolloquium Goslar, 18–20 June 1985*. VDI-Bericht 560, pp. 943–59.

de Bauer, L.I. and Krupa, S.V. (1990) The Valley of Mexico: summary of observational studies on its air quality and effects on vegetation. *Environmental Pollution* **65:** 109–18.

Baule, H. and Fricker, C. (1970) *The fertilizer treatment of forest trees*. Munich: BLV.

Beaubien, J. and Jobin, L. (1974) ERTS-1 imagery for broad mapping of forest damage and cover types of Anticosti Island. *Canadian Surveyor* **28,** 164–6.

Bechtold, W.A., Ruark, G.A. and Lloyd, F.T. (1991) Changing stand structure and regional growth reductions in Georgia's natural pine stands. *Forest Science* **37,** 703–17.

Becker, M. (1985) Le dépérissement du sapin dans les Vosges. Quelques facteurs liés à la détérioration des cimes. *Revue forestières françaises* **37,** 281–7.

Becker, M. (1987a) Bilan de santé actuel et rétrospectif du sapin (*Abies alba* Mill.) dans les Vosges. Étude écologique et dendrochronologique. *Annales des Sciences Forestières* **44:** 379–401.

Becker, M. (1987b) Santé de la forêt: le sapin témoignes. *Recherche* **18,** 1096–8.

Becker, M. (1989) The role of climate on present and past its vitality of silver fir forests in the Vosges Mountains of northeastern France. *Canadian Journal of Forest Research* **19,** 1110–7.

Becker, M. (1991) Impact of climate, soil and silviculture on forest growth and health. In: Landmann, G. (ed.), *French research into forest decline*. Nancy: École Nationale du Génie Rurual, des Eaux et des Forêts, pp. 23–38.

Becker, M., Bräker, O.U., Kenk, G., Schneider, O. and Schweingruber, F.H. (1990a) Aspect des houppiers et croissance des arbres au cours des dernières décennies dans les régions frontalières d'Allemagne, de France et de Suisse. *Revue forestières françaises* **42,** 284–300.

Becker, M., Bräker, O.U., Kenk, G., Schneider, O. and Schweingruber, F.H. (1990b) Kronenzustand und Wachstum von Waldbäumen im Dreiländereck Deutschland-Frankreich-Schweiz in den letzen Jahrzehnten. *Allgemeine Forst Zeitschrift* **45,** 263–6.

Becker, M., Landmann, G. and Lévy, G. (1989) Importance des facteurs climatiques, sylvicoles et stationnels dans le dépérissement du sapin dans le massif Vosgien (N.E. France). In: Bucher, J. and Bucher-Wallin, I. (eds), *Air pollution and forest decline.* Birmensdorf: Eidgenössische Anstalt für das forstliche Versuchswesen, pp. 277–82.

Becker, M. and Lévy, G. (1982) Le dépérissement du chêne en forêt de Tronçais. *Annales des Sciences Forestières* **39,** 439–44.

Becker, M. and Lévy, G. (1988) A propos du dépérissement des forêts: climat, sylviculture et vitalité de la sapinière Vosgienne. *Revue forestières françaises* **40,** 345–58.

Becker, M., Landmann, G. and Lévy, G. (1989) Silver fir decline in the Vosges Mountains (France): role of climate and silviculture. *Water, Air, and Soil Pollution* **48,** 77–86.

Beckett, P.H.T. and Webster, R. (1971) Soil variability: A review. *Soils and Fertilizers* **34,** 1–15.

Becquer, T., Merlet, D., Boudot, J.-P., Rouiller, J. and Gras, F. (1990) Nitrification and nitrate uptake: leaching balance in a declined forest ecosystem in eastern France. *Plant and Soil* **125,** 95–107.

Beevers, L. and Hageman, R.H. (1969) Nitrate reduction in higher plants. *Annual Review of Plant Physiology* **20,** 495–522.

Belanger, R.P. and Anderson, R.L. (1988) *A guide for visually assessing crown densities of Loblolly and Shortleaf pines.* Research Note SE-352. Asheville: United States Department of Agriculture Forest Service, South-eastern Forest Experiment Station.

Belkacem, S., Nys, C. and Gelhaye, D. (1992) Effets d'une fertilisation et d'un amendement sur l'immobilisation d'éléments dans la biomasse d'un peuplement adulte d'épicéa commun (*Picea abies* L. Karst.). *Annales des Sciences forestières* **49,** 235–52.

Bell, D.T. and Ward, S.C. (1984) Foliar and twig macronutrients (N, P, K, Ca and Mg) in selected species of *Eucalyptus* used in rehabilitation. Sources of variation. *Plant and Soil* **82,** 363–76.

Benarie, M. (1987) Lack of relationship between acidic rain and tree damage in urban areas of Europe: Part II. *Science of the Total Environment* **61,** 253–57.

Bender, J., Jäger, H.-J., Schweizer, B., Seufert, G. and Arndt, U. (1989) Long-term effects of air pollutants on forest trees in open-top chambers. In: Bucher, J. and Bucher-Wallin, I. (eds), *Air pollution and forest decline.* Birmensdorf: Eidgenössische Anstalt für das forstliche Versuchswesen, pp. 167–71.

Bender, J., Jäger, H.-J., Seufert, G. and Arndt, U. (1986) Untersuchungen zur Einzel- und Kombinationswirkung von SO_2 und O_3 auf den Stoffwechsel von Waldbäumen in Open-top-Kammern. *Angewandte Botanik* **60,** 461–79.

Bengtsson, B., Asp, H., Jensén, P. and Berggren, D. (1988) Influence of aluminium on phosphate and calcium uptake in beech (*Fagus sylvatica*) grown in nutrient solution and soil solution. *Physiologia Plantarum* **74,** 299–305.

Bengtsson, G. (1985) Approach and methodology for monitoring of the health of forests in Sweden. In: Schmid-Haas, P. (ed.), *Inventorying and monitoring endangered forests.* Birmensdorf: Eidgenössische Anstalt für das forstliche Versuchswesen, pp. 205–8.

Beniamino, F., Ponge, J.F. and Arpin, P. (1991) Soil acidification under the crown of oak trees 1. Spatial distribution. *Forest Ecology and Management* **40,** 221–32.

Benner, P., Sabel, P. and Wild, A. (1988) Photosynthesis and transpiration of healthy and diseased spruce trees in the course of three vegetation periods. *Trees: Structure and Function,* 223–32.

Benner, P. and Wild, A. (1987) Measurement of photosynthesis and transpiration in spruce trees with various degrees of damage. *Journal of Plant Physiology* **129,** 59–72.

Bennett, J.P., Anderson, R.L., Mielke, M.L. and Ebersole, J.J. (1989) An historical perspective on forest decline of eastern white pine. In: Bucher, J. and Bucher-Wallin, I. (eds), *Air pollution and forest decline.* Birmensdorf: Eidgenössische Anstalt für das forstliche Versuchswesen, pp. 69–75.

Benoit, L.F., Skelly, J.M., Moore, L.D. and Dochinger, L.S. (1982) Radial growth reductions in *Pinus strobus* L. correlated with foliar ozone sensitivity as an indicator of ozone-induced losses in eastern forests. *Canadian Journal of Forest Research* **12,** 673–8.

Benzian, B. (1966) Effects of nitrogen and potassium concentrations in conifer seedlings on frost damage. *Report of Rothamsted Experimental Station 1965,* 58–9.

Berg, B. (1986a) Nutrient release from litter and humus in coniferous forest soils – a mini review. *Scandinavian Journal of Forest Research* **1,** **3**359–69.

Berg, B. (1986b) The influence of experimental acidification on needle litter decomposition in a *Picea abies* L. forest. *Scandinavian Journal of Forest Research* **1,** **3**317–22.

Berg, B. (1986c) The influence of experimental acidification on nutrient release and decomposition rates of needle and root litter in the forest floor. *Forest Ecology and Management* **15,** 195–213.

Bergdahl, D.R., Grehan, J.R., Parker, B.L., Tobi, D.R. and Halik, S. (1992) Extent of injury and the fungi associated with feeding wounds created by conifer swift moth larvae on roots of red spruce, in, *52nd Meeting American Phytopathological Society, North-eastern Division, 28–30 October 1992, Portland, Maine, Program and Abstracts,* p. 30.

Bergkvist, B. (1987a) Leaching of metals from forest soils as influenced by tree species and management. *Forest Ecology and Management* **22,** 29–56.

Bergkvist, B. (1987b) Soil solution chemistry and metal budgets of spruce forest ecosystems in S. Sweden. *Water, Air, and Soil Pollution* **33,** 131–54.

Bergkvist, B., Folkeson, L. and Berggren, D. (1989) Fluxes of Cu, Zn, Pb, Cd, Cr, and Ni in temperate forest ecosystems. *Water, Air, and Soil Pollution* **47,** 217–86.

Bergmann, F. and Scholz, F. (1987) The impact of air pollution on the genetic structure of Norway spruce. *Silvae Genetica* **36,** 80–3.

Bergmann, W. (1927) *Heß-Beck, Forstschutz, 5.* Neudamm: Verlag J. Neumann.

Bernier, B. and Brazeau, M. (1986) Sugar maple decline in Quebec: the role of atmospheric pollution. In: *Maple decline*. Quebec: Ministère de l'Agriculture, des Pêcheries et de l'Alimentation, pp. 97–109.

Bernier, B. and Brazeau, M. (1988a) Foliar nutrient status in relation to sugar maple dieback and decline in the Quebec Appalachians. *Canadian Journal of Forest Research* **18,** 754–61.

Bernier, B. and Brazeau, M. (1988b) Nutrient deficiency symptoms associated with sugar maple dieback and decline in the Quebec Appalachians. *Canadian Journal of Forest Research* **18,** 762–7.

Bernier, B. and Brazeau, M. (1988c) Magnesium deficiency symptoms associated with sugar maple dieback in a Lower Laurentians site in southeastern Quebec. *Canadian Journal of Forest Research* **18,** 1265–9.

Bernier, B., Paré, D. and Brazeau, M. (1989) Natural stresses, nutrient imbalances and forest decline in southeastern Quebec. *Water, Air, and Soil Pollution* **48,** 239–50.

Berrang, P., Karnosky, D.F. and Bennett, J.P. (1991) Natural selection for ozone tolerance in *Populus tremuloides*: an evaluation of nationwide trends. *Canadian Journal of Forest Research* **21,** 1091–7.

Berry, C.R. (1961) White pine emergence tipburn, a physiogenic disturbance. Southeastern Forest Experiment Station Paper 130. Ashville: United States Department of Agriculture Forest Service, Southeastern Forest Experiment Station.

Berry, C.R. (1973) The differential sensitivity of Eastern White Pine to three types of air pollution. *Canadian Journal of Forest Research* **3,** 5543–47.

Berry, C.R. and Hepting, G.H. (1964) Injury to Eastern White Pine by unidentified atmospheric constituents. *Forest Science* **10,** 2–13.

Bert, G.D. (1992) Les principaux types de sapinières (*Abies alba* Mill.) dans le massif du Jura (France et Suisse). Étude phytoécologique. *Annales des Sciences Forestières* **49,** 161–83.

Bert, G.D. and Becker, M. (1990) Vitalité actuelle et passée du sapin (*Abies alba* Mill) dans le Jura. Étude dendroécologique. *Annales des Sciences Forestières* **47,** 395–412.

Bettany, J.R., Stewart, W.B. and Saggar, S. (1979) The nature and forms of sulfur in organic matter fractions of soils selected along an environmental gradient. *Soil Science Society of America Journal* **43,** 981–5.

Bewley, R.J.F. and Parkinson, D. (1984) Effects of sulphur dioxide pollution on forest soil microorganisms. *Canadian Journal of Microbiology* **30,** 179–85.

Bewley, R.J.F. and Stotzky, G. (1983a) Simulated acid rain (H_2SO_4) and microbial activity in soil. *Soil Biology and Biochemistry* **15,** 425–9.

Bewley, R.J.F. and Stotzky, G. (1983b) Anionic constituents of acid rain and microbial activity in soil. *Soil Biology and Biochemistry* **15,** 431–7.

Beyer, L., Blume, H.P. and Irmler, U. (1991) The humus of a 'Parabraunerde' (Orthic Luvisol) under *Fagus sylvatica* L and *Quercus robur* L and its modification in 25 years. *Annales des Sciences Forestières* **48,** 267–78.

Beyschlag, W., Wedler, M., Lange, O.L. and Heber, U. (1987) Einfluss einer Magnesiumdüngung auf Photosynthese und Transpiration von Fichten an einem Magnesium-Mangelstandort im Fichtelgebirge. *Allgemeine Forst Zeitschrift* **42,** 738–41.

Bialobok, S. (1989) Próba oceny uszkodzen drzew i krzewów iglastych poddanych dzialaniu dwutlenku siarki w otwartych komorach polowych. *Arboretum Kórnickie* **31,** 259–67.

Bian, Y.M. and Yu, S.W. (1992) Forest decline in Nanshan, China. *Forest Ecology and Management* **51,** 53–9.

Bicknell, S.H. (1989) Fog and rain chemistry in northern California. In: Olson, R.K. and Lefohn, A.S. (eds), *Effects of air pollution on western forests.* Pittsburgh: Air and Waste Management Association, pp. 147–67.

Billett, M.F., Fitzpatrick, E.A. and Cresser, M.S. (1988) Long-term changes in the acidity of forest soils in north-east Scotland. *Soil Use and Management* **4,** 102–7.

Billett, M.F., Fitzpatrick, E.A. and Cresser, M.S. (1990) Changes in the carbon and nitrogen status of forest soil organic horizons between 1949/50 and 1987. *Environmental Pollution* **66,** 67–79.

Billett, M.F., Fitzpatrick, E.A. and Cresser, M.S. (1991) Long-term changes in the Cu, Pb, and Zn content of forest soil organic horizons from North-east Scotland. *Water, Air, and Soil Pollution* **59,** 179–91.

Binkley, D., Driscoll, C.T., Allen, H.L., Schoeneberger, P. and McAvoy, D. (1989a) *Acidic deposition and forest soils. Context and case studies in the southeastern United States.* New York: Springer Verlag.

Binkley, D., Valentine, D., Wells, C. and Valentine, U. (1989b) An empirical analysis of the factors contributing to a 20-year decrease in soil pH in an old-field plantation of loblolly pine. *Biogeochemistry* **8,** 39–54.

Binns, W.O., Redfern, D.B., Boswell, R.C. and Betts, A.J.A. (1986) Forest health and air pollution 1985 survey. Forestry Commission Research and Development Paper 147. Edinburgh: Forestry Commission.

Binns, W.O., Redfern, D.B., Rennolls, K. and Betts, A.J.A. (1985) Forest health and air pollution 1984 survey. Forestry Commission Research and Development Paper 142. Edinburgh: Forestry Commission.

Birks, H.J.B. (1989) Holocene isochrone maps and patterns of tree spreading in the British Isles. *Journal of Biogeography* **16,** 503–40.

Björkdahl, G. and Eriksson, H. (1989) Effects of forest decline on increment in Norway spruce (*Picea abies* (L.) Karst.) in southern Sweden. *Meddelelser fra Norsk Institutt for Skogforskning* **42,** 19–36.

Björkman, E. (1970) The effect of fertilization on sulphur dioxide damage to conifers in industrial and built-up areas. *Studia Forestalia Suecica* **78.**

Black, V.J. (1984) The effects of air pollutants on apparent respiration. In: Koziol, M.J. and Whatley, F.R. (eds), *Gaseous air pollutants and plant metabolism.* London: Butterworths, pp. 231–48.

Black, V.J. (1985) SO_2 effects on stomatal behavior. In: Winner, W.E., Mooney, H.A. and Goldstein, R.A. (eds), *Sulfur dioxide and vegetation.* Stanford: Stanford University Press, pp. 96–117.

Blake, J., Webster, S.R. and Gessel, S.P. (1988) Soil sulfate-sulfur and growth responses of nitrogen-fertilized Douglas-fir to sulfur. *Soil Science Society of America Journal* **52,** 1141–7.

Blanchard, R.O., Baas, J. and Van Cotter, H. (1979) Oxidant damage to eastern white pine in New Hampshire. *Plant Disease Reporter* **63,** 177–82.

Blaschke, H. (1986) Einfluß von saurer Beregnung und Kalkung auf die Biomasse und Mykorrhizierung der Feinwurzeln von Fichten. *Forstwissenschaftliches Centralblatt* **105,** 324–9.

Blaser, P. (1982) Der Einfluss von Säureeinträgen auf den Boden. In: *Waldschadenforschung in der Schweiz: Stand der Kenntnisse.* Birmensdorf: Eidgenössische Forschungsanstalt für Wald, Schnee und Landschaft, pp. 103–12.

Boardman, R. (1978) Productivity under successive rotations of Radiata pine. *Australian Forestry* **41,** 177–9.

Boardman, R. and McGuire, D.O. (1990a) The role of zinc in forestry. I. Zinc in forest environments, ecosystems and tree nutrition. *Forest Ecology and Management* **37,** 167–205.

Boardman, R. and McGuire, D.O. (1990b) The role of zinc in forestry. II. Zinc deficiency and forest management: Effect on yield and silviculture of *Pinus radiata* plantations in South Australia. *Forest Ecology and Management* **37,** 207–18.

Bockheim, J.G., Leide, J.E. and Frelich, L.E. (1989) Red pine growth and chemical composition of foliage and forest floors across a precipitation–chemistry gradient in Wisconsin. *Canadian Journal of Forest Research* **19,** 1543–9.

Boehmer-Christiansen, S. (1988) The politics of dying forests. *Interdisciplinary Science Reviews* **13,** 5–8.

den Boer, W.M.J. and van den Tweel, P.A. (1985) The health condition of the Dutch forests in 1984. *Netherlands Journal of Agricultural Science* **33,** 167–74.

Böhm, M. (1989) A regional characterization of air quality and deposition in the coniferous forests of the western United States. In: Olson, R.K. and Lefohn, A.S. (eds), *Effects of air pollution on western forests.* Pittsburgh: Air and Waste Management Association, pp. 221–44.

Böhm, M. and Vandetta, T. (1989) *Ozone patterns in and around the coniferous forests of the western United States.* Report to the Forest Response Program, U.S. Environmental Protection Agency, Corvallis, Oregon.

Bondietti, E.A., Momoshima, N., Shortle, W.C. and Smith, K.T. (1990) A historical perspective on divalent cation trends in red spruce stemwood and the hypothetical relationship to acidic deposition. *Canadian Journal of Forest Research* **20,** 1850–8.

Bonneau, M. (1989) Étude de la nutrition minérale du sapin et d'épicéas jaunnissants du Jura central et méridional. In: *Journées de travail DEFORPA February-March 1989,* Volume 3. Nancy: INRA, pp. 6.4.1–16.

Bonneau, M. and Landmann, G. (1988) Forest dieback in Europe. *La Recherche* **19,** 1542–53.

Borchert, R. (1976) Differences in shoot growth patterns between juvenile and adult trees and their interpretation based on systems analysis of trees. *Acta Horticultura* **56,** 123–30.

Bordeleau, C. (1986) Sugar maple dieback: extent of damage in the Appalachians. In: *Maple decline.* Quebec: Ministère de l'Agriculture, des Pêcheries et de l'Alimentation, pp. 17–31.

Borie, L. (1987) Are sugar maples declining? *American Forests* November/December 1987, 26–8 and 66–70.

Bosch, C., Pfannkuch, E., Rehfuess, K.E., Runkel, K.H., Schramel, P. and Senser, M. (1986) Einfluß einer Düngung mit Magnesium und Calcium, von

Ozon und saurem Nebel auf Frosthärte, Ernährungszustand und Biomasseproduktion junger Fichten (*Picea abies* (L.) Karst.). *Forstwissenschaftliches Centralblatt* **105,** 218–29.

Bosch, C. and Rehfuess, K.E. (1988) Über die Rolle von Frostereignissen bei den 'neuartigen' Waldschäden. *Forstwissenschaftliches Centralblatt* **107,** 123–30.

Bosch, C., Pfannkuch, E., Baum, U. and Rehfuess, K.E. (1983) Über die Erkrankung der Fichte (*Picea abies* Karst.) in den Hochlagen des Bayerischen Waldes. *Forstwissenschaftliches Centralblatt* **102,** 167–81.

Botkin, D.B., Smith, W.H., Carlsen, R.W. and Smith, T.L. (1972) Effects of ozone on white pine saplings: Variation in inhibition and recovery of net photosynthesis. *Environmental Pollution* **3,** 273–89.

Bottacci, A. (1991) Effetti di deposizioni acide simulate su giovani pintine di abiete bianco (*Abies alba* Mill.) di due diverse provenienze. *Monti e Boschi* **42,** 43–8.

Bottacci, A., Brogi, L., Bussotti, F., Cenni, E., Clauser, F., Ferretti, M., Gellini, R., Grossoni, P. and Schiff, S. (1988) *Inquinamento ambientales e deperimento del bosco in Toscana.* Regione Toscana, Società Botanica Italiana, Firenze.

Boulton, T.W. and Flager, R.B. (1990) *Growth and water use efficiency of shortleaf pine as affected by ozone and acid rain.* Paper presented to the 83rd Annual Meeting of the Air and Waste Management Association, June 24–9, 1990, Pittsburgh.

Bouhot, L., Landmann, G. and Pierrat, J.C. (1992) *Visual assessment of the crown condition of forest trees by surveyors from six Mediterranean countries.* Unpublished manuscript available from G. Landmann, CRF-INRA, BP35, Champenoux, 54280 Seichamps, France.

Bouman, T. and Lohmann, H. (1987) Bodenchemische Bewertung von Forststandorten in Harz. *Mitteilungen der Deutschen Bodenkundlichen Gesellschaft* **55,** 719–21.

Bowden, R.D., Geballe, G.T. and Bowden, W.B. (1989) Foliar uptake of ^{15}N from simulated cloud water by red spruce (*Picea rubens*) seedlings. *Canadian Journal of Forest Research* **19,** 382–6.

Bower, J.S., Stevenson, K.J., Broughton, G.F.J., Lampert, J.E., Sweeney, B.P. and Wilken, J. (1991) *Ozone in the UK: A review of 1989/90 data from monitoring sites operated by Warren Spring Laboratory.* Report LR 793 (AP). Stevenage: Warren Spring Laboratory.

Bowling, P.J. and McLeod, D.E. (1968) A note on the presence of Armillaria in second growth eucalypt stands in southern Tasmania. *Australian Forest Research* **3,** 38–40.

Boxman, A.W. and Roelofs, J.G.M. (1988) Some effects of nitrate versus ammonium nutrition on the nutrient fluxes in *Pinus sylvestris* seedlings. Effects of mycorrhizal infection. *Canadian Journal of Botany* **66,** 1091–7.

Boyce, R.L. (1988) Wind direction and fir wave motion. *Canadian Journal of Forest Research* **18,** 461–6.

Boyce, R.L., Friedland, A.J., Webb, E.T. and Herrick, G.T. (1991) Modelling the effect of winter climate on high-elevation red spruce shoot water contents. *Forest Science* **37,** 1567–80.

Boyer, J.N., Houston, D.B. and Jensen, K.F. (1986) Impacts of chronic SO_2, O_3, and $SO_2 + O_3$ exposures on photosynthesis of *Pinus strobus* clones. *European Journal of Forest Pathology* **16,** 293–9.

Braanas, T. (1970) Fluorskader på furu i Vettismorki. *Tidsskrift for Skogbruk* **78,** 379–401.

Brabänder, H.D. and Bitter, A.W. (1991) Stand der Bewertung von Immissionsschäden in der Bundesrepublik Deutschland. In: Jöbstl, A. (ed.), *Ökonomische Bewertung von Waldschäden.* Vienna: Österreichischer Agrarverlag, pp. 63–74.

Bräker, O.U. (1990) Tree ring analysis of conifers in the Swiss forest decline study program. In: *Proceedings of the Lund Conference on tree rings.* In press.

Bräker, O.U. (1991) Der Radialzuwachs an unterschiedlich belaubten Buchen in zwei Beständen bei Zürich und Basel. *Schweizerische Zeitschrift für Forstwesen* **142,** 427–33.

Bräker, O.U. and Z'Graggen, S. (1987) Tree ring analysis in the Swiss forest decline study of 1984. In: Kairiukstis, L., Nilsson, S. and Straszak, A. (eds), *Forest decline and reproduction: Regional and global consequences.* Laxenburg: International Institute for Applied Systems Analysis, pp. 169–78.

Brand, D.G., Kehoe, P. and Connois, M. (1986) Coniferous afforestation leads to soil acidification in central Ontario. *Canadian Journal of Forest Research* **16,** 1389–91.

Brandl, H. (1985) Zur Bedeutung bestandesgeschichtlicher Untersuchungen in der Forstgeschichte am Bespiel des 'Tannensterbens' im Schwarzwald. *Allgemeine Forst- und Jagdzeitung* **156,** 142–6.

Brandl, H. and Matthies, F. (1984) Bewertungssystematik für Waldschäden durch Luftverunreinigungen. *Der Forst- und Holzwirt* **39,** 8–12.

Braun, S. and Schröter, H.J. (1991) Waldschadensbericht 1991 in Baden-Württemberg. *Forst und Holz* **46,** 637–40.

Breckle, S.-W. and Kahle, H. (1992) Effects of toxic heavy metals (Cd, Pb) on growth and mineral nutrition of beech (*Fagus sylvatica* L.). *Vegetatio* **101,** 43–53.

Van Breemen, N., Burrough, P.A., Velthorst, E.J., Van Dobben, H.F. De Wit, T., Ridder, T.B. and Reijnders, H.F.R. (1982) Soil acidification from atmospheric ammonium sulphate in forest canopy throughfall. *Nature* **299,** 548–50.

Van Breemen, N. and van Dijk, H.F.G. (1988) Ecosystem effects of atmospheric deposition of nitrogen in The Netherlands. *Environmental Pollution* **54,** 249–74.

Van Breemen, N., Driscoll, C.T. and Mulder, J. (1984) The role of acidic deposition and internal proton sources in acidification of soils and waters. *Nature* **307,** 599–604.

Van Breemen, N., Mulder, J. and Driscoll, C.T. (1983) Acidification and alkalinization of soils. *Plant and Soil* **75,** 283–308.

Van Breemen, N., Visser, W.F.J. and Pape, T. (1988) *Biogeochemistry of an oak-woodland ecosystem in The Netherlands affected by acid deposition.* Agricultural Research Reports 930. Wageningen: Pudoc.

Breeuwsma, A. and De Vries, W. (1984) The relative importance of natural production of H^+ in soil acidification. *Netherlands Journal of Agricultural Science* **32,** 161–3.

Brenninger, C. and Tranquillini, W. (1983) Photosynthese, Transpiration und Spaltöffnungsverhalten verschiedener Holzarten nach Begasung mit SO_2. *European Journal of Forest Pathology* **13,** 228–38.

Brewer, R.L., Gordon, R.J. and Shepard, L.S. (1983) Chemistry of mist and fog from the Los Angeles urban area. *Atmospheric Environment* **17,** 2267–70.

Bridge, M.C. and Ni Bhreartuin, C. (1988) A dendrochronological investigation of *Fagus sylvatica* in Co. Wicklow. *Irish Naturalists Journal* **22,** 407–11.

Briffa, K.R. (1991b) *Detection of any widespread and unprecedented changes in the growth of European conifers.* Report to the Forestry Commission, Alice Holt Lodge, Wrecclesham, Farnham, Surrey GU10 4LH, England.

Briffa, K.R. (1991) Increasing productivity of 'natural growth' conifers in Europe over the last century. In: *Proceedings of the International Dendrochronological Symposium 'Tree rings and environment'*, 3–9 September 1990, Lund, Sweden. In press.

Briffa, K.R. and Cook, E.R. (1989) Methods of response function analysis. In: Cook, E.R. and Kariukstis, L.A. (eds), *Methods of dendrochronology. Applications in the environmental sciences.* Dordrecht: Kluwer and Laxenburg: International Institute for Applied Systems Analysis, pp. 240–6.

Bringmark, E. (1989) Spatial variation in soil pH of beech forests in relation to buffering properties and soil depths. *Oikos* **54,** 165–77.

Brooks, R.T. (1989) An analysis of regional forest growth and atmospheric deposition patterns, Pennsylvania (USA). In: Bucher, J. and Bucher-Wallin, I. (eds), *Air pollution and forest decline.* Birmensdorf: Eidgenössische Anstalt für das forstliche Versuchswesen, pp. 283–88.

Brooks, R.T., Miller-Weeks, M. and Burkman, W. (1991) *Forest health monitoring New England 1990.* Report NE-INF-94-91. Northeastern Area Association of State Foresters and United States Department of Agriculture Forest Service.

Brown, B.N. (1976) *Phytopthora cinnamomi* associated with patch death in tropical rainforests in Queensland. *Australian Plant Pathology Society Newsletter* **5,** 1–4.

Brown, F.W. and Davis, F.W. (1991) Historical mortality of valley oak (*Quercus lobata* Née) in the Santa Ynez Valley, Santa Barbara County, 1938–1989. In: *Proceedings of the symposium on oak woodlands and hardwood rangeland management, October 31–November 2, 1990, Davis, California.* General Technical Report PSW-126. Berkeley: United States Department of Agriculture Forest Service, Pacific Southwest Research Station, pp. 202–7.

Brown, K.A. and Roberts, T.M. (1988) Effects of ozone on foliar leaching in Norway spruce (*Picea abies* (L.) Karst.): confounding factors due to NO_x production during ozone generation. *Environmental Pollution* **55,** 55–73.

Brown, K.A., Roberts, T.M. and Blank, L.W. (1987) Interaction between ozone and cold sensitivity in Norway spruce: a factor contributing to the forest decline in central Europe? *New Phytologist* **105,** 149–55.

Bruck, R.I. (1985) Boreal montane ecosystem decline in the southern Appalachian Mountains: potential role of anthropogenic pollution. In: Stubbs, H.S. (ed.), *Air pollution effects on forest ecosystems.* St. Paul, Minnesota: Acid Rain Foundation, pp. 137–55.

Bruck, R.I. and Robarge, W.P. (1988) Change in forest structure in the boreal montane ecosystem of Mount Mitchell, North Carolina. *European Journal of Forest Pathology* **18,** 357–66.

Bruck, R.I., Robarge, W.P. and McDaniel, A. (1989) Forest decline in the boreal montane ecosystem of the southern Appalachian Mountains. *Water, Air, and Soil Pollution* **48,** 161–80.

Bruns, H.-H. (1991) Die Waldschäden 1991 in Niedersachsen. *Forst und Holz* **46,** 615–19.

Bucher, J.B. (1988) Forest decline in Switzerland – critical remarks after four years of observation. Lecture presented to Joint Seminar on Global Habitability, Columbia University/University of Geneva, Geneva, 18–20 May, 1987.

Buchholzer, P. (1988) Buchenholz stark geschädigter Bäume als Rohstoff für die Spanplattenherstellung. *Holz-Zentralblatt* **114,** 225–6.

Buchholzer, P. and Harbs, C. (1986) Eignung von Fichtenholz aus Waldschadensgebieten zur Herstellung von Holzspanplatten. *Holz als Roh- und Werkstoff* **44,** 281–5.

Buck-Feucht, G. (1986) Vergleich alter und neuer Waldvegetationsaufnahmen im Forstbezirk Kirchheim unter Teck. *Mitteilungen des Vereins für Forstliche Standortskunde und Forstpflanzenzüchtung* **32,** 43–9.

Bues, C.T., Fengel, D., Grabner, I., Heindl, S., Schots-v.d. Zee, M. and Tang, J. (1989) Untersuchungen an kronengeschädigten Kiefern (*Pinus sylvestris* L.) in Nordost-Bayern. Teil II. Holzqualität. *Forstarchiv* **60,** 69–73.

Bues, C.T. and Schulz, H. (1988) Festigkeit und Feuchtegehalt von Kiefernholz aus Waldschadensgebieten. *Holz als Roh- und Werkstoff* **46,** 41–5.

Bues, C.T. and Schulz, H. (1990) Festigkeit und Feuchtegehalt von Eichenholz aus Waldschadensgebieten. *Holz als Roh- und Werkstoff* **48,** 85–9.

Bunce, H.W.F. (1984) Fluoride emissions and forest survival, growth and regeneration. *Environmental Pollution,* Series A, **35,** 169–88.

Bunce, H.W.F. (1989) The continuing effects of aluminium smelter emissions on coniferous forest growth. *Forestry* **62,** 223–31.

Van den Burg, J. (1988) Nitrogen deposition and mineral nutrient status of forests in The Netherlands on acid sandy soils. Paper read at the visit to The Netherlands of the EFMA Environmental Committee, Helmond, 14 June, 1988.

Van den Burg, J. (1991) Results and experiences from fertilization experiments in The Netherlands. *Fertilizer Research* **27,** 107–11.

Burgan, R.E. and Nelson, R.E. (1972) *Decline of 'ohi'a lehua forests in Hawaii.* General Technical Report PSW-3. Berkeley: United States Department of Agriculture Forest Service, Pacific Southwest Forest and Range Experimental Station.

Burke, M.K., Raynal, D.J. and Mitcell, M.J. (1992) Soil nitrogen availability influences seasonal carbon allocation patterns in sugar maple *(Acer saccharum). Canadian Journal of Forest Research* **22,** 447–56.

Burkman, W.G. (1990) Quality assurance aspects of the joint USA – Canada sugar maple decline project. In: Hart, D.R. (coordinator), *Proceedings of the third annual ecological quality assurance workshop.* Burlington, Ontario: Canada Center for Inland Waters, pp. 83–97.

Burkman, W.G. and Alerich, D.J. (1991) Forest health monitoring in New England: evaluation of data quality. Unpublished manuscript.

Burley, J. (1966) Provenance variation in growth of seedling apices of Sitka spruce. *Forest Science* **12,** 170–5.

Burns, B.S. (1990) *1985 Vermont sugarbush health survey.* Waterbury: Vermont Department of Forests, Parks, and Recreation.

Burschel, P. (1986) Waldbau, Waldzustand und neuartige Waldschäden. *Holz-Zentralblatt* **112,** 201–4.

Burton, A.J., Pregitzer, K.S. and Reed, D.D. (1991) Leaf area and foliar biomass relationships in northern hardwood forests located along an 800 km acid deposition gradient. *Forest Science* **37,** 1041–59.

Burton, K.W., Morgan, E. and Roig, A. (1983) The influence of heavy metals upon the growth of Sitka-spruce in South Wales forests. 1. Upper critical and foliar concentrations. *Plant and Soil* **73,** 327–36.

Burton, P.J. and Mueller-Dombois, D. (1984) Response of *Metrosideros* seedlings to experimental canopy openings. *Ecology* **64,** 779–91.

Buschmann, C., Rinderle, U. and Lichtenthaler, H.K. (1991) Detection of stress in coniferous trees with the VIRAF spectrometer. *IEEE Transactions on Geoscience and Remote Sensing* **29,** 96–100.

Busing, R.T. and Clebsch, E.E.C. (1988) Fraser fir mortality and the dynamics of a Great Smoky Mountains fir-spruce stand. *Castanea* **53,** 177–82.

Bussotti, F., Cenni, E., Ferretti, M., Brogi, L. and Mecci, A. (1991b) Danni forestali nei boschi della Toscana. Seconde parte – Sintomi e comportamento di singole specie. *Linea Ecologica* **23,** 42–6.

Bussotti, F., Ferretti, M., Cenni, E., Gellini, R., Clauser, F., Grossoni, P. and Barbolani, E. (1991a) New type of forest damage to mediterranean vegetation in southern Sardinian forests (Italy). *European Journal of Forest Pathology* **21,** 290–300.

Bussotti, F., Gellini, R., Ferretti, M., Cenni, E., Pietrini, R. and Sbrilli, G. (1992) Monitoring in 1989 of Mediterranean tree condition and nutritional status in southern Tuscany, Italy. *Forest Ecology and Management* **51,** 81–93.

Butzke, H. (1981) Versauern unsere Wälder? Erste Ergebnisse der Überprüfung 20 Jahre alter pH-Wert-Messungen in Waldböden Nordrhein-Westfalens. *Forst- und Holzwirtschaft* **36,** 542–8.

Butzke, H. (1984) Untersuchungsergebnisse aus Waldböden Nordrhein-Westfalens zur Frage der Bodenversauerung durch Immissionen. *Wissenschaft und Umwelt* **2/1984,** 80–8.

Bytnerowicz, A., Olszyk, D.M., Fox, C.A., Dawson, P.J., Kats, G., Morrison, C.L. and Wolf, J. (1988) Responses of desert annual plants to ozone and water stress in an *in situ* experiment. *Journal of the Air Pollution Control Association* **38,** 1145–51.

Bytnerowicz, A., Olszyk, D.M., Huttunen, S. and Takemoto, B. (1989) Effects of photochemical smog on growth, injury, and gas exchange of pine seedlings. *Canadian Journal of Botany* **67,** 2175–81.

Bytnerowicz, A., Poth, M. and Takemoto, B.K. (1990) Effects of photochemical smog and mineral nutrition on ponderosa pine seedlings. *Environmental Pollution* **67,** 233–48.

Bytnerowicz, A. and Takemoto, B.K. (1990) Effects of photochemical smog on the growth and physiology of ponderosa pine seedlings grown under nitrogen and magnesium deficiencies. In: Olson, R.K. and Lefohn, A.S. (eds), *Effects of air pollution on western forests.* Pittsburgh: Air and Waste Management Association, pp. 455–67.

Cadahia, D., Cobos, J.M., Soria, S., Clauser, F., Gellini, R., Grossoni, P. and Ferreira, M.C. (1992) *Observation of damages to mediterranean forest species.* Brussels. Commission of the European Communities.

Campbell, D.J. and Rudge, M.R. (1984) Vegetation changes induced over 10 years by goats and pigs at Port Ross, Auckland Islands (Subantarctic). *New Zealand Journal of Ecology* **7,** 103–18.

Canham, C.D. (1985) Suppression and release during canopy recruitment in *Acer saccharum. Bulletin of the Torrey Botanical Club* **112,** 134–45.

Cannell, M.G.R. (1978) Analysis of shoot apical growth of *Picea sitchensis* seedlings. *Annals of Botany* **42,** 1291–303.

Cannell, M.G.R. (1987) Photosynthesis, foliage development and productivity of Sitka spruce. *Proceedings of the Royal Society of Edinburgh* **93B,** 61–73.

Cannell, M.G.R. and Willett, S.C. (1975) Rates and times at which needles are initiated in buds on differing provenances of *Pinus contorta* and *Picea sitchensis* in Scotland. *Canadian Journal of Forest Research* **5,** 367–80.

Cape, J.N. (1988) Air pollutant effects on conifer leaf surfaces. In: Cape, J.N. and Mathy, P. (eds), *Scientific basis of forest decline symptomatology.* Air Pollution Report 15. Brussels: Commission of the European Communities, pp. 149–59.

Cape, J.N., Freer-Smith, P.H., Paterson, I.S., Parkinson, J.A. and Wolfenden, J. (1990) The nutritional status of *Picea abies* (L.) Karst. across Europe, and implications for 'forest decline'. *Trees: Structure and Function* **4,** 211–24.

Cape, J.N., Leith, I.D., Fowler, D., Murray, M.B., Sheppard, L.J., Eamus, D. and Wilson, R.H.F. (1991) Sulphate and ammonium in mist impair the frost hardening of red spruce seedlings. *New Phytologist* **118,** 119–26.

Cape, J.N., Paterson, I.S. and Wolfenden, J. (1989) Regional variation in surface properties of Norway spruce and Scots pine needles in relation to forest decline. *Environmental Pollution* **58,** 325–42.

Carlisle, A., Brown, A.H.F. and White, E.J. (1967) The nutrient content of tree stem flow and ground flora litter and leachates in a sessile oak (*Quercus petraea*) woodland. *Journal of Ecology* **55,** 615–27.

Carlson, C.E. (1972) *Monitoring fluoride pollution in Flathead National Forest and Glacier National Park.* Missoula: United States Department of Agriculture Forest Service, Division of State and Private Forestry.

Carlson, C.E. and Dewey, J.E. (1971) *Environmental pollution by fluorides in Flathead National Forest and Glacier National Park.* Missoula: United States Department of Agriculture Forest Service, Division of State and Private Forestry.

Carlson, C.E. and Gilligan, C.J. (1983) *Histological differentiation among abiotic causes of conifer needle necrosis.* Research Paper INT-298. Intermountain Forest and Range Experiment Station: United States Department of Agriculture Forest Service. 17 pp.

Carter, C.I. (1989) *The 1989 outbreak of green spruce aphid,* Elatobium abietinum. Forestry Commission Research Information Note 161. Edinburgh: Forestry Commission.

Carter, M.R. (1987) Seedling growth and mineral nutrition of Scots pine under acidic to calcareous soil conditions. *Soil Science* **144,** 175–80.

Cassens-Sasse, E. (1987) Witterungsbedingte saisonale Versauerungsschübe im Boden zweier Waldökosysteme. *Bericht Waldökosysteme/Waldsterben* **30,** 287 pp.

Castillo, F.J. (1986) Extracellular peroxidases as markers of stress? In: Greppin, H., Penel, C. and Gaspar, T. (eds), *Molecular and physiological aspects of plant peroxidase*. Geneva: Université de Genève – Centre de Botanique, pp. 419–26.

Castillo, F.J., Miller, P.R. and Greppin, H. (1987) Extracellular biochemical markers of photochemical oxidant air pollution damage to Norway spruce. *Experientia* **43,** 111–5.

Castillo, F.J., Ogier, G. and Greppin, H. (1989) Effect of ozone and sulfur dioxide on Norway spruce. Comparative biochemical analysis. In: Bucher, J.B. and Bucher-Wallin, I. (eds), *Air pollution and forest decline*. Birmensdorf: Eidgenössische Anstalt für das forstliche Versuchswesen, pp. 131–4.

Cejchan, S. (1985) Die forstlichen Sorgen der CSSR Mitte der achtziger Jahre. *Holz-Zentralblatt* **111,** 1368–9.

Cerny, A. (1989) Soucasny zdravotní stav jedle belokoré na území CSSR. *Lesnická práce* **68,** 402–7.

Chan, W.H. and Lusis, M.A. (1985) Post-superstack Sudbury smelter emissions and their fate in the atmosphere: an overview of the Sudbury environmental study. *Water, Air, and Soil Pollution* **20,** 211–30.

Chandler, J.W. and Dale, J.E. (1990) Needle growth in Sitka spruce (*Picea sitchensis*): effects of nutrient deficiency and needle position within shoots. *Tree Physiology* **6,** 41–56.

Chang, H.K. and Yong, M.C. (1987) Effects of simulated acid rain on the chlorophyll contents in the needles of *Pinus koraiensis* and *Ligustrum obtusifolium* seedlings. *Journal of the Korean Forestry Society* **76,** 11–6.

Chang, J. and Hanover, J.W. (1991) Geographic variation in the monoterpene composition of black spruce. *Canadian Journal of Forest Research* **21,** 1796–800.

Chang, M. and Aguilar, G.J.R. (1980) Effects of climate and soil on the radial growth of loblolly pine (*Pinus taeda* L.) in a humid environment of the southeastern USA. *Forest Ecology and Management* **3,** 141–50.

Chao, T.T., Harward, M.E. and Fang, S.C. (1962) Adsorption and desorption phenomena of sulfate ions in soils. *Soil Science Society of America Proceedings* **36,** 234–7.

Chappelka, A.H., Chevone, B.I. and Burk, T.E. (1985) Growth response of yellow-poplar (*Liriodendron tulipifera* L.) seedlings to ozone, sulfur dioxide, and simulated acidic precipitation, alone and in combination. *Environmental and Experimental Botany* **25,** 233–44.

Chappelka, A.H., Chevone, B.I. and Burk, T.E. (1988) Growth responses of green and white ash seedlings to ozone, sulfur dioxide, and simulated acid rain. *Forest Science* **34,** 1016–29.

Chappelka, A.H., Kush, J.S., Meldahl, R.S. and Lockaby, B.G. (1990) An ozone-low temperature interaction in loblolly pine (*Pinus taeda* L.). *New Phytologist* **114,** 721–6.

Cheret, C., Dagnac, J. and Fromard, F. (1987) Le dépérissement du sapin dans les Pyrénées luchonnaises. *Revue forestière française* **34,** 12–24.

Chevone, B.I., Yang, Y.S. and Reddick, G.S. (1984) Acidic precipitation and ozone effects on growth of loblolly and shortleaf pine seedlings. *Phytopathology* **74,** 756.

Christensen, N.L. and Peet, R.K. (1988) *Patterns of growth and mortality of loblolly pine (*Pinus taeda*) in permanent sample plots of the Duke Forest:*

1934–1988. Report to the Southern Commercial Forest Research Cooperative Program Manager.

Ciesla, W.M. and Hildebrandt, G. (1986) Forest decline inventory methods in West Germany: opportunities for application in North American forests. *United States Department of Agriculture Forest Service Report* **86–3**.

Ciesla, W.M., Lambert, H.L. and Franklin, R.T. (1963) *The status of the balsam wooly aphid in North Carolina and Tennessee.* Report 1-11-63. Asheville, North Carolina: United States Department of Agriculture Forest Service, Division of Forest Pest Management.

Ciesla, W.M. and Macias Samano, J.E. (1987) Desierto de los Leones: a forest in crisis. *American Forests* **93,** 29–31 and 72–4.

Clark, J. and Barter, G.W. (1958) Growth and climate in relation to dieback of yellow birch. *Forest Science* **4,** 343–64.

Clarkson, R.B. (1978) *Tumult on the mountain, lumbering in West Virginia, 1770–1920.* Parsons, West Virginia: McClain Printing Company.

Clauser, F. and Gellini, R. (1986) Rating of Waldsterben-symptoms (forest decline) in deciduous broadleaves during winter time. *European Journal of Forest Pathology* **16,** 250–3.

Clauser, F., Gellini, R., Bussotti, F., Cenni, E. and Bottacci, A. (1989) New types of damage to forest trees typical of the Mediterranean region. *European Journal of Forest Pathology* **19,** 78–83.

Cline, S.P. and Burkman, W.G. (1989) The role of quality assurance in ecological research programs. In: Bucher, J.B. and Bucher-Wallin, I. (eds), *Air pollution and forest decline.* Birmensdorf: Eidgenössische Anstalt für das forstliche Versuchswesen, pp. 361–5.

Cline, S.P., Burkman, W.G. and Geron, C.D. (1989) Use of quality control procedures for assessing variation in measurements of forest canopy condition. In: Olson, R.K. and Lefohn, A.S. (eds), *Effects of Air Pollution on western forests.* Air and Waste Management Association, Pittsburgh, 379–94.

Cloutier, A. (1985) Microdistribution des espèces végétales au pied des troncs d'*Acer saccharum* dans une érablière du sud du Québec. *Canadian Journal of Botany* **63,** 274–6.

Coile, T.S. and Schumacher, F.X. (1964) *Soil-site relations, stand structure, and yields of slash and loblolly pine plantations in the southern United States.* Durham: T. S. Coile, 296 p.

Cole, D.W. (1981) Nitrogen uptake and translocation by forest ecosystems. In: Clark, F.E. and Rosswall, T. (eds), *Terrestrial nitrogen cycles.* Ecological Bulletin, Stockholm, **33,** 319–32.

Coleman, J.D., Gillman, A. and Green, W.Q. (1980) Forest patterns and possum densities within podocarp/mixed hardwood forests on Mt Bryan O'Lynn, Westland. *New Zealand Journal of Ecology* **3,** 69–84.

Collins, W.E. (1978) Remote sensing of crop type and maturity using high resolution features in the near infra-red. *Photogrammetric Engineering and Remote Sensing* **44,** 43–55.

Committee on Biologic Markers of Air Pollution Damage in Trees (1989) *Biologic Markers of Air Pollution Stress and Damage in Forests.* Washington DC: National Academy Press, pp. 1–26.

Compton, O.C., Remmert, L.F. and Rudinsky, J.A. (1961) *Needle scorch and condition of ponderosa pine trees in The Dalles area.* Oregon Agricultural Experimental Station Miscellaneous Paper 120.

Conese, C., Maselli, F., Maracchi, G. and Falchi, M.A. (1989) Uso del telerilevamento per il monitoraggio del deperimento della vegetazione costiera in area protette della Toscana. *Monti e Boschi* **40,** 15–19.

Conkey, L.E. (1979) Response of tree ring density to climate in Maine, USA. *Tree Ring Bulletin* **39,** 29–38.

Conkey, L.E. (1988) Decline in old-growth red spruce stands in western Maine: an analysis of wood density and climate. *Canadian Journal of Forest Research* **18,** 1063–8.

Conway, M.J. (1952) The silviculture of red beech in Nelson and Westland. *New Zealand Journal of Forestry* **6,** 291–308.

Cook, E.R. (1986) *The use of climatic response models of tree rings in the analysis and prediction of forest decline.* Paper presented to the Task Force Meeting 'Methodology of Dendrochronology', Krakow, Poland, 2–6 June 1986. IIASA, Laxenburg.

Cook, E.R. (1987) The use and limitations of dendrochronology in studying effects of air pollution on forests. In: Hutchinson, T.C. and Meema, K.A. (eds), *Effects of atmospheric pollutants on forests, wetlands and agricultural ecosystems.* New York: Springer-Verlag, pp. 277–90.

Cook, E.R. (1988) A tree ring analysis of red spruce in the Southern Appalachian Mountains. In: Van Deusen, P.C. (ed.), *Analyses of Great Smoky Mountain red spruce tree ring data.* General Technical Report SO-69. New Orleans: United States Department of Agriculture Forest Service, Southern Forest Experiment Station, pp. 6–20.

Cook, E.R. (1990) Bootstrap confidence intervals for red spruce ring-width chronologies and an assessment of age-related bias in recent growth trends. *Canadian Journal of Forest Research* **20,** 1326–31.

Cook, E.R. and Innes, J.L. (1989) Tree ring analysis as an aid to evaluating the effects of air pollution on tree growth. In: Committee on Biologic Markers of Air-pollution Damage in Trees (ed.), *Biologic markers of air pollution stress and damage in forests.* Washington DC: National Academy Press, pp. 157–68.

Cook, E.R. and Johnson, A.H. (1988) Threshold responses and climatic change: possible contributors to the decline of red spruce. In: *Proceedings of the US/FRG research symposium: Effects of atmospheric pollutants on the spruce–fir forests of the eastern United States and the Federal Republic of Germany.* General Technical Report NE-120. Broomall: United States Department of Agriculture Forest Service, Northeastern Forest Experiment Station, pp. 11–322.

Cook, E.R. and Johnson, A.H. (1989) Climate change and forest decline: a review of the red spruce case. *Water, Air, and Soil Pollution* **48,** 127–40.

Cook, E.R., Johnson, A.H. and Blasing, T.J. (1987) Forest decline: modelling the effect of climate in tree rings. *Tree Physiology* **3,** 27–40.

Cooley, D.R. and Manning, W.J. (1987) The impact of ozone on assimilate partitioning in plants: a review. *Environmental Pollution* **47,** 95–113.

Cooper, R.J., Dodge, K.M. and Whitmore, R.C. (1987) Estimating defoliation using stratified point intercept sampling. *Forest Science* **33,** 157–63.

Cornic, G. (1987) Interaction between sublethal pollution by sulfur dioxide and drought stress. The effect on photosynthetic capacity. *Physiologia Plantarum* **71,** 115–19.

Cornic, G. and Pierre, M. (1987) Interaction between sublethal pollution by sulfur dioxide and drought stress. The effect on photosynthetic capacity. In: Bervaes, J., Mathy, P. and Evers, P. (eds), *Relationship between above and below ground influences of air pollutants on forest trees.* Air Pollution Research Report No. 16. Brussels: Commission of the European Communities, pp. 102–8.

Cosby, B.J., Hornberger, G.M., Wright, R.F. and Galloway, J.N. (1986) Modelling the effects of acid deposition: control of long-term sulfate dynamics by soil sulfate adsorption. *Water Resources Research* **22,** 1283–91.

Courchesne, F. and Hendershot, W.H. (1990) The role of basic aluminium sulfate minerals in controlling sulfate retention in the mineral horizons of two spodosols. *Soil Science* **150,** 571–8.

Cowling, E.B. (1989) Recent changes in chemical climate and related effects on forests in North America and Europe. *Ambio* **18,** 167–71.

Cox, R.M., Spavold-Tims, J. and Hughes, R.N. (1989) Acid fog and ozone: their possible role in birch deterioration around the Bay of Fundy, Canada. *Water, Air, and Soil Pollution* **48,** 263–76.

Coyne, P.I. and Bingham, G.E. (1982) Variation in photosynthesis and stomatal conductance in an ozone-stressed ponderosa pine stand: Light response. *Forest Science* **28,** 257–73.

Craig, B.W. and Friedland, A.J. (1990) Spatial patterns in forest composition and standing dead spruce in montane forests of the Adirondacks and Northern Appalachians. *Environmental Monitoring and Assessment,* in press. (Reported in NAPAP, 1990).

Cramer, H.H. (1984) On the predisposition to disorders of Middle European forests. *Pflanzenschutz-Nachrichten Bayer* **37,** 97–207.

Cramer, H.H. and Cramer-Middendorf, M. (1984) Studies on the relationships between periods of damage and factors of climate in the Middle European forests since 1851. *Pflanzenschutz Nachrichten Bayer* **37,** 208–334.

Crampton, C.B. (1980) Variations of pH with annual cumulative precipitation in acid forest soils. *Canadian Journal of Soil Science* **60,** 385–7.

Crang, R.E. and McQuattie, C.J. (1986) Qualitative and quantitative effects of acid misting and two air pollutants on foliar structures of *Liriodendron tulipifera. Canadian Journal of Botany* **64,** 1237–43.

Crang, R.E., Robinson, S.G., Noble, R.D. and Dochinger, L.S. (1983) Acid fog effects on yellow poplar leaf morphology. *Micron* **14,** 75–6.

Crawley, M.J. and Akhteruzzaman, M. (1988) Individual variation in the phenology of oak trees and its consequences for herbivorous insects. *Functional Ecology* **2,** 409–15.

Cronan, C.S. (1980) Controls on leaching from coniferous forest floor microcosms. *Plant and Soil* **56,** 301–22.

Cronan, C.S. (1985) Biogeochemical influence of vegetation and soils in the ILWAS watersheds. *Water, Air, and Soil Pollution* **26,** 355–71.

Cronan, C.S., April, R., Bartlett, R.J., Bloom, P.R., Driscoll, C.T., Gherini, S.A., Henderson, G.S., Joslin, J.D., Kelly, J.M., Newton, R.M., Parnell, R.A., Patterson, H.H., Raynal, D.J., Schaedle, M., Schofield, C.L., Sucoff, E.I., Tepper, H.B. and Thornton, F.C. (1989) Aluminium toxicity in forests exposed to acidic deposition: the ALBIOS results. *Water, Air, and Soil Pollution* **48,** 181–92.

Cronan, C.S. and Goldstein, R.A. (1989) ALBIOS: a comparison of aluminium biogeochemistry in forested watersheds exposed to acidic deposition. In: Adriano, D.C. and Havas, M. (eds), *Acidic precipitation. Volume 1. Case studies*. New York: Springer Verlag, pp. 113–35.

Cronan, C.S., Kelly, J.M., Schofield, C.L. and Goldstein, R.A. (1987) Aluminium geochemistry and tree toxicity in forests exposed to acidic deposition. In: Perry, R., Harrison, R.M., Bell, J.N.B. and Lester, J.N. (eds), *Acid rain: scientific and technical advances*. London: Selper, pp. 649–56.

Cronan, C.S. and Schofield, C.L. (1979) Aluminium leaching response to acid precipitation effects on high-elevation watersheds in the Northeast. *Science* **204,** 304–5.

Cropper, W.P. and Gholz, H.L. (1991) *In situ* needle and fine root respiration in mature slash pine (*Pinus elliottii*) trees. *Canadian Journal of Forest Research* **21,** 1589–95.

Crossley, A. and Fowler, D. (1986) The weathering of Scots pine epicuticular wax in polluted and clean air. *New Phytologist* **103,** 207–18.

Curry, J. and Church, T. (1952) Observations of winter drying of conifers in the Adirondacks. *Journal of Forestry* **50,** 114–16.

Dale, V.H., Gardner, R.H., DeAngelis, D.L., Eagar, C.C. and Webb, J.W. (1991) Elevation-mediated effects of balsam wooly adelgid on southern Appalachian spruce-fir forests. *Canadian Journal of Forest Research* **21,** 1639–48.

Dässler, H.G. and Ranft, H. (1989) Zur Änderung der Aziditätsverhältnisse im Oberboden sowie in Waldbächen. *Allgemeine Forst Zeitschrift* **45,** 176–8, 180.

Davey, B.G. and Conyers, M.K. (1988) Determining the pH of acid soils. *Soil Science* **146,** 141–50.

David, M.B., Fuller, R.D., Fernandez, I.J., Mitchell, M.J., Rustad, L.E., Vance, G.F., Stam, A.C. and Nodvin, S.C. (1990) Spodosol variability and assessment of response to acidic deposition. *Soil Science Society of America Journal* **54,** 541–8.

David, M.B., Grigal, D.F., Ohmann, L.F. and Gertner, G.Z. (1988) Sulfur, carbon, and nitrogen relationships in forest soils across the northern Great Lakes States as affected by atmospheric deposition and vegetation. *Canadian Journal of Forest Research* **18,** 1386–91.

Davis, F.W., Borchert, M., Harvey, L.E. and Michaelsen, J.C. (1991) Factors affecting seedling survivorship of blue oak (*Quercus douglasii* H. & A.) in central California. In: *Proceedings of the symposium on oak woodlands and hardwood rangeland management, October 31 – November 2, 1990, Davis, California*. General Technical Report PSW-126. Berkeley: United States Department of Agriculture Forest Service, Pacific Southwest Research Station, pp. 81–6.

Davis, M.R. (1990) Chemical composition of soil solutions extracted from New Zealand beech forests and West German beech and spruce forests. *Plant and Soil* **126,** 237–46.

Dean, T.J. and Johnson, J.D. (1992) Growth response of young slash pine trees to simulated acid rain and ozone stress. *Canadian Journal of Forest Research* **22,** 839–48.

Deans, J.D., Leith, I.D., Sheppard, L.J., Cape, J.N., Fowler, D., Murray, M.B. and Mason, P.A. (1990) The influence of acid mists on growth, dry matter partitioning, nutrient concentrations and mycorrhizal fruiting bodies in red spruce seedlings. *New Phytologist* **115,** 459–64.

Debus, R. (1988) Histological changes and contents of carbohydrates in *Picea abies.* In: Cape, J.N. and Mathy, P. (eds), *Scientific basis of forest decline symptomatology.* Air Pollution Report 15. Brussels: Commission of the European Communities, pp. 252–61.

Decker, B. and Gürth, P. (1985) Das Waldsterben im Forstbezirk Müllheim/Südlicher Schwarzwald. *Allgemeine Forst- und Jagdzeitung* **156,** 233–40.

DeHayes, D.H. and Hawley, G.J. (1992) Genetic implications in the decline of red spruce. *Water, Air, and Soil Pollution* **62,** 233–48.

DeHayes, D.H., Ingle, M.A. and Waite, C.E. (1989) Nitrogen fertilization enhances cold tolerance of red spruce seedlings. *Canadian Journal of Forest Research* **19,** 1037–43.

DeHayes, D.H., Waite, C.E., Ingle, M.A. and Willimas, M.W. (1990) Winter injury susceptibility and cold tolerance of current and year-old needles of red spruce trees from several provenances. *Forest Science* **36,** 982–94.

DeHayes, D.H., Thornton, F.C., Waite, C.E. and Ingle, M.A. (1991) Ambient cloud deposition reduces cold tolerance of red spruce seedlings. *Canadian Journal of Forest Research* **21,** 1292–5.

Delatour, C. (1983) Les dépérissements de chênes en Europe. *Revue forestière française* **4,** 265–82.

Denstorf, O, Heeschen, G. and Kenneweg, H. (1984) Ergebnisse der grossräumigen Inventur von Waldschäden 1983 mit Farb-Infrarot-Luftbildern im südlichen Schleswig-Holstein. *Allgemeine Forst- und Jagdzeitung* **155,** 126–31.

Derome, J., Kukkola, M. and Mälkönen, E. (1986) Forest liming on mineral soils. Results of Finnish experiments. *National Swedish Environmental Protection Board Report* **3084,** 1–107.

Derome, J. and Pätilä, A. (1989) The liming of forest soils in Finland. *Meddelser fran der Norsk Institut for Skogforskning* **42,** 147–55.

Derome, J. and Pätilä, A. (1990) Alleviation of forest soil acidification through liming. In: Kauppi, P., Anttilla, P. and Kenttämies, K. (eds), *Acidification in Finland.* Springer Verlag, Berlin, pp. 1093–115.

DeWalle, D.R., Swistock, B.R., Sayre, R.G. and Sharpe, W.E. (1991) Spatial variations of sapwood chemistry with soil acidity in Appalachian forests. *Journal of Environmental Quality* **20,** 486–91.

Diamandis, S. (1978) 'Top dying' of Norway spruce, *Picea abies* (L.) Karst., with special reference to *Rhizosphaera Kalkhoffi* Bubak. I. Development of foliar symptoms. *European Journal of Forest Pathology* **8,** 337–45.

Diamandis, S. (1979) 'Top-dying' of Norway spruce, *Picea abies* (L.) Karst. with special reference to *Rhizosphaera Kalkhoffi* Bubak. VI. Evidence related to the primary cause of top dying. *European Journal of Forest Pathology* **9,** 183–91.

Dick, J. McP., Jarvis, P.G. and Leakey, R.R.B. (1990) Influence of male cones on early season vegetative growth of *Pinus contorta* trees. *Tree Physiology* **6,** 105–17.

Van Dijk, H.F.G. and Roelofs, J.G.M. (1988) Effects of excessive ammonium deposition on the nutritional status and condition of pine needles. *Physiologia Plantarum* **73,** 494–501.

Dittrich, A.P.M., Yin, Z.-H., Slovik, S. and Heber, U. (1991) Wirkung von Schadgasen auf Blätter. In: *Proceedings 2. Statusseminar der PBWU zum Forschungsschwerpunkt 'Waldschäden'. 4–6 February, 1991, GSF-Forschungszentrum Neuherberg*. Neuherberg: GSF, 439–51.

Dizengremel, P. and Citerne, A. (1988) Air pollutant effects on mitochondria and respiration. In: Schulte-Hostede, S., Darrall, N.M., Blnak, L.W. and Wellburn, A.R. (eds), *Air pollution and plant metabolism*. London: Elsevier, pp. 169–88.

Dmyterko, E. (1991) Monitoring des Einflusses der Luftverschmutzung mittels 'Forsteinrichtungsmethode' auf die ständige Probefläche. In: *Proceedings of the IUFRO Workshop 'Monitoring air pollution impact on permanent sample plots, data processing and results interpretation'*, Prachatice, Czechoslovakia: IUFRO, pp. 84–90.

Dobler, D., Hohloch, K., Lisbach, B. and Saliari, M. (1988) Trieblängen-Messungen an Buchen. *Allgemeine Forst Zeitschrift* **43,** 811–12.

Dobson, M.C. (1991) *De-icing salt damage to trees and shrubs*. Forestry Commission Bulletin 101. London: HMSO, 64 pp.

Dobson, M.C., Taylor, G. and Freer-Smith, P.H. (1990) The control of ozone uptake by *Picea abies* (L.) Karst. and *P. sitchensis* (Bong.) Carr. during drought and interacting effects on shoot water relations. *New Phytologist* **116,** 465–74.

Dochinger, L.S. and Jensen, K.F. (1985) Effect of acid mist and air pollutants on yellow-poplar seedling height and leaf growth. Research Paper NE-572. Broomall: United States Department of Agriculture Forest Service, North-eastern Forest Experiment Station.

Dochinger, L.S. and Seliskar, C.E. (1965) Results from grafting chlorotic dwarf and healthy eastern white pine. *Phytopathology* **55,** 404–7.

Dochinger, L.S. and Seliskar, C.E. (1970) Air pollution and chlorotic dwarf disease of eastern white pine. *Forest Science* **16,** 46–55.

Dohrenbusch, A. (1990) Die Verjüngungsentwicklung der Buche (*Fagus sylvatica* L.). Bericht einer langfristigen Beobachtung im Solling. *Schriften aus der Forstlichen Fakultät der Universität Göttingen und der Niedersächsischen Forstlichen Versuchsanstalt* **97,** 70 pp.

Doley, D. (1988) Fluoride-induced enhancement and inhibition of photosynthesis in four taxa of Pinus. *New Phytologist* **110,** 21–31.

Donaubauer, E. (1987) Auftreten von Krankheiten und Schädlingen der Eiche und ihr Bezug zum Eichensterben. *Österreichische Forstzeitung* **98,** 46–8.

Dong, P.H. and Kramer, H. (1986) Auswirkungen von Umweltbelastungen auf das Wuchsverhalten verschiedener Nadelbaumarten im nordwestdeutschen Küstenraum. *Der Forst- und Holzwirt* **41,** 286–90.

Dong, P.H. and Kramer, H. (1987) Zuwachsverlust in erkrankten Fichtenbeständen. *Allgemeine Forst- und Jagdzeitung* **158,** 122–5.

Donnelly, J.R., Shane, J.B. and Schaberg, P.G. (1990) Lead mobility within the xylem of red spruce seedlings: implications for the development of pollution histories. *Journal of Environmental Quality* **19,** 268–71.

Donoso Z., C., Escobar, B. and Urrutia, J. (1985) Estructura y estrategias regenerativas de un bosque virgen de Ulmo (*Eucryphia cordifolia* Cav.), Tepa (*Laurelia philippiana* Phil) Looser en Chiloé, Chile. *Revista Chilena Historica Naturales* **58,** 171–86.

Donoso Z., C., Grez Z., R., Escobar, B. and Real, P. (1984) Estructura y dinamica de bosques del tipo forestal siempreverde en un sector de Chiloé Insular. *Bosque* **5,** 82–104.

Dottavio, C.L. and Williams, D.L. (1983) Satellite technology: an improved means for monitoring forest insect defoliation. *Journal of Forestry* **81,** 30–4.

Doty, R.D. (1982) Annual precipitation on the island of Hawaii between 1890 and 1977. *Pacific Science* **36,** 421–5.

Doty, R.D. (1983) Stream flow in relation to ohia forest decline on the island of Hawaii. *Water Resources Bulletin* **19,** 217–21.

Dowding, P. (1988) Air pollutant effects on plant pathogens. In: Schulte-Hostede, S., Darrall, N.M., Blank, L.W. and Wellburn, A.R. (eds), *Air pollution and plant metabolism.* London: Elsevier, pp. 329–55.

Downing, G.L. (1960) A cedar bark beetle outbreak. *United States Department of Agriculture Forest Service Research Centre for Forest Insects Survey Report* 6.

Van den Driessche, R. (1974) Prediction of mineral nutrient status of trees by foliar analysis. *Botanical Reviews* **40,** 347–94.

Van den Driessen, R. (1984) Nutrient storage, retranslocation and relationship of stress to nutrition. In: Bowen, G.D. and Nambiar, E.K.S. (eds), *Nutrition of plantation forests.* London: Academic Press, pp. 181–210.

Drillias, M., Carlson, J.E.C. and Rezabek, C.L. (1990) *Investigation of sugar maple (*Acer saccharum*) health in Wisconsin sugarbushes.* Report No. PUBL-AM-036-90. Madison: Bureau of Air Management, Wisconsin Department of Natural Resources.

Duckworth, C.M.S. and Cresser, M.S. (1991) Factors influencing nitrogen retention in forest soils. *Environmental Pollution* **72,** 1–21.

Dull, C.W., Ward, J.D., Brown, H.D., Ryan, G.W., Clarke, W.H. and Uhler, R.J. (1988) *Evaluation of spruce and fir mortality in the southern Appalachian Mountains.* Protection Report R8-PR13. Atlanta, Georgia: United States Department of Agriculture Forest Service, Southern Region. 92 pp.

Dunkl, M. and Knoppik, D. (1991) Freilandexperimente zur Untersuchung eines direkten Effektes von Ozon und anderen Photooxidantien auf den Gaswechsel von Fichten am Wank. In: *Proceedings 2. Statusseminar der PBWU zum Forschungsschwerpunkt 'Waldschäden'. 4–6 February, 1991, GSF-Forschungszentrum Neuherberg.* Neuherberg: GSF, pp. 125–37.

Durand, P., Gelpe, J., Lemoine, B., Riom, J. and Timbal, J. (1983) Le dépérissement du chêne pédonculé dans les Pyrénées Atlantiques. *Revue forestières françaises* **35,** 357–68.

Duriscoe, D.M. (1987) *Evaluation of ozone injury to selected tree species in Sequoia and Kings Canyon National Parks.* Final Report No. 122. Indianapolis, Indiana: Holcomb Research Institute, Butler University.

Duriscoe, D.M. and Stolte, K.W. (1989) Photochemical oxidant injury to ponderosa pine (*Pinus ponderosa* Laws.) and Jeffrey pine (*Pinus Jeffreyi* Grev. and Balf.) in the national parks of the Sierra Nevada of California. In: Olson, R.K. and Lefohn, A.S. (eds), *Effects of air pollution on western forests.* Anaheim, California: Air and Waste Management Association, pp. 261–78.

Dustin, C.D. and Raynal, D.J. (1988) Effects of simulated acid rain on sugar maple seedling growth. *Environmental and Experimental Botany* **28,** 207–13.

Dwire, K., Huntley, B., Miller-Weeks, M. and Brooks, R. (compilers) (1990) *New England forest health monitoring. Methods manual for field measurements.* Durham, New Hampshire: United States Department of Agriculture Forest Service.

Dzialuk, R. (1987) Threat to forests in Poland. Especially to forests in mountainous regions. In: Kairiukstis, L., Nilsson, S. and Straszak, A. (eds), *Forest decline and reproduction: Regional and global consequences.* Laxemburg: International Institute for Applied Systems Analysis, pp. 369–73.

Eamus, D. and Fowler, D. (1990) Photosynthetic and stomatal conductance responses to acid mist of red spruce seedlings. *Plant, Cell and Environment* **13,** 349–57.

Eamus, D., Leith, I. and Fowler, D. (1989) Water relations of red spruce seedlings treated with acid mist. *Tree Physiology* **5,** 387–97.

Eberhardt, J., Brennan, E., Kuser, J. and Harkov, R. (1988) Ozone tolerance of field-grown eastern white pine. *Journal of Arboriculture* **14,** 185–92.

Ebermayer, E. (1876) *Die gesamte Lehre von der Waldstreu.* Berlin: Springer.

Eckert, R.T. (1989) Genetic variation in red spruce and its relation to forest decline in the northeastern United Sates. In: Bucher, J. and Bucher-Wallin, I. (eds), *Air pollution and forest decline.* Birmensdorf: Eidgenössische Anstalt für das forstliche Versuchswesen, pp. 319–24.

Eckert, R.T. and O'Malley, D.M. (1988) Population genetic variation in New England red spruce. In: *Proceedings of the US/FRG Research Symposium: Effects of atmospheric pollutants on the spruce-fir forests of the eastern United States and the Federal Republic of Germany.* General Technical Report NE-120. Broomall: United States Department of Agriculture Forest Service, Northeastern Forest Experiment Station, pp. 323–27.

Eckstein, D. (1985) On the application of dendrochronology for the evaluation of forest damage. In: Schmid-Haas, P. (ed.), *Inventorying and monitoring endangered forests.* Birmensdorf: Eidgenössische Anstalt für das forstliche Versuchswesen, pp. 287–90.

Eckstein, D., Aniol, R.W. and Bauch, J. (1983) Dendroklimatologische Untersuchungen zum Tannensterben. *European Journal of Forest Pathology* **13,** 279–88.

Eckstein, D., Hooggesteger, J. and Holmes, R.L. (1991) Insect-related differences in growth of birch and pine at northern treeline in Swedish Lapland. *Holarctic Ecology* **14,** 18–23.

Eckstein, D. and Krause, C. (1989) Dendroecological studies on spruce trees to monitor environmental changes around Hamburg. *IAWA Bulletin* **10,** 175–82.

Eckstein, D., Richter, K., Aniol, R.W. and Quiehl, F. (1984) Dendroklimatologische Untersuchungen zum Buchensterben im südwestlichen Vogelsberg. *Forstwissenschaftliches Centralblatt* **103,** 274–90.

Eckstein, D. and Saß, U. (1989) Dendroecological assessment of decline and recovery of fir and spruce in the Bavarian forest. In: Bucher, J.B. and Bucher-Wallin, I. (eds), *Air pollution and Forest Decline.* Birmensdorf: Eidgenössische Anstalt für das forstliche Versuchswesen, pp. 255–60.

Edmonds, R.L. (1988) Pollution gradient studies in the Puget Sound region and evaluation of existing conifer plot data for correlative studies. *Forest Response Program Annual Meeting* Vol. II. Raleigh, North Carolina: North Carolina State University, pp. 289–93.

Edmunds, G.F. and Alstad, D.N. (1985) Incidence of basal needle necrosis of pines in relation to environmental factors. *Environmental Pollution* **37A,** 217–27.

Edwards, G.S., Friend, A.L., O'Neill, E.G. and Tomlinson, P.T. (1992) Seasonal patterns of biomass accumulation and carbon allocation in *Pinus taeda* seedlings exposed to ozone, acidic precipitation, and reduced soil Mg. *Canadian Journal of Forest Research* **22,** 640–6.

Edwards, G.S. and Kelly, J.M. (1992) Ectomycorrhizal colonization of loblolly pine seedlings during three growing seasons in response to ozone, acidic precipitation, and soil Mg status. *Environmental Pollution* **76,** 71–7.

Edwards, G.S., Pier, P.A. and Kelly, J.M. (1990) Influence of ozone and soil magnesium status on the cold hardiness of loblolly pine (*Pinus taeda* L.) seedlings. *New Phytologist* **115,** 157–64.

Edwards, G.S., Wullschleger, S.D. and Kelly, J.M. (1993) Growth and physiology of northern red oak: Comparison of mature tree and seedling responses to ozone. *Environmental Pollution*, in press.

Van der Eerden (1982) Toxicity of ammonia to plants. *Agriculture and Environment* **7,** 223–35.

Van der Eerden, Lekkerkerk, L.J.A., Smeulders, S.M. and Jansen, A.E. (1992) Effects of atmospheric ammonia and ammonium sulphate on Douglas fir (*Pseudotsuga menziesii*). *Environmental Pollution* **76,** 1–9.

Van der Eerden, L.J.M. and Perez-Soba, M.G.F.J. (1992) Physiological responses of *Pinus sylvestris* to atmospheric ammonia. *Trees: Structure and Function* **6,** 48–53.

Eichholz, U. (1985) Sterben von Eichenjungbeständen in Südhessen. *Allgemeine Forst Zeitschrift* **40,** 47–8.

Eichkorn, T. (1986) Wachstumsanalysen an Fichten in Südwestdeutschland. *Allgemeine Forst- und Jagdzeitung* **157,** 125–39.

Eickmeier, W.G. (1988) Ten years of forest dynamics at Radnor Lake, Tennessee. *Bulletin of the Torrey Botanical Club* **115,** 100–7.

Eiden, R., Peters, K., Trautner, F., Herterich, R. and Gietl, G. (1989) Air pollution and deposition. In: Schulze, E.-D., Lange, O.L. and Oren, R. (eds), *Forest decline and air pollution. A study of spruce (*Picea abies*) on acid soils.* Berlin: Springer-Verlag, pp. 57–106.

Eisenhauer, D.R. (1989) Untersuchung zur Entwicklung der ökologischen Stabilität von Eichenbeständen im nordöstlichen Harzvorland. *Beiträge für die Forstwirtschaft* **23,** 55–62.

Eklund, B. (1957) Om granens årsringsvariationer inom mellersta Norrland och deras samband med klimatet. *Meddelanden från Statens Skogsforskningsinstitut* **47,** Part 1.

Elagin, I.N. and Cherkashin, V.P. (1987) Structure and dynamics of litter in iris-crowberry pine forest. *Soviet Journal of Ecology* **18,** 196–9.

Elder, N.L. (1962) Vegetation of the Kaimanawa Ranges. *Transactions of the Royal Society of New Zealand (Botany)* **2,** 1–37.

Elder, N.L. (1963) Evidence of climatic change from the vegetation of the North Island. *Proceedings of the New Zealand Ecological Society* **10,** 45–8.

Eldhuset, T.D. (1988) Virkninger av aluminium på hyere planter ved lav pH:

En litteraturoversikt. *Meddelelser frå Norsk Institutt for Skogforskning* **40(8)** 19 pp.

Ellenberg, H. (1979) Veränderungen der Flora Mitteleuropas unter dem Einfluss von Düngung und Immissionen. *Schweizerische Zeitschrift für Forstwesen* **136,** 19–39.

Elliott, C.L., Eberhardt, J.C. and Brennan, E.G. (1987) The effect of ambient ozone pollution and acidic rain on the growth and chlorophyll content of green and white ash. *Environmental Pollution* **44,** 61–70.

Ellis, R.C. (1964) Dieback of alpine ash in north eastern Tasmania. *Australian Forestry* **28,** 75–90.

Ellis, R.C. (1971) Dieback of alpine ash as related to changes in soil temperature. *Australian Forestry* **35,** 152–63.

Ellis, R.C., Mount, A.B. and Mattay, J.P. (1980) Recovery of *Eucalyptus delegatensis* from high altitude dieback after felling and burning the understorey. *Australian Forestry* **43,** 29–35.

Ellsworth, D.S. and Reich, P.B. (1992) Water relations and gas exchange of *Acer saccharum* seedlings in contrasting natural light and water regimes. *Tree Physiology* **10,** 1–20.

Elstner, E.F. and Osswald, W. (1988) Yellowing associated with forest decline. In: Cape, J.N. and Mathy, P. (eds), *Scientific basis of forest decline symptomatology.* Air Pollution Report 15. Brussels: Commission of the European Communities, pp. 126–32.

Elvidge, C.D. (1988) Vegetation reflectance features in AVIRIS data. *Proceedings of the International Symposium on Remote Sensing of Environment, 6th Thematic Conference, Remote Sensing for Exploration Geology.* Environmental Research Institute of Michigan, p. 169.

Elvidge, C.D., Vogelmann, J.E. and Fox, C.A. (1989) Comparisons of aircraft and satellite sensors for monitoring air pollution effects on forests. In: Olson, R.K. and Lefohn, A.S. (eds), *Effects of air pollution on western forests.* Pittsburgh: Air and Waste Management Association, pp. 337–49.

Emanuelsson, U. (1988) The relationship of different agricultural systems to the forest and woodlands of Europe. In: Sabiano, F. (ed.) *Human influence on forest ecosystems development in Europe.* Bologna: Pitagora Editrice, pp. 169–78.

Emmer, I.M. and Tietema, A. (1990) Temperature-dependent nitrogen transformations in acid oak-beech forest litter in the Netherlands. *Plant and Soil* **122,** 193–6.

Enright, N.J. (1984) Principal components analysis of tree-ring/climate relationships in white spruce (*Picea glauca*) from Schefferville, Canada. *Journal of Biogeography* **11,** 353–61.

Ericsson, A., Hellqvist, C., Lángström, B., Larsson, S. and Tenow, O. (1985) Effects on growth of simulated and induced shoot pruning by *Tomicus piniperda* as related to carbohydrate and nitrogen dynamics in Scots pine. *Journal of Applied Ecology* **22,** 105–24.

Escudero, A., del Arco, J.M., Sanz, I.C. and Ayala, J. (1992) Effects of leaf longevity and retranslocation efficiency on the retention time of nutrients in the leaf biomass of different woody species. *Oecologia* **90,** 80–7.

Estivalet, D., Perrin, R., Le Tacon, F. and Bouchard, D. (1990) Nutritional and microbiological aspects of decline in the Vosges Forest area (France). *Forest Ecology and Management* **37,** 233–48.

Euteneuer, T., Steubing, L. and Debus, R. (1988a) Quantitative Beurteilung der Morphologie epikutikulare Wachse von *Picea abies* (L.) Karst. *Angewandte Botanik* **62,** 63–72.

Euteneuer, T., Steubing, L. and Debus, R. (1988b) Alkane distribution pattern and wettability of *Picea abies* epicuticular waxes in regions with low immission loads. In: Cape, J.N. and Mathy, P. (eds), *Scientific basis of forest decline symptomatology.* Air Pollution Report 15. Brussels: Commission of the European Communities, pp. 262–72.

Evans, A. and Zelazny, L.W. (1987) Effects of sulfate additions on the status of exchangeable aluminium in a Cecil soil. *Soil Science* **143,** 410–17.

Evans, J. (1990) Long-term productivity of forest plantations – status in 1990. In: *Proceedings, IUFRO XIX World Congress, 5–11 August 1990, Montreal, Canada.* Montreal: IUFRO, pp. 165–80.

Evans, L.S. (1986) Proposed mechanisms of initial injury-causing apical dieback in red spruce at high elevation in eastern North America. *Canadian Journal of Forest Research* **16,** 1113–16.

Evans, L.S. and Biesemeyer, P.T. (1988) Initial twig injury to high elevation *Picea rubens* in eastern North America caused by inadequate periderm formation. *American Journal of Botany* **75,** 1378–86.

Evans, L.S. and Miller, P.R. (1972) Comparative needle anatomy and relative ozone sensitivity of four pine species. *Canadian Journal of Botany* **50,** 1067–71.

Evans, L.S. and Miller, P.R. (1972) Ozone damage to ponderosa pine: a histological and histochemical appraisal. *American Journal of Botany* **59,** 297–304.

Evenson, W.E. (1983) Climate analysis in 'ohi'a dieback area on the island of Hawai'i. *Pacific Science* **37,** 375–84.

Everard, J. (1973) Foliar analysis. Sampling methods, interpretation and application of the results. *Quarterly Journal of Forestry* **67,** 51–66.

Everett, R.L. and Thran, D.F. (1992) Nutrient dynamics in singleleaf pinyon pine (*Pinus monophylla* Torr. & Frem.) needles. *Tree Physiology* **10,** 59–68.

Evers, F.H. (1972) Die Jahrweisen Fluktuationen der Nährelementkonzentrationen in Fichtennadeln und ihre Bedeutung für die Interpretation nadelanalytischer Befunde. *Allgemeine Forst- und Jagdzeitung* **143,** 68–74.

Evers, F.H. (1984) Bericht über die Arbeit der Abteilung Botanik und Standortskunde. *Mitteilungen der Forstlichen Versuchs- und Forschungsanstalt Baden-Württemberg* **108,** 83–92.

Evers, F.H. (1989) Düngung im Wald – Möglichkeiten und Grenzen. *Bericht 5, PEF Statuskolloquium 7-9.3.1989 im Kernforschungszentrum Karlsruhe.* Volume 1. Karlsruhe: PEF, pp. 35–60.

Evers, F.H. (1991) Forest fertilization – present state and history with special reference to South German conditions. *Fertilizer Research* **27,** 71–86.

Evers, F.H. and Schöpfer, W. (1988) *Ergebnisse der Belastungsinventur Baden-Württemberg 1983 – Darstellung der Ernährungs- und Belastungsverhältnisse bei Fichte und Tanne.* Forstliche Versuchs- und Forschungsanstalt Baden-Württemberg. 44 pp.

Evers, F.H., Hildebrand, E.E., Kenk, G. and Kremer, W.L. (1986) Boden-, ernährungs- und ertragskundliche Untersuchungen in einem stark geschädigten Fichtenbestand des Buntsandstein-Schwarzwaldes. *Mitteilungen des Vereins für Forstliche Standortskunde und Forstpflanzenzüchtung* **32,** 72–80.

Ewell, D.M., Mazzu, L.C. and Duriscoe, D.M. (1990) Specific leaf weight and other characteristics of ponderosa pine as related to visible ozone injury. In: Olson, R.K. and Lefohn, A.S. (eds), *Effects of air pollution on western forests.* Pittsburgh: Air and Waste Management Association, pp. 411–18.

Ewers, F.W. and Skied, R. (1981) Longevity of needle fascicles of *Pinus longaeva* (bristlecone pine) and other north American pines. *Oecologia* **51,** 107–15.

Fabijanowski, J. (1986) Hodowla lasu wobec zagrozenia imisjami drzewostanów w górach. *Sylwan* **130,** 53–66.

Facelli, J.M. and Pickett, S.T.A. (1991) Plant litter: its dynamics and effects on plant community structure. *Botanical Review* **57,** 1–32.

Facteau, T.J., Wang, S.Y. and Rowe, K.E. (1973) The effect of hydrogen fluoride on pollen germination and pollen tube growth in *Prunus avium* L. cv. 'Royal Ann'. *American Society for Horticultural Science* **98,** 234–6.

Falkengren-Grerup, U. (1986) Soil acidification and vegetation changes in deciduous forests in southern Sweden. *Oecologia* **70,** 339–47.

Falkengren-Grerup, U. (1987) Long-term changes in pH of forest soils in southern Sweden. *Environmental Pollution* **43,** 79–90.

Falkengren-Grerup, U. (1989a) Soil acidification and its impact on ground vegetation. *Ambio* **18,** 179–83.

Falkengren-Grerup, U. (1989b) Effect of stemflow on beech forest soils and vegetation in southern Sweden. *Journal of Applied Ecology* **26,** 341–52.

Falkengren-Grerup, U. and Eriksson, H. (1990) Changes in soil, vegetation and forest yield between 1947 and 1988 in beech and oak sites of southern Sweden. *Forest Ecology and Management* **38,** 37–53.

Falkengren-Grerup, U. and Tyler, G. (1991a) Changes of cation pools of the topsoil in south Swedish beech forests between 1979 and 1989. *Scandinavian Journal of Forest Research* **6,** 145–52.

Falkengren-Grerup, U. and Tyler, G. (1991b) Dynamic floristic changes of Swedish beech forest in relation to soil acidity and stand management. *Vegetatio* **95,** 149–58.

Fangmeier, A., Kress, L.W., Lepper, P. and Heck, W.W. (1990) Ozone effects on the fatty acid composition of loblolly pine needles (*Pinus taeda* L.). *New Phytologist* **115,** 639–47.

Farrar, J.F., Relton, J. and Rutter, A.J. (1977) Sulphur dioxide and the scarcity of *Pinus sylvestris* in the industrial Pennines. *Environmental Pollution* **14,** 63–8.

Farrell, E.P. (1990) Aspects of the nitrogen cycle in peatland and plantation forest ecosystems in western Ireland. *Plant and Soil* **128,** 13–20.

Faulds, W. (1977) A pathogenic fungus associated with *Platypus* attack on New Zealand *Nothofagus* species. *New Zealand Journal of Forestry Science* **7,** 384–96.

Fayle, D.C.F. and MacDonald, G.B. (1977) Growth and development of sugar maple as revealed by stem analysis. *Canadian Journal of Forest Research* **7,** 526–36.

Federer, C.A. and Hornbeck, J.W. (1985) The buffer capacity of forest soils in New England. *Water, Air, and Soil Pollution* **26,** 163–73.

Federer, C.A. and Hornbeck, J.W. (1987a) Expected decrease in diameter growth of even-aged red spruce. *Canadian Journal of Forest Research* **17,** 266–9.

Federer, C.A. and Hornbeck, J.W. (1987b) Red spruce diameter growth and Weibull functions for even-aged stand development. In: Jacoby, G.C. and Hornbeck, J.W. (compilers) *Proceedings, International symposium on ecological aspects of treering analysis,* Marymount College, Tarrytown, New York, 17-21 Aug. 1986. Report No. ER-12 GTN. Washington DC: Department of Energy, pp. 18–25.

Federer, C.A., Hornbeck, J.W. and Smith, R.B. (1988) Regional dendrochronologies of red spruce and other species in New England. In: *Proceedings of the US/FRG Research Symposium: Effects of atmospheric pollutants on the spruce–fir forests of the eastern United States and the Federal Republic of Germany.* General Technical Report NE-120. Northeastern Forest Experiment Station, Broomall: United States Department of Agriculture Forest Service, pp. 285–90.

Federer, C.A., Hornbeck, J.W., Tritton, L.M., Martin, C.W., Pierce, R.S. and Smith, C.T. (1989a) Long-term depletion of calcium and other nutrients in eastern US forests. *Environmental Management* **13,** 593–601.

Federer, C.A., Tritton, L.M., Hornbeck, J.W. and Smith, R.B. (1989b) Physiologically based dendroclimate models for effects of weather on red spruce basal-area growth. *Agricultural and Forest Meteorology* **46,** 159–72.

Feger, K.H. and Raspe, S. (1992) Ernährungszustand von Fichtennadeln und -wurzeln in Abhängigkeit vom Nährstoffangebot im Boden. *Forstwissenscahftliches Centralblatt* **111,** 73–86.

Feger, K.H., Raspe, S., Schmid, M. and Zöttl, H.W. (1991) Verteilung der Elementvorräte in einem schlechtwüchsigen 100jährigen Fichtenbestand auf Buntsandstein. *Forstwissenschaftliches Centralblatt* **110,** 248–62.

Feiler, S. (1985) Einflüsse von Schwefeldioxid auf die Membranpermeabilität und Folgen für die Frostempfindlichkeit der Fichte (*Picea abies* (L.) Karst.). *Flora* **177,** 217–26.

Felten, V., Giot, P., and Marechal, P. (1985) Application of airborne digital multispectral scanner data and testing of spaceborne data for the evaluation of forest degradation. In: Schmid-Haas, P. (ed.), *Inventorying and monitoring endangered forests.* Birmensdorf: Eidgenössische Anstalt für das forstliche Versuchswesen, pp. 147–9.

Fengel, D. (1987) Chemischanalytische Untersuchungen am Holz erkrankter Bäume. Gelöste Substanzen im frischen Splintholz. *Holz als Roh- und Werkstoff* **45,** 501–7.

Fengel, D. (1988) Chemischanalytische Untersuchungen am Holz erkrankter Bäume. Löslichkeit der Zellwandbestandteile. *Holz als Roh- und Werkstoff* **46,** 1–6.

Fenn, M. (1991) Increased site fertility and litter decomposition rate in high-pollution sites in the San Bernardino Mountains. *Forest Science* **37,** 1163–81.

Fenn, M.E., Dunn, P.H. and Wilborn, R. (1990) Black stain root disease in ozone-stressed ponderosa pine. *Plant Disease* **74,** 426–30.

Feret, P.P., Diebel, K.E. and Sharik, T.L. (1990) Effect of simulated acid rain on reproductive attributes of red spruce (*Picea rubens* Sarg.). *Environmental and Experimental Botany* **30,** 309–12.

Ferm, A., Hytönen, J., Kolari, K.K. and Veijalainen, H. (1988) Metsäpuiden kasvuhäiriöt turkistarhojen läheisyydessä. *Metsäntutkimuslaitoksen Tiedonantoja, Finland 320, 77 p.*

Fernandez, I.J. and Rustad, L.E. (1990) Soil response to S and N treatments in a northern New England low elevation coniferous forest. *Water, Air and Soil Pollution* **52,** 23–39.

Ferretti, M., Barbolani, E., Bussotti, F. and Gellini, R. (1990) Variazioni stagionali di nutienti minerali ed elementi in traccia nelle foglie di *Fagus sylvatica* L. a Vallombrosa (FI). *Giornale Botanico Italiano* **124,** 589–99.

Ferretti, M., Udisti, R. and Barbolani, E. (1992) A new approach to experimental data evaluation from tree ring analysis. *Fresenius Journal of Analytical Chemistry* 1992, in press.

Fiedler, H.J. (1986) Forstdüngung gegen negative Immissionswirkungen im Mittelgebirge. *Archiv für Naturschutz und Landschaftsforschung Berlin* **26,** 117–31.

Fiedler, H.J. (1988) Zur Zinkausstattung der Fichtenökosysteme in Gebieten mit 'neuartigen' Waldschäden. *Beiträge für die Forstwirtschaft* **22,** 61–6.

Fiedler, H.J., Baronius, G. and Ehrig, F. (1990) Rasterelektronenmikroskopische und chemische Untersuchungen grüner und chlorotischer Nadeln eines immissionsgeschädigten Kiefernbestandes. *Flora* **184,** 91–101.

Fife, D.N. and Nambiar, E.K.S. (1984) Movement of nutrients in Radiata pine needles in relation to the growth of shoots. *Annals of Botany* **54,** 303–14.

Fife, D.N. and Nambiar, E.K.S. (1987) Growth and nutrient retranslocation in needles of Radiata pine in relation to nitrogen supply. *Annals of Botany* **60,** 147–56.

Fincher, J. and Alscher, R.G. (1992) The effect of long-term ozone exposure on injury in seedlings of red spruce (*Picea rubens* Sarg.). *New Phytologist* **120,** 49–59.

Fincher, J., Cumming, J.R., Alscher, R.G., Rubin, G. and Weinstein, L. (1989) Long-term ozone exposure affects winter hardiness of red spruce (*Picea rubens* Sarg.) seedlings. *New Phytologist* **113,** 85–96.

Fink, S. (1986a) Anatomische und biochemische Untersuchungen zur Klärung der Beteiligung biotischer Schadensverursacher am Waldsterben. *Bericht 2, Statuskolloquium der PEF, 4–7.3.1986 im Kernforschungszentrum Karlsruhe.* Volume 1. Karlsruhe: PEF, 55–64.

Fink, S. (1986b) Microscopical investigations on wood formation and function in diseased trees. *IAWA Bulletin* **7,** 351–5.

Fink, S. (1988) Histological and cytological changes caused by air pollutants and other abiotic factors. In: Schulte-Hostede, S., Darrall, N.M., Blank, L.W. and Wellburn, A.R. (eds), *Air pollution and plant metabolism.* London: Elsevier Applied Science, pp. 36–54.

Fink, S. (1991) Structural changes in conifer needles due to Mg and K deficiency. *Fertilizer Research* **27,** 23–7.

Fischer, H. and Rommel, W.-D. (1989) Jahrringbreiten und Höhentrieblängen von Buchen mit unterschiedlicher Belaubungsdichte in Baden-Württemberg. *Allgemeine Forst Zeitschrift* **44,** 264–5.

Fisher, W.R. (1908) *Report on planting trees in watersheds of Bury and District Joint Board.* Bury.

Fitzgerald, A.E. and Wardle, P. (1979) Food of the opossum (*Trichosurus vulpecula* Kerr) in the Waiho Valley, South Westland. *New Zealand Journal of Zoology* **6,** 339–45.

Fitzgerald, J.W., Strickland, T.C. and Swank, W.T. (1982) Metabolic fate of inorganic sulphate in soil samples from undisturbed and managed forest ecosystems. *Soil Biology and Biochemistry* **14,** 529–36.

Florence, R.G. and Chuong, P.H. (1974) The influence of soil type on foliar nutrients in *Pinus radiata* plantations. *Australian Forest Research* **6,** 1–8.

Flower-Ellis, J.G.K. (1985) Litterfall in an age series of Scots pine stands: summary of results for the period 1973–1983. In: Lindroth, A. (ed.), *Climate, photosynthesis and litterfall in pine forest on sandy soil – basic ecological measurement at Jädraås.* Report 19. Uppsala: Swedish University of Agricultural Sciences, Department of Ecology and Environmental Research, pp. 65–94.

Flückiger, W., and Braun, S. (1992) *Waldschadensbericht. Untersuchungen in Buchenbeobachtungsflächen 1984–1988.* Schönenbuch: Institut für Angewandte.

Flückiger, W., Braun, S. and Bolsinger, M. (1988) Air pollution: effect on host plant-insect relationships. In: Schulte-Hostede, S., Darrall, N.M., Blank, L.W. and Wellburn, A.R. (eds), *Air pollution and plant metabolism.* London: Elsevier, pp. 366–80.

Flückiger, W., Braun, S., Flückiger-Keller, H., Leonardi, S., Asche, N., Bühler, U. and Lier, M. (1986a) Untersuchungen über Waldschäden in festen Buchenbeobachtungsflächen der Kantone Basel-Landschaft, Basel-Stadt, Aargau, Solothurn, Bern, Zürich und Zug. *Schweizerische Zeitschrift für Forstwesen* **137,** 917–1010.

Flückiger, W., Braun, S., Leonardi, S., Asche, N. and Flückiger-Keller, H. (1986b) Factors contributing to forest decline in northwestern Switzerland. *Tree Physiology* **1,** 177–84.

Flückiger, W., Braun, S., Leonardi, S., Förderer, L. and Bühler, U. (1989) Untersuchungen an Buchen in festen Waldbeobachtungsflächen des Kantons Zürich. *Schweizerische Zeitschrift für Forstwesen* **140,** 536–50.

Flühler, H. (1981) Eine Übersicht über die laufenden Untersuchungen und bisherigen Ergebnisse. *Mitteilungen, Eidgenössische Anstalt für das Forstliche Versuchswesen, Birmensdorf* **57,** 367–72.

Flühler, H. (1983) Longterm fluoride pollution of a forest ecosystem: time, the dimension of pitfalls and limitations. In: Ulrich, B. and Pankrath, J. (eds), *Effects of accumulation of air pollutants in forest ecosystems.* Dordrecht: D. Reidel, pp. 303–17.

Flühler, H., Keller, T. and Schwager, H. (1981) Die Immissionsbelastung der Föhrenwälder im Walliser Rhonetal. *Mitteilungen, Eidgenössische Anstalt für das forstliche Versuchswesen, Birmensdorf* **57,** 399–414.

Fölster, H. (1985) Proton consumption rates in Holocene and present-day weathering of acid forest soils. In: Drever, J.I. (ed.) *The chemistry of weathering.* Dordrecht: D. Reidel, pp. 197–209.

Ford, E.D., Deans, J.D. and Milne, R. (1987a) Shoot extension in *Picea sitchensis.* 1. Seasonal variation within a forest canopy. *Annals of Botany* **60,** 531–42.

Ford, E.D., Milne, R. and Deans, J.D. (1987b) Shoot extension in *Picea sitchensis.* 2. Weather influences on daily growth rate. *Annals of Botany* **60,** 543–52.

Forschner, W. and Wild, A. (1988) Comparative investigation on the nutrient composition of healthy and injured spruces of different locations. In: Mathy, P. (ed.) *Air pollution and ecosystems.* Dordrecht: D. Reidel, pp. 604–8.

Forschner, W., Schmitt, V. and Wild, A. (1989) Investigations on the starch content and ultrastructure of spruce needles relative to the occurrence of novel forest decline. *Botanica Acta* **102,** 208–21.

Forschungsbeirat Waldschäden (1986) 2. *Bericht.* Karlsruhe, Germany: Forschungsbeirat Waldschäden/Luftverunreinigungen der Bundesregierung und der Länder.

Förster, H. (1990) Über die Magnesiumverarmung der Böden in den Hochlagen des inneren Bayerischen Waldes. *Zeitschrift für Pflanzenernährung und Bodenkunde* **153,** 81–7.

Fosberg, F.R. (1961) *Guide to Excursion III.* Tenth Pacific Science Congress. Honolulu: Tenth Pacific Science Congress and University of Hawaii.

Fourchy, P. (1951) Sécheresse, variations climatiques et végétation. *Revue forestières françaises* **3,** 47–55.

Fowells, H.A. (1965) *Silvics of forest trees of the United States.* United States Department of Agriculture Forest Service Agricultural Handbook No. 271. Washington, DC: United States Government Printing Office.

Fowler, D., Cape, J.N., Deans, J.D., Leith, I.D., Murray, M.B., Smith, R.I., Sheppard, L.J. and Unsworth, M.H. (1989a) Effects of acid mist on the frost hardiness of red spruce seedlings. *New Phytologist* **113,** 321–35.

Fox, C.A., Kincaid, W.B., Nash, T.H., Young, D.L. and Fritts, H.C. (1986) Tree ring variation in western larch (*Larix occidentalis*) exposed to sulphur dioxide emissions. *Canadian Journal of Forest Research* **16,** 283–92.

France, E.A., Binkley, D. and Valentine, D. (1989) Soil chemistry changes after 27 years under four tree species in southern Ontario. *Canadian Journal of Forest Research* **19,** 1648–50.

Franich, R.A., Wells, L.G. and Barnett, J.R. (1977) Variation with tree age of needle cuticle topography and stomatal structure in *Pinus radiata* needles. *Phytochemistry* **17,** 1617–23.

Franklin, D.A. (1967) The synecology of the Tararua indigenous forests. *New Zealand Forest Service Technical Paper* No. 53.

Frear, S.T. (1982) What's killing the Alaska yellow-cedar? *American Forests* **88,** 41–3 and 62–3.

Freedman, B. and Hutchinson, T.C. (1980a) Pollutant inputs from the atmosphere and accumulations in soils and vegetation near a nickel-copper smelter at Sudbury, Ontario, Canada. *Canadian Journal of Botany* **58,** 108–32.

Freedman, B. and Hutchinson, T.C. (1980b) Effects of smelter pollutants on forest leaf litter decomposition near a nickel-copper smelter at Sudbury, Ontario, Canada. *Canadian Journal of Botany* **58,** 1722–36.

Freedman, B. and Hutchinson, T.C. (1980c) Long-term effects of smelter pollution at Sudbury, Ontario, on forest community composition. *Canadian Journal of Botany* **58,** 2123–40.

von Freiesleben, N.E. (1988) Sulfate mobility in an acid Danish soil. *Journal of Environmental Quality* **17,** 278–84.

Frelich, L.E., Bockheim, J.G. and Leide, J.E. (1989) Historical trends in tree-growth and chemistry across an air-quality gradient in Wisconsin. *Canadian Journal of Forest Research* **19,** 113–21.

Friedland, A.J. (1989a) Recent changes in the montane spruce–fir forests of the northeastern United States. *Environmental Monitoring and Assessment* **12,** 237–44.

Friedland, A.J. (1989b) *1988 Northern spruce regional assessment.* Forest Response Program, Spruce–fir Research Cooperative Report. Radnor, Pennsylvania: United States Department of Agriculture Forest Service.

Friedland, A.J. and Battles, J.J. (1987) Red spruce (*Picea rubens* Sarg.) decline in the northeastern United States; review and recent data from Whiteface Mountain. In: Kairiukstis, L., Nilsson, S. and Straszak (eds), *Forest decline and reproduction: regional and global consequences.* Laxenburg: IIASA, pp. 287–96.

Friedland, A.J., Gregory, R.A., Kärenlampi, L. and Johnson, A.H. (1984) Winter damage to foliage as a factor in red spruce decline. *Canadian Journal of Forest Research* **14,** 963–5.

Friedland, A.J., Hawley, G.J. and Gregory, R.A. (1985) Investigations of nitrogen as a possible contributor to red spruce (*Picea rubens* Sarg.) decline. In: *Air pollutants effects on forest ecosystems.* Minneapolis, Minnesota: University of Minnesota Press, pp. 95–106.

Friedland, A.J., Hawley, G.J. and Gregory, R.A. (1988) Red spruce (*Picea rubens* Sarg.) foliar chemistry in northern Vermont and New York, USA. *Plant and Soil* **105,** 189–93.

Friend, A.L. and Hafley, W.L. (1989) Climatic limitations to growth in loblolly and shortleaf pine (*Pinus taeda* and *P. echinata*): a dendroclimatological approach. *Forest Ecology and Management* **26,** 113–22.

Friend, D.J.C. (1980) Effect of different photon flux densities (PAR) on seedling growth and morphology of *Metrosideros collina* (Forst.) Gray. *Pacific Science* **34,** 93–100.

Fromard, F., Dagnac, J., Gauquelin, T. and Cheret, V. (1991) Results of research into decay of the fir (*Abies alba* Mill.) in the Pyrenees. New data about nutritional and physiological disturbances. In: Longhurst, J.W.S. (ed.), *Acid deposition. Origins, impacts and abatement strategies.* Berlin: Springer Verlag, pp. 109–22.

Frühwald, A. (1986) Technological properties of wood from trees in polluted regions. *IAWA Bulletin* **7,** 389–97.

Frühwald, A., Bauch, J. and Göttsche-Kühn, H. (1984a) Holzeigenschaften von Fichten aus Waldschadensgebieten. Untersuchungen am frisch gefällten Holz. *Holz als Roh- und Werkstoff* **42,** 441–49.

Frühwald, A., Bauch, J. and Gottsche-Kuhn, H. (1984b) Über die Holzeigenschaften von Fichten aus Waldschadensgebieten. *Holz als Roh- und Werkstoff* **44,** 399–406.

Frühwald, A., Schwab, E., Mehringer, H. and Krause, H.A. (1986) Zuwachs, Dichte und mechanische Eigenschaften des Holzes von Kiefern aus Waldschadensgebieten. *Holz als Roh- und Werkstoff* **44,** 399–406.

Führer, E.G. (1992) Der Zusammenhang zwischen der Dürre und der Erkrankung der Traubeneichenbestände in Ungarn. *Forstwissenschaftliches Centralblatt* **111,** 129–36.

Fuller, R.D., Mitchell, M.J., Schindler, S.C., Schecher, W.D. and Driscoll, C.T. (1985) Sulfur transformations in forested spodosols of the northeastern US. Paper presented at the International Symposium on Acidic Precipitation, Muskoka, Ontario, September 15–20, 1985.

Fuller, R.D., Driscoll, C.T., Lawrence, G.B. and Nodvin, S.C. (1987) Processes regulating sulphate flux after whole-tree harvesting. *Nature* **325,** 707–10.

Gagnon, G., Roy, G., Gravel, C. and Gagné, J. (1986) State of dieback research at the Ministère de l'Énergie et des Resources. In: *Maple decline.* Quebec: Ministère de l'Agriculture, des Pêcheries et de l'Alimentation, pp. 43–80.

Gansner, D.A, Birch, T.W. and Frieswyk, T.S. (1987) What's up with sugar maple. *National Woodlands* **10,** 5–6.

Gao, R.T., Qin, X.X., Li, J.Z., Hao, W.Q. and Wang, X.Z. (1985) A preliminary study of the relationships between artificial defoliation of poplar trees and their growth. *Scientia Silvae Sinicae* **21,** 199–205.

Garrec, J.-P. (1991) Effects of atmospheric pollution on the above-ground part of the trees. In: Landmann, G. (ed.), *French research into forest decline.* Nancy: École Nationale du Génie Rural, des Eaux et des Forêts, pp. 59–86.

Garrec, J.-P. and Kerfourn, C. (1989) Effets des pluies acides et de l'ozone sur la perméabilité à l'eau et aux ions de cuticules isolées. Implication dans le phénomène de dépérissement des forêts. *Environmental and Experimental Botany* **29,** 215–28.

Garrec, J.-P., Richardin, I., Le Maout, L., Rose, C., Gérard, B. and Mathieu, Y. (1989) Étude du pH et du pouvoir tampon des aiguilles des sapins (*Abies alba* L.) et d'épicéas (*Picea abies* L.) dépérissants. Application en symptomatologie? *Annales des Sciences Forestières* **46,** 55–69.

Garrec, J.-P., Rose, C. and Le Maout, L. (1988) Diagnosis of forest decline – a study of four physiological tests. In: Cape, J.N. and Mathy, P. (eds), *Scientific basis of forest decline symptomatology.* Air Pollution Report 15. Brussels: Commission of the European Communities, pp. 288–94.

Gärtner, E. (1985) Mangangehalte in Altfichten, Boden und Kronendurchlass an jeweils gleichen Standorten. In: Stratmann, H. (ed.), *Waldschäden. Einflussfaktoren und ihre Bewertung. Kolloquium Goslar, 18. bis 20. Juni 1985.* VDI-Bericht 560, pp. 559–73.

Gärtner, E.J. (1988) Operational forest decline symtomatology. In: Cape, J.N. and Mathy, P. (eds), *Scientific basis of forest decline symptomatology.* Air Pollution Report 15. Brussels: Commission of the European Communities, pp. 295–306.

Gärtner, R. and Hradetzky, J. (1988) Gesundheitszustand und Zuwachs der Fichte. *Forst und Holz* **43,** 131–2.

Gasch, G., Grünhage, L., Jäger, H.-J. and Wentzel, K.-F. (1988) Das Verhältnis der Schwefelfraktionen in Fichtennadeln als Indikator für Immissionsbelastungen durch Schwefeldioxid. *Angewandte Botanik* **62,** 73–84.

Geballe, G.T., Smith, W.H. and Wargo, P.M. (1990) Red spruce seedling health: an assessment of acid fog deposition and heavy metal soil contamination as interactive stress factors. *Canadian Journal of Forest Research* **20,** 1680–3.

Gehrig, R. (1992) Immissionssituation im Schweizer Wald. In: *Waldschadensforschung in der Sceweiz: Stand der Kenntnisse.* Birmensdorf: Eidgenössische Forschungsanstalt für Wald, Schnee und Landschaft, pp. 81–6.

Gehrmann, J., Büttner, G. and Ulrich, B. (1987) Untersuchungen zum Stand der Bodenversauerung wichtiger Waldstandorte im Land Nordrhein-Westfalen. *Bericht des Forschungszentrums Waldökosysteme der Universität Göttingen* **B4,** 233 pp.

Genys, J.B. and Heggestad, H.E. (1978) Susceptibility of different species, clones and strains of pines to acute injury caused by ozone and sulphur dioxide. *Plant Disease Reporter* **62,** 687–91.

Genys, J.B. and Heggestad, H.E. (1983) Relative sensitivity of various types of eastern white pine, *Pinus strobus*, to sulphur dioxide. *Canadian Journal of Forest Research* **13,** 1262–5.

Gerecke, K.-L. (1986) Zuwachsuntersuchungen an vorherrschenden Tannen aus Baden-Württemberg. *Allgemeine Forst- und Jagdzeitung* **157,** 59–68.

Gerecke, K.-L. (1990) 'Tannensterben' und 'Neuartige Waldschäden' – Ein Beitrag aus der Sicht der Waldwachstumskunde. *Allgemeine Forst- und Jagdzeitung* **161,** 81–96.

Gerhold, H.D. (1977) *Effects of air pollution on* Pinus strobus *L. and genetic resistance.* Report EPA-600/3-77-002. Corvallis: Corvallis Environmental Research Laboratory, Office of Research and Development, US Environmental Protection Agency. 44 pp.

Gerrish, G. (1990) Relating carbon allocation patterns to the senescence in *Metrosideros* forest. *Ecology* **71,** 1176–84.

Gerrish, G. and Bridges, K.W. (1984) A thinning and fertilization experiment in *Metrosideros* dieback stands in Hawaii. *Hawaii Botanical Science Paper* 43. Honolulu: University of Hawaii.

Gerrish, G., Mueller-Dombois, D. and Bridges, K.W. (1988) Nutrient limitation and *Metrosideros* forest dieback in Hawaii. *Ecology* **69,** 723–7.

Gersper, P.L. and Holowaychuk, N. (1971) Some effects of stemflow from forest canopy trees on chemical properties of soils. *Ecology* **52,** 691–702.

Gertner, G. (1991) Comment II. *Forest Science* **37,** 723–5.

Gholz, H.L. and Cropper, W.P. (1991) Carbohydrate dynamics in mature *Pinus elliottii* var. *elliottii* trees. *Canadian Journal of Forest Research* **21,** 1742–7.

Giersberg, F. (1901) *Künstliche Düngung im forstlichen Betriebe.* Berlin: Patz and Gerleb.

Giese, R.L., Houston, D.R., Benjamin, D.M., Kuntz, J.E., Kapler, J.E. and Skilling, D.D. (1964) Studies of maple blight. *University of Wisconsin Research Bulletin* 250.

Giese, K. (1991) Die Entwicklung der immissionsbedingten Waldschäden und ihre Auswirkungen auf die ökonomische Situation der Forstbetriebe. In: *Impact of the new types of forest damage on the economic situation of forestry enterprises.* Bonn: Bundesministerium für Ernährung, Landwirtschaft und Forsten pp. 41–51.

Gilbert, O.L. (1975) Effects of air pollution on landscape and land-use around Norwegian aluminium smelters. *Environmental Pollution* **8,** 113–21.

Gignac, L.D. and Beckett, P.J. (1986) The effect of smelting operations on peatlands near Sudbury, Ontario, Canada. *Canadian Journal of Botany* **64,** 1138–47.

Di Giulio, R.T. and Richardson, C.J. (1987) *Effects of atmospheric deposition on red spruce: a free-radical based approach.* Report to the Spruce–Fir Cooperative, Broomall, Pennsylvania.

Glass, G.E. and Loucks, O.L. (1986) Implications of a gradient in acid and ion deposition across the northern Great Lakes States. *Environmental Science and Technology* **20,** 35–43.

Glatzel, G. (1985) Schwermetallbelastung von Wäldern in der Umgebung eines

Hüttenwerkes in Brixlegg – Tirol. II. Wachstum und Mineralstoffernährung von Fichtenpflanzen (*Picea abies*) auf schwermetallbelastetem Waldbodenhumus. *Centralblatt für das Gesamte Forstwesen* **102,** 41–51.

Glatzel, G. (1990) The nitrogen status of Austrian forest ecosystems as influenced by atmospheric deposition, biomass harvesting and lateral organomass exchange. *Plant and Soil* **128,** 67–74.

Glatzel, G. (1991) The impact of historic landuse and modern forestry on nutrient relations of Central European forest ecosystems. *Fertilizer Research* **27,** 1–8.

Glatzel, G., Kazda, M., Grill, D., Halbwachs, G. and Katzensteiner, K. (1987) Ernährungsstörungen bei Fichte als Komplexwirkung von Nadelschäden und erhöhter Stickstoffdeposition – ein Wirkungsmechanismus des Waldsterbens? *Allgemeine Forst- und Jagdzeitung* **158,** 91–7.

Glavac, V. (1988) Blatt- Verkleinerungen in immissionsgeschädigten Buchenwäldern. *Allgemeine Forst Zeitschrift* **43,** 813–14.

Glavac, V., Jochheim, H., Koenies, H., Rheinstädter, R. and Schäfer, H. (1985) Einfluss des Stammablaufwassers auf den Boden im Stammfussbereich von Altbuchen in unterschiedlich immissionsbelasteten Gebieten. *Allgemeine Forst Zeitschrift* **40,** 1397–8.

Glavac, V., Koenies, H. and Ebben, U. (1990a) Seasonal variation of calcium, magnesium, potassium, and manganese contents in xylem sap of beech (*Fagus sylvatica* L.) in a 35-year-old limestone beech forest stand. *Trees: Structure and Function* **4,** 75–80.

Glavac, V., Koenies, H. and Ebben, U. (1990b) Seasonal variations in mineral concentrations in the trunk xylem sap of beech (*Fagus sylvatica* L.) in a 42-year-old beech forest stand. *New Phytologist* **116,** 47–54.

Glos, P. and Schulz, H. (1986) Qualität und Festigkeit von Bauschnittholz aus Waldschadensgebieten. *Holz als Roh- und Werkstoff* **44,** 293–8.

Gobran, G.R. and Nilsson, S.I. (1988) Effects of forest floor leachate on sulfate retention in a spodosol soil. *Journal of Environmental Quality* **17,** 235–9.

Godbold, D.L., Dictus, K. and Hüttermann, A. (1988a) Influence of aluminium and nitrate on root growth and mineral nutrition of Norway spruce (*Picea abies*) seedlings. *Canadian Journal of Forest Research* **18,** 1167–71.

Godbold, D.L., Fritz, E. and Hüttermann, A. (1988b) Aluminium toxicity and forest decline. *Proceedings of the National Academy of Science of the USA* **85,** 3888–92.

van Goor, C.P. (1985) The impact of tree species on soil productivity. *Netherlands Journal of Agricultural Science* **33,** 133–40.

Göransson, A. and Eldhuset, T.D. (1987) Effects of aluminium on growth and nutrient uptake of *Betula pendula* seedlings. *Physiologia Plantarum* **69,** 193–9.

Gorham, E. and Gordon, A.G. (1960a) Some effects of smelter pollution northeast of Falconbridge, Ontario. *Canadian Journal of Botany* **38,** 307–12.

Gorham, E. and Gordon, A.G. (1960b) The influence of smelter fumes upon the chemical composition of lake water near Sudbury, Ontario, and upon the surrounding vegetation. *Canadian Journal of Botany* **38,** 477–87.

Gosz, J.R. (1981) Nitrogen cycling in coniferous ecosystems. In: Clark, F.E. and Rosswall, T. (eds), *Terrestrial nitrogen cycles. Ecological Bulletin* **33,** 405–26.

Göttsche-Kuhn, H. (1988) Bildung und Eigenschaften des Holzes von Fichten

(*Picea abies* (L.) Karst.) aus Waldschadensgebieten. *Mitteilungen der Bundesforschungsanstalt für Forst- und Holzwirtschaft, Reinbeck bei Hamburg* **157.** 171 pp.

Göttsche-Kuhn, H. and Frühwald, A. (1986) Holzeigenschaften von Fichten aus Waldschadensgebieten. Untersuchungen an gelagertem Holz. *Holz als Roh- und Werkstoff* **44,** 313–18.

Goulding, K.W.T., Johnston, A.E. and Poulton, P.R. (1988) The effect of atmospheric deposition, especially of nitrogen, on grassland and woodland at Rothamsted Experimental Station, England, measured over more than 100 years. In: Mathy, P. (ed.), *Air pollution and ecosystems.* Dordrecht, D. Reidel, pp. 841–6.

Grabherr, G. and Mucina, L. (1989) Waldsterben und Waldmelioration in den Alpen aus der Sicht der Vegetationsökologie. *Allgemeine Forst- und Jagdzeitung* **160,** 216–22.

Grammel, R.H., Becker, G., Gross, M. and Höwecke, B. (1986) Einige Holzeigenschaften erkrankter Fichten und Tannen aus Baden-Württemberg, als Ergebnis einer flächendeckenden Erhebung. *Holz als Roh- und Werkstoff* **44,** 287–92.

Grant, I.F., Bancroft, K. and Alexander, M. (1979a) SO_2 and NO_2 effects on microbial activity in an acid forest soil. *Microbial Ecology* **5,** 85–9.

Grant, I.F., Bancroft, K. and Alexander, M. (1979b) Effect of SO_2 and bisulphite on heterotrophic activity in acid soil. *Applied and Environmental Microbiology* **38,** 78–83.

Grant, P.J. (1984) Drought effect on high altitude forests, Ruahine Range, North Island, New Zealand. *New Zealand Journal of Botany* **22,** 15–27.

Graumlich, L.J. and Brubaker, L.B. (1987) Increasing net primary productivity in Washington (USA) forests during the last 100 years. In: Jacoby, G.C. and Hornbeck, J.W. (eds), *Proceedings of the International Symposium on Ecological Aspects of Tree Ring Analysis.* Springfield: US Department of Commerce, National Technical Information Service, pp. 59–69.

Gray, T.R.G. and Ineson, P. (1981) The effects of sulphur dioxide and acid rain on the decomposition of leaf litter. *Journal of the Science of Food and Agriculture* **32,** 624–5.

Graybill, D.A. (1986) The reconstruction of prehistoric Salt River stream flow. In: Heathington, D.A. and Gregory, D.A. (eds), *The 1982–1984 Excavations at las Colinas, Environment and Subsistence.* Archaeological Series 162, Volume 5, Part 1. Tucson: Cultural Resource Management Division, Arizona State Museum, University of Arizona.

Graybill, D.A. (1987) A network of high elevation conifers in the western US for detection of tree ring growth response to increasing atmospheric carbon dioxide. In: Jacoby, G.C. and Hornbeck, J.W. (eds), *Proceedings of the International Symposium on Ecological Aspects of Tree Ring Analysis.* Springfield: US Department of Commerce, National Technical Information Service, pp. 463–74.

Graybill, D.A. and Rose, M.R. (1989) Analysis of growth trends and variation in conifers from Arizona and New Mexico. In: Olson, R.K. and Lefohn, A.S. (eds), *Effects of air pollution on western forests.* Pittsburgh: Air and Waste Management Association, pp. 395–407.

Gregory, S.C. and Redfern, D.B. (1987) The pathology of Sitka spruce in northern Britain. *Proceedings of the Royal Society of Edinburgh* **93B,** 145–56.

Grehan, J.R., Paricer, B.L., Wagner, D.L., Rosovsky, J. and Aleong, J. (1992) Root damage by the conifer swift moth: a mortality factor in montane red spruce regeneration. *Forest Science* **38,** 611–22.

Greig, B.J.W. (1992) *Occurrence of decline and dieback of oak in Great Britain.* Research Information Note 214. Farnham: Forestry Commission.

Von Grenzius, R. (1984) Starke Versauerung der Waldböden Berlins. *Forstwissenschaftliches Centralblatt* **103,** 131–9.

Greszta, J., Barszcz, J. and Niemtur, S. (1989) Evaluation of damage to montane forests in southern Poland. In: Bucher, J.B. and Bucher-Wallin, I. (eds), *Air pollution and forest decline.* Birmensdorf: Eidgenössische Anstalt für das forstliche Versuchswesen, pp. 41–50.

Greszta, J. and Maczynski, M. (1989) Silviculture methods in Polish forests affected by industrial pollutants. *Working paper – International Institute for Applied Systems Analysis. Publication No. 102 of the Biosphere Dynamics Project.* Laxenburg: IIASA.

Greve, U., Eckstein, D. and Scholz, F. (1983) Holzbiologische Untersuchungen an Fichten unterschiedlicher Immissionstoleranz. *Aquilo Serio Botanico* **19,** 90–6.

Greve, U., Eckstein, D., Aniol, R.W. and Scholz, F. (1986) Dendroklimatologische Untersuchungen an Fichten unterschiedlicher Immissionsbelastung in Nordostbayern. *Allgemeine Forst- und Jagdzeitung* **157,** 174–9.

Greve, U., Eckstein, D., Scholz, F. and Schweingruber, F.H. (1985) Holzbiologische Untersuchungen an Fichtenklonen unterschiedlicher Empfindlichkeit gegen eine HF-Begasung. *Angewandte Botanik* **59,** 81–93.

Grier, C.C. (1988) Foliage loss due to snow, wind and winter drying damage: its effect on leaf biomass of some western conifer forests. *Canadian Journal of Forest Research* **18,** 1097–102.

Grieve, I. (1978) Some effects of the plantation of conifers on a freely drained lowland soil, Forest of Dean, UK. *Forestry* **51,** 21–8.

Griffin, H.D. (1965) Maple dieback in Ontario. *Forestry Chronicle* **41,** 295–300.

Griffin, J.R. (1971) Oak regeneration in the upper Carmel Valley, California. *Ecology* **52,** 862–8.

Griffin, J.R. (1973) Valley oaks, the end of an era? *Fremontia* **1,** 5–9.

Griffin, J.R. (1976) Regeneration of *Quercus lobata* savannas, Santa Lucia mountains, California. *American Midland Naturalist* **95,** 422–35.

Griffin, J.R. (1980) Animal damage to valley oak and seedlings, Carmel Valley. In: Plumb, T.R. (coordinator), *Proceedings of the symposium on the ecology, management, and utilization of California oaks, June 26–28, 1979, Claremont, California.* General Technical Report PSW-44. Berkeley: United States Department of Agriculture Forest Service, Pacific Southwest Forest and Range Experiment Station, pp. 242–5.

Grigal, D.F., McRoberts, R.E. and Ohmann, L.F. (1991) Spatial variation in chemical properties of forest floor and surface mineral soil in the North Central United Sates. *Soil Science* **151,** 282–90.

Griggs, R.F. (1946) The timberlines of North America and their interpretations. *Ecology* **27,** 275–89.

Grill, D. (1971) Pufferkapazität gesunder und rauchgeschädigter Fichtennadeln. *Zeitschrift für Pflanzenkrankheit* **78,** 612–22.

Grill, D., Pfeifhofer, H., Halbwachs, G. and Waltinger, H. (1987) Investigations on epicuticular waxes of differently damaged spruce needles. *European Journal of Forest Pathology* **17,** 246–55.

Grill, D., Pfeifhofer, H., Tschulik, A., Hellig, K. and Holzer, K. (1988) Thiol content of spruce needles at forest limits. *Oecologia* **76,** 294–7.

Grimm, R. and Rehfuess, K. (1986) Kurzfristige Veränderungen von Bodenreaktion und Kationenaustauscheigenschaften in einem Meliorationsversuch zu Kiefer (*Pinus syl.* L.) auf Podsol-Pseudogley in der Oberpfalz. *Allgemeine Forst-und Jagdzeitung* **157,** 205–13.

Gross, H.L. (1975) Crown deterioration and reduced growth associated with excessive seed production by birch. *Canadian Journal of Botany* **50,** 2431–8.

Gross, H.L. and Harden, A.A. (1968) *Dieback and abnormal growth of yellow birch induced by heavy fruiting.* Information Report 0-X-79. Sault Sainte Marie: Forest Research Laboratory.

Grossman, D.M. (1988) *Invertebrates associated with spruce-fir stands on Mt. Moosilauke, NH and at Howland, ME.* M.Sc. Thesis, University of Maine, Orono.

Groton, E. and Eagar, C. (1988) Southern red spruce-Fraser fir forests. In: Van Deusen, P.C. (ed.), *Analyses of Great Smoky Mountain Red Spruce Tree ring Data.* General Technical report SO-69. New Orleans: United States Department of Agriculture Forest Service, Southern Forest Experimental Station, pp. 6–20.

Gruber, F. (1986) *Beiträge zum morphogenetischen Zyklus der Knospe, zur Phyllotaxis und zum Triebwachstum der Fichte (*Picea abies *(L.) Karst) auf unterschiedlichen Standorten.* Beiträge Forschungszentrum Waldökosystemen/Waldschäden **A25.** (215 pp.).

Gruber, F. (1987) Zur Präventivtriebbildung der Fichte. *Allgemeine Forst Zeitschrift* **49,** 1285–8.

Gruber, F. (1989) Phänotypen der Fichte (*Picea abies* (L.) Karst.) I. Verzweigungsphänotypen: Genotyp und Modifikation. *Allgemeine Forst- und Jagdzeitung* **160,** 157–65.

Gruber, F. (1990) *Verzweigungssystem, Benadelung und Nadelfall der Fichte (*Picea abies*).* Basel: Birkhäuser Verlag. 135 pp.

Grulke, N.E., Miller, P.R., Wilborn, R.D. and Hahn, S. (1990) Photosynthetic response of giant sequoia seedlings and rooted branchlets of mature foliage to ozone fumigation. In: Olson, R.K. and Lefohn, A.S. (eds), *Effects of air pollution on western forests.* Pittsburgh: Air and Waste Management Association, pp. 429–41.

Grünhage, L. and Jäger, H.J. (1988) Entwicklung der Nährstoff- und Schwermetallgehalte in Fichtennadeln aus dem Rhein-Main-Gebiet. *Angewandte Botanik* **62,** 85–91.

Guderian, R., Tingey, D.T. and Rabe, R. (1985a) Effects of photochemical oxidants on plants. In: Guderian, R. (ed.), *Air pollution by photochemical oxidants.* Berlin: Springer Verlag, pp. 129–333.

Guerrero, M.G., Vega, J.M. and Losada, M. (1981) The assimilatory nitrate-reducing system and its regulation. *Annual Reviews of Plant Physiology* **32,** 169–204.

Guillaumin, J.-J., Bernard, C., Delatour, C. and Belgrand, M. (1983) Le dépérissement du chêne à Tronçais: pathologie racinaire. *Revue forestière françaises* **35,** 415–24.

Gullvåg, B. (1986) Kan volslaget på barnålene gi informasjon om forurensningsbelastning? *Norsk Skogbruk* **32,** 16–17.

Gundersen, P. (1991) Nitrogen deposition and the forest nitrogen cycle: role of denitrification. *Forest Ecology and Management* **44,** 15–28.

Gundersen, P. and Beier, C. (1988) Aluminium sulfate solubility in acid forest soils in Denmark. *Water, Air, and Soil Pollution* **39,** 247–61

Gundersen, P. and Rasmussen, L. (1990) Nitrification in forest soils: effects from nitrogen deposition on soil acidification and aluminium release. *Reviews of Environmental Contamination and Toxicology* **113,** 1–45.

Günthardt-Georg, M.S. (1989) Site-dependent changes in the epicuticular wax structure of spruce. In: Bucher, J. and Bucher-Wallin, I. (eds), *Air pollution and forest decline.* Birmensdorf: Eidgenössische Anstalt für das forstliche Versuchswesen, pp. 416–18.

Günthardt-Goerg, M.S. and Keller, Th. (1987) Some effects of long-term ozone fumigation on Norway spruce. II. Epicuticular wax and stomata. *Trees: Structure and Function* **1,** 145–50.

Günthardt-Goerg, M.S. and Keller, Th. (1988) Cuticular features and conductivity of needle diffusate in spruce. In: Cape, J.N. and Mathy, P. (eds), *Scientific basis for forest decline symptomatology.* Air Pollution Report 15. Brussels: Commission of the European Communities, pp. 316–22.

Guyon, J.P. (1991) Dépérissement du pin maritime (*Pinus pinaster* Aït) en Vendée. Les causes écologiques. *Annales des Sciences Forestères* **48,** 333–46.

Hadfield, J.S. (1968) *Evaluation of diseases of red spruce on the Chamberlain Hill sale, Rochester Ranger District, Green Mt. Nat. Forest.* File Report A-68-8 5230. Amherst, Maryland: United States Department of Agriculture Forest Service Northeastern Area, State and Private Forestry, Amherst FPC Field Office.

Hadley, J.L. and Smith, W.K. (1989) Wind erosion of leaf surface wax in alpine timberline conifers. *Arctic and Alpine Research* **21,** 392–8.

Hafley, W.L. (1985) *A review of Forest Service growth loss data in the Southeast.* Report to the US Department of Agriculture Forest Service, Southeast Forest Experiment Station, Asheville.

Hager, H. and Kazda, M. (1985) The influence of stand density and canopy position on sulfur content in needles of Norway spruce (*Picea abies* L. Karst.). *Water, Air and Soil Pollution* **25,** 321–9.

Hägi, K. (1989) *Terrestrische Waldschadeninventur.* Bericht 314. Birmensdorf: Eidgenössische Anstalt für das forstliche Versuchswesen.

Hågvar, S. and Kjøndal, B.R. (1981a) Decomposition of birch leaves: Dry weight loss, chemical changes and effects of artificial acid rain. *Pedobiologia* **22,** 232–45.

Hågvar, S. and Kjøndal, B.R. (1981b) Effects of artificial acid rain on the microarthropod fauna in decomposing birch leaves. *Pedobiologia* **22,** 409–22.

Haines, B.L. and Carlson, C.L. (1989) *Effects of acidic precipitation on trees.* In: Adriano, D.C. and Johnson, A.H. (eds), Acidic precipitation. Volume 2: Biological and ecological effects. New York: Springer Verlag, pp. 1–27.

Hall, J.P. (1991) *ARNEWS annual report 1990*. Information Report ST-X–1. Ottawa: Forestry Canada.

Hall, R.J., Still, G.N. and Crown, P.H. (1983) Mapping the distribution of aspen defoliation using Landsat color composites. *Canadian Journal of Remote Sensing* **9,** 86–91.

Hallbäcken, L. (1992) Long term changes of base cation pools in soil and biomass in a beech and a spruce forest of southern Sweden. *Zeitschrift für Pflanzenernährung und Bodenkunde* **155,** 51–60.

Hallbäcken, L. and Tamm, C.O. (1986) Changes in soil acidity from 1927 to 1982–1984 in a forest area of south-west Sweden. *Scandinavian Journal of Forest Research* **1,** 219–32.

Hambuckers, A. and Remacle, J. (1991) Nutritional status of declining spruce (*Picea abies* (L.) Karst.): effect of soil organic matter turnover rate. *Water, Air and Soil Pollution* **59,** 95–106.

Hamburg, S.P. and Cogbill, C.V. (1988) Historical decline of red spruce populations and climatic warming. *Nature* **331,** 428–31.

Häni, H. (1975) Wechselwirkung von Fluorid mit einem mineralischen, illithaltigen Boden und Veränderung in den darauf wachsenden Maispflanzen. *Schweizerische landwirtschaftliche Forschung* **14,** 189–201.

Hanson, D.W., Norton, S.A. and Williams, J.S. (1982) Modern and paleolimnological evidence for accelerated leaching and metal accumulation in soils in New England, caused by atmospheric deposition. *Water, Air, and Soil Pollution* **18,** 227–39.

Hanson, P.J., McLaughlin, S.B. and Edwards, N.T. (1988) Net CO_2 exchange of *Pinus taeda* shoots exposed to variable ozone levels and rain chemistries in field and laboratory settings. *Physiologia Plantarum* **74,** 635–42.

Hantschel, R., Kaupenjohann, M., Horn, R. and Zech, W. (1988) Acid rain studies in the Fichtelgebirge (NE-Bavaria). In: Mathy, P. (ed.), *Air pollution and ecosystems.* Reidel, Dordrecht, 881–6.

Hapla, F. and Kottwitz, K. (1987) Holzqualität von Kiefern aus einem Waldschadensgebiet. *Holz-Zentralblatt* **113,** 1333–5.

Härdtle, W. (1990) Damage in beech forests of Schleswig-Holstein (Northern Germany) in relation to forest structure and soil conditions. *Vegetatio* **90,** 117–23.

Hari, P. and Arovaara, H. (1988) Detecting CO_2 induced enhancement in the radial increment of trees. Evidence from northern timberline. *Scandinavian Journal of Forest Research* **3,** 67–74.

Harmer, R. (1989) The effect of mineral nutrients on growth, flushing, apical dominance and branching in *Quercus petraea* (Matt.) Liebl. *Forestry* **62,** 383–95.

Harr, D.R. (1982) Fog drip in the Bull Run municipal watershed, Oregon. *Water Resources Bulletin* **18,** 785–9.

Harrington, T.C. (1986) Growth decline of wind-exposed red spruce and balsam fir in the White Mountains. *Canadian Journal of Forest Research* **16,** 232–8.

Harrington, T.C., Rizzo, D.M. and Marchand, P.J. (1984) Wind, rocks, root disease and mortality of subalpine red spruce and balsam fir. *Phytopathology* **74,** 823.

Harris, A.S. and Farr, W.A. (1974) *The forest ecosystem of southeast Alaska. 7. Forest ecology and timber management.* General Technical Report PNW-25.

Portland: US Department of Agriculture Forest Service, Pacific Northwest Research Station.

Harrison, R.B., Johnson, D.W. and Todd, D.E. (1989) Sulfate adsorption and desorption reversibility in a variety of forest soils. *Journal of Environmental Quality* **18,** 419–26.

Härtel, O. and Cerny, M. (1981) Veränderungen in Fichtenwaldböden durch Langzeiteinwirkung von SO_2. *Mitteilungen der Forstlichen Bundesversuchsanstalt, Wien* 137/II, 233–40.

Hartley, G. and Merrill, T.G. (1915) Storm and drought injury to foliage of ornamental trees. *Phytopathology* **5,** 20–29.

Hartmann, G. and Blank, R. (1992) Winterfrost, Kahlfrass und Prachtkäferbefall als Faktoren im Ursachenkomplex des Eichensterbens in Norddeutschland. *Forst und Holz* **47,** 443–52.

Hartmann, G., Blank, R. and Lewark, S. (1989) Eichensterben in Norddeutschland – Verbreitung, Schadbilder, mögliche Ursachen. *Forst und Holz* **44,** 475–87.

Hartmann, G., Saborowski, J., Uebel, R. and Voretzsch, A. (1986) Entwicklung und Verteilung von Waldschäden an Fichte im Harz. Ergebnisse und methodische Aspekte der Luftbild-Waldschadenserhebungen 1983 bis 1985. *Der Forst- und Holzwirt* **41,** 413–20.

Hartmann, G., Uebel, R. and Stock, R. (1985) Zur Verbreitung der Nadelvergilbung an Fichte im Harz. Vorläufige Mitteilung von Luftbildauswertungsergebnissen. *Der Forst- und Holzwirt* **40,** 286–92.

Hartmann, G. and Voretzsch, A. (1985) Ergebnisse der Luftbild-Waldschadenserhebung Niedersachsen 1985. *Der Forst- und Holzwirt* **40,** 271–8.

Hartmann, P., Schneider, O., Petter, D.A. and Schlaepfer, R. (1987) Étude des relations, chez l'épicéa, entre la largeur du cerne et l'état sanitaire de la cime. *Schweizerische Zeitschrift für Forstwesen* **138,** 923–43.

Hasemann, G., Jung, G. and Wild, A. (1990) The loss of structural integrity in damaged spruce needles from locations exposed to air pollution. II. Epidermis and stomata (dermal tissue). *Journal of Phytopathology* **128,** 33–45.

Hasemann, G. and Wild, A. (1990) The loss of structural integrity in damaged spruce needles from locations exposed to air pollution. I. Mesophyll and central cylinder. *Journal of Phytopathology* **128,** 15–32.

Häsler, R. (1991) Photosynthese und stomatäre Leitfähigkeit der Fichte unter dem Einfluss von Witterung und Luftschadstoffen. In: Stark, M. (ed.), *Luftschadstoffe und Wald.* Zürich: Verlag der Fachvereine, ETH, pp. 143–68.

Haug, I. and Feger, K.H. (1990) Effects of fertilization with $MgSO_4$ and $(NH_4)_2SO_4$ on soil solution chemistry, mycorrhizae and nutrient content of fine roots in a Norway spruce stand. *Water, Air, and Soil Pollution* **54,** in press.

Hauhs, M. (1985) Wasser- und Stoffhaushalt im Einzugsgebiet der Langen Bramke (Harz). *Bericht des Forschungszentrums Waldökosysteme/Waldsterben* 17.

Hauhs, M. (1989) Lange Bramke: an ecosystem study of a forested catchment. In: Adriano, D.C. and Havas, M. (eds), *Acidic Precipitation.* Volume 1. New York: Springer-Verlag, pp. 275–305.

Hauhs, M., Rost-Siebert, K., Raben, G., Paces, T. and Vigerust, B. (1989) Summary of the European data. In: Malanchuk, J.L. and Nilsson, J. (eds),

The role of nitrogen in the acidification of soils and surface waters. Copenhagen: Nordic Council of Ministers.

Hauhs, M. and Wright, R.F. (1986) Regional pattern of acid deposition and forest decline along a cross section through Europe. *Water, Air, and Soil Pollution* **31,** 463–74.

Haupolter, R. (1987) Baumsterben in Mitteleuropa. Eine Literaturübersicht. Teil 1: Fichtensterben. *FBVA Berichte* **19**, 55–73.

Hausladen, A., Madamanchi, N.R., Fellows, S., Alscher, R.G. and Amundson, R.G. (1990) Seasonal changes in antioxidants in red spruce as affected by ozone. *New Phytologist* **115,** 447–58.

Havranek, W.M., Pfeifhofer, H. and Grill, D. (1990) Pigmentgehalte und Gaswechsel von Tief- und Hochlagenfichten nach chronischer Ozonbelastung. *Forstwissenschaftliches Centralblatt* **109,** 200–209.

Hawley, G.J. and DeHayes, D.H. (1992) Degree of red spruce–black spruce introgression in montane ecosystems. In: DeHayes, D.H. and Hawley, G.J. (eds), *Proceedings of the First Northern Forest Genetics Association Conference, Burlington, Vermont, July 23–25, 1991.* University of Vermont, Burlington, p.25.

Hayes, E.M. and Skelly, J.M. (1977) Transport of ozone from the Northeast US into Virginia and its effect on eastern white pines. *Plant Disease Reporter* **61,** 778–82.

Haynes, R.J. (1986) *Mineral nitrogen in the plant-soil system.* London: Academic Press. 483 pp.

Haynes, R.J. and Swift, R.S. (1986) Effects of soil acidification and subsequent leaching on levels of extractable nutrients in a soil. *Plant and Soil* **95,** 327–36.

Haywood, J.K. (1910) *Injury to vegetation and animal life by smelter wastes.* Bulletin No. 113. Washington, DC: US Department of Agriculture, Bureau of Chemistry.

Hearman, J. (1938) Minor elements and the pine. *Australian Forestry* **3,** 24–7.

Heath, R.L. (1989) Alteration of chlorophyll in plants upon air pollutant exposure. In: Committee on Biologic Markers of Air-Pollution Damage in Trees (eds), *Biologic Markers of Air-Pollution Stress and Damage in Forests.* Washington DC: National Academy Press, pp. 347–56.

Hedgecock, G.G. (1914) Injuries by smelter smoke in southeastern Tennessee. *Journal of the Washington Academy of Science* **4,** 70–71.

Heij, G.T. and Schneider, T. (eds) (1991) *Acidification research in The Netherlands.* Amsterdam: Elsevier.

Heikkinen, O. (1984a) Climatic changes during recent centuries as indicated by dendrochronological studies, Mount Baker, Washington, USA. In: Mörner, N.-A. and Karlén, W. (eds), *Climatic changes on a yearly to millennial basis.* Dordrecht: D. Reidel, pp. 353–61.

Heilman, P.E. and Gessel, S.P. (1963) Nitrogen requirements and the biological cycling of nitrogen in Douglas-fir stands in relationship to the effects of nitrogen fertilization. *Plant and Soil* **18,** 386–402.

Heinsdorf, M. (1991) Einfluß der Emission N-haltiger Abprodukte auf Ernährungszustand und Mykorrhizaentwicklung von benachbarten Kiefernstangenhölzern. *Beiträge für die Forstwirtschaft* **25,** 62–5.

Heinsdorf, D., Krauß, H.H. and Hippeli, P. (1988) Ernährungs- und bodenkundliche Untersuchungen in Fichtenbeständen des mittleren Thüringer Waldes unter Berücksichtigung der in den letzten Jahren aufgetretenen Umweltbelastungen. *Beiträge für die Forstwirtschaft* **22,** 160–7.

Heliotis, F.D., Karandinos, M.G. and Whiton, J.C. (1988) Air pollution and the decline of the fir forest in Parnis National Park, near Athens, Greece. *Environmental Pollution* **54,** 29–40.

Helmisaari, H.-S. (1990) Temporal variation in nutrient concentrations of *Pinus sylvestris* needles. *Scandinavian Journal of Forest Research* **5,** 177–93.

Hendershot, W.H. (1991) Fertilization of sugar maple showing dieback symptoms in the Quebec Appalachians, Canada. *Fertilizer Research* **27,** 63–70.

Hennessey, T.C., Dougherty, P.M., Cregg, B.M. and Wittwer, R.F. (1992) Annual variation in needle fall of a loblolly pine stand in relation to climate and stand density. *Forest Ecology and Management* **51,** 329–38.

Henning, E. (1895) Studier öfver vegetationsförhållandena i Jemtland ur forstlig, agronomisk och geologisk synpunkt. *Sveriges Geologiska Undersökning* Series C **145,** 1–75.

Hennon, P.E., Hansen, E.M. and Shaw, C.G. (1990a) Dynamics of decline and mortality of *Chamaecyparis nootkatensis* in southeast Alaska. *Canadian Journal of Botany* **68,** 651–62.

Hennon, P.E., Shaw, C.G. and Hansen, E.M. (1990b) Dating decline and mortality of *Chamaecyparis nootkatensis* in Southeast Alaska. *Forest Science* **36,** 502–15.

Hennon, P.E., Hansen, E.M. and Shaw, C.G. (1990c) Causes of basal scars on *Chamecyparis nootkatensis. Northwest Science* **64,** 45–54.

Hern, J.A., Rutherford, G.K. and Vanloon, G.W. (1985) Chemical and pedogenetic effects of simulated acidic precipitation on two eastern Canadian forest soils. I. Nonmetals. *Canadian Journal of Forest Research* **15,** 839–47.

Hernandez Tejeda, T., de Bauer, M. de L. de la I and Krupa, S. (1982) Daño por gases oxidantes en pinos y avena, reconocimiento y evaluación en el Ajusco, D.F. *Revista Chapingo* 33–4, 19–28.

Herrick, G.T. and Friedland, A.J. (1991) Winter desiccation and injury of subalpine red spruce. *Tree Physiology* **8,** 23–36.

Herrmann, K., Rock, B.N., Ammer, U. and Paley, H.N. (1988) Preliminary assessment of airborne imaging spectrometer and airborne thematic mapper data acquired for forest decline areas in the Federal Republic of Germany. *Remote Sensing of the Environment* **24,** 129–49.

Herwitz, S.R., Peterson, D.L. and Eastman, J.R. (1989) Thematic mapper detection of changes in leaf area of closed canopy pine plantations in central Massachussetts. *Remote Sensing of the Environment* **30,** 129–40.

Heyer, C. and Hess, R. (1906) *Der Waldbau oder die Forstproduktezucht.* Berlin: BG Teubner.

Hibben, C.R. (1969) Ozone toxicity to sugar maple. *Phytopathology* **59,** 1423.

Hibbs, D.E. (1982) Gap dynamics in a hemlock-hardwood forest. *Canadian Journal of Forest Research* **12,** 522–7.

Higashi, T. (1990) Extraction of expanded area of damaged stands by pine wilt disease using two Landsat TM data. *Journal of the Remote Sensing Society of Japan* **10,** 389–95.

Hildebrand, E. (1986) Zustand und Entwicklung der Austauschereigenschaften von Mineralböden aus Standorten mit erkrankten Waldbeständen. *Forstwissenschaftliches Centralblatt* **105,** 60–76.

Hildebrand, E.E. (1988) Strukturgebundene chemische Ungleichgewichte in Waldböden. *Bulletin Bodenkundliche Gesellschaft der Schweiz* **12,** 67–86.

Hildebrandt, G. (1980) Voraussetzungen und Praxis der Inventur von Vegetationsschäden durch Fernerkundung. *Allgemeine Forst Zeitschrift* **27,** 720–3.

Hildebrandt, G. and Kadro, A. (1984) Aspects of a countrywide inventory and monitoring of actual forest damages in Germany. *Bildmessung und Luftbildwesen* **52,** 201–16.

Hiley, W.E. and Cuncliffe, N. (1922) An investigation into the relation between height growth of trees and meteorological conditions. *Oxford Forestry Memoirs* **1,** 1–19.

Hiley, W.E. and Cuncliffe, N. (1923) Further observations on the relation of the height growth of trees to meteorological conditions. *Annals of Applied Biology* **10,** 442–52.

Hiltbrunner, E. and Flückiger, W. (1992) Altered feeding preference of beech weevil *Rhynchaenus fagi* L. for beech foliage under ambient air pollution. *Environmental Pollution* **75,** 333–6.

Hilton, G.M., Packham, J.R. and Willis, A.J. (1987) Effects of experimental defoliation on a population of pedunculate oak (*Quercus robur* L.). *New Phytologist* **107,** 603–12.

Hippeli, P. and Branse, C. (1992) Veränderungen der Nährelementkonzentrationen in den Nadeln mittelalter Kiefernbestände auf pleistozänen Sandstandorten Brandenburgs in den Jahren 1964 bis 1988. *Forstwissenschaftliches Centralblatt* **111,** 44–60.

Hippoliti, G. (1989) Sulle cause del deperimento delle abetine toscane. *Annali – Accademia Italiana di Scienze Forestali* **38,** 517–39.

L'Hirondelle, S.J., Jacobson, J.S. and Lassoie, J.P. (1992) Acidic mist and nitrogen fertilization effects on growth, nitrate reductase activity, gas exchange, and frost hardiness of red spruce seedlings. *New Phytologist* **121,** 611–22.

Hitschold, J. (1934) Das Fichtensterben in Ostpreußen. *Der Deutsche Forstwirt* **16,** 845–7 and 853–6.

Hocevar, M. (1990) Poskodovanost in rast smrekovega gorskega gozda na Pokljusko-Jeloviski Planoti. *Zbornik gozdarstva in lesarstva, Ljubljana* **36,** 27–68.

Hocking, G.H. (1946) Drought in Hawke's Bay. *New Zealand Journal of Forestry* **5,** 230–1.

Hoffmann, G., Hemsdorf, D. and Krauss, H.H. (1990) Wirkung atmogener Stickstoffeinträge auf Produktivität und Stabilität von Kiefern-Forstökosystemen. *Beiträge für die Forstwirtschaft* **24,** 59–73.

Hofmann-Schiele, C. (1991) Beziehungen zwischen den Böden und der Intensität der Fichtenerkrankung in den Hochlagen des Inneren Bayerischen Waldes. *Forstwissenschaftliches Centralblatt* **110,** 228–39.

Hogsett, W.E., Plocher, M., Wildman, V., Tingey, D.T. and Bennett, J.P. (1985) Growth response of two varieties of slash pine seedlings to chronic ozone exposures. *Canadian Journal of Botany* **63,** 2369–76.

Hogsett, W.E., Tingey, D.T., Hendricks, C. and Rossi, D. (1990) Sensitivity of western conifers to SO_2 and seasonal interaction of acid fog and ozone. In: Olson, R.K. and Lefohn, A.S. (eds), *Effects of air pollution on western forests*. Pittsburgh: Air and Waste Management Association, pp. 469–91.

Holdaway, M.R. (1988) The effects of climate, acid deposition and their interaction on Lake States forests. In: *Healthy forests, healthy world. Proceedings of the Society of American Foresters, October 16–19, Rochester, New York*. Bethesda: Society of American Foresters, pp. 67–71.

Holloway, J.T. (1954) Forests and climate in the South Island of New Zealand. *Transactions of the Royal Society of New Zealand* **82,** 329–410.

Holmgren, B. and Wastenson, L. (1985) Small scale aerial infrared photographs for inventory of forest damage – experiences from Swedish test areas. In: Schmid-Haas, P. (ed.), *Inventorying and monitoring endangered forests*. Birmensdorf: Eidgenössische Anstalt für das forstliche Versuchswesen, pp. 123–8.

Holzwirth, G.C. and Fischer, R.W. (1979) *Climatological summaries of the lower few kilometers of rawinsonde observations*. US Environmental Protection Agency Report No. EPA-600/4-79-026.

Hopkins, A.D. (1899) *Report on investigations to determine the cause of unhealthy conditions of spruce and pine from 1880–1893*. West Virginia Agricultural Station Bulletin 56. Morgantown, West Virginia.

Hopkins, A.D. (1901) Insect enemies of the spruce in the northeast. *US Department of Agriculture, Division of Entomology Bulletin* **28,** 15–29.

Horler, D.N.H., Dockray, M., Barber, J. and Barringer, A.R. (1983) Red-edge measurements for remotely sensing plant chlorophyll content. *Advances in Space Research* **3,** 273–7.

Hornbeck, J.W. (1987) Growth patterns of red oak and red and sugar maple relative to atmospheric deposition. In: *Proceedings of the 6th Central Hardwood Conference, Knoxville, TN, February 24–26, 1987,* pp. 277–82.

Hornbeck, J.W. and Smith, R.B. (1985) Documentation of red spruce growth decline. *Canadian Journal of Forest Research* **15,** 1199–1201.

Hornbeck, J.W., Smith, R.B. and Federer, C.A. (1986) Growth decline in red spruce and balsam fir relative to natural processes. *Water, Air, and Soil Pollution* **31,** 425–30.

Hornbeck, J.W., Smith, R.B. and Federer, C.A. (1987) Extended growth decreases in New England are limited to red spruce and balsam fir. In: Jacoby, G.C. and Hornbeck, J.W. (compilers) *Proceedings, International symposium on ecological aspects of tree ring analysis,* Marymount College, Tarrytown, New York, 17–21 Aug. 1986. Report No ER–12 GTN. Washington DC: Department of Energy, pp. 38–44.

Hornbeck, J.W., Smith, R.B. and Federer, C.A. (1988) Growth trends in 10 species of trees in New England, 1950–1980. *Canadian Journal of Forest Research* **18,** 1337–40.

Hornung, M. (1985) Acidification of soils by trees and forests. *Soil Use and Management* **1,** 24–8.

Horton, S.J., Reinart, R.A. and Heck, W.W. (1990) Effects of ozone on three open-pollinated families of *Pinus taeda* L. grown in two substrates. *Environmental Pollution* **65,** 279–92.

Hoshizaki, T., Rock, B.N. and Wong, S.K.S. (1988) Pigment analysis and spectral assessment of spruce trees undergoing forest decline in the NE United States and Germany. *GeoJournal* **17,** 173–8.

Hosking, G.P. (1989) Beech forest health – implications for management. *New Zealand Journal of Forestry Science* **19,** 290–3.

Hosking, G.P. and Hutcheson, J.A. (1986) Hard beech decline on the Mamaku Plateau, North Island, New Zealand. *New Zealand Journal of Botany* **24,** 263–9.

Hosking, G.P. and Hutcheson, J.A. (1988) Mountain beech (*Nothofagus solandri* var. *cliffortioides*) decline in the Kaweka Range, North Island, New Zealand. *New Zealand Journal of Botany* **26,** 393–400.

Hosking, G.P. and Kershaw, D.J. (1985) Beech mortality in the Maruia Valley, South Island, New Zealand. *New Zealand Journal of Botany* **23,** 201–11.

Hotz, A., Bilger, W., Kaiser, W.M. and Lange, O.L. (1990) Chlorophyll-Fluoreszenz, Photosynthesekapazität, Chlorophyll- und Nährstoffgehalte der Blätter 'ungeschädigter' und als 'geschädigt' eingestufter Buchen (*Fagus sylvatica* L.) im Gebiet um Würzberg. *Forstwissenschaftliches Centralblatt* **109,** 40–54.

Houle, G. (1990) Growth patterns of sugar maple seedlings and mature trees in healthy and declining hardwood stands. *Canadian Journal of Forest Research* **20,** 894–901.

Houston, D.B. and Stairs, G.R. (1973) Genetic control of sulfur dioxide and ozone tolerance in eastern white pine. *Forest Science* **19,** 267–71.

Houston, D.B. (1974) Responses of selected *Pinus strobus* L. clones to fumigation with sulfur dioxide and ozone. *Canadian Journal of Forest Research* **4,** 65–8.

Houston, D.R. (1973) Diebacks and declines: diseases initiated by stress, including defoliation. *International Shade Tree Conference Proceedings* **49,** 73–6.

Houston, D.R. (1981) *Stress-triggered diseases: the diebacks and declines.* Report NE-INF-41-81. Broomall: United States Department of Agriculture Forest Service, Northeastern Forest Experiment Station.

Houston, D.R. (1984) Stress related to diseases. *Arboricultural Journal* **8,** 137–49.

Houston, D.R., Allen, D.C. and Lachance, D. (1990) *Sugarbush management: A guide to maintaining tree health.* General Technical Report NE–129. Broomall: US Department of Agriculture Forest Service, Northeastern Forest Experiment Station.

Hovland, J., Abrahamsen, G. and Ogner, G. (1980) Effects of artificial rain on decomposition of spruce needles and on mobilization and leaching of elements. *Plant and Soil* **56,** 365–78.

Hoy, J.M. (1958) Coccids associated with rata and kamahi in New Zealand. *New Zealand Journal of Science* **1,** 170–200.

Humphreys, F.R. and Truman, R. (1964) Aluminium and phosphorus requirements of *Pinus radiata. Plant and Soil* **20,** 131–4.

Hunger, W. (1989) Neue Beobachtungen zu Waldschäden in der Nähe einer Schweinemastanlage. *Beiträge für die Forstwirtschaft* **23,** 64–8.

Hunt, R.E., Rock, B.N. and Nobel, P.S. (1987) Measurement of leaf relative water content by infra-red reflectance. *Remote Sensing of the Environment* **22,** 429–36.

Huntington, T.G. and Ryan, D.F. (1988) Soil chemical properties in the spruce–fir zone on Mount Moosilauke in New Hampshire.In: *Proceedings of the US/FRG*

Research Symposium: Effects of atmospheric pollutants on the spruce-fir forests of the eastern United States and the Federal Republic of Germany. General Technical Report NE–120. Broomall: US Department of Agriculture Forest Service, Northeastern Forest Experiment Station, pp. 191–8.

Huntington, T.G., Peart, D.R., Hornig, J., Ryan, D.F. and Russo-Savage, S. (1990) Relationships between soil chemistry, foliar chemistry, and condition of red spruce at Mount Moosilauke, New Hampshire. *Canadian Journal of Forest Research* **20,** 1219–27.

Hunziker, P., Mandallaz, D. and Schlaepfer, R. (1988) Zeitliche Entwicklung der Waldschäden – Möglichkeiten und Grenzen der statistischen Auswertung. *Forstarchiv* **59,** 43–7.

Hursh, C.R. (1948) *Local climate in the Copper basin of Tennessee as modified by the removal of vegetation.* Circular No. 774. Washington, DC: US Department of Agriculture.

Hüser, R. (1983) Forstdüngung mit Blickrichtung auf die Immissionsbelastungen. *Allgemeine Forstzeitung* **38,** 1089–92.

Hüttl, R.F. (1985a) Jüngste Waldschäden, Ernährungsstörungen und diagnostische Düngung. *VDI-Berichte* **560,** 863–86.

Hüttl, R.F. (1985b) 'Neuartige' Waldschäden und Nährelementversorgung von Fichtenbeständen (*Picea abies* Karst.) in Südwestdeutschland. *Freiburger Bodenkundliche Abhandlungen* **16,** 195 pp.

Hüttl, R.F. (1989) Vergleichende Analyse von Düngungsversuchen in der Bundesrepublik Deutschland und in den USA. *KfK-PEF-Berichte* **55,** 221–47.

Hüttl, R.F. (1990) Nutrient supply and fertilizer experiments in view of N saturation. *Plant and Soil* **128,** 45–58.

Hüttl, R.F. and Fink, S. (1988) Diagnostische Düngungsversuche zur Revitalisierung geschädigter Fichtenbestände (*Picea abies* Karst.) in Südwestdeutschland. *Forstwissenschaftliches Centralblatt* **107,** 173–83.

Hüttl, R.F. and Mehne, B.M. (1988) New type of 'forest decline', nutrient deficiencies and the 'Virus' hypothesis. In: Mathy, P. (ed.), *Air pollution and ecosystems.* Dordrecht: D. Reidel, pp. 870–5.

Hüttl, R.F. and Wisniewski, J. (1987) Fertilisation as a tool to mitigate forest decline associated with nutrient deficiencies. *Water, Air, and Soil Pollution* **32,** 265–76.

Hüttl, R.F. and Zöttl, H.W. (1985) Ernährungszustand von Tannenbeständen in Süddeutschland – ein historischer Vergleich. *Allgemeine Forst Zeitschrift* **41,** 1011–16.

Huttunen, S. (1973) Studies on tree damage due to air pollution in Oulu: the toxins contained in pine needles as assayed by a microanalyser. *Aquilo Serio Botanico* **12,** 1–11.

Huttunen, S. (1980) Seasonal variation of air pollution stresses in conifers. *Mitteilungen der forstlichen Versuchsantalt Wien* **37**/1, 103–13.

Huttunen, S., Reinikainen, J. and Turunen, M. (1990) Wintering response of conifers to acid rain treatment under northern conditions. In: Kauppi, P., Antilla, P. and Kenttamies, K. (eds), *Acidification in Finland.* Berlin: Springer Verlag, pp. 607–33.

Huttunen, S. and Törmälehto, H. (1982) Air pollution resistance of some Finnish *Pinus sylvestris* L. provenances. *Aquilo Serio Botanica* **18,** 1–9.

Huttunen, S., Laine, K. and Torvela, H. (1985) Seasonal sulphur contents of pine needles as indices of air pollution. *Annales Botanici Fennici* **22,** 343–59.

Hwang, S.H. (1977) *'Ohi'a decline: the role of Phytophthora cinnamomi.* Technical Report 12. Honolulu: University of Hawaii, Cooperative National Park Service Resources Study Unit.

Hyink, D.M. (1991) Comment 1. *Forest Science* **37,** 718–22.

Hyink, D.M. and Zedaker, S.M. (1987) Stand dynamics and the evaluation of forest decline. *Tree Physiology* **3,** 17–26.

Ilvesniemi, H. (1991) Spatial and temporal variation of soil chemical characteristics in pine sites in southern Finland. *Silva Fennica* **25,** 99–108.

Innes, J.L. (1983a) Landuse changes in the Scottish Highlands during the 19th century: the role of pasture degeneration. *Scottish Geographical Magazine* **99,** 141–9.

Innes, J.L. (1983b) Lichenometric dating of debris-flow deposits in the Scottish Highlands. *Earth Surface Processes and Landforms* **8,** 399–403.

Innes, J.L. (1983c) Stratigraphic evidence of episodic talus accumulation on the Isle of Skye, Scotland. *Earth Surface Processes and Landforms* **8,** 399–403.

Innes, J.L. (1986) Textural properties of regoliths on steep slopes. *Transactions of the Royal Society of Edinburgh: Earth Sciences* **77,** 241–50.

Innes, J.L. (1987) The interpretation of international forest health data. In: Perry, R., Harrison, R.M., Bell, J.N.B. and Lester, J.N. (eds), *Acid rain: technical and scientific advances.* London: Selper, pp. 633–40.

Innes, J.L. (1988a) Forest health surveys – a critique. *Environmental Pollution* **54,** 1–15.

Innes, J.L. (1988b) Forest health surveys: problems in assessing observer objectivity. *Canadian Journal of Forest Research* **18,** 560–5.

Innes, J.L. (1988c) Forest decline in urban areas – a comment. *Science of the Total Environment* **72,** 227–33.

Innes, J.L. (1989a) Rapid yellowing of needles of *Picea abies. European Journal of Forest Pathology* **19,** 21–8.

Innes, J.L. (1989b) Rapid mass movements in upland Britain: a review with particular reference to debris flows. *Studia Geomorphologica Carpatho-Balcanica* **23,** 53–67.

Innes, J.L. (1990a) Plants and air pollution. In: Clouston, B. (ed.), *Landscape design with plants.* Oxford: Heinemann Newnes, pp. 199–211.

Innes, J.L. (1990b) Some problems with the interpretation of international assessments of forest damage. In: *Proceedings, IUFRO XIX World Congress, 5–11 August 1990, Montreal, Canada.* Volume 2. Montreal: IUFRO, pp. 380–7.

Innes, J.L. (1990c) *Assessment of tree condition.* Forestry Commission Fieldbook 12. London: HMSO.

Innes, J.L. (1990d) The role of quality assurance in the assessment of forest condition in Europe. In: Hart, D.R. (coordinator), *Proceedings of the third annual ecological quality assurance workshop.* Burlington, Ontario: Canada Center for Inland Waters, pp. 99–112.

Innes, J.L. (1991a) High-altitude and high-latitude tree growth in relation to past, present and future climate change. *The Holocene* **1,** 168–73.

Innes, J.L. (1991b) The application of cause–effect criteria to the relationship between air pollution and forest decline in Europe. In: Longhurst, J.W.S.

(ed.), *Acid deposition. Origins, impacts and abatement strategies.* Berlin: Springer Verlag, pp. 63–74.

Innes, J.L. (ed.) (1991c) *Interim report on cause-effect relationships in forest decline.* Geneva: United Nations Economic Commission for Europe.

Innes, J.L. (1992a) Observations on the condition of beech (*Fagus sylvatica* L.) in Britain in 1990. *Forestry* **65,** 35–60.

Innes, J.L. (1992c) Forest decline. *Progress in Physical Geography* **16,** 1–64.

Innes, J.L. (1992d) Some factors affecting the crown density of trees in Great Britain based on recent annual surveys of forest condition. In: Hüttl, R.F. and Mueller-Dombois, D. (eds), *Forest decline in the Atlantic and Pacific region.* Berlin: Springer Verlag, pp. 40–53.

Innes, J.L. (1993) The climatic sensitivity of temperate forests. *Environmental Pollution,* in press.

Innes, J.L. and Boswell, R.C. (1988) *Forest health surveys 1987. Part 2: Analysis and interpretation.* Forestry Commission Bulletin 79. London: HMSO.

Innes, J.L. and Boswell, R.C. (1989a) *Monitoring of forest condition in the United Kingdom 1988.* Forestry Commission Bulletin 88. London: HMSO.

Innes, J.L. and Boswell, R.C. (1989b) Sulphur contents of conifer needles in Great Britain. *GeoJournal* **19,** 63–6.

Innes, J.L. and Boswell, R.C. (1990a) *Monitoring of forest condition in Great Britain 1989.* Forestry Commission Bulletin 94. London: HMSO.

Innes, J.L. and Boswell, R.C. (1990b) Reliability, presentation, and relationships among data from inventories of forest condition. *Canadian Journal of Forest Research* **20,** 790–9.

Innes, J.L. and Boswell, R.C. (1991a) *Monitoring of forest condition in Great Britain in 1990.* Forestry Commission Bulletin 98. London: HMSO.

Innes, J.L. and Boswell, R.C. (1991b) *Forest monitoring programme. 1991 results.* Forestry Commission Research Information Note 209. London: HMSO.

Innes, J.L., Boswell, R.C., Binns, W.O. and Redfern, D.B. (1986) Forest health and air pollution 1986 survey. Forestry Commission Research and Development Paper 150. Edinburgh: Forestry Commission.

Innes, J.L. and Cook, E.R. (1989) Tree ring analysis as an aid to evaluating the effects of pollution on tree growth. *Canadian Journal of Forest Research* **19,** 1174–89.

Innes, J.L., Landmann, G. and Mettendorf, B. (1992) Consistency of observations of forest tree defoliation in three European countries. *Environmental Monitoring and Assessment* in press.

Innes, J.L., Moffat, A.J. and Lonsdale, D. (1989) *Weather conditions during the summer of 1989 and their effect on trees.* Forestry Commission Research Information Note 162. Edinburgh: Forestry Commission.

Innes, J.L. and Neumann, H. (1991) Past growth variations in *Picea sitchensis* with differing crown densities. *Scandinavian Journal of Forest Research* **6,** 395–405.

Innes, J.L. and Whittaker, R.J. (1993) Relationships between forest condition and the environment in Great Britain. *Journal of Applied Ecology,* in press.

Isaac, L.A. (1965) Fog drip and rain interception in coastal forests. Research Note 34. Portland: US Department of Agriculture Forest Service, Pacific Northwest Forest and Range Experiment Station.

Isermann, K. (1985) Diagnose und Therapie der 'neuartigen Waldschäden' aus der

Sicht der Waldernährung. In: Stratmann, H. (ed.), *Waldschäden. Einflussfaktoren und ihre Bewertung. Kolloquium Goslar, 18. bis 20. Juni 1985*. VDI-Berichte 560, pp. 897–920.

Isermann, K. (1987) Revitalisierung geschädigter Fichten-Altbestände durch Mineraldüngung. *Allgemeine Forstzeitung* **42,** 1–4.

Iwaki, H. and Totsuka, T. (1959) Ecological and physiological studies on the vegetation of Mt. Shimagare II. On the crescent-shaped 'dead trees strips' in the Yatsugatake and ChiChibu Mountains. *Botanical Magazine Tokyo* **72,** 255–60.

Jacks, G., Andersson, S. and Segman, B. (1989) pH-changes in forested and open land in Sweden. In: Szabolcs, T. (ed.), *Ecological impact of acidification.* Budapest: Hungarian Academy of Sciences, pp. 103–4.

Jacobi, J.D. (1983) Metrosideros dieback in Hawaii: A comparison of adjacent dieback and non-dieback rain forest stands. *New Zealand Journal of Ecology* **6,** 79–97.

Jacobi, J.D., Gerrish, G., Mueller-Dombois, D. and Whittaker, L. (1988) Stand-level dieback and *Metrosideros* regeneration in the montane rain forest of Hawaii. *GeoJournal* **17,** 193–200.

Jacobson, J.S., Bethard, T., Heller, L.I. and Lassoie, J.P. (1990) Response of *Picea rubens* seedlings to intermittent mist varying in acidity, and in concentrations of sulfur-, and nitrogen-containing pollutants. *Physiologia Plantarum* **78,** 595–601.

Jacobson, J.S. and Lassoie, J.P. (1988) Response of red spruce to sulfur- and nitrogen-containing contaminants in simulated acidic mist. In: *Proceedings of the US/FRG Research Symposium: Effects of atmospheric pollutants on the spruce-fir forests of the eastern United States and the Federal Republic of Germany.* General Technical Report NE-120. Broomall: United States Department of Agriculture Forest Service, Northeastern Forest Experiment Station, pp. 371–9.

Jacobson, J.S., Lassoie, J.P., Osmeloski, J. and Yamada, K. (1989) Changes in foliar elements in red spruce seedlings after exposure to sulfuric and nitric acid mist. *Water, Air, and Soil Pollution* **48,** 141–59.

Jacoby, G.C. (1986) Long-term temperature trends and a positive departure from the climate-growth response since the 1950s in high elevation lodgepole pine from California. In: Rozenzweig, C. and Dickinson, R. (eds), *Proceedings of the NASA Conference on Climate-Vegetation Interactions,* Report OIES-2. Boulder, Colorado: University Corporation for Atmospheric Research, pp. 81–3.

Jagels, R. (1986) Acid fog, ozone and low elevation spruce decline. *IAWA Bulletin* **7,** 299–307.

Jagels, R., Carlisle, J., Cronan, C., Cunningham, R., Gordon, G., Piatek, K.B., Serreze, S., Stubbs, C. (1988) Coastal red spruce health along an acidic fog/ozone gradient. In: *Proceedings of the US/FRG Research Symposium: Effects of atmospheric pollutants on the spruce-fir forests of the eastern United States and the Federal Republic of Germany.* General Technical Report NE-120. Broomall: US Department of Agriculture Forest Service, Northeastern Forest Experiment Station, pp. 229–33.

Jäger, H.J. (1982) Biochemical indication of an effect of air pollution on plants. In:

Steubing, L. and Jäger, H.J. (eds), *Monitoring of air pollutants by plants. Methods and problems*. The Hague: W. Junk, pp. 99–107.

Jäger, H.J., Weigel, H.J. and Grünhage, L. (1986) Physiologische und biochemische Aspekte der Wirkung von Immissionen auf Waldbäume. *European Journal of Forest Pathology* **16,** 98–109.

Jakucs, P. (1988) Ecological approach to forest decay in Hungary. *Ambio* **17,** 267–74.

Jalkanen, R. and Kurkela, T. (1990) Needle retention, age, shedding and budget, and growth of Scots pine between 1865 and 1988. In: Kauppi, P., Anttila, P. and Kenttämies, K. (eds), *Acidification in Finland*. Berlin: Springer-Verlag, pp. 691–7.

James, B.R. and Riha, S.J. (1989) Aluminium leaching by mineral acids in forest soils: I. Nitric-sulfuric acid differences. *Soil Science Society of America Journal* **53,** 259–64.

James, R.L., Cobb, F.W., Miller, P.R. and Parmeter, J.R. (1980) Effects of oxidant air pollution on susceptibility of pine roots to *Fomes annosus*. *Phytopathology* **70,** 560–3.

Jane, G.T. (1986) Wind damage as an ecological process in mountain beech forests of Canterbury, New Zealand. *New Zealand Journal of Ecology* **9,** 25–39.

Jane, G.T. and Green, T.G.A. (1986) Etiology of forest dieback areas within the Kaimai Range, North Island, New Zealand. *New Zealand Journal of Botany* **24,** 513–27.

Jansen, A.E. (1988) Air pollutant effects on ectomycorrhizal fungi. In: Cape, J.N. and Mathy, P. (eds), *Scientific basis of forest decline symptomatology*. Air Pollution Report 15. Brussels: Commission of the European Communities, pp. 182–9.

Jansen, A.E., Dighton, J. and Bresser, A.H.M. (eds), (1988) *Ectomycorrhiza and acid rain*. Air Pollution Research Report 12. Brussels: Commission of the European Communities.

Jarva, M. and Tervahauta, A. (1989) Neulasten rikkianalysivertailu. *Silva Fennica* **23,** 117–24.

Jaworski, A. and Skrzyszewski, J. (1987) Ocena zywotnosci drzewostanów jodlowych w Beskidzie Sadeckim na podstawie wybranych cech przyrostowych i morfologicznych. *Sylwan* **131,** 1–14.

Jensen, K.F. and Dochinger, L.S. (1989) Response of eastern hardwood species to ozone, sulfur dioxide and acid precipitation. *Journal of the Air Pollution Control Association* **39,** 852–5.

Jensen, K.F. and Patton, R.L. (1990) Response of yellow-poplar (*Liriodendron tulipifera* L.) seedlings to simulated acid rain and ozone – 1. Growth modifications. *Environmental and Experimental Botany* **30,** 59–66.

Jensen, K.F., Patton, R.L. and Schier, G.A. (1988) Effect of simulated acid rain and ozone on red spruce seedlings: an interim report, in *Proceedings of the United States – Federal Republic of Germany Research Symposium: The effects of atmospheric pollutants on the spruce-fir forests of the eastern United States and the Federal Republic of Germany*. General Technical Report NE-120. Broomall: United States Department of Agriculture Forest Service, Northeastern Forest Experimental Staton, pp. 365–370.

Jepson, W.L. (1910) The silva of California. *University of California Memoirs* **2,** 1–480.

Johnson, A.H. (1983) Red spruce decline in the northeastern U.S.: hypotheses regarding the role of acid rain. *Environmental Science and Technology* **17,** 294A–305A.

Johnson, A.H., Cook, E.R. and Siccama, T.G. (1988a) Climate and red spruce growth and decline in the northern Appalachians. *Proceedings of the National Academy of Science of the USA* **85,** 5369–73.

Johnson, A.H., Cook, E.R. and Siccama, T.G. (1988b) Tree rings, climate, and red spruce decline. In: *Proceedings of the US/FRG Research Symposium: Effects of atmospheric pollutants on the spruce-fir forests of the eastern United States and the Federal Republic of Germany.* General Technical Report NE-120. Broomall: US Department of Agriculture Forest Service, Northeastern Forest Experiment Station, pp. 299–309.

Johnson, A.H., Friedland, A.J. and Dushoff, J.G. (1986) Recent and historic red spruce mortality: evidence of climatic influence. *Water, Air, and Soil Pollution* **30,** 319–30.

Johnson, A.H. and McLaughlin, S.B. (1986) The nature and timing of the deterioration of red spruce in the Northern Appalachian mountains. In: *Acid deposition long-term trends.* Washington DC: National Academy Press, pp. 200–30.

Johnson, A.H., Schwartzman, T.N. and Miller, R.W. (1989b) *Relationships between foliar chemistry, soil chemistry, and forest condition at Whiteface Mt., NY.* MPO Contribution SF08-3. US Department of Agriculture Forest Service, Spruce–fir Cooperative, Radnor Corporate Center, Radnor, Pennsylvania.

Johnson, A.H. and Siccama, T.G. (1983a) Acid deposition and forest decline. *Environmental Science and Technology* **17,** 294A–305A.

Johnson, A.H. and Siccama, T.G. (1983b) Spruce decline in the northern Appalachians: evaluating acid deposition as a possible cause. *Proceedings of the Technical Association of the Pulp and Paper Industry.* Atlanta, Georgia, pp. 301–10.

Johnson, A.H., Siccama, T.G., Silver, T.L. and Battles, J.J. (1989a) Decline of red spruce in high-elevation forests of New York and New England. In: Adriano, D.C. and Havas, M. (eds), *Acidic Precipitation. Volume 1. Case studies.* New York: Springer Verlag, pp. 85–112.

Johnson, D.W. (1992) Nitrogen retention in forest soils. *Journal of Environmental Quality* **21,** 1–12.

Johnson, D.W., Cole, D.W. Van Miegroet, H. and Horng, F.W. (1986) Factors affecting anion movement and retention in four forest soils. *Soil Science Society of America Journal* **50,** 776–83.

Johnson, D.W., Cresser, M.S., Nilsson, S.I., Turner, J., Ulrich, B., Binkley, D. and Cole, D.W. (1991b) Soil changes in forest ecosystems: evidence for and probable causes. *Proceedings of the Royal Society of Edinburgh* **978,** 81–116.

Johnson, D.W., Van Miegroet, H., Lindberg, S.E., Todd, D.E. and Harrison, R.B. (1991a) Nutrient cycling in red spruce forests of the Great Smoky Mountains. *Canadian Journal of Forest Research* **21**; 769–87.

Johnson, D.W. and Reuss, J.O. (1984) Soil-mediated effects of atmospherically deposited sulphur and nitrogen. *Philosophical Transactions of the Royal Society of London,* Series B, **305,** 383–92.

Johnson, D.W., Richter, D.D., Lovett, G.M. and Lindberg, S.E. (1985) The effects of atmospheric deposition on potassium, calcium, and magnesium cycling in two deciduous forests. *Canadian Journal of Forest Research* **15,** 773–82.

Johnson, D.W. and Taylor, G.E. (1989) Role of air pollution in forest decline in Eastern North America. *Water, Air, and Soil Pollution* **48,** 21–43.

Johnson, D.W. and Todd, D.E. (1990) Nutrient cycling in forests of Walker Branch Watershed, Tennessee: roles of uptake and leaching in causing soil changes. *Journal of Environmental Quality* **19,** 97–104.

Johnston, A.E., Goulding, K.W.T. and Poulton, P.R. (1986) Soil acidification during more than 100 years under permanent grassland and woodland at Rothamsted. *Soil Use and Management* **2,** 3–10.

Jönsson, P. (1991) Can climate trends explain the damage in central Swedish lowland forests? *Scandinavian Journal of Forest Research* **6,** 153–60.

Joria, P.E., Ahearn, S.C. and Connor, M. (1991) A comparison of the SPOT and Landsat Thematic Mapper satellite systems for detecting gypsy moth defoliation in Michigan. *Photogrammetric Engineering and Remote Sensing* **57,** 1605–12.

Jorns, A. and Hecht-Buchholz, C. (1985) Aluminium induzierter Magnesium- und Calciummangel im Laborversuch bei Fichtensämlingen. *Allgemeine Forst Zeitschrift* **46,** 1248–52.

Joslin, J.D. and Wolfe, M.H. (1989) Aluminium effects on northern red oak seedling growth in six forest soil horizons. *Soil Science Society of America Journal* **53,** 274–81.

Joslin, J.D. and Wolfe, M.H. (1992) Red spruce soil solution chemistry and root distribution across a cloud water deposition gradient. *Canadian Journal of Forest Research* **22,** 893–904.

Jukola-Sulonen, E.-L., Mikkola, K., Nevalainen, S. and Yli-Kojola, H. (1987) Havupuiden elinvoimaisuus Suomessa 1985–6. *Metsäntutkimuslaitoksen tiedonantoja* **256,** 1–92.

Jukola-Sulonen, E.-L., Mikkola, K. and Salemaa, M. (1990) The vitality of conifers in Finland, 1986–88. In: Kauppi, P., Anttila, P. and Kenttämies, K. (eds), *Acidification in Finland.* Berlin: Springer Verlag, pp. 523–60.

June, S.R. and Ogden, J. (1975) Studies on the vegetation of Mount Colenso, New Zealand. 3. The population dynamics of red beech seedlings. *Proceedings of the New Zealand Ecological Society* **22,** 61–6.

June, S.R. and Ogden, J. (1978) Studies on the vegetation of Mount Colenso, New Zealand. 4. An assessment of the processes of canopy maintenance and regeneration strategy in a red beech (*Nothofagus fusca*) forest. *New Zealand Journal of Ecology* **1,** 7–15.

Jung, E. (1968) Bestandesumwandlungen im Rauchschadengebiet von ranshofen. *Proceedings, Wplyw Zanieczyszen Powietrza na Lasy,* **6,** pp. 407–13. Katowice.

Jung, I. and Winter, K. (1992) Mineral nutrient deficiency increases the sensitivity of photosynthesis to sulphur dioxide in needles of a coniferous tree, *Abies nordmanniana. Oecologia* **90,** 70–3.

Junga, U. (1984) Sterilkultur als Modellsystem zur Untersuchung des Mechanismus der Aluminium-Toxizität bei Fichtenkeimlingen (*Picea abies* L. Karst.). *Berichte Waldökosysteme/Waldsterben* **5,** 173 pp.

Kadro, A. and Kuntz, S. (1985) Results of computer-aided investigations into forestry damage with multispectral scanner data. In: Schmid-Haas, P. (ed.), *Inventorying and monitoring endangered forests.* Birmensdorf: Eidgenössische Anstalt für das forstliche Versuchswesen, pp. 179–82.

Kadro, A. and Kuntz, S. (1986) Experiences in application of multispectral scanner data for forest damage inventory. *Symposium on remote sensing, resources development and environmental management, Enschede,* pp. 469–72.

Kalela, E.K. (1941) Über die Holzarten und die durch die klimatischen Verhältnisse verursachten Holzartenwechsel in den Wäldern Ostpatagoniens. *Annales Academiae Scientiarum Fennicae,* Series A, **2,** 5–151.

Kandler, O. (1988) Epidemiologische Bewertung der Waldschadenserhebungen 1983 bis 1987 in der Bundesrepublik Deutschland. *Allgemeine Forst- und Jagdzeitung* **159,** 179–94.

Kandler, O. (1989) Epidemiological evaluation of the course of 'Waldsterben' from 1983 to 1987. In: Bucher, J. and Bucher-Wallin, I. (eds), *Air pollution and forest decline.* Birmensdorf, Eidgenössische Anstalt für das forstliche Versuchswesen, pp. 297–302.

Kandler, O. (1992) Historical declines and diebacks of European forests and present conditions. **11** 1077–94 *Environmental Toxicology and Chemistry.*

Kandler, O., Miller, W. and Ostner, R. (1987) Dynamik der 'akuten Vergilbung' der Fichte. Epidemiologische und physiologische Befunde. *Allgemeine Forst Zeitschrift* **42,** 715–23.

Kärenlampi, L. (1986) Relationships between macroscopic symptoms of injury and cell structural changes in needles of ponderosa pine exposed to air pollution in California. *Annales Botanici Fennici* **23,** 255–64.

Kärenlampi, L. and Houpis, J.L. (1986) Structural conditions of mesophyll cells of *Pinus ponderosa* var. *scopulorum* after SO_2 fumigation. *Canadian Journal of Forest Research* **16,** 1381–5.

Karhu, M. and Huttunen, S. (1986) Erosion effects of air pollution on needle surfaces. *Water, Air, and Soil Pollution* **31,** 417–23.

Karl, T.R., Livezey, R.E. and Epstein, E.S. (1984) Recent unusual mean winter temperatures across the contiguous United States. *Bulletin of the American Meteorological Society* **65,** 1302–9.

Karnosky, D.F. (1989) Air pollution induced population changes in North American forests. In: Bucher, J. and Bucher-Wallin, I. (eds), *Air pollution and forest decline.* Birmensdorf: Eidgenössische Anstalt für das forstliche Versuchswesen, pp. 315–17.

Karnosky, D.F., Berrang, P.C., Scholz, F. and Bennett, J.P. (1989) Variation in and natural selection for air pollution tolerances in trees. In: Scholz, F., Gregorius, H.-R. and Rudin, D. (eds), *Genetic effects of air pollutants in forest tree populations.* Berlin: Springer Verlag, pp. 29–37.

Kashimura, T. (1981) On the dehydration resistance and the frost hardiness of *Abies mariesii.* In: IIzumi, S. (ed.), *Ecological studies on the* Abies mariesii *forest.* Sendai: The Mount Hokkoda Botanical Laboratory of Tohoku University, pp. 15–24.

Kaupenjohann, M., Döhler, H. and Bauer, M. (1989) Effects of N-immissions on nutrient status and vitality of *Pinus sylvestris* near a hen-house. *Plant and Soil* **113,** 279–82.

Kaupenjohann, M., Zech, W., Hantschel, R. and Horn, R. (1987) Ergebnisse von Düngungsversuchen mit Magnesium an vermutlich immissionsgeschädigten Fichten (*Picea abies* Karst.) im Fichtelgebirge. *Forstwissenschaftliches Centralblatt* **106,** 78–84.

Kazda, M. (1990) Indications of unbalanced nitrogen nutrition of Norway spruce stands. *Plant and Soil* **128,** 97–101.

Kazda, M. and Weilgony, P. (1988) Seasonal dynamics of major cations in xylem sap and needles of Norway spruce (*Picea abies* L. Karst.) in stands with different soil solution chemistry. *Plant and Soil* **110,** 91–100.

Kazda, M. and Zvacek, L. (1989) Aluminium and manganese and their relation to calcium in soil solution and needles in three Norway spruce (*Picea abies,* L. Karst.) stands of Upper Austria. *Plant and Soil* **114,** 257–67.

Ke, J. and Skelly, J.M. (1989) An evaluation of Norway spruce in northeastern United States. In: Bucher, J.B. and Bucher-Wallin, I. (eds), *Air pollution and forest decline.* Birmensdorf: Eidgenössische Anstalt für das forstliche Versuchswesen, pp. 55–60.

Keeves, A. (1966) Some evidence of loss of productivity with successive rotations of *Pinus radiata* in the south east of S. Australia. *Australian Forestry* **30,** 51–63.

Keil, P., Hecht-Buchholz, C. and Ortmann, U. (1986) Zum Einfluß von erhöhtem Manganangebot auf Fichtensämlinge. *Allgemeine Forst Zeitschrift* **41,** 855–8.

Keil, R., Benz, T., Einig, W. and Hampp, R. (1988) Adenine and pyridine nucleotides, and carbohydrates in healthy and deceased needles: alterations in relation to vegetation period (May to October 1985) and needle age (1979 to 1985). In: Mathy, P. (ed.), *Air pollution and ecosystems.* Dordrecht, D. Reidel, pp. 584–9.

Keller, T. (1973) Report on the IUFRO meeting 'Air pollution effects on forests', Sopron, Hungary, October 9–14, 1972. *European Journal of Forest Pathology* **3,** 56–60.

Keller, T. (1981a) Die Beeinflussung physiologischer Prozesse der Fichte durch eine Winterbegasung mit SO_2. *Mitteilungen der Forstlichen Bundesversuchsanstalt Wien* 137/I, 115–20.

Keller, T. (1981b) Winter uptake of airborne SO_2 by shoots of deciduous species. *Environmental Pollution* **26A,** 313–17.

Keller, T. (1984) The influence of SO_2 on CO_2 uptake and peroxidase activity. *European Journal of Forest Pathology* **14,** 354–9.

Keller, T. (1986) The electrical conductivity of Norway spruce needle diffusate as affected by certain air pollutants. *Tree Physiology* **1,** 85–94.

Keller, T. (1991) Air pollutants. In: Raghavendra, A.S. (ed.), *Physiology of trees.* New York: Wiley, pp. 337–57.

Keller, T. and Häsler, R. (1984) The influence of a fall fumigation with ozone on the stomatal behaviour of spruce and fir. *Oecologia* **64,** 284–6.

Keller, T. and Häsler, R. (1986) The influence of a prolonged SO_2-fumigation on the stomatal reaction of spruce. *European Journal of Forest Pathology* **16,** 110–15.

Keller, T. and Matyssek, R. (1990) Limited compensation of ozone stress by potassium in Norway spruce. *Environmental Pollution* **67,** 1–14.

Keller, T. and Schwager, H. (1977) Air pollution and ascorbic acid. *European Journal of Forest Pathology* **7,** 338–50.

Keller, W. and Imhof, P. (1987) Zum Einfluss der Durchforstung auf die

Waldschäden. I. Teil: Erste Ergebnisse von Waldschadenuntersuchungen in Buchen-Durchforstungsflächen der EAFV. *Schweizerische Zeitschrift für Forstwesen* **138,** 39–54.

Kelley, R.S. and Eav, B.B. (1987) *Vermont hardwood tree health survey - 1986.* Waterbury, Vermont: Vermont Department of Forests, Parks and Recreation.

Kelley, R.S. and Eav, B.B. (1988) Vermont hardwood tree health survey with color infra-red photography. In: *Healthy forests, healthy world. Proceedings of the Society of American Foresters, October 16–19, Rochester, New York.* Bethesda: Society of American Foresters, pp. 58–61.

Kelly, J.M. and Mays, P.A. (1989) Root zone physical and chemical characteristics in southeastern spruce–fir stands. *Soil Science Society of America Journal* **53,** 1248–55.

Kelly, J.M. and Strickland, R.C. (1986) Throughfall and plant nutrient concentration response to simulated acid rain treatment. *Water, Air, and Soil Pollution* **29,** 219–31.

Kelly, J.M., Schaedle, M., Thornton, F.C. and Joslin, J.D. (1990) Sensitivity of tree seedlings to aluminium: II. Red oak, sugar maple, and European beech. *Journal of Environmental Quality* **19,** 172–9.

Kelly, P.M., Munro, M.A.R., Hughes, M.K. and Goodess, C.M. (1989) Climate and signature years in west European oaks. *Nature* **340,** 57–60.

Keltjens, W.G. and Van Loenen, E. (1989) Effects of aluminium and mineral nutrition on growth and chemical composition of hydroponically grown seedlings of five different forest tree species. *Plant and Soil* **119,** 39–50.

Kenk, G. (1983) Zuwachsuntersuchungen in geschädigten Tannenbeständen in Baden-Württemberg. *Allgemeine Forst Zeitschrift* **38,** 650–2.

Kenk, G. (1986) Wachstumsanalysen im Zusammenhang mit den gegenwärtigen Waldschäden: Ergebnisse und Folgerungen für Nadelbaumbestände in Baden-Württemberg. DVFA, Sektion Ertragskunde, *Bericht über die Tagung 12–14 Mai 1986,* pp. 10/1–14.

Kenk, G. (1990) Effects of air pollution on forest growth in southwest Germany – Hunting for a phantom? In: *Proceedings of the IUFRO XIX World Congress, 5–11 August, Montréal, Canada.* Montreal: IUFRO, pp. 388–95.

Kenk, G., Evers, F., Unfried, P. and Schröter, H. (1983) Düngung als Therapie gegen Immissionswirkungen in Tannen-Fichten-Beständen? *Allgemeine Forst- und Jagdzeitung* **154,** 153–70.

Kenk, G. and Fischer, H. (1988) Evidence from nitrogen fertilization in forests of Germany. *Environmental Pollution* **54,** 199–218.

Kenk, G., Kremer, N., Bonaventura, D. and Gallus, M. (1984b) *Jahrring- und zuwachsanalytische Untersuchungen in erkrankten Tannenbeständen des Landes Baden-Württemberg.* Mitteilungen, Forstliche Versuchs- und Forschungsanstalt Baden-Württemberg 112.

Kenk, G. and Spiecker, H. (1988) Einige Ergebnisse zum aktuellen und früheren Wachstumsverhalten von Fichten. *4 Statuskolloquien des PEF, Karlsruhe, March, 1988.* Karlsruhe: PEF, pp. 371–81.

Kenk, G., Unfried, P., Evers, F.H. and Hildebrand, E.E. (1984a) Düngung zur Minderung der neuartigen Waldschäden. Auswertung eines alten Düngungsversuchs zu Fichte im Buntsandstein-Odenwald. *Forstwissenschaftliches Centralblatt* **103,** 307–20.

Kenk, G., Unfried, P., Evers, F.H. and Hildebrand, E. (1985) Ergebnisse von Düngungsversuchen mit Magnesium an vermutlich immissionsgeschädigten Fichten (Picea abies (L.) Karst.) im Fichtelgebirge. *Forstwissenschaftliches Centralblatt* **106,** 78–84.

Kennel, E. (1986) *Entwicklung der Waldschäden auf den Dauerbeobachtungsflächen in Bayern von 1981 bis 1985*. Bayerische Forstliche Versuchs- und Forschungsanstalt Sachgebiet II, Munich.

Kenneweg, H. (1973) Interpretationen von Luftaufnahmen für die Erforschung und Gestaltung von Vegetationsbeständen in westdeutschen Ballungsräumen. In: *Proceedings of the IUFRO Symposium 'Remote Sensing including aerial photography'*, Freiburg, Germany, pp. 87–116.

Kenneweg, H. (1989) Remote sensing approaches to assessment and diagnosis of forest decline. In: Bucher, J.B. and Bucher-Wallin, I. (eds), *Air Pollution and Forest Decline*. Birmensdorf: Eidgenössische Anstalt für das forstliche Versuchswesen, pp. 217–21.

Kenneweg, H., Förster, B., Runkel, M. and Winter, R. (1989) Satellitenbilder zur Waldschadenserfassung – wo liegen die Probleme? *Allgemeine Forst- und Jagdzeitung* **160,** 73–6.

Kenward, R.E. (1989) Bark-stripping by grey squirrels in Britain and North America: why does the damage differ? In: Putman, R.J. (ed.), *Mammals as pests. Papers presented at a symposium of the Mammal Society, London, November 1987*. London: Chapman and Hall, pp. 144–54.

Keslo, E.G. (1965) Memorandum 5220, 2480 July 23, 1965. Amherst, Massachusetts: US Forest Service, Northern FPC Zone.

Kessler, K.J. (1965) Dieback of managed, old-growth northern hardwoods in upper Michigan. *Plant Disease Reporter* **49,** 483–6.

Keyser, P., Kirk, T.K. and Zeikus, J.G. (1978) Lignolytic enzyme of *Phanaerochaete chrysosporium* synthesized in the absence of lignin in response to nitrogen starvation. *Journal of Bacteriology* **135,** 780–97.

Khanna, P.K., Prenzel, J., Meiwes, K.J., Ulrich, B. and Matzner, E. (1987) Dynamics of sulfate retention by acid forest soils in an acidic deposition environment. *Soil Science Society of America Journal* **51,** 446–52.

Khanna, P.K., Weaver, G.T. and Beese, F. (1986) Effect of sulfate on ionic transport and balance in a slightly acidic forest soil. *Soil Science Society of America Journal* **50,** 770–6.

Kienast, F. (1982) Jahrringanalytische Untersuchungen in immissionsgefährdeten Waldschadengebieten des Walliser Rhonetales. *Geographica Helvetica* 1982, 143–8.

Kienast, F. (1985) Tree ring analysis, forest damage and air pollution in the Swiss Rhône Valley. *Land Use Policy* January 1985, 74–7.

Kienast, F., Flühler, H. and Schweingruber, F.H. (1981) Jahrringanalysen an Föhren (*Pinus sylvestris* L.) aus immissionsgefährdeten Waldbeständen des Mittelwallis (Saxon, Schweiz). *Mitteilungen, Eidgenössische Anstalt für das forstliche Versuchswesen, Birmensdorf* **57,** 415–32.

Kile, G.A. (1974) Insect defoliation in the eucalypt regrowth forests of southern Tasmania. *Australian Forest Research* **6,** 9–18.

Kile, G.A. (1980) Behaviour of an Armillaria in some *Eucalyptus obliqua – Eucalyptus regnans* forests in Tasmania and its role in their decline. *European Journal of Forest Pathology* **10,** 278–96.

Kile, G.A., Turnbull, C.R.A. and Podger, F.D. (1981) Effect of 'regrowth dieback' on some properties of *Eucalyptus obliqua* trees. *Australian Forest Research* **11,** 55–62.

Kimball, K.D. and Keifer, M. (1988) Climatic comparisons with tree ring data from montane forests: are the climatic data appropriate? *Canadian Journal of Forest Research* **18,** 385–90.

Kimber, P.C. (1976) Increased girth increment associated with crown scorch in jarrah. *Western Australian Forests Department Research Paper* **37,** 4pp.

Kimmins, J.P. (1990) Monitoring the condition of the Canadian forest environment: The relevance of the concept of 'ecological indicators'. *Environmental Monitoring and Assessment* **15,** 231–40.

Kimura, B.Y. and Nagata, K.M. (1980) *Hawaii's vanishing flora.* Honolulu: Oriental Publishing Co.

Kincaid, D.T. and Lyons, E.E. (1981) Winter water relations of red spruce on Mount Monadnock, New Hampshire. *Ecology* **62,** 1155–61.

Kinloch, B.B., Westfall, R.D. and Forrest, G.I. (1986) Caledonian Scots pine: origins and genetic structure. *New Phytologist* **104,** 703–29.

Klein, H. and Jäger, H.J. (1976) Einfluss der Nährstoffversorgung auf die SO_2-Empfindlichkeit von Erbsenpflanzen. *Zeitschrift für Pflanzenpathologie und Pflanzenschutz* **83,** 555–68.

Klein, R.M. and Perkins, T.D. (1988) Primary and secondary causes and consequences of contemporary forest decline. *Botanical Reviews* **54,** 1–43.

Klein, R.M., Perkins, T.D. and Myers, H.L. (1989) Nutrient status and winter hardiness of red spruce foliage. *Canadian Journal of Forest Research* **19,** 754–8.

Kleman, J. (1985) The spectral signatures of damaged spruce stands measured from a helicopter. In: Schmid-Haas, P. (ed.), *Inventorying and monitoring endangered forests.* Birmensdorf: Eidgenössische Anstalt für das forstliche Versuchswesen, pp. 135–40.

Klemm, O. (co-ordinator) (1989) Leaching and uptake of ions through aboveground Norway spruce tree parts. In: Schulze, E.-D., Lange, O.-L. and Oren, R. (eds), *Forest decline and air pollution. A study of spruce (*Picea abies*) on acid soils.* Berlin: Springer Verlag, pp. 210–37.

Kliejunas, J.T. and Ko, W.H. (1973) Root rot of 'ohi'a (*Metrosideros collina* subsp. *polymorpha*) caused by *Phytophthora cinnamomi. Plant Disease Reporter* **57,** 383–4.

Kliejunas, J.T. and Ko, W.H. (1974) Deficiency of inorganic nutrients as a contributing factor to 'ohi'a decline. *Phytopathology* **64,** 891–6.

Klumpp, A., Klumpp, G. and Guderian, R. (1988) Wuchsleistung und äussere Schädigungsmerkmale bei Buchen nach Einwirkung von Ozon, Schwefeldioxid und Stickstoffdioxid. *Allgemeine Forst Zeitschrift* **43,** 731–4.

Klumpp, A., Küppers, K. and Guderian, R. (1989) Nitrate reductase activity of needles of Norway spruce fumigated with different mixtures of ozone, sulfur dioxide, and nitrogen dioxide. *Environmental Pollution* **58,** 261–71.

Knabe, W., Pohlmann, H. and Godde, D. (1989) The role of silicon in detecting and explaining forest decline. In: Bucher, J.B. and Bucher-Wallin, I. (eds), *Air pollution and forest decline.* Birmensdorf: Eidgenössische Anstalt für das forstliche Versuchswesen, pp. 111–17.

Knight, H.A. (1987) The pine decline. *Journal of Forestry* **85,** 25–8.

Koch, B. (1987) *Quantifizierung von Schäden an Waldbäumen auf der Grundlage spektraler Rückstrahlungseigenschaften.* Munich: Forschungsbericht, Lehrstuhl für Landschaftstechnik der Ludwig-Maximilians-Universität. 195 pp.

Koch, B. and Kritikos, G. (1984) Integrative investigation on forest damage detection based on airborne multispectral scanner data. *Proceedings of the Symposium on integrated approaches in remote sensing,* Guildford, Surrey. pp. 109–13.

Koch, B., Ammer, U., Kritikos, G. and Kubler, D. (1984) Untersuchungen zur Beurteilung der Vitalität von Fichten anhand multispektraler Scannerdaten. *Forstwissenschaftliches Centralblatt* **103,** 214–31.

Koch, B., Schneider, T. and Wittmeier, H. (1989) Teil III: Quantifizierung der Schäden anhand spektraler Rückstrahlungsmessungen. *Forstarchiv* **60,** 123–5.

Koch, R. (1876) Untersuchung über Bacterien. V. Die etiologie der Milzbrand–Krankheit, begründet auf die Entwicklungsgeschichte des Bacillus Anthracis. *Beiträge zur Biologie der Pflanzen* **2**, 277–310.

Koch, W. (1989) Der Reinluft/Standortsluft – Vergleich an Fichte. *Forstwissenschaftliches Centralblatt* **108,** 73–82.

Koch, W. and Lautenschlager, K. (1988) Photosynthesis and transpiration in the upper crown of a mature spruce in purified and ambient atmosphere in a natural stand. *Trees: Structure and Function* **2,** 213–22.

Köhl, M. (1987) Gegenwärtiger Stand der Methoden zur Inventur und Überwachung gefährdeter Wälder. *Allgemeine Forst Zeitschrift* **42,** 573–7.

Köhl, M. (1991) Waldschadeninventuren: mögliche Ursachen der Variation der Nadel-/Blattverlustschätzung zwischen Beobachtern und Folgerungen für Kontrollaufnahmen. *Allgemeine Forst- und Jagdzeitung* **162,** 210–21.

Kohut, R.J., Laurence, J.A., Amundson, R.G., Raba, R.M. and Melkonian, J.J. (1990) Effects of ozone and acidic precipitation on the growth and photosynthesis of red spruce after two years of exposure. *Water, Air, and Soil Pollution* **51,** 277–86.

Kohyama, H. and Fujita, N. (1981) Studies on the *Abies* population of Mt. Shimagare I. Survivorship curve. *Botanical Magazine Tokyo* **94,** 55–8.

Kohyama, T. (1988) Etiology of 'Shimagare' dieback and regeneration in subalpine *Abies* forests of Japan. *GeoJournal* **17,** 201–8.

Koltzenburg, C. and Knigge, W. (1987a) Holzeigenschaften von Buchen aus immissionsgeschädigten Beständen. *Holz als Roh- und Werkstoff* **45,** 81–7.

Koltzenburg, C. and Knigge, W. (1987b) Holzeigenschaften von Buchen aus immissionsgeschädigten Beständen. Anatomische Holzeigenschaften. *Holz als Roh- und Werkstoff* **45,** 267–72.

Konnert, V. and Mettendorf, B. (1990) Beobachtungsflächen zu den 'neuartigen Waldschäden' an Tanne in Baden-Württemberg: Nadelverlust, Klima, Zuwachs und Ernährungssituation. *Allgemeine Forst- und Jagdzeitung* **161,** 116–23.

Kontic, R. (1987) Comparative studies on the annual ring pattern and crown condition of conifers: The situtation in Switzerland and southern Germany: General conclusions. In: Kairiukstis, L., Nilsson, S. and Straszak, A. (eds), *Forest decline and reproduction: Regional and global consequences.* Laxenburg: International Institute for Applied Systems Analysis, pp. 184–7.

Kontic, R., Niederer, M., Nippel, C.-A. and Winkler-Seifert, A. (1986)

Jahrringanalysen an Nadelbäumen zur Darstellung und Interpretation von Waldschäden (Wallis, Schweiz). Bericht 283. Birmensdorf: Eidgenössische Anstalt für das forstliche Versuchswesen, 46 pp.

van Kooten, O. and van Hove, L.W.A. (1988) Fluorescence as a means of diagnosing the effect of pollutant induced stress in plants. In: Mathy, P. (ed.), *Air pollution and ecosystems.* Dordrecht: D. Reidel, pp. 596–601.

Korstian, C.F. (1937) Perpetuation of spruce on cut-over and burned lands in the higher southern Appalachian Mountains. *Ecological Monographs* **7,** 125–67.

de Kort, I. (1986) Wood structure and growth ring width of vital and non-vital Douglas fir (*Pseudotsuga menziesii*) from a single stand in The Netherlands. *IAWA Bulletin* **7,** 309–18.

de Kort, I. and Baas, P. (1986) Diktegroei en houtstructuur van vitale en niet-vitale douglas in de Peel. *Nederlandes Bosbouwtijdschrift* **58,** 52–7.

de Kort, I., Loeffen, V. and Baas, P. (1991) Ring width, density and wood anatomy of Douglas fir with different crown vitality. *IAWA Bulletin* **12,** 453–65.

Köstner, B., Czygan, F.-C. and Lange, O.-L. (1990) An analysis of needle yellowing in healthy and chlorotic Norway spruce (*Picea abies*) in a forest decline area of the Fichtelgebirge (N.E. Bavaria). 1. Annual time-course changes in chloroplast pigments for five different needle age classes. *Trees: Structure and Function* **4,** 55–67.

Kouki, J. and Hokkanen, T. (1992) Long-term needle litterfall of a Scots pine, *Pinus sylvestris*, stand: relation to temperature. *Oecologia* **89,** 176–81.

Kovbassa, N.P. (1988) (Vulnerability of Norway spruce during tending fellings). *Lesovedenie i Lesnoe Khozyaistvo* **23,** 94–7.

Koziol, M.J. (1984) Interactions of gaseous pollutants with carbohydrate metabolism. In: Koziol, M.J. and Whatley, R.F. (eds), *Gaseous air pollutants and plant metabolism.* London: Butterworths, pp. 251–73.

Koziol, M.J., Whatley, F.R. and Shelvey, J.D. (1988) An integrated view of the effects of gaseous air pollutants on plant carbohydrate metabolism. In: Schulte-Hostede, S., Darrall, N.M., Blank, L.W. and Wellburn, A.R. (eds), *Air pollution and plant metabolism.* London: Elsevier, pp. 148–68.

Kozlowski, T.T. and Constantinidou, H.A. (1986a) Responses of woody plants to environmental pollution. Part 1. Sources and types of pollutants and plant responses. *Forestry Abstracts* **47,** 1–51.

Kozlowski, T.T. and Constantinidou, H.A. (1986b) Responses of woody plants to environmental pollution. Part 2. Factors affecting responses to pollution and alleviation of pollution effects. *Forestry Abstracts* **47,** 105–32.

Kral, F. and Mayer, H. (1985) Ergebnisse vergleichender Resistenz-Untersuchungen an Tannenherkünften. *Schweizerische Zeitschrift für Forstwesen* **136,** 41–8.

Kramer, H. (1986a) Relation between crown parameters and volume increment of Picea abies stands damaged by environmental pollution. *Scandinavian Journal of Forest Research* **1,** 251–63.

Kramer, H. (1986b) Dégâts dans les couronnes et accroisements dans les peuplements forestiers résineux frappés par le récent 'dépérissement des forêts'. *Revue forestières française* **38,** 124–33.

Krämer, P. and Suda, M. (1987) Das Waldsterben in den Massenmedien. *Allgemeine Forst Zeitschrift* **49,** 1291–4.

Krapfenbauer, A. (1983) Von der Streunutzung zur Ganzbaumnutzung. *Centralblatt für das gesamte Forstwesen* **100,** 143–74.

Kratochvílová, I., Janous, D., Marek, M., Barták, M. and Ríha, L. (1989) Production activity of mountain cultivated Norway spruce stands under the impact of air pollution. *Ekológia (CSSR)* **8,** 407–19.

Krauss, H. (1957) Weißtannen-Vorkommen und Weißtannen-Sterben in Thüringen. *Forst und Jagd* (Berlin) **7:** 102–9 and 151–3.

Krehan, H. (1989) Das Tannensterben in Europa. Eine Literaturstudie mit kritischer Stellungnahme. *FBVA Berichte* **39,** 56 p.

Kress, L.W., Allen, H.L., Mudano, J.E. and Heck, W.W. (1989) Ozone effects on growth of loblolly pine. In: Bucher, J.B. and Bucher-Wallin, I. (eds), *Air pollution and forest decline.* Birmensdorf, Eidgenössische Anstalt für das forstliche Versuchswesen, pp. 153–8.

Kress, L.W. and Skelly, J.M. (1982) Response of several eastern forest tree species to chronic doses of ozone and nitrogen dioxide. *Plant Disease* **66,** 1149–52.

Kress, L.W., Skelly, J.M. and Hinkelmann, J.H. (1982a) Growth impact of O_3, NO_2, and/or SO_2 on *Pinus taeda. Environmental Monitoring and Assessment* **1,** 229–39.

Kress, L.W., Skelly, J.M. and Hinkelmann, J.H. (1982b) Relative sensitivity of 18 full-sib families of *Pinus taeda* L. to ozone. *Canadian Journal of Forest Research* **12,** 203–9.

Kreutzer, K. (1972) Über den Einfluß der Streunutzung auf den Stickstoffhaushalt von Kiefernbeständen. *Forstwissenschaftliches Centralblatt* **91,** 263–70.

Kreutzer, K. (1989) Änderungen im Stickstoffhaushalt der Wälder und die dadurch verursachten Auswirkungen auf die Qualität des Sickerwassers. *DVWK-Mitteilungen* **17,** 121–39.

Kreutzer, K. and Bittersohl, J. (1986) Stoffauswaschung aus Fichtenkronen (*Picea abies* Karst.) durch saure Beregnung. *Forstwissenschaftliches Centralblatt* **105,** 357–63.

Kreutzer, K., Knorr, A., Brosinger, F. and Kretzschmar, P. (1983) Scots pine-dying within the neighbourhood of an industrial area. In: Ulrich, B. and Pankrath, J. (eds), *Effects of accumulation of air pollutants in forest ecosystems.* Dordrecht, D. Riedel, pp. 343–57.

Kriebel, H.B. and Leben, C. (1981) The impact of air pollution on the gene pool of eastern white pine. *XVII IUFRO World Congress, Kyoto, Japan. Proceedings of Division 2,* pp. 185–9.

Kriebitzsch, W.U. (1978) Stickstoffnachlieferung in sauren Waldböden Nordwestdeutschlands. *Scripta Botanica* **14,** 1–66.

Krieter, M. (1986) Untersuchungen von Restabilisierungsmaßnahmen in immissionsgeschädigten Wäldern des südlichen Taunus. *IMA-Querschnittseminar 'Restabilisierungsmaßnahmen – Düngung'.* Karlsruhe: Berichtsband KfK-PEF, July 1986, pp. 23–33.

Kritikos, G., Kubler, D. and Herrmann, K. (1985) Comparative classification of forestry damage with scanner data from different altitudes. In: Schmid-Haas, P. (ed.), *Inventorying and monitoring endangered forests.* Birmensdorf: Eidgenössische Anstalt für das forstliche Versuchswesen, pp. 183–7.

Krivan, V., Lüttge, U. and Schaldach, G. (1986) Profile von Makro- und Mikromineralnährstoffen in gesunden und kranken Fichten (*Picea abies* (L.)

Karst.) auf verschiedenen Standorten in Südwestdeutschland. *Angewandte Botanik* **60,** 373–89.

Król, A. (1983) Owadyszkodniki wtórne i techniczne debu szypulkowego (*Quercus robur* L.) w rejonie eksploatacji siarki w Tarnobrzegu. *Acta Agraria et Silvestria, Silvestris* **21,** 3–13.

Kroth, W., Grottker, T. and Bartelheimer, P. (1989) Forstwirtschaftliche Massnahme bei neuartigen Waldschäden. *Angewandte Wissenschaft* 368. Bonn: Schriftenreihe des Bundesministers für Ernährung, Landwirtschaft und Forsten.

Kryuchkov, V.V. (1991) Maximal anthropogenic loading and condition of northern ecosystems. *Soviet Journal of Ecology* **22,** 155–66.

Krzak, J., Dong, P.H., Büttner, G., Hütterman, A., Kramer, H. and Ulrich, B. (1988) Photosynthesis, nutrient, growth and soil investigations of a declining Norway spruce (*Picea abies*) stand in the coastal region of northern Germany. *Forest Ecology and Management* **24,** 263–81.

Kubiková, J. (1991) Forest dieback in Czechoslovakia. *Vegetatio* **93,** 101–8.

Kufner, M. and Schulz, H. (1986) Biegefestigkeit von Fichten- und Kiefernholz aus Waldschadensgebieten in Bayern. *Holz als Roh- und Werkstoff* **44,** 409–14.

Kuhn, N., Amiet, R. and Hufschmid, N. (1987) Veränderungen in der Waldvegetation der Schweiz infolge Nährstoffanreicherungen aus der Atmosphäre. *Allgemeine Forst- und Jagdzeitung* **158,** 77–84.

Kuhns, M.R., Garret, H.E., Teskey, R.O. and Hinckley, T.M. (1985) Root growth of black walnut trees related to soil temperature, soil water potential, and leaf water potential. *Forest Science* **31,** 617–29.

Kuivinen, K.C. and Lawson, M.P. (1982) Dendroclimatic analysis of birch in South Greenland. *Arctic and Alpine Research* **14,** 243–50.

Kullman, L. (1988) Subalpine *Picea abies* decline in the Swedish Scandes. *Mountain Research and Development* **8,** 33–42.

Kullman, L. (1989) Tree-limit history during the Holocene in the Scandes Mountains, Sweden, inferred from subfossil wood. *Reviews of Palaeobotany and Palynology* **58,** 163–71.

Kullman, L. and Högberg, N. (1989) Rapid natural decline of upper montane forests in the Swedish Scandes. *Arctic* **42,** 217–26.

Kunert, K.J. and Hofer, G. (1987) Geben Veränderungen des antioxidativen Systems von Pflanzen Hinweise auf die Wirkung von Luftschadstoffen. *Allgemeine Forst Zeitschrift* **42,** 697–700.

Küppers, M., Zech, W., Schulze, E.-D. and Beck, E. (1985) CO_2 - Assimilation, Transpiration und Wachstum von *Pinus sylvestris* L. bei unterschiedlicher Magnesiumversorgung. *Forstwissenschaftliches Centralblatt* **104,** 23–36.

Kurkela, T. and Jalkanen, R. (1990) Revealing past needle retention in *Pinus* spp. *Scandinavian Journal of Forest Research* **5,** 481–5.

Kurz, W.A. and Kimmins, J.P. (1987) Analysis of some sources of error in methods used to determine fine root production in forest ecosystems: a simulation approach. *Canadian Journal of Forest Research* **17,** 909–12.

Kuser, J.E., Eberhardt, J.C. and Brennan, E.G. (1989) Genecological variation in *Pinus strobus*: growth rate, date of fall needle-shedding and second-winter needle retention. *Silvae Genetica* **38,** 121–5.

Kuylenstierna, J.C.I. and Chadwick, M.J. (1991) Increases in soil acidity between 1957 and 1990. *Ambio* **20,** 118–19.

Laamanen, A., Markkanen, K. and Riekkinen, A.-L. (1972) Kokkolan malminjalostus-, peruskemikaali- ja lannoiteteollisuuden vaikutus yhdyskuntailmaan. *Työterveyslaitoksen tutkimuksia* **80,** 1–78.

Lacasse, N.L. and Treshow, M. (eds) (1978) *Diagnosing vegetation injury caused by air pollution.* Research Triangle Park, North Carolina: United States Environmental Protection Agency.

Lachance, D. (1985) Répartition géographique et intensité de dépérissement de l'érable à sucre dans les érablières au Québec. *Phytoprotection* **66,** 83–90.

Lachance, D., Benoit, P., LaFlamme, G., Bonneau, G. and Piché, R. (1984) Insectes et maladies des arbres, Québec, 1983. *Forêt-Conservation* 50. Supplement.

Lackner, A.L. and Alexander, S.A. (1983) Root disease and insect infestations on air-pollution-sensitive *Pinus strobus* and studies of pathogenicity of *Verticicladiella procera. Plant Disease* **67,** 679–81.

LaMarche, V.C., Graybill, D.A., Fritts, H.C. and Rose, M.R. (1984) Increasing atmospheric carbon dioxide: tree ring evidence for growth enhancement in natural vegetation. *Science* **225,** 1019–20.

LaMarche, V.C. and Stockton, C.W. (1974) Chronologies from temperature-sensitive bristlecone pines at upper treeline in Western United States. *Tree Ring Bulletin* **34,** 21–45.

Lambert, M.J. and Turner, J. (1977) Dieback in high site quality *Pinus radiata* stands – the role of sulphur and boron deficiencies. *New Zealand Journal of Forest Science* **7,** 333–48.

Landmann, G. (1988a) Evaluation de l'incidence des conditions ecologiques sur le dépérissement et premiers resultats d'une approche symptomatologique. *Journals DEFORPA* 1988, **1,** 2–48.

Landmann, G. (1988b) Observation au sol de l'état sanitaire des forêts. Problems liés a l'interpretation des données, evaluation retrospective sommaire des resultats. *Journals DEFORPA* 1988, **2,** 1–30.

Landmann, G. (1988c) The main symptoms of forest decline in France and a tentative classification. In: Moisl, K. and Zlinska, J. (eds), *Workshop on Methodologies.* Prague: Institute for Environment and Utilization of Natural Resources, pp. 51–8.

Landmann, G. (1991) Ground monitoring of forest vitality: an initial evaluation of the French results (for the period 1983–1988) and a comparison with the results from neighbouring countries. Unpublished manuscript.

Landmann, G. (1992) Role of climate, stand dynamics and past management in the 'novel forest decline': a critical review of ten years of field ecology in France. In: Hültl, R.F. and Mueller-Dombois, D. (eds), Forest decline in the Atlantic and Pacific regions. Berlin: Springer-Verlag, pp. 18–39.

Landmann, G. and Fabbri, B. (1989) Relations entre l'aspect des cimes, la croissance, les paramètres sylvicoles et stationnels. Le cas du sapin dans les Vosges. In: *Journées de travail DEFORPA February-March 1989.* Vol. 1. INRA, Nancy, pp. 3.6.1–17.

Landolt, W., Bucher, J.B. and Kaufmann, E. (1984) Waldschäden in der Schweiz – 1983 (II. Teil). Interpretation der Sanasilva-Umfrage und der Fichtennadelanalysen aus der Sicht der forstlichen Ernährungslehre. *Schweizerische Zeitschrift für Forstwesen* **135,** 637–53.

Landolt, W., Guecheva, M. and Bucher, J.B. (1989) The spatial distribution of different elements in and on the foliage of Norway spruce growing in Switzerland. *Environmental Pollution* **56,** 155–67.

Landolt, W., Lüthy, B. and Pfenninger, I. (1988) The effect of ozone on the cyclite and starch content in Scotch pine (*Pinus sylvestris*) and Norway spruce (*Picea abies*) needles. In Cape, J.N. and Mathy, P. (eds), *Scientific basis of forest decline symptomatology.* Air Pollution Report 15. Brussels: Commission of the European Communities, pp. 332–41.

Landsberg, J. and Ohmart, C. (1989) Levels of insect defoliation in forests: patterns and concepts. *Trends in Ecology and Evolution* **4,** 96–100.

Lang, E. and Jagnow, G. (1986) Fungi of a forest soil nitrifying at low pH values. *FEMS Microbiology Ecology* **38,** 257–65.

Lange, O.L., Heber, U., Schulze, E.-D. and Ziegler, H. (1989b) Atmospheric pollutants and plant metabolism. In: Schulze, E.-D., Lange, O.L. and Oren, R. (eds), *Forest decline and air pollution. A study of spruce (*Picea abies*) on acid soils.* Berlin: Springer-Verlag, pp. 238–73.

Lange, O.L., Weikert, R.M., Wedler, M., Gebel, J. and Heber, U. (1989a) Photosynthese und Nährstoffversorgung von Fichten aus einem Waldschadensgebiet auf basenarmen Untergrund. *Allgemeine Forst Zeitschrift* **44,** 55–64.

Lange, O.L., Zellner, H., Gebel, J., Schramel, P., Kostner, B. and Czygan, F.C. (1987) Photosynthetic capacity, chloroplast pigments, and mineral content of the previous year's spruce needles with and without the new flush: analysis of the forest decline phenomenon of needle bleaching. *Oecologia* **73,** 351–7.

Långström, B. and Hellqvist, C. (1991) Shoot damage and growth losses following three years of *Tomicus*-attacks in Scots pine stands close to a timber storage site. *Silva Fennica* **25,** 133–45.

Långström, B. and Hellqvist, C. (1992) Height growth recovery and crown development in top-damaged pine trees. *Scandinavian Journal of Forest Research* 7, in press.

Långström, B., Tenow, O., Ericsson, A., Hellqvist, C. and Larsson, S. (1990) Effects of shoot pruning on stem growth, needle biomass, and dynamics of carbohydrates and nitrogen in Scots pine as related to season and tree age. *Canadian Journal of Forest Research* **20,** 514–23.

Lappi, J. and Bailey, R.L. (1987) Estimation of tree diameter increment function or other tree relations using angle-count samples. *Forest Science* **33,** 725–39.

Larsen, J.B. (1986a) Das Tannensterben: Eine neue Hypothese zur Klärung des Hintergrundes dieser rätselhaften Komplexkrankheit der Weißtanne (*Abies alba* Mill.). *Forstwissenschaftliches Centralblatt* **105,** 381–96.

Larsen, J.B. (1986b) Die geographische Variation der Weisstanne (*Abies alba* Mill.). Wachstumsentwicklung und Frostresistenz. *Forstwissenschaftliches Centralblatt* **105,** 396–406.

Laurence, J.A., Kohut, R.J. and Amundson, R.G. (1989) Response of red spruce seedlings exposed to ozone and simulated acidic precipitation in the field. *Archives for Environmental Contamination and Toxicology* **18,** 285–90.

Laurent, T.H. (1982) Cedar dieback – is it the problem? In *Proceedings of the 2nd Alaska Integrated Pest Management Conference, Anchorage, Alaska.* Anchorage: University of Alaska, Cooperative Extension Service, pp. 104–7.

Lautenschlager, K. (1984) Osmotic potential and gas exchange in healthy and diseased silver firs (*Abies alba* Mill.). *European Journal of Forest Pathology* **14,** 359–72.

Lavalle, A. and Bard, G. (1978) Comportment of cytospora canker on black spruce in natural forests. *Phytoprotection* **59,** 132–6.

Lawesson, J.E. (1988) Stand-level dieback and regeneration of forests in the Galápagos Islands. *Vegetatio* **77,** 87–93.

Lawrence, G.B. and Fernandez, I.J. (1991) Biogeochemical interactions between acidic deposition and a low-elevation spruce-fir stand in Howland, Maine. *Canadian Journal of Forest Research* **21,** 867–75.

Leak, W.B. (1988) Comparison of standard and actual tree-growth trends for deciduous and coniferous species in New Hampshire. *Canadian Journal of Forest Research* **17,** 1297–300.

LeBlanc, D.C. (1990a) Red spruce decline on Whiteface Mountain, New York. 1. Relationships with elevation, tree age, and competition. *Canadian Journal of Forest Research* **20,** 1408–14.

LeBlanc, D.C. (1990b) Relationships between breast-height and whole-stem growth indices on Whiteface Mountain, New York. *Canadian Journal of Forest Research* **20,** 1399–1407.

LeBlanc, D.C., Nicholas, N.S. and Zedaker, S.M. (1992) Prevalence of individual-tree growth decline in red spruce populations of the southern Appalachian Mountains. *Canadian Journal of Forest Research* **22,** 905–14.

LeBlanc, D.C. and Raynal, D.J. (1990) Red spruce decline on Whiteface Mountain, New York. II. Relationships between apical and radial growth decline. *Canadian Journal of Forest Research* **20,** 1415–21.

Lechowicz, M.J. (1987) Resource allocation by plants under air pollution stress: Implications for plant-pest-pathogen interactions. *Botanical Review* **53,** 281–300.

Leckie, D.G. and Ostaff, D.P. (1988) Classification of airborne multispectral scanner data for mapping current defoliation caused by the spruce budworm. *Forest Science* **34,** 259–75.

Lee, E.H., Tingey, D.T. and Hogsett, W.E. (1988) Evaluation of ozone exposure indices in exposure-response modelling. *Environmental Pollution* **53,** 43–62.

Lee, K.J. and Kim, T.W. (1990) Air pollution and forest declines in Far-East Asia. *Proceedings of the 19th World Congress, IUFRO, Montreal, 5–11 August, 1990.* Montreal: IUFRO, pp. 37–48.

Lee, W.S. and Chevone, B.I. (1990) Growth and gas exchange of loblolly pine seedlings as influenced by drought and air pollutants. *Water, Air, and Soil Pollution* **51,** 105–16.

Lee, W.S., Chevone, B.I. and Seiler, J.R. (1990) Growth response and drought susceptibility of red spruce seedlings exposed to simulated acidic rain and ozone. *Forest Science* **36,** 265–75.

Lees, J.A. (1991) *A dendrochronological investigation into factors affecting 'top dying' of Norway spruce (*Picea abies *(L.) Karst.) in England and Wales.* Dissertation, Department of Geography, Coventry Polytechnic.

Leffef, F. (1988) De gezondheid van de zomereik in relatie tot de aantasting door insekten. *Nederlands Bosbouwtijdschrift* **12,** 414–20.

Lefohn, A.S., Knudsen, H.P. and McEvoy, L.R. (1988) The use of kriging to estimate monthly ozone exposure parameters for the southeastern United States. *Environmental Pollution* **53,** 27–42.

Lefohn, A.S. and Mohnen, V.A. (1986) The characterization of ozone, sulphur dioxide and nitrogen dioxide for selected monitoring sites in the Federal Republic of Germany. *Journal of the Air Pollution Control Association* **36,** 1329–37.

Legge, A.H., Bogner, J.C. and Krupa, S.V. (1988) Foliar sulphur species in pine: a new indicator of a forest ecosystem under air pollution stress. *Environmental Pollution* **55,** 15–27.

Leininger, T.D., Winner, W.E. and Alexander, S.A. (1990) Root disease incidence in eastern white pine plantations with and without symptoms of ozone injury in the Coweeta Basin of North Carolina. *Plant Disease* **74,** 552–4.

Leith, I.D., Murray, M.B., Sheppard, L.J., Cape, J.N., Deans, J.D., Smith, R.I. and Fowler, D. (1989a) Visible foliar injury of red spruce seedlings subjected to simulated acid mist. *New Phytologist* **113,** 313–20.

Leith, I.D., Cape, J.N., Sheppard, L.J., Murray, M.B., Deans, F.D. and Fowler, D. (1989b) Frost hardiness and visible injury of spruce seedlings subjected to simulated acid mist. In: Bucher, J.B. and Bucher-Wallin, I. (eds), *Forest decline and air pollution.* Birmensdorf, Eidgenössische Anstalt für das forstliche Versuchswesen, pp. 175–80.

Lenz, O., Nogler, P. and Bräker, O.U. (1988) *L'évolution du temps et le dépérissement du Sapin blanc dans la région de Berne.* Bericht 303. Birmensdorf: Eidgenössische Anstalt für das forstliche Versuchswesen.

Lenz, R. and Haber, W. (1990) Longterm assessment of spruce vitality in the Fichtelgebirge (West Germany) under ongoing acid deposition. *Vegetatio* **89,** 121–35.

Leonard, J.G., Tobi, D.R. and Parker, B.L. (1991) Spatial and temporal distribution of *Korscheltellus gracilis* larvae (Lepidoptera: Hepialidae) in the Green Mountains, Vermont. *Environmental Entomology* **20** pp. 371–6.

Lesinski, J.A. (1989) Dynamics of injury symptoms in Norway spruce. In: *International congress on forest decline research: State of knowledge and perspectives,* Poster Abstracts Vol. 1. Bonn: Forschungsbeirat Waldschäden/Luftverunreinigungen der Bundesregierung und der Länder, pp. 55–6.

Lesinski, J.A. (1991a) Development of injury symptoms in Norway spruce on permanent plots in Sweden. In: *Proceedings of the IUFRO Workshop 'Monitoring air pollution impact on permanent sample plots, data processing and results interpretation', Prachatice, Czechoslovakia, 1991.* Prachatice: IUFRO, pp. 213–18.

Lesinski, J.A. (1991b) Morphological vigour indicators in Norway spruce. In: *Proceedings of the IUFRO Workshop on Ecological Modelling, Usti nad Labem, Czechoslovakia, 1991.* Usti nad Labem: IUFRO, pp. 38–46.

Lesinski, J.A., Dmyterko, E. and Grzyb, M. (1991) Comparison of two methods when classifying injury symptoms in Norway spruce from the Sudety Mountains, Poland. In: *Proceedings of the IUFRO Workshop 'Monitoring air pollution impact on permanent sample plots, data processing and results interpretation',* Prachatice, Czechoslovakia, 1991. Prachatice: IUFRO, pp. 208–12.

Lesinski, J.A. and Landmann, G. (1988) Crown and branch malformation in conifers related to forest decline. In: Cape, J.N. and Mathy, P. (eds), *Scientific basis of forest decline methodology.* Air Pollution Report 15. Brussels: Commission of the European Communities, pp. 92–107.

Lesinski, J.A. and Westman, L. (1987) Crown injury types in Norway spruce and their applicability for forest inventory. In: Perry, R., Harrison, R.M., Bell, J.N.B. and Lester, J.N. (eds), *Acid rain: scientific and technical advances.* London: Selper, pp. 657–62.

Levin, M.J. (1985) Ergebnisse und waldbaumliche Konsequenzen einer Immissionsschadenserhebung im Privatforstbetrieb Sachsenwald. *Waldhygiene* **16,** 19–22.

Lévy, G. and Becker, M. (1987) Le dépérissement du sapin dans les Vosges: rôle primordial de déficits d'alimentation en eau. *Annales des Sciences Forestières* **44,** 403–16.

Lewton-Brain, L. (1909) The Maui forest trouble. *Hawaiian Planter's Record* **1,** 92–5.

Li, X. and Zhang, Y.B. (1989) Relationship of sulphur content in tree leaves to SO_2 concentration in air. *Bulletin of Environmental Contamination and Toxicology* **42,** 878–83.

Li, Z. and Aneja, V.P. (1992) Regional analysis of cloud chemistry at high elevations in the eastern United States. *Atmospheric Environment* **26A,** 2001–17.

Lick, E. and Krapfenbauer, A. (1986) Terrestrische Waldzustandsinventuren und Probleme einer Objektivierung. *Centralblatt für das gesamte Forstwesen* **103,** 227–41.

Liedeker, H. and Klein, R.M. (1989) Characteristic discolorations of coniferous foliage related to spruce decline in Europe and North America. *Canadian Journal of Forest Research* **19,** 145–8.

Liese, W., Schneider, M. and Eckstein, D. (1975) Histometrische Untersuchungen am Holz einer rauchgeschädigten Fichte. *European Journal of Forest Pathology* **5,** 152–61.

Like, D.E. and Klein, R.M. (1985) The effect of simulated acid rain on nitrate and ammonium production in soils from three ecosystems of Camels Hump Mountain, Vermont. *Soil Science* **140,** 352–5.

Lindberg, S.E., Lovett, G.M., Richter, D.D. and Johnson, D.W. (1986) Atmospheric deposition and canopy interactions in a forest. *Science* **231,** 141–5.

Lines, R. (1984) Species and seed origin trials in the industrial Pennines. *Quarterly Journal of Forestry* **78,** 9–23.

Lines, R. (1987) *Choice of seed origins for the main forest species in Britain.* Forestry Commission Bulletin 66. London: HMSO.

Lines, R. and Mitchell, A.F. (1965) Results of some older Scots pine provenance experiments. *Forestry Commission Report on Forest Research 1964.* London: HMSO, pp. 172–94.

Linzon, S.N. (1971) Economic effects of sulfur dioxide on forest growth. *Journal of the Air Pollution Control Association* **21,** 81–6.

Lipas, E., Levula, T. and Väliknga, P. (1983) Eräitä metsänlannoitustuloksia Lapista. *Metsäntutkimuslaitoksen tiedonantoja* 114.

Liu, J.-C. and Hüttl, R.F. (1991) Relations between damage symptoms and nutritional status of Norway spruce stands (*Picea abies* Karst.) in southwestern Germany. *Fertilizer Research* **27,** 9–22.

Liu; J.-C., Pfirrmann, T., Payer, H.-D., Runkel, K.H. and Schramel, P. (1991)

Wasser- und Stoffdynamik der Böden am Wank und ihr möglicher Einfluß auf die Fichten. In: *Proceedings 2. Statusseminar der PBWU zum Forschungsschwerpunkt 'Waldschäden'. 4–6 February, 1991, GSF-Forschungszentrum Neuherberg.* Neuherberg: GSF, pp. 237–51.

Liu, J.-C. and Trüby, P. (1989) Bodenanalytische Diagnose von K- und Mg-Mangel in Fichtenbeständen (*Picea abies* Karst.). *Zeitschrift für Pflanzenernährung und Bodenkunde* **152,** 307–11.

Ljungström, M., Gyllin, M. and Nihlgård, B. (1990) Effects of liming on soil acidity and beech (*Fagus sylvatica* L.) regeneration on acid soils in south Swedish beech forests. *Scandinavian Journal of Forest Research* **5,** 243–54.

Loehle, C. (1988) Forest decline: endogenous dynamics, tree defenses, and the elimination of spurious correlation. *Vegetatio* **77,** 65–78.

Lonsdale, D. (1986a) *Beech health study 1985.* Forestry Commission Research and Development Paper 146. Edinburgh: Forestry Commission.

Lonsdale, D. (1986b) *Beech health study 1986.* Forestry Commission Research and Development Paper 149. Edinburgh: Forestry Commission.

Lonsdale, D., Hickman, I.T., Mobbs, I.D. and Matthews, R.W. (1989) A quantitative analysis of beech health and pollution across southern Britain. *Naturwissenschaften* **76,** 571–3.

Lonsdale, D. and Wainhouse, D. (1987) *Beech bark disease.* Forestry Commission Bulletin 69. London: HMSO.

Lorenc-Plucinska, G. (1986) Effect of sulphur dioxide on the partitioning of assimilates in Scots pine (*Pinus sylvestris* L.) seedlings of different susceptibility to this gas. *European Journal of Forest Pathology* **16,** 266–73.

Lorenz, M. and Eckstein, D. (1988) Wachstumsreaktion von Einzelbäumen in Douglasien-, Fichten- und Kiefernbeständen in norddeutschen Waldschadensgebieten. *Forst und Holz* **43,** 8–12.

Lorimer, C.G. (1980) Age structure and disturbance history of a southern Appalachian virgin forest. *Ecology* **61,** 1169–84.

Lovett, G.M. and Hubbell, J.G. (1991) Effects of ozone and acid mist on foliar leaching from eastern white pine and sugar maple. *Canadian Journal of Forest Research* **21,** 794–802.

Lucas, P.W. (1989) The development of needle chlorosis following long-term exposure of Sitka spruce to ozone. In: *International congress on forest decline research: state of knowledge and perspectives,* Poster abstract, vol. 1. Bonn: Forschungsbeirat Waldschäden/Luftverunreinigungen der Bundesregierung und der Länder, pp. 331–2.

Lucier, A.A. (1988) Pine growth-rate changes in the Southeast: A summary of key issues for forest managers. *Journal of Applied Forestry* **12,** 84–9.

Lucier, A.A. and Haines, S.G. (eds) (1990) *Mechanisms of forest response to acidic deposition.* New York: Springer Verlag.

Ludewig, M., Naumann, R., Fenner, R., Lalk, I., Döring, M., Hartmann, A. and Dörffling, K. (1989) Peroxidase-Aktivität in Fichtennadeln unter Schadgasbelastung im Freiland und in Expositionskammern. *Forstwissenschaftliches Centralblatt* **108,** 289–302.

Lutz, H.J. and Breiniger, M.T. (1986) Erste Ergebnisse von Auswaschungsversuchen mit Fichten (*Picea abies*) in Wasserkultur. *Kali-Briefe* **18,** 1–7.

Lyle, E.S. and Adams, F. (1971) Effect of available soil calcium on taproot elongation of loblolly pine (*Pinus taeda* L.) seedlings. *Soil Science Society of America Proceedings* **35,** 800–805.

Lynch, D.F. (1951) Diameter growth of ponderosa pine in relation to the Spokane pine-blight problem. *Northwestern Science* **25,** 157–63.

Lyon, H.L. (1909) The forest disease on Maui. *Hawaiian Planter's Record* **1,** 151–9.

Ma Guangjing (1989) La pollution atmospherique et le dépérissement des forêts. In Bucher, J. and Bucher-Wallin, I. (eds), *Air pollution and forest decline.* Birmensdorf, Eidgenössische Anstalt für das forstliche Versuchswesen, pp. 51–4.

Madgwick, H.A.I., Beets, P.N., Sandberg, A.M. and Jackson, D.S. (1983) Nitrogen concentration in foliage of *Pinus radiata* as affected by nitrogen nutrition, thinning, needle age and position in crown. *New Zealand Journal of Forestry Science* **13,** 197–204.

Magasi, L.P. (1988) *Acid rain national early warning system. Manual on plot establishment and monitoring.* Canadian Forestry Service Report DPC-X-25. Ottawa: Forestry Canada.

Magasi, L.P. (1989) *White birch deterioration in the Bay of Fundy region, New Brunswick 1979–1988.* Information Report M-X-175. Fredericton: Forestry Canada – Maritimes Region.

Magasi, L.P. (1990) *Forest pest conditions in the Maritimes in 1989.* Information Report M-X-177. Fredericton Forestry Canada – Maritimes Region.

Magasi, L.P. (1992) *Forest pest conditions in the Maritimes in 1991.* Information Report M-X-181. Fredericton Forestry Canada – Maritimes Region.

Mahler, G., Klebes, J. and Höwecke, B. (1988a) Holzkundliche Untersuchungen an Buchen mit neuartigen Waldschäden. *Allgemeine Forst- und Jagdzeitung* **159,** 121–5.

Mahler, G., Klebes, J. and Höwecke, B. (1988b) Holztechnologische Eigenschaften von Buchen mit neuartigen Waldschäden. *Holz-Zentralblatt* **114,** 402–4.

Mahrer, F. (1989) Problems in the determination and interpretation of needle and leaf loss. In: Bucher, J. and Bucher-Wallin, I. (eds), *Air pollution and forest decline.* Birmensdorf, Eidgenössische Anstalt für das forstliche Versuchswesen, pp. 229–31.

Maier, G. (1985) Entwicklung der Waldschäden auf den Fichten-Dauerbeobachtungsflächen in Bayern von 1981 bis 1984. *Allgemeine Forstzeitschrift* **40,** 500–1.

Maier-Maercker, U. (1986) Histological studies on the bundle sheath in needles of *Picea abies* (L.) Karst., diseased or fumigated with SO_2. *European Journal of Forest Pathology* **16,** 352–9.

Maier-Maercker, U. (1991) Verminderung des stomatären Regelvermögens durch Luftschadstoffe. In: *Proceedings 2. Statusseminar der PBWU zum Forschungsschwerpunkt 'Waldschäden'. 4–6 February, 1991, GSF-Forschungszentrum Neuherberg.* Neuherberg: GSF, 139–49.

Majumdar, S.K., Cline, S.W. and Zelnick, R.W. (1989) Chemical analyses of soils and oak tree tissues from two forest habitats in Pennsylvania differing in their sensitivity to acid precipitation. *Environmental Technology Letters* **10,** 1019–26.

Malcolm, D.C. and Garforth, M.F. (1977) The sulphur : nitrogen ratio of conifer foliage in relation to atmospheric pollution with sulphur dioxide. *Plant and Soil* **47,** 89–102.

Mälkönen, E. (1974) Annual primary production and nutrient cycle in some Scots pine stands. *Communicationes Institutale Forestales Fennica* **84,** 1–87.

Mälkönen, E. (1990) Estimation of nitrogen saturation on the basis of long-term fertilization experiments. *Plant and Soil* **128,** 75–82.

Mälkonen, E. and Kukkola, M. (1991) Effect of long-term fertilization on the biomass production and nutrient status of Scots pine stands. *Fertilizer Research* **27,** 113–27.

Mälkonen, E., Derome, J. and Kukkola, M. (1990) Effects of nitrogen inputs on forest ecosystems. Estimation based on long-term fertilization experiments. In: Kauppi, P., Anttilla, P. and Kenttämies, K. (eds), *Acidification in Finland*. Berlin: Springer Verlag, pp. 325–47.

Mandallaz, D., Schlaepfer, R. and Arnould, J. (1986) Dépérissement des forêts: essai d'analyse des dépendances. *Annales des Sciences Forestières* **43,** 441–58.

Manderscheid, R. and Jäger, H.-J. (1990) Comparative investigations on antioxidative components in needles of spruce (*Picea abies* (L.) Karst.) kept in open-top chambers with charcoal-filtered and non-filtered air. *Angewandte Botanik* **64,** 489–502.

Manderscheid, R. and Jäger, H.-J. (1992) Biochemisch-physiologische Untersuchungen an Fichten. (I) Nachweis von Schadgaswirkungen in Open-Top-Kammern. (II) Antioxidantien in unterschiedlich geschädigten Bäumen vom Standort Schöllkopf (Nordschwarzwald). In: *8. Statuskolloquium des PEF vom 17. bis 19. März 1992 im Kernforschungszentrum Karlsruhe*. Karlsruhe: Kernforschungszentrum, pp. 156–64.

Manderscheid, R., Jäger, H.-J. and Kress, L.W. (1992) Effects of ozone on foliar nitrogen metabolism of *Pinus taeda* L. and implications for carbohydrate metabolism. *New Phytologist* **121,** 623–33.

Manes, F., Altieri, A., Angelini, R., Bruno, F., Cortiello, M., Del Caldo, L. and Frederico, R. (1988) Micromorphological and biochemical changes in *Pinus pinea* L., *Pinus pinaster* Aiton, *Nicotiana tabacum* L. in relation to atmospheric pollutants. In: Cape, J.N. and Mathy, P. (eds), *Scientific basis for forest decline symptomatology*. Air Pollution Report 15. Brussels: Commission of the European Communities, pp. 342–53.

Manion, P.D. (1981) *Tree disease concepts*. New York: Prentice Hall. 324 pp.

Manion, P.D. (1985) Factors contributing to the decline of forests, a conceptual overview. In: Stubbs, H.S. (ed.), *Air pollutants effects on forest ecosystems*. St. Paul, Minnesota: Acid Rain Foundation, pp. 63–73.

Manion, P.D. (1989) Hardwood forest declines – concepts and management. In: *Proceedings of the 1988 SAF, Healthy forests, healthy world, 16–19 October, 1988, Rochester, New York*. Washington DC: Society of American Foresters, pp. 127–30.

Mankovská, B. (1988) The accumulation of atmospheric pollutants by *Picea abies* Karst. *Ekológia (CSSR)* **7,** 95–108.

Mann, L.K., Johnson, D.W., West, D.C., Cole, D.W., Hornbeck, J.W., Martin, C.W., Riekerk, H., Cmith, C.T., Swank, W.T., Tritton, L.M. and Van Lear, D.H. (1988) Effects of whole-tree and stem-only clearcutting on postharvest hydrologic losses, nutrient capital, and regrowth. *Forest Science* **34,** 412–28.

Mann, L.K., McLaughlin, S.B. and Shriner, D.S. (1980) Seasonal physiological responses of white pine under chronic air pollution stress. *Environmental and Experimental Botany* **20,** 99–105.

Marchand, P.J. (1984) Dendrochronology of a fir wave. *Canadian Journal of Forest Research* **14,** 51–6.

Marchand, P.J. and Chabot, B.F. (1978) Winter water relations of tree-line plant species on Mt. Washington, New Hampshire. *Arctic and Alpine Research* **10,** 105–16.

Marchand, P.J. and Goulet, F.L. (1984) Wave mortality in subalpine forest ecosystems with particular reference to foliage and root attrition in *Abies balsamea*, White Mountains, New Hampshire (USA). In: Turner, H. and Tranquillini, W. (eds), *Establishment and tending of subalpine forests: research and management.* Bericht 270. Birmensdorf, Eidgenössische Anstalt für das forstliche Versuchswesen, pp. 173–7.

Marchand, P.J., Goulet, F.L. and Harrington, T.C. (1986) Death by attrition: a hypothesis for wave mortality of subalpine *Abies balsamea. Canadian Journal of Forest Research* **16,** 591–6.

Markkola, A.M., Ohtonen, R. and Tarvainen, O. (1990) Peroxidase activity as an indicator of pollution stress in the fine roots of *Pinus sylvestris. Water, Air, and Soil Pollution* **52,** 149–56.

Marschner, B., Stahr, K. and Renger, M. (1989) Potential hazards of lime application in a damaged pine forest ecosystem in Berlin, Germany. *Water, Air, and Soil Pollution* **48,** 45–57.

Marschner, B. and Wilczynski, A.W. (1991) The effect of liming on quantity and chemical composition of soil organic matter in a pine forest in Berlin, Germany. *Plant and Soil* **137,** 229–36.

Marschner, H., Häussling, M. and George, E. (1991) Ammonium and nitrate uptake rates and rhizosphere pH in non-mycorrhizal roots of Norway spruce (*Picea abies* (L.) Karst.). *Trees: Structure and Function* **5,** 14–21.

Marx, D.H. and Shafer, S.R. (1989) Fungal and bacterial symbionts as potential biological markers of effects of atmospheric deposition on forest health. In: Committee on Biologic Markers of Air-Pollution Damage in Trees (eds), *Biologic Markers of Air-Pollution Stress and Damage in Forests.* Washington DC: National Academy Press, pp. 217–32.

Maser, C. (1988) *The redesigned forest.* San Pedro: R. & E. Miles. 234 pp.

Masuch, G., Kicinski, H.G., Dälme, W. and Kettrup, A. (1989) Hydrogen peroxide dissolved in acidic fog as air pollutant – effects on spruce needles. *International Journal of Environmental and Analytical Chemistry* **37,** 161–85.

Materna, J. (1962) Auswertung von Düngungsversuchen in rauchgeschädigten Fichtenbeständen. *Wissenschaftliche Zeitschrift der Technischen Universität Dresden* 11, 589–93.

Materna, J. (1982) Concentration of sulphur dioxide in the air and sulphur content in Norway spruce seedlings (*Picea abies* Karst.). *Communicationes Institutale Forestalis Cechosloveniae* **12,** 137–46.

Materna, J. (1987) Maßnahmen der CSSR gegen das Waldsterben. *Allgemeine Forstzeitung, Wien* **98,** 36–7.

Materna, J. (1988) Monitoring of air pollution effects on forests in the Czech Socialist Republic. In: Moisl, K. and Zlinska, J. (eds), *Workshop on*

Methodologies. Prague: Institute for Environment and Utilization of Natural Resources, pp. 5–10.

Materna, J. (1989) Mineral nutrition of Norway spruce stands in the western part of Czechoslovakia. *Lesnictvi* **35,** 975–82.

Matson, P.A. and Boone, R.D. (1984) Natural disturbance and nitrogen mineralization: wave-form dieback of mountain hemlock in the Oregon Cascades. *Ecology* **65,** 1511–16.

Matson, P.A. and Waring, R.H. (1984) Effects of nutrient and light limitation on mountain hemlock: Susceptibility to laminated root rot. *Ecology* **65,** 1517–24.

Matthews, J. (1955) The influence of weather on the frequency of beech mast years in England. *Forestry* **28,** 107–16.

Mattson, K.G., Arnaut, L.Y., Reams, G.A., Cline, S.P., Peterson, C.E. and Vong, R.J. (1990) *Response of forest trees to sulfur, nitrogen, and associated pollutants.* US Environmental Protection Agency Report Nr. EPA/600/3-90/074. Corvallis: United States Environmental Protection Agency.

Matyssek, R., Günthardt-Georg, M.S., Keller, T. and Scheidegger, C. (1991) Impairment of gas exchange and structure in birch leaves (*Betula pendula*) caused by low ozone concentrations. *Trees: Structure and Function* **5,** 5–13.

Matyssek, R., Günthardt-Georg, M.S., Saurer, M. and Keller, T. (1992) Seasonal growth, $\delta^{13}C$ in leaves and stem, and phloem structure of birch (*Betula pendula*) under low ozone concentrations. *Trees: Structure and Function* **6,** 69–76.

Matzner, E. (1985) Auswirkungen von Düngung und Kalkung auf den Elementumsatz und die Elementverteilung in zwei Waldökosystemen im Solling. *Allgemeine Forst Zeitung* **41,** 1143–7.

Matzner, E. (1992) Factors controlling Al-activity in soil solutions in an acid forest soil of the German Solling area. *Zeitschrift für Pflanzenernährung und Bodenkunde* **155,** 333–8.

Matzner, E., Blanck, K., Hartmann, G. and Stock, R. (1989) Needle chlorosis pattern in relation to soil chemical properties in two Norway spruce (*Picea abies* Karst.) forests of the German Harz Mountains. In: Bucher, J. and Bucher-Wallin, I. (eds), *Air pollution and forest decline.* Birmensdorf: Eidgenössische Anstalt für das forstliche Versuchswesen, pp. 195–9.

Matzner, E., Murach, D. and Fortmann, H. (1986) Soil acidity and its relationship to root growth in declining forest stands in Germany. *Water, Air, and Soil Pollution* **31,** 273–82.

Matzner, E. and Thoma, E. (1983) Auswirkungen eines saisonalen Versauerungsschubes im Sommer/Herbst 1982 auf den chemischen Bodenzustand verschiedener Waldökosysteme. *Allgemeine Forst Zeitschrift* **38,** 677–83.

Matzner, E. and Ulrich, B. (1981) Bilanzierung jährlicher Elementflüsse in Waldökosystemen im Solling. *Zeitschrift für Pflanzenernährung und Bodenkunde* **144,** 660–81.

Matzner, E. and Ulrich, B. (1985) Implications of the chemical soil conditions for forest decline. *Experientia* **41,** 578–84.

Maurice, C.G. and Crang, R.E. (1986) Increase in *Pinus strobus* needle transectional areas in response to acid misting. *Archives for Environmental Contamination and Toxicology* **15,** 77–82.

Maurice, C.G. and Crang, R.E. (1989) Foliar injury and shoot growth in response to prolonged acidic misting of *Pinus strobus* seedlings *Archives for Environmental Contamination and Toxicology* **18,** 277–84.

Mauser, H. (1991) Homogenität und Stabilität des Interpretationsverhaltens bei der einzelbaumweisen Kronenbeurteilung aus Farbinfrarotluftbildern. *Centralblatt für das gesamte Forstwesen* **108,** 315–29.

Mayer, H., König, C. and Rall, A. (1988) Identifikation von Witterungsereignissen mit pflanzenphysiologischer Streßwirkung für Waldbäume. *Forstwissenschaftliches Centralblatt* **107,** 131–40.

Mayhead, G.J. (1973) The effect of altitude above sea level on the Yield Class of Sitka spruce. *Scottish Forestry* **27,** 231–7.

Mazurski, K.R. (1986) The destruction of forests in the Polish Sudetes Mountains by industrial emissions. *Forest Ecology and Management* **17,** 303–15.

Mazurski, K.R. (1990) Industrial pollution: the threat to Polish forests. *Ambio* **19,** 70–4.

McBride, J.R. (1974) Plant succession in the Berkeley Hills. *Madrono* **22,** 317–29.

McCauley, K.J. and Cook, S.A. (1980) *Phellinus weirii* infestation of two mountain hemlock forests in the Oregon Cascades. *Forest Science* **26,** 23–9.

McClaugherty, C.A., Aber, J.D. and Melillo, J.M. (1982) The role of fine roots in the organic matter and nitrogen budgets of two forested ecosystems. *Ecology* **63,** 1481–90.

McClenahen, J.R. (1978) Community changes in a deciduous forest exposed to air pollution. *Canadian Journal of Forest Research* **8,** 432–8.

McColl, J.G. and Firestone, M.K. (1987) Cumulative effects of simulated acid rain on soil chemical and microbial characteristics and conifer seedling growth. *Soil Science Society of America Journal* **51,** 794–800.

McColl, J.G., Monette, L.K. and Bush, D.S. (1982) Chemical characteristics of wet and dry atmospheric deposition in northern California. *Journal of Environmental Quality* **11,** 585–90.

McCool, P.M. (1988) Effect of air pollutants on mycorrhizae. In: Schulte-Hostede, S., Darrall, N.M., Blank, L.W. and Wellburn, A.R. (eds), *Air pollution and plant metabolism.* London: Elsevier, pp. 356–65.

McCormick, L.H. and Steiner, K.C. (1978) Variation in aluminium tolerance among six genera of trees. *Forest Science* **24,** 565–8.

McIlveen, W.D., Rutherford, S.T. and Linzon, S.N. (1986) *A historical perspective of sugar maple decline within Ontario and outside of Ontario.* Report No. ARB-141-86-Phyto. Toronto: Ministry of the Environment. 40 pp.

McIntosh, R. (1984) *Fertiliser experiments in established conifer stands.* Forestry Commission Forest Record 127. Edinburgh: Forestry Commission.

McKay, H.M. (1988) Non-pollutant abiotic factors affecting needle loss. In Cape, J.N. and Mathy, P. (eds), *Scientific basis of forest decline symptomatology.* Air Pollution Report 15. Brussels: Commission of the European Communities, pp. 31–8.

McLaughlin, D.L., Linzon, S.N., Dimma, D.E. and McIlven, W.D. (1985) Sugar maple decline in Ontario. Report No. ARB-144-85-Phyto. Toronto: Ontario Ministry of the Environment.

McLaughlin, S.B., Adams, M.B., Edwards, N.T., Hanson, P.J., Layton, P.A., O'Neill, E.G. and Roy, W.K. (1988) Comparative sensitivity, mechanisms, and whole plant physiological implications of responses of loblolly pine genotypes to ozone and acid deposition. Environmental Sciences Division Publication No. 3105, Oak Ridge National Laboratory (ORNL/TM-10777). Oak Ridge, Tennessee: Oak Ridge National Laboratory.

McLaughlin, S.B., Andersen, C.P., Edwards, N.T., Roy, W.K. and Layton, P.A. (1990) Seasonal patterns of photosynthesis and respiration of red spruce saplings from two elevations in declining southern Appalachian stands. *Canadian Journal of Forest Research* **20,** 485–95.

McLaughlin, S.B., Andersen, C.P., Hanson, P.J., Tjoelker, M.G. and Roy, W.K. (1991) Increased dark respiration and calcium deficiency of red spruce in relation to acidic deposition at high-elevation southern Appalachian Mountain sites. *Canadian Journal of Forest Research* **21,** 1234–44.

McLaughlin, S.B., Blasing, T.J., Mann, L.K. and Duvick, D.N. (1983) Effects of acid rain and gaseous pollutants on forest productivity: a regional scale approach. *Journal of the Air Pollution Control Association* **33,** 1042–9.

McLaughlin, S.B., Downing, D.J., Blasing, T.J., Cook, E.R. and Adams, H.S. (1987) An analysis of climate and competition as contributors to decline of red spruce in high elevation Appalachian forests of the eastern United States. *Oecologia* **72,** 487–501.

McLaughlin, S.B., McConathy, R.K., Duvick, D. and Mann, L.K. (1982) Effects of chronic air pollution stress on photosynthesis, carbon allocation, and growth of white pine trees. *Forest Science* **28,** 60–70.

McLaughlin, S.B. and Tjoelker, M.J. (1992) Growth and physiological changes in red spruce saplings associated with acidic deposition at high elevations in the southern Appalachians, USA. *Forest Ecology and Management* **51,** 43–51.

McLeod, A.R., Holland, M.R., Shaw, P.J.A., Sutherland, P.M., Darrall, N.M. and Skeffington, R.A. (1990) Enhancement of nitrogen deposition to forest trees exposed to SO_2. *Nature* **347,** 277–9.

McQuattie, C.J. and Schier, G.A. (1990) Response of red spruce seedlings to aluminium toxicity in nutrient solution: alterations in root anatomy. *Canadian Journal of Forest Research* **20,** 1001–11.

Mead, D.J. and Will, G.M. (1976) Seasonal and between-tree variation in the nutrient levels in *Pinus radiata* foliage. *New Zealand Journal of Forest Science* **6,** 3–13.

Medlarz, S. (1988) The role of quality assurance in the Spruce–Fir Forest Research Cooperative. In: *Proceedings of the US/FRG Research Symposium: Effects of atmospheric pollutants on the spruce-fir forests of the eastern US and the Federal Republic of Germany, Burlington, Vermont, October 19–23, 1987.* General Technical Report NE-120. Broomall: US Department of Agriculture Forest Service, Northeastern Forest Experiment Station, pp. 61–5.

Medwecka-Kornas, A. and Gawronski, S. (1990) The dieback of fir *Abies alba* Mill. and changes in the Pino-Quercetum stands in the Ojców National Park (Southern Poland). *Vegetatio* **87,** 175–86.

Mehlhorn, H., Francis, B.J. and Wellburn, A.R. (1988) Prediction of the probability of forest decline damage to Norway spruce using three simple site-independent diagnostic parameters. *New Phytologist* **110,** 525–34.

Mehlhorn, H., Seufert, G., Schmidt, A. and Kuhnert, K.J. (1986) Effect of SO_2 and O_3 on production of antioxidants in conifers. *Plant Physiology* **82,** 336–8.

Mehringer, H., Bauch, J. and Frühwald, A. (1988) Holzbiologische Untersuchungen an Buchen aus Waldschadensgebieten. *Holz als Roh- und Werkstoff* **46,** 447–55.

Meier, S., Grand, L.F., Schoeneberger, M.M., Reinert, R.A. and Bruck, R.I. (1990) Growth, ectomycorrhizae and nonstructural carbohydrates of loblolly pine seedlings exposed to ozone and soil water deficit. *Environmental Pollution* **64,** 11–27.

Meier, S., Robarge, W.P., Bruck, R.I. and Grand, L.F. (1989) Effects of simulated rain acidity on ectomycorrhizae of red spruce seedlings potted in natural soil. *Environmental Pollution* **59,** 315–24.

Meiwes, K.J., Khanna, P.K. and Ulrich, B. (1980) Retention of sulphate by an acid brown earth and its relationship with the atmospheric input of sulphur to forest vegetation. *Zeitschrift für Pflanzenernährung und Bodenkunde* **13,** 402–11.

Meiwes, K.J., Khanna, P.K. and Ulrich, B. (1986) Parameters for describing soil acidification and their relevance to the stability of forest ecosystems. *Forest Ecology and Management* **15,** 161–79.

Mejnartowicz, L.E. (1984) Enzymatic investigations on tolerance in forest trees. In: Koziol, M.J. and Whatley, F.R. (eds), *Gaseous pollutants and plant metabolism.* London: Butterworths, pp. 381–98.

Mejstrik, V. (1987) The air SO_2 concentrations, pH of soils and response of spruce forest in Czechoslovakia. In: Perry, R., Harrison, R.M., Bell, J.N.B. and Lester, J.N. (eds), *Acid rain: Scientific and technical advances.* London: Selper, pp. 669–73.

Melillo, J.M. (1981) Nitrogen cycling in deciduous forests. In: Clark, F.E. and Rosswall, T. (eds), *Terrestrial nitrogen cycles. Ecological Bulletin, Stockholm,* **33,** 427–42.

Melin, J. (1986) *Omsättning och fördelning av gödselkväve i tre barrskogsekosystem i Mellansverige.* Swedish University of Agricultural Sciences, Department of Forest Soils, Reports in Forest Ecology and Forest Soils **55,** 122pp.

Melin, J. and Nömmik, H. (1988) Fertilizer nitrogen distribution in a *Pinus sylvestris/Picea abies* ecosystem, central Sweden. *Scandinavian Journal of Forest Research* **3,** 3–15.

Mengel, K., Lutz, H.J. and Breiniger, T.H. (1986) Mineralstoff und Kohlenhydratauswaschung durch saure Nebel aus jungen Fichten. *Kernforschungsanlage Jülich* **369,** 292–8.

Mengel, K., Lutz, H.J. and Breiniger, T.H. (1987) Auswaschung von Nährstoffen durch sauren Nebel aus jungen intakten Fichten (*Picea abies*). *Zeitschrift für Pflanzenernährung und Bodenkunde* **150**, 61–8.

Merk, L., Kloos, M., Schönwitz, R. and Ziegler, H. (1988) Influence of various factors on quantitative composition of leaf monoterpenes of *Picea abies* (L.) Karst. *Trees: Structure and Function* **2**, 45–51.

Messenger, A.S. (1980) Spruce plantation effects on soil moisture and chemical element distribution. *Forest Ecology and Management* **3,** 113–25.

Mettendorf, B., Schröter, H. and Hradetsky, J. (1988) Analysenergebnisse zur Schadensentwicklung auf Tannen- und Fichten-Dauerbeobachtungsflächen in Baden-Württemberg. *Allgemeine Forst- und Jagdzeitung* **159,** 171–7.

Meyberg, M., Lockhausen, J. and Kristen, U. (1988) Ultrastructural changes in mesophyll cells of spruce needles from a declining forest in northern Germany. *European Journal of Forest Pathology* **18,** 169–75.

Meyer, F.H. (1987) Das Wurzelsystem geschädigter Waldbestände. *Allgemeine Forst Zeitschrift* **42,** 754–7.

Meyer, W.H. (1929) *Yields of second-growth spruce and fir in the Northeast.* Technical Bulletin 142. Washington DC: United States Department of Agriculture. 53 pp.

Mickler, R.A. and Medlarz, S.A. (1987) The role of quality assurance in national acid rain research in the United States. *Environmental Technology Letters* **8,** 459–66.

van Miegroet, H. and Cole, D.W. (1985) Acidification sources in red alder and Douglas fir soils – importance of nitrification. *Soil Science Society of America Journal* **49,** 1274–9.

Mielikäinen, K. and Kauppi, P. (1991) Forest resources in Finland and findings on the impacts of air pollution on forests. In: *Impact of the new types of forest damage on the economic situation of forestry enterprises.* Bonn: Bundesministerium für Ernährung, Landwirtschaft und Forsten, pp. 32–40.

Mielke, M.E. (1988) Condition of red spruce in West Virginia. In: *Proceedings of the US/FRG Research Symposium: Effects of atmospheric pollutants on the spruce-fir forests of the eastern United States and the Federal Republic of Germany.* General Technical Report NE-120. Broomall: US Department of Agriculture Forest Service, Northeastern Forest Experiment Station, pp. 219–28.

Mielke, M.E., Rezabek, C.L., Carlson, J.C. and Gillespie, A.J.R. (undated) *Survey to assess the health of sugar maple and other northern hardwoods across a pH gradient in Wisconsin, 1987–1988.* St. Paul: Wisconsin Department of Natural Resources, Madison and Fort Collins: US Department of Agriculture Forest Service.

Mielke, M.E., Soctomach, M.A., Marsden, M.A. and Ciesla, W.M. (1986) *Decline and mortality of red spruce in West Virginia.* Report 86-4. Fort Collins, Colorado: United States Department of Agriculture Forest Service, Forest Pest Management/Methods Application Group.

Mies, E. and Zöttl, H.W. (1985) Zeitliche Änderung der Chlorophyll- und Elementgehalte in den Nadeln eines gelb-chlorotischen Fichtenbestandes. *Forstwissenschaftliches Centralblatt* **104,** 1–8.

Miller, E.K., Huntington, T.G., Johnson, A.H. and Friedland, A.J. (1992) Aluminium in soil solutions from a subalpine spruce-fir forest at Whiteface Mountain, New York. *Journal of Environmental Quality* **21,** 345–52.

Miller, H.G. and Cooper, J.M. (1973) Changes in amount and distribution of stem growth in pole-stage Corsican pine following the application of nitrogen fertilizer. *Forestry* **46,** 157–90.

Miller, H.G. and Miller, J.D. (1988) Response to heavy nitrogen applications in fertilizer experiments in British forests. *Environmental Pollution* **54,** 219–31.

Miller, H.G., Miller, J.D. and Binns, W.O. (1977) Growth of Scots pine under nutritional and climatic stress. *Plant and Soil* **48,** 103–14.

Miller, H.G., Miller, J.D. and Cooper, J.M. (1981) Optimum foliar nitrogen concentration in pine and its change with stand age. *Canadian Journal of Forest Research* **11,** 563–72.

Miller, H.G., Miller, J.D. and Pauline, O.J.L. (1976) Effect of nitrogen supply on nutrient uptake in Corsican pine. *Journal of Applied Ecology* **13,** 955–66.

Miller, P.R. (1973) Oxidant-induced community change in a mixed conifer forest. *Advances in Chemistry Series* **122,** 101–17.

Miller, P.R. (1983) Ozone effects in the San Bernardino National Forest. In: *Proceedings of the Symposium on Air Pollution and the Productivity of the Forest.* Arlington, Virginia: Izaak Walton League, pp. 161–97.

Miller, P.R. (1988) Root and shoot growth during early development of *Sequoiadendron giganteum* seedlings stressed by ozone. *XIV International Botanical Congress, Berlin, Abstract* 4-29-2.

Miller, P.R. (1989) Biomarkers for defining air pollution effects in western coniferous forests. In: Committee on Biologic Markers of Air Pollution Damage in Trees (eds), *Biologic Markers of Air Pollution Stress and Damage in Forests.* Washington DC: National Academy Press, pp. 111–18.

Miller, P.R. and McBride, J.R. (1975) Effects of air pollutants on forests. In: Mudd, J.B. and Kozlowski, T.T. (eds), *Responses of plants to air pollution.* New York: Academic Press, pp. 195–235.

Miller, P.R. and McBride, J.R. (1989) Trends of ozone damage to conifer forests in the western United States, particularly southern California. In: Bucher, J. and Bucher-Wallin, I. (eds), *Air pollution and forest decline.* Birmensdorf: Eidgenössische Anstalt für das forstliche Versuchswesen, pp. 61–8.

Miller, P.R. and Millecan, A.A. (1971) Extent of air pollution damage to some pines and other conifers in California. *Plant Disease Reports* **55,** 555–9.

Miller, P.R., Dunn, P.H., Leininger, T.D., Schilling, S.L., Larson, D.A., Herring, D.C., Heckmann, B.A. and Gomez, A.P. (1988) *Testing the sensitivity of five western conifer species to sulfur dioxide alone, ozone alone, and ozone followed by acid fog.* Report to the Forest Response Program, US Environmental Protection Agency, Corvallis, Oregon.

Miller, P.R., Longbotham, G.J., Van Doren, R.E. and Thomas, M.A. (1979) Effect of chronic oxidant air pollution exposures on California black oak in the San Bernardino Mountains. In: *Proceedings of the Symposium on Ecology, Management and Utilization of California oaks, June 26–28, 1979, Claremont, California.* Forest Service Research Paper. Berkeley: US Department of Agriculture, Pacific Southwest Forest and Range Experiment Station, pp. 220–9.

Miller, P.R., McBride, J.R., Schilling, S.L. and Gomez, A.P. (1989) Trend of ozone damage to conifer forests between 1974 and 1988 in the San Bernardino Mountains of Southern California. In: Olson, R.K. and Lefohn, A.S. (eds), *Effects of air pollution on western forests.* Pittsburgh: Air and Waste Management Association, pp. 309–23.

Miller, P.R., McCutchan, M.H. and Milligan, H.P. (1972) Oxidant air pollution in the Central Valley, Sierra Nevada foothills, and Mineral King Valley of California. *Atmospheric Environment* **6,** 623–33.

Miller, P.R. and Parmeter, J.R. (1967) Effects of ozone injury to ponderosa pine. *Phytopathology* **57,** 822.

Miller, P.R., Parmeter, J.R., Flick, B.H. and Martinez, C.W. (1969) Ozone dosage response to ponderosa pine seedlings. *Journal of the Air Pollution Control Association* **19,** 435–8.

Miller, P.R., Parmeter, J.R., Taylor, O.C. and Cardiff, E.A. (1963) Ozone injury to the foliage of *Pinus ponderosa. Phytopathology* **53,** 1072–6.

Miller, P.R., Taylor, O.C. and Poe, M.P. (1986) Spatial variation of summer ozone concentrations in the San Bernardino Mountains, Paper 86-39.2. In: *Proceedings of the 79th Annual Meeting of the Air Pollution Control Association, June 22–24, Minneapolis, Minnesota,* pp. 1–14.

Miller, P.R. and Van Doren, R.E. (1981) Ponderosa and Jeffrey pine foliage retention indicates ozone dose response. In: *Proceedings of the Symposium on the Dynamics and Management of Mediterranean-type Ecosystems, San Diego, California.* Berkeley: US Department of Agriculture Forest Service, p. 621.

Miller-Weeks, M. and Cooke, R. (1989) *Symptomatology and trend of tree condition of red spruce and balsam fir in the northeastern U.S.* Report to the US Department of Agriculture Forest Service Forest Response Program. Northeastern Forest Experiment Station, Radnor.

Miller-Weeks, M., Millers, I. and Cooke, R. (1988a) Description of crown symptoms on red spruce and balsam fir in the northeastern United States: a progress report. In: *Proceedings of the US/FRG Research Symposium: Effects of atmospheric pollutants on the spruce-fir forests of the eastern United States and the Federal Republic of Germany.* General Technical Report NE-120. Broomall: US Department of Agriculture Forest Service, Northeastern Forest Experiment Station, pp. 199–210.

Miller-Weeks, M., Spruce, J. and Smoronk, D. (1988b) Red spruce and balsam fir mortality mapping in New York, Vermont, New Hampshire and Western Maine: a progress report. In: *Proceedings of the US/FRG Research Symposium: Effects of atmospheric pollutants on the spruce-fir forests of the eastern United States and the Federal Republic of Germany.* General Technical Report NE-120. Broomall: US Department of Agriculture Forest Service, Northeastern Forest Experiment Station, pp. 211–18.

Millers, I., Shriner, D.S. and Rizzo, D. (1989) History of hardwood decline in the eastern United States. General Technical Report NE-126. Broomall: US Department of Agriculture Forest Service, Northeastern Forest Experiment Station.

Millers, I., Allen, D.C. and Lachance, D. (1992) *Sugar maple crown conditions improve between 1988 and 1990.* Radnor: US Department of Agriculture Forest Service, Northeastern Area State and Private Forestry and Ottawa: Forestry Canada.

Milligan, R.H. (1974) Insects damaging beech (*Nothofagus*) forests. *Proceedings of the New Zealand Ecological Society* **21,** 32–40.

Mitchell, A.F., Hallett, V.E. and White, J.E.J. (1990) *Champion trees in the British Isles.* Forestry Commission Field Book 10. London: HMSO.

Mitchell, H.L. and Chandler, R.F. (1939) The nitrogen nutrition and growth of certain deciduous trees of northeastern United States. *Black Rock Forestry Bulletin* 11.

Mitscherlich, G. (1983) Untersuchungen über die Reaktion kranker Alttannen auf die Witterung. *Allgemeine Forst- und Jagdzeitung* **154,** 101–5.

Moehring, K. (1987) Kritische Anmerkung zur Waldschadenserhebung 1986 des Bundesministeriums für Ernährung, Landwirtschaft und Forsten. *Forstarchiv* **58,** 10–11.

Moffat, A.J. and Boswell, R.C. (1990) Effect of tree species and species mixtures on soil properties at Gisburn Forest, Yorkshire. *Soil Use and Management* **6,** 46–51.

Mohnen, V.A., Aneja, V., Bailey, B., Cowling, E., Goltz, S.M., Healey, J.,

Hornig, J., Kadlecek, J.A., Meagher, J., Mueller, S. and Sigmon, J.T. (1990) *An assessment of atmospheric exposure and deposition to high elevation forests in the eastern United States.* Mountain Cloud Chemistry Program, Final Report to the US Environmental Protection Agency, Research Triangle Park, North Carolina.

Mohren, G.M.J., van den Burg, J. and Burger, F.W. (1986) Phosphorus deficiency induced by nitrogen input in Douglas fir in The Netherlands. *Plant and Soil* **95,** 191–200.

Moll, W. (1961) Vergleichende Betrachtungen von Feinwurzel intensität Bodenanalyse und Nadelanalyse eines Standortes. *Der Forst- und Holzwirt* **16,** 477–8.

Möller, G. (1983) Variation of boron concentration in pine needles from trees growing on mineral soil in Sweden and response to nitrogen fertilization. *Communicationes Instituti Forestales Fennica* **116,** 111–15.

Moloney, K.A., Stratton, L.J. and Klein, R.M. (1983) Effects of simulated acidic, metal-containing precipitation on coniferous litter decomposition. *Canadian Journal of Botany* **61,** 3337–42.

Molski, B. and Dmuchowski, W. (1984) Evaluation of air pollution pressure on forest and agricultural areas (mapping on macro regions scale) on the basis of sulphur and fluorine accumulation by the needles of Scots pine. In: *Proceedings of the 13th International Meeting of Specialists in Air Pollution Damages in Forests.* Jiloviste-Strnady: Forestry and Game Management Research Institute.

Momoshima, N. and Bondietti, E.A. (1990) Cation binding in wood: applications to understanding historical changes in divalent cation availability to red spruce. *Canadian Journal of Forest Research* **20,** 1840–9.

Monchaux, P. and Thivolle-Gazat, A. (1990/1991) Correction of magnesium deficiency in a young stand of Norway spruce. *Water, Air, and Soil Pollution* **54,** 595–605.

Moody, B.H. (compiler) (1988) *Forest insect and disease conditions in Canada, 1987.* Ottawa: Forestry Canada.

Moore, R., Warrington, S. and Whittaker, J.B. (1991) Herbivory by insects on oak trees in pure stands compared with paired mixtures. *Journal of Applied Ecology* **28,** 290–304.

Morgenstern, E.K. (1969) Winter drying of red spruce provenances related to introgressive hybridization with black spruce. *Bi-monthly Research Nores, Canadian Forest Service* **25,** 34–6.

Morison, J.I.L. and Spence, R.A. (1989) Warm spring conditions and aphid infestations. *Weather* **44,** 374–80.

Morrison. I.K. (1984) Acid rain. A review of literature on acid deposition effects in forest ecosystems. *Forestry Abstracts* **45,** 483–506.

Mößmer, R. and Mayer, F.-J. (1991) Ergebnisse der Waldzustandserhebung 1991 in Bayern. *Forst und Holz* **46,** 640–3.

Mößmer, R., Tränkner, H., Maier, G., Burgis, M., Grad, M. and Marin, K. (1990) Schwerpunktprogramm 'Erforschung der Wechselwirkungen von Klima- und Waldschäden am Schwerpunktstandort Kalkalpen'. In: *Waldschadenserhebung 1989 – Luftbildauswertung – terrestriche Schadansprache.* Munich: Bayerische Forstliche Versuchs- und Forschungsanstalt.

Mößmer, R., Tränkner, H. and Schuster, M. (1991) Waldschadenserhebungen am Schwerpunktstandort Wank in den Jahren 1987 bis 1990. In: *Proceedings 2. Statusseminar der PBWU zum Forschungsschwerpunkt 'Waldschäden'. 4–6 February, 1991, GSF-Forschungszentrum Neuherberg*. Neuherberg: GSF, pp. 65–76.

Mosteller, F. and Tukey, J.W. (1977) *Data analysis and regression: A second course in statistics*. Reading, MS: Addison-Wesley.

Motta, R. and Quaglino, A. (1989) Sui danni causati dalla fauna selvatica ai popolamenti forestali in alta valle di Susa (TO). *Italia Forestale e Montana* **44,** 241–60.

Mudano, J.E., Allen, H.L. and Kress, L.W. (1992) Stem and foliage elongation of young loblolly pine as affected by ozone. *Forest Science* **38,** 324–35.

Mueller-Dombois, D. (1980) The 'ohi'a dieback phenomenon in the Hawaiian rain forest. In: Cairns, J. (ed.), *The recovery process in damaged ecosystems*. Ann Arbor: Ann Arbor Science, pp. 153–61.

Mueller-Dombois, D. (1981) Spatial variation and succession in tropical island rain forests: a progress report. *Hawaii Botanical Science Paper* 41. Honolulu: University of Hawaii.

Mueller-Dombois, D. (1982) Canopy dieback in indigenous forests of Pacific islands. *Hawaiian Botanical Society Newsletter* **21,** 2–6.

Mueller-Dombois, D. (1983) Canopy dieback and successional processes in Pacific forests. *Pacific Science* **37,** 317–25.

Mueller-Dombois, D. (1985) Ohi'a dieback in Hawaii: 1984 synthesis and evaluation. *Pacific Science* **39,** 150–70.

Mueller-Dombois, D. (1987) Natural dieback in forests. *BioScience* **37,** 575–83.

Mueller-Dombois, D. (1988) Forest decline and dieback – a global perspective. *Trends in Ecology and Evolution* **3,** 310–12.

Mueller-Dombois, D. and Krajina, V.J. (1968) Comparison of east-flank vegetation on Mauna Loa and Mauna Kea, Hawaii. *Recent Advances in Tropical Ecology* **2,** 508–20.

Mueller-Dombois, D., Canfield, J.E., Holt, R.A. and Buelow, G.P. (1983) Tree-group death in North American and Hawaiian forests: a pathological problem or a new problem for vegetation ecology? *Phytocoenologia* **11,** 117–37.

Mueller-Dombois, D., Vitousek, P.M. and Bridges, K.W. (1984) Canopy dieback and ecosystem processes in Pacific forests: A progress report and research proposal. *Hawaiian Botanical Science Paper* 44.

Mueller-Starck, G. (1985) Genetic differences between 'tolerant' and 'sensitive' beeches (*Fagus sylvatica* L.) in an environmentally stressed adult forest stand. *Silva Gentica* **34,** 241–7.

Mueller-Starck, G. (1989) Genetic implications of environmental stress in adult forest stands of *Fagus sylvatica* L. In: Scholz, F., Gregorius, H.-R. and Rudin, D. (eds), *Genetic effects of air pollutants in forest tree populations*. Berlin: Springer Verlag, pp. 127–42.

Muir, P.S. (1992) Effects of ambient acidic rain on cation concentrations in leaves of *Acer saccharum* and *Cornus florida*. *Canadian Journal of Forest Research* **22,** 490–6.

Muir, P.S. and Armentano, T.V. (1988) Evaluating oxidant injury to foliage of *Pinus ponderosa*: a comparison of methods. *Canadian Journal of Forest Research* **18,** 498–505.

Muir, P.S. and Böhm, M. (1989) Cloud chemistry and occurrence in the western United States: A synopsis of current information. In: Olson, R.K. and Lefohn, A.S. (eds), *Effects of air pollution on western forests*. Pittsburgh: Air and Waste Management Association, pp. 73–101.

Muir, P.S. and McCune, B. (1987) Index construction for foliar symptoms of air pollution injury. *Plant Disease* **71,** 558–65.

Mülder, D. (1934) Erfahrungen über das Fichtensterben in einem ostpreußen Revier. *Der Deutsche Forstbeamte* **2,** 736–7.

Mulder, J., van Breemen, N. and Eijck, H.C. (1989) Depletion of soil aluminium by acid deposition and implications for acid neutralization. *Nature* **337,** 247–9.

Müller, E. and Stierlin, H.R. (1990) *Sanasilva-Kronenbilder mit Nadel- und Blattverlustprozenten*. Birmensdorf: Eidgenössische Forschungsanstalt für Wald, Schnee und Landschaft.

Müller-Starck, G. (1985) Genetic differences between 'tolerant' and 'sensitive' beeches (*Fagus sylvatica* L.) in an environmentally stressed adult forest stand. *Silvae Genetica* **34,** 241–7.

Muller-Using, B. and Schlegel, F. (1980) Die Entwicklung von Sämlingen chilenischer Nothofagusarten in einer Beschattungsanlage der Universität Valdivia. *Allgemeine Forst- und Jagdzeitung* **151,** 79–96.

Mullins, J.A. (1983) Heritabilities and gains in volume, diameter, and height for open-pollinated progeny of eastern white pine from the Cumberland Mountains. *Proceedings, North-Central Tree Improvement Conference,* 95–102.

Münch, D., Feger, K.H. and Zöttl, H.W. (1990) Nadelvergilbung, Elementverteilung und Wachstum von Fichten eines Hochlagenstandortes im Südschwarzwald. *Allgemeine Forst- und Jagdzeitung* **161,** 210–18.

Münster-Swendsen, M. (1987a) Sammenhæng imellem nedbr, tilvækst og viklerangreb hos rdgran. *Dansk Skovforenings Tidsskrift* **72,** 41–50.

Münster-Swendsen, M. (1987b) Index of vigour in Norway spruce (*Picea abies* Karst.). *Journal of Applied Ecology* **24,** 551–61.

Murach, D. (1984) Die Reaktion der Feinwurzeln von Fichten (*Picea abies* L. Karst.) auf zunehmende Bodenversauerung. *Göttingen Bodenkundliche Berichte* **77,** 126 pp.

Murach, D. (1988) Effects of soil acidification on root growth in beech (*Fagus sylv.* L.) and spruce (*Picea abies* Karst.) stands and its possible consequences for the shoots. In: Bervaes, J., Mathy, P. and Evers, P. (eds), *Relationship between above and below ground influences of air pollutants on forest trees*. Air Pollution Research Report 16. Brussels: Commission of the European Communities, pp. 148–67.

Murray, J.S. (1954) Two diseases of spruce under investigation in Great Britain. *Forestry* **27,** 54–62.

Murray, J.S. (1957) Top dying of Norway spruce in Great Britain. *Proceedings of the British Commonwealth Forestry Conference,* 2–9.

Murri, M. and Schlaepfer, R. (1987) Zusammenhänge von Kroneneigenschaften und Durchmesser bzw. Grundflächenzuwachs von Fichte auf zwei Gebirgsstandorten. *Forstwissenschaftliches Centralblatt* **106,** 328–40.

Murtha, P.A. (1983) Some air-photo scale effects on Douglas fir damage type interpretation. *Photogrammetric Engineering and Remote Sensing* **49,** 327–35.

Muys, B. (1990) N-excess in the forest: effects and possible measures. *Silva Gandavensis* **55,** 35–42.

NAPAP (National Acid Precipitation Assessment Program) (1990) *Changes in forest health and productivity in the United States and Canada.* Acidic deposition: State of Science and Technology Report 16. Washington DC: NAPAP.

NAPAP (National Acid Precipitation Assessment Program) (1991) *1990 Integrated assessment report.* Washington DC: NAPAP Office of the Director.

Nadelhoffer, K.J., Aber, J.D. and Melillo, J.M. (1984) Seasonal patterns of ammonium and nitrate uptake in nine temperate forest ecosystems. *Plant and Soil* **80,** 321–35.

Nagel, W.P. (1959) *Forest insect conditions in the Southeast during 1958.* Research Paper SE-100. Asheville: US Department of Agriculture Forest Service.

Nageleisen, L.M., Hartmann, G. and Landmann, G. (1991) Dépérissement d'essences feuillues en Europe occidentale: cas particulier des Chênes rouvre et pédonculé. *Proceedings, 10th World Forestry Congress, Paris, 17–26 September 1991.*

Nakane, K. and Kimura, Y. (1992) Assessment of pine forest damage by blight based on Landsat TM data and correlation with environmental factors. *Ecological Research* **7,** 9–18.

Namias, J. (1970) Climatic anomaly over the United States during the 1960s. *Science* **170,** 741–3.

Nassauer, K.-G. (1991) Waldschadenserhebung 1991 in Hessen. *Forst und Holz* **46,** 624–6.

Nebe, W. (1987) Zur Manganernährung der Fichte. *Archiv für Forstwesen* **16,** 109–18.

Nebe, W. and Lucas, A. (1990) Standörtliche Bindung von Immissionsschäden an Fichte im Staatlichen Forstwirtschaftsbetrieb Gotha. *Beiträge für die Forstwirtschaft* **24,** 22–6.

Nebe, W. and Rossbach, T. (1990) Zur Beurteilung von Immissionsbelastungen im Harz durch Nadelanalysen in Fichtenaufwüchsen. *Hercynia* **27,** 82–91.

Needham, R.H. (1960) Problems associated with regeneration of *Eucalyptus gigantea* in the Surrey Hills area. *APPITA* 13, No. 4.

Neger, F.W. (1908) Das Tannensterben in den sächsischen und anderen deutschen Mittelgebirgen. *Tharandter Forstliches Jahrbuch* **58,** 201–25.

Neite, H., Kazda, M. and Paulissen, D. (1992) Schwermetallgehalte in Waldböden Nordrhein-Westfalen. Klassifizierung und kartographische Auswertung. *Zeitschrift für Pflanzenernährung und Bodenkunde* **155,** 217–22.

Nelson, R.F. (1983) Detecting forest canopy change due to insect activity using Landsat MSS. *Photogrammetric Engineering* **50,** 607–17.

Nelson, T.C. and Zillgitt, W.M. (1969) *A forest atlas of the South.* New Orleans: US Department of Agriculture Forest Service, Southern Forest Experiment Station, and Asheville: US Department of Agriculture Forest Service, Southeastern Forest Experiment Station.

Neumann, M. (1989a) Einfluss von Standortsfaktoren auf den Kronenzustand. In: Bucher, J. and Bucher-Wallin, I. (eds), *Air pollution and forest decline.* Birmensdorf: Eidgenössische Anstalt für das forstliche Versuchswesen, pp. 209–14.

Neumann, M. (1989b) Zu Fragen der Waldzustandserfassung durch großräumige Inventuren. *Centralblatt für das gesamte Forstwesen* **106,** 161–78.

Neumann, M. and Stowasser, S. (1987) Waldzustandsinventur: zur Objektivität von Kronenklassifizierungen. In: *Forstliche Bundesversuchsanstalt Wien, Jahresbericht 1986.* Vienna: Forstliche Bundesversuchsanstalt, pp. 101–8.

Neuvonen, S. and Suomela, J. (1990) The effects of simulated acid rain on pine needle and birch leaf litter decomposition. *Journal of Applied Ecology* **27,** 857–72.

Nicholas, N.S., Gregoire, T.G. and Zedaker, S.M. (1991) The reliability of tree crown position classification. *Canadian Journal of Forest Research* **21,** 698–701.

Nicholas, N.S. and Zedaker, S.M. (1989) Ice damage in spruce-fir forests of the Black Mountains, North Carolina. *Canadian Journal of Forest Research* **19,** 1487–91.

Nicholas, N.S., Zedaker, S.M., Eagar, C.C. and Bonner, F.T. (1989a) *Spruce–fir regeneration in the southern Appalachian Mountains, U.S.A.* US Department of Agriculture Forest Service internal report, March 1989.

Nicholas, N.S., Zedaker, S.M. and Eagear, C. (1989b) *A comparison of overstory community structure in three south Appalachian spruce–fir forests.* Report to the US Department of Agriculture Forest Service. Northeastern Forest Experiment Station, Radnor.

Nichols, P.H. (1981) *Spatial analysis of forest growth.* Forestry Commission Occasional Paper 12. Forestry Commission, Edinburgh.

Nieboer, E., MacFarlane, J.D. and Richardson, D.H.S. (1984) Modification of cell buffering capacities by gaseous air pollutants. In: Koziol, M.J. and Whatley, F.R. (eds), *Gaseous air pollutants and plant metabolism.* London: Butterworths, pp. 313–30.

Niehaus, G. (1986) Zur Verbreitung der Kiefern-Chlorosen im niedersächsischen Tiefland. *Der Forst- und Holzwirt* **41,** 435–8.

Niehaus, G. (1989) Terrestrische Zustandserfassung von Kiefernkronen (*Pinus sylvestris*) – ein differenzierter Anspracheschlüssel. *Forst und Holz* **44,** 200–206.

Niehaus, G. (1990) CIR-Luftbild-gestützte Differentialdiagnose spezifischer Symptome an Kiefernkronen. *Forst und Holz* **45,** 83–90.

Nihlgård, B. (1989) Nutrients and structural dynamics of conifer needles in South Sweden 1985–1987. *Meddelelser fra Norsk Institut for Skogforskning* **42,** 157–65.

Nihlgård, B. (1990) Relationship of forest damage to air pollution in the Nordic countries. *Agricultural and Forest Meteorology* **50,** 87–98.

Nihlgård, B., Schlyter, P. and Stjernquist, I. (1985) Luftförorreningsskador på bok i Sydsverige. *Skogen* **7,** 20–2.

Nilsson, S. (1974) Tillväxtförluster hos tall vid angrepp av märgborrar. Institutionen för skogsteknik, Skogshögskolan, Garpenberg, Sweden. Report No. 78.

Nilsson, J. (1986) *Critical loads for nitrogen and sulphur.* Report 1986: 11. Nordic Council of Ministers, Copenhagen.

Nilsson, S. (1991) *Possible market effects of forest decline attributed to air pollutants in Europe.* Report ECE/TIM/57, 29 pp. Geneva: United Nations Economic

Commission for Europe, and Rome: Food and Agriculture Organization of the United Nations.

Nilsson, S. (1992) Economic impacts of forest decline caused by air pollutants in Europe. In: De Steiguer, J.E. (ed.), *The economic impact of air pollution on timber markets. Studies from North America and Europe.* General Technical Report SE-75. Asheville: US Department of Agriculture Forest Service, pp. 27–38.

Nilsson, S.I. (1983) Effects on soil chemistry as a consequence of proton input. In: Ulrich, B. and Pankrath, J. (eds), *Effects of accumulation of air pollutants in forest ecosystems.* Dordrecht: D. Reidel, pp. 105–11.

Nilsson, J. and Grennfelt, P. (1988) *Critical loads for sulphur and nitrogen.* Report 1988: 15. Nordic Council of Ministers, Copenhagen.

Nilsson, S.I., Miller, H.G. and Miller, J.D. (1982) Forest growth as a possible cause of soil and water acidification: an examination of the concepts. *Oikos* **39,** 40–9.

Nimman, B. and Knigge, W. (1989) Anatomische Holzeigenschaften und Lagerungsverhalten von Kiefern aus immissionsbelasteten Standorten der Norddeutschen Tiefebene. *Forstarchiv* **60,** 78–83.

Nodvin, S.C., Driscoll, C.T. and Likens, G.E. (1986) The effect of pH on sulphate adsorption by a forest soil. *Soil Science* **142,** 69–75.

Nohrstedt, H.-Ö. (1989) Some changes of water and chemical properties following forest fertilization. *Meddelelser fra Norsk Institut for Skogforskning* **42,** 167–76.

Nohrstedt, H.Ö. (1990) Effects of repeated nitrogen fertilization with different doses on soil properties in a *Pinus sylvestris* stand. *Scandinavian Journal of Forest Research* **5,** 3–15.

Norby, R.J. (1989) Foliar nitrate reductase: A marker for assimilation of atmospheric nitrogen oxides. In: Committee on Biologic Markers of Air Pollution Damage in Trees (eds), *Biologic Markers of Air Pollution Stress and Damage in Forests.* Washington DC: National Academy Press, pp. 245–50.

Norby, R.J., Weerasuriya, Y. and Hanson, P.J. (1989) Induction of nitrate reductase activity in red spruce needles by NO_2 and HNO_3 vapour. *Canadian Journal of Forest Research* **19,** 889–96.

Norby, R.J., Gunderson, C.A., Wullschleger, S.D., O'Neill, E.G. and McCracken, M.K. (1992) Productivity and compensatory responses of yellow poplar trees in elevated CO_2. *Nature* **357,** 322–4.

Nordstrom, D.K. (1982) The effect of sulfate on aluminium concentrations in natural waters: Some stability relations in the systems Al_2O_3-SO_3-H_2O at 298 K. *Geochimica et Cosmochimica Acta* **46,** 681–92.

Norton, S.A. (1976) Changes in chemical processes in soils caused by acid precipitation. In: Dochinger, L.S. and Seliga, T.A. (eds), *Proceedings of the first international symposium on acid precipitation and the forest ecosystem.* General Technical Report NE-23. Upper Darby, Pennsylvania: US Department of Agriculture Forest Service, pp. 711–24.

Norton, D.A. (1984a) Tree-growth–climate relationships in subalpine *Nothofagus* forests, South Island, New Zealand. *New Zealand Journal of Botany* **22,** 471–81.

Norton, D.A. (1984b) Phenological growth characteristics of *Nothofagus solandri* trees at three altitudes in the Craigieburn Range, New Zealand. *New Zealand Journal of Botany* **22,** 411–22.

Norton, D.A. (1984c) A dendrochronological study of *Nothofagus solandri* tree growth along an elevational gradient, South Island, New Zealand. In: Turner, H. and Tranquillini, W. (eds), *Establishment and Tending of subalpine forest: research and management.* Birmensdorf: Eidgenössische Anstalt für das forstliche Versuchswesen, pp. 159–71.

Nykvist, N. and Rosén, K. (1985) Effect of clear-felling and slash removal on the acidity of northern coniferous soils. *Forest Ecology and Management* **11,** 157–69.

Nykvist, N. and Skyllberg, U. (1989) The spatial variation of pH in the mor layer of some coniferous forest stands in northern Sweden. *Scandinavian Journal of Forest Research* **4,** 3–11.

Nys, C. (1981) Modifications des characteristiques physico-chemiques d'un sol brun acide des Ardennes primaires par la monoculture d'Epicéa commun. *Annales des Sciences Forestières* **38,** 237–58.

Nys, C. (1989) Fertilisation, dépérissement et production de l'épicéa commun (*Picea abies*) dans les Ardennes. *Revue forestière française* **41,** 336–47.

Oberlander, G.T. (1956) Summer fog precipitation on the San Francisco peninsula. *Ecology* **37,** 851–2.

Ogden, J. (1970) Studies on the vegetation of Mount Colenso, New Zealand. 2. The population dynamics of red beech. *Proceedings of the New Zealand Ecological Society* **18,** 66–75.

Ogden, J. (1976) Notes on the influence of drought on the bush remnants of the Manawatu Lowlands. *Proceedings of the New Zealand Ecological Society* **23,** 92–8.

Ogden, J. (1985a) An introduction to plant demography with special reference to New Zealand trees. *New Zealand Journal of Botany* **23,** 751–72.

Ogden, J. (1985b) Past, present and future: studies on the population dynamics of some long lived trees. In: White, J. (ed.), *Studies in plant demography: A festschrift for John L. Harper.* New York: Academic Press, pp. 3–16.

Ogden, J. (1988) Forest dynamics and stand-level dieback in New Zealand's *Nothofagus* forests. *GeoJournal* **17,** 225–30.

Ohman, J.H. (1969) Diseases. In: *Proceedings, Sugar Maple Conference, 20–22 August 1968, Houghton, Michigan.* Washington DC: US Department of Agriculture Forest Service, pp. 37–51.

Ohmann, L.F. and Grigal, D.F. (1990) Spatial and temporal patterns of sulfur and nitrogen in wood of trees across the north central United States. *Canadian Journal of Forest Research* **20,** 508–13.

Oka, S. (1983) Reconstruction on the distribution of the Shimagare phenomenon in Japan. *Journal of Geography (Tokyo)* **92,** 219–34.

Okubo, K. (1933) On Mt. Shimagare. *Japanese Journal of Forestry* **15,** 652–68.

Oleksyn, J. (1988) Height growth of different European Scots pine *Pinus sylvestris* L. provenances in a heavily polluted and a control environment. *Environmental Pollution* **55,** 289–99.

Oleksyn, J. (1989) Provenance differentiation as a factor in susceptibility of Scots pine to air pollution. In: Bucher, J.B. and Bucher-Wallin, I. (eds), *Air pollution and forest decline.* Birmensdorf: Eidgenössische Anstalt für das forstliche Versuchswesen, pp. 329–35.

Oleksyn, J. and Bialobok, S. (1986) Net photosynthesis, dark respiration and susceptibility to air pollution of 20 European provenances of Scots pine *Pinus sylvestris* L. *Environmental Pollution* Series A, **40,** 287–302.

Oliver, C.D. (1981) Forest development in North America following major disturbances. *Forest Ecology and Management* **3,** 153–68.

Olsson, M.T. (1986) Long-term impact of lime on the humus form in a beech stand on sandy till. *Studia Forestalia Suecica* **124,** 12 pp.

Oosterbaan, A. and Borgesius, J.J. (1986) Sterfte bij zomereik 1984/1985. *Rapport Rijksinstitut voor Onderzoek in de Bos- en Landschapsbouw 'De Dorschkamp', Wageningen,* Nr. 464.

Oosterbaan, A. and Nabuurs, G.J. (1991) Relationships between oak decline and groundwater class in The Netherlands. *Plant and Soil* **136,** 87–93.

Oosting, H.J. and Billings, W.D. (1951) A comparison of virgin spruce–fir forests in the northern and southern Appalachian system. *Ecology* **32,** 84–103.

Oosting, H.J. and Reed, J.F. (1944) Ecological composition of pulpwood forests in Northwestern Maine. *American Midland Naturalist* **31,** 182–210.

Oren, R., Schulze, E.-D., Matyssek, R. and Zimmermann, R. (1986) Estimating photosynthetic rate and annual carbon gain in conifers from specific leaf weight and leaf biomass. *Oecologia* **70,** 187–93.

Oren, R., Schulze, E.-D., Schneider, U., Werk, K.S. and Meyer, J. (1988a) Carbon assimilation and nutrition of trees in two *Picea abies* (Karst.) stands in Bavaria exhibiting apparent decline differences. In: Mathy, P. (ed.), *Air pollution and ecosystems.* Dordrecht, D. Riedel, pp. 919–22. V

Oren, R., Schulze, E.-D., Werk, K.S. and Meyer, J. (1988b) Performance of two *Picea abies* (L.) Karst. stands at different stages of decline. VII. Nutrient relations and growth. *Oecologia* **77,** 163–73. V

Oren, R., Werk, K.S., Schulze, E.-D., Meyer, J., Schneider, B.U. and Schramel, P. (1988c) Performance of two *Picea abies* (L.) Karst. stands at different stages of decline. VI. Nutrient concentration. *Oecologia* **77,** 151–62.

Oren, R., Schulze, E.-D., Werk, K.S., Meyer, J., Schneider, B.U. and Heilmeier, H. (1988d) Performance of two *Picea abies* (L.) Karst. stands at different stages of decline. I. Carbon relations and stand growth. *Oecologia* **75,** 25–37.

Oren, R. and Zimmermann, R. (1989) CO_2 assimilation and the carbon balance of healthy and declining Norway spruce stands. In: Schulze, E.-D., Lange, O.L. and Oren, R. (eds), *Forest decline and air pollution. A study of spruce (*Picea abies*) on acid soils.* Berlin: Springer Verlag, pp. 352–69.

Orlov, A.Y. (1980) Dynamics of needle mass in pine stands. *Lesovedenie* 1980, 34–41.

Orthofer, R. and Kienzl, K. (1990) Acidic precipitation and forest damage research in Austria. In: Bresser, A.H.M. and Salomons, W. (eds), *Acidic precipitation. Volume 5: International overview and assessment.* New York: Springer Verlag, pp. 107–38.

Osonubi, O., Oren, R., Werk, K.S., Schulze, E.D., Heilmeier, H. (1988) Performances of two *Picea abies* (L.) Karst. stands at different stages of decline. IV. Xylem sap concentrations of magnesium, calcium, potassium and nitrogen. *Oecologia* **77,** 1–6.

Oßwald, W.F. and Elstner, E.F. (1987) Vergleichende Untersuchungen der

Fichtenerkrankungen in den bayerischen Mittelgebirgen. *Allgemeine Forst Zeitschrift* **42,** 693–6.

Oßwald, W.F., Senger, H. and Elstner, E.F. (1987) Ascorbic acid and glutathione contents of spruce needles from different locations in Bavaria. *Zeitschrift für Naturforschung* **42c,** 879–84.

Oßwald, W.E., Senger, H. and Schrammel, P. (1990) Vergleich der Ascorbinsäure-, Glutathion-, Zucker- und Ionengehalte gesunder und geschädigter Fichten des Standortes Laugna nordwestlich von Augsburg. *Forstwissenschaftliches Centralblatt* **109,** 181–9.

Ouimet, R. and Fortin, J.-M. (1992) Growth and foliar nutrient status of sugar maple: incidence of forest decline and reaction to fertilization. *Canadian Journal of Forest Research* **22,** 699–706.

Paffrath, D. and Peters, W. (1988) Betrachtung der Ozonvertikalverteilung im Zusammenhang mit den neuartigen Waldschäden. *Forstwissenschaftliches Centralblatt* **107,** 152–9.

Paganelli, D.J., Seiler, J.R. and Feret, P.P. (1987) Root regeneration as an indicator of aluminium toxicity in loblolly pine. *Plant and Soil* **102,** 115–18.

Paganucci, L. (1989) Sperimentazione sulla densità di impianto delle abetine di Vallombrosa. *Italia Forestale e Montana* **44,** 30–44.

Page, G. (1968) Some effects of conifer crops on soil properties. *Commonwealth Forestry Review* **47,** 52–62.

Påhlsson, A.M.B. (1989) Mineral nutrients, carbohydrates and phenolic compounds in leaves of beech (*Fagus sylvatica* L.) in southern Sweden as related to environmental factors. *Tree Physiology* **5,** 485–95.

Pain, C.F. (1971) Rapid mass movement under forest and grass in the Hunua Ranges, New Zealand. *Australian Geographical Studies* **9,** 77–84.

Pain, C.F. (1972) Characteristics and geomorphic effects of earthquake-initiated landslides in the Adelbert Range, Papua-New Guinea. *Engineering Geology* **6,** 261–74.

Pain, C.F. and Bowler, J.M. (1973) Denudation following the November 1970 earthquake at Madang, Papua-New Guinea. *Zeitschrift für Geomorphologie, Supplementband* **18,** 92–104.

Päivinen, R., Nousiainen, M. and Korhonen, K.T. (1992) *Puutunnusten mittaamisen luotettavuus.* Folia Forestalia 787.

Pallant, E. and Riha, S.J. (1990) Surface soil acidification under red pine and Norway spruce. *Soil Science Society of America Journal* **54,** 1124–30.

Palta, J. and Li, P. (1978) Cell membrane properties in relation to freezing injury. In: Li, P. and Sakai, A. (eds), *Plant cold hardiness and freezing stress.* New York: Academic Press, pp. 93–115.

Palzer, C. (1985) Crown symptoms of regrowth dieback. *Pacific Science* **34,** 465–70.

Papp, R.P., Kliejunas, J.T., Smith, R.S. and Scharpf, R.F. (1979) Association of *Plagithmysus bilineatus* (Coleoptera: Cerambycidae) and *Phytophthora cinnamomi* with the decline of ohia-lehua forests on the islands of Hawaii. *Forest Science* **25,** 187–96.

Paré, D. and Brazeau, B. (1989) Origin of the phosphorus deficiency observed in declining sugar maple stands in the Quebec Appalachians. *Canadian Journal of Forest Research* **19,** 24–34.

Parker, G.G. (1990) Evaluation of dry deposition, pollutant damage, and forest health with throughfall studies. In: Lucier, A.A. and Haines, S.G. (eds), *Mechanisms of forest response to acidic deposition.* New York: Springer Verlag, pp. 10–61.

Parker, M.L. (1987) Recent abnormal increase in tree ring widths: a possible effect of elevated atmospheric carbon dioxide. In: Jacoby, G.C. and Hornbeck, R.W. (eds), *Proceedings of the international symposium on ecological aspects of tree ring analysis.* Springfield: US Department of Commerce, National Technical Information Service, pp. 511–21.

Parmeter, J.R., Bega, R.V. and Neff, T. (1962) A chlorotic decline of ponderosa pine in southern California. *Plant Disease Reporter* **46,** 269–73.

Parmeter, J.R. and Miller, P.R. (1968) Studies relating to the cause of decline and death of ponderosa pine in southern California. *Plant Disease Reporter* **52,** 707–11.

Pastor, J., Gardner, R.H., Dale, V.H. and Post, W.M. (1987) Successional changes in nitrogen availability as a potential factor contributing to spruce declines in boreal North America. *Canadian Journal of Forest Research* **17,** 1394–400.

Pasuthová, J. (1981) Pufracní hapacita jehlici smrku jako indikatór rezistence k pusobení SO_2. *Práce Vyzkumného Ustavu Lesního Hospodárství a Myslivosti* **58,** 139–51.

Patel, J.D. and Devi, G.S. (1986) Variations in chloroplasts of leaf mesophyll cells of *Syzygium cumini* L. and *Tamarindus indica* L. growing under pollution stress of a fertilizer complex. *Indian Journal of Ecology* **13,** 1–4.

Paternak, P.S., Voron, V.P., Pristupa, G.K. and Mazepa, V.G. (1987) Trials of mineral fertilizers in Scots pine stands suffering from atmospheric pollution in the western Poles'e of the Ukraine. *Lesovodstvo i Agrolesomelioratsiya* **75,** 10–14.

Paterson, E., Goodman, B.A. and Farmer, V.C. (1991) The chemistry of aluminium, iron and manganese oxides in acid soil. In: Ulrich, B. and Sumner, M.E. (eds), *Soil acidity.* Berlin: Springer-Verlag, pp. 97–124.

Patterson, M.T. and Rundel, P.W. (1989) Seasonal physiological responses of ozone stressed Jeffrey pine in Sequoia National Park, California. In: Olson, R.K. and Lefohn, A.S. (eds), *Effects of air pollution on western forests.* Pittsburgh: Air and Waste Management Association, pp. 419–27.

Patton, R.L. and Garraway, M.O. (1986) Ozone-induced necrosis and increased peroxidase activity in hybrid poplar (*Populus* sp.) leaves. *Environmental and Experimental Botany* **26,** 137–41.

Patton, R.L., Jensen, K.F. and Schier, G.A. (1991) Responses of red spruce seedlings to ozone and acid deposition. *Canadian Journal of Forest Research* **21,** 1354–59.

Pauley, E.F. and Clebsch, E.E.C. (1990) Patterns of *Abies fraseri* regeneration in a Great Smoky Mountains spruce–fir forest. *Bulletin of the Torrey Botanical Club* **117,** 375–81.

Pavan, M.A. and Bingham, F.T. (1982) Toxicity of aluminium to coffee seedlings grown in nutrient solution. *Soil Science Society of America Journal* **46,** 993–7.

Pavan, M.A., Bingham, F.T. and Pratt, P.F. (1982) Toxicity of aluminium to coffee in ultisols and oxisols amended with $CaCO_3$, $MgCO_3$ and $CaSO_4.2H_2O$. *Soil Science Society of America Journal* **46,** 1201–7.

Payette, S., Filion, L., Gauthier, L. and Boutin, Y. (1985) Secular climatic change in old-growth tree-line vegetation of northern Quebec. *Nature* **315,** 135–8.

Payton, I.J. (1989) Fungal (*Sporothrix*) induced mortality of kamahi (*Weinmannia racemosa*) after attack by pinhole borer (*Platypus* spp.). *New Zealand Journal of Botany* **27,** 359–68.

Peart, D.R., Conkey, L.E., Smith, W.H., Knight, F.B., Keifer, M. and Grosman, D.M. (1988) Condition of the spruce–fir forest at Mount Moosilauke, New Hampshire. In: *Proceedings of the US/FRG Research Symposium: Effects of atmospheric pollutants on the spruce–fir forests of the eastern United States and the Federal Republic of Germany.* General Technical Report NE-120. Broomall: US Department of Agriculture Forest Service, Northeastern Forest Experiment Station, pp. 173–82.

Peart, D.R., Jones, M.B. and Palmiotto, P.A. (1991) Winter injury to red spruce at Mount Moosilauke, New Hampshire. *Canadian Journal of Forest Research* **21,** 1380–9.

Peart, D.R., Poage, N.J. and Jones, M.B. (1992) Winter injury to subalpine spruce: influence of prior vigor and effects on subsequent growth. *Canadian Journal of Forest Research* **22,** 888–92.

Pedersen, B.S. (1989) Ozone injury to Jeffrey and ponderosa pines surrounding Lake Tahoe, California and Nevada. In: Olson, R.K. and Lefohn, A.S. (eds), *Effects of air pollution on western forests.* Pittsburgh: Air and Waste Management Association, pp. 279–92.

Pedersen, B.S. and Cahill, T.A. (1989) Ozone at a remote, high-altitude site in Sequoia National Park, California. In: Olson, R.K. and Lefohn, A.S. (eds), *Effects of air pollution on western forests.* Pittsburgh: Air and Waste Management Association, pp. 207–20.

Pekelharing, C.J. (1979) Fluctuation in opossum populations along the north bank of the Taramakau catchment and its effect on the forest canopy. *New Zealand Journal of Forestry Science* **9,** 212–24.

Pekelharing, C.J. and Reynolds, R.N. (1983) Distribution and abundance of browsing mammals in Westland National Park in 1978, and some observations on their impact on the vegetation. *New Zealand Journal of Forestry Science* **13,** 247–65.

Pellett, N. and White, D.B. (1969a) Effect of soil nitrogen and soil moisture levels on the cold acclimation of container grown *Juniperus chinensis* 'Hetzi'. *Journal of the American Horticultural Society* **94,** 457–9.

Pellett, N. and White, D.B. (1969b) Relationship of seasonal tissue changes to cold acclimation of *Juniperus chinensis* 'Hetzi'. *Journal of the American Horticultural Society* **94,** 460–62.

Penka, M. and Cervená, M. (1985) Influence of SO_2 on peroxidase activity and chlorophyll content in adult individuals of Norway spruce (*Picea abies* (L.) Karsten). *Acta Universitatis Agriculturae (Brno) Series C (Facultas Silviculturae)* **54,** 35–45.

Percy, K.E. and Baker, E.A. (1990) Effects of simulated acid rain on epicuticular wax production, morphology, chemical composition and on cuticular membrane thickness in two clones of Sitka spruce [*Picea sitchensis* (Bong.) Carr.] *New Phytologist* **116,** 79–87.

Perera, W.R.H. (1978) Thotupolakanda – an environmental disaster? *Sri Lanka Forester* **13,** 53–5.

Perkins, T.D. (1988) Changes in the montane spruce–fir forest of Camels Hump, Vermont. In: *Proceedings of the US/FRG Research Symposium: Effects of atmospheric pollutants on the spruce-fir forests of the eastern United States and the Federal Republic of Germany.* General Technical Report NE-120. Broomall: US Department of Agriculture Forest Service, Northeastern Forest Experiment Station, pp. 279–84.

Perkins, T.D., Klein, R.M., Badger, G.J. and Easter, M.J. (1992) Spruce–fir decline and gap dynamics on Camels Hump, Vermont. *Canadian Journal of Forest Research* **22,** 413–22.

Perkins, T.D., Vogelmann, H.W. and Klein, R.M. (1987) Changes in light intensity and soil temperature as a result of forest decline on Camels Hump, Vermont. *Canadian Journal of Forest Research* **17,** 565–8.

Perpet, M. (1988) Zur Differentialdiagnose bei der Waldschadenserhebung auf Buchenbeobachtungsflächen. *Allgemeine Forst- und Jagdzeitung* **159,** 108–13.

Perring, F.H. and Walters, S.N. (1962) *Atlas of the British flora.* London: Nelson.

Persson, H.A. (1983) The distribution and productivity of fine roots in boreal forests. *Plant and Soil* **71,** 87–101.

Petersen, L. (1986) Effects of acid deposition on soil and sensitivity of the soil to acidification. *Experientia* **42,** 340–4.

Peterson, D.L. and Arbaugh, M.J. (1988a) An evaluation of the effects of ozone injury on radial growth of Ponderosa pine (*Pinus ponderosa*) in the southern Sierra Nevada. *Journal of the Air Pollution Control Association* **38,** 921–7.

Peterson, D.L. and Arbaugh, M.J. (1988b) *Growth trends in the mixed conifer forest of the Sierra Nevada.* Final report to the Western Conifers Research Cooperative of the Forest Response Program. Corvallis: US Environmental Protection Agency. 100 pp.

Peterson, D.L. and Arbaugh, M.J. (1990) Growth trends of whitebark pine and lodgepole pine in a subalpine Sierra Nevada forest. *Arctic and Alpine Research* **22,** 233–43.

Peterson, D.L., Arbaugh, M.J. and Robinson, L.J. (1989) The effects of ozone stress on tree growth and vigour in the Sierra Nevada of California, USA. In: Bucher, J.B. and Bucher-Wallin, I. (eds), *Air pollution and forest decline.* Birmensdorf: Eidgenössische Anstalt für das forstliche Versuchswesen, pp. 289–94.

Peterson, D.L., Arbaugh, M.J. and Robinson, L.J. (1991) Regional growth trends of ozone-stressed ponderosa pine (*Pinus ponderosa*) in the Sierra Nevada, California, USA. *The Holocene* **1,** 50–61.

Peterson, D.L., Arbaugh, M.J., Wakefield, V.A. and Miller, P.R. (1987a) Evidence of growth reduction in ozone-injured Jeffrey pine (*Pinus jeffreyi* Grev. and Balf.) in Sequoia and Kings Canyon National Parks. *Journal of the Air Pollution Control Association* **37,** 906–12.

Peterson, D.L., Spanner, M.A., Running, S.W. and Teuber, K.B. (1987b) Relationships of Thematic Mapper simulator data to leaf area index of temperate coniferous forests. *Remote Sensing of the Environment* **22,** 323–41.

Petteys, E.Q.P., Burgan, R.E. and Nelson, R.E. (1975) *Ohia forest decline: its spread and severity in Hawaii.* Research Paper PSW-105. Berkeley, California:

US Department of Agriculture Forest Service, Pacific South West Forest and Range Experiment Station. 11 pp.

Philipson, J.J. (1990) Prospects for enhancing flowering of conifers and broadleaves of potential silvicultural importance in Britain. *Forestry* **63,** 223–40.

Phillips, S.O., Skelly, J.M. and Burkhart, H.E. (1977) Eastern white pine exhibits growth retardation by fluctuating air pollutant levels: interaction of rainfall, age, and symptom expression. *Phytopathology* **67,** 721–5.

Phillipson, J. and Thompson, D.J. (1983) Phenology and intensity of pyllophage attack on *Fagus sylvatica* in Wytham Woods, Oxford. *Ecological Entomology* **8,** 315–30.

Phipps, R.L. and Whiton, J.C. (1988) Decline in long-term growth trends of white oak. *Canadian Journal of Forest Research* **18,** 24–32.

Pielou, E.C. (1963) Runs of healthy and diseased trees in transects through an infected forest. *Biometrics* **19,** 603–14.

Piene, H. (1989) Spruce budworm defoliation and growth loss in young balsam fir: recovery of growth in spaced stands. *Canadian Journal of Forest Research* **19,** 1616–24.

Pietiläinen, P., Poikolainen, J. and Lähdesmäki, P. (1991) Long-term monitoring of nitrate reductase activity in the needles of *Pinus sylvestris* in the context of environmental temperature and ground frost as an indication of nitrogen balance in N. Finland. *Annales Botanici Fennici* **28,** 131–4.

Pinkerton, J.E. and Lefohn, A.S. (1987) The characterization of ozone data for sites located in forested areas of the eastern US. *Journal of the Air Pollution Control Association* **37,** 1005–10.

Pisano, V.E. (1975) Caracteristicas de la biota magallánica derivadas de factores especiales. *Anales del Instituto de Patagonia* **6,** 123–37.

Pitblado, J.R. and Amiro, B.D. (1982) LANDSAT mapping of the industrially disturbed vegetation communities of Sudbury, Canada. *Canadian Journal of Remote Sensing* **8,** 17–28.

Podger, F.D. (1972) *Phytopthora cinnamomi* – a cause of lethal disease in indigenous plant communities in western Australia. *Phytopathology* **62,** 972–81.

Podger, F.D., Kile, G.A., Bird, T., Turnbull, C.R.A. and McLeod, D.E. (1980) An unexplained decline in some forests of *Eucalyptus obliqua* and *E. regnans* in southern Tasmania. *Australian Forest Research* **10,** 53–70.

Polle, A., Chakrabarti, K., Schürmann, W. and Rennenberg, H. (1990) Composition and properties of hydrogen peroxide decomposing systems in extracellular and total extracts from needles of Norway spruce (*Picea abies* L., Karst.). *Plant Physiology* **94,** 312–19.

Polle, A., Krings, B. and Rennenberg, H. (1989) Superoxide dismutase activity in needles of Norwegian spruce trees (*Picea abies* L.). *Plant Physiology* **90,** 1310–15.

Polle, A. and Rennenberg, H. (1989) Defense against oxidative damage in needles of spruce trees (*Picea abies,* L.) growing in an alpine mountain forest (Wank). In: *International congress on forest decline research: State of knowledge and perspectives,* Poster abstracts, vol. 1. Bonn: Forschungsbeirat Waldschäden/Luftverunreinigungen der Bundesregierung und der Länder, pp. 421–2.

Polle, A. and Rennenberg, H. (1992) Field studies on Norway spruce trees at high altitudes: II. Defence systems against oxidative stress in needles. *New Phytologist* **121,** 635–42.

Polomski, J., Flühler, H. and Blaser, P. (1982a) Fluoride-induced mobilization and leaching of organic matter, iron, and aluminium. *Journal of Environmental Quality* **11,** 452–6.

Polomski, J., Flühler, H. and Blaser, P. (1982b) Accumulation of airborne fluoride in soils. *Journal of Environmental Quality* **11,** 457–61.

Pomerleau, R. (1944) Observations sur quelques maladies non-parasitaires des arbres dans le Québec. *Canadian Journal of Research,* Section C **22,** 171–89.

Pomerleau, R. (1962) *Severe winter browning of red spruce in southeastern Quebec.* Bi-monthly Reports of the Canadian Department of Forests: Entomology and Pathology 18, 3. C.

Powell, D.S. and Dickson, D.R. (1984) *Forest statistics for Maine – 1971 and 1982.* Bulletin NE-81. Broomall: US Department of Agriculture Forest Service.

van Praag, H.J., Sougnez-Remy, S., Weissen, F. and Carletti, G. (1988) Root turnover in a beech and a spruce stand of the Belgian Ardennes. *Plant and Soil* **105,** 87–103.

van Praag, H.J. and Weissen, F. (1986) Foliar mineral composition, fertilization and dieback of Norway spruce in the Belgian Ardennes. *Tree Physiology* **1,** 169–76.

Pretzsch, H. (1989) Untersuchungen an kronengeschädigten Kiefern (*Pinus sylvestris* L.) in Nordost-Bayern. Teil 1: Krankheitsbild und Wuchsverhalten. *Forstarchiv* **60,** 62–9.

Priehäusser, G. (1943) Über Fichtenwurzelfäule, Kronenform und Standort. Beitrag zur Kenntis der Fichtenrotfäule. *Forstwissenschaftliches Centralblatt* **65,** 259–73.

Prinz, B. and Krause, G.H.M. (1989) State of scientific discussion about the causes of the novel forest decline in the Federal Republic of Germany and surrounding countries. In: Bucher, J. and Bucher-Wallin, I. (eds), *Air pollution and forest decline.* Birmensdorf: Eidgenössische Anstalt für das forstliche Versuchswesen, pp. 27–34.

Prinz, B., Krause, G.H.M. and Jung, K.D. (1985) Untersuchungen der LIS Essen zur Problematik der Waldschäden. In: Kortzfleisch, V. (ed.), *Waldschäden – Theorie und Praxis auf der Suche nach Antworten.* Munich: Oldenbourg, pp. 143–94.

Prinz, B., Krause, G.H.M. and Stratmann, H. (1982) *Waldschäden in der Bundesrepublik Deutschland.* LIS-Berichte **28,** 54 pp.

Pronos, J. and Vogler, D.R. (1981) *Assessment of ozone injury to pines in the southern Sierra Nevada, 1979/1980.* US Department of Agriculture Forest Service Pacific and Southwest Region Forest Pest Management Report 81–20, 13 pp.

Pronos, J., Vogler, D.R. and Smith, R.S. (1978) *An evaluation of ozone injury to pines in the southern Sierra Nevada.* US Department of Agriculture Forest Service Pacific and Southwest Region Forest Pest Management Report 78–1, 17 pp.

Puls, J. and Rademacher, P. (1986) Chemische Untersuchungen an Fichten aus Waldschadensgebieten. *Holz als Roh- und Werkstoff* **44,** 307–12.

Pyle, C. and Schafale, M.P. (1988) Land use history of three spruce–fir forest sites in southern Appalachia. *Journal of Forest History* **32,** 4–21.

Pyle, C., Schafale, M.P., Wentworth, T.R. and White, P.S. (1985) *History of disturbance in spruce–fir forests of the SARRMC intensive study sites – Mt. Rogers, National Recreation Area, Black Mountains, and Great Smoky Mountains.* SARRMC-Southern Appalachian Spruce–Fir Ecosystem Assessment Project. 67 pp.

Qiu, Z., Chappelka, A.H., Somers, G.L., Lockaby, B.G. and Meldahl, R.S. (1992) Effects of ozone and simulated acidic precipitation on above- and below-ground growth of loblolly pine *(Pinus taeda). Canadian Journal of Forest Research* **22,** 582–7.

Raddi, P. and Rinallo, C. (1989) Variation in needle wax degradation in two silver fir provenances differentiated by tolerance to air pollution. In: Scholz, F., Gregorius, H.-R. and Rudin, D. (eds), *Genetic effects of air pollutants in forest tree populations.* Berlin: Springer Verlag, pp. 67–76.

Rademacher, P. (1986) *Morphologische und physiologische Eigenschaften von Fichten, Tannen, Kiefern und Buchen gesunder und erkrankter Waldstandorte.* GKSS Forschungszentrum Geesthacht 86/E/10.

Rademacher, P., Bauch, J. and Puls, J. (1986) Biological and chemical investigations of the wood from pollution-affected spruce (*Picea abies* (L.) Karst.). *Holzforschung* **40,** 331–8.

Raison, R.J., Khanna, P.K., Connell, M.J. and Falkner, R.A. (1990) Effects of water availability and fertilization on nitrogen cycling in a stand of *Pinus radiata. Forest Ecology and Management* **30,** 31–43.

Raitio, H. (1987) Neulasvuosikertojen merkitys neulasanalyysin tulkinnassa. *Silva Fennica* **21,** 11–16.

Rajan, S.S.S. (1979) Adsorption and desorption of sulfate and charge relationship to allophanic clays. *Soil Science Society of America Journal* **43,** 65–9.

Ramann, E. (1883) Die Einwirkung der Streuentnahme auf Sandböden. *Zeitschrift für Forst- und Jagdwesen* **11,** 577 and **12,** 633.

Rampazzo, N. (1991) *Physikalisch-chemische und mineralogische Zustandsänderungen von Waldböden durch Luftimmissionen.* Dissertationen der Universität für Bodenkultur in Wien. 159 pp.

Ranft, H. (1984) Zwei Jahrzehnte geführte Düngeversuche unter Immissionsbedingungen im Erzgebirge. *Berichte XIII Internationale Arbeitstagung forstliche Rauchschadenssachverständiger, 27.8 – 1.9, Most, CSSR.* Jiloviste-Strnady: Forschungsanstalt für Forst- und Jagdwesen, Czechoslovakia, pp. 86–9.

Ratkowsky, D.A., Myers, B.J. and Bird, T. (1980) Analysis of pattern of crown damage in forests. *Forest Ecology and Management* **3,** 245–53.

Rawlings, G.B. (1953) Insect epidemics on forest trees in New Zealand. *New Zealand Journal of Forestry* **6,** 405–12.

Raymond, C.A. and Cotterill, P.P. (1990) Methods of assessing crown form of *Pinus radiata. Silva Genetica* **39,** 67–71.

Raynal, D.J., Joslin, J.D., Thonton, F.C., Schaedle, M. and Henderson, G.S. (1990) Sensitivity of tree seedlings to aluminium: III. Red spruce and loblolly pine. *Journal of Environmental Quality* **19,** 180–7.

Raynal, D.J., Roman, J.R. and Eichenlaub, W.M. (1982a) Response of tree seedlings to acid precipitation-1. Effect of substrate acidity on seed germination. *Environmental and Experimental Botany* **22,** 377–83.

Raynal, D.J., Roman, J.R. and Eichenlaub, W.M. (1982b) Response of tree seedlings to acid precipitation-II. Effect of simulated acidified canopy throughfall on sugar maple seedling growth. *Environmental and Experimental Botany* **22,** 385–92.

Read, J. and Hill, R.S. (1985) Photosynthetic response to light of Australian and Chilean species of *Nothofagus* and their relevance to the rain-forest dynamics. *New Phytologist* **101,** 731–42.

Reams, G.A. and Huso, M.M.P. (1990) Stand history: an alternative explanation of red spruce radial growth reduction. *Canadian Journal of Forest Research* **20,** 250–3.

Reams, G.A. and Peterson, C.E. (1992) Evaluating changes in forest condition potentially related to acidic deposition: an example using red spruce. *Forest Ecology and Management* **51**, 5–16.

Reams, G.A., Vong, R.J., Böhm, M., Mattson, K.G. and Arnaut L.Y. (1990) *Extent and magnitude of recent change in forest condition and the role of air pollution and non-air pollution factors.* Forest Response Program Major Program Output Nos. 1 and 2. Document No. EPA/600/3-90/042. Corvallis: US Environmental Protection Agency.

Rebbeck, J., Jensen, K.F. and Greenwood, M.S. (1992) Ozone effects on the growth of grafted mature and juvenile red spruce. *Canadian Journal of Forest Research* **22,** 756–60.

Redfern, D.B., Gregory, S.C., Macaskill, G.A. and Pratt, J.E. (1987) Pathology. Advisory services. Scotland and northern England. In: *Report on Forest Research 1987.* London: HMSO, pp. 42–3.

Redfern, D.B., Gregory, S.C., Pratt, J.E. and Macaskill, G.A. (1986) Foliage browning and shoot death in Sitka spruce and other conifers in northern Britain during the winter 1983–1984. *European Journal of Forest Pathology* **17,** 166–80.

Reemtsma, J.B. (1986) Der Magnesium-Gehalt von Nadeln niedersächsischer Fichtenbestände und seine Beurteilung. *Allgemeine Forst- und Jagdzeitung* **157,** 196–200.

Reemtsma, J.B. (1989) Ernährung der Fichte in Niedersachsen und erste Ergebnisse. *KfK-PEF-Berichte* **55,** 249–60.

Rehfuess, C. and Rehfuess, K.E. (1988) Ersatztriebe an Fichten – Entwicklung, Nährstoffversorgung und Bedeutung für die Kronenmorphologie. *Allgemeine Forst- und Jagdzeitung* **159,** 20–6.

Rehfuess, K.E. (1981) *Waldböden – Entwicklung, Eigenschaften und Nutzung.* Hamburg: Verlag Paul Parey.

Rehfuess, K.E. (1981b) Über die Wirkungen der sauren Niederschläge in Waldökosystemen. *Forstwissenschaftliches Centralblatt* **100,** 363–81.

Rehfuess, K.E. (1983) Waldkrankungen und Immissionen – eine Zwischenbilanz. *Allgemeine Forst Zeitschrift* **38,** 601–10.

Rehfuess, K.E. (1985a) On the causes of decline of Norway spruce (*Picea abies Karst.*) in Central Europe. *Soil Use and Management* **1,** 30–2.

Rehfuess, K.E. (1985b) Walderkrankungen und Immissionen – erläutert am Beispiel der Fichtenerkrankungen. *Düsseldorfer Geobotanisches Kolloquieum* **2,** 25–9.

Rehfuess, K.E. (1987) Perceptions on forest diseases in Central Europe. *Forestry* **60,** 1–11.

Rehfuess, K.E. (1988a) Übersicht über die bodenkundliche Forschung in Zusammenhang mit den neuartigen Waldschäden. In: *PEF-Statuskolloquium, March 1988*. Karlsruhe: PEF, pp. 1–26.

Rehfuess, K.E. (1988b) Disease types in Norway spruce stands in southern Germany – extent and causes. In: *Proceedings of the US/FRG research symposium: effects of atmospheric pollutants on the spruce-fir forests of the eastern United States and the Federal Republic of Germany*. General Technical Report NE-120. Broomall: US Department of Agriculture Forest Service, Northeastern Forest Experiment Station, pp. 79–86.

Rehfuess, K.E. (1989) Acidic deposition – extent and impact on forest soils, nutrition, growth and disease phenomena in Central Europe: a review. *Water, Air, and Soil Pollution* **48,** 1–20.

Rehfuess, K.E. and Rodenkirchen, H. (1984) Über die Nadelröte-Erkrankung der Fichte (*Picea abies* Karst.) in Süddeutschland. *Forstwissenschaftliches Centralblatt* **103,** 248–62.

Rehfuess, K.E. and Rodenkirchen, H. (1985) Über die Nadelröte-Erkrankung der Fichte (*Picea abies* Karst.) in Süddeutschland - Entgegnung auf kritische Stellungnahmen von P. Schütt sowie H. Butin und Chr. Wagner. *Forstwissenschaftliches Centralblatt* **104,** 388–90.

Rehfuess, K.E., Flurl, H., Franz, F. and Raunecker, E. (1983) Growth patterns, phloem nutrient contents and root characteristics of beech (*Fagus sylv.* L.) on soils of different reaction. In: Ulrich, B. and Pankrath, J. (eds), *Effects of accumulation of air pollutants in forest ecosystems*. Reidel, Dordrecht, pp. 359–75.

Rehfuess, K.E., Makeschin, F. and Rodenkirchen, H. (1991) Results and experience from amelioration trials in Scots pine (*Pinus sylvestris* L.) forests of Northeastern Bavaria. *Fertilizer Research* **27,** 95–105.

Reich, P.B., Schoettle, A.W. and Amundson, R.G. (1986) Effects of O_3 and acidic rain on photosynthesis and growth in sugar maple and northern red oak seedlings. *Environmental Pollution* (Series A) **40,** 1–15.

Reich, P.B., Schoettle, A.W., Stroo, H.F., Troiana, J. and Amundson, R.G. (1987) Effects of ozone and acid rain on white pine (*Pinus strobus*) seedlings grown in five soils. I. Net photosynthesis and growth. *Canadian Journal of Botany* **65,** 977–87.

Reich, P.B., Teskey, R.O., Johnson, P.S. and Hinckley, T.M. (1980) Periodic root growth and shoot growth in oak. *Forest Science* **26,** 590–8.

Reichelt, G. (1983) Untersuchungen zum Nadelbaumsterben in der Region Schwarzwald-Baar-Heuberg. *Allgemeine Forst- und Jagdzeitung* **154,** 66–75.

Reicosky, D.A. and Hannover, J.W. (1976) Seasonal changes in leaf surface waxes of *Picea pungens*. *American Journal of Botany* **63,** 449–56.

Reid, N.J. (1987) Remote sensing and forest damage. *Environmental Science and Technology* **21,** 428–9.

Reif, A. and Allen, R.B. (1988) Plant communities of the steepland conifer-broadleaved hardwood forests of central Westland, South Island, New Zealand. *Phytocoenologia* **16,** 145–224.

Reiners, W.A. and Lang, G.E. (1979) Vegetational patterns and processes in the balsam fir zone, White Mountains, New Hampshire. *Ecology* **60,** 403–17.

Renaud, J.-P. and Mauffette, Y. (1991) The relationships of crown dieback with carbohydrate content and growth of sugar maple (*Acer saccharum*). *Canadian Journal of Forest Research* **21,** 1111–18.

Reuss, J.O. and Johnson, D.W. (1986) *Acid deposition and the acidification of soils and waters.* New York: Springer Verlag. 119 pp.

Rey Balmaceda, R.C. (1985) La modificación del paisaje y su apreciacíon cientifica. In: Boelcke, O., Moore, D.M. and Roig, F.A. (eds), *Transecta botánica de la Patagonia austral.* Buenos Aires: Consejo Nacional de Ciencia y Tecnología, pp. 95–104.

Richards, B.L., Taylor, O.C. and Edmunds, G.F. (1968) Ozone mottle of pine in southern California. *Journal of the Air Pollution Control Association* **18,** 73–7.

Richards, J.H. (1989) Evaluation of root growth and functioning of trees exposed to air pollutants. In: Committee on Biologic Markers of Air Pollution Damage in Trees (eds), *Biologic Markers of Air Pollution Stress and Damage in Forests.* Washington DC: National Academy Press, pp. 169–81.

Richards, L.W. (1989) Airborne chemical measurements in nighttime stratus clouds in the Los Angeles Basin. In: Olson, R.K. and Lefohn, A.S. (eds), *Effects of air pollution on western forests.* Pittsburgh: Air and Waste Management Association, pp. 51–72.

Richardson, C.J., Di Giulio, R.T. and Tandy, N.E. (1989) Free-radical mediated processes as markers of air pollution stress in trees. In: Committee on Biologic Markers of Air Pollution Damage in Trees (eds), *Biologic Markers of Air Pollution Stress and Damage in Forests.* Washington DC: National Academy Press, pp. 251–60.

Riederer, M., Kurbasik, K., Steinbrecher, R. and Voss, A. (1988) Surface areas, lengths and volumes of *Picea abies* (L.) Karst. needles: determination, biological variability and effect of environmental factors. *Trees: Structure and Function* **2,** 165–72.

Riha, S.J., James, B.R., Senesac, G.P. and Pallant, E. (1986a) Spatial variability of soil pH and organic matter in forest plantations. *Soil Science Society of America Journal* **50,** 1347–52.

Riha, S.J., Senesac, G. and Pallant, E. (1986b) Effects of forest vegetation on spatial variability of surface mineral soil pH, soluble aluminium and carbon. *Water, Air, and Soil Pollution* **31,** 929–40.

Rinallo, C., Raddi, P., Gellini, R. and di Lonardi, V. (1986) Effects of simulated acid deposition on the surface structure of Norway spruce and silver fir needles. *European Journal of Forest Pathology* **16,** 440–6.

Robak, H. (1969) Litt om fluoropptaket: nålene hos barskogen omkring aluminiumsverk. *Tidsskrift for Skogbruk* **77,** 269–90.

Robak, H. (1969) Aluminium plants and conifers in Norway. In: *Air Pollution. Proceedings of the 1st European Congress on the Influence of Air Pollution on Plants and Animals.* Wageningen: Centre for Agricultural Publishing and Documentation, pp. 27–31.

Robarge, W.P., Pye, J.M. and Bruck, R.I. (1989) Foliar elemental composition of spruce-fir in the southern Blue Ridge province. *Plant and Soil* **114,** 19–34.

Roberge, M.R. (1987) *Evolution of the pH of various soils over the past twenty years.* Information Note LAU-X-77. Sainte-Foy, Quebec: Canadian Forest Service, Laurentian Forestry Research Centre.

Roberge, M.R. (1988) *Twenty years of group management in a yellow birch-sugar maple stand*. Information Note LAU-X-81B. Sainte Foy, Quebec: Canadian Forest Service, Laurentian Forestry Research Centre.

Roberts, B.R. (1990) Physiological response of yellow-poplar seedlings to simulated acid rain, ozone fumigation, and drought. *Forest Ecology and Management* **31,** 215–24.

Roberts, B.R. and Cannon, W.N. (1989) Changes in xylem pressure potential of red spruce seedlings treated with ozone and simulated acid rain. *Canadian Journal of Forest Research* **19,** 1200–203.

Roberts, T.M., Skeffington, R.A. and Blank, L.W. (1989) Causes of Type 1 spruce decline in Europe. *Forestry* **62,** 179–222.

Robertson, A. (1987) The centroid of tree crowns as an indicator of abiotic processes in a balsam fir wave forest. *Canadian Journal of Forest Research* **17,** 746–55.

Rock, B.N., Defeo, N. and Vogelmann, J.E. (1987) Vegetation survey pilot study: detection and quantification of forest decline damage using remote sensing techniques. *Proceedings of the NAPAP Terrestrial Effects Peer Review.* Atlanta, Georgia: NAPAP, pp. 67–74.

Rock, B.N., Hoshizaki, T. and Miller, J.R. (1988b) Comparison of in situ and airborne multispectral measurements of the blue shift associated with forest decline. *Remote Sensing of Environment* **24,** 109–27.

Rock, B.N. and Vogelmann, J.E. (1986) The use of TMS/TM data for mapping of forest decline damage in the northeastern United States. In: *Proceedings of the International Geoscience and Remote Sensing Symposium (IGARSS'86), Zurich, Switzerland.* European Space Agency Document SP–254, pp. 1405–10.

Rock, B.N., Vogelmann, J.E. and Defeo, N.J. (1988a) Vegetation survey pilot study: Assessing forest damage on Whiteface Mountain, NY using remote sensing techniques. In: *Proceedings of the US/FRG Research Symposium: Effects of atmospheric pollutants on the spruce-fir forests of the eastern United States and the Federal Republic of Germany.* General Technical Report NE-120. Broomall: US Department of Agriculture Forest Service, Northeastern Forest Experiment Station, pp. 263–9.

Rock, B.N., Vogelmann, J.E. and Defeo, N.J. (1989) The use of remote sensing for the study of air pollution effects in forests. In: Committee on Biologic Markers of Air Pollution Damage in Trees (eds), *Biologic markers of air pollution stress and damage in forests,* Washington DC: National Academy Press, pp. 183–94.

Rock, B.N., Vogelmann, J.E., Williams, D.L., Vogelmann, A.F. and Hoshizaki, T. (1986) Remote detection of forest damage. *BioScience* **36,** 439–45.

Roelofs, J.G.M., Boxman, A.W., van Dijk, H.F.G. (1987) Effects of airborne ammonium on natural vegetation and forests. In: *Ammonia and acidification, Proceedings of a EURASAP Symposium,* pp. 266–76.

Roelofs, J.G.M., Kempers, A.J., Houdjik, A.L.F.M. and Jansen, J. (1985) The effect of airborne ammonium sulphate on *Pinus nigra* var. *maritima* in The Netherlands. *Plant and Soil* **84,** 45–56.

Röhle, H. (1986) Waldschäden und Zuwachsreaktion – dargestellt am Beispiel geschädigter Fichtenbestände im Nationalpark Bayerischer Wald. *Forstwissenschaftliches Centralblatt* **105,** 115–22.

Rohmeder, E. (1960) Die Wirkung von Staub und Russ auf den Zuwachs der Fichte. *Der Forst- und Holzwirt* **15,** 245–8.

Rol, R. (1951) Le dépérissement des chênes. *Revue forestières françaises* **3,** 707–9.

Roloff, A. (1985a) Schadstufen bei der Buche. *Der Forst- und Holzwirt* **40,** 131–4.

Roloff, A. (1985b) Untersuchungen zum vorzeitigen Laubfall und zur Diagnose von Trockenschäden in Buchenbeständen. *Allgemeine Forst Zeitschrift* **40:** 157–60.

Roloff, A. (1988) Branching structure in hardwoods related to forest decline. In: Cape, J.N. and Mathy, P. (eds), *Scientific basis of forest decline symptomatology.* Air Pollution Report 15. Brussels: Commission of the European Communities, pp. 73–91.

Roloff, A. (1989) *Kronenentwicklung und Vitalitätsbeurteilung ausgewählter Baumarten der gemäßigten Breiten.* Schriften aus der Forstlichen Fakultät der Universität Göttingen und der Niedersächsischen Forstlichen Versuchsanstalt, 93. 258 pp.

Roloff, A. (1991) Crown structure and tree vitality. In: Longhurst, J.W.S. (ed.), *Acid deposition. Origins, impacts and abatement strategies.* Berlin: Springer Verlag, pp. 193–213.

Rommel, W.D. (1989) Absterbende Tannenwälder 1784 im Schwäbisch-Fränkischen Wald. *Allgemeine Forst Zeitschrift* **45,** 269–70.

Ronse, A., de Temmerman, L., Guns, M. and de Borger, R. (1988) Evolution of acidity, organic matter content, and CEC in uncultivated soils on North Belgium during the past 25 years. *Soil Science* **146,** 453–60.

Rose, M.R., Robinson, W.J. and Dean, J.S. (1982) *Dendroclimatic reconstruction for the southeastern Colorado Plateau.* Final Report to Division of Chaco Research, National Park Service, Order No. PX7486-7-0121.

Rosén, K. (1989) Observed and predicted changes in exchangeable calcium, magnesium and potassium in Swedish forest soils. In: Szabolcs, I. (ed.) *Ecological impact of acidification.* Budapest: Hungarian Academy of Sciences, pp. 115–22.

Rost-Siebert, K. (1983) Aluminium-Toxizität und -Toleranz an Keimpflanzen von Fichte (*Picea abies* Karst.) und Buche (*Fagus sylvatica* L.). *Allgemeine Forst- und Holzwirtschaftliche Zeitung* **26/27,** 686–680.

Rost-Siebert, K. (1985) Untersuchungen zur H- und Al-Ionentoxizität an Keimpflanzen von Fichte (*Picea abies* L. Karst.) und Buche (*Fagus sylvatica* L.) in Lösungskultur. *Berichte Waldökosysteme/Waldsterben* **12,** 219 pp.

Rost-Siebert, K. (1988) Ergebnisse vegetationskundlicher und bodenchemischer Vergleichsuntersuchungen zur Feststellung immissionsbedingter Veränderungen während der letzten Jahrzehnte. *Berichte des Forschungszentrums Waldökosystemen der Universität Göttingen* **B 8,** 158 pp.

Rost-Siebert, K. and Jahn, G. (1988) Veränderungen der Waldbodenvegetation während der letzten Jahrzehnte – Eignung zur Bioindikation von Immissionswirkungen? *Forst und Holz* **43,** 75–81.

Rothe, G.M., Weil, H., Geider, M., Pfennig, P., Wilhelmi, V. and Maurer, W.D. (1988) Nutrient element and carbohydrate status of Norway spruce at Mt. Kleiner Feldberg in Taunus exposed to air pollution and soil acidification. *European Journal of Forest Pathology* **18,** 98–111.

Rowland, A., Murray, A.J.S. and Wellburn, A.R. (1985) Oxides of nitrogen and their impact upon vegetation. *Reviews on Environmental Health* **5,** 295–342.

Roy, A.K., Sharma, A. and Talukder, G. (1988) Some aspects of aluminium toxicity in plants. *Botanical Review* **54,** 145–78.

Ruark, G.A., Thomas, C.E., Bechtold, W.A. and May, D.M. (1991) Growth reductions in naturally regenerated pine stands in Alabama and Georgia. *Southern Journal of Applied Forestry* **15,** 73–9.

Rubner, K. (1934) *Die pflanzengeographisch-ökologischen Grundlagen des Waldbaus.* 3. Neudamm, Germany: Verlag J. Neumann.

Ruiter, J.H. (1969) Suspected copper deficiency in radiata pine. *Plant and Soil* **31,** 197–200.

Runkel, M. and Kenneweg, H. (1986) Waldschadens- und Waldstrukturanalyse in Schleswig-Holstein. *Landschaftsentwicklung und Umweltforschung,* Berlin, **36,** 125 pp.

Salemaa, M. and Jukola-Sulonen, E.-L. (1990) Vitality rating of *Picea abies* by defoliation class and other vigour indicators. *Scandinavian Journal of Forest Research* **5,** 413–26.

Salemaa, M., Jukola-Sulonen, E.-L. and Lindgren, M. (1991) Forest condition in Finland, 1986–1990. *Silva Fennica* **25,** 147–75.

Salonius, P.O. (1990) Respiration rates in forest soil organic horizon materials treated with simulated acid rain. *Canadian Journal of Forest Research* **20,** 910–13.

Sanasilva (1986) *Kronenbilder.* Birmensdorf: Eidgenössische Anstalt für das forstliche Versuchswesen.

Sanasilva (1991) *Sanasilva-Waldschadenbericht 1991.* Berne: BUWAL, and Birmensdorf: Eidgenössische Forschungsanstalt für Wald, Schnee und Landschaft.

Santerre, A., Mermet, J.M. and Villaneuva, V.R. (1990) Comparative time-course mineral content study between healthy and diseased *Picea* trees from polluted areas. *Water, Air, and Soil Pollution* **52,** 157–74.

Sasek, T.W. and Richardson, C.J. (1989) Effects of chronic doses of ozone on loblolly pine: photosynthetic characteristics in the third growing season. *Forest Science* **35,** 745–55.

Sasek, T.W., Richardson, C.J., Fendick, E.A., Bevington, S.R. and Kress, L.W. (1991) Carryover effects of acid rain and ozone on the physiology of multiple flushes of loblolly pine seedlings. *Forest Science* **37,** 1078–98.

Sauter, J.J. and Voß, J.-U (1986) SEM-observations on the structural degradation of epistomatal waxes in *Picea abies* (L.) Karst. – and its possible role in the 'Fichtensterben'. *European Journal of Forest Pathology* **16,** 408–23.

Schadauer, K. (1991) Die Ermittlung von Genauigkeitsmaßen terrestrischer Kronenzustandsinventuren im Rahmen der Österreichischen 'Waldzustandsinventur'. *Centralblatt für das gesamte Forstwesen* **108,** 253–82.

Schäffner, G. and Gürth, P. (1992) Untersuchungen über Naturverjüngung und Vorbau in immissionsgeschädigten Hochlagen des Südschwarzwaldes. *Allgemeine Forst- und Jagdzeitung* **163,** 31–9.

Scheidter, F. (1919) Das Tannensterben im Frankenwalde. *Naturwissenschaftliche Zeitschrift für Forst- und Landwirtschaft* **17,** 69–90.

Scherbatskoy, T. and Klein, R.M. (1983) Response of spruce and birch foliage to leaching by acidic mists. *Journal of Environmental Quality* **12,** 189–95.

Scherbatskoy, T., Klein, R.M. and Badger, G.J. (1987) Germination responses of forest tree seed to acidity and metal ions. *Environmental and Experimental Botany* **27,** 157–64.

Scheromm, P. and Plassard, C. (1988) Nitrogen nutrition of non-mycorrhizal pine (*Pinus pinaster*) grown on nitrate or ammonium. *Plant Physiology and Biochemistry* **26,** 261–69.

Schier, G.A. (1985) Response of red spruce and balsam fir seedlings to aluminium toxicity in nutrient solution. *Canadian Journal of Forest Research* **15,** 29–33.

Schier, G.A. (1986) Seedling growth and nutrient relationships in a New Jersey Pine Barrens soil treated with 'acid rain'. *Canadian Journal of Forest Research* **16,** 136–42.

Schier, G.A. (1987) Throughfall chemistry in a red maple provenance plantation sprayed with 'acid rain'. *Canadian Journal of Forest Research* **17,** 660–5.

Schierl, R. and Kreutzer, K. (1991) Effects of liming in a Norway spruce stand (*Picea abies* (L.) Karst.). *Fertilizer Research* **27,** 39–47.

Schlaepfer, R. (1985) Problems in the planning of forest damage inventories. In: Schmid-Haas, P. (ed.), *Inventorying and monitoring endangered forests.* Birmensdorf: Eidgenössische Anstalt für das forstliche Versuchswesen, pp. 339–42.

Schlaepfer, R. (1986) Relationship between diameter growth and crown parameters, in particular foliage loss, of *Picea abies* and *Abies alba:* results of a pilot study in Switzerland. In: Fujimori, T. and Whitehead, D. (eds), *Crown and canopy structure in relation to productivity.* Ibaraki, Japan: Forestry and Forest Products Research Institute, pp. 70–92.

Schlaepfer, R. (1992) Forest vegetation and acidification: a critical review. In: Schneider, T. (ed.), *Acidification research: Evaluation and policy applications.* Amsterdam: Elsevier, pp. 27–44.

Schlaepfer, R. and Haemmerli, F. (1990) Das 'Waldsterben' in der Schweiz aus heutiger Sicht. *Schweizerische Zeitschrift für Forstwesen* **141,** 163–88.

Schlatter, J. and Gerding, V. (1985) Deficiencia de boro en plantaciones de *Pinus radiata* D.Don en Chile. I. Distribucion y origen. *Bosque* **6,** 24–31.

Schmeink, B. and Wild, A. (1990) Studies on the content of free amino acids in needles of undamaged and damaged spruce trees at a natural habitat. *Journal of Plant Physiology* **136,** 66–71.

Schmid-Haas, P. (1987) Inventur und Überwachung des Gesundheitszustandes des Waldes. *Schweizerische Zeitschrift für Forstwesen* **138,** 837–53.

Schmid-Haas, P. (1989a) Der Nachweis der Ursache des Tannensterbens an einem Beispiel. *Forstwissenschaftliches Centralblatt* **108,** 244–54.

Schmid-Haas, P. (1989b) Do the observed needle losses reduce increments? In: Bucher, J.B. and Bucher-Wallin, I. (eds), *Air pollution and forest decline.* Birmensdorf: Eidgenössische Anstalt für das forstliche Versuchswesen, pp. 271–5.

Schmid-Haas, P. (1991a) Kronenverlichtung und Waldwachstum. *Schweizerische Zeitschrift für Forstwesen* **141,** 189–209.

Schmid-Haas, P. (1991b) Monitoring the state of health as a tool for causal research in forest decline. In: *IUFRO and ICP-Forests workshop on monitoring air pollution impact on permanent sample plots, data processing and results interpretation.* Prachatice: IUFRO, pp. 94–101.

Schmid-Haas, P., Masumy, S.A., Niederer, M. and Schweingruber, F.H. (1986)

Zuwachs- und Kronenanalysen an geschwächten Tannen. *Schweizerische Zeitschrift für Forstwesen* **137,** 811–32.

Schmidt, M. (1985) Die trockenen und warmen Sommer 1976, 1982 und 1983 in Deutschland – Stressfaktoren für den Wald. In: Stratmann, H. (ed.), *Waldschäden. Einflussfaktoren und ihre Bewertung. Kolloquium Goslar, 18. bis 20. Juni 1985.* VDI-Bericht 560, pp. 527–44.

Schmidtke, H. (1987) Zur Definition von Schadniveaus für Waldschadensinventuren. *Allgemeine Forst Zeitschrift* **42,** 930–2.

Schmitt, U., Liese, W. and Ruetze, M. (1986) Ultrastrukturelle Veränderungen in grünen Nadeln geschädigter Fichten. *Angewandte Botanik* **60,** 441–50.

Schneckenburger, H. and Frenz, M. (1986) Time-resolved fluorescence of conifers exposed to environmental pollutants. *Radiation and Environmental Biophysics* **25,** 289–95.

Schneider, B.U. (1989) Wachstum und Ernährung von Feinwurzeln geschädigter und ungeschädigter Waldbestände im Fichtelgebirge. *Bayreuther Bodenkundliche Berichte* 13.

Schneider, B.U., Meyer, J., Schulze, E.-D. and Zech, W. (1989) Root and mycorrhizal development in healthy and declining Norway spruce stands. In: Schulze, E.-D., Lange, O.L. and Oren, R. (eds), *Forest decline and air pollution. A study of spruce (*Picea abies*) on acid soils.* Berlin: Springer Verlag, pp. 370–91.

Schneider, O., Hartmann, P., Schlaepfer, R. and Peter, D.A. (1988) Relationship between tree ring width and crown transparency of spruce (*Picea abies* Karst.). *Dendrochronologia* 1988 (6), 9–31.

Schönwitz, R., Kloos, M., Merk, L. and Ziegler, H. (1990a) Patterns of monoterpenes stored in needles of *Picea abies* (L.) Karst. from several locations in mountainous regions of southern Germany. *Trees: Structure and Function* **4,** 27–33.

Schönwitz, R., Lohwasser, K., Kloos, M. and Ziegler, H. (1990b) Seasonal variation in the monoterpenes in needles of *Picea abies* (L.) Karst. *Trees: Structure and Function* **4,** 34–40.

Schöpfer, W. (1985) Das Schulungs- und Kontrollsystem der terrestrischen Waldschadensinventuren. *Allgemeine Forst Zeitschrift* **40,** 1353–7.

Schöpfer, W. (1986) Zusammenhang zwischen Wuchsraum und Zuwachs in erkrankten Fichten- und Tannen-Beständen. *Der Forst- und Holzwirt* **41,** 315–19.

Schreiber, U., Vidaver, W., Runeckles, V.C. and Rosen, P. (1978) Chlorophyll fluorescence assay for ozone injury in intact plants. *Plant Physiology* **61,** 80–4.

Schreuder, H.T. and Thomas, C.E. (1991) Establishing cause-effect relationships using forest survey data. *Forest Science* **37,** 1497–512.

Schröder, W. and Vetter, L. (1989) Synthetisierende Darstellung standörtlicher Randbedingungen der Waldschäden in der Bundesrepublik Deutschland. Ein systemtheoretischer und geostatistischer Ansatz. *Allgemeine Forst- und Jagdzeitung* **160,** 144–50.

Schröter, H. (1985) Gesundheitszustand von Tannen, Fichten und Buchen auf Dauerbeobachtungsflächen der FVA in Baden-Württemberg. *Allgemeine Forst- und Jagdzeitung* **156,** 123–32.

Schröter, H., Achstetter, L. and Holzapfel, W. (1985) Gesundheitszustand von

Tannen, Fichten und Buchen auf Dauerbeobachtungsflächen der FVA in Baden-Württemberg. *Allgemeine Forst- und Jagdzeitung* **156,** 123–32.

Schulte-Hostede, S., Darrall, N.M., Blank, L.W. and Wellburn, A.R. (1988) (eds), *Air pollution and plant metabolism.* London: Elsevier.

Schulz, H. (1984) Immissionen-Waldschäden und Holzqualität. *Holz-Zentralblatt* **110,** 1493–5.

Schulz, H. (1986) Biochemische und faktoranalytische Untersuchungen zur Interpretation von SO_2-Indikationen in Nadeln von *Pinus sylvestris. Biochemische Physiologische Pflanzen* **181,** 241–56.

Schulz, H. (1989) Biochemische Indikation mit Koniferennadeln – Ein Verfahren zur Früherkennung von Immissionswirkungen. *Biochemische Physiologische Pflanzen* **184,** 419–32.

Schulze, E.-D. (1989) Air pollution and forest decline in a spruce (*Picea abies*) forest. *Science* **244,** 776–83.

Schulze, E.-D. and Küppers, M. (1985) Responses of *Pinus sylvestris* to magnesium deficiency. In: Turner, H. and Tranquillini, W. (eds), *Establishment and tending of subalpine forest: research and management.* Proceedings of the 3rd International Workshop, IUFRO Project Group P 1.07-00, Ecology of Subalpine Zones, 3–5 Sept. 1984, Riederalp, Switzerland. Report 270. Birmensdorf: Eidgenössische Anstalt für das forstliche Versuchswesen, pp. 193–6.

Schulze, E.-D., Lange, O.L. and Oren, R. (eds) (1989) *Forest decline and air pollution. A study of spruce (*Picea abies*) on acid soils.* Berlin: Springer-Verlag. 475 pp.

Schulze, E.-D., McCracken, I., Zimmermann, R. and Benecke, U. (1990) Performance of two *Picea abies* (L.) Karst. stands at different stages of decline. VIII. Photosynthesis and nutrition under polluted and clean air conditions of Germany and New Zealand. *Oecologia* **82,** 158–61.

Schulze, E.-D., Oren, R. and Zimmermann, R. (1987) Die Wirkung von Immissionen auf 30jährige Fichten in mittleren Höhenlagen des Fichtelgebirges auf Phyllit. *Allgemeine Forst Zeitschrift* **42,** 725–30.

Schütt, P. (1985) Das Waldsterben – eine Pilzkrankheit? *Forstwissenschaftliches Centralblatt* **104,** 169–77.

Schütt, P. (1988) Forest decline in Germany. In: *Proceedings of the US/FRG research symposium: Effects of atmospheric pollutants on the spruce–fir forests of the eastern United States and the Federal Republic of Germany.* General Technical Report NE-120. Broomall: US Department of Agriculture Forest Service, Northeastern Forest Experiment Station, pp.87–8.

Schütt, P. (1989) Symptoms of bioindicators of decline in European forests. Committee on Biologic Markers of Air Pollution Damage in Trees (eds), *Biologic markers of air pollution stress and damage in forests.* Washington DC: National Academy Press, pp. 119–24.

Schütt, P. and Cowling, E.B. (1985) Waldsterben, a general decline of forests in central Europe: symptoms, development, and possible causes. *Plant Disease* **69,** 548–58.

Schütz, J.-P. (1985) Forest decay in a continental-wide polluted environment: control by silvicultural measures. *Experientia* **41,** 320–5.

Schwappach, A. (1889) Absterben der Fichte im norddeutschen Küstengebiete. *Zeitschrift für Forst- und Jagdwesen* **21,** 608–11.

Schweingruber, F.H. (1985) Abrupt changes in growth reflected in tree ring sequences as an expression of biotic and abiotic influence. In: Schmid-Haas, P. (ed.), *Inventorying and monitoring endangered forests*. Birmensdorf: Eidgenössische Anstalt für das forsliche Versuchswesen, pp. 291–5.

Schweingruber, F.H. (1986) Abrupt growth changes in conifers. *IAWA Bulletin* **7,** 277–83.

Schweingruber, F.H. (1987) Potentials and limitations of dendrochronology in pollution research. In: Jacoby, G.C. and Hornbeck, J.W. (eds), *Proceedings of the international symposium on ecological aspects of tree-ring analysis*. Washington DC: US Department of Energy, US Environmental Protection Agency and US Department of Agriculture Forest Service, pp. 344–52.

Schweingruber, F.H. (1989) Lässt sich fehlendes Datenmaterial zur Waldschadenssituation anhand von Postkarten ergänzen? *Allgemeine Forst Zeitschrift* **44,** 266–8.

Schweingruber, F.H., Aellen-Rumo, K., Weber, U. and Wehrli, U. (1990) Rhythmic growth reductions in forest trees of Central Europe and the Front Range in Colorado. *Trees: Structure and Function* **4,** 99–106.

Schweingruber, F.H., Albrecht, H., Beck, M., Hessel, J., Joos, K., Keller, D., Kontic, R., Lange, K., Niederer, M., Nippel, C., Spang, S., Spinnler, A., Steiner, B. and Winkler-Siefert, A. (1986). Abrupte Zuwachsschwankungen in Jahrringabfolgen als oekologische Indikatoren. *Dendrochronologia* **4,** 125–83.

Schweingruber, F.H., Kontic, R., Niederer, M., Nippel, C.A. and Winkler-Siefert, A. (1985) Diagnosis and distribution of conifer decay in the Swiss Rhone Valley. In: Turner, H. and Tranquillini, W. (eds), *Establishment and tending of subalpine forest: research and management*. Proceedings of the 3rd International Workshop, IUFRO Project Group P 1.07-00, Ecology of Subalpine Zones, 3–5 Sept. 1984, Riederalp, Switzerland. Report 270. Birmensdorf: Eidgenössische Anstalt für das forstliche Versuchswesen, pp. 189–92.

Schweingruber, F.H., Kontic, R. and Winkler-Siefert, A. (1983) *Eine jahrring-analytische Studie zum Nadelbaumsterben in der Schweiz*. Bericht 253. Birmensdorf: Eidgenössiche Anstalt für das forstliche Versuchswesen. 29 pp.

Scott, J.T., Siccama, T.G., Johnson, A.H. and Breisch, A.R. (1984) Decline of red spruce in the Adirondacks, New York. *Bulletin of the Torrey Botanical Club* **111,** 438–44.

Seiler, M. (1991) Zeitlicher Trend der troposphärischen Ozone im Bereich der bayerischen Kalkalpen. In: *Proceedings 2. Statusseminar der PBWU zum Forschungsschwerpunkt 'Waldschäden'. 4–6 February, 1991, GSF-Forschungszentrum Neuherberg*. Neuherberg: GSF, pp. 41–63.

Seitschek, O. (1982) Die gegenwärtige Waldschutz-Situation in Bayern. *Allgemeine Forst Zeitschrift* **37,** 423–8.

Selinger, H.D., Knoppik, D. and Ziegler-Jöns, A. (1986) Einfluß von Mineralstoffernährung, Ozon und saurem Nebel auf Photosynthese-Parameter und stomatäre Leitfähigkeit von *Picea abies* (L.) Karst. *Forstwissenschaftliches Centralblatt* **105,** 239–42.

Sell, J., Schnell, G. and Arnold, M. (1988) Holzqualität gesunder und geschädigter Fichten und Tannen von 6 schweizerischen Standorten. *Dübendorf Eidgenössische Materialprüfungs- und Forschungsanstalt* Report **217,** 112 pp.

Sell, J., Schnell, G. and Arnold, M. (1989) Vergleichende Untersuchung der

Holzqualität gesunder und kranker Fichten und Tannen von 6 schweizerischen Standorten. *Holz als Roh- und Werkstoff* **47,** 87–92.

Selling, O.H. (1948) Studies in Hawaiian pollen statistics. Part III. On the late Quaternary history of the Hawaiian vegetation. *Bernice P. Bishop Museum Special Publication* 39.

Senoo, T. and Honjyo, T. (1988) Assessment of tree stress by airborne multi-spectral data. *Journal of the Japanese Forestry Society* **70,** 45–56.

Senser, M., Hopker, K., Peuker, A. and Glashagen, B. (1987) Wirkungen extremer Ozonkonzentrationen auf Koniferen. *Allgemeine Forst Zeitschrift* **42,** 709–14.

Seppälä, H. (1992) Western European air pollution and the Finnish forest sector. In: De Steiguer, J.E. (ed.), *The economic impact of air pollution on timber markets. Studies from North America and Europe.* General Technical Report SE-75. Asheville: US Department of Agriculture Forest Service, Southeastern Forest Experimental Station, pp. 39–48.

Serre-Bachet, F. (1987) Tree-growth, ARMA modelling and pollution. In: Jacoby, G.C. and Hornbeck, J.W. (eds), *Proceedings of the international symposium on ecological aspects of tree ring analysis.* Washington DC: US Department of Energy, US Environmental Protection Agency and US Department of Agriculture Forest Service, pp. 591–8.

Seufert, G. and Arndt, U. (1986) Beobachtungen in definiert belasteten Modell-ökosystemen mit jungen Bäumen. *Allgemeine Forst Zeitschrift* **41,** 545–9.

Seymour, R.S. (1985) Forecasting growth and yield of budworm-infested forests. Part 1: eastern North America. In: Sanders, C.J., Stark, R.W., Mullins, E.J. and Murphy, J. (eds), *Recent advances in spruce budworm research.* Ottawa: Minister of Supply and Services, pp. 200–13.

Shafer, S.R. and Heagle, A.S. (1989) Growth responses of field-grown loblolly pine to chronic doses of ozone during multiple growing seasons. *Canadian Journal of Forest Research* **19,** 821–31.

Shafer, S.R., Heagle, A.S. and Camberato, D.M. (1987) Effects of chronic doses of ozone on field-grown loblolly pine: Seedlings response in the first year. *Journal of the Air Pollution Control Association* **37,** 1179–84.

Shaw, C.G., Fischer, G.W., Adams, D.F., Adams, M.F. and Lynch, D.W. (1951) Fluorine injury to ponderosa pine: A summary. *Northwestern Science* **15,** 156.

Shaw, C.G., Laurent, T.H., Eglitis, A. and Hennon, P.E. (1985) Decline and mortality of *Chamaecyparis nootkatensis* in southeastern Alaska, a problem of long duration but unknown cause. *Plant Disease* **69,** 13–17.

Shaw, W.B. (1983) Tropical cyclones: determinants of pattern and structure in New Zealand's indigenous forests. *Pacific Science*37, 405–13.

Sheffield, R.M. and Cost, N.D. (1987) Behind the decline. *Journal of Forestry* **85,** 29–33.

Sheffield, R.M., Cost, N.D., Bechtold, W.A. and McClure, J.P. (1985) *Pine growth reductions in the Southeast.* Resource Bulletin SE-83. Asheville: US Department of Agriculture Forest Service, Southeastern Forest Experimental Station. 112p.

Shelbourne, C.J.A. and Low, C.B. (1980) Multi-trait index selection and associated genetic gains of *Pinus radiata* at five sites. *New Zealand Journal of Forest Science* **10,** 307–24.

Sheldon, C. (1912) *The wilderness of the North Pacific Coast islands: A hunter's experience while searching for wapiti, bears, and caribou on the larger islands of British Columbia and Alaska.* New York: Scribner's Sons. 246 pp.

Sheppard, L.J., Smith, R.I. and Cannell, M.G.R. (1989) Frost hardiness of *Picea rubens* growing in spruce decline regions of the Appalachians. *Tree Physiology* **5,** 25–37.

Shigo, A.L. and Marx, H.G. (1977) *Compartmentalization of decay in trees* (CODIT). Information Bulletin 405. Washington DC: US Department of Agriculture.

Shigo, A.L., Dorn, D. and Lee, H.C. (1983) Selections of maple and birch trees with high resistance to spread of decay associated with wounds. In: *Proceedings of the 28th Northeastern Forest Tree Improvement Group, Durham, New Hampshire.* pp. 110–17.

Shortle, W.C. and Smith, K.T. (1988) Aluminium-induced calcium deficiency syndrome in declining red spruce. *Science* **240,** 1017–18.

Siccama, T.G., Bliss, M. and Vogelmann, H.W. (1982) Decline of red spruce in the Green Mountains of Vermont. *Bulletin of the Torrey Botanical Club* **109,** 162–8.

Sidhu, S.S. and Staniforth, R.J. (1986) Effects of atmospheric fluorides on foliage, and cone and seed production in balsam fir, black spruce, and larch. *Canadian Journal of Botany* **64,** 923–31.

Siegel, B.Z., Nachbar-Hapai, M. and Siegel, S.M. (1990) The contribution of sulfate to rainfall pH around Kilauea Volcano, Hawaii. *Water, Air and Soil Pollution* **52,** 227–35.

Sigal, L.L., Eversman, S. and Berglund, D.L. (1988) Isolation of protoplasts from loblolly pine needles and their flow-cytometric analysis for air pollution effects. *Environmental and Experimental Botany* **28,** 151–61.

Silveira Lopes, M.I.M., Garrido, M.A. de O. and de Mello, F. de A.F. (1985) Influência do cultivo de *Pinus sobre* a manta orgânica e propriedades químicas em um Latossolo Vermelho Escuro primitivamente sob a vegetaçao de cerrado. II. Effeitos sobre o pH e teores de H^+ e Al^{3+} do solo. *Silvicultura em Sao Paulo* 17–19, 25–29.

Silver, W.L., Johnson, A.H., Johnson, C.E. and Siccama, T.G. (1988) *Changes in red spruce populations in montane forests of the eastern U.S., 1982–1987.* Forest Response Program Report. Corvallis: US Environmental Protection Agency.

Simmleit, N. and Schulten, H.R. (1987) On the impact of acidic precipitation on the epicuticular wax of spruce needles. In: Perry, R., Harrison, R.M., Bell, J.N.B. and Lester, J.N. (eds), *Acid rain: Scientific and technical advances.* London: Selper, pp. 546–53.

Simonett, D.S. (1967) Landslide distribution and earthquakes in the Bewani and Torricelli Mountains, New Guinea. In: Jennings, J.N. and Mabbutt, J.A. (eds), *Landform studies from Australia and New Guinea.* Canberra: Australian National University Press, pp. 64–84.

Sinclair, W.A. (1964) Comparison of recent declines of white ash, oaks, and sugar maple in northeastern woodlands. *Cornell Plantations* **20,** 62–7.

Singh, B.R. and Johnson, D.W. (1986) Sulfate content and adsorption in soils of two forest watersheds in southern Norway. *Water, Air, and Soil Pollution* **31,** 847–56.

Singh, B.R., Abrahamsen, G. and Stuanes, A. (1980) Effect of simulated acid rain on sulphate movement in acid forest soils. *Soil Science Society of America Journal* **44,** 75–80.

Singh, J.S., Lauenroth, W.K., Hunt, H.W. and Swift, D.M. (1984) Bias and random errors in estimators of net root production: a simulation approach. *Ecology* **65,** 1760–4.

Skalski, J.R. (1990) A design for long-term status and trends monitoring. *Journal of Environmental Management* **30,** 139–44.

Skeffington, R.A. (1983) Soil properties under three species of tree in southern England in relation to acid deposition in throughfall. In: Ulrich, B. and Pankrath, J. (eds), *Effects of accumulation of air pollutants in forest ecosystems.* Dordrecht: D. Reidel, pp. 219–31.

Skeffington, R.A. and Wilson, E.J. (1988) Excess nitrogen deposition: issues for consideration. *Environmental Pollution* **54,** 169–84.

Skelly, J.M. (1989) Forest decline versus tree decline – pathological considerations. *Environmental Monitoring and Assessment* **12,** 23–7.

Skelly, J.M., Davis, D.D., Merrill, W., Cameron, E.A., Brown, H.D., Drummond, D.B. and Dochinger, L.S. (1987) *Diagnosing injury to eastern forest trees.* Atlanta, Georgia: US Department of Agriculture Forest Service, Forest Pest Management and University Park, Pennsylvania: Pennsylvania State University.

Skelly, J.M., Moore, L.D. and Stone, L.L. (1972) Symptom expression of eastern white pine located near a source of oxides of nitrogen and sulfur dioxide. *Plant Disease Reporter* **56,** 3–6.

Skelly, J.M., Yang, Y.S., Chevone, B.I., Long, S.J., Nellesen, J.E. and Winner, W.E. (1983) Ozone concentrations and their influence on forest species in the Blue Ridge Mountains of Virginia. In: Davis, D.D., Miller, A.A. and Dochinger, L. (eds), *Proceedings air pollution and productivity of the forests.* Washington DC: Izaak Walton League, pp. 143–59.

Skiba, U., Cresser, M.S., Derwent, R.G. and Futty, D.W. (1988) Peat acidification in Scotland. *Nature* **337,** 68–9.

Skipworth, J.P. (1981) Mountain beech mortality in the West Ruapehu forests. *Wellington Botanical Society Bulletin* **41,** 26–34.

Skuhravy, V. (1990) Das Vorhandensein von Nadeln an den Trieben der gemeinen Kiefer, *Pinus sylvestris* L. als Indikator für den Grad des Umwelteinflusses. *Anzeiger für Schädlingskunde, Pflanzenschutz, Umweltschutz* **63,** 145–7.

Skyllberg, U. (1990) Correlation between pH and depth in the mor layer of a *Picea abies* (L.) Karst. stand on till soils in northern Sweden. *Scandinavian Journal of Forest Research* **5,** 143–53.

Smith, J.M. (1988) *The role of remote sensing in the study of tree damage caused by air pollution.* Report to the Department of the Environment, Marsham Street, London. 43 pp.

Smith, M.E. and Bayliss, N.S. (1942) The necessity of zinc for *Pinus radiata. Plant Physiology* **17,** 303–10.

Smith, R.B., Hornbeck, J.W., Federer, C.A. and Krusic, P.J. (1989) *Regionally averaged diameter growth in New England forests.* Research Paper NE-637. Broomall: US Department of Agriculture Forest Service, Northeastern Forest Experiment Station.

Smith, W.H. (1981) *Air pollution and forests. Interactions between air contaminants and forest ecosystems.* New York: Springer Verlag.

Smith, W.H. (1990) *Air pollution and forests. Interactions between air contaminants and forest ecosystems.* New York: Springer Verlag. 2nd edition.

Smykala, J. (1985) Aktualne zagrozenie lasów w Polsce. *Las Polski* **17,** 12–13.

Sohet, K., Herbauts, J. and Gruber, W. (1988) Changes caused by Norway spruce in an ochreous brown earth, assessed by the isoquartz method. *Journal of Soil Science* **39,** 549–61.

Soikkeli, S. and Kärenlampi, L. (1984) The effects of nitrogen fertilization on the ultrastructure of mesophyll cells of conifer needles in northern Finland. *European Journal of Forest Pathology* **14,** 129–36.

Soikkeli, S. and Paakkunainen, T. (1981) The effect of air pollution on the ultrastructure of the developing and current year needles of Norway spruce. *Mitteilungen, Forstliche Bundesversuchsanstalt Wien* 137/I, 159–64.

Soikkeli, S. and Tuovinen, T. (1979) Damage in mesophyll ultrastructure of needles of Norway spruce in two industrial environments in central Finland. *Annales Botanici Fennici* **16,** 50–60.

Solomon, D.S., Hosmer, R.A., Zarnoch, S.J. and Droessler, T.D. (1989) *A spatial and temporal assessment of forest growth decline for tree species in the Northeast.* Report to the US Department of Agriculture Forest Service, Forest Response Program. Northeastern Forest Experiment Station, Radnor.

Souto, D. and McCreery, L. (1988) Spruce beetle survey on red spruce in New York, Vermont, New Hampshire, and Western Maine. In: *Proceedings of the US/FRG Research Symposium: Effects of atmospheric pollutants on the spruce–fir forests of the eastern United States and the Federal Republic of Germany.* General Technical Report NE-120. Broomall: US Department of Agriculture Forest Service, Northeastern Forest Experiment Station, pp. 343–5.

Spanner, M.A., Pierce, L.L., Peterson, D.L. and Running, S.W. (1990) Remote sensing of temperate coniferous forest leaf area index. The influence of canopy closure, understory vegetation and background reflectance. *International Journal of Remote Sensing* **11,** 95–111.

Speers, C.F. (1958) The balsam wooly adelgid in the Southeast. *Journal of Forestry* **56,** 515–16.

Spelsberg, G. (1987) Zum Problem der Beurteilung des Zuwachses in geschädigten Beständen. *Allgemeine Forst- und Jagdzeitung* **158,** 205–10.

Spelsberg, G. (1988) Zur Frage der Wachstumsdifferenzierung bei vorherrschenden Fichten unter besonderer Berücksichtigung der Waldschäden. *Forst und Holz* **43,** 538–42.

Spence, R.D., Rykiel, E.J. and Sharpe, P.J.H. (1990) Ozone alters carbon allocation in loblolly pine: assessment with Carbon-11 labeling. *Environmental Pollution* **64,** 93–106.

Spiecker, H. (1986) Das Wachstum der Tannen und Fichten auf Plenterwald-Versuchsflächen des Schwarzwaldes in der Zeit von 1950 bis 1984. *Allgemeine Forst- und Jagdzeitung* **157,** 152–64.

Spiecker, H. (1987) Düngung Niederschlag und der jährliche Volumenzuwachs einiger Fichtenbestände Südwestdeutschlands. *Allgemeine Forst- und Jagdzeitung* **158,** 70–6.

Spiecker, H. (1991) Liming, nitrogen and phosphorus fertilization and the annual volume increment of Norway spruce stands on long-term permanent plots in Southwestern Germany. *Fertilizer Research* **27,** 87–93.

Sprugel, D.G. (1976) Dynamic structure of wave-generated *Abies balsamea* forests in the Northeastern United States. *Journal of Ecology* **64,** 889–911.

Sprugel, D.G. and Bormann, F.H. (1981) Natural disturbance and the steady state in high-altitude balsam fir forests. *Science* **211,** 390–3.

Staland, P. (1985) Skogsgödsling – vad vet vi idag? *Institutet för skogsförbättrig, Sammanfattningar* Nr. 1, 1985/1986, Uppsala. 6 pp.

Standiford, R., McDougald, N., Phillips, R. and Nelson, A. (1991) South Sierra oak regeneration weak in sapling stage. *California Agriculture* **45,** 12–14.

Stanturf, J.A., Stone, E.L. and McKittrick, R.C. (1989) Effects of added nitrogen on growth of hardwood trees in southern New York. *Canadian Journal of Forest Research* **19,** 279–84.

Stark, R.W., Miller, P.R., Cobb, F.W., Wood, D.W. and Parmeter, J.R. (1968) Incidence of bark beetle infestation in injured trees. *Hilgardia* **39,** 121–6.

Steele, M.J., Coutts, M.P. and Yeoman, M.M. (1989) Developmental changes in Sitka spruce as indices of physiological age. I. Changes in needle morphology. *New Phytologist* **113,** 367–75.

Stefan, K. (1989) Schwefel- und Nährstoffgehalte in Pflanzenproben des österreichischen Bioindikatorennetzes. In: Bucher, J.B. and Bucher-Wallin, I. (eds), *Air pollution and forest decline.* Birmensdorf: Eidgenössische Anstalt für das forstliche Versuchswesen, pp. 99–104.

Van der Stegen, J. and Myttenaere, C. (1991) Status of sulphur in the foliage of Norway spruce (*Picea abies* (L.) Karst.) in relation to the mode of contamination. *Environmental Pollution* **69,** 327–36.

Sterba, H. and Eckmüllner, O. (1988) Nadelverlust – Zuwachsrückgang: Doch eine Beziehung. *Österreichische Forstzeitung* **99,** 52–3.

Stewart, G.H. (1989) Ecological considerations of dieback in New Zealand's indigenous forests. *New Zealand Journal of Forestry Science* **19,** 243–9.

Stewart, G.H. and Rose, A.B. (1988) Factors predisposing rata-kamahi (*Metrosideros umbellata-Weinmannia racemosa*) forests to canopy dieback, Westland, NZ. *GeoJournal* **17,** 217–24.

Stienen, H. and Bauch, J. (1988) Element content in tissues of spruce seedlings from hydroponic cultures simulating acidification and deacidification. *Plant and Soil* **106,** 231–8.

Stock, R. (1988) Aspekte der regionalen Verbreitung 'Neuartiger Waldschäden' an Fichte im Harz. *Forst und Holz* **43,** 283–6.

Stone, E.L. (1953) Magnesium deficiency of some northeastern pines. *Soil Science Society of America* **17,** 297–300.

Stone, E.L. (1990) Boron deficiency and excess in forest trees: A review. *Forest Ecology and Management* **37,** 49–75.

Stone, L.L. and Skelly, J.M. (1974) The growth of two forest tree species adjacent to a periodic source of air pollution. *Phytopathology* **64,** 773–8.

Stow, T.K., Allen, H.L. and Kress, L.W. (1992) Ozone impacts on seasonal foliage dynamics of young loblolly pine. *Forest Science* **38,** 102–19.

Strader, R.H., Binkley, D. and Wells, C.G. (1989) Nitrogen mineralization in high elevation forests of the Appalachians. 1. Regional patterns in southern spruce–fir forests. *Biogeochemistry* **7,** 131–45.

Strand, L. (1980a) *Acid precipitation and regional tree ring analyses.* Project Report IR 73/80. Oslo: SNSF. 36 pp.

Strand, L. (1980b) The effect of acid precipitation on tree growth. In: Drabloes, D. and Tollan, A. (eds), *Ecological impact of acid precipitation.* Oslo: SNSF, pp. 64–7.

Strand, M. and Lundmark, T. (1987) Effects of low night temperature and light on chlorophyll fluorescence of field-grown seedlings of Scots pine. *Tree Physiology* **3,** 211–24.

Strand, M. and Öquist, G. (1988) Effects of frost hardening, dehardening and freezing stress on in vivo chlorophyll fluorescence of seedlings of Scots pine (*Pinus sylvestris* L.). *Plant, Cell and Environment* **11,** 231–8.

Strickland, T.C., Fitzgerald, J.W. and Swank, W.T. (1986) *In situ* measurements of sulfate incorporation into forest floor and soil organic matter. *Canadian Journal of Forest Research* **16,** 549–53.

Sucoff, E., Thornton, F.C. and Joslin, J.D. (1990) Sensitivity of tree seedlings to aluminium: I. Honeylocust. *Journal of Environmental Quality* **19,** 163–71.

Sumner, M.E., Fey, M.V. and Noble, A.D. (1991) Nutrient status and toxicity problems in acid soils. In: Ulrich, B. and Sumner, M.E. (eds) *Soil acidity.* Berlin: Springer-Verlag, pp. 149–82.

Süsser, P. and Schwertmann, U. (1991) Proton buffering in mineral horizons of some acid forest soils. *Geoderma* **49,** 63–76.

Sutinen, S. (1990) Structural changes in needle tissue of spruce trees from a damaged stand in southern Finland. *Scandinavian Journal of Forest Research* **5,** 403–12.

Sverdrup, H., Warfvinge, P. and Rosén, K. (1992) A model for the impact of soil solution chemistry Ca: Al ratio, soil moisture and temperature on tree base cation uptake. *Water, Air, and Soil Pollution* **61,** 365–83.

Swain, E.H.F. (1935) Pinus radiata *plantations in the S.E. of South Australia.* Woods and Forests Department, South Australia, Volume 2, pp. 69–121.

Swank, W.T., Fitzgerald, J.W. and Ash, J.T. (1984) Microbial transformation of sulfate in forest soils. *Science* **223,** 182–4.

Takahashi, K. (1974) Meteorological damage of subalpine forests through clear cutting. *Japanese Journal of Forestry* **56,** 337–40.

Takahashi, K. and Takahashi, K. (1974) Wind damage of subalpine forests through road construction I. *Japanese Journal of Forestry* **56,** 288–93.

Tamm, O. (1950) *Northern coniferous forest soils.* Oxford: Scrivenor Press.

Tamm, C.O. (1955) Studies on forest nutrition. I. Seasonal variation in the nutrient content of conifer needles. *Meddelanden fran Statens skogsforskningsinstitut* **45,** 1–34.

Tamm, C.O. and Hallbäcken, L. (1988) Changes in soil acidity in two forest areas with different acid deposition: 1920s to 1980s. *Ambio* **17,** 56–61.

Taylor, G.E., Norby, R.J., McLauglin, S.B., Johnosn, A.H. and Turner, R.S. (1986) Carbon dioxide assimilation and growth of red spruce (*Picea rubens* Sarg.) seedlings in response to ozone, precipitation chemistry, and soil type. *Oecologia* **70,** 163–71.

Taylor, O.C., Miller, P.R., Page, A.L. and Lund, L.J. (1986) *Effects of ozone and sulfur dioxide mixtures on forest vegetation of the Southern Sierra Nevada,* Final report, California Air Resources Board, Contract No. AO-135-33.

Tegethoff, A.C. (1964) *High elevation spruce mortality. Memorandum 5220,*

September 25, 1964. Amherst, Maryland, US Forest Service Northern FPC Zone.

de Temmerman, L. and Coosemans, P. (1989) Ammonia and ammonium deposition on pine stands in Belgium. In: Bucher, J. and Bucher-Wallin, I. (eds), *Air pollution and forest decline.* Birmensdorf: Eidgenössische Anstalt für das forstliche Versuchswesen, pp. 91–6.

Temple, P.J. (1988) Injury and growth of Jeffrey pine and giant sequoia in response to ozone and acidic mist. *Environmental and Experimental Botany* **28,** 323–33.

Temple, P.J. (1989) Oxidant air pollution effects on plants of Joshua Tree National Monument. *Environmental Pollution* **57,** 35–47.

Tepper, H.B., Yang, C.S. and Schaedle, M. (1989) Effect of aluminium on growth of root tips of honey locust and loblolly pine. *Environmental and Experimental Botany* **29,** 165–73.

Terborgh, J. (1989) *Where have all the birds gone?* Princeton: Princeton University Press. 207 pp.

Tesar, V. and Temmlová, B. (1982) Content of silicon and sulphur in spruce needles as a demonstration of an accelerated aging of the organism by action of immissions. *Communicationes Instituti Forestalis Cechosloveniae* **12,** 147–57.

Teskey, R.O. and Hinckley, T.M. (1981) Influence of temperature and water potential on root growth of white oak. *Physiologia Plantarum* **52,** 363–9.

Tessier, L. (1986) Chronologie de melèzes des Alpes et petit âge glaciaire. *Dendrochronologia* **4,** 97–113.

Tessier, L., Serre-Bachet, F. and Guiot, J. (1990) Pollution fluorée et croissance radiale des conifères en Maurienne (Savoie, France). *Annales des Sciences Forestières* **47,** 309–23.

Thiébaut, B. (1988) Tree growth, morphology and architecture, the case of beech: *Fagus sylvatica* L. In: Cape, J.N. and Mathy, P. (eds), *Scientific basis for forest decline symptomatology.* Air Pollution Report 15. Brussels: Commission of the European Communities, pp. 49–72.

Thijsse, G. and Baas, P. (1990) 'Natural' and NH_3-induced variation in epicuticular needle wax morphology of *Pseudotsuga menziesii* (Mirb.) Franco. *Trees: Structure and Function* **4,** 111–19.

Thomas, C.E. and Parresol, B.R. (1989) Comparing basal area growth rates in repeated inventories: Simpson's paradox in forestry. *Forest Science* **35,** 1029–39.

Thompson, C.R., Olszyk, D.M., Kats, G., Bytnerowicz, A., Dawson, P.J. and Wolf, J. (1984) Effects of ozone and sulfur dioxide on annual plants in the Mojave Desert. *Journal of the Air Pollution Control Association* **34,** 1017–27.

Thompson, T.R.E. and Loveland, P.J. (1985) The acidity of Welsh soils. *Soil Use and Management* **1,** 21–4.

Thomsen, M. (1991) Long-term monitoring sites. In: Innes, J.L. (ed.), *Interim report on cause-effect relationships in forest decline.* Geneva: United Nations Economic Commission for Europe.

Thor, E. and Gall, W.R. (1978) Variation in air pollution tolerance and growth rate among progenies of southern Appalachian white pine. *Metropolitan Tree Improvement Alliance (METRIA) Proceedings* **1,** 80–6.

Thornton, F.C., Pier, P.A. and McDuffie, C. (1990) Response of growth, photosynthesis, and mineral nutrition of red spruce seedlings to ozone and acidic cloud deposition. *Environmental and Experimental Botany* **30,** 313–23.

Thornton, F.C., Pier, P.A. and McDuffie, C. (1992) Red spruce response to ozone and cloudwater after three years exposure. *Journal of Environmental Quality* **21,** 196–202.

Thornton, F.C., Schaedle, M. and Raynal, D.J. (1987) Effects of aluminium on red spruce seedlings in solution culture. *Environmental and Experimental Botany* **27,** 489–98.

Thornton, F.C., Schaedle, M. and Raynal, D.J. (1989) Tolerance of red oak and American and European beech seedlings to aluminium. *Journal of Environmental Quality* **18,** 541–5.

Tippett, J.T. and Shigo, A.L. (1981) Barrier zone formation: a mechanism of tree defense against vascular pathogens. *IAWA Bulletin* **2,** 163–8.

Tjoelker, M.G., McLaughlin, S.B., DiCosty, R.J., Lindberg, S.E. and Norby, R.J. (1992) Seasonal Variation in nitrate reductase activity in needles of high-elevation red spruce trees. *Canadian Journal of Forest Research* **22,** 375–80.

Torelli, N., Cufar, K. and Robic, D. (1986) Some wood anatomical, physiological, and silvicultural aspects of Silver fir dieback in Slovenia (NW Yugoslavia). *IAWA Bulletin* **7,** 343–50.

Tovar, D.C. (1989) Air pollution and forest decline near Mexico City. *Environmental Monitoring and Assessment* **12,** 49–58.

Treshow, M. (1975) Interaction of air pollutants and plant diseases. In: Mudd, J.B. and Kozlowski, T.T. (eds), *Responses of plants to air pollution. New York: Academic Press, pp. 307–34.*

Treshow, M. and Anderson, F.K. (1989) Plant stress from air pollution. New York: Wiley.

von Tubeuf, C. (1906) Pathologische Erscheinungen beim Absterben von Fichten im Sommer 1904. *Zeitschrift für Forst- und Landwirtschaft* **4,** 449–66 and 511–12.

Tuomisto, H. (1988) Use of Picea abies needles as indicators of air pollution: epicuticular wax morphology. *Annales Botanici Fennici* **25,** 351–64.

Turc, L. (1927) Note sur le dépérissement du chêne pédonculé dans les forêts du plateau nivernais. *Revue des Eaux et Forêts* **65,** 561–5.

Turner, R.S., Olson, R.J. and Brandt, C.C. (1986) *Areas having soil characteristics that may indicate sensitivity to acidic deposition under alternate forest damage hypotheses.* Report no. ORNLTM-9917. Oak Ridge: Oak Ridge National Laboratory.

Turvey, N.D. (1984) Copper deficiency in *Pinus radiata* planted in a podzol in Victoria, Australia. *Plant and Soil* **77**, 73–86.

Turvey, N.D., Carlyle, C. and Downes, G.M. (1992) Effects of micronutrients on the growth form of two families of *Pinus radiata* (D. Don) seedlings. *Plant and Soil* **139**, 59–65.

Tyler, G. (1987a) Probable effects of soil acidification and nitrogen deposition on the floristic composition of oak (*Quercus robur* L.) forest. *Flora* **179,** 165–70.

Tyler, G. (1987b) Acidification and chemical properties of *Fagus sylvatica* L. forest soils. *Scandinavian Journal of Forest Research* **2,** 263–71.

Tyree, M.T., Flanagan, L.B. and Adamson, N. (1987) Response of trees to drought. In: Hutchinson, T.C. and Meema, K. (eds), *Effects of atmospheric pollutants on forests, wetlands, and agricultural ecosystems*. Berlin: Springer-Verlag, pp. 201–16.

Tzschacksch, O. and Weiss, M. (1972) Die Variation der SO_2-Resistenz von Provenienzen der Baumart Fichte (*Picea abies* (L.) Karst.). *Beiträge für die Forstwirtschaft* **3,** 21–3.

Udvardny, M.D.F. (1969) *Dynamic zoogeography with special reference to land animals*. New York: Van Nostrand Reinhold.

Uebelhör, K. (1984) *Struktur und Dynamik von Nothofagus-Urwäldern in den Mittellagen der valdivianischen Anden Chiles*. Forstliche Forschungsberichte München **58.** 230 pp.

Ukleja-Dobrowolska, D. (1989) Zamieranie jodly – wciaz nie wyjasnione zjawisko. *Sylwan* **133,** 59–67.

Ulrich, B. (1981) Eine ökosystemare Hypothese über die Ursachen des Tannensterbens (*Abies alba* Mill.). *Forstwissenschaftliches Centralblatt* **100,** 228–36.

Ulrich, B. (1983a) A concept of forest ecosystem stability and of acid deposition as driving force for destabilization. In: Ulrich, B. and Pankrath, J. (eds), *Effects of accumulation of air pollutants on forest ecosystems*. Dordrecht: D. Reidel, pp. 1–32.

Ulrich, B. (1983b) Soil acidity and its relations to acid deposition. In: Ulrich, B. and Pankrath, J. (eds), *Effects of accumulation of air pollutants on forest ecosystems*. Dordrecht: D. Reidel, pp. 127–46.

Ulrich, B. (1986) Natural and anthropogenic components of soil acidification. *Zeitschrift für Pflanzenernährung und Bodenkunde* **149,** 702–17.

Ulrich, B. (1991) An ecosystem approach to soil acidification. In: Ulrich, B. and Sumner, M.E. (eds), *Soil acidity*. Berlin: Springer Verlag, pp. 28–79.

Ulrich, B. and Matzner, E. (1983) *Abiotische Folgewirkungen der weiträumigen Ausbreitung von Luftverunreinigungen*. Umweltforschungsplan des Bundesministers des Innern. Luftreinhaltung Forschungsbeirat 10402615.

Ulrich, B., Mayer, R. and Khanna, P.K. (1979) Deposition von Luftverunreinigungen und ihre Auswirkungen in Waldökosystemen im Solling. *Schriften aus der Forstlichen Fakultät, Universität Göttingen und der Niedersächsischen Forstlichen Versuchsanstalt* **58,** 1–291.

Ulrich, B., Mayer, R. and Khanna, P.K. (1980) Chemical changes due to acid precipitation in a loess-derived soil in central Europe. *Soil Science* **130,** 193–9.

Ulrich, B., Meyer, H., Jänich, K. and Büttner, G. (1989) Basenverluste in den Böden von Hainisimsen-Buchenwäldern in Südniedersachsen zwischen 1954 und 1986. *Forst und Holz* **44,** 251–3.

Ulrich, B. and Sumner, M.E. (eds) (1991) *Soil Acidity*. Berlin: Springer Verlag.

Ulrich, E. (1991) The pollution climate (especially in the Vosges Mountains). In: Landmann, G. (ed.), *French research into forest decline*. Nancy: École Nationale du Génie Rurual, des Eaux et des Forêts, pp. 39–58.

United States Department of Agriculture, Forest Service (1978) *Forest statistics of the U.S., 1977*. Washington DC: US Department of Agriculture Forest Service.

Usher, R.W. and Williams, W.T. (1982) Air pollution toxicity to eastern white pine in Indiana and Wisconsin. *Plant Disease* **66,** 199–204.

Ustin, S.L. and Curtiss, B. (1990) Spectral characteristics of ozone-treated conifers. *Environmental and Experimental Botany* **30,** 293–308.

Utschig, H. (1989) Waldwachstumskundliche Untersuchungen im Zusammenhang mit Waldschäden. Auswertung der Zuwachstrendanalyseflächen des Lehrstuhles für Waldwachstumskunde für die Fichte (*Picea abies* (L.) Karst.) in Bayern. *Forstliche Forschungsberichte München* **97,** 185 pp.

Väisänen, S. (1986) Effects of air pollution by metal, chemical and fertilizer plants on forest vegetation at Kokkola, w Finland. *Annales Botanici Fennici* **23,** 305–15.

Van Deusen, P.C. (1987) Testing for stand dynamics effects on red spruce growth trends. *Canadian Journal of Forest Research* **17,** 1487–95.

Van Deusen, P.C. (1988) Red spruce tree ring analysis using a Kalman filter. In: Van Deusen, P.C. (ed.), *Analyses of Great Smoky Mountain red spruce tree ring data.* General Technical Report 20-69. New Orleans: US Department of Agriculture Forest Service, Southern Forest Experiment Station, pp. 57–67.

Van Deusen, P.C. (1989) Indicator plants in forest health surveys. In: *Proceedings, Advanced Forest Inventory Methods, Syracuse, New York, August 1989.* Publishers unknown.

Van Deusen, P.C. (1990) Stand dynamics and red spruce decline. *Canadian Journal of Forest Research* **20,** 743–49.

Van Deusen, P.C. (1992) Growth trends and stand dynamics in natural loblolly pine in the southeastern United States. *Canadian Journal of Forest Research* **22,** 660–6.

Van Deusen, P.C., Reams, G.A. and Cook, E.R. (1991) Possible red spruce decline. Contributions of tree ring analysis. *Journal of Forestry* **89,** 20–4.

Vann, D.R., Strimbeck, G.R. and Johnson, A.H. (1992) Effects of ambient levels of airborne chemicals on the freezing resistance of red spruce foliage. *Forest Ecology and Management* **51,** 69–79.

Verein Deutsche Ingenieure – Kommission Reinhaltung der Luft (1987) *Acidic precipitation. Formation and impact on terrestrial ecosystems.* Düsseldorf: Verein Deutscher Ingenieure.

Veblen, T.T. (1979) Structure and dynamics of *Nothofagus* forests near timberline in south-central Chile. *Ecology* **60,** 937–45.

Veblen, T.T. (1989a) Tree regeneration responses to gaps along a Transandean gradient. *Ecology* **70,** 541–3.

Veblen, T.T. (1989b) *Nothofagus* regeneration in treefall gaps in northern Patagonia. *Canadian Journal of Forest Research* **19,** 365–71.

Veblen, T.T. and Ashton, D.H. (1978) Catastrophic influences on the vegetation of the Valdivian Andes, Chile. *Vegetatio* **36,** 149–67.

Veblen, T.T. and Lorenz, D.C. (1988) Recent vegetation changes along the forest/steppe ecotone of northern Patagonia. *Annals of the Association of American Geographers* **78,** 93–111.

Veblen, T.T. and Stewart, G.H. (1980) Comparison of forest structure and regeneration on Bench and Stewart Islands, New Zealand. *New Zealand Journal of Ecology* **3,** 50–68.

Veblen, T.T. and Stewart, G.H. (1982a) The effects of introduced wild animals on New Zealand forests. *Annals of the Association of American Geographers* **72,** 372–97.

Veblen, T.T. and Stewart, G.H. (1982b) On the conifer regeneration gap in New Zealand: The dynamics of *Librocedrus bidwillii* stands on South Island. *Journal of Ecology* **70,** 415–36.

Veblen, T.T., Ashton, D.H., Rubulis, S., Lorenz, D.C. and Cortes, M. (1989) *Nothofagus* stand development on in-transit moraines, Casa Pangue Glacier, Chile. *Arctic and Alpine Research* **21,** 144–55.

Veblen, T.T., Ashton, D.H., Schlegel, F.M. and Veblen, A.T. (1977b) Distribution and dominance of species in the understorey of a mixed evergreen-deciduous *Nothofagus* forest in south-central Chile. *Journal of Ecology* **65,** 815–30.

Veblen, T.T., Donoso, C., Schlegel, F.M. and Escobar, B. (1981) Forest dynamics in south-central Chile. *Journal of Biogeography* **8,** 211–47.

Veblen, T.T., Schlegel, F.M. and Escobar R., B. (1980) Structure and dynamics of old-growth *Nothofagus* forests in the Valdivian Andes, Chile. *Journal of Ecology* **68,** 1–31.

Veblen, T.T., Schlegel, F.M. and Oltremari, J.V. (1983) Temperate broad-leaved evergreen forests of South America. In: Ovington, J.D. (ed.), *Ecosystems of the World 10. Temperate broad-leaved evergreen forests.* Berlin: Springer Verlag, pp. 5–31.

Villanueva, V.R., Mardon, M. and le Goff, M.Th. (1986) Comparison of amino-compounds in the needles of healthy and damaged *Picea* trees in air polluted areas. Preliminary results. *International Journal of Environmental and Analytical Chemistry* **25,** 127–34.

Villanueva, V.R. and Santerre, A. (1989) On the mechanism of adaptive metabolism of healthy-resistant trees from forest polluted areas. *Water, Air, and Soil Pollution* **48,** 59–75.

Villanueva, V.R. and Santerre, A. (1991) Contribution to forest decline studies; biochemical effects of atmospheric pollution (acid rain) in *Picea* trees. In: Longhurst, J.W.S. (ed.), *Acid deposition. Origins, impacts and abatement strategies.* Berlin: Springer Verlag, pp. 141–60.

Villanueva, V.R., Santerre, A. and Mardon, M. (1988) Comparative study of cellular metabolism in healthy and diseased trees from polluted areas. In: Cape, J.N. and Mathy, P. (eds), *Scientific basis of forest decline symptomatology.* Air Pollution Report 15. Brussels: Commission of the European Communities, pp. 358–70.

Viro, P.J. (1955) Investigations on forest litter. *Communicationes Institutales Forestales Fennica* **45,** Part 6.

Visser, H. (1986) Analysis of tree ring data using the Kalman filter technique. *IAWA Bulletin* **7,** 289–97.

Visser, H. and Molenaar, J. (1986) *Time-dependent responses of trees to weather variations: an application of the Kalman filter.* KEMA Report 50385-MOA 86–3041. De Dorschkamp: N.V. Kema.

Vithayasai, C. (1971) Runs of healthy and diseased trees in transects through an infected forest: statistical and computational methods. *Biometrics* **27,** 1003–16.

Vithayasai, C. (1972) Statistical analysis of the ordering of diseased and healthy trees in a transect. *Forest Science* **18,** 87–92.

Vitousek, P.M., Van Cleve, K., Balakrishnan, N. and Mueller-Dombois, D. (1983) Soil development and nitrogen turnover in montane rainforest soils on Hawai'i. *Biotropica* **15,** 268–74.

Vogel, M. (1989) Heuristisches Modell zur Schätzung des Radialzuwachses aus den meteorologischen Elementen Niederschlag und Temperatur. *Beiträge für die Forstwirtschaft* **23,** 178–81.

Vogelei, A. and Rothe, G.M. (1988) Die Wirkung von Säure und Aluminiumionen auf den Nährelementgehalt und den histologischen Zustand nichtmykorrizierter Fichtenwurzeln (*Picea abies* (L.) Karst.). *Forstwissenschaftliches Centralblatt* **107,** 348–57.

Vogelmann, H.W., Badger, G.J., Bliss, M. and Klein, R.M. (1985) Forest decline on Camels Hump, Vermont. *Bulletin of the Torrey Botanical Club* **112,** 274–87.

Vogelmann, H.W., Perkins, T.D., Badger, G.J. and Klein, R.M. (1988) A 21-year record of forest decline on Camels Hump, Vermont, U.S.A. *European Journal of Forest Pathology* **18,** 240–9.

Vogelmann, J.E. and Rock, B.N. (1986) Assessing forest decline in coniferous forests of Vermont using NS-001 thematic mapper simulator data. *International Journal of Remote Sensing* **7,** 1303–21.

Vogelmann, J.E. and Rock, B.N. (1988) Assessing forest damage in high elevation coniferous forests in Vermont and New Hampshire using thematic mapper data. *Remote Sensing of Environment* **24,** 227–46.

Vogler, D.R. (1982) *Ozone monitoring in the southern Sierra Nevada, 1976–1981.* Report No. 82–17. Berkeley: US Department of Agriculture Forest Service, Pacific Southwest Region, Forest Pest Management.

Vogt, K.A., Grier, C.C., Gower, S.T., Sprugel, D.G. and Vogt, D.J. (1986a) Overestimation of net root production: a real or imaginary problem? *Ecology* **67,** 577–9.

Vogt, K.A., Grier, C.C. and Vogt, D.J. (1986b) Production, turnover, and nutrient dynamics of above- and belowground detritus of world forests. *Advances in Ecological Research* **15,** 303–77.

Vong, R.J. and Guttorp, P. (1991) Co-occurrence of ozone and acidic cloudwater in high-elevation forests. *Environmental Science and Technology* **25,** 1325–9.

Vonk, J.W., Barug, D. and Bosma, T.N.P. (1988) Fate of mineral nitrogen in acid heathland and forest soils. In: Mathy, P. (ed.), *Air pollution and ecosystems.* Dordrecht: D. Reidel, pp. 835–40.

Voretzsch, A., Herms, P. and Hartmann, G. (1986) Standardisierung von CIR-Luftbildern. Gegenwärtiger Stand der Möglichkeiten. *Der Forst- und Holzwirt* **41,** 420–6.

Vose, J.M. and Allen, H.L. (1988) Leaf area, stemwood growth, and nutrient relationships in loblolly pine. *Forest Science* **34,** 547–63.

De Vries, W. (1988) Critical deposition level for nitrogen and sulphur on Dutch forest ecosystems. *Water, Air, and Soil Pollution* **42,** 221–39.

De Vries, W. and Breeuwsma, A. (1984) Causes of soil acidification. *Netherlands Journal of Agricultural Science* **32,** 159–61.

Wachter, A. (1978) Deutschsprachige Literatur zum Weißtannensterben (1930–1978). *Zeitschrift für Pflanzenpathologie und Pflanzenschutz* **85,** 361–81.

Wachter, H. (1986) Beobachtungen zur Benadelung in einem Fichtenstangenholz des Eggegebirges 1983–86. *Der Forst- und Holzwirt* **41,** 146–50.

Wagner, A. (1986) *Waldbaurichtlinien für den Staatswald des Saarlandes. 1. Teil: Standortsökologische Grundlagen.* Saarbrücken: Ministerium für Wirtschaft.

Wahlmann, B., Braun, E. and Lewark, S. (1986) Radial increment in different tree heights in beech stands affected by air pollution. *IAWA Bulletin* **7,** 285–8.

Wallner, D.W. and Fong, M. (1982) *Analysis of ozone injury to ponderosa and Jeffrey pines in Sequoia and Kings Canyon National Parks. Final survey results 1980–1982.* Sequoia National Park: Resource Management, Sequoia National Park, California.

Wardlaw, T.J. (1989) Management of Tasmanian forests affected by regrowth dieback. *New Zealand Journal of Forestry Science* **19,** 265–76.

Wardle, J.A. (1983) Disturbance and the ecology of beeches with emphasis on mountain beech. In: Silvester, W.B. and Watt, V. (eds), *The future of Tongariro National Park beech forests.* Department of Lands and Survey, Wellington, New Zealand. Wellington: New Zealand Forest Service, pp. 13–18.

Wardle, J.A. and Allen, R.B. (1983) Dieback in New Zealand *Nothofagus* forests. *Pacific Science* **37,** 397–404.

Wardle, P. (1970) The ecology of *Nothofagus solandri,* Parts 1–4. *New Zealand Journal of Botany* **8,** 494–646.

Wardle, P. (1981) Winter desiccation of conifer needles simulated by artificial freezing. *Arctic and Alpine Research* **13,** 419–23.

Wargo, P.M., Bergdahl, D.R., Olson, C.W. and Tobi, D.R. (1989) *Root vitality and decline of red spruce.* Report to the US Department of Agriculture Forest Service, Forest Response Program. Northeastern Forest Experiment Station, Radnor.

Warner, T.E., Wallner, D.W. and Vogler, D.R. (1982) Ozone injury to ponderosa and Jeffrey pines in the Sequoia-Kings Canyon National Parks. In: van Riper, C., Whittig, L.D. and Murphy, P.L. (eds), *Proceedings of the conference on research in California's National Parks.* Davis: University of California, pp. 1–7.

Warren, W.G. (1990) Some novel statistical analyses relevant to the reported growth decline of pine species in the Southeast. *Forest Science* **36,** 448–63.

Waterman, A.M. (1955) The relation of *Valsa kunzei* to cankers on conifers. *Phytopathology* **45,** 686–92.

Watson, W.Y. and Richardson, D.H. (1972) Appreciating the potential of a devastated land. *Forestry Chronicle* **48,** 312–15.

Wehrmann, J. (1960) Methodische Untersuchungen zur Durchführung von Nadelanalysen in Kiefernbeständen. *Forstwissenschaftliches Centralblatt* **78,** 77–97.

Weidner, M. and Kraus, M. (1987) Ribulose-1,5-bisphosphate carboxylase activity and influence of air pollution in spruce. *Physiologia Plantarum* **70,** 664–72.

Weikert, R.M. (1986) Die photosynthetische Leistungsfähigkeit verschieden stark geschädigter Fichten – Vergleich zwischen einem 'belasteten' und 'unbelasteten' Standort. Diploma thesis, University of Würzberg.

Weikert, R.M., Wedler, M., Lippert, M., Schramel, P. and Lange, O.L. (1989) Photosynthetic performance, chloroplast pigments, and mineral content of various needle age classes of spruce (*Picea abies*) with and without the new flush: an experimental approach for analysing forest decline phenomena. *Trees: Structure and Function* **3,** 161–72.

Weiss, M.J. and Millers, I. (1988) Historical impacts on red spruce and balsam fir in the northeastern United States. In: *Proceedings of the US/FRG Research Symposium: Effects of atmospheric pollutants on the spruce-fir forests of the eastern United States and the Federal Republic of Germany.* General Technical Report NE-120. Broomall: US Department of Agriculture Forest Service, Northeastern Forest Experiment Station, pp. 271–7.

Weiss, M.J. and Rizzo, D.M. (1987) Forest declines in major forest types of the eastern United States. In: Kairiukstis, L., Nilsson, S. and Straszak, A. (eds), *Forest decline and reproduction: Regional and global consequences.* Laxenburg: International Institute of Applied Systems Analysis, pp. 297–306.

Weiss, M.J. (1985) *Red spruce and balsam fir decline and mortality.* Interim Report, US Department of Agriculture Forest Service. Northeastern Area, Broomall, Pennsylvania.

Weiss, V. (1991) Waldschadenserhebung 1991 in Schleswig-Holstein. *Forst und Holz* **46,** 622–3.

Weiss, W.J. and Knigge, W. (1989) Auswirkung von Immissionsbelastungen auf einige technologische Eigenschaften und das Lagerungsverhalten von Stammholz der Fichte (*Picea abies* (L.) Karst.). *Forstarchiv* **60,** 22–7.

Weissen, F., Hambuckers, A., van Praag, H.J. and Remacle, J. (1990) A decennial control of N-cycle in the Belgian Ardenne forest ecosystems. *Plant and Soil* **128,** 59–66.

Weissen, F., van Praag, H., Marechal, P., Delecour, F. and Farcy, Ch. (1988) Les causes de la dégradation sanitaire de forêts en Wallonie: le point de la situation. *Bulletin de la Société Royale Forestière de Belgique* **95,** 57–68.

Wells, C.G. (1980) *Effects of spacing and age on nutrients in needles and soil properties of loblolly pine.* Progress report. FS-SE-1103-3(3). Asheville: US Department of Agriculture Forest Service, Southeastern Forest Experiment Station.

Wells, C.G. and Metz, L.J. (1963) Variation in nutrient content of loblolly pine needles with season, age, soil and position on the crown. *Soil Science Society of America Proceedings* **27,** 90–3.

Wentworth, T.R., White, P.S., Pyle, C. and Schafale, M.P. (1988) Compilation and interpretation of the vegetation data base and disturbance history of southern Appalachian spruce-fir. In: *Proceedings of the US/FRG Research Symposium: Effects of atmospheric pollutants on the spruce-fir forests of the eastern United States and the Federal Republic of Germany.* General Technical Report NE-120. Broomall: US Department of Agriculture Forest Service, Northeastern Forest Experiment Station, pp. 145–9.

Werk, K.S., Oren, R., Schulze, E.-D., Zimmermann, R. and Meyer, J. (1988) Performance of two *Picea abies* (L.) Karst. stands at different stages of decline. III. Canopy transpiration of green trees. *Oecologia* **76,** 519–24.

Werner, W.L. (1984) *Die Höhen- und Nebelwälder auf der Insel Ceylon (Sri Lanka).* Tropical und subtropical Pflanzenwelt 46. Wiesbaden: Steiner.

Werner, W.L. (1988) Canopy dieback in the upper montane rain forests of Sri Lanka. *GeoJournal* **17,** 245–8.

West, P.W. (1979) Date of onset of regrowth dieback and its relation to summer drought in eucalypt forest of southern Tasmania. *Annals of Applied Biology* **93,** 337–50.

West, P.W. and Podger, F.D. (1980) Loss in timber volume and value due to regrowth dieback of eucalypts in southern Tasmania. *Australian Forestry* **43,** 20–8.

Westing, A.H. (1966) Sugar maple decline: an evaluation. *Economic Botany* **20,** 196–212.

Westman, L. (1989) A new method for assessment of visible damage to birch and other deciduous trees. In: Bucher, J.B. and Bucher-Wallin, I. (eds), *Air pollution and forest decline.* Birmensdorf: Eidgenössische Anstalt für das forstliche Versuchswesen, pp. 223–8.

Westman, L. and Lesínski, J. (1985) Thinning out of the tree crown – what is hidden in that integrated measure of forest damage? In: Schmid-Haas, P. (ed.), *Inventorying and monitoring endangered forests.* Birmensdorf: Eidgenössische Anstalt für das forstliche Versuchswesen, pp. 223–8.

Westman, L. and Lesínski, J. (1986) *Kronutglesning och andra förändringar i grankronan. Morfologisk beskrivning.* Stockholm: Naturvårdsverket Rapport 3262. 96 pp.

Westman, W.E. and Price, C.V. (1988a) Detecting air pollution stress in southern California vegetation using Landsat Thematic Mapper band data. *Photogrammetric Engineering and Remote Sensing* **54,** 1305–11.

Westman, W.E. and Price, C.V. (1988b) Spectral changes in conifers subjected to air pollution and water stress: experimental studies. *IEEE Transactions on Geoscience and Remote Sensing* **26,** 11–21.

Westveld, M. (1931) *Reproduction of pulpwood lands in the Northeast.* Technical Bulletin 223. Washington DC: US Department of Agriculture. 52pp.

Wharton, E.H. and Powell, D.S. (1985) Eastern white pine: Inventory and dynamics. In: Funk, D. (ed.), *Eastern white pine: Today and tomorrow.* General Technical Report WO-51. Washington DC: US Department of Agriculture Forest Service, pp. 16–21.

Wheeler, G.S. (1965) Memorandum 2400, 5100 July 1, 1965. Laconia, New Hampshire. US Forest Service Northern FPC Zone.

White, K.L. (1966) Structure and composition of foothill woodland in central coastal California. *Ecology* **47,** 229–37.

Whyte, A.G.D. (1973) Productivity of first and second crops of *Pinus radiata* on the Moutere gravel soils of Nelson. *New Zealand Journal of Forestry* **18,** 87–103.

Widmann, H. and Avemark, W. (1988) Bestimmung von Weiserjahren mit Hilfe des Sensitivitätsindexes. *Forst und Holz* **43,** 128–31.

Wiedemann, E. (1925) *Zuwachsrückgang und Wuchsstockungen der Fichte in den mittleren und unteren Höhenlagen der sächsischen Staatsforsten.* Tharandt: Akademische Buchhandlung Walter Laux.

Wiedemann, E. (1927) Untersuchungen über das Tannensterben. *Forstwissenschaftliches Centralblatt* **49,** 759–80, 815–27 and 845–53.

Wiesner, J. (1904) Über den Hitzelaubfall. *Berichte der Deutschen Botanischen Gesellschaft, Berlin* **22:** 501–505.

Wild, A. (1987) Physiologische und cytomorphologische Charakterisierung von immissionsbelasteten Fichten. *Allgemeine Forst Zeitschrift* **42,** 734–7.

Wilkinson, R.C. (1988) Geographic variation in needle morphology of red spruce in relation to winter injury and decline. In: *Proceedings of the US/FRG Research Symposium: Effects of atmospheric pollutants on the spruce-fir forests of the eastern United States and the Federal Republic of Germany.* General

Technical Report NE-120. Broomall: US Department of Agriculture Forest Service, Northeastern Forest Experiment Station, pp. 507–14.

Wilkinson, R.C. (1990) Effects of winter injury on basal area and height growth of 30-year-old red spruce from 12 provenances growing in northern New Hampshire. *Canadian Journal of Forest Research* **20,** 1616–22.

Will, G.M. (1957) Variation in the mineral content of Radiata pine needles with age and position in the tree crown. *New Zealand Journal of Science and Technology* **38B,** 699–706.

Will, G.M. (1978) Nutrient deficiencies in *Pinus radiata* in New Zealand. *New Zealand Journal of Forest Science* **8,** 4–14.

Will, G.M. (1985) *Nutrient deficiency and fertilizer use in New Zealand exotic forests.* Forest Research Institute Bulletin 97. Rotorua: New Zealand Forest Service.

Williams, B.L., Cooper, J.M. and Pyatt, D.G. (1979) Effects of afforestation with *Pinus contorta* on nutrient content, acidity, and exchangeable cations in peat. *Forestry* **51,** 29–35.

Williams, K., Davis, S.D., Gartner, B.L. and Karlsson, S. (1991) Factors affecting the establishment of a chapparal oak, *Quercus durata* Jeps., in grassland. In: *Proceedings of the symposium on oak woodlands and hardwood rangeland management, October 31 – November 2, 1990, Davis, California.* General Technical Report PSW-126. Berkeley: US Department of Agriculture Forest Service, Pacific Southwest Research Station, pp. 70–3.

Williams, K.A. (1982) *Tolerances of four species of southern pine to aluminium in solution cultures.* MSc thesis, University of Florida, Gainesville.

Williams, W.T. (1980) Air pollution disease in the Californian forests. A base line for smog disease on ponderosa and Jeffrey pines in Sequoia and Los Padres National Forests, California. *Environmental Science and Technology* **14,** 179–82.

Wilmanns, O. (1989) Zur Frage der Reaktion der Waldboden-Vegetation auf Stoffeintrag durch Regen – eine Studie auf der Schwäbischen Alb. *Allgemeine Forst- und Jagdzeitung* **160,** 165–75.

Wilson, J. (1984) Microscopic features of wind damage to leaves of *Acer pseudoplatanus* L. *Annals of Botany* **53,** 73–82.

Winner, W.E., Gillespie, C., Shen, W.-S. and Mooney, H.A. (1988) Stomatal responses to SO_2 and O_3. In: Schulte-Hostede, S., Darrall, N.M., Blank, L.W. and Wellburn, A.R. (eds), *Air pollution and plant metabolism.* London: Elsevier, pp. 255–71.

Winner, W.E., Lefohn, A.S., Cotter, I.S., Greitner, C.S., Nellesen, J., McEvoy, L.R., Olson, R.L., Atkinson, C.J. and Moore, L.D. (1989) Plant responses to elevational gradients of O_3 exposures in Virginia. *Proceedings of the National Academy of Sciences of the USA* **86,** 8828–32.

Winner, W.E. and Mooney, H.A. (1985) Ecology of SO_2 resistance V. Effects of volcanic SO_2 on native Hawaiian plants. *Oecologia* **66,** 387–93.

Winterhalder, K. (1988) Trigger factors initiating revegetation processes on barren, acid, metal-toxic soils near Sudbury, Ontario smelters. *Proceedings, 1988 Annual Meeting of American Society for Surface Mining and Reclamation, April 17–22, 1988.* Pittsburgh: American Society for Surface Mining and Reclamation.

Wittich, W. (1951) Der Einfluss der Streunutzung auf den Boden. *Forstwissenschaftliches Centralblatt* **70,** 65–92.

Wittig, R. (1986) Acidification phenomena in beech (*Fagus sylvatica*) forests of Europe. Water, Air, and Soil Pollution **31,** 317–23.

Wittig, R. and Neite, H. (1985) Acid indicators around the trunk base of *Fagus sylvatica* in limestone and loess beechwoods: distribution pattern and phytosociological problems. *Vegetatio* **64,** 113–19.

Wittig, R., Ballach, H.-J. and Brandt, C.J. (1985) Increase of number of acid indicators in the herb layer of the Millet Grass-Beech forest of the Westphalian Bight. *Angewandte Botanik* **59,** 219–32.

Wittmann, O. and Fetzer, K.D. (1982) *Aktuelle Bodenversauerung in Bayern.* Materialien Nr. 20, Bayerischer Staatsministerium für Landesentwicklung und Weltfragen, 71 pp.

Wolfe, M.H., Kelly, J.M. and Wolt, J.D. (1987) Soil pH and extractable sulfate-sulfur distribution as influenced by tree species and distance from the stem. *Soil Science Society of America Journal* **51,** 1042–6.

Wolfenden, J., Robinson, D.C., Cape, J.N., Paterson, I.S., Francis, B.J., Mehlhorn, H. and Wellburn, A.R. (1988) Use of carotenoid ratios, ethylene emissions and buffer capacities for the early diagnosis of forest decline. *New Phytologist* **109,** 85–95.

Wood, T. and Bormann, F.H. (1975) Increases in foliar leaching caused by acidification of an artificial mist. *Ambio* **4,** 169–71.

Woods, R.V. (1990/1991) Second rotation decline in *P. radiata* plantations in South Australia has been corrected. *Water, Air and Soil Pollution* **54,** 607–19.

Worbes, M. (1989) Abrupte Zuwachsreduktionen an Fichten aus dem Harz. *Forst und Holz* **44,** 254–9.

Worrall, J.J. and Harrington, T.C. (1988) Etiology of canopy gaps in spruce-fir forests at Crawford Notch, NH. *Canadian Journal of Forest Research* **18,** 1463–9.

Worrell, R. (1987) *Predicting the productivity of Sitka spruce on upland sites in northern Britain.* Forestry Commission Bulletin 72. London: HMSO.

Worrell, R. (1992) A comparison between European continental and British provenances of some British native trees: growth, survival and stem form. *Forestry* **65,** 253–80.

Worrell, R. and Malcolm, D.C. (1990a) Productivity of Sitka spruce in northern Britain 1. The effects of elevation and climate. *Forestry* **63,** 105–18.

Worrell, R. and Malcolm, D.C. (1990b) Productivity of Sitka spruce in northern Britain 2. Prediction from site factors. *Forestry* **63,** 119–28.

Wright, J.W. and Bull, W.I. (1963) Geographic variation in Scotch pine. *Silvae Genetica* **12,** 1–25.

Wyttenbach, A. and Tobler, L. (1988) The seasonal variation of 20 elements in 1st and 2nd year needles of Norway spruce, *Picea abies* (L.) Karst. *Trees: Structure and Function* **2,** 52–64.

Wyttenbach, A., Bajo, S., Tobler, L. and Keller, T. (1985) Major and trace element concentrations in needles of *Picea abies:* levels, distribution functions, correlations and environmental influences. *Plant and Soil* **85,** 313–25.

Wyttenbach, A., Tobler, L. and Bajo, S. (1989a) Nadelinhaltstoffe und Ablagerungen auf Nadeloberflächen von Fichten (*Picea abies* Karst.). *Forstwissenschaftliches Centralblatt* **108,** 233–43.

Wyttenbach, A., Tobler, L. and Bajo, S. (1989b) Na, Cl and Br in needles of Norway spruce (*P. abies*) and in the aerosol adhering to the needles. *Toxicological and Environmental Chemistry* **19,** 25–33.

Xiaozhu, L. and Youbiao, Z. (1989) Relationship of sulphur content in the leaves to SO_2 concentration in air. *Bulletin of Environmental Contamination and Toxicology* **42,** 878–83.

Yang, Y.S., Skelly, J.M. and Chevone, B.I. (1982) Clonal response of eastern white pine clones to acute doses of ozone, sulphur dioxide, or nitrogen dioxide. *Phytopathology* **73,** 1234–7.

Yang, Y.S., Skelly, J.M., Chevone, B.I. and Birch, J.B. (1983) Effects of long-term ozone exposure on photosynthesis and dark respiration of eastern white pine. *Environmental Science and Technology* **17,** 371–3.

Yarmishko, V.T. (1989) (Formation of needle biomass in young Scots pine forests on the Kola peninsula). *Botanicheskii Zhurnal* **74,** 1376–86.

Yoshino, M. (1976) Geographical distribution of 'dead trees strips' in Japanese subalpine forests. In: Kato, T., Nakano, S. and Umesao, T. (eds), *Mountains, forests, and ecology*. Tokyo: Chuokoron-sha, pp. 183–202.

Yost, R.S., Uehara, G. and Fox, R.L. (1982) Geostatistical analysis of soil chemical properties of large land areas. I. Semi-variograms. *Soil Science Society of America Journal* **46,** 1028–32.

Young, C.W.T. (1965) *Death of pedunculate oak and variations in annual radial increments related to climate.* Forestry Commission Forest Record 55. London: HMSO.

Yu, S.W., Yu, Z.W., Ma, G.J., Zhu, C.L., Bian, Y.M., Chen, S.Y., Zhu, X.Y. (1988) (Preliminary investigation on the causes of masson pine forest decline in the Nanshan region of Chongqing, Sichuan Province). *Chinese Journal of Environmental Science* **9,** 77–81.

Zahner, R. (1988) A model for tree-ring time series to detect regional growth changes in young, evenaged forest stands. *Tree Ring Bulletin* **48,** 13–20.

Zahner, R. and Saucier, J.R. (1989) Twenty-five years of climate-related growth trends for natural stands of loblolly pine in Central Georgia. In: *Fifth Biennial Silvicultural Resources Conference, November 1–3, Memphis Tennessee.*

Zahner, R., Saucier, J.R. and Myers, R.K. (1989) Tree ring model interprets growth decline in natural stands of loblolly pine in the southeastern United States. *Canadian Journal of Forest Research* **19,** 612–21.

Zareba, R. (1984) Mozliwosci reintrodukcji w lasach i utrzymania w rezerwatach cisa pospolitego – *Taxus baccata* L. *Sylwan* **128,** 31–4.

Zech, W., Suttner, T. and Kotschenreuther, R. (1983) Mineralversorgung vermutlich immissionsgeschädigter Bäume in NO Bayern. *KALI – Briefe* **16,** 565–71.

Zech, W., Suttner, T. and Popp, E. (1985) Elemental analyses and physiological responses of forest trees in SO_2-polluted areas of NE-Bavaria. *Water, Air, and Soil Pollution* **25,** 175–83.

Zedaker, S.M., Hyink, D.M. and Smith, D.W. (1987) Growth declines in red spruce. *Journal of Forestry* **85,** 34–6.

Zedaker, S.M., Nicholas, N.S. and Eager, C. (1989) Assessment of forest decline in the southern Appalachian spruce–fir forest, U.S.A. In: Bucher, J. and Bucher-Wallin, I. (eds), *Air pollution and forest decline.* Birmensdorf: Eidgenössische Anstalt für das forstliche Versuchswesen, pp. 239–44.

Zedaker, S.M., Nicholas, N.S., Eagar, C., White, P.S. and Burk, T.E. (1988) Stand characteristics associated with potential decline of spruce–fir forests in the Southern Appalachians. In: *Proceedings of the US/FRG Research Symposium: Effects of atmospheric pollutants on the spruce-fir forests of the eastern United States and the Federal Republic of Germany.* General Technical Report NE-120. Broomall: US Department of Agriculture Forest Service, Northeastern Forest Experiment Station, pp. 123–31.

Zedler, B., Plarre, R. and Rothe, G.M. (1986) Impact of atmospheric pollution on the protein and amino acid metabolism of spruce *Picea abies* trees. *Environmental Pollution* **40A,** 193–212.

Zellnig, G., Bermadinger, E., Gailhofer, M. and Grill, D. (1989a) Structural investigations (TEM, SEM) of spruce needles from inversion-zones. In: Bucher, J.B. and Bucher-Wallin, I. (eds), *Air pollution and forest decline.* Birmensdorf: Eidgenössische Anstalt für das forstliche Versuchswesen, pp. 553–5.

Zellnig, G., Gailhofer, M., Pfeifhofer, H.W. and Grill, D. (1989b) Ultrastructure and pigment composition of chloroplasts of differently damaged green and yellowed needles of *Picea abies. Phyton* **29,** 213–25.

von Zezschwitz, E. (1982) Akute Bodenversauerung in den Kammlagen des Rothaargebirges. *Der Forst- und Holzwirt* **37,** 275–6.

von Zezschwitz, E. (1985) Immissionsbedingte Änderungen analytischer Kennwerte nordwestdeutscher Mittelgebirgsböden. *Geologische Jahrbuche* **F20,** 3–41.

Zimmermann, R.-D. (1989) Schadstoff- und Nährelementverteilung im Baumkronenbereich der Rotbuche. *Allgemeine Forstzeitschrift* **44,** 260–3.

Zimmermann, R.-D., Oren, R., Schulze, E.-D. and Werk, K.S. (1988) Performance of two *Picea abies* (L.) Karst. stands at different stages of decline. II. Photosynthesis and leaf conductance. *Oecologia* **76,** 513–18.

Zinke, P.J. (1962) The pattern of influence of individual trees on soil properties. *Ecology* **43,** 130–3.

Zöttl, H.W. (1983) Zur Frage der toxischen Wirkung von Aluminium auf Pflanzen. *Allgemeine Forst Zeitschrift* **38,** 206–8.

Zöttl, H.W. (1985a) Waldschäden und Nährelementversorgung. *Düsseldorfer Geobotanisches Kolloquium* **2,** 31–41.

Zöttl, H.W. (1985b) Die Rolle der Nährelementversorgung bei der Entwicklung 'neuartiger' Waldschäden. *VDI-Berichte* **560,** 887–96.

Zöttl, H.W. (1987) Stoffumsätze in Ökosystemen des Schwarzwaldes. *Forstwissenschaftliches Centralblatt* **106,** 105–14.

Zöttl, H.W. and Hüttl, R.F. (1985) Schadsymptome und Ernährungszustand von Fichtenbeständen im südwestdeutschen Alpenraum. *Allgemeine Forst Zeitschrift* **40,** 197–200.

Zöttl, H.W. and Hüttl, R.F. (1986) Nutrient supply and forest decline in southwest Germany. *Water, Air, and Soil Pollution* **31,** 449–62.

Zöttl, H.W., Hüttl, R.F., Fink, S., Tomlinson, G.H. and Wisniewski, J. (1989) Nutritional disturbances and histological changes in declining forests. *Water, Air, and Soil Pollution* **48,** 87–109.

Zöttl, H.W. and Mies, E. (1983) Nährelementversorgung und Schadstoffbelastung von Fichtenökosystemen im Südschwarzwald unter Immissionseinfluß. *Mitteilungen der Deutschen Bodenkundlichen Gesellschaft* **38,** 429–34.

Zvacek, L. (1988) *Mikronährstoffe und toxische Metalle an Waldstandorten.* Dissertation, Universität Wien, Austria, 237 pp.

Zwiazek, J.J. and Shay, J.M. (1987) Fluoride- and drought-inducing structural alterations of mesophyll and guard cells in cotyledons of jack pine (*Pinus banksiana*). *Canadian Journal of Forest Research* **65,** 2310–17.

Index